NEW ADVANCES IN STEM CELL TRANSPLANTATION

Edited by **Taner Demirer**

New Advances in Stem Cell Transplantation
Edited by Taner Demirer

Published by InTech

Janeza Trdine 9, 51000 Rijeka, Croatia

Publishing Process Manager Masa Vidovic

Technical Editor Teodora Smiljanic

Cover Designer InTech Design Team

Additional hard copies can be obtained from orders@intechopen.com

New Advances in Stem Cell Transplantation, Edited by Taner Demirer
p. cm.
ISBN 978-953-51-0013-3

For EU product safety concerns:
IN TECH d.o.o., Prolaz Marije Krucifikse Kozulić 3, 51000 Rijeka, Croatia,
info@intechopen.com or visit our website at intechopen.com.

Meet the editor

Dr Demirer graduated from Ankara University Medical School in Turkey in 1984. He then went on to train in the USA from 1987-1997. He was given the title of 'Fellow of American College of Physicians (FACP)' by ACP in July of 1996. During his career, he has written many papers in the medical journals and books in regard to stem cell mobilization kinetics, factors influencing the stem collection and engraftment, as well as HDC in patients with multiple myeloma, breast, and ovarian cancer. He was Chair of the EBMT Solid Tumors Working Party (STWP) between 2001-2007. Dr. Demirer is currently a professor of Medicine and Hematology/Oncology at the Ankara University Medical School in Ankara, Turkey.

Contents

Preface

This book documents the increased number of stem cell-related research, clinical applications, and views for the future. The book covers a wide range of issues in cell-based therapy and regenerative medicine, and includes clinical and preclinical chapters from the respected authors involved with stem cell studies and research from around the world. It complements and extends the basics of stem cell physiology, hematopoietic stem cells, issues related to clinical problems, tissue typing, cryopreservation, dendritic cells, mesenchymal cells, neuroscience, endovascular cells and other tissues. In addition, tissue engineering that employs novel methods with stem cells is explored. Clearly, the continued use of biomedical engineering will depend heavily on stem cells, and this book is well positioned to provide comprehensive coverage of these developments.

This book will be the the main source for clinical and preclinical publications for scientists working toward cell transplantation therapies with the goal of replacing diseased cells with donor cells of various organs, and transplanting those cells close to the injured or diseased target. With the increased number of publications related to stem cells and *Cell Transplantation*, we feel it is important to take this opportunity to share these new developments and innovations describing stem cell research in the cell transplantation field with our worldwide readers.

Stem cells have a unique ability. They are able to self renew with no limit, allowing them to replenish themselves, as well as other cells. Another ability of stem cells is that they are able to differentiate to any cell type. A stem cell does not differentiate directly to a specialized cell however- there are often multiple intermediate stages. A stem cell will first differentiate to a progenitor cell. A progenitor cell is similar to a stem cell, although they are limited in the number of times they can replicate, and they are also restricted in which cells they can further differentiate to. Serving as a sort of repair system for the body, they can theoretically divide without limit in order to replenish other cells for the rest of the person or animal's natural life. When a stem cell divides, each new cell has the potential to either remain a stem cell, or become another type of cell with a more specialized function, such as a muscle cell, a red blood cell, or a brain cell.

Because of the unique abilities of stem cells, as opposed to a typical somatic cell, they are currently the target of ongoing research. Research on stem cells is advancing in the knowledge about how an organism develops from a single cell and how healthy cells replace damaged cells in adult organisms. This promising area of science is also leading scientists to investigate the possibility of cell-based therapies to treat disease such as diabetes or heart disease. It is often referred to as regenerative medicine or reparative medicine.

During this last decade, the number of published articles or books investigating the role of stem cells in cell transplantation or regenerative medicine increased remarkably across all sections of the stem cell related journals. The largest number of stem cell articles was published mainly in the field of neuroscience, followed by the bone, muscle, cartilage, and hepatocytes. Interestingly, in recent years, the number of stem cell articles describing the potential use of stem cell therapy and islet transplantation in diabetes is also slowly increasing, even though this field of endeavor could have one of the greatest clinical and societal impacts.

Stem cells could have the potential to diminish the problem of the availability of transplantable organs that, today, limits the number of successful large-scale organ replacements. Several different methods using stem cells are currently used, but there are still several obstacles that need to be resolved before attempting to use stem cells in the clinic. Regarding the transplantation of differentiated cells derived from stem cells, one can argue that there are several regulatory, scientific, and technical issues, such as cell manufacturing procedures, regulatory mechanisms for differentiation, and developing screening methods to avoid developing inappropriate differentiated cells. One of the next steps in stem cell therapy is the development of treatments that will function not only at an early stage of transplantation, but will also remain intact throughout the life of the host recipient.

It will be exciting and interesting for our readers to follow the recent developments in the field of stem cells and cell transplantation, via this book, such as authors' search for the clues to what pathways are used by stem cells to repair tissue, or what can trigger wound healing, bone growth, and brain repair. Although we are close to finding pathways for stem cell therapies in many medical conditions, scientists need to be careful how they use stem cells ethically, and should not rush into clinical trials without carefully investigating the side effects. Focus must be on Good Manufacturing Procedures (GMP) and careful monitoring of the long-term effects of transplanted stem cells in the host.

In conclusion, *Cell Transplantation* is bridging cell transplantation research in a multitude of disease models as methods and technology continue to be refined. The use of stem cells in many therapeutic areas will bring hope to many patients awaiting replacement of malfunctioning organs, or repairing of damaged tissues. We hope that this book will be an important tool and reference guide for all scientists worldwide who work in the field of stem cells and cell transplantation. Additionally, we hope that it will shed a light upon many important debatable issues in this field.

I would like to thank all authors who contributed this book with excellent up to date chapters relaying the recent developments in the field of stem cell transplantation to our readers. I would like to give special thanks to Masa Vidovic, Publishing Process Manager, and all InTech workers for their valuable contribution in order to make this book available.

Taner Demirer, MD, FACP
Professor of Medicine, Hematology/Oncology
Dept. of Hematology
Ankara University Medical School
Ankara
Turkey

Part 1

Basic Aspects of Stem Cell Transplantation

1

Generation of Patient Specific Stem Cells: A Human Model System

Stina Simonsson, Cecilia Borestrom and Julia Asp
Department of Clinical Chemistry and Transfusion Medicine, Institute of Biomedicine, University of Gothenburg, Gothenburg Sweden

1. Introduction

In 2006, Shinya Yamanaka and colleagues reported that only four transcription factors were needed to reprogram mouse fibroblasts back in development into cells similar to embryonic stem cells (ESCs). These reprogrammed cells were called induced pluripotent stem cells (iPSCs). The year after, iPSCs were successfully produced from human fibroblasts and in 2008 reprogramming cells were chosen as the breakthrough of the year by Science magazine. In particular, this was due to the establishment of patient-specific cell lines from patients with various diseases using the induced pluripotent stem cell (iPSC) technique. IPSCs can be patient specific and therefore may prove useful in several applications, such as; screens for potential drugs, regenerative medicine, models for specific human diseases and in models for patient specific diseases. When using iPSCs in academics, drug development, and industry, it is important to determine whether the derived cells faithfully capture biological processes and relevant disease phenotypes. This chapter provides a summary of cell types of human origin that have been transformed into iPSCs and of different iPSC procedures that exist. Furthermore we discuss advantages and disadvantages of procedures, potential medical applications and implications that may arise in the iPSC field.

1.1 Preface

For the last three decades investigation of embryonic stem (ES) cells has resulted in better understanding of the molecular mechanisms involved in the differentiation process of ES cells to somatic cells. Under specific *in vitro* culture conditions, ES cells can proliferate indefinitely and are able to differentiate into almost all tissue specific cell lineages, if the appropriate extrinsic and intrinsic stimuli are provided. These properties make ES cells an attractive source for cell replacement therapy in the treatment of neurodegenerative diseases, blood disorders and diabetes. Before proceeding to a clinical setting, some problems still need to be overcome, like tumour formation and immunological rejection of the transplanted cells. To avoid the latter problem, the generation of induced pluripotent stem (iPS) cells have exposed the possibility to create patient specific ES-like cells whose differentiated progeny could be used in an autologous manner. An adult differentiated cell has been considered very stable, this concept has however been proven wrong experimentally, during the past decades. One ultimate experimental proof has been cloning

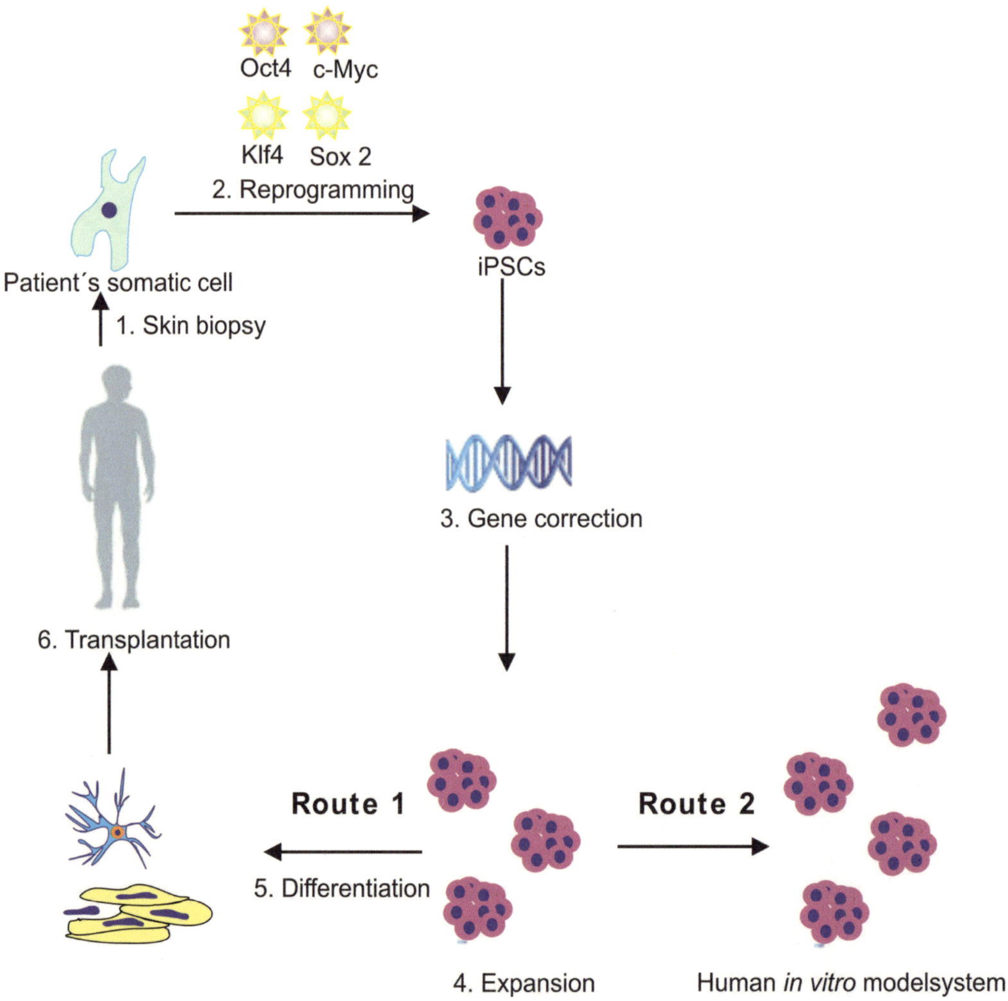

Fig. 1. Schematic picture of establishment of patient-specific induced pluripotent stem cells (iPSCs), from which two prospective routes emerge1) *in vivo* transplantation 2) *in vitro* human model system. Patient-specific induced pluripotent stem cells that are similar to embryonic stem cells (ESCs) are produced by first 1) collecting adult somatic cells from the patient, for example skin fibroblasts by a skin biopsy, 2) and reprogramming by retroviral transduction of defined transcription factors (Oct4, c-Myc, Klf4 and Sox 2 or other combinations) in those somatic fibroblast cells. Reprogrammed cells are selected by the detection of endogenous expression of a reprogramming marker, for example Oct4. 3) Generated patient-specific iPSCs can be genetically corrected of a known mutation that causes the disease. 4) Expansion of genetically corrected patient-specific iPSCs theoretically in eternity. First prospective route (Route 1): 5) upon external signals (or internal) iPSCs can theoretically be stimulated to differentiate into any cell type in the body. 6) In this way patient-specific dopamine producing nerve cells or skin cells can be generated and transplanted into individuals suffering from Parkinson´s disease or Melanoma respectively. Second route (Route 2): Generated disease-specific iPSCs can be used as a human *in vitro* system to study degenerative disorders or any disease, cause of disease, screening for drugs or recapitulate development.

animals using somatic cell nuclear transfer (SCNT) to eggs. Such experiments can result in a new individual from one differentiated somatic cell. The much more recent method to reprogram cells was the fascinating finding that mouse embryonic fibroblasts (MEFs) can be converted into induced pluripotent stem cells (iPSCs) by retroviral expression of four transcription factors: Oct4, c-Myc, Sox2 and Klf4. iPSCs are a type of pluripotent stem cell derived from a differentiated somatic cell by overexpression of a set of proteins. Nowadays, several ways of generating iPSCs have been developed and includes 1) overexpression of different combinations of transcription factors most efficiently in combination with retroviruses (step 2 in Figure 1), 2) exposure to chemical compounds in combination with the transcription factors Oct4, Klf4 and retroviruses, 3) retroviruses alone, 4) recombinant proteins or 5) mRNA. The iPSCs are named pluripotent because of their ability to differentiate into all different differentiation pathways. Generation of patient-specific iPSC lines capable of giving rise to any desired cell type provides great opportunities to treat many disorders either as therapeutic treatment or discovery of patient specific medicines in human iPSC model systems (Figure 1). Here, some of this field's fast progress and results mostly concerning human cells are summarized.

2. Reprogramming-Induced Pluripotent Stem Cells (iPSCs)

Reprogramming is the process by which induced pluripotent stem cells (iPSCs) are generated and is the conversion of adult differentiated somatic cells to an embryonic-like state. Takahashi and Yamanaka demonstrated that retrovirus-mediated delivery of Oct4, Sox2, c-Myc and Klf4 is capable of inducing pluripotency in mouse fibroblasts (Takahashi and Yamanaka, 2006) and one year later was reported the successful reprogramming of human somatic fibroblast cells into iPSCs using the same transcription factors (Takahashi et al., 2007). Takahashi and Yamanaka came up with those four reprogramming proteins after a search for regulators of pluripotency among 24 cherry picked pluripotency-associated genes. These initial mouse iPSC lines differed from ESCs in that they had a diverse global gene expression pattern compared to ESCs and failed to produce adult chimeric mice. Later iPSCs were shown to have the ability to form live chimeric mice and were transmitted through the germ line to offspring when using Oct4 or Nanog as selection marker for reprogramming instead of Fbx15, which was used in the initial experiments (Meissner et al., 2007; Okita et al., 2007; Wernig et al., 2007). Various combinations of the genes listed in table 1 have been used to obtain the induced pluripotent state in human somatic cells. The first human iPSC lines were successfully generated by Oct4 and Sox2 combined with either, Klf4 and c-Myc, as used earlier in the mouse model, or Nanog and Lin28 (Lowry et al., 2008; Nakagawa et al., 2008; Park et al., 2008b; Takahashi et al., 2007; Yu et al., 2007). Subsequent reports have demonstrated that Sox2 can be replaced by Sox1, Klf4 by Klf2 and c-Myc by N-myc or L-myc indicating that they are not fundamentally required for generation of iPSCs (Yamanaka, 2009). Oct4 has not yet been successfully replaced by another member of the Oct family to generate iPSCs which is logical due to the necessity of Oct4 in early development. However, Blx-01294 an inhibitor of G9a histone methyl transferase, which is involved in switching off Oct4 during differentiation, enables neural progenitor cells to be reprogrammed without exogenous Oct4, although transduction of Klf4, c-Myc and Sox2 together with endogenous Oct4 was required (Shi et al., 2008). Recently, Oct4 has been replaced with steroidogenic factor 1, which controls Oct4 expression in ESCs by binding the

Oct4 proximal promoter, and iPSCs were produced without exogenous Oct4 (Heng et al., 2010). Remarkably, exogenous expression of E-cadherin was reported to be able to replace the requirement for Oct4 during reprogramming in the mouse system (Redmer et al., 2011). iPSCs are similar to embryonic stem cells (ESCs) in morphology, proliferation and ability to form teratomas. In mice, pluripotency of iPSCs has been proven by tetraploid complementation (Zhao et al., 2009). Both ESCs and iPSCs can be used as the pluripotent starting cells for the generation of differentiated cells or tissues in regenerative medicine. However, the ethical dilemma associated with ESCs is avoided when using iPSCs since no embryos are destroyed when iPSCs are obtained. Moreover, iPSCs can be patient-specific and as such patient-specific drugs can be screened and in personalized regenerative medicine therapies immune rejection could be circumvented. However the question surrounding the potential immunogenicity remains unclear due to recent reports that iPSCs do not form teratomas probably because iPSCs are rejected by the immune system (Zhao et al., 2011).

Genes	Description
Oct4	Transcription factor expressed in undifferentiated pluripotent embryonic stem cells and germ cells during normal development. Together with Nanog and Sox 2, is required for the maintenance of pluripotent potential.
Sox2	Transcription factor expressed in undifferentiated pluripotent embryonic stem cells and germ cells during development. Together with Oct4 and Nanog, is necessary for the maintenance of pluripotent potential.
Myc family	Proto-oncogenes, including c-Myc, first used for generation of human and mouse iPSCs.
Klf family	Zinc-finger-containing transcription factor Kruppel-like factor 4 (KLF4) was first used for generation of human and mouse iPSCs
Nanog	Homeodomain-containing transcription factor essential for maintenance of pluripotency and self-renewal in embryonic stem cells. Expression is controlled by a network of factors including the key pluripotency regulator Oct4.
Lin 28	Conserved RNA binding protein and stem cell marker. Inhibitor of microRNA processing in embryonic stem (ES) and carcinoma (EC) cells.

Table 1. Combinations of the genes that have been used to obtain the induced pluripotent state in human somatic cells

2.1 Differentiation of iPSCs into cells of the heart

After the cells have been reprogrammed, it will be possible to differentiate them towards a wide range of specialized cells, using existing protocols for differentiation of hESCs. Differentiation of beating heart cells, the cardiomyocytes, from hESCs has now been achievable through various protocols for a decade (Kehat et al., 2001; Mummery et al., 2002). In 2007, human iPSCs were first reported to differentiate into cardiomyocytes (Takahashi et al., 2007), using a protocol including activin A and BMP4 which was described for differentiation of hESCs the same year (Laflamme et al., 2007). A comparison between the cardiac differentiation potential of hESCs and iPSCs concluded that the difference between the two cell sources were no greater than the known differences between different hESC lines and that iPSCs thus should be a viable alternative as an autologous cell source (Zhang et al., 2009). Furthermore, a recent study demonstrated that reprogramming excluding c-MYC yielded iPSCs which efficiently up-regulated a cardiac gene expression pattern and showed spontaneous beating in contrast to iPSCs reprogrammed with four factors including c-MYC (Martinez-Fernandez et al., 2010). On the transcriptional level, beating clusters from both iPSCs and hESCs were found to be similarly enriched for cardiac genes, although a small difference in their global gene expression profile was noted (Gupta et al., 2010). Taken together, these results indicate that cardiomyocytes differentiated from both hESCs and iPSCs are highly similar, although differences exist.

2.2 Additional methods to achieve reprogramming- 1.cloning = Somatic Cell Nuclear Transfer (SCNT) 2.cell fusion 3.egg extract

In addition to the iPSC procedure other ways exist to reprogram somatic cells including: 1) somatic cell nuclear transfer (SCNT), 2) cell fusion of somatic adult cells with pluripotent ESCs to generate hybrid cells and 3) cell extract from ESCs or embryo carcinoma cells (ECs). From the time when successful SCNT experiments, more commonly known as cloning, in the frog *Xenopus Laevis* (Gurdon et al., 1958) to the creation of the sheep Dolly (Wilmut et al., 1997), it has been proven that an adult cell nucleus transplanted into an unfertilized egg can support development of a new individual, and researchers have focused on identifying the molecular mechanisms that take place during this remarkable process. Even though SCNT has been around for 50 years, the molecular mechanisms that take place inside the egg remain largely unknown. The gigantic egg cell receiving a tiny nucleus is extremely difficult to study. Single cell analysis are required and gene knock-out of egg proteins is very challenging. In 2007 a report that the first primate ESCs were isolated from SCNT blastula embryos of the species Rhesus Monkey was published (Byrne et al., 2007). The reason why it took so long to perform successful SCNT in Rhesus Monkey was a technical issue; to enucleate the egg, modified polarized light was used instead of traditional methods using either mechanical removal of DNA or UV light mediated DNA destruction. The first reliable publication of successful human SCNT reported generation of a single cloned blastocyst (Stojkovic et al., 2005). Unfortunately, the dramatic advances in human SCNT reported by Hwang and colleagues in South Korea were largely a product of fraud (Cho et al., 2006). In human SCNT reports, left over eggs from IVF (*in vitro* fertilization) that failed to fertilize have been used, indicating poor egg quality. However, human SCNT using 29 donated eggs (oocytes) of good quality, and not leftovers from IVF, from three young women were reported to develop into cloned blastocysts, at a frequency as high as 23% (French et al., 2008). Theoretically, hESC lines can be derived *in vitro* from SCNT generated blastocysts. However, so far no established hESC line using the SCNT procedure has been reported. The shortage of donated high quality human eggs for research is a significant impediment for this field.

Other methods that have been used to elucidate the molecular mechanism of reprogramming are 2) fusion of somatic adult cells with pluripotent ESCs to generate hybrid cells or 3) cell extract from ESCs or ECs (Bhutani et al., 2010; Cowan et al., 2005; Freberg et al., 2007; Taranger et al., 2005; Yamanaka and Blau, 2010).

3. Molecular mechanisms of reprogramming

The mechanisms of nuclear reprogramming are not yet completely understood. The crucial event during reprogramming is the activation of ES- and the silencing of differentiation markers, while the genetic code remains intact. Major reprogramming of gene expression takes place inside the egg and genes that have been silenced during embryo development are awakened. In contrast, genes that are expressed in, and are specific for, the donated cell nucleus become inactivated most of the time, however some SCNT embryos remember their heritage and fail to inactivate somatic-specific genes (Ng and Gurdon, 2008). It has been reported that reprogramming involves changes in chromatin structure and chromatin components (Jullien et al., 2010; Kikyo et al., 2000). Importantly, initiation of Oct4 expression has been found to be crucial for successful nuclear transfers (Boiani et al., 2002; Byrne et al., 2003) and important for iPSC creation; all other reprogramming iPSC transcription factors have been replaced with other factors or chemical compounds, but only one report so far could exclude Oct4. In murine ES cells, Oct4 must hold a precise level to maintain them as just ES cells (Niwa et al., 2000) and therefore understanding the control of the Oct4 level will be key if one wants to understand pluripotency and reprogramming at the molecular level. A recent report demonstrated that Oct4 expression is regulated by scaffold attachment factor A (SAF-A). SAF-A was found on the Oct4 promoter only when the gene is actively transcribed in murine ESCs, depending on LIF, and gene silencing of SAF-A in ESCs resulted in down regulation of Oct4 (Vizlin-Hodzic et al., 2011). Other Oct4 modulators have been reported that in similarity with SAF-A are in complex with RNA polymerase II (Ding et al., 2009; Ponnusamy et al., 2009). Post-translational modifications have been shown to be able to modify the activity of Oct4, such as sumoylation (Wei et al., 2007) and ubiquitination (Xu et al., 2004). During the reprogramming process epigenetic marks are changed such as the removal of methyl groups on DNA (DNA demethylation) of the Oct4 promoter which has been shown during SCNT (Simonsson and Gurdon, 2004) and has also been observed in mouse (Yamazaki et al., 2006). The growth arrest and DNA damage inducible protein Gadd45a and deaminase Aid was shown to promote DNA demethylation of the Oct4 and Nanog promoters (Barreto et al., 2007; Bhutani et al., 2010). Consistent with those findings is that Aid together with Gadd45 and Mbd4 has been shown to promote DNA demethylation in zebrafish (Rai et al., 2008). Translational tumor protein (Tpt1) has been proposed to control Oct4 and shown to interact with nucleophosmin (Npm1) during mitosis of ESCs and such complexes are involved in cell proliferation (Johansson et al., 2010b; Koziol et al., 2007). Furthermore, phosphorylated nucleolin (Ncl-P) interacts with Oct4 during interphase in both murine and human ESCs (Johansson et al., 2010a). Core transcription factors, Oct4, Sox2 and Nanog, were shown to individually form complexes with nucleophosmin (Npm1) to control ESCs (Johansson and Simonsson, 2010). ESCs also display high levels of telomerase activity which maintain the length of the telomeres. The telomerase activity or Tert gene expression is rapidly down regulated during differentiation and are much lower or absent in somatic cells. Therefore, reestablishment of high telomerase activity (or reactivation of Tert gene) is important for reprogramming. In SCNT animals, telomere length in somatic cells has been reported to be comparable to that in normally fertilized animals (Betts et al., 2001; Lanza et al., 2000; Tian et al., 2000). A telomere length-resetting mechanism has been identified in the *Xenopus* egg (Vizlin-Hodzic et al., 2009).

When iPSCs first were introduced many thought that the molecular mechanism of reprogramming was solved once and for all. It was soon shown that to generate iPSC colonies one could use different combinations of transcription factors most efficiently together with retroviruses or more recently, exposure to chemical compounds together with the transcription factors, Oct4 and Klf4, and with retroviruses (Zhu et al., 2010) or retroviruses alone (Kane et al., 2010). What retroviruses do for the reprogramming process is unknown and the efficiency by which the egg reprograms the somatic cells is far more efficient than the iPSC procedure. Moreover, mutagenic effects have been documented in both laboratory and clinical gene therapy studies, principally as a result of a dysregulated host gene expression in the proximity of gene integration sites. So the first question to ask is whether all iPSC experiments so far forgot the obvious control of using only virus. The answer is probably no because the efficiency is very low with viruses alone as compared to using transcription factors combined with virus or identified reprogramming compounds. Reprogramming an adult somatic frog cell nucleus to generate a normal "clonal" new individual is far less efficient (0.1-3%) than reprogramming to create a blastocyst, from which ESCs are isolated (efficiency 20-40%) (Gurdon, 2008) and is comparable with blastula formation after human SCNT (23%). This number could be compared with iPSC procedure that has reported 0.5 % success rate at most with human cells (table 1). The low efficiency and slow kinetics of iPSC derivation suggest that there are other procedures that are more efficient, yet to decipher. There is a belief that there are different levels of pluripotency when it comes to ESC and also that reprogramming follows an organized sequence of events, beginning with downregulation of somatic markers and activation of pluripotency markers alkaline phosphatase, SSEA-4, and Fbxo15 before pluripotency endogenous genes such as Oct4, Nanog, Tra1-60 and Tra-1-80 become expressed and cells gain independence from exogenous transcription factor expression (Brambrink et al., 2008; Stadtfeld et al., 2008a). Only a small subset of somatic cells expressing the reprogramming factors down-regulates somatic markers and activates pluripotency genes (Wernig et al., 2008a).

3.1 History of reprogramming

SCNT has been around for more than fifty years although it was already proposed in 1938 by Hans Spemann (Spemann, 1938), an embryologist who received the Nobel Prize in Medicine for his development of new embryological micro surgery techniques. Spemann anticipated that "transplanting an older nucleus into an egg would be a fantastic experiment". Later on, Robert Briggs and Thomas King were the first to put the nuclear transfer technique into practice. However, they only managed to obtain viable offspring through nuclear transfer of undifferentiated cells in the frog species *Rana pipiens* (Briggs and King, 1952). During the 1950s to the 1970s a series of pioneering somatic nuclear transfer experiments performed by John Gurdon showed that nuclei from differentiated amphibian cells, for example tadpole intestinal or adult skin cells could generate cloned tadpoles (Gurdon, 1962; Gurdon et al., 1958; Gurdon et al., 1975). In 1997, the successful cloning of a mammal was first achieved. The sheep Dolly was produced by using the nuclei of cells cultured from an adult mammary gland (Wilmut et al., 1997). Following the cloning of Dolly, researchers have reported successful cloning of a number of species including cow, pig, mouse, rabbit, cat (named Copycat) and monkey. In 2006, reprogrammed murine iPSCs were reported by Takahashi and Yamanaka (Takahashi and Yamanaka, 2006) and in 2007 human iPSCs were reported (Takahashi et al., 2007; Yu et al., 2009).

4. Producing iPSCs from other cell types than fibroblasts

The most studied somatic cell type that has been reprogrammed into iPSCs is fibroblasts. The different human somatic cell types that have been transformed into iPSCs so far are summarized in table 2. The efficiency of fibroblast reprogramming does not exceed 1-5% but generally is extremely inefficient (0.001-0.1%) and occurs at a slow speed (> 2 weeks). In order to use iPSCs in clinical applications, improved efficiency, suitable factor delivery techniques and identification of true reprogrammed cells are crucial. In the fast growing field of regenerative medicine, patient-specific iPSCs offer a unique source of autologous cells for clinical applications. Although promising, using somatic cells of an adult individual as starting material for reprogramming in this context has also raised concern. Acquired somatic mutations that have been accumulated during an individual's life time will be transferred to the iPSCs, and there is a fear that these mutations may be associated with adverse events such as cancer development. As an alternative, iPSCs have been generated from human cord blood. These cells have been shown to differentiate into all three germ layers including spontaneous beating cardiomyocytes (Haase et al., 2009). Reprogrammed cells from cord blood have not only the advantage to come from a juvenescent cell source. In addition, cord blood is already routinely harvested for clinical use.

Another issue that has been raised in this field is a wish to harvest cells for reprogramming without surgical intervention. Therefore, reprogramming experiments have also been performed using plucked human hair follicle keratinocytes. These iPSCs were also able to differentiate into cells from all three germ layers including cardiomyocytes (Novak et al., 2010).

Human Origin Somatic Cell type	Efficiency	Reprogramming Factors	Reference
Fibroblasts	0.02%	OKSM	(Takahashi et al., 2007)
	0.02%	OSLN	(Yu et al., 2007)
	0.002%	OKS	(Nakagawa et al., 2008)
Hepatocytes	0.1%	OKSM	(Liu et al., 2010)
Keratinocytes	ND	OKSM	(Aasen et al., 2008)
	ND	OKS	(Aasen et al., 2008)
Neural stem cells	<0.004%	O	(Kim et al., 2008)
Amniotic cells	0.05-1.5%	OKSM	(Li et al., 2009)
	0.1%	OSN	(Zhao et al., 2010)
Adipose-derived stem cells	0.5%	OKSM	(Sugii et al., 2010)
	<0.1%	OKS	(Aoki et al., 2010)
Cord blood stem cells	ND	OKSM	(Eminli et al., 2009)
	<0.01%	OS	(Giorgetti et al., 2009)
Cord blood endothelial cells	<0.01%	OSLN	(Haase et al., 2009)
Mobilized peripheral blood	0.01%	OKSM	(Loh et al., 2009)

Table 2. Different somatic cell types that human iPSCs have been generated from

4.1 iPSC as a disease model

The introduction of iPSC technology holds a great promise for disease modelling. By differentiating iPSCs from patients into various cell lineages there is hope to be able to follow the disease progression and to identify new prognostic markers as well as to use the differentiated cells for drug screening in both toxicological testing and the development of new treatment. This approach has already been tested for monogenic diseases using genetically modified hESCs or hESCs from embryos carrying these diseases (reviewed in (Stephenson et al., 2009)). However, diseases with a more complex genetic background involving several or unknown genes have not been able to be studied in this way before iPSCs became available. An additional advantage with iPSCs is that since many diseases differ in both clinical symptoms and penetrance between patients, iPSCs derived from patients will offer the opportunity to reveal a clinical history as well. It could also provide a model for late-onset degenerative diseases such as Alzheimer's disease or osteoarthritis.

Recent work on cardiac arrhythmias has fully shown the potential of disease modelling using iPSCs. Long QT syndrome (LQTS) is characterized by rapid irregular heart beats due to abnormal ion channel function and the condition can lead to sudden death. So far, various mutations in at least 12 different genes have been associated with LQTS and the disease is subdivided into different types depending on which gene is affected (reviewed in (Bokil et al., 2010)). Fibroblasts from patients with LQTS1 (Moretti et al., 2010) and LQTS2 (Itzhaki et al., 2011; Matsa et al., 2011) were reprogrammed and differentiated into the cardiac lineage. These cells displayed the electrophysiological pattern characteristic to the disease. Moreover, the cells responded appropriately when treated with pharmacological compounds, which further extends the usability of these cells.

iPSCs have also been generated from fibroblasts from patients suffering from the LEOPARD syndrome, an autosomal-dominant developmental disorder where one of the major disease phenotypes includes hyperthropic cardiomyopathy. The authors showed that cardiomyocytes derived from those iPSCs were larger with another intracellular organization compared to cardiomyocytes derived from hESCs or iPSCs generated from a healthy sibling (Carvajal-Vergara et al., 2010). Today many laboratories and hospitals worldwide are producing iPSC lines from patients with various diseases. Patient-specific iPSC lines can be used as 1) a human modelling system for studying the molecular cause of, and in the long run for 2) the treatment of, degenerative diseases with autologous transplantation, which refers to the transplantation to a patient of his/her own cells. The therapeutic potential of iPSCs in combination with genetic repair has already been successfully shown in mouse models of sickle cell anemia (Hanna et al., 2007), Duchenne muscular dystrophy (DMD) (Kazuki et al., 2010), hemophilia A (Xu et al., 2009) and, in a rat model, Parkinson's disease (Wernig et al., 2008c). For diseases where animal and human physiology differ, disease-specific iPSC lines capable of differentiation into the tissue affected by the disease could recapitulate tissue formation and thereby enable determination of the cause of the disease and could provide cues to drug targets. Therefore iPSC lines from patients suffering from a variety of genetic diseases with either Mendelian or complex inheritance have been secured for future research, and include deaminase deficiency-related severe combined immunodeficiency (ADA-SCID), Shwachman-Bodian-Diamond syndrome (SBDS), Gaucher disease (GD) type III, Duchenne (DMD) and Becker muscular dystrophy (BMD), Parkinson disease (PD), Huntington disease (HD), juvenile-onset (type1) diabetes mellitus (JDM), Downs syndrome (DS)/trisomy21 and Lesch-Nyhan syndrome (Park et al., 2008a). Furthermore, iPSCs derived from amyotrophic lateral sclerosis (ALS) patients were terminally differentiated into motor neurons (Dimos et al., 2008).

4.2 Procedures to produce iPSCs

In the first iPSC reprogramming studies, retroviral or lentiviral vectors were used to introduce the transcription factors into somatic cells. By using these viral delivery systems,

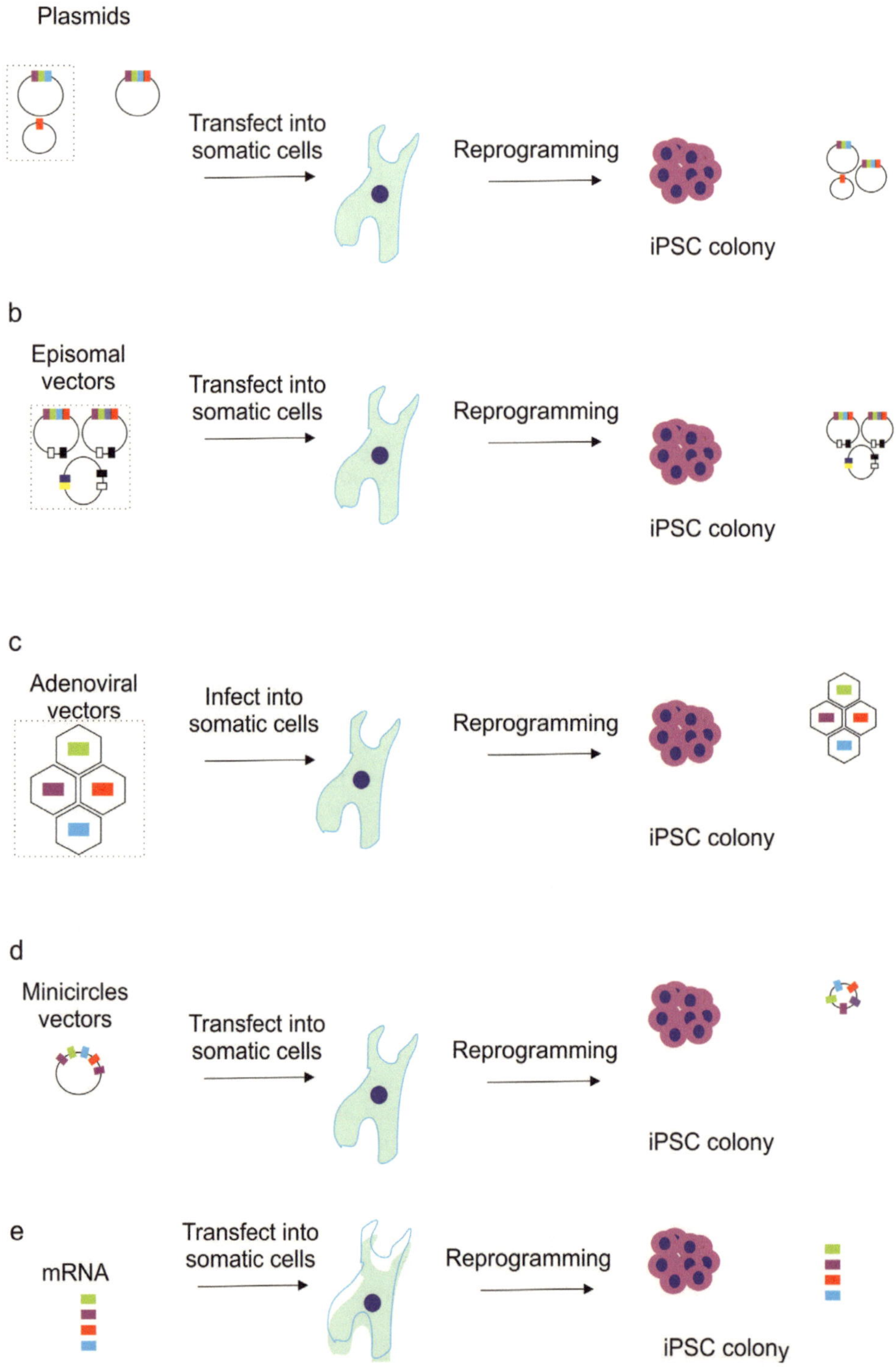

Fig. 2. Methods for producing induced pluripotent stem cells (iPSCs) by non-integrating vectors. Several different methods exist to generate iPSCs by non-integrating vectors: for

example by plasmid, episomal, adenoviral minicircle vectors and mRNA. a) A combination of expression plasmid vectors for defined reprogramming factors is transfected into somatic cells. Plasmid vectors are not integrated into the genome of transfected cells and are gradually lost during reprogramming. This method therefore requires multiple transfection steps. b) Somatic cells can be transfected by episomal vectors expressing defined reprogramming factors. These vectors can replicate themselves autonomously in cells during reprogramming under drug selection and are not integrated into the genome. Upon withdrawal of drug selection, the episomal vectors are lost. c) Adenovirus carrying defined reprogramming factors can be infected into somatic cells to transiently express these factors. This method requires multiple transductions since adenoviral vectors are lost upon celldivision. d) The minicircle vector method is based on PhiC31-vector intra molecular recombinant system that allows the bacterial elements of the vector to be degraded in bacteria. Minicircle vector containing only defined reprogramming factors is not degraded and is delivered into somatic cells by nucleofection. This strategy requires multiple transfection steps since minicircle vectors are lost upon cell division. e) Reprogramming using mRNA reprogramming factors have been achieved.

the transduced viral vectors and transgenes are randomly and permanently integrated into the genome of infected somatic cells and remains in the iPSCs. The vector integration into the host genome is a limitation of this technology if it is going to be used in human therapeutic applications due to increased risk of tumor formation (Okita et al., 2007). Approaches to derive transgene-free iPSCs are therefore critical. The first strategy was by using non-integrating (Figure 2) vectors. Efforts have been made to derive iPSCs by repeated plasmid transfections (Gonzalez et al., 2009; Okita et al., 2008) (Figure 2a), adenoviral (Stadtfeld et al., 2008b) (Figure 2b) and episomal vectors (Yu et al., 2009) (Figure 2c). Recently, minicircle vectors (Figure 2d) have been used to generate iPSCs (Jia et al., 2010). Unfortunately, reprogramming with these techniques has extremely low efficiency as compared to integrating viral vectors. Another promising alternative is the use of excisable integrating vectors, allowing for the generation of transgene-free iPSCs. A classical expression-excision system uses vectors with inserts flanked with recognition sites, loxP sites, for Cre-recombinase (Figure 3a). Consequently, DNA is excised upon Cre-recombinase expression in the cells. Cre-loxP-based approaches have been used to reprogram human somatic cells from individuals with Parkinson's disease by four different vectors (Soldner et al., 2009) or by a single, polycistronic lentiviral vector encoding reprogramming factors (Chang et al., 2009). Though, a potential limitation of Cre-loxP-based approaches is that a long terminal repeat (LTR) will remain after Cre-mediated excision which may interfere with the expression of endogenous genes. An alternative integration-free strategy is based on the piggy-Bac transposon (Figure 3b), a mobile genetic element from insects that integrates into the genome of mammalian cells and, most importantly, can be entirely removed by a transposase. Two research teams generated iPSCs using this system to deliver a single polycistron encoding four reprogramming factors into somatic cells (Woltjen et al., 2009; Yusa et al., 2009). Interestingly, the latest development indicates that gene transfection may not even be needed for the generation of iPSCs and that direct delivery of four recombinant reprogramming proteins that can penetrate the plasma membrane of somatic cells is sufficient (Zhou et al., 2009), or mRNA (Angel &Yanik, 2010; Plews et al., 2010; Warren et al. 2010; Yakoba et al., 2010; Zhou et al.,2009).

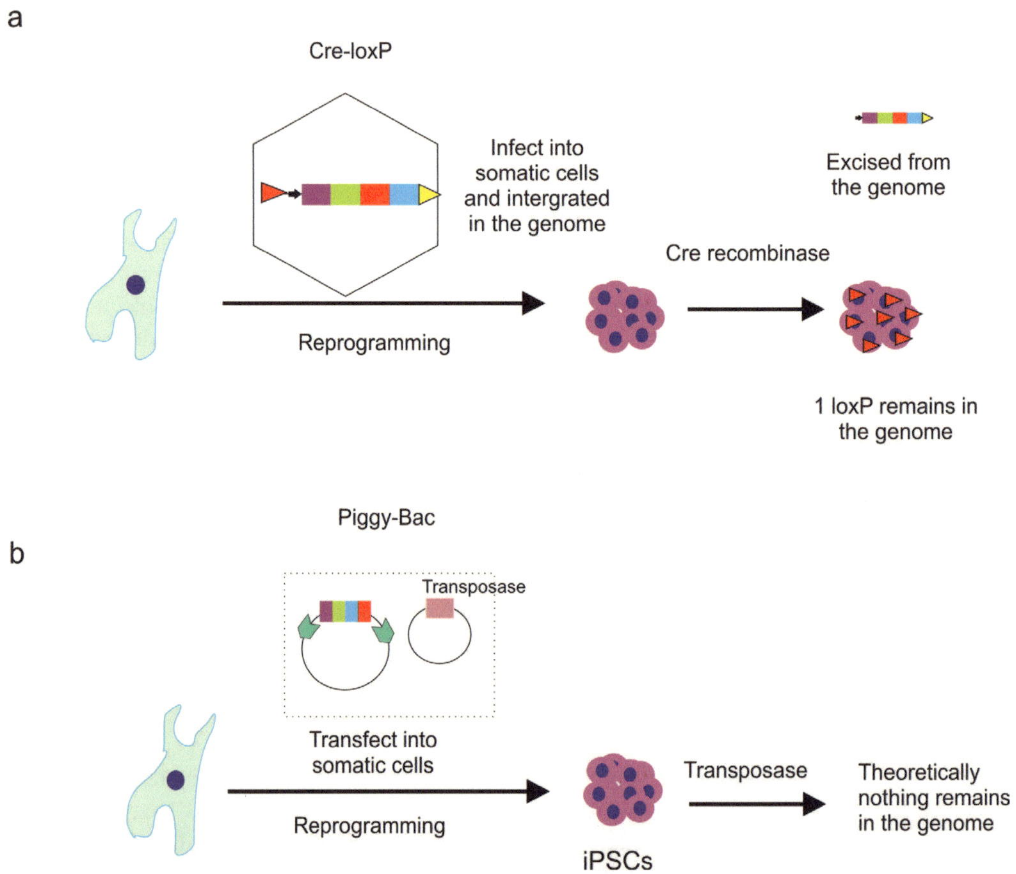

Fig. 3. Methods for production of induced pluripotent stem cells (iPSCs) by excisable integrating vectors. Two different methods exist today to generate iPSCs by excisable integrating vectors: by Cre-loxP and Piggy-Bac vectors. a) In the Cre-loxP viral delivery system, defined reprogramming factors are cloned into vectors flanked by recognition sites, loxP sites, for Cre-recombinase. Upon transduction into somatic cells, the loxP site is duplicated and reprogramming factos are stably integrated into the genome flanked by loxP sites. When Cre-recombinase is expressed, the integrated reprogramming factors are excised from the genome but one loxP site is left behind integrated into the genome of iPSCs. b) The Piggy-Bac transposon gene delivery system is based on a mobile genetic element that efficiently integrates into the genome of mammalian cells. When fusion gene encoding defined reprogramming factors in the transposon expression vector as well as transposase expression vector are transfected into somatic cells, the fusion gene is stably integrated into the genome. When transposase is expressed, the interated genetic material is excised from the genome resulting in transgene- and vector free iPSCs.

The therapeutic application of iPSCs is limited by another concern due to the use of potential oncogenes when iPSCs are produced. C-Myc is an oncogene and as such causes tumor formation, which has been observed in iPSC-derived chimeric mice (Okita et al., 2007). As a major step towards solving this issue, several studies have demonstrated that mouse and human iPSCs can be derived without C-Myc but the efficiency of reprogramming is reduced (Nakagawa et al., 2008; Wernig et al., 2008b; Yu et al., 2007). Although the oncogenic potential of C-Myc is mostly discussed, Oct4, Sox2 and Klf4 are also associated with multiple types of cancer (Bass et al., 2009; Gidekel et al., 2003; Wei et al., 2006). To circumvent this problem, a recent trend is to avoid the transduction of some of the oncogenes by 1) reprogramming somatic cells which already endogenously express sufficient levels of some of the reprogramming factors (Tsai et al., 2010), 2) replacing one or more reprogramming factors by small molecules like histone deacetylase inhibitor vaporic acid, the DNA methyltransferase inhibitor 5-aza-cytidine, the Wnt signaling component WNT3a, the L-channel calcium channel agonist Bayk8644 (Huangfu et al., 2008a; Huangfu et al., 2008b), or 3) dual inhibition of mitogen activated protein kinase signaling and glycogen synthase kinase-3 (Silva et al., 2008). It has been reported that Sox2 can be replaced by Sox1, Klf4 by Klf2 and c-Myc by N-myc or L-myc indicating that they are not fundamentally required for generation of iPSCs (Yamanaka, 2009). Tet-on™ technology has been used to express exogenously reprogramming factors in presence of Doxycycline. Removal of Doxycycline results in that iPSC colonies that endogenously express pluripotent genes and colonies that are truly reprogrammed remains.

5. Transplanting cells

In order to make cell therapy (route 1 in Figure 1) using iPSCs a reality in medicine many obstacles need to be overcome. Organ transplantation between individuals is complicated due to the limited availability of matched tissues and consequently the requirement for life-long treatment with immunosuppressive drugs that can cause serious side effects. The hope is that iPSCs that are already genetically matched with the patient would circumvent these issues. Another advantage of iPSCs over current transplantation approaches is the opportunity of repairing mutations that cause the disease by homologous recombination, which has not been very successful in adult stem cells due to difficulties in propagating those cells *in vitro*. In mouse, iPSC technology combined with correction of a known disease-causing mutation has been proven successful. In human autologous cell therapy has been used since the mid 90´s for the treatment of focal cartilage lesions, using the patient's chondrocytes transplanted into the injured knee (Brittberg et al., 1994), thereby alleviated osteoarthritic symptoms and induction of tissue repair. The cell therapy gives stable long-term results up to 20 years after surgery in some patients but is less successful in others (Lindahl et al., 2003; Peterson et al., 2010). One drawback with this technique is the supply of cells. Large injuries require large amounts of cells, and there is a limit of the size of the biopsies that can be taken out from the patient. Introducing the iPSC technique in such system might improve the process. Since the iPSCs have theoretically an unlimited proliferation capacity, these cells can be used to reach larger quantities of cells. When sufficient numbers have been produced, the iPSCs are differentiated into chondrocytes and transplanted to the lesion. In this case, no biopsy would need to be harvested, since iPSCs can be made from a regular skin fibroblast. Before this somewhat futuristic scenario can come true, rigorous characterization of the iPSC is needed, since these cells, as all stem cells, can form teratoma *in vivo* (Fairchild, 2010). The iPSCs have however, been shown to retain their epigenic memory from the tissue from which they originate. It would therefore be easier to differentiate an iPSC to a chondrocyte if the donor cell was a chondrocyte (Kim et al., 2010), and maybe terminally so, thus avoiding risk for terratoma formation. A biopsy would thus be needed, but a relatively small cell harvest could with the iPSC technique result in the treatment of larger injuries. The iPSC procedure could also lead to a therapy-outcome that is more predicted and constant due to that chondrogenic differentiation of iPSC probably result in a more homogeneous cell-population. Since cartilage lacks vascularisation and thus is immunoprivileged the derivation of a universal donor chondrocytes cell line based on the iPSC technology could be an interesting option. If such cells are combined with a suitable matrix scaffold a cartilage regeneration therapy could potentially have a much wider application and be more cost effective than current autologous procedures.

5.1 Directprogramming of somatic cells into another cell type

Switching from one somatic cell type into another cell type, not necessarily via a pluripotent cell state was first demonstrated when fibroblasts formed myofibers after transduction with retroviral vectors expressing the skeletal muscle factor MyoD (Davis et al., 1987). Further, it has been reported that pancreatic acinar cells could be transformed into insulin-producing β cells by overexpression of the pancreatic factors Pdx1, MafA and Ngn3 *in vivo* (Zhou et al., 2008) as well as that ESCs could be directly differentiated into specific dopamine neurons by overexpression of only one factor, Lmx1 (Friling et al., 2009). These experiments proved that transdifferentiation do not require reprogramming into a pluripotent state, although all such experiments have used some kind of retroviruses and if only virus in itself can contribute to pluripotency as has recently been shown one cannot completely rule out that the switch hasn't passed via a pluripotent state.

6. Final remarks

To date, clinically valid iPSCs do not yet exist, but are under development worldwide. Some will argue that the complexity of reprogramming is solved by the iPSC technology, however apart from the defined reprogramming factors, retroviruses help in the reprogramming process in an unknown way, and is still inefficient compared to SCNT which argues for that more can be learnt about reprogramming. Also the fact that different combinations of reprogramming factors, or replacement with chemicals, have been used successfully indicates that there exist reprogramming molecules yet to be discovered. Therefore, further investigations are needed to learn more about the molecular mechanisms of iPSCs and how to prevent tumor formation following *in vivo* transplantation. Awaiting *in vivo* safety, these techniques offer exciting possibilities for mapping mechanisms of different diseases and screening for patient-specific therapies and drugs. To derive iPSCs from the patient's own cells following differentiation into the disease-causing cells means recapitulating the disease in a test tube for genomic, proteomic and epigenomic analysis. The iPSC as a human *in vitro* disease modeling system is a new promising and fast expanding research area.

7. References

Aasen, T., Raya, A., Barrero, M.J., Garreta, E., Consiglio, A., Gonzalez, F., Vassena, R., Bilic, J., Pekarik, V., Tiscornia, G., *et al.* (2008). Efficient and rapid generation of induced pluripotent stem cells from human keratinocytes. Nat Biotechnol *26*, 1276-1284.

Angel, M. & Yanik, M. F. (2010). Innate immune suppression enables frequent transfection with RNA encoding reprogramming proteins. *PLoS One* 5, e11756.

Aoki, T., Ohnishi, H., Oda, Y., Tadokoro, M., Sasao, M., Kato, H., Hattori, K., and Ohgushi, H. (2010). Generation of induced pluripotent stem cells from human adipose-derived stem cells without c-MYC. Tissue Eng Part A *16*, 2197-2206.

Barreto, G., Schafer, A., Marhold, J., Stach, D., Swaminathan, S.K., Handa, V., Doderlein, G., Maltry, N., Wu, W., Lyko, F., *et al.* (2007). Gadd45a promotes epigenetic gene activation by repair-mediated DNA demethylation. Nature *445*, 671-675.

Bass, A.J., Watanabe, H., Mermel, C.H., Yu, S., Perner, S., Verhaak, R.G., Kim, S.Y., Wardwell, L., Tamayo, P., Gat-Viks, I., *et al.* (2009). SOX2 is an amplified lineage-survival oncogene in lung and esophageal squamous cell carcinomas. Nat Genet *41*, 1238-1242.

Betts, D., Bordignon, V., Hill, J., Winger, Q., Westhusin, M., Smith, L., and King, W. (2001). Reprogramming of telomerase activity and rebuilding of telomere length in cloned cattle. Proc Natl Acad Sci U S A *98*, 1077-1082.

Bhutani, N., Brady, J.J., Damian, M., Sacco, A., Corbel, S.Y., and Blau, H.M. (2010). Reprogramming towards pluripotency requires AID-dependent DNA demethylation. Nature *463*, 1042-1047.

Boiani, M., Eckardt, S., Scholer, H.R., and McLaughlin, K.J. (2002). Oct4 distribution and level in mouse clones: consequences for pluripotency. Genes Dev *16*, 1209-1219.

Bokil, N.J., Baisden, J.M., Radford, D.J., and Summers, K.M. (2010). Molecular genetics of long QT syndrome. Mol Genet Metab *101*, 1-8.

Brambrink, T., Foreman, R., Welstead, G.G., Lengner, C.J., Wernig, M., Suh, H., and Jaenisch, R. (2008). Sequential expression of pluripotency markers during direct reprogramming of mouse somatic cells. Cell Stem Cell *2*, 151-159.

Briggs, R., and King, T.J. (1952). Transplantation of Living Nuclei From Blastula Cells into Enucleated Frogs' Eggs. Proc Natl Acad Sci U S A *38*, 455-463.

Brittberg, M., Lindahl, A., Nilsson, A., Ohlsson, C., Isaksson, O., and Peterson, L. (1994). Treatment of deep cartilage defects in the knee with autologous chondrocyte transplantation. N Engl J Med *331*, 889-895.

Byrne, J.A., Pedersen, D.A., Clepper, L.L., Nelson, M., Sanger, W.G., Gokhale, S., Wolf, D.P., and Mitalipov, S.M. (2007). Producing primate embryonic stem cells by somatic cell nuclear transfer. Nature *450*, 497-502.

Byrne, J.A., Simonsson, S., Western, P.S., and Gurdon, J.B. (2003). Nuclei of adult mammalian somatic cells are directly reprogrammed to oct-4 stem cell gene expression by amphibian oocytes. Curr Biol *13*, 1206-1213.

Carvajal-Vergara, X., Sevilla, A., D'Souza, S.L., Ang, Y.S., Schaniel, C., Lee, D.F., Yang, L., Kaplan, A.D., Adler, E.D., Rozov, R., *et al.* (2010). Patient-specific induced pluripotent stem-cell-derived models of LEOPARD syndrome. Nature *465*, 808-812.

Chang, C.W., Lai, Y.S., Pawlik, K.M., Liu, K., Sun, C.W., Li, C., Schoeb, T.R., and Townes, T.M. (2009). Polycistronic lentiviral vector for "hit and run" reprogramming of adult skin fibroblasts to induced pluripotent stem cells. Stem Cells *27*, 1042-1049.

Cho, M.K., McGee, G., and Magnus, D. (2006). Research conduct. Lessons of the stem cell scandal. Science *311*, 614-615.

Cowan, C.A., Atienza, J., Melton, D.A., and Eggan, K. (2005). Nuclear reprogramming of somatic cells after fusion with human embryonic stem cells. Science *309*, 1369-1373.

Davis, R.L., Weintraub, H., and Lassar, A.B. (1987). Expression of a single transfected cDNA converts fibroblasts to myoblasts. Cell *51*, 987-1000.

Dimos, J.T., Rodolfa, K.T., Niakan, K.K., Weisenthal, L.M., Mitsumoto, H., Chung, W., Croft, G.F., Saphier, G., Leibel, R., Goland, R., *et al.* (2008). Induced pluripotent stem cells generated from patients with ALS can be differentiated into motor neurons. Science *321*, 1218-1221.

Ding, L., Paszkowski-Rogacz, M., Nitzsche, A., Slabicki, M.M., Heninger, A.K., de Vries, I., Kittler, R., Junqueira, M., Shevchenko, A., Schulz, H., *et al.* (2009). A genome-scale RNAi screen for Oct4 modulators defines a role of the Paf1 complex for embryonic stem cell identity. Cell Stem Cell *4*, 403-415.

Eminli, S., Foudi, A., Stadtfeld, M., Maherali, N., Ahfeldt, T., Mostoslavsky, G., Hock, H., and Hochedlinger, K. (2009). Differentiation stage determines potential of hematopoietic cells for reprogramming into induced pluripotent stem cells. Nat Genet *41*, 968-976.

Fairchild, P.J. (2010). The challenge of immunogenicity in the quest for induced pluripotency. Nat Rev Immunol *10*, 868-875.

Freberg, C.T., Dahl, J.A., Timoskainen, S., and Collas, P. (2007). Epigenetic reprogramming of OCT4 and NANOG regulatory regions by embryonal carcinoma cell extract. Mol Biol Cell *18*, 1543-1553.

French, A.J., Adams, C.A., Anderson, L.S., Kitchen, J.R., Hughes, M.R., and Wood, S.H. (2008). Development of human cloned blastocysts following somatic cell nuclear transfer with adult fibroblasts. Stem Cells *26*, 485-493.

Friling, S., Andersson, E., Thompson, L.H., Jonsson, M.E., Hebsgaard, J.B., Nanou, E., Alekseenko, Z., Marklund, U., Kjellander, S., Volakakis, N., *et al.* (2009). Efficient production of mesencephalic dopamine neurons by Lmx1a expression in embryonic stem cells. Proc Natl Acad Sci U S A *106*, 7613-7618.

Gidekel, S., Pizov, G., Bergman, Y., and Pikarsky, E. (2003). Oct-3/4 is a dose-dependent oncogenic fate determinant. Cancer Cell *4*, 361-370.

Giorgetti, A., Montserrat, N., Aasen, T., Gonzalez, F., Rodriguez-Piza, I., Vassena, R., Raya, A., Boue, S., Barrero, M.J., Corbella, B.A., *et al.* (2009). Generation of induced pluripotent stem cells from human cord blood using OCT4 and SOX2. Cell Stem Cell *5*, 353-357.

Gonzalez, F., Barragan Monasterio, M., Tiscornia, G., Montserrat Pulido, N., Vassena, R., Batlle Morera, L., Rodriguez Piza, I., and Izpisua Belmonte, J.C. (2009). Generation of mouse-induced pluripotent stem cells by transient expression of a single nonviral polycistronic vector. Proc Natl Acad Sci U S A *106*, 8918-8922.

Gupta, M.K., Illich, D.J., Gaarz, A., Matzkies, M., Nguemo, F., Pfannkuche, K., Liang, H., Classen, S., Reppel, M., Schultze, J.L., *et al.* (2010). Global transcriptional profiles of beating clusters derived from human induced pluripotent stem cells and embryonic stem cells are highly similar. BMC Dev Biol *10*, 98.

Gurdon, J. (2008). Primate therapeutic cloning in practice. Nat Biotechnol *26*, 64-65.

Gurdon, J.B. (1962). The developmental capacity of nuclei taken from intestinal epithelium cells of feeding tadpoles. J Embryol Exp Morphol *10*, 622-640.

Gurdon, J.B., Elsdale, T.R., and Fischberg, M. (1958). Sexually mature individuals of Xenopus laevis from the transplantation of single somatic nuclei. Nature *182*, 64-65.

Gurdon, J.B., Laskey, R.A., and Reeves, O.R. (1975). The developmental capacity of nuclei transplanted from keratinized skin cells of adult frogs. J Embryol Exp Morphol *34*, 93-112.

Haase, A., Olmer, R., Schwanke, K., Wunderlich, S., Merkert, S., Hess, C., Zweigerdt, R., Gruh, I., Meyer, J., Wagner, S., *et al.* (2009). Generation of induced pluripotent stem cells from human cord blood. Cell Stem Cell *5*, 434-441.

Hanna, J., Wernig, M., Markoulaki, S., Sun, C.W., Meissner, A., Cassady, J.P., Beard, C., Brambrink, T., Wu, L.C., Townes, T.M., *et al.* (2007). Treatment of sickle cell anemia mouse model with iPS cells generated from autologous skin. Science *318*, 1920-1923.

Heng, J.C., Feng, B., Han, J., Jiang, J., Kraus, P., Ng, J.H., Orlov, Y.L., Huss, M., Yang, L., Lufkin, T., *et al.* (2010). The nuclear receptor Nr5a2 can replace Oct4 in the reprogramming of murine somatic cells to pluripotent cells. Cell Stem Cell *6*, 167-174.

Huangfu, D., Maehr, R., Guo, W., Eijkelenboom, A., Snitow, M., Chen, A.E., and Melton, D.A. (2008a). Induction of pluripotent stem cells by defined factors is greatly improved by small-molecule compounds. Nat Biotechnol *26*, 795-797.

Huangfu, D., Osafune, K., Maehr, R., Guo, W., Eijkelenboom, A., Chen, S., Muhlestein, W., and Melton, D.A. (2008b). Induction of pluripotent stem cells from primary human fibroblasts with only Oct4 and Sox2. Nat Biotechnol *26*, 1269-1275.

Itzhaki, I., Maizels, L., Huber, I., Zwi-Dantsis, L., Caspi, O., Winterstern, A., Feldman, O., Gepstein, A., Arbel, G., Hammerman, H., *et al.* (2011). Modelling the long QT syndrome with induced pluripotent stem cells. Nature *471*, 225-229.

Jia, F., Wilson, K.D., Sun, N., Gupta, D.M., Huang, M., Li, Z., Panetta, N.J., Chen, Z.Y., Robbins, R.C., Kay, M.A., *et al.* (2010). A nonviral minicircle vector for deriving human iPS cells. Nat Methods *7*, 197-199.

Johansson, H., and Simonsson, S. (2010). Core transcription factors, Oct4, Sox2 and Nanog, individually form complexes with nucleophosmin (Npm1) to control embryonic stem (ES) cell fate determination. Aging (Albany NY) *2*, 815-822.

Johansson, H., Svensson, F., Runnberg, R., Simonsson, T., and Simonsson, S. (2010a). Phosphorylated nucleolin interacts with translationally controlled tumor protein during mitosis and with Oct4 during interphase in ES cells. PLoS One *5*, e13678.

Johansson, H., Vizlin-Hodzic, D., Simonsson, T., and Simonsson, S. (2010b). Translationally controlled tumor protein interacts with nucleophosmin during mitosis in ES cells. Cell Cycle *9*.

Jullien, J., Astrand, C., Halley-Stott, R.P., Garrett, N., and Gurdon, J.B. (2010). Characterization of somatic cell nuclear reprogramming by oocytes in which a linker histone is required for pluripotency gene reactivation. Proc Natl Acad Sci U S A *107*, 5483-5488.

Kane, N.M., Nowrouzi, A., Mukherjee, S., Blundell, M.P., Greig, J.A., Lee, W.K., Houslay, M.D., Milligan, G., Mountford, J.C., von Kalle, C., *et al.* (2010). Lentivirus-mediated reprogramming of somatic cells in the absence of transgenic transcription factors. Mol Ther *18*, 2139-2145.

Kazuki, Y., Hiratsuka, M., Takiguchi, M., Osaki, M., Kajitani, N., Hoshiya, H., Hiramatsu, K., Yoshino, T., Kazuki, K., Ishihara, C., *et al.* (2010). Complete genetic correction of ips cells from Duchenne muscular dystrophy. Mol Ther *18*, 386-393.

Kehat, I., Kenyagin-Karsenti, D., Snir, M., Segev, H., Amit, M., Gepstein, A., Livne, E., Binah, O., Itskovitz-Eldor, J., and Gepstein, L. (2001). Human embryonic stem cells can differentiate into myocytes with structural and functional properties of cardiomyocytes. J Clin Invest *108*, 407-414.

Kikyo, N., Wade, P.A., Guschin, D., Ge, H., and Wolffe, A.P. (2000). Active remodeling of somatic nuclei in egg cytoplasm by the nucleosomal ATPase ISWI. Science *289*, 2360-2362.

Kim, J.B., Zaehres, H., Wu, G., Gentile, L., Ko, K., Sebastiano, V., Arauzo-Bravo, M.J., Ruau, D., Han, D.W., Zenke, M., *et al.* (2008). Pluripotent stem cells induced from adult neural stem cells by reprogramming with two factors. Nature *454*, 646-650.

Kim, K., Doi, A., Wen, B., Ng, K., Zhao, R., Cahan, P., Kim, J., Aryee, M.J., Ji, H., Ehrlich, L.I., *et al.* (2010). Epigenetic memory in induced pluripotent stem cells. Nature *467*, 285-290.

Koziol, M.J., Garrett, N., and Gurdon, J.B. (2007). Tpt1 activates transcription of oct4 and nanog in transplanted somatic nuclei. Curr Biol *17*, 801-807.

Laflamme, M.A., Chen, K.Y., Naumova, A.V., Muskheli, V., Fugate, J.A., Dupras, S.K., Reinecke, H., Xu, C., Hassanipour, M., Police, S., *et al.* (2007). Cardiomyocytes derived from human embryonic stem cells in pro-survival factors enhance function of infarcted rat hearts. Nat Biotechnol *25*, 1015-1024.

Lanza, R.P., Cibelli, J.B., Blackwell, C., Cristofalo, V.J., Francis, M.K., Baerlocher, G.M., Mak, J., Schertzer, M., Chavez, E.A., Sawyer, N., *et al.* (2000). Extension of cell life-span and telomere length in animals cloned from senescent somatic cells. Science *288*, 665-669.

Li, C., Zhou, J., Shi, G., Ma, Y., Yang, Y., Gu, J., Yu, H., Jin, S., Wei, Z., Chen, F., *et al.* (2009). Pluripotency can be rapidly and efficiently induced in human amniotic fluid-derived cells. Hum Mol Genet *18*, 4340-4349.

Lindahl, A., Brittberg, M., and Peterson, L. (2003). Cartilage repair with chondrocytes: clinical and cellular aspects. Novartis Found Symp *249*, 175-186; discussion 186-179, 234-178, 239-141.

Liu, H., Ye, Z., Kim, Y., Sharkis, S., and Jang, Y.Y. (2010). Generation of endoderm-derived human induced pluripotent stem cells from primary hepatocytes. Hepatology *51*, 1810-1819.

Loh, Y.H., Agarwal, S., Park, I.H., Urbach, A., Huo, H., Heffner, G.C., Kim, K., Miller, J.D., Ng, K., and Daley, G.Q. (2009). Generation of induced pluripotent stem cells from human blood. Blood *113*, 5476-5479.

Lowry, W.E., Richter, L., Yachechko, R., Pyle, A.D., Tchieu, J., Sridharan, R., Clark, A.T., and Plath, K. (2008). Generation of human induced pluripotent stem cells from dermal fibroblasts. Proc Natl Acad Sci U S A *105*, 2883-2888.

Martinez-Fernandez, A., Nelson, T.J., Ikeda, Y., and Terzic, A. (2010). c-MYC independent nuclear reprogramming favors cardiogenic potential of induced pluripotent stem cells. J Cardiovasc Transl Res *3*, 13-23.

Matsa, E., Rajamohan, D., Dick, E., Young, L., Mellor, I., Staniforth, A., and Denning, C. (2011). Drug evaluation in cardiomyocytes derived from human induced pluripotent stem cells carrying a long QT syndrome type 2 mutation. Eur Heart J *32*, 952-962.

Meissner, A., Wernig, M., and Jaenisch, R. (2007). Direct reprogramming of genetically unmodified fibroblasts into pluripotent stem cells. Nat Biotechnol *25*, 1177-1181.

Moretti, A., Bellin, M., Welling, A., Jung, C.B., Lam, J.T., Bott-Flugel, L., Dorn, T., Goedel, A., Hohnke, C., Hofmann, F., *et al.* (2010). Patient-specific induced pluripotent stem-cell models for long-QT syndrome. N Engl J Med *363*, 1397-1409.

Mummery, C., Ward, D., van den Brink, C.E., Bird, S.D., Doevendans, P.A., Opthof, T., Brutel de la Riviere, A., Tertoolen, L., van der Heyden, M., and Pera, M. (2002). Cardiomyocyte differentiation of mouse and human embryonic stem cells. J Anat *200*, 233-242.

Nakagawa, M., Koyanagi, M., Tanabe, K., Takahashi, K., Ichisaka, T., Aoi, T., Okita, K., Mochiduki, Y., Takizawa, N., and Yamanaka, S. (2008). Generation of induced pluripotent stem cells without Myc from mouse and human fibroblasts. Nat Biotechnol *26*, 101-106.

Ng, R.K., and Gurdon, J.B. (2008). Epigenetic memory of an active gene state depends on histone H3.3 incorporation into chromatin in the absence of transcription. Nat Cell Biol *10*, 102-109.

Niwa, H., Miyazaki, J., and Smith, A.G. (2000). Quantitative expression of Oct-3/4 defines differentiation, dedifferentiation or self-renewal of ES cells. Nat Genet *24*, 372-376.

Novak, A., Shtrichman, R., Germanguz, I., Segev, H., Zeevi-Levin, N., Fishman, B., Mandel, Y.E., Barad, L., Domev, H., Kotton, D., *et al.* (2010). Enhanced reprogramming and cardiac differentiation of human keratinocytes derived from plucked hair follicles, using a single excisable lentivirus. Cell Reprogram *12*, 665-678.

Okita, K., Ichisaka, T., and Yamanaka, S. (2007). Generation of germline-competent induced pluripotent stem cells. Nature *448*, 313-317.

Okita, K., Nakagawa, M., Hyenjong, H., Ichisaka, T., and Yamanaka, S. (2008). Generation of mouse induced pluripotent stem cells without viral vectors. Science *322*, 949-953.

Park, I.H., Arora, N., Huo, H., Maherali, N., Ahfeldt, T., Shimamura, A., Lensch, M.W., Cowan, C., Hochedlinger, K., and Daley, G.Q. (2008a). Disease-specific induced pluripotent stem cells. Cell *134*, 877-886.

Park, I.H., Zhao, R., West, J.A., Yabuuchi, A., Huo, H., Ince, T.A., Lerou, P.H., Lensch, M.W., and Daley, G.Q. (2008b). Reprogramming of human somatic cells to pluripotency with defined factors. Nature *451*, 141-146.

Peterson, L., Vasiliadis, H.S., Brittberg, M., and Lindahl, A. (2010). Autologous chondrocyte implantation: a long-term follow-up. Am J Sports Med *38*, 1117-1124.

Plews, J. R., Li, J., Jones, M., Moore, H. D., Mason, C., Andrews, P. W. & Na, J. (2010). Activation of pluripotency genes in human fibroblast cells by a novel mRNA based approach. *PLoS One* 5, e14397.

Ponnusamy, M.P., Deb, S., Dey, P., Chakraborty, S., Rachagani, S., Senapati, S., and Batra, S.K. (2009). RNA polymerase II associated factor 1/PD2 maintains self-renewal by its interaction with Oct3/4 in mouse embryonic stem cells. Stem Cells *27*, 3001-3011.

Rai, K., Huggins, I.J., James, S.R., Karpf, A.R., Jones, D.A., and Cairns, B.R. (2008). DNA demethylation in zebrafish involves the coupling of a deaminase, a glycosylase, and gadd45. Cell *135*, 1201-1212.

Redmer, T., Diecke, S., Grigoryan, T., Quiroga-Negreira, A., Birchmeier, W., and Besser, D. (2011). E-cadherin is crucial for embryonic stem cell pluripotency and can replace OCT4 during somatic cell reprogramming. EMBO Rep *12*, 720-726.

Shi, Y., Do, J.T., Desponts, C., Hahm, H.S., Scholer, H.R., and Ding, S. (2008). A combined chemical and genetic approach for the generation of induced pluripotent stem cells. Cell Stem Cell *2*, 525-528.

Silva, J., Barrandon, O., Nichols, J., Kawaguchi, J., Theunissen, T.W., and Smith, A. (2008). Promotion of reprogramming to ground state pluripotency by signal inhibition. PLoS Biol *6*, e253.

Simonsson, S., and Gurdon, J. (2004). DNA demethylation is necessary for the epigenetic reprogramming of somatic cell nuclei. Nat Cell Biol *6*, 984-990.

Soldner, F., Hockemeyer, D., Beard, C., Gao, Q., Bell, G.W., Cook, E.G., Hargus, G., Blak, A., Cooper, O., Mitalipova, M., *et al.* (2009). Parkinson's disease patient-derived induced pluripotent stem cells free of viral reprogramming factors. Cell *136*, 964-977.

Spemann, H. (1938). Embryonic Development and Induction. New Haven: Yale University Press.

Stadtfeld, M., Maherali, N., Breault, D.T., and Hochedlinger, K. (2008a). Defining molecular cornerstones during fibroblast to iPS cell reprogramming in mouse. Cell Stem Cell *2*, 230-240.

Stadtfeld, M., Nagaya, M., Utikal, J., Weir, G., and Hochedlinger, K. (2008b). Induced pluripotent stem cells generated without viral integration. Science *322*, 945-949.

Stephenson, E.L., Mason, C., and Braude, P.R. (2009). Preimplantation genetic diagnosis as a source of human embryonic stem cells for disease research and drug discovery. BJOG *116*, 158-165.

Stojkovic, M., Stojkovic, P., Leary, C., Hall, V.J., Armstrong, L., Herbert, M., Nesbitt, M., Lako, M., and Murdoch, A. (2005). Derivation of a human blastocyst after heterologous nuclear transfer to donated oocytes. Reprod Biomed Online *11*, 226-231.

Sugii, S., Kida, Y., Kawamura, T., Suzuki, J., Vassena, R., Yin, Y.Q., Lutz, M.K., Berggren, W.T., Izpisua Belmonte, J.C., and Evans, R.M. (2010). Human and mouse adipose-derived cells support feeder-independent induction of pluripotent stem cells. Proc Natl Acad Sci U S A *107*, 3558-3563.

Takahashi, K., Tanabe, K., Ohnuki, M., Narita, M., Ichisaka, T., Tomoda, K., and Yamanaka, S. (2007). Induction of pluripotent stem cells from adult human fibroblasts by defined factors. Cell *131*, 861-872.

Takahashi, K., and Yamanaka, S. (2006). Induction of pluripotent stem cells from mouse embryonic and adult fibroblast cultures by defined factors. Cell *126*, 663-676.

Taranger, C.K., Noer, A., Sorensen, A.L., Hakelien, A.M., Boquest, A.C., and Collas, P. (2005). Induction of dedifferentiation, genomewide transcriptional programming, and epigenetic reprogramming by extracts of carcinoma and embryonic stem cells. Mol Biol Cell *16*, 5719-5735.

Tian, X.C., Xu, J., and Yang, X. (2000). Normal telomere lengths found in cloned cattle. Nat Genet *26*, 272-273.

Tsai, S.Y., Clavel, C., Kim, S., Ang, Y.S., Grisanti, L., Lee, D.F., Kelley, K., and Rendl, M. (2010). Oct4 and klf4 reprogram dermal papilla cells into induced pluripotent stem cells. Stem Cells *28*, 221-228.

Vizlin-Hodzic, D., Johansson, H., Ryme, J., Simonsson, T., and Simonsson, S. (2011). SAF-A Has a Role in Transcriptional Regulation of Oct4 in ES Cells Through Promoter Binding. Cell Reprogram.

Vizlin-Hodzic, D., Ryme, J., Simonsson, S., and Simonsson, T. (2009). Developmental studies of Xenopus shelterin complexes: the message to reset telomere length is already present in the egg. FASEB J *23*, 2587-2594.

Warren, L., Manos, P. D., Ahfeldt, T., Loh, Y. H., Li, H., Lau, F., Ebina, W., Mandal, P. K., Smith, Z. D. & other authors (2010). Highly efficient reprogramming to pluripotency and directed differentiation of human cells with synthetic modified mRNA. *Cell Stem Cell* 7, 618-630.

Wei, D., Kanai, M., Huang, S., and Xie, K. (2006). Emerging role of KLF4 in human gastrointestinal cancer. Carcinogenesis *27*, 23-31.

Wei, F., Scholer, H.R., and Atchison, M.L. (2007). Sumoylation of Oct4 enhances its stability, DNA binding, and transactivation. J Biol Chem *282*, 21551-21560.

Wernig, M., Lengner, C.J., Hanna, J., Lodato, M.A., Steine, E., Foreman, R., Staerk, J., Markoulaki, S., and Jaenisch, R. (2008a). A drug-inducible transgenic system for direct reprogramming of multiple somatic cell types. Nat Biotechnol *26*, 916-924.

Wernig, M., Meissner, A., Cassady, J.P., and Jaenisch, R. (2008b). c-Myc is dispensable for direct reprogramming of mouse fibroblasts. Cell Stem Cell *2*, 10-12.

Wernig, M., Meissner, A., Foreman, R., Brambrink, T., Ku, M., Hochedlinger, K., Bernstein, B.E., and Jaenisch, R. (2007). In vitro reprogramming of fibroblasts into a pluripotent ES-cell-like state. Nature *448*, 318-324.

Wernig, M., Zhao, J.P., Pruszak, J., Hedlund, E., Fu, D., Soldner, F., Broccoli, V., Constantine-Paton, M., Isacson, O., and Jaenisch, R. (2008c). Neurons derived from reprogrammed fibroblasts functionally integrate into the fetal brain and improve symptoms of rats with Parkinson's disease. Proc Natl Acad Sci U S A *105*, 5856-5861.

Wilmut, I., Schnieke, A.E., McWhir, J., Kind, A.J., and Campbell, K.H. (1997). Viable offspring derived from fetal and adult mammalian cells. Nature *385*, 810-813.

Woltjen, K., Michael, I.P., Mohseni, P., Desai, R., Mileikovsky, M., Hamalainen, R., Cowling, R., Wang, W., Liu, P., Gertsenstein, M., *et al.* (2009). piggyBac transposition reprograms fibroblasts to induced pluripotent stem cells. Nature *458*, 766-770.

Xu, D., Alipio, Z., Fink, L.M., Adcock, D.M., Yang, J., Ward, D.C., and Ma, Y. (2009). Phenotypic correction of murine hemophilia A using an iPS cell-based therapy. Proc Natl Acad Sci U S A *106*, 808-813.

Xu, H.M., Liao, B., Zhang, Q.J., Wang, B.B., Li, H., Zhong, X.M., Sheng, H.Z., Zhao, Y.X., Zhao, Y.M., and Jin, Y. (2004). Wwp2, an E3 ubiquitin ligase that targets transcription factor Oct-4 for ubiquitination. J Biol Chem *279*, 23495-23503.

Yakubov, E., Rechavi, G., Rozenblatt, S. & Givol, D. (2010). Reprogramming of human fibroblasts to pluripotent stem cells using mRNA of four transcription factors. *Biochem Biophys Res Commun* 394, 189-193.

Yamanaka, S. (2009). A fresh look at iPS cells. Cell *137*, 13-17.
Yamanaka, S., and Blau, H.M. (2010). Nuclear reprogramming to a pluripotent state by three approaches. Nature *465*, 704-712.
Yamazaki, Y., Fujita, T.C., Low, E.W., Alarcon, V.B., Yanagimachi, R., and Marikawa, Y. (2006). Gradual DNA demethylation of the Oct4 promoter in cloned mouse embryos. Mol Reprod Dev *73*, 180-188.
Yu, J., Hu, K., Smuga-Otto, K., Tian, S., Stewart, R., Slukvin, II, and Thomson, J.A. (2009). Human induced pluripotent stem cells free of vector and transgene sequences. Science *324*, 797-801.
Yu, J., Vodyanik, M.A., Smuga-Otto, K., Antosiewicz-Bourget, J., Frane, J.L., Tian, S., Nie, J., Jonsdottir, G.A., Ruotti, V., Stewart, R., *et al.* (2007). Induced pluripotent stem cell lines derived from human somatic cells. Science *318*, 1917-1920.
Yusa, K., Rad, R., Takeda, J., and Bradley, A. (2009). Generation of transgene-free induced pluripotent mouse stem cells by the piggyBac transposon. Nat Methods *6*, 363-369.
Zhang, J., Wilson, G.F., Soerens, A.G., Koonce, C.H., Yu, J., Palecek, S.P., Thomson, J.A., and Kamp, T.J. (2009). Functional cardiomyocytes derived from human induced pluripotent stem cells. Circ Res *104*, e30-41.
Zhao, H.X., Li, Y., Jin, H.F., Xie, L., Liu, C., Jiang, F., Luo, Y.N., Yin, G.W., Wang, J., Li, L.S., *et al.* (2010). Rapid and efficient reprogramming of human amnion-derived cells into pluripotency by three factors OCT4/SOX2/NANOG. Differentiation *80*, 123-129.
Zhao, T., Zhang, Z.N., Rong, Z., and Xu, Y. (2011). Immunogenicity of induced pluripotent stem cells. Nature *474*, 212-215.
Zhao, X.Y., Li, W., Lv, Z., Liu, L., Tong, M., Hai, T., Hao, J., Guo, C.L., Ma, Q.W., Wang, L., *et al.* (2009). iPS cells produce viable mice through tetraploid complementation. Nature *461*, 86-90.
Zhou, H., Wu, S., Joo, J.Y., Zhu, S., Han, D.W., Lin, T., Trauger, S., Bien, G., Yao, S., Zhu, Y., *et al.* (2009). Generation of induced pluripotent stem cells using recombinant proteins. Cell Stem Cell *4*, 381-384.
Zhou, Q., Brown, J., Kanarek, A., Rajagopal, J., and Melton, D.A. (2008). In vivo reprogramming of adult pancreatic exocrine cells to beta-cells. Nature *455*, 627-632.
Zhu, S., Li, W., Zhou, H., Wei, W., Ambasudhan, R., Lin, T., Kim, J., Zhang, K., and Ding, S. (2010). Reprogramming of human primary somatic cells by OCT4 and chemical compounds. Cell Stem Cell *7*, 651-655.

2

Importance of Non-HLA Gene Polymorphisms in Hematopoietic Stem Cell Transplantation

Jeane Visentainer and Ana Sell
Maringa State University
Brazil

1. Introduction

In the last 10 years, non-HLA genotypes have been investigated for their potential roles in the occurrence and severity of graft-versus-host disease (GvHD) as well as for their contribution to overall transplant-related mortality, infectious episodes, and overall survival.

These non-HLA-encoded genes include polymorphisms within the regulatory sequences of the cytokine genes, or genes associating with innate immunity: *KIR* (killer immunoglobulin-like receptor) genes, *MIC* (MHC class I chain-related) genes, and others.

The first studied non-HLA genes were polymorphisms in regulatory cytokine genes because of cytokine role in GvHD immunopathogenesis. Single nucleotide polymorphisms in several regions of cytokine genes were correlated with the transplant overcome in several studies (Kim et al., 2005; Laguila Visentainer et al., 2005; Lin et al., 2003; Mlynarczewska et al., 2004; Viel et al., 2007; reviewed in Dickinson, 2008).

2. Role of cytokines in graft-versus-host disease after allogeneic stem cell transplantation

The pathophysiology of acute GvHD can be considered a cytokine storm (Ferrara, 2000), initializing with the transplant conditioning regimen that damages and activates host tissues. Activated host cells secrete inflammatory cytokines, such as tumor necrosis factor (TNF)-α and interleukin (IL)-1. This initial cytokine release is further amplified in the second phase by presentation of host antigens to donor T cells and the subsequent proliferation and differentiation of these activated T cells. These cells secrete a variety of cytokines, such as IL-2, TNF-α, interferon (IFN)-γ, IL-4, IL-6, IL-10, and transforming growth factor-beta (TGF)-β1. Several reports have demonstrated the increase of these cytokines in the serum from patients with acute GvHD (Kayaba et al., 2000; Liem et al., 1998; Sakata et al., 2001; Visentainer et al., 2003).

Although chronic GvHD remains a frequent complication of hematopoietic stem cell transplantation (HSCT), the pathogenesis is still unclear. However, it is known that cytokines also play an important role in its development (Iwasaki, 2004; Letterio & Roberts, 1998; Liem et al., 1999; Margolis & Vogelsang, 2000; Zhang et al., 2006). Chronic GvHD is a multisystem alloimmune and autoimmune disorder characterized by immune dysregulation, immunodeficiency, impaired organ function and decreased survival (Baird & Pavletic, 2006). It starts with the expansion of donor T cells in response to allo or auto-antigens that escape assessment thymus and the mechanisms of deletion. T cells induce damage in target organs by attacking cytolytic, inflammatory cytokines and fibrosis by activating B cells, with production of autoantibodies (Pérez-Simón et al., 2006).

Thus multiple cytokines are important in GvHD pathogenesis and regulation (Ferrara & Krenger, 1998; Jung et al., 2006; Kappel et al., 2009; Reddy et al., 2003; Tawara et al., 2008; Visentainer et al., 2005; Yi et al., 2008). Furthermore, the timing and duration of cytokine expression may be a critical factor determining the induction of the graft-versus-host (GvH) reaction, and cytokine dysregulation could potentially contribute to the severity of GvHD.

Recently, Choi et al. (2010) and Paczesny et al. (2010) reviewed the biology of acute GvHD, and concluded that the underlying mechanisms of GvHD have emerged as a complex network of immune interactions where the key players are the naive T cells, the host and donor APCs, CTLs and regulatory T cells, along with new players such as Plasmacytoid DCs (pDCs), B cells and Th17 cells.

2.1 Cytokine gene polymorphisms

The production of some cytokines is under genetic control. Polymorphisms in the regulatory regions of several cytokine genes may cause inter-individual differences in cytokine production (Wilson et al., 1997; Turner et al., 1997; Awad et al., 1998; Fishman et al., 1998; Pravica et al., 1999). As these polymorphisms segregate independently, each person is a mosaic of high-, intermediate-, and low-producing phenotypes. These cytokine polymorphisms are known to have functional relevance in post-transplant outcome, rejection and GvHD, following solid organ (Benza et al., 2009; Fernandes et al., 2002; Hahn et al., 2001; Karimi et al., 2011; Reviron et al., 2001) and hematopoietic stem cell transplantation (Ambruzova et al., 2009; Karimi et al., 2010; Laguila Visentainer et al., 2005; Leffell et al., 2001; Takahashi et al., 2000; Tambur et al., 2001), respectively.

2.2 Impact of cytokine gene polymorphisms on graft-vs-host disease

Many studies in recent years have focused on correlating donor and/or recipient genotype with GvHD risk. Table 1 summarizes the various polymorphisms in genes encoding both pro- and anti-inflammatory factors and their receptors that have been studied in GvHD.

3. Killer immunoglobulin-like receptors and hematopoietic stem cell transplantation

Natural killer (NK) cell effector function is regulated by a balance between activating receptors and inhibitory receptors for major histocompatibility complex (MHC) class I molecules (Joyce & Sun, 2011; Parham et al., 2006; Yokoyama et al., 2006). In the setting of allogeneic HSCT, donor NK cells may attack recipient cells that lack the appropriate HLA class I ligands for the donor KIR. Several studies have shown that certain combinations of killer immunoglobulin-like receptors and human leukocyte antigens (in both donors and recipients) can affect the chances of survival of transplant patients, particularly in relation to the graft-versus-leukemia effect, which may be associated to decreased relapse rates in certain groups (reviewed in Franceschi et al., 2011).

GvHD		
Acute	Chronic	
SNP/Genotype		References
	*IL1A-889*2*	Cullup et al. (2003)
IL6-174 G/C	*IL6-174 GG*	Lin et al. (2003)
	IL6-174 CC	Laguila Visentainer et al. (2005)
TNF-308 GG/GA		Takahashi et al. (2000)
	TNF-238 GA	Viel et al. (2007)
IL2-330 GT		Macmillan et al. (2003)
	IL10-1082,-819,-592 ATA/ATA	Kim et al. (2005)
IL10-592 A/A		Lin et al. (2003)
IL-10RB A/A	*IL-10RB A/A*	Sivula et al. (2009)
TGFB1+869,+915 TG/GG		Leffell et al. (2001)
TGFB1+869 T		Hattori et al. (2002)
TGF- beta1 codon 25 GG		Rashidi-Nezhad et al. (2010)
IFN-γ T+874A		Karimi et al. (2010)
IL-7RA		Shamim et al. (2011)

Table 1. Polymorphisms in genes encoding both pro- and anti-inflammatory factors and their receptors in GvHD

3.1 Killer immunoglobulin-like receptors

The group of *KIR* genes comprises a region of approximately 150 Kb in the leukocyte receptor complex (LRC) on chromosome 19q13.4 (Uhrberg et al., 1997). KIRs are members of a group of regulatory molecules on the surface of NK cells, in subgroups of Tγδ+ lymphocytes, effector Tαβ+ lymphocytes and memory lymphocytes (Rajagopalan & Long, 2005). The KIR family includes activating and inhibitory molecules. Inhibitory KIRs (2DL and 3DL) have a long cytoplasmic tail containing tyrosine-based inhibitory motifs (ITIMs) that trigger inhibitory events of cytotoxicity. In contrast, activating KIRs (2DS and 3DS) interact with the DAP12 molecule, which has tyrosine-based activation motifs (ITAMs) that cause a cascade that results in an increase in cytoplasmic granulation and the production of cytokines and chemokines, thereby initiating immune response (McVicar et al., 2001).

The balance between activation and inhibition of NK cells occurs through the binding of KIRs with HLA class I molecules present in all nucleated cells of an individual. Most KIRs bind to HLA-C molecules. It is worth remembering the importance of the dimorphism of amino acids, such as residue 80 of α-helix-1, in the definition of this HLA receptor. On this basis, HLA-C alleles may be defined as "Group 1" or "Group 2": C1 - HLA-Cw*01, *03, *07, *08, *12, *13, *14, and *16 and C2 - HLA-Cw*02, *04, *05, *06, *07, *15, *17, and *18, which are specific for KIR2DL2/2DL3/2DS2 and KIR2DL1/2DS1, respectively (Boyton & Altmann, 2007). Evidence suggests that HLA-Cw4 is a receptor for KIR2DS4 (Katz et al., 2001). The KIR2DL4, for example, specificity binds to the HLA-G molecule (Rajagopalan & Long, 1999), while the KIR3DL1 receptor binds to a subset of HLA molecules with the Bw4 epitope, present in approximately one third of all HLA-B molecules. The KIR3DS1 is highly homologous with 3DL1 and seems to share the Bw4 epitope as ligand, although this needs to be experimentally verified. The KIR3DL2 receptor is still being discussed, but studies suggest that HLA-A3 and HLA-A11 perform this role (O'Connor et al., 2006).

Based on the genetic content and pattern of segregation at the population level, *KIR* haplotypes are divided into two groups, A and B, varying in the type and number of genes present. The *KIR* group A haplotype is uniform in terms of gene content (*3DL3, 2DL3, 2DL1, 2DP1, 3DP1, 2DL4, 3DL1, 2DS4,* and *3DL2*), of which all but 1 encode inhibitory receptors. In contrast, the *KIR* group B haplotype is more diverse in the *KIR* genes it contains, has more activating receptors, and is characterized by the *2DL2, 2DS1, 2DS2, 2DS3,* and *2DS5* genes (Uhrberg et al., 1997).

3.2 Impact of killer immunoglobulin-like receptors and hematopoietic stem cell transplantation

Previous studies have examined the effect of donor and recipient *KIR* genotypes on the outcome of allogeneic HSCT (Bishara et al., 2004; Gagne et al., 2002; Sun et al., 2005). One study found a 100% risk of GvHD after unrelated donor BMT, when the donor contained *KIR* genes absent in the recipient, compared to a 60% risk of GvHD with other combinations (Gagne et al., 2002).

In 2004, one study carried out *KIR-HLA* genotyping of 220 related HLA identical donor-recipient pairs (112 for myeloid diseases and 108 lymphoid diseases) (Cook et al., 2004). For patients with myeloid diseases, survival was lower in those homozygous for Group 2 (C2) HLA-C compared to patients with Group 1 (C1). This effect was observed only when the donor had the *KIR2DS2* gene. As *KIR2DS2* is in strong linkage disequilibrium with *KIR2DL2* (receptor inhibited by C1), this would indirectly indicate lower survival in patients who do not have the receptor for *KIR2DL2*, an opposite result to the model in which this lack of inhibition could result in NK cell alloreactivity with a consequent elimination of residual leukemic cells (Witt et al., 2006). In 178 patients with AML, CML, ALL and primary myelodysplastic syndrome (MDS) who received HSCT with T cell depletion from HLA-identical related donors, some authors observed that the disease-free survival was significantly higher in patients with AML and MDS that did not have the HLA ligand for the inhibitory KIR of the donor (Hsu et al., 2005). Moreover, the relapse rate was lower in these individuals, which may be related to higher survival rates. The results differ from a study in which T cell depletion was not performed (Cook et al., 2004). In another study (Schellekens et al., 2008) involving 83 patients with different types of hematologic malignancies who received HSCT from related HLA-identical donors without T cell depletion, a high relapse rate was found when high numbers of activating KIRs were present in both the patient and donor. According to the authors, a consequence of this finding may be an increased alloreactivity of the host against graft, impairing the response of donor cells resulting in an insufficient graft-versus-leukemia effect and increased risk of leukemic relapse.

Nowadays, there is no unequivocal evidence that polymorphic genes for KIR involved in innate immunity sufficiently influence GvHD and transplant outcome to change clinical practice (Davies et al., 2002; Cooley et al., 2009; Giebel et al., 2003; Hsu et al., 2005; Ludajic et al., 2009; Miller et al., 2007; Moretta et al., 2009; Schellekens et al., 2008; Symons et al., 2010; Witt et al., 2006).

Using a large cohort of patients, Venstrom et al. (2010) demonstrated that individual donor activating KIR, recipient HLA class I ligands, and donor *KIR* gene copy number all impact KIR-driven NK effects. They also showed that not all *KIR* B haplotypes have equivalent clinical impact, and they proposed that future studies consider specific B haplotype subsets or individual *KIR* genes in their analyses.

However, there are conflicting results in many studies, which may be due to the heterogeneity in HSCT protocols employed, differences in inclusion criteria, the HSCT preparative regimen and graft content, the degree of donor HLA-incompatibility, and posttransplant immunosuppression. Beside of this, according to early studies, Symons et al. (2010) have described 4 models of NK cell alloreactivity to predict HSCT outcomes: 1) KIR ligand incompatibility; 2) receptor-ligand model; 3) missing ligand model; and 4) *KIR* gene-gene model. And, contradictory results obtained from these models have made it difficult to conclude which model is most predictive of transplant outcome.

4. *MICA* and *MICB* matching in bone marrow transplantation

Retrospective and prospective studies have shown that matching donors and recipients for non-HLA DNA sequences in the MHC (beta and delta block matching) can result in improved patient survival and less severe GvHD (Tay et al., 1995; Witt et al., 2000). One of these blocks, the beta block, spans about 300 kb and contains the immunologically relevant *HLA-B*, *HLA-C*, *MICA*, and *MICB* genes (Kitcharoen et al., 2006). The polymorphic MICA molecule likely may be a target for specific antibodies and T cells in solid organ grafts or in GvHD (Zhang & Stastny, 2006).

4.1 *MICA* and *MICB* genes

In 1994, two new polymorphic families of MHC class I related genes, termed MHC class I-related chain A (*MICA*) and B (*MICB*) were described (Bahram et al., 1994). These genes are highly polymorphic with at least 76 alleles for *MICA* and 31 alleles for *MICB* (IMGT/HLA database; http://www.ebi.ac.uk/imgt/hla/stats.html), and are located in the MHC classical class I region (Horton et al., 2004), 46.4 and 141.2 kb centromeric to *HLA-B*, respectively (Bahram et al., 1994; Bahram et al., 2000; Leelayuwat et al., 1994). They encode cell surface glycoproteins that do not associate with β_2-microglobulin. These molecules function as restriction elements for intestinal γδ T cells and they behave as cell stress molecules. MICA is expressed in endothelial cells, keratinocytes and monocytes, but not in CD4+, CD8+ or CD19+ lymphocytes (Zwirner et al., 1999).

The MICA gene products have been shown to play a role in some aspects of antigen presentation and T-cell recognition, and appear to be important in innate immunity as ligands to NKG2D receptor expressed on most γδ T cells, CD8 αβ T cells, and NK cells (Tieng et al., 2002).

4.2 *MICA* and *MICB* and relevance to stem cell transplantation outcome

Several studies have shown that the highly polymorphic MIC antigens are expressed in transplanted organs and may cause early graft rejection (Hankey et al., 2002; Mizutani et al., 2006; Narayan et al., 2011; Panigrahi et al., 2007; Sumitran-Holgersson, 2008; Terasaki et al., 2007). The polymorphisms of MICA and MICB may be involved in allogeneic BMT and GvHD (Gannage et al., 2008; Murai et al., 2003; Parmar et al. 2009; Przepiorka et al., 1995) because they are augmented by stress in epithelia (Groh et al., 1996) and are recognized by a subpopulation of intestinal γδ T cells (Zou et al. 2007). In addition to classical HLA class I and II matching, matches at *MICA* and *MICB* loci have been shown to increase patient survival (Kitcharoen et al., 2006).

Recent review has discussed the genetics and biology of the *MICA* gene and its products, and their importance in disease related to NK activity and allograft rejection or GvHD

(Choy & Phipps, 2010). According to Parmar et al. (2009), some HSCT cases with matched *HLA* but mismatched *MICA* showed an increased incidence of GvHD, and according to Boukouaci et al. (2009), MICA-129 valine and soluble MICA are risk factors for chronic GvHD, whereas the presence of anti-MICA antibodies that can neutralize soluble MICA confers protection.

A methionine to valine change at position 129 of the α2-heavy chain domain categorized the *MICA* alleles into strong (MICA-129 met) and weak (MICA-129 val) binders of NKG2D receptor (Steinle et al., 2001). Varying affinities of *MICA* alleles for NKG2D may affect thresholds of NK-cell triggering and T-cell modulation. According to Boukouaci et al. (2009), in the context of cGVHD, the weak engagement of NKG2D receptors by the weak binder MICA-129 val allele may impair NK/cytotoxic T lymphocyte cell activation/costimulation, possibly skewing the TH1 pathway toward TH2 with consequent B-cell activation and Ab production.

5. Conclusion

Analysis of non-HLA genetics may permit more accurate assessment of transplant-related complications, improve donor selection and individualized prophylaxis, and aid in the development of a prognostic risk index. Overall, this type of analysis could potentially define high- and low-risk patient groups, and to result in effective therapeutic strategies for GvHD.

6. References

Ambruzova, Z.; Mrazek, F.; Raida, L.; Jindra, P.; Vidan-Jeras, B.; Faber, E.; Pretnar, J.; Indrak, K. & Petrek, M. (2009). Association of IL6 and CCL2 gene polymorphisms with the outcome of allogeneic haematopoietic stem cell transplantation. *Bone Marrow Transplantation*, Vol.44, No.4, (August 2009), pp.227-235, ISSN 0268-3369

Awad, M.R.; El-Gamel, A.; Hasleton, P.; Turner, D.M.; Sinnott, P.J. & Hutchinson, I.V. (1998). Genotypic variation in the transforming growth factorb1 gene. *Transplantation*, Vol.66, No.8, (October 1998), pp.1014-1020, ISSN 0041-1337

Bahram, S. (2000). MIC genes: from genetics to biology. *Advances in Immunology*, Vol.76, (2000), pp.1-60, ISSN 0065-2776

Bahram, S.; Bresnahan, M.; Geraghty, D.E. & Spies, T. (1994). A second lineage of mammalian major histocompatibility complex class I genes. *Proceedings of the National Academy of Sciences of the United States of America*, Vol.91, No.14, (Jul 1994), pp. 6259-6263, ISSN 0027-8424

Baird, K. & Pavletic, S.Z. Chronic graft versus host disease. (2006). *Current Opinion in Hematology*, Vol.13, No.6, (November 2006), pp. 426-435, ISSN 1065-6251

Benza, R.L.; Coffey, C.S.; Pekarek, D.M.; Barchue J.P.; Tallaj, J.A.; Passineau, M.J. & Grenett, H.E. (2009). Transforming growth factor-beta polymorphisms and cardiac allograft rejection. *The Journal of Heart and Lung Transplantation*, Vol.28, No.10, (October 2009), pp.1057-1062, ISSN 1053-2498

Bishara, A; De Santis, D.; Witt C.C.; Brautbar, C.; Christiansen, F.T.; Or, R.& Nagler, A., Slavin, S. (2004). The beneficial role of inhibitory KIR genes of HLA class I NK epitopes in haploidentically mismatched stem cell allografts may be masked by residual donor-alloreactive T cells causing GVHD. *Tissue Antigens*, Vol.63, (March 2004), pp.204-211, ISSN 0001-2815

Boukouaci, W.; Busson, M.; Peffault de Latour, R.; Rocha, V.; Suberbielle, C.; Bengoufa, D.; Dulphy, N.; Haas, P.; Scieux, C.; Amroun, H.; Gluckman, E.; Krishnamoorthy, R.; Toubert, A.; Charron, D.; Socié, G. & Tamouza, R. (2009). MICA-129 genotype, soluble MICA, and anti-MICA antibodies as biomarkers of chronic graft-versus-host disease. *Blood*, Vol.114, No.25, (December 2009), pp.5216–5224, ISSN 0006-4971

Boyton, R.J. & Altmann, D.M. (2007). Natural killer cells, killer immunoglobulin like receptors and human leukocyte antigen class I in disease. *Clinical of Experimental Immunology*, Vol.149, No.1, (January 2007), pp.1-8, ISSN 0009-9104

Choi, S.W.; Levine, J.E. & Ferrara, J.L. (2010). Pathogenesis and Management of Graft-versus-Host Disease. *Immunology and Allergy Clinics of North America*, Vol.30, No.1, (February 2010), pp.75–101, ISSN 0889-8561

Choy, M.K. & Phipps, M.E. (2010). MICA polymorphism: biology and importance in immunity and disease. *Trends in Molecular Medicine*, Vol.16, No.3, (March 2010), pp.97-106, ISSN 1471-4914

Cook, M.A.; Milligan, D.W.; Fegan, C.D.; Darbyshire, P.J.; Mahendra, P.; Craddock, C.F.; Moss, P.A. & Briggs, D.C. (2004). The impact of donor KIR and patient HLA-C genotypes on outcome following HLA-identical sibling hematopoietic stem cell transplantation for myeloid leukemia. *Blood*, Vol.103, No.4, (February 2004), pp.1521-1526, ISSN 0006-4971

Cullup, H.; Dickinson, A.M.; Cavet, J.; Jackson, G.H. & Middleton, P.G. (2003). Polymorphisms of interleukin-1alpha constitute independent risk factors for chronic graft-versus-host disease after allogeneic bone marrow transplantation. *British Journal of Haematology*, Vol.122, No.5, (September 2003), pp.778-787, ISSN 0007-1048

Davies, S.M.; Ruggieri, L.; DeFor, T.; Wagner, J.E.; Weisdorf, D.J.; Miller, J.S.; Velardi, A. & Blazar, B.R. (2002). Evaluation of KIR ligand incompatibility in mismatched unrelated donor hematopoietic transplants. *Blood*, Vol.100, No.10, (November 2002), pp.3825-3827, ISSN 0006-4971

Fernandes, H.; Koneru, B.; Fernandes, N.; Hameed, M.; Cohen, MC.; Raveche, E. & Cohen, S. (2002). Investigation of promoter polymorphisms in the tumor necrosis factor-a and interleukin-10 genes in liver transplant patients. *Transplantation*, Vol.73, No.12, (June 2002), pp.1886-1891, ISSN 0041-1337

Ferrara, J.L. & Krenger, W. (1998). Graft-versus-host disease: the influence of type 1 and type 2 T cell cytokines. *Transfusion Medicine Reviews*, Vol.12, No.1, (January 1998), pp.1–17, ISSN 0887-7963

Ferrara, J.L. (2000). Pathogenesis of acute graft-versus-host disease: cytokines and cellular effectors. *Journal of Hematotherapy & Stem Cell Research*, Vol.9, No.3, (June 2000), pp. 299–306, ISSN 1061-6128

Fishman, D.; Faulds, G.; Jeffery, R.; Mohamedali, V.; Yudkin, J.S.; Humphries, S. & Woo, P. (1998). The effect of novel polymorphisms in the interleukin-6 (IL-6) gene on IL-6 transcription and plasma IL-6 levels, and an association with systemic-onset juvenile chronic arthritis. *The Journal of Clinical Investigation*, Vol.102, No.7, (October 1998), pp.1369-1376, ISSN 0021-9738

Franceschi, D.S.;de Souza, C.A.; Aranha, F.J.; Cardozo, D.M.; Sell, A.M. & Visentainer, J.E. (2011). Importance of killer immunoglobulin-like receptors in allogeneic hematopoietic stem cell transplantation. *Revista Brasileira de Hematologia e Hemoterapia*, Vol.33, No.2, (April 2011), ISSN 1516-8484

Gagne, K.; Brizard, G.; Gueglio. B.; Milpied, N. ; Herry, P.; Bonneville, F.; Chéneau, M.L.; Schleinitz, N.; Cesbron, A.; Folléa, G.; Harrousseau, J.L & Bignon, J.D. (2002). Relevance of KIR gene polymorphisms in bone marrow transplantation outcome. *Human Immunology*, Vol.63, (April 2002), pp.271-280, ISSN 0198-8859

Gannage, M.; Buzyn, A. ; Bogiatzi, S.I.; Lambert, M.; Soumelis, V; Dal Cortivo, L.; Cavazzana-Calvo, M.; Brousse, N. & Caillat-Zucman, S. (2008). Induction of NKG2D ligands by gamma radiation and tumor necrosis factor-alpha may participate in the tissue damage during acute graft-versus-host disease. *Transplantation*, Vol.85, No.6, (March 2008), pp.911-915, ISSN 0041-1337

Gasser, S. & Raulet, D.H. (2006). Activation and self-tolerance of natural killer cells. *Immunological Reviews*, Vol.214, No.1, (December 2006), pp.130-142, ISSN: 0105-2896

Giebel, S.; Locatelli, F.; Lamparelli, T.; Velardi, A.; Davies, S.; Frumento, G.; Maccario, R.; Bonetti, F.; Wojnar, J.; Martinetti, M.; Frassoni, F.; Giorgiani, G.; Bacigalupo, A. & Holowiecki, J. (2003). Survival advantage with KIR ligand incompatibility in hematopoietic stem cell transplantation from unrelated donors. *Blood*, Vol.102, No.3, (August 2003), pp.814-819, ISSN 0006-4971

Groh, V.; Bahram, S; Bauer, S.; Herman, A.; Beauchamp, M. & Spies T. (1996). Cell stress-regulated human major histocompatibility complex class I gene expressed in gastrointestinal epithelium. *Proceedings of the National Academy of Sciences of the United States of America*, Vol.93, No.22 (October 1996), pp. 12445-12450, ISSN 0027-8424

Hahn, A.B.; Kasten-Jolly, J.C.; Constantino, D.M.; Graffunder, E. & Conti, D.J. (2001). TNF-alpha, IL-6, IFN-gamma, and IL-10 gene expression polymorphisms and the IL-4 receptor a-chain variant Q576R: effects on renal allograft outcome. *Transplantation*, Vol.72, No.4, (August 2001), pp.660-665, ISSN 0041-1337

Hankey, K.G.; Drachenberg, C.B.; Papadimitriou, J.C.; Klassen, D.K.; Philosophe, B.; Bartlett, S.T.; Groh, V.; Spies, T. & Mann, D.L. (2002). MIC expression in renal and pancreatic allografts. *Transplantation* Vol.73, No.2, (January 2002), pp. 304-306, ISSN 0041-1337

Hattori, H.; Matsuzaki, A.; Suminoe, A.; Ihara, K.; Nagatoshi, Y.; Sakata, N.; Kawa, K.; Okamura, J. & Hara, T. (2002). Polymorphisms of transforming growth factor-beta1 and transforming growth factor-beta1 type II receptor genes are associated with acute graft-versus-host disease in children with HLA matched sibling bone marrow transplantation. *Bone Marrow Transplanation*, Vol.30, No.10, (November 2002), pp.665-671, ISSN 0268-3369

Horton, R.; Wilming, L.; Rand, V.; Lovering R.C.; Bruford, E.A.; Khodiyar, V.K.; Lush, M.; Povey, S.; Talbot, C.C. Jr.; Wright, M.W.; Wain, H.M.; Trowsdale, J.; Ziegler, A. & Beck, S. (2004). Gene map of the extended human MHC. *Nature Reviews Genetics*, Vol.5, No.12, (December 2004), pp.889-899, ISSN 1471-0056

Hsu, K.C.; Keever-Taylor, C.A.; Wilton, A.; Pinto, C.; Heller, G.; Arkun, K.; O'Reilly, R.J.; Horowitz, M.M. & Dupont, B. (2005). Improved outcome in HLA-identical sibling hematopoietic stem-cell transplantation for acute myelogenous leukemia predicted by KIR and HLA genotypes. *Blood*, Vol.105, No.12, (June 2005), pp.4878-4884, ISSN 0006-4971

Iwasaki, T. (2004). Recent advances in the treatment of graft-versus-host disease. *Clinical Medicine & Research*, Vol.2, No.4, (November 2004), pp.243-252, ISSN 1539-4182

Joyce, M.G. & Sun, P.D. (2011). The structural basis of ligand recognition by natural killer cell receptors. *Journal of Biomedicine and Biotechnology*, Vol.2011, (May 2011), pp.1-15, ISSN 1110-7243

Jung, U.; Foley, J.E.; Erdmann, A.A.; Toda, Y.; Borenstein, T.; Mariotti, J. & Fowler, D.H. (2006). Ex vivo rapamycin generates Th1/Tc1 or Th2/Tc2 effector T cells with enhanced in vivo function and differential sensitivity to post-transplant rapamycin therapy. *Biology of Blood and Marrow Transplantation*, Vol.12, No.9, (September 2006), pp.905–918, ISSN 1083-8791

Kappel, L.W.; Goldberg, G.L.; King, C.G.; Suh, D.Y.; Smith, O.M.; Ligh, C.; Holland, A.M.; Grubin, J.; Mark, N.M.; Liu, C.; Iwakura, Y.; Heller, G. & van den Brink, M.R. (2009). IL-17 contributes to CD4-mediated graft-versushost disease. *Blood*, Vol.113, No.4, (January 2009), pp.945–952, ISSN 0006-4971

Karimi, M.H.; Daneshmandi, S.; Pourfathollah, A.A.; Geramizadeh, B.; Ramzi, M.; Yaghobi, R. & Ebadi, P. (2010). The IFN-γ Allele Correlated to Moderate-to-Severe Acute Graft-Versus-Host Disease After Allogeneic Stem Cell Transplant. *Experimental and Clinical Transplantation*, Vol.8, No.2, (June 2010), pp.125-129, ISSN 1304-0855

Karimi, M.H.; Daneshmandi, S.; Pourfathollah, A.A.; Geramizadeh, B.; Yaghobi, R.; Rais-Jalali, G.A.; Roozbeh, J. & Bolandparvaz, S. (2011). A study of the impact of cytokine gene polymorphism in acute rejection of renal transplant recipients. *Molecular Biology Reports*, [Epub ahead of print], (May 2011), ISSN 0301-4851

Katz, G.; Markel, G.; Mizrahi, S.; Arnon, T.I. & Mandelboim, O. (2001). Recognition of HLA-Cw4 but not HLA-Cw6 by the NK cell receptor killer cell Ig-like receptor two-domain short tail number 4. *Journal of Immunology*, Vol.166, No.12, (June 2001), pp.7260-7267, ISSN 0022-1767

Kayaba, H.; Hirokawa, M.; Watanabe, A.; Saitoh, N.; Changhao, C.; Yamada, Y.; Honda, K., Kobayashi, Y.; Urayama, O. & Chihara, J. (2000). Serum markers of graft-versus-host disease after bone marrow transplantation. *The Journal of Allergy and Clinical Immunology*, Vol.106, No.2, (July 2000), pp. S40-S44, ISSN 00917-6749

Kim, D.H.; Lee, N.Y.; Sohn, S.K.; Baek, J.H.; Kim, J.G.; Suh, J.S.; Lee, K.B. & Shin, I.H. (2005). IL-10 promoter gene polymorphism associated with the occurrence of chronic GVHD and its clinical course during systemic immunosuppressive treatment for chronic GVHD after allogeneic peripheral blood stem cell transplantation. *Transplantation*, Vol.79, No.11, (June 2005), pp.1615-1622, ISSN 0041-1337

Kitcharoen, K.; Witt, C.S.; Romphruk, A.V.; Christiansen, F.T. & Leelayuwat, C. (2006). MICA, MICB, and MHC beta block matching in bone marrow transplantation: relevance to transplantation outcome. *Human Immunology*, Vol.67, No.3, (March 2006), pp.238–246, ISSN 0198-8859

Laguila Visentainer, J.E.; Lieber, S.R.; Lopes Persoli, L.B.; Dutra Marques, S.B.; Vigorito, A.C.; Penteado Aranha, F.J.; de Brito Eid, K.A.; Oliveira, G.B.; Martins Miranda, E.C.; Bragotto, L. & de Souza, C.A. (2005). Relationship between cytokine gene polymorphisms and graft-versus-host disease after allogeneic stem cell transplantation in a Brazilian population. *Cytokine*, Vol.32, No.3-4, (November 2005), pp.171-177, ISSN: 1043-4666

Lanier, L.L. (2008). Up on the tightrope: natural killer cell activation and inhibition. *Nature Immunology*, Vol.9, No.5, (May 2008), pp.495-502, ISSN 1529-2908

Leelayuwat, C.; Townend, D.C.; Degli-Esposti, M.A.; Abraham, L.J. & Dawkins, R.L. (1994). A new polymorphic and multicopy RL: A new polymorphic and multicopy MHC gene family related to non mammalian class I. *Immunogenetics,* Vol.40, No.51, (1994), pp. 339-351, ISSN 0093-7711

Leffell, M.S.; Vogelsang, G.B.; Lucas, D.P.; Delaney, N.L. & Zachary, A.A. (2001). Association between TGF-beta expression and severe GVHD in allogeneic bone marrow transplantation. *Transplantation Proceedings*, Vol.33, No.1-2, (February-March 2001), pp.485-486, ISSN 0041-1345

Letterio, J.J. & Roberts, A.B. (1998). Regulation of immune responses by TGFbeta. *Annual Review of Immunology*, Vol.16, (April 1998), pp.137-161, ISSN 15453278

Liem, L.M.; Fibbe, W.E.; van Houwelingen, H.C. & Golmy, E. (1999). Serum transforming growth factor-b1 levels in bone marrow transplant recipients correlate with blood cell counts and chronic graft-versus-host disease. *Transplantation*, Vol.67, No.1, (January 1999), pp. 59-65, ISSN 0041-1337

Liem, L.M.; van Houwelingen, H.C. & Goulmy, E. (1998). Serum cytokine levels after HLA-identical bone marrow transplantation. *Transplantation*, Vol.66, No.7, (October 1998), pp.863-871, ISSN 0041-1337

Lin, M.T., Storer, B., Martin, P.J., Tseng, L.H., Gooley, T., Chen, P.J. & Hansen, J.A. (2003). Relation of an interleukin-10 promoter polymorphism to graft-versus-host disease and survival after hematopoietic-cell transplantation. *The New England Journal of Medicine*, Vol.349, No.23, (December 2003), pp.2201-2210, ISSN 0028-4793

Ludajic, K.; Balavarca, Y.; Bickeböller, H.; Rosenmayr, A.; Fae, I.; Fischer, G.F.; Kouba, M.; Pohlreich, D.; Kalhs, P. & Greinix, H.T. (2009). KIR genes and KIR ligands affect occurrence of acute GVHD after unrelated, 12/12 HLA matched, hematopoietic stem cell transplantation. *Bone Marrow Transplantation*, Vol.44, No.2, (July 2009), pp.97–103, ISSN 0268-3369

Macmillan, M.L.; Radloff, G.A.; Kiffmeyer, W.R.; Defor, T.E.; Weisdorf, D.J. & Davies, S.M. (2003). High-producer interleukin-2 genotype increases risk for acute graft-versus-host disease after unrelated donor bone marrow transplantation. *Transplantation*, Vol.76, No.12, (December 2003), pp.1758-1762, ISSN 0041-1337

Margolis, J. & Vogelsang, G. (2000). Chronic graft-versus-host disease. *Journal of Hematotherapy & Stem Cell Research*, Vol.9, No.3, (June 2000), pp.339-346, ISSN 1061-6128

McVicar, D.W. & Burshtyn, D.N. (2001). Intracellular signaling the killer immunoglobulin-like receptors and Ly49. *Sciences´s STKE: signal transduction knowledge environment*, Vol.2001, No.75, (March 2001), pp. re1, ISSN 1945-0877

Mlynarczewska, A.; Wysoczanska, B.; Karabon, L.; Bogunia-Kubik, K. & Lange, A. (2004) Lack of IFN-gamma 2/2 homozygous genotype independently of recipient age and intensity of conditioning regimen influences the risk of aGVHD manifestation after HLA-matched sibling haematopoietic stem cell transplantation. *Bone Marrow Transplantation*, Vol.34, No.4, (August 2004), pp.339-344, ISSN 0268-3369

Miller, J.S.; Cooley, S.; Parham, P.; Farag, S.S.; Verneris, M.R.; McQueen, K.L.; Guethlein, L.A.; Trachtenberg, E.A.; Haagenson, M.; Horowitz, M.M.; Klein, J.P. & Weisdorf, D.J. (2007). Missing KIR-ligands are associated with less relapse and increased graft versus host disease (GVHD) following unrelated donor allogeneic HCT. *Blood*, Vol.109, No.11, (June 2007), pp.5058–5061, ISSN 0006-4971

Mizutani, K.; Terasaki, P.I.; Shih, R.N.; Pei, R. Ozawa, M. & Lee, J. (2006). Frequency of MIC Antibody in Rejected Renal Transplant Patients without HLA Antibody. *Human Immunology*, Vol.67, No.3, (March 2006) pp. 223–229, ISSN 0198-8859

Moretta, A.; Pende, D.; Locatelli, F. & Moretta, L. (2009). Activating and inhibitory killer immunoglobulin-like receptors (KIR) in haploidentical haemopoietic stem cell transplantation to cure high-risk leukaemias. *Clinical Experimental of Immunology*, Vol.157, No.3, (September 2009), pp.325–331, ISSN 0093-9104

Murai, M.; Yoneyama, H.; Ezaki, T.; Suematsu, M.; Terashima, Y.; Harada, A.; Hamada, H.; Asakura, H.; Ishikawa, H & Matsushima, K. (2003). Peyer's patch is the essential site in initiating murine acute and lethal graft-versus-host reaction. *Nature Immunology*, Vol.4, No.2, (February 2003), pp. 3154-3160, ISSN 1529-2908

Narayan, S.; Tsai, E.W.; Zhang, Q.; Wallace, W.D.; Reed, E.F. & Ettenger, R.B. (2011). Acute rejection associated with donor specific anti-MICA antibody in a highly sensitized pediatric renal transplant recipient. *Pediatric Transplantation*, Vol.15, No.1, (February 2011), pp.E1-7, ISSN 1397-3142

O'Connor, G.M.; Hart, O.M. & Gardiner, C.M. (2006). Putting the natural killer cell in its place. *Immunology*, Vol.117, No.1, (January 2006), pp.1-10, ISSN 0019-2805

Paczesny, S.; Hanauer, D.; Sun, Y. & Reddy, P. (2010). New perspectives on the biology of acute GVHD. *Bone Marrow Transplantation*, Vol.45, No.1, (January 2010), pp.1–11, ISSN 0268-3369

Panigrahi, A.; Gupta, N.; Siddiqui, J.A.; Margoob, A.; Bhowmik, D.; Guleria, S. & Mehra, N.K. (2007). Post Transplant Development of MICA and Anti-HLA Antibodies is Associated with Acute Rejection Episodes and Renal Allograft Loss. *Human Immunology*, Vol.68, No.5, (May 2007), pp.362–367, ISSN 0198-8859

Parham, P. (2006). Taking license with natural killer cell maturation and repertoire development. *Immunological Reviews*, Vol.214, No.1, (December 2006), pp.155-160, ISSN 1529-2908

Parmar, S.; Del Lima, M.; Zou, Y.; Patah, P.A.; Liu, P.; Cano, P.; Rondon, G.; Pesoa, S.; de Padua Silva, L.; Qazilbash, M.H.; Hosing, C.; Popat, U.; Kebriaei, P.; Shpall, E.J.; Giralt, S.; Champlin, R.E.; Stastny, P. & Fernandez-Vina, M. (2009). Donor-recipient mismatches in MHC class I chain-related gene A in unrelated donor transplantation lead to increased incidence of acute graft-versus-host disease. *Blood*, (October 2009), Vol.114, No.14, pp.2884–2887, ISSN 0006-4971

Pérez-Simón, J.A.; Sánchez-Abarca, I.; Díezcampelo, M.; Caballero, D. & San Miguel, J. (2006). Chronic Graft-Versus-Host Disease Pathogenesis and Clinical Management. *Drugs*, Vol.66, No.8, (2006), pp. 1041-1057, ISSN 0012-6667

Pravica, V.; Asderakis, A.; Perrey, C.; Hajeer, A.; Sinnott, P.J. & Hutchinson, I.V. (1999). In vitro production of IFN-gamma correlates with CA repeat polymorphism in the human IFN-gamma gene. *European Journal of Immunogenetics*, Vol.26, No.1, (February 1999), pp.1-3, ISSN 0960-7420

Przepiorka, D.; Weisdorf, D.; Martin, P.; Klingemann, H.G.; Beatty, P.; Hows, J & Thomas, E.D. (1995). 1994 Consensus Conference on Acute GVHD Grading. Bone Marrow Transplantation, Vol.15, No.6, (June 1995), pp. 825-828, ISSN 0268-3369

Rajagopalan, S. & Long, E.O. (1999). A human histocompatibility leukocyte antigen (HLA)-G-specific receptor expressed on all natural killer cells. *Journal of Experimental Medicine*, Vol.189, No.7, (1999), pp.1093-1100. ISSN 0022-1007

Rajagopalan, S. & Long, E.O. (2005). Understanding how combinations of HLA and KIR genes influence disease. *The Journal of Experimental Medicine,* Vol.201, No.7, (April 2005), pp.1025-1029, ISSN 0022-1007

Rashidi-Nezhad, A.; Azimi, C.; Alimoghaddam, K.; Ghavamzadeh, A.; Hossein-Nezhad, A.; Izadi, P.; Sobhani, M.; Noori-Daloii, A.R. & Noori-Daloii, M.R. (2010). TGF-Beta codon 25 polymorphism and the risk of graft-versus-host disease after allogenic hematopoietic stem cell transplantation. *Iranian Journal of Allergy, Asthma and Immunology*, Vol.9, No.1, (March 2010), pp.1-6, ISSN 1735-1502

Reddy, P. & Ferrara, J.L. (2003). Immunobiology of acute graft-versus-host disease. *Blood Reviews*, (December 2003), Vol.17, No.4, pp.187–194, ISSN 0268-960X

Reviron, D.; Dussol, B.; Andre, M.; Brunet, P.; Mercier, P. & Berland, Y. (2001). TNF-alpha and IL-6 gene polymorphism and rejection in kidney transplantation recipients. *Transplantation Proceedings*, Vol.33, No.1-2, (February-March 2001), pp.350-351, ISSN 0041-1345

Sakata, N; Yasui, M.; Okamura, T.; Inoue, M.; Yumura-Yagi, K. & Kawa, K. (2001). Kinetics of plasma cytokines after hematopoietic stem cell transplantation from unrelated donors: the ratio of plasma IL-10/sTNFR level as a potential prognostic marker in severe acute graft-versus-host disease. *Bone Marrow Transplantation*, Vol.27, No.11, (June 2001), pp. 1153-1161, ISSN 0268-3369

Schellekens, J.; Rozemuller, E.H.; Petersen, E.J.; van den Tweel, J.G.; Verdonck, L.F. & Tilanus, M.G. (2008). Activating KIRs exert a crucial role on relapse and overall survival after HLA-identical sibling transplantation. *Molecular Immunology*, Vol.45, No.8, (April 2008), pp.2255-2261, ISSN 0161-5890

Shamim, Z.; Ryder, L.P.; Christensen, I.J.; Toubert, A.; Norden, J.; Collin, M.; Jackson, G.; Dickinson, A.M. & Müller, K. (2011). Prognostic Significance of Interleukin-7 Receptor-α Gene Polymorphisms in Allogeneic Stem-Cell Transplantation: A Confirmatory Study. *Transplantation*, (April 2011), Vol.91, No.7, pp.731–736, ISSN 0041-1337

Sivula, J.; Turpeinen, H.; Volin, L. & Partanen, J. (2009). Association of IL-10 and IL-10Rβ gene polymorphisms with graft-versus-host disease after haematopoietic stem cell transplantation from an HLA-identical sibling donor. *BMC Immunology*, Vol.10, (May 2009), pp.24-30, ISSN 1471-2172

Steinle, A.; Li, P.; Morris, D.L.; Groh, V.; Lanier, L.L.; Strong, R.K. & Spies, T. (2001). Interactions of human NKG2D with its ligands MICA, MICB, and homologs of the mouse RAE-1 protein family. *Immunogenetics*, Vol.53, No.4, (May-June 2001), pp.279-287, ISSN: 0093-7711

Sumitran-Holgersson, S. (2008). Relevance of MICA and other non- HLA antibodies in clinical transplantation. *Current Opinion of Immunology*, Vol.20, No.5 (October 2008), pp.607–613, ISSN 0952-7915

Sun, J.Y.; Gaidulis, L.; Dagis, A.; Palmer, J.; Rodriguez, R.; Miller, M.M.; Forman, S.J. & Senitzer, D. (2005). Killer Ig-like receptor (KIR) compatibility plays a role in the prevalence of acute GVHD in unrelated hematopoietic cell transplants for AML. *Bone Marrow Transplantation*, Vol.36, No.6, (September 2005), pp.525-530, ISSN 0268-3369

Symons, H.J.; Leffell, M.S.; Rossiter, N.D.; Zahurak, M.; Jones, R.J. & Fuchs, E.J. (2010). Improved Survival with Inhibitory Killer Immunoglobulin Receptor (KIR) Gene Mismatches and KIR Haplotype B Donors after Nonmyeloablative, HLA-Haploidentical Bone Marrow Transplantation. *Biology of Blood and Marrow Transplantation*, Vol.16, No.4, (April 2010), pp.533-542, ISSN 1083-8791

Takahashi, H.; Furukawa, T.; Hashimoto, S.; Susuki, N.; Yamazaki, F.; Inano, K.; Takahashi, M.; Aizawa, Y. & Koike, T. (2000). Contribution of TNF-alpha and IL-10 gene polymorphisms to graftversus-host disease following allo-hematopoietic stem cell transplantation. *Bone Marrow Transplantation*, Vol.26, No.12, (December 2000), pp.1317-1323, ISSN 0268-3369

Tambur, A.R.; Yaniv, I.; Stein, J.; Lapidot, M.; Shabtai, E.; Kfir, B. & Klein, T. (2001). Cytokine gene polymorphism in patients with graft-versus-host disease. *Transplantation Proceedings*, Vol.33, No.1-2, (February-March 2001), pp.502-503, ISSN 0041-1345

Tawara, I.; Maeda, Y.; Sun, Y.; Lowler, K.P.; Liu, C.; Toubai, T.; McKenzie, A.N & Reddy, P. (2008). Combined Th2 cytokine deficiency in donor T cells aggravates experimental acute graft-vs-host disease. *Experimental Hematology*, Vol.36, No.8, (August 2008), pp.988–996, ISSN 0301-472X

Tay, G.K.; Witt, C.S.; Christiansen, F.T.; Charron, D.; Baker, D.; Herrmann, R.; Smith, L.K.; Diepeveen, D.; Mallal, S. & McCluskey, J. (1995). Matching for MHC haplotypes results in improved survival following unrelated bone marrow transplantation. *Bone Marrow Transplantation*, Vol.15, No.3, (March 1995), pp.381–385, ISSN 0268-3369

Terasaki, P.I.; Ozawa, M. & Castro R. (2007). Four-year follow-up of a prospective trial of HLA an MICA antibodies on kidney graft survival. *American Journal of Transplantation*, Vol.7, No.2, (February 2007), pp. 408–415, ISSN 1600-6135

Tieng, V.; Le Bouguenec, C. ; du Merle, L.; Bertheau, P.; Desreumaux, P.; Janin, A.; Charron, D. & Toubert, A. (2002). Binding of Escherichia coli adhesin AfaE to CD55 triggers cell-surface expression of the MHC class I-related molecule MICA. *Proceedings of the National Academy of Sciences of the United States of America*, Vol.99, No.5, (March 2002), pp.2977-2982, ISSN 0027-8424

Turner, D.M.; Williams, D.M.; Sankaran, D.; Lazarus, M.; Sinnott, P.J. & Hutchinson, I.V. (1997). An investigation of polymorphism in the interleukin-10 gene promoter. *European Journal of Immunogenetics*, Vol.24, No.1, (February 1997), pp.1-8, ISSN 0960-7420

Uhrberg, M.; Valiante, N.M.; Shum, B.P.; Shilling, H.G.; Lienert-Weidenbach, K.; Corliss, B.; Tyan, D.; Lanier, L.L. & Parham, P. (1997). Human diversity in killer cell inhibitory receptor genes. *Immunity*, Vol.7, No.6, (December 1997), pp.753-763, ISSN 1074-7613

Venstrom, J.M., Ted, A. Gooley, T.A.; Spellman, S.; Pring, J.; Malkki, M.; Dupont, B.; Petersdorf, E. & Hsu, K.C. (2010). Donor activating KIR3DS1 is associated with decreased acute GVHD in unrelated allogeneic hematopoietic stem cell transplantation. *Blood*, Vol.115, No.15, (April 2010), pp.3162-3165, ISSN 0006-4971

Viel, D.O.; Tsuneto, L.T.; Sossai, C.R.; Lieber, S.R.; Marques, S.B.; Vigorito, A.C.; Aranha, F.J.; De Brito Eid, K.A.; Oliveira, G.B.; Miranda, E.C.; De Souza, C.A. & Visentainer, J.E. (2007). IL2 and TNFA gene polymorphisms and the risk of graft-versus-host disease after allogeneic stem cell transplantation. *Scandinavian Journal of Immunology*, Vol.66, No.6, (December 2007), pp.703-710, ISSN 0300-9475

Visentainer, J.E.; Lieber, S.R.; Persoli, L.B.; Vigorito, A.C.; Aranha, F.J.; Eid, K.A.; Oliveira, G.B.; Miranda, E.C. & de Souza, C.A. (2003). Serum cytokine levels and acute graft-versus-host disease after HLA-identical hematopoietic stem cell transplantation. *Experimental Hematology*, Vol.31, No.11, (November 2003), pp. 1044-1050, ISSN 0301-472X

Wilson, A.G.; Symons, J.A.; Mcdowell, T.L.; Mcdevitt, H.O. & Duff, G.W. (1997). Effects of a polymorphism in the human tumor necrosis factor alpha promoter on transcriptional activation. *Proceedings of the National Academy of Sciences of the United States of America*, Vol.94, No.7, (April 1997), pp.3195-319, ISSN 1091-6490

Witt, C.; Sayer, D.; Trimboli, F.; Saw, M.; Herrmann, R.; Cannell, P.; Baker, D. & Christiansen, F. (2000). Unrelated donors selected prospectively by block-matching have superior bone marrow transplant outcome. *Human Immunology*, Vol. 61, No.2, (February 2000), pp.85-91, ISSN 0198-8859

Witt, C.S. & Christiansen, F.T. (2006). The relevance of natural killer cell human leukocyte antigen epitopes and killer cell immunoglobulinlike receptors in bone marrow transplantation. *Vox Sanguinis*, Vol.90, No.1, (January 2006), pp.10-20, ISSN 0042-9007

Yi, T.; Zhao, D.; Lin, C.L.; Zhang, C.; Chen, Y.; Todorov, I.; LeBon, T.; Kandeel, F.; Forman, S. & Zeng, D. (2008). Absence of donor Th17 leads to augmented Th1 differentiation and exacerbated acute graft-versus-host disease. *Blood*, Vol.112, No.5, (September 2008), pp.2101–2110, ISSN 0006-4971

Yokoyama, W.M. & Kim, S. (2006). Licensing of natural killer cells by self major histocompatibility complex class I. *Immunological Reviews*, Vol.214, No.1, (December 2006), pp.143-154, ISSN 1529-2908

Zhang, C.; Todorov, I.; Zhang, Z.; Liu, Y.; Kandeel, F.; Forman, S.; Strober, S. & Zeng, D. (2006). Donor CD4+ T and B cells in transplants induce chronic graft-versus-host disease with autoimmune manifestations. *Blood*, Vol.107, No.7, (April 2006), pp.2993-3001, ISSN 0006-4971

Zhang, Y. & Stastny, P. (2006). MICA antigens stimulate T cell proliferation and cell-mediated cytotoxicity. *Human Immuno*logy, Vol.67, No.3, (March 2006), pp.215–222, ISSN 0198-8859

Zou, Y.; Stastny, P.; Susal, C.; Dohler, B. & Opelz G. (2007). Antibodies against MICA antigens and kidneytransplant rejection. *The New England Journal of Medicine*, Vol.357, No.13, (September 2007), pp. 1293-1300, ISSN 0028-4793

Zwirner, N.W.; Dole, K. & Stastny, P. (1999). Differential surface expression of MICA by endothelial cells, fibroblasts, keratinocytes, and monocytes. *Human Immuno*logy, Vol.60, No.4, (April 1999), pp.323–330, ISSN 0198-8859

3

Relevance of HLA Expression Variants in Stem Cell Transplantation

Britta Eiz-Vesper and Rainer Blasczyk
Institute for Transfusion Medicine, Hannover Medical School, Hannover Germany

1. Introduction

Matching the donor and recipient for class I and II human leukocyte antigens (HLA) is pivotal to the success of allogeneic hematopoietic stem cell transplantation (HSCT). Transplantation across HLA barriers will lead to the development of T-cell responses to the mismatched HLA molecules, resulting in T-cell–mediated graft-versus-host disease (GvHD) or graft rejection in patients with insufficient immune suppression. The accuracy of testing and matching criteria has an important impact on the transplant outcome, but exact matching across multiple HLA loci (e.g., HLA-A, HLA-B, HLA-C, and HLA-DRB1) is a challenging task. Today, serological HLA diagnostic tests are being replaced by DNA-based typing methods considering only selected regions of the genes. Therefore, HLA null alleles or expression variants bearing their variation outside of these regions may be misdiagnosed as normally expressed variants, resulting in HLA mismatches that are highly likely to stimulate allogeneic T cells and trigger GvHD. This chapter will address the relevance, genetics, prevalence and diagnosis of HLA expression, variants of HLA class I loci and will discuss their clinical implications for transplantation.

2. The human major histocompatibility complex

The human major histocompatibility complex (MHC), also referred to as the human leukocyte antigen (HLA) complex, is encoded on the short arm of chromosome 6 (6p21) and is extremely polymorphic (Parham et al. 1988). HLA class I molecules are expressed on most nucleated cells. The HLA class I region comprises the gene loci for the heavy chains of the three classical human leukocyte antigens, HLA-A, -B, and -C. They consist of a heavy chain (44 kDa) and a non-covalently bound β2 microglobulin (β2m) light chain (12 kDa) encoded by chromosome 15. The heavy chain is made up of three extracellular domains: α1, α2, and α3. The highly polymorphic region of HLA class I molecules is located in the DNA and amino acid sequences of the α1 and α 2 domains, which form the peptide-binding groove. Endogenous 8 to 12 amino acid peptides are presented to CD8+ cytotoxic T lymphocytes (CTLs) (Natarajan et al. 1999). The α3 domain is mainly invariant and contains the binding site for the co-receptor CD8. Because of the MHC's role in recognizing pathogenic and cancerous peptides, these genes are under high environmental pressure to be very polymorphic. A total of 4,946 HLA class I alleles have been identified to date (http://www.ebi.ac uk/imgt/hla; released April 2011).

HLA class I molecules are stabilized by disulfide bonds located in the α2 and α3 domains between cysteine (C) residues at amino acid positions 101/164 and 203/259. These bonds are essential for the correct processing and function of the molecules (Solheim 1999). Amino acid substitutions in these crucial C residues are likely to cause aberrant expression of the respective HLA class I molecules and may also change the affinity of the peptide-binding groove towards endogenous peptides (Warburton et al. 1994; Hirv et al. 2006; Hinrichs et al. 2009; Hinrichs et al. 2010).

HLA class II molecules (DR, DQ, DP) are mainly expressed on hematopoietic cells (macrophages, dendritic cells, T cells and B cells). The heterodimers are formed by two membrane-bound chains (α and β), each consisting of two domains (α1/α2 or β1/β2, respectively) encoded by two genes co-located in the centromeric part of the MHC. The antigenic peptide (up to 30 amino acids) is presented to CD4+ T helper cells (T_h cells) in a cleft formed by the outermost α1 and β1 domains. Nearly all of the polymorphisms occur at exon 2 of the respective A or B genes. Peptides presented by HLA class II molecules are derived from exogenous proteins as well as from epitopes of plasma membranes or endosomes (Rudensky et al. 1991; Chicz et al. 1993; Sant 1994). The nonpolymorphic β2 domain contains the binding site for the T cell co-receptor CD4. More than 1,457 HLA class II alleles have been identified to date (http://www.ebi.ac uk/imgt/hla; released April 2011).

2.1 Peptide presentation by HLA

The ability to recognize and distinguish between self and non-self is primarily mediated by T lymphocytes, which survey the protein environment of cell surfaces for binding partners, i.e. for signs of foreign invasion. T cells do not recognize proteins directly; instead, they recognize imprints of ongoing protein metabolism in the form of peptides presented by HLA molecules. This phenomenon is called MHC restriction. The biological function of HLA molecules is to present antigenic peptides to T cells. Therefore, HLA molecules play a central role in T cell-mediated adoptive immunity. MHC class I molecules present peptides from endogenously synthezised proteins, whereas MHC class II molecules present peptides from incorporated exogenous proteins. All of these peptides originate from foreign or host cell proteins and are generated by proteasomal cleavage (class I pathway) or lysosomal processing (class II pathway). It has been estimated that about 0.5% of presented peptides are bound to MHC molecules, whereas more than 99% are ignored. Consequently, peptide binding to HLA is the single most selective event involved in antigen processing and presentation (Yewdell, Norbury, and Bennink 1999; Yewdell and Bennink 2001). A T cell-mediated immune response occurs when the T-cell receptor recognizes a specific peptide-MHC complex and thus identifies cells that have been infected by intracellular parasites or viruses or cells containing abnormal proteins (e.g., tumor cells). The peptides beeing part of a certain peptide-MHC complex triggering T-cell recognition are important tools for diagnosis and treatment of infectious, autoimmune, allergic and neoplastic diseases (Ferrari et al. 2000; Haselden, Kay, and Larche 2000; Singh 2000; Wang, Phan, and Marincola 2001).

Different polymorphic HLA molecules have different peptide binding specificities (Falk et al. 1991; Sette et al. 1994; Bade-Doeding et al. 2007; Bade-Doeding et al. 2011). Peptides presented by MHC class I molecules are derived from cytoplasmic proteins by proteolytic degradation in the proteasome. Therefore, the MHC class I presentation pathway is often called the cytosolic or endogenous pathway. The MHC class I crystal structure features a unique peptide-binding groove at the outer polymorphic α2 and α3 domains (Bjorkman et al. 1987, Madden et al. 1991). This groove can be subdivided into six pockets (A-F) of different size, shape, and function (Garrett et al. 1989; Matsumura et al. 1992). A pocket is defined as a unit having an affinity for a certain peptide side chain (e.g., affinity of pocket A for peptide position P1 and pocket B for P2). Some pockets have a well-shaped structure with an affinity for only one side chain, whereas others have an affinity for a group of side chains. In some cases, the boundaries between pockets are unclear. The most important residues and positions of a peptide are known as anchor residues and anchor positions. The identity and spacing of these primary anchors constitutes the peptide motif of an HLA specificity (Sette et al. 1987; Sette et al. 1989; Jardetzky et al. 1991; Ruppert et al. 1994; Rammensee, Friede, and Stevanoviic 1995). A typical peptide is 8 to 12 amino acids in length and binds in the peptide-binding groove, exhibiting an extended conformation with its terminal amino group bound to a pocket at one end of the groove and its terminal carboxyl group bound to a pocket at the other end of the groove.

Peptide binding motifs generally contain two to three anchor positions (Rammensee, Friede, and Stevanoviic 1995). Other features such as secondary anchors and disfavored residues have also been described as playing an important role in defining the peptide-MHC interaction (Ruppert et al. 1993). The peptide-binding cleft of HLA class II molecules is formed by the outer α1 and β1 domains. Since it does not narrow at the ends, it can accommodate longer peptides containing up to 30 but usually 13 to 17 amino acids. The peptides presented by class II molecules are derived from extracellular proteins internalized by endophagocytosis and degraded in an endocytic compartment. Hence, the MHC class II-dependent pathway of antigen presentation is called the endocytic or exogenous pathway.

2.2 HLA nomenclature and typing methods

According to the World Health Organization (WHO) Committee on Nomenclature for Factors of the HLA System (Holdsworth et al. 2009), each HLA allele name has a unique number corresponding to up to four sets of digits separated by colons. The 2-digits before the first colon describe the type, which often corresponds to the serological antigen carried by an allotype. The next set of digits are used to list the subtypes, numbers being assigned in the order in which DNA sequences have been determined. Broad families of alleles are clustered into serotypes (e.g., HLA-A1).

There are two levels of typing: low-resolution (2-digits) and high-resolution (at least 4-digits). Low-resolution typing delivers results equivalent to serological typing and can be achieved by serological (microlymphocytotoxicity test) or molecular techniques. Due to its simplicity and low cost, serologic typing is still used in some laboratories. High-resolution typing can only be achieved by DNA-based techniques allowing classification of the individual alleles within each serotype (e.g., HLA-A*01:01). A number of HLA typing methods based on PCR technology have been developed. PCR with sequence-specific primers (PCR-SSP), PCR followed by sequence-specific oligonucleotide probing (PCR-SSO) and PCR followed by sequencing-based typing (PCR-SBT) are currently the most commonly used molecular methods for low- and high-resolution HLA typing. These methods have displaced serology in most laboratories because of a much greater accuracy.

2.2.1 Types and nomenclature of HLA expression variants

To label HLA alleles with an alternative expression pattern the WHO Nomenclature Committee for Factors of the HLA System defined suffixes ('N', 'L', 'S', 'C', 'A', 'Q'), that are added to an allele name to indicate its expression status (Holdsworth et al. 2009). Alleles shown to be not expressed ('Null' alleles) are given the suffix 'N'. The alteration does not necessarily imply the lack of production of an internal partial product which might be a T-cell target (Elsner and Blasczyk 2004). HLA alleles with 'Low' cell surface expression of an intact antigen compared to normal levels are indicated using the suffix 'L'. The suffix 'S' is used to denote an allele specifying a protein which is exclusively expressed as a 'Secreted' molecule but not as a cell surface protein. A 'Q' suffix is used when the expression of an allele is 'Questionable' given that the mutation seen in the allele has previously been shown to affect normal expression levels. The suffix 'C' is used to denote an allele product found in the 'Cytoplasm' but not on the cell surface, and the suffix 'A' indicates 'Aberrant' expression.

Currently, 197 HLA class I alleles (168 N, 5 L, 24 Q and 1 S allele) and 21 HLA class II alleles (all null alleles) with variant expression are listed in the IMGT/HLA database on the HLA nomenclature website (www.ebi.ac.uk/imgt/hla; released April 2011). As of April 2011, no alleles have been named with a 'C' or 'A' suffix. Most of these alleles carry mutations causing stop codons, leaving no doubt about their non-expression. Examples include HLA-A*02:82N, HLA-A*23:08N, HLA-A*24:132N, HLA-B*14:07N, HLA-B*39:40N, HLA-B*46:07N, HLA-B*56:190N, or HLA-C*06:49N. In the case of HLA-A*03:03N, a frame deletion is responsible for non-expression (Lienert et al. 1996).

Only four HLA-A alleles (HLA-A*01:01:38L, HLA-A*02:01:01:02L, HLA-A*24:02:01:02L, HLA-A*30:14L) and one HLA-B allele (HLA-B*39:01:01:02L) with low-expression patterns have been identified up to now (Balas et al. 1994; Magor et al. 1997; Laforet et al. 1997; Dunn et al. 2004; Hirv et al. 2006; Perrier et al. 2006). Low expression of these alleles is usually associated with a low expression of the corresponding mRNA. However, the alteration causing the low expression of HLA*A-30:14L is not associated with a reduced mRNA level, but rather seems to result from the loss of the disulfide bond between the cysteine residues at positions 101 and 164 in the α2 domain (Hirv et al. 2006; Hinrichs et al. 2009).

The only soluble secreted allele (S) known so far is HLA-B*44:02:01:02S (Dubois et al. 2004). This HLA-B44 variant was typed as a null allele by microlymphocytotoxicity, whereas the B*44:02:01:01 allele was identified by PCR-SSP. DNA sequencing revealed a single nucleotide difference at the end of intron 4 in the acceptor splicing site, leading to a splicing error characterized by the deletion of exon 5 (transmembrane domain of the HLA antigen).

All known HLA class I Q alleles (7 HLA-A, 9 HLA-B and 8 HLA-C) and the HLA-A*30:14L allele have cysteine residue mutations at amino acid position 101 or 164 affecting the 101/164 disulfide bridge in the α2 domain. Point mutations altering codon 101 have been described for HLA-C*02:25Q and HLA-C*03:22Q (Middleton et al. 2006). In the case of HLA-A*02:293Q, HLA-A*11:50Q, HLA-A*30:14L (Hirv et al. 2006), HLA-A*32:11Q (Tang et al. 2006), HLA-B*15:218Q, HLA-B*35:65Q (Elsner et al. 2006), HLA-B*37:16Q, HLA-B*39:38Q (Tang et al. 2006), HLA-B*40:133Q, HLA-C*04:59Q, HLA-C*07:121Q, HLA-C*12:42Q, HLA-C*15:32Q and HLA-C*16:16Q, point mutations in codon 164 result in a replacement of the Cys residue, causing disruption of the disulfide bond in the α2 domain. HLA-A*30:14L is the only one of these alleles described as having a low expression pattern not affecting the corresponding mRNA levels (Hirv et al. 2006; Hinrichs et al. 2009). There are no known alleles with an amino acid mutation at positions 203 or 259 affecting the bridge in the α3 domain.

2.3 HLA in transplantation

The best donor is an HLA genotypically matched sibling identified by family typing. When no identical sibling donor is available, transplantation of stem cells from an HLA-matched unrelated donor can result in comparable disease-free survival, particularly for good-risk patients (Petersdorf et al. 2004; Petersdorf 2007; 2008). Nevertheless, unrelated transplantation is associated with a higher frequency of post-transplant complications than in genotypically matched sibling HSCT, mainly because of undefined HLA incompatibilities. The negative impact of HLA mismatches on the outcome of hematopoietic stem cell transplantation has been demonstrated in a variety of studies (Mickelson et al. 2000; Ottinger et al. 2003; Schaffer et al. 2003). Most allele mismatches affect differences in the T-cell receptor contact area of the heavy chain or the peptide-binding site causing a change in the peptide binding repertoire both leading to a T cell-mediated allorecognition.

HistoCheck (www.histocheck.org) is an online tool which helps clinicians and researchers visualize the amino acid substitutions of HLA alleles so that they can make informed judgments about their functional similarity (Elsner et al. 2004). Because exact HLA matching is often not possible, it is important to understand which alleles are the most similar. *HistoCheck* provides crystallography-based 3-dimensional (3D) visualizations of the allelic mismatches by highlighting amino acid mismatches, positions, and functions. The user is provided with dissimilarity scores (DSSs) for the amino acids involved as well as an over-all DSS for the two alleles. However, scoring HLA mismatches by HistoCheck has not been shown to predict clinical outcome in unrelated hematopoietic stem cell transplantation.

Several large-scale studies have shown that high-resolution matching of patients and unrelated donors significantly improves post-transplant survival (Bray et al. 2008), the incidence and severity of acute and chronic GVHD (Morishima et al. 2002; Morishima et al. 2007), and engraftment (Petersdorf et al. 2001; Flomenberg et al. 2004; Lee et al. 2007; Petersdorf 2008). Regarding cord blood transplantation, several studies have shown that the degree of HLA match is important as well, but a large cell dose may be at least equally important (Laughlin et al. 2004; Rocha, Sanz, and Gluckman 2004; Arcese et al. 2006; Eapen et al. 2007).

The National Marrow Donor Program (NMDP¸ www.marrow.org) proposed minimum HLA matching requirements for adult donors for HLA-A, -B, -C and -DRB1 (8/8) typed, at high resolution by DNA-based methods and cord blood units (CBU) for HLA-A, -B, (low resolution) and -DRB1 (high resolution) (Table 1) (Bray et al. 2008; Kamani et al. 2008).

Considering HLA allele and haplotype frequencies can be very useful when interpreting typing results and finding appropriate donors. Simply knowing that a patient's haplotype is extremely rare can prevent futile registry searches. Considering allele frequency alone is insufficient, because a rare allele can be acceptable when it is found in its most common haplotype. Being aware of rare alleles and haplotypes is also an important factor in quality control. Furthermore, typing results in registries are often incomplete. In the case where there are two matching donors, but each donor typing is incomplete with respect to different alleles, then haplotype frequencies can help choose the donor who is most likely to be an exact match. To overcome these limitations the new matching algorithm *HapLogic* (www.marrow.org) and *Haplocheck* (www.haplocheck.org) were developed. *HapLogic* a new enhanced matching algorithm that automatically identifies the donors or CBUs with the highest potential to match the patient, was established by the NMDP to accelerate and improve the efficiency of searches. The new matching algorithm analyzes the haplotypes of millions of donors on NMDP's *Be The Match Registry*. HapLogic uses advanced logic to predict a donor's or CBU's high-resolution match and builds upon mathematical formulas that predict DR match in AB donors (Hurley et al. 2006).

HLA locus	Tissue type patient?	Match donor and patient?
A	Yes, allele level	Yes
B	Yes, allele level	Yes
C	Yes, allele level	Yes
DRA	No	No
DRB1	Yes, allele level	Yes
DRB3, 4, and 5	Yes (DRB1 association)	Unknown
DQA1	No	No
DQB1	Yes (DRB1 association)	Uncertain
DPA1	No	No
DPB1	No	Uncertain

Table 1. HLA tissue typing recommended by the NMDP (from www.marrow.org) (Bray et al. 2008; Kamani et al. 2008)

The web tool *HaploCheck* is addressing this chance by ranking typing results based upon haplotype frequencies. The user enters the typing results for a patient, for which the cis/trans phase is unknown. The result is a list of separated haplotypes, ordered by frequency. Very rare alleles and associations are highlighted to inform the user of potential problems when searching registries, or to identify potential typing errors. For the case that a single mismatch is unavoidable, the user is presented with a list of mismatch-containing haplotypes and their frequencies. This can not only prevent futile registry searches, but also enable the clinician to make decisions about accepted mismatches before initiating a registry search.

3. Prevalence and allogenicity of HLA class I expression variants

Few investigators have systematically addressed the prevalence of HLA null and alternatively expressed alleles, which has been shown to be about 0.003% and 0.3%, respectively (Noreen et al. 2001; Elsner and Blasczyk 2004; Smith et al. 2005). Considering that most studies indicate that the prevalence of these alleles is around 1 per 1000 individuals, these alleles are not particularly rare. Consequently, it was recommended that laboratories typing unrelated bone marrow patients and donors should have a strategy to identify these expression variants (Elsner and Blasczyk 2004).

HLA null and expression variants are typically identified by the discrepancy between serological and molecular typing results. As molecular typing techniques have nearly displaced serological methods and are focusing on selected regions of the HLA genes many expression variants are likely to be overlooked. In solid organ transplantation, HLA expression variants are not considered in the matching procedure. In allogeneic HSCT, expression variants make an essential difference and can strongly affect transplant-related mortality since HLA mismatches are the major cause of severe GvHD or graft rejection. Thus, in contrast to solid organ transplantation, excluding HLA expression variants is required in the matching process for HSCT (Elsner and Blasczyk 2004; Hirv et al. 2006; Hinrichs et al. 2009).

Overlooking an HLA null allele in the donor would result in a T cell-mediated allorecognition of the recipient's HLA and may lead to the development of acute severe GvHD (Elsner and Blasczyk 2004). In the reverse setting (recipient null allele, donor expressed allele), allogeneic recognition of the recipient's stem cells may lead to their destruction and subsequent graft failure. Accordingly, mismatches between expressed and non-expressed HLA variants should be avoided in HSCT. In case of a recipient with an HLA null allele having no HLA-identical donor with the same null variant, matching must be performed as if the patient would be homozygous for the expressed allele of the respective HLA locus (Figure 1).

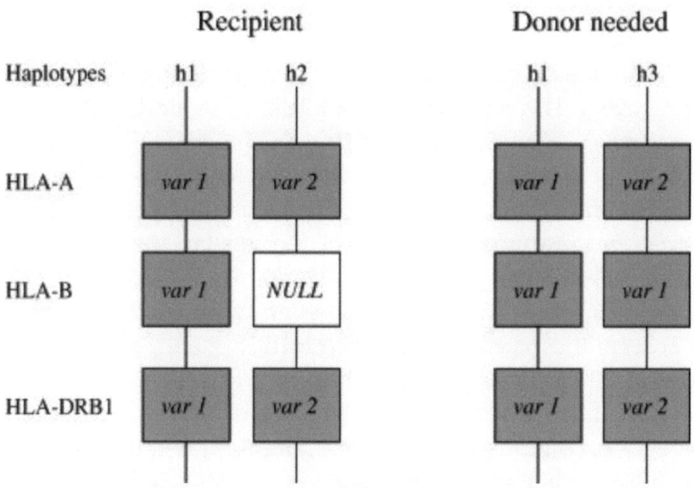

Fig. 1. Haplotypes in a recipient-donor combination with a null allele in one of the recipient's haplotypes. Shadowed boxes indicate normally expressed variants. The recipient carries an HLA-B null allele (white box). In the donor search the recipient's haplotype h2 has to be 'replaced' by haplotype (h3) containing the expressed HLA-B allele (var1). However, such a haplotype may be rare and a matching donor hard to find (Elsner and Blasczyk 2004).

On the other hand, an incomplete HLA molecule may be generated, as has been shown for HLA-B*44:02:01:02S, which might be presented via the indirect allogeneic recognition pathway (Magor et al. 1997; Dubois et al. 2004). Provided that the HLA-derived peptides fit into the peptide-binding groove and are capable of triggering a strong T-cell response, they may act as minor histocompatibility antigens (mHags). This could also apply to those HLA expression variants where the transcription of a truncated mRNA is known and/or translation is probable. It also shows that premature stop codons do not automatically lead to the interruption of transcription (Balas et al. 1994; Laforet et al. 1997; Magor et al. 1997; Dunn et al. 2004; Hirv et al. 2006; Perrier et al. 2006; Eiz-Vesper, Blasczyk, and Horn 2007). In the light of countless non-HLA mHags this is probably of inferior importance.

4. Characterization of HLA expression variants by cytokine-induced HLA secretion

Because of the clinical importance of expresion variants an HLA secretion assay was designed capable of discriminating between low-expression (L) and non-expressed (N) HLA variant alleles and assigning questionably expressed (Q) alleles to either group (Hinrichs et al. 2009).

All of the aforementioned HLA class I alleles with an unknown expression profile (Q alleles; 7 HLA-A, 9 HLA-B and 8 HLA-C) and HLA-A*30:14L, have a mutation of cysteine residue 101 or 164 affecting disulfide bridge 101/164 in the α2 domain. Because HLA-A*30:14L is the only one of these alleles described to have a low expression pattern with no effect on mRNA levels (Hirv et al. 2006; Hinrichs et al. 2009), A*30:14L was used as an expression model. HLA-A*30:14L was reported to be non-expressed under normal conditions and to show weak aberrant expression after cultivation of the corresponding B-lymphoblastoid cell line at 30°C (Hirv et al. 2006).

HLA-A*30:14L was originally identified in a patient suffering from chronic myeloid leukemia (Hirv et al. 2006). The sequence of this allele is identical to that of HLA-A*30:01 except for a transversion at nucleotide position 563 in exon 3 (guanine to cytosine substitution), resulting in a replacement of cysteine by serine at position 164, impairing disulfide bridge formation in the α2 domain of the mature polypeptide. This alteration of the secondary structure presumably decreases expression, rendering HLA-A*30:14L basically undetectable by serology.

Human cell lines (HEK293, C1R and K562) expressing recombinant soluble HLA (sHLA) molecules (Table 2) were incubated with interferon (IFN)-γ and/or tumor necrosis factor (TNF)-α (Hinrichs et al. 2009). These pro-inflammatory cytokines are known to enhance the expression of HLA molecules by affecting the interaction of DNA-binding proteins with the HLA-A promoter regions, resulting in the increased transcription of heavy and light chain genes (Girdlestone 1996; Gobin et al. 1997; Gobin et al. 1998; Gobin et al. 1999; Johnson 2003). In addition, these cytokines induce the transcription of proteasome subunits, peptide transporters and chaperones that promote the expresson of HLA class I molecules by providing peptides for presentation (Ma et al. 1997; Lankat-Buttgereit and Tampe 2002).

Expression of soluble HLA-A*30:14L and HLA-A*30:01 was measured in the supernatants of transfected and untransfected cells incubated with or without IFN-γ and/or TNF-α using a W6/32 and anti-β2-microglobulin-based sandwich ELISA (Figure 1) (Bade-Doeding et al. 2007). HLA-A*30:14L was not detected in the supernatant of unstimulated transfectants. Stimulation with IFN-γ and/or TNF-α increased HLA-A*30:14L secretion to detectable levels and increased HLA-A*30:01 expression up to 8-fold, but did not result in any difference between mRNA levels of HLA-A*30:14L and A*30:01 (Figure 2).

Day		Expression level (ng/ml)		
		HEK293	C1R	K562
1	HLA-A*30:01	31.3 ± 10.3	98.0 ± 14.7	155.8 ± 73.9
	HLA-A*30:14L	0	3.2 ± 1.7	0
3	HLA-A*30:01	383.2 ± 56.5	225.8 ± 177.5	143.5 ± 40.8
	HLA-A*30:14L	1.8 ± 1.4	0	0
7	HLA-A*30:01	160.4 ± 3.2	253.6 ± 16.1	175.9 ± 74.7
	HLA-A*30:14L	0	9.7 ± 4.9	0

Table 2. Soluble HLA-A*30:01 and HLA-A*30:14L expression levels (ng/ml) in the supernatant of three transfected cell lines (HEK293, C1R and K562) after 1, 3 and 7 days of incubation.

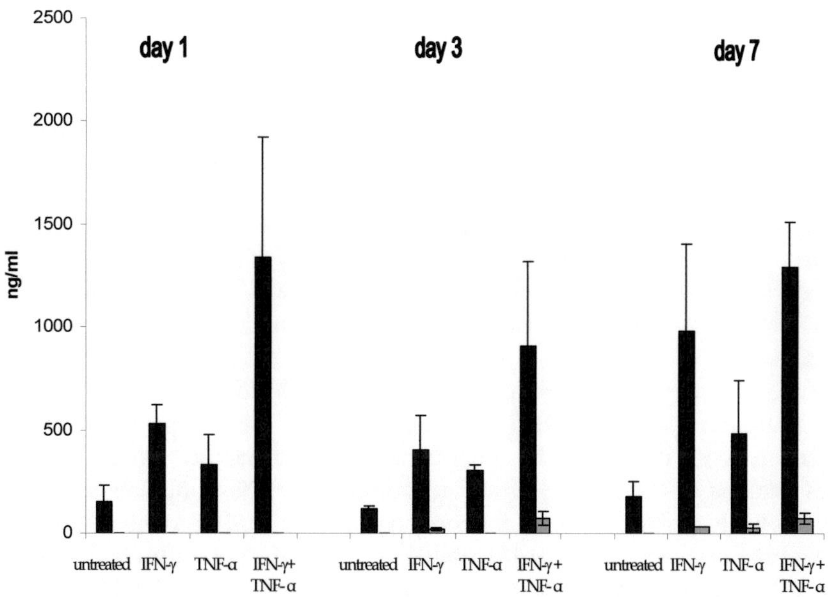

Fig. 2. Secretion of soluble HLA-A*30:01 and HLA-A*30:14L by transfected K562 cells

Expression of mRNA transcripts of both alleles was determined by real-time PCR. For control, Epstein-Barr virus (EBV)-transformed B-lymphoblastoid cell line (B-LCL) expressing HLA-A*30:14L was established from cells of the patient's mother (genotype HLA-A*30:14L,*02:01) (Hirv et al. 2006). The positive control was a B-LCL expressing HLA-A*30:01 (genotype HLA-A*30:01,*02:01). In both B-LCLs and HEK293 cells, the mRNA level of HLA-A*30:14L was nearly identical to that of HLA-A*30:01 (Figure 3). This finding suggests that the mRNA transcription rate of sHLA-A*30:14L is not affected by the mutation at nucleotide position 563 (G->C). The mRNA levels of both alleles clearly increased in response to combined stimulation with IFN-γ and TNF-α. In view of this lack of any difference in mRNA transcription, the protein expression defect is most likely caused by the missing disulfide bond in the α2 domain.

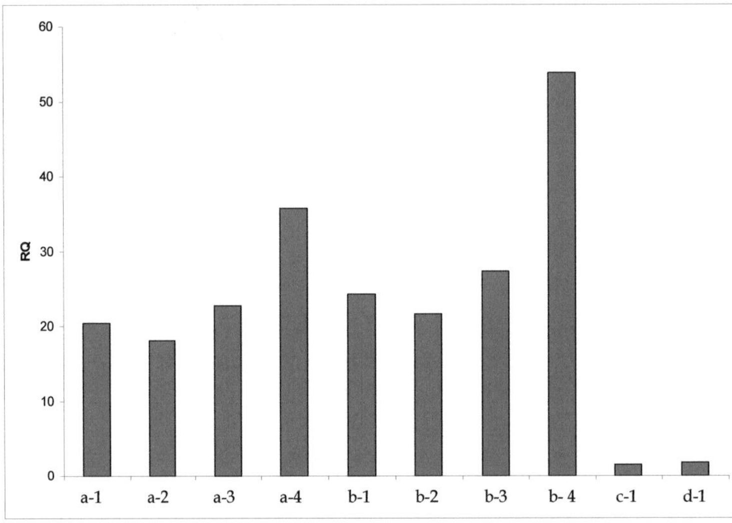

Fig. 3. Detection of mRNA levels of HLA-A*30 alleles in B-LCLs and HEK293 cells

mRNA expression of the HLA-A*30 alleles in transfected cell lines and B-LCLs was determined by real-time PCR. Shown are representative results for HEK293 cells measured after 3 days of culture in the presence ("treated") or absence ("untreated") of the cytokines IFN-γ and/or TNF-α. Data were acquired using a probe specific for the HLA-A30 sequence. Similar results were achieved for all transfected cell lines. Lanes: a) sHLA-A*30:01-transfected HEK293, b) sHLA-A*30:14L-transfected HEK293, c) EBRCC-256 (HLA-A*30:01), d) EBRCC-1818 (HLA-A*30:14L), 1 untreated, 2 IFN-γ-treated, 3 TNF-α-treated, 4 IFN-γ plus TNF-α-treated

The observation that HLA-A*30:14L protein accumulates inside the cells indicates that HLA-A*30:14L translation is not affected. Consequently, the lack of protein secretion in the supernatant is best explained by post-translational instability of the HLA-A*30:14L molecules because of the missing disulfide bridge (Hinrichs et al. 2009). Based on these findings, it is likely that the intracellular enriched HLA-A*30:14L protein is a major substrate for proteasomal cleavage and that it provides a flood of peptide fragments presented to cytotoxic T lymphocytes. As a result of this indirect surface expression by the presentation of peptide fragments, it is possible that GvHD or graft rejection might be promoted in the event of mismatching (Benichou 1999). Consequently, considering HLA-A*30:14L as null allele is, in case of a mismatch with any other HLA-A allele, potentially more dangerous in terms of GvHD and graft rejection than a mismatch with its most related allele HLA-A*30:01. Indeed, mistyping HLA-A*30:14L as an N allele has led to a severe GvHD in a patient transplanted with hematopoietic stem cells from an HLA-A*02:01 homozygous donor (Hirv et al. 2006).

In recent studies, the cytokine-based HLA secretion assay was used to classify the expression patterns of HLA-A*32:11Q (Tang et al. 2006) and HLA-B*35:65Q (Elsner et al. 2006). Both alleles undergo cysteine substitution at amino acid position 164 and thus lack the disulfide bond between the cysteine residues at amino acid positions 101 and 164 in the α2 domain of the mature protein. This interferes with HLA maturation inside the ER and therefore impairs cell surface expression. In concordance with the results of Hinrichs et al. (Hinrichs et al. 2009), IFN-γ and TNF-α increased the expression of the HLA expression variants, making HLA-A*32:11Q and HLA-B*35:65Q distinctly detectable. Compared to HLA-A*32:01 and HLA-B*35:01, the variants have very weak protein levels, indicating a low expression status. Consequently, they should be handled as low expression variants (L alleles).

5. The nature of peptides presented by HLA class I expression variants

The functional integrity of HLA low-expression variants is a prerequisite for considering them as essential in hematopoietic stem cell donor and recipient matching to diminish the risk of serious complications such as GvHD or graft rejection. HLA class I molecules present endogenous peptides 8-12 amino acids in length to CD8+ cytotoxic T lymphocytes (Natarajan et al. 1999). Most amino acid polymorphisms of different HLA class I molecules are located in the peptide-binding region shaped by parts of the α1 and α2 domains; these polymorphisms determine the characteristics of presented peptides. Peptide motifs have been reported for the most common HLA-A and B alleles and for some rare variants. Importantly, differences in peptide binding among the alleles of a serological group have also been described (Prilliman et al. 1999; Bade-Doeding et al. 2011,). Identification and comparison of allele-specific peptide-binding motifs provide important information for donor-recipient matching and prediction of HLA subtype allogenicity in allogeneic HSCT.

In order to determine the functionality of HLA low-expression alleles, peptides from recombinant truncated HLA-A*30:14L molecules secreted in the supernatant of a human cell line were eluted and sequenced (Hinrichs et al 2010). The suitability of the monoclonal anti-HLA class I antibody W6/32 for purifying recombinant HLA-A*30:14L molecules suggested its proper folding and assembly. Presumably, more soluble HLA-A*30:14L is produced and secreted into the supernatant that might not be correctly folded because of the lack of a disulfide bridge in the α2 domain.

Edman pool sequencing of eluted peptides corroborated the hypothesis that peptides are presented by HLA low expression variants and showed idential peptide motifs in HLA-A*30:01 and HLA-A*30:14L confirming the previously described peptide motif of A*30:01 (Lamberth et al. 2008; Sidney et al. 2008). The C-terminal position (PΩ) was identified as a primary anchor position. The preferred residues of the HLA-A*30 peptide epitopes at this position are lysine (K), valine (V) or arginine (R). The preference for lysine as the top amino acid at the PΩ position of the bound peptides, like described by positional scanning combinatorial peptide libraries (PSCPL) analysis, could be consolidated by the obtained peptide sequence data (Lamberth et al. 2008; Sidney et al. 2008). Position P3 of the peptides was identified as a primary-secondary anchor showing a high preference for the basic amino acids K and R. Six amino acids are reportedly favored at position P2: phenylalanine (F), serine (S), threonine (T), valine (V), isoleucine (I) or leucine (L).

The size of the obtained peptides ranged from 8 to 14 amino acids, but most had a length of 9 to 10 aa. The sequences of 200 HLA-A*30:01 ligands and of 100 HLA-A*30:14L ligands were identified. The following three peptide epitopes (3%) were presented by both HLA-A*30:01 and HLA-A*30:14L: 1) VLDTPGPPV, a nonameric peptide derived from titin (isoform N2-A, aa position 19783-19791), a protein of human muscle ultrastructure and elasticity; 2) EITALAPSTMK, an 11-mer peptide derived from human muscle protein ACTA1 (actin, alpha 1, skeletal muscle; aa position 301-311); and 3) DNIQGITKPAIR, a 12-mer peptide derived from a histone protein (HIST2H4A; aa position 25-36) (Table 3).

Peptide position	1	2	**3**	4	5	6	7	8	**9**	**10**	**11**	**12**	Source
Ligand	V	L	D	T	P	G	P	P	**V**				Titin (TTN titin isoform N2-A)
	E	I	T	A	L	A	P	S	T	M	**K**		Actin (ACTA1)
	D	N	I	Q	G	I	T	K	P	A	I	**R**	Histone (HIST2H4A)

Table 3. Shared peptide epitopes of HLA-A*30:14L and HLA-A*30:01

To verify the presentation of naturally presented peptides from recombinant HLA-A*30:01/30:14L molecules, peptide binding was analyzed by flow cytometry (Storkus et al. 1993; Zeh et al. 1994; Maeurer et al. 1996) in three EBV-transformed B-LCLs expressing either HLA-A*30:14L,*02:01 (Ulm-241539), HLA-A*30:01,*02:01 (EBRCC-256) or HLA-A*02:01 (EBRCC-2296) (Warburton et al. 1994; Hirv et al. 2006; Hinrichs et al. 2009; 2010). Acid treatment of the cell lines resulted in the dissociation of the naturally bound peptides and the release of β2 microglobulin from the HLA class I heavy chain. The HLA class I molecules were then reconstituted by adding fluorescein isothiocyanate (FITC)-labeled HLA peptide ligands and recombinant β2 microglobulin. The synthetic FITC-labeled peptide EITALAK(FITC)PSTMK (HLA-A*30:01/30:14L) and the immunodominant HLA-A*02:01-restricted CMVpp65$_{495\text{-}503}$ peptide (NLVPMK(FITC)VATV) were used. Reconstitution of HLA with the HLA-A*30 ligand mounted up to 51% (Ulm-241539) and 74% (EBRCC-256), respectively, compared to 25% for the HLA-A*02:01 homozygous cell line (EBRCC-2296). Binding on cells expressing the normal HLA-A*30:01 allele was higher than on those expressing HLA-A*30:14L, the low expression variant (Figure 4). The results confirm that the A*30 peptide previously isolated binds to HLA-A*30 on the cell surface. Peptide binding was found for the A*30:01 specific peptide on the HLA-A*30:14L-expressing cell line (Ulm-241539), indicating the stability of HLA-A*30:14L cell surface expression.

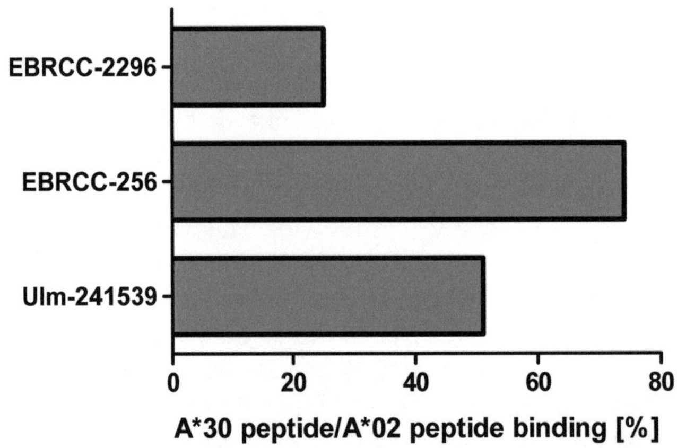

Fig. 4. Relative A*30/A*02:01 peptide-binding intensities for different HLA-expressing B-LCLs

Homology-based modeling for each HLA-A*30 alleles with the shared 9, 11 and 12-mer peptide epitopes revealed only marginal differences between the two HLA-A*30 alleles. The HLA-A*30:01 and HLA-A*30:14L models were essentially identical with the Cys164 Ser substitution, but simply adopted an alternate rotamer conformation upon breakage of the disulfide bond. Therefore, only the HLA-A*30:14L model is illustrated (Figure 5).
Although the models look identical and the alleles appear to bind identical peptides, the Cys164Ser variation could potentially generate additional flexibility within the peptide-binding groove, thereby influencing binding kinetics, particularly in peptides of lower affinity. Such an effect could stimulate a T-cell immune response and have serious implications in allogeneic HSCT.

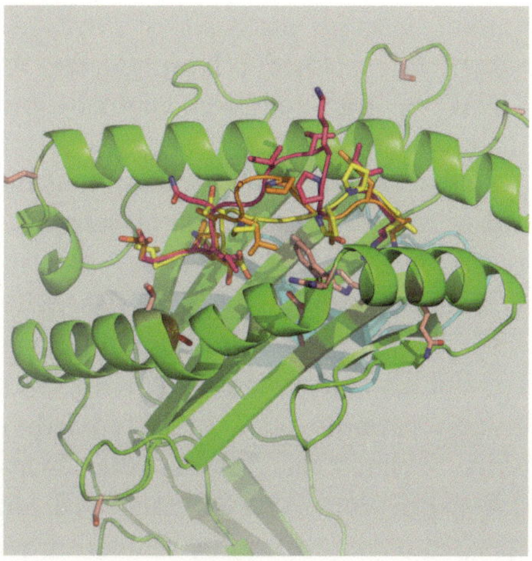

Fig. 5. Homology-based model of HLA-A*30:14L with the three shared peptide ligands

Modeling of the HLA-A*30:01 and HLA-A*30:14L structures was carried out using the SCWRL homology-based modeling server (Wang, Canutescu, and Dunbrack 2008) while employing the crystal structure of the closely related HLA-A*11:01 (1Q94) as a template. Peptide templates for 9-mer (1Q94), 11-mer (2BVO) and 12-mer (3BW9) were superimposed and merged with the HLA-A*30:14L model. Peptide mutagenesis was then performed using DeepView (Guex and Peitsch 1997) and the rotamer library to find the best side chain orientations with minimum steric clashes. Each model was then subjected to energy minimization using DeepView software. The graphics program PyMOL (http://www.pymol.org) was used to generate the structural models.

6. Conclusions

Since HLA mismatches are the main cause of severe GvHD and graft rejection, misinterpretation of HLA null alleles and expression variants as irrelevant could strongly affect transplant-related mortality.
The cytokine-based HLA secretion assay can be used to distinguish between low-expressed and non-expressed HLA alleles in order to classify alleles with a currently undefined expression status (questionable alleles, Q) as well as to re-classify certain alleles which have been assigned as null variants (N). Additionally, discrimination between cytokine inducible and non-inducible defect alleles may be important in allotransplant settings in which a

cytokine storm usually occurs following pre-transplant myeloablative conditioning or post-transplant immunosuppressive therapy.

The fact that the monoclonal anti-HLA class I antibody W6/32 is a conformational antibody implies that only correctly folded, β2 microglobulin-assembled and peptide-loaded MHC complexes can be detected. This suggests that parts of soluble HLA-A*30:14L molecules are assembled correctly and secreted by transfectants. Presumably, more soluble HLA-A*30:14L is produced and secreted into the supernatant, but it might not be correctly folded because of the lack of the disulfide bridge in the α2 domain. Therefore, these molecules are not detected by the conformational anti-HLA-ABC mAb. This assumption arose after comparing mRNA and associated protein levels of HLA-A*30:14L and HLA-A*30:01 alleles (Hinrichs et al. 2009). Additionally, it was found that HLA-A*30:14L accumulates inside the cells; therefore, it might be a major substrate for proteasomal cleavage and could provide a flood of peptide fragments presented to cytotoxic T lymphocytes. As a result of this indirect allorecognition pathway, GvHD or graft rejection might be promoted in the event of a severe mismatch.

It was shown for the first time that an HLA low expression allele (HLA-A*30:14L) presents peptides with identical features to those of its most closely related relative, HLA-A*30:01 (Hinrichs et al. 2010). The results indicate that a mismatch at amino acid position 164 might be permissive. Therefore, mismatching of these alleles will presumably be of low allogenicity in allogeneic HSCT. The fact that a low expression variant is not only functional and able to present peptides, but also shares epitopes with its related variant leads to the conclusion that low expression variants need to be considered in donor selection as permissive or non-permissive mismatches, respectively. Increasing knowledge of the expression behavior of HLA expression variants, such as L and Q alleles, will help to improve HLA allogenicity prediction algorithms by delivering proof that these variants are fully functional. Taking all relevant factors into account, the results shown allow to predict the immunogenicity of aberrantly expressed alleles in a transplant setting.

In the case of HLA-A*30:14L misinterpreting it as a null allele is, in case of a mismatch with any other HLA-A allele, potentially more dangerous in terms of GvHD and graft rejection according to the direct and indirect allo-recognition pathway than a mismatch with its most related allele HLA-A*30:01. Indeed, mistyping HLA-A*30:14L as an N allele has led to a severe GvHD in a patient transplanted with hematopoietic stem cells from an HLA-A*02:01 homozygous donor (Hirv et al. 2006).

In order to predict the relevance of similar alleles with disulfide bridge rearrangements (e.g., HLA-A*32:11Q and B*35:65Q) in allogeneic HSCT, it is important to know their surface expression as well as their peptide binding of HLA variants. From a clinical perspective, HLA variants with similar disulfide bridge variations need to be considered as functionally active in an allogeneic HSCT setting as long as the opposite has not been shown.

7. Acknowledgments

The authors would like to thank Jan Hinrichs and Daniel Föll for performing the experiments and contributing helpful discussions. The authors are grateful to Christina Bade-Döding, Nektarios Ladas and Trevor Huyton for help in peptide data analysis and homology-based modeling. This work is supported in part by funding from the Deutsche Forschungsgemeinschaft (DFG, German Research Foundation) for the Cluster of Excellence

REBIRTH (From Regenerative Biology to Reconstructive Therapy) and by the German Federal Ministry of Education and Research (reference number: 01EO0802).

8. References

Arcese, W., V. Rocha, M. Labopin, G. Sanz, A. P. Iori, M. de Lima, A. Sirvent, A. Busca, S. Asano, I. Ionescu, P. Wernet, and E. Gluckman. 2006. Unrelated cord blood transplants in adults with hematologic malignancies. *Haematologica* 91 (2):223-30.

Bade-Doeding, C., A. Theodossis, S. Gras, L. Kjer-Nielsen, B. Eiz-Vesper, A. Seltsam, T. Huyton, J. Rossjohn, J. McCluskey, and R. Blasczyk. 2011. The impact of human leukocyte antigen (HLA) micropolymorphism on ligand specificity within the HLA-B*41 allotypic family. *Haematologica* 96 (1):110-8.

Bade-Doeding, C., D. S. DeLuca, A. Seltsam, R. Blasczyk, and B. Eiz-Vesper. 2007. Amino acid 95 causes strong alteration of peptide position Pomega in HLA-B*41 variants. *Immunogenetics* 59 (4):253-9.

Balas, A., F. Garcia-Sanchez, F. Gomez-Reino, and J. L. Vicario. 1994. HLA class I allele (HLA-A2) expression defect associated with a mutation in its enhancer B inverted CAT box in two families. *Hum Immunol* 41 (1):69-73.

Benichou, G. 1999. Direct and indirect antigen recognition: the pathways to allograft immune rejection. *Front Biosci* 4:D476-80.

Bjorkman, P. J., M. A. Saper, B. Samraoui, W. S. Bennett, J. L. Strominger, and D. C. Wiley. 1987. The foreign antigen binding site and T cell recognition regions of class I histocompatibility antigens. *Nature* 329 (6139):512-8.

Bjorkman P.J., M. A. Saper , B. Samraoui , W. S. Bennett, J. L.Strominger, and D. C. Wiley. 1987. Structure of the human class I histocompatibility antigen, HLA-A2. *Nature* 329 (6139):506-12.

Bray, R. A., C. K. Hurley, N. R. Kamani, A. Woolfrey, C. Muller, S. Spellman, M. Setterholm, and D. L. Confer. 2008. National marrow donor program HLA matching guidelines for unrelated adult donor hematopoietic cell transplants. *Biol Blood Marrow Transplant* 14 (9 Suppl):45-53.

Chicz, R. M., R. G. Urban, J. C. Gorga, D. A. Vignali, W. S. Lane, and J. L. Strominger. 1993. Specificity and promiscuity among naturally processed peptides bound to HLA-DR alleles. *J Exp Med* 178 (1):27-47.

Dubois, V., J. M. Tiercy, M. P. Labonne, A. Dormoy, and L. Gebuhrer. 2004. A new HLA-B44 allele (B*44020102S) with a splicing mutation leading to a complete deletion of exon 5. *Tissue Antigens* 63 (2):173-80.

Dunn, P. P., J. R. Turton, J. Downing, S. Williams, C. V. Navarrete, and C. Darke. 2004. HLA-A*24020102L in the UK blood donor population. *Tissue Antigens* 63 (6):589-91.

Eapen, M., P. Rubinstein, M. J. Zhang, C. Stevens, J. Kurtzberg, A. Scaradavou, F. R. Loberiza, R. E. Champlin, J. P. Klein, M. M. Horowitz, and J. E. Wagner. 2007. Outcomes of transplantation of unrelated donor umbilical cord blood and bone marrow in children with acute leukaemia: a comparison study. *Lancet* 369 (9577):1947-54.

Eiz-Vesper, B., R. Blasczyk, and P. A. Horn. 2007. Description of the first HLA-DRB1 null allele. *Immunogenetics* 59 (6):507-10.

Elsner, H. A., and R. Blasczyk. 2004. Immunogenetics of HLA null alleles: implications for blood stem cell transplantation. *Tissue Antigens* 64 (6):687-95.

Elsner, H. A., D. DeLuca, J. Strub, and R. Blasczyk. 2004. HistoCheck: rating of HLA class I and II mismatches by an internet-based software tool. *Bone Marrow Transplant* 33 (2):165-9.

Elsner, H. A., P. A. Horn, C. Schoenemann, W. W. Altermann, and R. Blasczyk. 2006. Aberrant expression of HLA-B*3565Q is associated with a disrupted disulfide bond. *Immunogenetics* 58 (11):929-31.

Falk, K., O. Rotzschke, S. Stevanovic, G. Jung, and H. G. Rammensee. 1991. Allele-specific motifs revealed by sequencing of self-peptides eluted from MHC molecules. *Nature* 351 (6324):290-6.

Ferrari, G., D. D. Kostyu, J. Cox, D. V. Dawson, J. Flores, K. J. Weinhold, and S. Osmanov. 2000. Identification of highly conserved and broadly cross-reactive HIV type 1 cytotoxic T lymphocyte epitopes as candidate immunogens for inclusion in Mycobacterium bovis BCG-vectored HIV vaccines. *AIDS Res Hum Retroviruses* 16 (14):1433-43.

Flomenberg, N., L. A. Baxter-Lowe, D. Confer, M. Fernandez-Vina, A. Filipovich, M. Horowitz, C. Hurley, C. Kollman, C. Anasetti, H. Noreen, A. Begovich, W. Hildebrand, E. Petersdorf, B. Schmeckpeper, M. Setterholm, E. Trachtenberg, T. Williams, E. Yunis, and D. Weisdorf. 2004. Impact of HLA class I and class II high-resolution matching on outcomes of unrelated donor bone marrow transplantation: HLA-C mismatching is associated with a strong adverse effect on transplantation outcome. *Blood* 104 (7):1923-30.

Garrett, T. P., M. A. Saper, P. J. Bjorkman, J. L. Strominger, and D. C. Wiley. 1989. Specificity pockets for the side chains of peptide antigens in HLA-Aw68. *Nature* 342 (6250):692-6.

Girdlestone, J. 1996. Transcriptional regulation of MHC class I genes. *Eur J Immunogenet* 23 (5):395-413.

Gobin, S. J., V. Keijsers, M. van Zutphen, and P. J. van den Elsen. 1998. The role of enhancer A in the locus-specific transactivation of classical and nonclassical HLA class I genes by nuclear factor kappa B. *J Immunol* 161 (5):2276-83.

Gobin, S. J., A. Peijnenburg, V. Keijsers, and P. J. van den Elsen. 1997. Site alpha is crucial for two routes of IFN gamma-induced MHC class I transactivation: the ISRE-mediated route and a novel pathway involving CIITA. *Immunity* 6 (5):601-11.

Gobin, S. J., M. van Zutphen, A. M. Woltman, and P. J. van den Elsen. 1999. Transactivation of classical and nonclassical HLA class I genes through the IFN-stimulated response element. *J Immunol* 163 (3):1428-34.

Guex, N., and M. C. Peitsch. 1997. SWISS-MODEL and the Swiss-PdbViewer: an environment for comparative protein modeling. *Electrophoresis* 18 (15):2714-23.

Haselden, B. M., A. B. Kay, and M. Larche. 2000. Peptide-mediated immune responses in specific immunotherapy. *Int Arch Allergy Immunol* 122 (4):229-37.

Hinrichs, J., C. Figueiredo, K. Hirv, J. Mytilineos, R. Blasczyk, P. A. Horn, and B. Eiz-Vesper. 2009. Discrimination of HLA null and low expression alleles by cytokine-induced secretion of recombinant soluble HLA. *Mol Immunol* 46 (7):1451-7.

Hinrichs, J., D. Foll, C. Bade-Doeding, T. Huyton, R. Blasczyk, and B. Eiz-Vesper. 2010. The nature of peptides presented by an HLA class I low expression allele. *Haematologica* 95 (8):1373-80.

Hirv, K., U. Pannicke, J. Mytilineos, and K. Schwarz. 2006. Disulfide bridge disruption in the alpha2 domain of the HLA class I molecule leads to low expression of the corresponding antigen. *Hum Immunol* 67 (8):589-96.

Holdsworth, R., C. K. Hurley, S. G. Marsh, M. Lau, H. J. Noreen, J. H. Kempenich, M. Setterholm, and M. Maiers. 2009. The HLA dictionary 2008: a summary of HLA-A, -B, -C, -DRB1/3/4/5, and -DQB1 alleles and their association with serologically defined HLA-A, -B, -C, -DR, and -DQ antigens. *Tissue Antigens* 73 (2):95-170.

Hurley, C. K., J. E. Wagner, M. I. Setterholm, and D. L. Confer. 2006. Advances in HLA: practical implications for selecting adult donors and cord blood units. *Biol Blood Marrow Transplant* 12 (1 Suppl 1):28-33.

Jardetzky, T. S., W. S. Lane, R. A. Robinson, D. R. Madden, and D. C. Wiley. 1991. Identification of self peptides bound to purified HLA-B27. *Nature* 353 (6342):326-9.

Johnson, D. R. 2003. Locus-specific constitutive and cytokine-induced HLA class I gene expression. *J Immunol* 170 (4):1894-902.

Kamani, N., S. Spellman, C. K. Hurley, J. N. Barker, F. O. Smith, M. Oudshoorn, R. Bray, A. Smith, T. M. Williams, B. Logan, M. Eapen, C. Anasetti, M. Setterholm, and D. L. Confer. 2008. State of the art review: HLA matching and outcome of unrelated donor umbilical cord blood transplants. *Biol Blood Marrow Transplant* 14 (1):1-6.

Laforet, M., N. Froelich, A. Parissiadis, H. Bausinger, B. Pfeiffer, and M. M. Tongio. 1997. An intronic mutation responsible for a low level of expression of an HLA-A*24 allele. *Tissue Antigens* 50 (4):340-6.

Lamberth, K., G. Roder, M. Harndahl, M. Nielsen, C. Lundegaard, C. Schafer-Nielsen, O. Lund, and S. Buus. 2008. The peptide-binding specificity of HLA-A*3001 demonstrates membership of the HLA-A3 supertype. *Immunogenetics* 60 (11):633-43.

Lankat-Buttgereit, B., and R. Tampe. 2002. The transporter associated with antigen processing: function and implications in human diseases. *Physiol Rev* 82 (1):187-204.

Laughlin, M. J., M. Eapen, P. Rubinstein, J. E. Wagner, M. J. Zhang, R. E. Champlin, C. Stevens, J. N. Barker, R. P. Gale, H. M. Lazarus, D. I. Marks, J. J. van Rood, A. Scaradavou, and M. M. Horowitz. 2004. Outcomes after transplantation of cord blood or bone marrow from unrelated donors in adults with leukemia. *N Engl J Med* 351 (22):2265-75.

Lee, S. J., J. Klein, M. Haagenson, L. A. Baxter-Lowe, D. L. Confer, M. Eapen, M. Fernandez-Vina, N. Flomenberg, M. Horowitz, C. K. Hurley, H. Noreen, M. Oudshoorn, E. Petersdorf, M. Setterholm, S. Spellman, D. Weisdorf, T. M. Williams, and C. Anasetti. 2007. High-resolution donor-recipient HLA matching contributes to the success of unrelated donor marrow transplantation. *Blood* 110 (13):4576-83.

Lienert, K., G. Russ, S. Lester, G. Bennett, X. Gao, and J. McCluskey. 1996. Stable inheritance of an HLA-"blank" phenotype associated with a structural mutation in the HLA-A*0301 gene. *Tissue Antigens* 48 (3):187-91.

Ma, W., P. J. Lehner, P. Cresswell, J. S. Pober, and D. R. Johnson. 1997. Interferon-gamma rapidly increases peptide transporter (TAP) subunit expression and peptide transport capacity in endothelial cells. *J Biol Chem* 272 (26):16585-90.

Madden, D. R., J. C. Gorga, J. L. Strominger, and D. C. Wiley. 1991. The structure of HLA-B27 reveals nonamer self-peptides bound in an extended conformation. *Nature* 353 (6342):321-5.

Maeurer, M. J., D. Martin, E. Elder, W. J. Storkus, and M. T. Lotze. 1996. Detection of naturally processed and HLA-A1-presented melanoma T-cell epitopes defined by CD8(+) T-cells' release of granulocyte-macrophage colony-stimulating factor but not by cytolysis. *Clin Cancer Res* 2 (1):87-95.

Magor, K. E., E. J. Taylor, S. Y. Shen, E. Martinez-Naves, N. M. Valiante, R. S. Wells, J. E. Gumperz, E. J. Adams, A. M. Little, F. Williams, D. Middleton, X. Gao, J. McCluskey, P. Parham, and K. Lienert-Weidenbach. 1997. Natural inactivation of a common HLA allele (A*2402) has occurred on at least three separate occasions. *J Immunol* 158 (11):5242-50.

Matsumura, M., D. H. Fremont, P. A. Peterson, and I. A. Wilson. 1992. Emerging principles for the recognition of peptide antigens by MHC class I molecules. *Science* 257 (5072):927-34.

Mickelson, E. M., E. Petersdorf, C. Anasetti, P. Martin, A. Woolfrey, and J. A. Hansen. 2000. HLA matching in hematopoietic cell transplantation. *Hum Immunol* 61 (2):92-100.

Middleton, D., A. Meenagh, S. G. Marsh, and J. Martin. 2006. A HLA-Cw*03 allele, Cw*0322Q with limited or no expression. *Tissue Antigens* 67 (4):343-5.

Morishima, Y., T. Kawase, M. Malkki, and E. W. Petersdorf. 2007. Effect of HLA-A2 allele disparity on clinical outcome in hematopoietic cell transplantation from unrelated donors. *Tissue Antigens* 69 Suppl 1:31-5.

Morishima, Y., T. Sasazuki, H. Inoko, T. Juji, T. Akaza, K. Yamamoto, Y. Ishikawa, S. Kato, H. Sao, H. Sakamaki, K. Kawa, N. Hamajima, S. Asano, and Y. Kodera. 2002. The clinical significance of human leukocyte antigen (HLA) allele compatibility in patients receiving a marrow transplant from serologically HLA-A, HLA-B, and HLA-DR matched unrelated donors. *Blood* 99 (11):4200-6.

Natarajan, K., H. Li, R. A. Mariuzza, and D. H. Margulies. 1999. MHC class I molecules, structure and function. *Rev Immunogenet* 1 (1):32-46.

Noreen, H. J., N. Yu, M. Setterholm, M. Ohashi, J. Baisch, R. Endres, M. Fernandez-Vina, U. Heine, S. Hsu, M. Kamoun, Y. Mitsuishi, D. Monos, L. Perlee, S. Rodriguez-Marino, A. Smith, S. Y. Yang, K. Shipp, J. Hegland, and C. K. Hurley. 2001. Validation of DNA-based HLA-A and HLA-B testing of volunteers for a bone marrow registry through parallel testing with serology. *Tissue Antigens* 57 (3):221-9.

Ottinger, H. D., S. Ferencik, D. W. Beelen, M. Lindemann, R. Peceny, A. H. Elmaagacli, J. Husing, and H. Grosse-Wilde. 2003. Hematopoietic stem cell transplantation: contrasting the outcome of transplantations from HLA-identical siblings, partially HLA-mismatched related donors, and HLA-matched unrelated donors. *Blood* 102 (3):1131-7.

Parham, P., C. E. Lomen, D. A. Lawlor, J. P. Ways, N. Holmes, H. L. Coppin, R. D. Salter, A. M. Wan, and P. D. Ennis. 1988. Nature of polymorphism in HLA-A, -B, and -C molecules. *Proc Natl Acad Sci U S A* 85 (11):4005-9.

Perrier, P., A. Dormoy, C. Andre-Botte, and N. Froelich. 2006. HLA-A*02010102L: a laborious assignment. *Tissue Antigens* 68 (5):442-5.

Petersdorf, E. W. 2007. Risk assessment in haematopoietic stem cell transplantation: histocompatibility. *Best Pract Res Clin Haematol* 20 (2):155-70.

Petersdorf, E. W. 2008. Optimal HLA matching in hematopoietic cell transplantation. *Curr Opin Immunol* 20 (5):588-93.

Petersdorf, E. W., C. Anasetti, P. J. Martin, T. Gooley, J. Radich, M. Malkki, A. Woolfrey, A. Smith, E. Mickelson, and J. A. Hansen. 2004. Limits of HLA mismatching in unrelated hematopoietic cell transplantation. *Blood* 104 (9):2976-80.

Petersdorf, E. W., J. A. Hansen, P. J. Martin, A. Woolfrey, M. Malkki, T. Gooley, B. Storer, E. Mickelson, A. Smith, and C. Anasetti. 2001. Major-histocompatibility-complex class I alleles and antigens in hematopoietic-cell transplantation. *N Engl J Med* 345 (25):1794-800.

Prilliman, K. R., K. W. Jackson, M. Lindsey, J. Wang, D. Crawford, and W. H. Hildebrand. 1999. HLA-B15 peptide ligands are preferentially anchored at their C termini. *J Immunol* 162 (12):7277-84.

Rammensee, H. G., T. Friede, and S. Stevanoviic. 1995. MHC ligands and peptide motifs: first listing. *Immunogenetics* 41 (4):178-228.

Rocha, V., G. Sanz, and E. Gluckman. 2004. Umbilical cord blood transplantation. *Curr Opin Hematol* 11 (6):375-85.

Rudensky, AYu, P. Preston-Hurlburt, S. C. Hong, A. Barlow, and C. A. Janeway, Jr. 1991. Sequence analysis of peptides bound to MHC class II molecules. *Nature* 353 (6345):622-7.

Ruppert, J., R. T. Kubo, J. Sidney, H. M. Grey, and A. Sette. 1994. Class I MHC-peptide interaction: structural and functional aspects. *Behring Inst Mitt* (94):48-60.

Ruppert, J., J. Sidney, E. Celis, R. T. Kubo, H. M. Grey, and A. Sette. 1993. Prominent role of secondary anchor residues in peptide binding to HLA-A2.1 molecules. *Cell* 74 (5):929-37.

Sant, A. J. 1994. Endogenous antigen presentation by MHC class II molecules. *Immunol Res* 13 (4):253-67.

Schaffer, M., A. Aldener-Cannava, M. Remberger, O. Ringden, and O. Olerup. 2003. Roles of HLA-B, HLA-C and HLA-DPA1 incompatibilities in the outcome of unrelated stem-cell transplantation. *Tissue Antigens* 62 (3):243-50.

Sette, A., S. Buus, E. Appella, J. A. Smith, R. Chesnut, C. Miles, S. M. Colon, and H. M. Grey. 1989. Prediction of major histocompatibility complex binding regions of protein antigens by sequence pattern analysis. *Proc Natl Acad Sci U S A* 86 (9):3296-300.

Sette, A., S. Buus, S. Colon, J. A. Smith, C. Miles, and H. M. Grey. 1987. Structural characteristics of an antigen required for its interaction with Ia and recognition by T cells. *Nature* 328 (6129):395-9.

Sette, A., J. Sidney, M. F. del Guercio, S. Southwood, J. Ruppert, C. Dahlberg, H. M. Grey, and R. T. Kubo. 1994. Peptide binding to the most frequent HLA-A class I alleles measured by quantitative molecular binding assays. *Mol Immunol* 31 (11):813-22.

Sidney, J., E. Assarsson, C. Moore, S. Ngo, C. Pinilla, A. Sette, and B. Peters. 2008. Quantitative peptide binding motifs for 19 human and mouse MHC class I molecules derived using positional scanning combinatorial peptide libraries. *Immunome Res* 4:2.

Singh, R. R. 2000. The potential use of peptides and vaccination to treat systemic lupus erythematosus. *Curr Opin Rheumatol* 12 (5):399-406.

Smith, D. M., J. E. Baker, W. B. Gardner, G. W. Martens, and E. D. Agura. 2005. HLA class I null alleles and new alleles affect unrelated bone marrow donor searches. *Tissue Antigens* 66 (2):93-8.

Solheim, J. C. 1999. Class I MHC molecules: assembly and antigen presentation. *Immunol Rev* 172:11-9.

Storkus, W. J., H. J. Zeh, 3rd, R. D. Salter, and M. T. Lotze. 1993. Identification of T-cell epitopes: rapid isolation of class I-presented peptides from viable cells by mild acid elution. *J Immunother Emphasis Tumor Immunol* 14 (2):94-103.

Tang, T. F., L. Hou, B. Tu, W. Y. Hwang, A. E. Yeoh, J. Ng, and C. K. Hurley. 2006. Identification of nine new HLA class I alleles in volunteers from the Singapore stem cell donor registries. *Tissue Antigens* 68 (6):518-20.

Wang, E., G. Q. Phan, and F. M. Marincola. 2001. T-cell-directed cancer vaccines: the melanoma model. *Expert Opin Biol Ther* 1 (2):277-90.

Wang, Q., A. A. Canutescu, and R. L. Dunbrack, Jr. 2008. SCWRL and MolIDE: computer programs for side-chain conformation prediction and homology modeling. *Nat Protoc* 3 (12):1832-47.

Warburton, R. J., M. Matsui, S. L. Rowland-Jones, M. C. Gammon, G. E. Katzenstein, T. Wei, M. Edidin, H. J. Zweerink, A. J. McMichael, and J. A. Frelinger. 1994. Mutation of the alpha 2 domain disulfide bridge of the class I molecule HLA-A*0201. Effect on maturation and peptide presentation. *Hum Immunol* 39 (4):261-71.

Yewdell, J. W., and J. R. Bennink. 2001. Cut and trim: generating MHC class I peptide ligands. *Curr Opin Immunol* 13 (1):13-8.

Yewdell, J. W., C. C. Norbury, and J. R. Bennink. 1999. Mechanisms of exogenous antigen presentation by MHC class I molecules in vitro and in vivo: implications for generating CD8+ T cell responses to infectious agents, tumors, transplants, and vaccines. *Adv Immunol* 73:1-77.

Zeh, H. J., 3rd, G. H. Leder, M. T. Lotze, R. D. Salter, M. Tector, G. Stuber, S. Modrow, and W. J. Storkus. 1994. Flow-cytometric determination of peptide-class I complex formation. Identification of p53 peptides that bind to HLA-A2. *Hum Immunol* 39 (2):79-86.

4

The T-Cells' Role in Antileukemic Reactions - Perspectives for Future Therapies'

Helga Maria Schmetzer[1] and Christoph Schmid[2]
[1]*Department of Medicine III, José Carreras Unit for Hematopoietic Stem Cell Transplantation, Ludwig Maximilian University of Munich*
[2]*Stem Cell Transplantation Unit, Klinikum Augsburg, Ludwig Maximilian University of Munich*
Germany

1. Introduction

1.1 Background (Schmid, Schmetzer)

Highly specialized and sensitive defence against infections as well as tumors is provided in great part by the adaptive, T-cell-mediated part of our immune system. Those T-cell specialists arise out of a cell pool of 25-100 million distinct naïve T-cell clones, after a very efficient priming phase by dendritic cells (DC), that present antigens in the context of major histocompatibility complexes (MHC), T-cell receptors (TCR) and costimulatory signals. It is well known, that T-cells are most important effectors of cellular tumor immunity and carry long lasting memory. Therefore recent tumor research has focused on the development and improvement of T-cell based immunotherapies (Barrett & Le Blanc, 2010; Smits et al., 2011).

The T-lymphocyte pool includes naïve (T_{naive}), effector (T_{eff}), effector memory (T_{em}) and central memory (T_{cm}), either CD8 or CD4 T-cells. After an antigen driven TCR engagement T_{naive} proliferate and give rise to large numbers of T_{eff}. Most of them die after depleting target cells. Some 'memory T-cells' of either T_{cm} or T_{em} phenotype remain, persist and can differentiate to T_{eff} after a re-challenge with antigens. It has been demonstrated that memory T-cells are able to self-renewal and differentiation (Sallusto et al., 2005). However antitumor immunity is limited by regulatory T-cells (T_{reg}) that are in general responsible for the prevention of autoimmunity, regulation of inflammatory antimicrobial or antitumor reactions and are regarded as the mediators of tolerance (Vignali et al., 2008). T_{reg} reactions can be either mediated by inhibitory cytokines (IL-10, IL-35, TGF-ß), Granzyme A/B dependent cytolysis, CD25 dependent IL-2 deprivation-mediated apoptosis, adenosine receptor mediated immunosuppression of DCs' maturation (Vignali et al., 2008). T_{reg} subtypes can be identified by their expression profiles and be subdivided in resting or activated T_{regs} of either CD4 or CD8 subtype (Miyora et al., 2009; Jordanova et al., 2008; Schick et al., 2011). Resting T_{regs} can convert to activated T_{regs} after proliferation. The functional repertoire of T-cells depends on their survival and their cooperation via cellular contact and the secretion of humoral factors. Improved knowledge of these properties should contribute to detect or select T-cells with defined markers or functional profiles for adoptive therapy.

In the past the potential of adoptively transferred T-cells to eradicate tumor cells has been intensively studied in patients with melanoma, EBV-associated tumors and especially in the context of a transplantation of allogeneic hematopoietic stem cells including immunocompetent cells in patients with lymphoid- or myeloid-derived malignant diseases (Kahl et al., 2007; Kolb et al., 2004; Schmid et al., 2011; Parmar et al., 2011; Moosmann et al., 2010).
In patients with acute myeloid leukemia (AML), a malignant clonal disease of the hematopoietic stem cells, conventional chemotherapy induces remission in 60- 80% of patients. However, relapse occurs in the majority of patients in the following two years (Buechner et al., 2003). Allogeneic stem cell transplantation (SCT) is considered as a curative therapy for patients with AML: Following engraftment of donor cells, the established hematopoietic chimerism persists even after discontinuation of immunsuppressive therapy, reflecting tolerance both in Graft-versus-Host and Host-versus-Graft direction (Kolb et al., 2004; Schmid et al., 2011). T-cells of the healthy donor mediate the 'graft versus leukemia' (GvL), effect which was perceived in principle as early as in the 1960ies. Clinical evidence for the efficacy of GvL reactions in AML came from the observation, that leukemia relapse after allogeneic SCT, the most frequent reason of treatment failure in AML, occurred most frequently in patients who had either been transplanted from a syngeneic twin, or had received a graft which had been depleted from donor T-cells prior to transfusion. In contrast, relapse incidence was lowest in patients developing acute or chronic Graft-versus-Host Disease (GvHD), which represents the second clinical manifestation of the allogeneic immune reaction mediated by donor T-cells (Hemmati et al., 2011; Schmid et al., 2007; Schmid et al., 2011; van den Brink et al., 2010). Based on these observations, the infusion of donor lymphocytes (DLI) was developed as a form of immunotherapy for relapsed disease after allogeneic SCT. However, up to now, not every patient, in particular those with rapidly proliferating AML, responds to or permanently benefits from those T-cell based immunotherapy.

1.2 Aims (Schmetzer)

The ability of T-cells to eliminate tumor cells and even to cure tumors has been demonstrated in experimental animal models. In man the development of effective T-cell therapies to treat human tumors remains still a challenge. Tumor antigens that elicit curative responses have been identified in animal models; in man tumor associated antigens are also known, but their use is limited due to HLA-restriction and limited duration of responses. This applies also to hematological malignancies as AML. In the last years several approaches have been made to further characterize T-cells and to explore the role of soluble and cellular factors on the regulation and mediation of antitumor reactions in AML-patients. Known tumor antigens may help to identify specific T-cells and their subtypes. This could be monitored in the course of treatment and they could be prepared for further treatment. However the majority of potential tumor antigens is unknown. Their presence may be deduced from T-cell reactions initiated and mediated by leukemia-derived dendritic cells (DC), presenting the whole antigenic repertoire of the leukemic cell. In analogy to T-cell reactions against known tumor antigens DC-stimulated T-cells reacting against unknown tumor antigens may be analysed against healthy cells before further use. In addition the identification of possible immune escape phenomena in cases without successful SCT or DLI or without antileukemic functions *ex vivo* could contribute to develop strategies to overcome those immunological barriers.

In this chapter we want to present experimental and clinical results of our group with a special focus on the following topics and discuss perspectives for future therapies:

- T-cells addressing known (leukemia-) specific antigens
- T-cells addressing unknown leukemia-specific antigens
- T-cell profiles to predict antileukemic reactions and prognosis
- Clinical use of donor T-cells for prevention and treatment of AML relapse after allogeneic SCT
- Perspectives for future therapies: adoptive transfer or *in vivo* activation of antileukemic T-cells?

2. Research methods (Schmetzer)

Cellular characterizations (especially of T-cells, leukemic blasts and dendritic cells) were performed by Flow Cytometric Analyses applying a panel of marker-specific, fluorochrome labelled monoclonal antibodies: **T-cells**: Naïve T-cells (T_{naive}) CD45RO-CCR7$^+$; non-naïve T-cells ($T_{non\text{-}naïve}$) CD45RO$^+$; central memory T-cells (T_{cm}) CD45RO$^+$CCR7$^+$CD8$^+$; Effector memory T-cells (T_{em}) CD45RO$^+$CCR7$^-$CD27$^+$; Effector T-cells (T_{eff}) CD45RO$^+$CCR7$^-$CD27$^-$; Regulatory T-cells (CD8$^+$$T_{reg}$) CD8$^+$CD25$^+$CD127$_{low}$; (CD4$^+$$T_{reg}$) CD4$^+$CD25$^+$CD127$_{low}$; ($T_{naive\ reg}$) CD25$^+$CD127$_{low}$CCR7$^+$CD45RO$^-$; ($T_{cm\ reg}$) CD25$^+$CD127$_{low}$CCR7$^+$CD45RO$^+$; ($T_{eff/em\ reg}$) CD25$^+$CD127$_{low}$CCR7$^-$CD45RO$^+$ (Liepert et al., 2010; Vogt et al., 2011; Schick et al., 2011). **Blasts**: Myeloid cells co-expressing patient-specific markers (e.g. CD34, CD117, CD56, CD65); **DC:** DC co-expressing DC-antigens; Mature DC: DC coexpressing CD83; DC$_{leu}$: DC coexpressing DC-antigens (e.g. CD80, CD86, CD40) with blast-markers; (Schmetzer et al., 2007; Kremser et al., 2010). Untouched or touched CD4$^+$, CD8$^+$ or CD3$^+$ T-cells were isolated by Magnetic labelled cell sorting (MACS) (Schick et al., 2011; Vogt et al., 2011; Grabrucker et al., 2010), leukemia-antigen specific T-cells by Interferon gamma (IFN-γ) capture assay (Neudorfer et al., 2007) or by MHC-multimer (Streptamers) staining for several described leukemia associated antigens (Knabel et al., 2002). In some cases spectratyping analyses were performed to observe clonal restriction among T-cells characterized by defined T-cell receptor-Vß chains (Schuster et al., 2008). The expression of leukemia-associated antigens (LAA; e.g. WT1, PRAME, PR1) was evaluated by PCR-technology (Steger et al., 2011). 'Taq man low density arrays' were used to study expressions of most of the known protein-coding Y-chromosome genes (Liu et al., 2005). To predict the HLA-A0201 binding potential of selected peptides a HLA-A0201 peptide binding assay was performed by using the HLA-A0201 positive TAP-deficient T2 cell line system (Nijman et al., 1993; Saller et al., 1985). Leukemic blasts were cultured in 'DC-media' containing a cocktail of immune-modulators and cytokines, (thereby converting blasts to leukemia-derived DC (DC$_{leu}$), theoretically presenting the whole leukemic antigen spectrum (Kremser et al., 2010; Schmetzer et al., 2007).

Alternatively LAA/HA1 or 'male specific' antigens were loaded as peptides or full length proteins on either unmanipulated or irradiated antigen presenting cells (APC; e.g. MNCs or CD4 cell depleted cell fraction (Adhikary et al., 2008), EBV-transformed B-cells 'mini-LCL' or DC (Moosmann et al., 2002) were irradiated with 45/80 Gray and used for T-cell stimulations as given in figure 3.1.3-1 (Steger et al., 2012; Bund et al., 2011). Antileukemic reactivity of T-effector cells was measured by chromium release-, IFN-γ ELISPOT assays or non-radioactive Fluorolysis assays (Bund et al., 2011; Kremser et al., 2010). Intracellular cytokine staining (ICS) and cytokine release profiles were performed by FACS (Cytometric Bead arrays (CBA)), ELISA or ELISPOT (Elbaz & Shaltout 2001; Schmittel 2000; Fischbacher

et al., 2011; Merle et al., 2011; Bund et al., 2011). In a dog model we performed in addition an *in vivo* immunisation with antigen positive cells (Bund et al., 2011).
Using these methods we could evaluate antigen expression profiles on T-cells, antigen presenting cells or blast cells as well as cytokine secretion profiles assigning these profiles to cellular subtypes and correlate them with antileukemic reaction profiles or the clinical response to immunotherapies. Moreover we could contribute and quantify reaction profiles of 'antigen stimulated' or '-unstimulated' as well as of (specifically) selected, enriched or cloned T-cells against blast targets. Statistical evaluations were performed with standard excel programmes or SPSS software.

3. Experimental key results

In the following chapter the most important experimental results generated by our group are summarized. Responsible co-workers and cooperation partners are given and their contributions listed below.

3.1 T-cells addressing known (leukemia-) specific antigens

3.1.1 Y-chromosome encoded proteins overexpressed in acute myeloid leukemia and CD8+ T-cell reactions (Groupleaders: Kolb, Adamski; Scientists: Bund, Gallo)

In haploidentical SCT we and others observed that female-donors (especially mothers) show a higher GvL reactivity against male-patients (particularly sons) compared to all other haploidentical donor-recipient combinations for AML patients (Stern et al., 2008). These effects could be due to Y-chromosome-encoded male minor histocompatibility antigens (minor-H-Ags, mHAs) recognized by female alloimmune effector (memory) T-cells, immunized during pregnancy, in the context of a GvL-reaction (Ofran et al., 2010). We studied the expression profiles of Y-chromosome genes in healthy male stem cell donors compared to male AML patients in order to possibly detect new AML-typical Y-restricted expression patterns. Blasts from male patients with acute (myelo) monocytic leukemia and monocytes from healthy male donors as the healthy control counterpart were used to determine and compare the expression profiles of Y-chromosome genes. We could detect several genes being up-regulated in male AML-cells. Among those, we focused on *PCDH11*, *VCY*, *TGIF2LY*, known to be expressed only in male tissues and its X-chromosome-encoded homolog *TGIF2LX* (Blanco-Arias et al., 2002; Skaletsky et al., 2003; Gallo et al., 2011). In a next step, we studied the immunological impact of the identified Y-encoded genes. We analyzed the proteins encoded by those four genes for the presence of nonameric peptides that could potentially bind HLA-A0201 molecules. Such analysis was performed with the help of publicly available peptide-motif scoring systems (*http://bimas.dcrt.nih.gov/molbio/hla_bind/* and *http://www.syfpeithi.de;* Rammensee et al., 1995). High-scoring peptides were not found for *PCDH11*, whereas HLA-A0201-binding peptides could be identified *in silico* for the *VCY*, *TGIF2LY* and *TGIF2LX* genes. Their effective binding efficacy was determined in a standard HLA-A0201-T2 binding assay: Two peptides derived from VCY as well as from TGIF2LX and one derived from TGIF2LY were able to bind to the HLA-A0201-molecules of the T2 cells. Furthermore these peptides were tested for their ability to induce a CD8+ T-cell response: we selected CD3+ T-cells from female volunteers and cultured them in the presence of T2 cells loaded with the different Y-chromosome-encoded peptides for four weeks. We could show that the *VCY*-encoded peptide G54 stimulated female effector T-cells as shown by a specific lysis of G54-loaded T2 target cells in a chromium release assay

(Figure 3.1.1-1). Further research will focus on the immunogenic G54-peptid particularly with respect to our assumption that mothers (who had given birth to a son) bear CD8+ T-cells being reactive to male-specific antigens, leading to a strong GvL effect. Moreover, peptides restricted to other HLA-types will also be investigated.

In conclusion we identified three new potentially antileukemic, male-specific human genes being upregulated in blasts of male AML-patients. These might be valuable targets for T-cells.

As a rule Y-chromosome coded antigens are expressed in all cells of the organism. However some data have been reported, that differentially spliced forms of the Y-linked UTY (ubiquitously-transcribed tetratricopeptide-repeat-gene, Skaletsky et al., 2003) may be restricted in a tissue restricted fashion (Warren et al., 2000). UTY is highly conserved in man and animal species. We worked out in a dog model *in vitro* and *in vivo* that the Y-chromosome coded mHA UTY might be a promising candidate target-structure to improve GvL immune reactions after SCT: female T-cells were stimulated with either autologous (female) DCs, loaded with three different UTY-derived (male) peptides or with allogeneic, donor-matched male cells (PBMCs) endogenously expressing UTY. We identified 3 out of 15 identified UTY-encoded peptides bearing immunological potential to stimulate 'antimale'-(i.e. anti-UTY-) directed immune reactions. Amongst those, W248 showed highest immunogenic potential in both *in vitro* and *in vivo* settings: *In vitro* expanded CTLs specifically recognized mainly bone-marrow (BM) from DLA-identical male littermates in an MHC-I-restricted manner (Figure 3.1.1-2; *in vitro*). *In vivo*, comparable W248-(UTY-) specific reactivity against BM was also obtained after stimulation and immunization of a female dog with DLA-identical male PBMCs (Table 3.1.1-1; *in vivo* (Bund et al., 2001)).

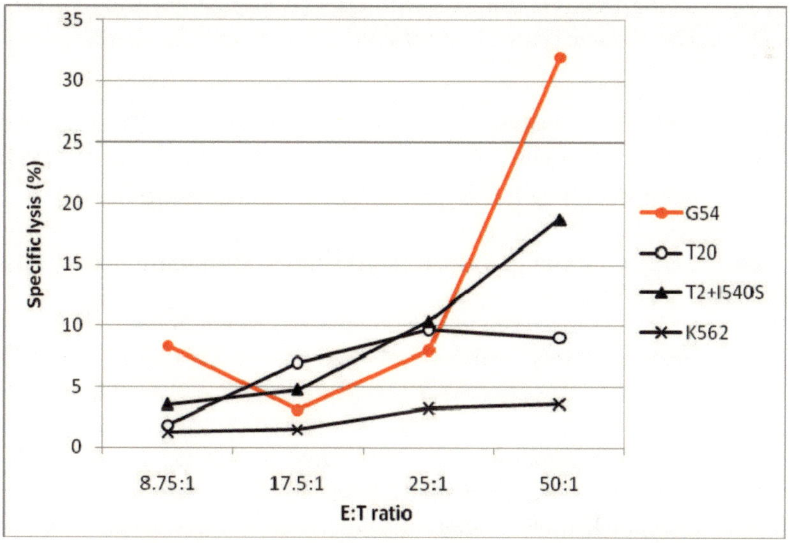

T2-cells were loaded with the VCY-derived G54-peptide and incubated with female HLA-A0201+ T-cells for 4 weeks. Cytotoxic activity of the generated G54-specific CTLs were tested in Chromium release assays (E:T = 8.75:1 to 50:1). T-cells specifically lysed T2-cells loaded with the cognate peptides G54 (G54, • red). T2-cells alone (T20, ○), T2-cells loaded with I540S (non-HLA-A0201 binding; HFLLWKLIA; T2+I540S, ▲) and K562-cells (NK-cell target, ×) were not recognized or only to a low extent

Fig. 3.1.1-1. Female T-cells stimulated with the Y-chromosomally encoded G54-peptid can specifically lyse T2-cells loaded with G54 *in vitro*.

	E:T*	male DLA-identical target cells	W248-specific spots/100,000 T-cells*
In vitro	80:1	BM	85
		BM + <anti MHC-I antibody>	24
In vivo	20:1	BM	45
		BM + <anti MHC-I antibody>	25
		BM + W248-peptide	35
		BM + W248-peptide + <anti MHC-I antibody>	15

BM bone marrow; <MHC-I>=<MHC-I>-antibody.
* E:T= Effector-to-target-ratio
**number of UTY-specific spots per 100,000 T-cells

T-cells from female dog(s) were expanded using autologous DCs pulsed with the UTY-encoded peptide W248 (in vitro; n=3/6) or male DLA-identical PBMCs (in vivo; n=1). Female UTY-specific T-cell recognition was determined in IFN-γ-ELISPOT-assays (day 21-28 (E:T = 80:1) and day 49 (E:T = 20:1), respectively). Female T-cells mainly recognized male DLA-identical BM in vitro and in vivo verifying the male-specific UTY-presentation. T-cells` MHC-I-restriction was shown by an anti-MHC-class-I-mAb.

Table 3.1.1-1 Female dog T-cells stimulated with UTY (W248)-peptides loaded on autologous DC *in vitro* or male DLA-identical PBMC *in vivo* specifically recognized 'male' target cells

Taken together we could demonstrate, that female dog effector T-cells could be specifically stimulated against male, *UTY*-gene product specific cells meaning in turns, that UTY seems to be a promising candidate antigen to improve GvL reactions after SCT.

3.1.2 LAA-specific CD8+ T-cells (Groupleaders: Busch, Borkhardt, Kolb; scientists: Doessinger, Steger, Schuster)

Between 60-90% of AML cases overexpress leukemia-associated antigens (LAA), that means antigens that are absent or only weakly expressed in normal tissues (e.g. WT1, PR1 or PRAME (Greiner et al., 2006; Steger et al., 2011)). Therefore T-cell based immunotherapeutic strategies addressing those LAA-expressing cells could be promising. Alternatively mHAs, preferentially expressed on hematopoietic cells, could qualify as T-cell targets in a GvL reaction (Mutis & Goulmy, 2002). Principally T-cell based strategies could be based on a vaccination with LAA/mHAs or by identification, selection and transfer of LAA/mHAs specific T-cells already present in the donor. LAA/mHAs specific T-cells can be found at a low frequency in normal persons implying a low level of immunity. Vice versa a long lasting immunity against leukemic cells overexpressing LAA or mHAs should imply the presence of specific T-cells. Therefore our experimental approach was to detect LAA-specific T-cells by MHC-multimer technology in AML patients after SCT. We constructed human HLA-A2 peptide multimers (Knabel et al., 2002) and tested CD8+ T-cells in 5 AML- and 2 MDS-patients after SCT for antigen specificity, as given in table 3.1.2-1

Patients (pt)	Dgn.	stage of the disease at T-cell aquisition	Cytogen. Marker at first dgn.	Blasts in PB at sample aquisi- tion	IC blast phenotype (CD) in acute phases	IC mono- cytes (%)	IC B- cells (%)	IC T- cells (%)	IC NK- cells (%)	performed analyses
pt 1147	AML-M2	CR2 after SCT and DLI	+21,+21 , +21	0	7,33,34,117	nd	nd	nd	nd	MHC- multimer staining
pt 1148	AML-M2	CR after 3rd SCT	inv (3qq)	0	13,33,34,11 7	nd	nd	nd	nd	MHC- multimer staining, ICS
pt 1149	AML-M4	CR after SCT and DLI	46, XX	0	13,15,33,34 , 117	7	6	14	6	MHC- multimer staining, ICS
pt 1150	MDS- RAEB I	CR after SCT and DLI	-5, +1 (at relapse)	0	13,33,34,11 7	11	6	13	7	MHC- multimer staining, ICS
pt 1151	AML-M4	CR after SCT	46, XX	5	15,33,7,65, 64,4	nd	nd	28	12	MHC- multimer staining, ICS
pt 1152	AML-M5	CR after SCT	46, XX	0	nd	nd	nd	nd	nd	MHC- multimer staining, Spectraty- ping
pt 1153	MDS- CMML	Pers after SCT and DLI	46, XX	8	nd	nd	nd	nd	nd	MHC- multimer staining, ICS, LAA, CD4+ exp.
pt 1154	MDS- RAEB II	CR after SCT and DLI	46, XY	0	nd	10	3	34	8	ICS, LAA, CD4+ exp.
pt 1155	MPS-atyp. CML	CR after SCT and DLI	t (8;22)	0	nd	nd	nd	nd	nd	ICS, LAA, CD4+ exp.

nd not done; CR complete remission; Rel. relapse; MDS myelodysplastic syndrome; RAEB Refractory Anaemia with Excess Blasts; AML-M3 acute myeloid leukemia FAB M2; CMML Chronic Myelomonocytic Leukemia; LAA leukemia associated antigen, overexpression analyses compared to healthy controls, detected by the RQ= $2^{-\Delta\Delta ct}$ method in MNCs; dgn. diagnosis; ICS intracellular staining; CD4+ exp. CD4+ experiments

Table 3.1.2-1 Patients'characteristics I:

In all of these 7 cases we could detect LAA-specific CD8+ T-cells (that means more than 0.1% T-cells with LAA specificity) –although not directed against all given LAA, suggesting the persistence of T-cells with antileukemic potential (table 3.1.2-1 and fig 3.1.2-2). Four of five cases with two different types of LAA- specific CD8+ T-cells (pt 1149, 1151, 1152, 1153) were characterized by long-lasting clinical remissions for more than 2 years. One patient with two different types of LAA-specific CD8+ T-cells relapsed after 9 months (pt 1151) and in addition the patient with only one type of LAA-specific T-cells, who relapsed after one year (pt 1147). One patient could not be analysed under a clinical point of view since he died one month after sample acquisition of an infect (pt 1148, table 3.1.2-2, Steger et al., 2012).

Patients (pt)	**LAA over-expression in MNCs (at sample preparation)**	**Peak response MHC-Multimer staining (% of specific CD8+ T-cells)**	**Cytokine profile (IFN-γ, IL-2) after LAA-stimulation** **CD 4+/C8+ T-cells**	**Time to relapse or last follow up**
pt 1147	nd	A2/WT1 0.11 A2/PR1 0.13 A2/PRAME 0.034	NR/NR	04/07 local rel.; 08/09 systemic rel., 05/11 CR after 2nd SCT, DLI and 3th rel.
pt 1148	nd	A2/WT1 0.37 A2/PR1 0.4 A2/PRAME nd	NR/NR	03/06 death with infect in CR after SCT
pt 1149	nd	A2/WT1 0.61 A2/PR1 0.39 A2/PRAME 0.4	NR/NR	until 02/11 in CR after SCT
pt 1150	nd	A2/WT1 0.27 A2/PR1 0.13 A2/PRAME 0.081	NR/NR	until 03/11 in CR after SCT
pt 1151	nd	A2/WT1 0.16 A2/PR1 0.088 A2/PRAME 0.42	NR/NR	11/06 rel; 04/07 death after 3th DLI, rel. and 2nd SCT
pt 1152*	nd	A2/WT1 0.16 A2/PR1 0.074 A2/PRAME 0.38	NR/NR	until 04/11 in CR after SCT
pt 1153	WT1 (-), PRAME (+), PR1 (-)	A2/WT1 0.1 A2/PR1 0.14 A2/PRAME 0.13	*PRAME 0.32*/NR	until 02/11 in CR after SCT
pt 1154	WT1 (-), PRAME (-), PR1 (-)	nd	NR/NR	until 07/11 in CR after SCT
pt 1155	WT1 (-), PRAME (-), PR1 (-)	nd	NR/NR	until 07/11 in CR after SCT

No LAA overexpression (-), 1–10x LAA overexpression (+), compared to healthy controls, analysed by the RQ= $2^{-\Delta\Delta ct}$ method; nd not done

Bold: > 0.1% CD8+ T-cells defined as 'LAA-specific T-cells present'; * restricted T-cells by spectratyping in *PRAME*- MHC multimer selected T-cells detectable

Table 3.1.2-2 Anti-LAA-peptide reactive T-cells are detectable by MHC-Multimer staining in all given PB samples from AML/MDS patients after allogeneic SCT

In one case (pt 1152) we performed spectratyping in T-cells selected for PRAME-specificity and could demonstrate a highly restricted TCR-repertoire. This points to a specific clonal expansion of CD8+ cells after (*in vivo*) PRAME challenge by residual PRAME overexpressing blasts (Figure 3.1.2-1).

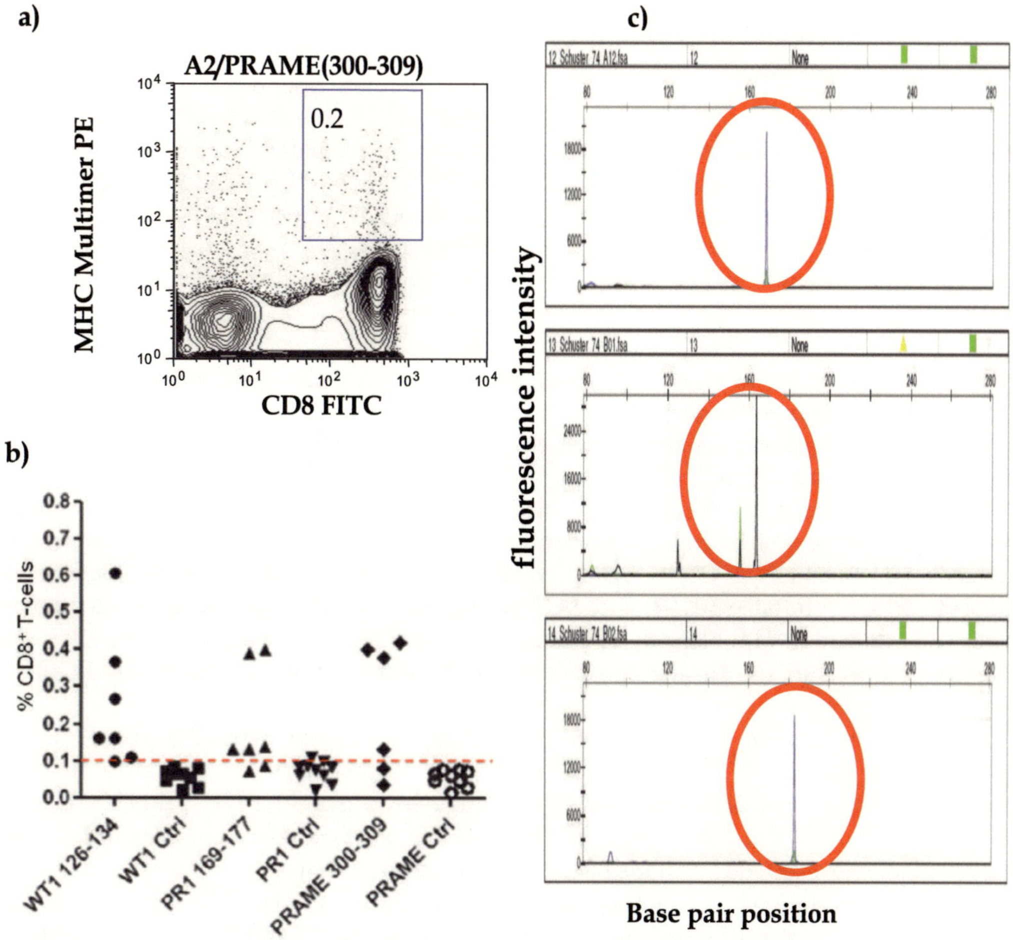

a) representative FACS-plot shows MHC-Multimer staining from peripheral blood of an AML-patient after SCT.

b) LAA-specific T-cells from PBMCs from AML-pts after SCT were labelled with MHC multimers binding T-cells, that recognize frequently overexpressed LAA antigens. PBMCs were analyzed by FACS. Healthy donors served as controls and multimer-staining revealed a threshold value of 0.11% multimer-positive cells in the CD8 cell fraction.

c) MHC-multimer positive $CD8^+$ T-cells from peripheral blood of an AML-patient after SCT were enriched by FACS sorting. RNA was extracted and Spectratyping of the Vβ-composition reveals strong clonal restriction among TCR-species (1rd row: Vß3-Jß1.5; 2nd row: Vß3-Jß2.1; 3rd row: Vß3-Jß2.5)

Fig. 3.1.2-1. LAA-specific T-cells can be isolated from AML-patients after SCT by MHC multimer technique. Selected T-cells are highly Vß restricted

Upon in vitro peptide-challenge we could not detect secretion of IFN-γ or IL-2 in T-cells by intracellular cytokine staining in 7 out of 8 patients tested (table 3.1.2-2). This however does not exclude a cytokine-independent functionality of multimer positive T-cells. Interestingly we could detect an IFN-γ response in CD4+ T-cells in one case (pt 1153) with PRAME overexpressing leukemic blasts detectable during persisting relapse after SCT. This is very surprising as the peptide triggering this response is MHC class I restricted and could indicate a CD8 coreceptor independent binding of a MHC I directed TCR and an unusual involvement of CD4+ T-cells in the leukemia directed immune response.
In summary that means, that in general LAA-specific, HLA-A2-restricted CD8+ T-cells can be prepared by MHC multimer technology from most of the patients after SCT at various time points. Possibly the simultaneous detection of two different LAA-specific CD8+ T-cells correlates with a higher chance of longlasting remissions. Optimal time points have to be evaluated for the preparation of (sufficient) LAA-specific CD8+ T-cells that could be used for adoptive transfer and concerning minimal amounts needed for maintenance of remission.

3.1.3 LAA-specific CD4+ T-cells (Groupleaders: Buhmann, Milosevic, Kolb, Schmetzer; scientists: Steger)

CD8+ T-cells recognize HLA-class I restricted peptides, therefore they can mediate strong cytotoxic, antileukemic reactions, but also severe graft versus host (GvH) disease. In contrast CD4+ T-cells recognize HLA class II restricted peptides that are mainly expressed by cells of the haematopoetic system and absent from other organs. In order to further analyse the antileukemic function of CD4+ T-cells we prepared (untouched) CD4+ T-cells from 6 patients, as given in table 3.1.3-1, after SCT or DLI immunotherapies and stimulated them with LAA-proteins (WT1, PRAME, PR1 and the mHA HA-1), that were loaded on the CD4 depleted cell fraction (containing monocytes and DC as antigen presenting cells (APCs)) or on 'mini-LCL'.
During the stimulation phase the cells lost their naïve (and central memory) T-cell phenotype and gained an effector memory or effector cell phenotype (data not shown).
As already shown for CD8 selected T-cells cytokine release assays for IFN-γ (ELISPOT, ICS) or GM-CSF (ELISA) did not reveal clear and specific cytokine release profiles of LAA or HA-1 stimulated, expanded or cloned CD4+ T-cells (data not shown).
Therefore we performed a functional Fluorolysis assay in case pt 1158: *blast cells* of the patient, that were characterized by a overexpression of WT1, PRAME and PR1 at first diagnosis served as leukemic target and *fibroblasts, effector cells* or *non-blast cells* of the patient as negative controls. *Effector cells* (E) used for these assays were: unstimulated MNC, CD3+ or CD4+ T-cells obtained in different stages of the disease, LAA or HA-1 stimulated CD4+ T-cells in different stimulation phases, enriched proliferating and CD40L+ CD4+ cells before or after single cell cloning (table 3.1.3-2).
We could demonstrate that most of the CD4+ T- cell containing effector cell fractions (except MNCs before SCT, CD4+ T-cells after 16 stimulation rounds (E11) and two of the CD4+ T-cell clones (E13, E15)) were able to regularly and specifically lyse leukemic blasts, but not fibroblasts or non-blast-cells of the patient. Highest antileukemic activity was demonstrated for enriched proliferating and CD40L+CD4+ T-cells followed by one of the three tested clones. Unfortunately despite of provable (CD4+) immunity against leukemic cells the patient died two years after SCT in hematological CR from a myocardial chloroma.

Patients (pt)	Dgn.	T-cell source (stage of disease)	Blasts in PB at sample aquisition	Cell source	Analyses performed in addition to CD4+experiments
pt 1153	MDS-CMML	Pers after SCT	8	CD4+, CD8+, MNCs	ICS, IFN-γ ELISPOT, MHC-Multimer staining, LAA analyses
pt 1154	MDS-RAEB II	CR after SCT	0	CD4+, CD8+, MNCs	ICS, IFN-γ ELISPOT, LAA analyses
pt 1155a and 1155 b	MPS-atyp. CML	CR after SCT	0	CD4+, CD8+, MNCs	ICS, IFN-γ ELISPOT, GM-CSF ELISA, LAA analyses
pt 1156	Biphen. ALLL/ AML	CR after SCT	0	CD4+	IFN-γ ELISPOT
pt 1157	AML-M4	CR after SCT	0	CD4+	IFN-γ ELISPOT
pt 1158 a ('E1')	AML-M4	CR before SCT	0	MNCs	LAA analyses, Fluorolysis assay
pt 1158 b ('E2')	AML-M4	CR after 1st SCT and 1st DLI	0	CD4+	Fluorolysis assay
pt 1158 c ('E3') pt 1158 d ('E12': after P5) pt 1158 e ('E11': after P16) pt 1158 f ('E6': after P17,FACS-Sort, P14) pt 1158 g, h, i ('E13, 14, 15' clones, after FACS-Sort, P14)	AML-M4	CR after 1st SCT and 1st DLI	0	CD4+	ICS, IFN-γ ELISPOT, GM-CSF ELISA , Fluorolysis assay
pt 1158 p ('blast targets')	AML-M4	Rel. after 1st SCT and 2nd DLI	34	CD3 depleted blasts	Fluorolysis assay
pt 1158 j ('E4')	AML-M4	Rel. after 1st SCT and 2nd DLI	nd	CD4+	Fluorolysis assay
pt 1158 k ('E5')	AML-M4	CR after 2nd SCT and 4th DLI	0	CD4+	Fluorolysis assay
pt 1158 l ('E16')	AML-M4	CR after 2nd SCT and 4th DLI	0	CD4+	Fluorolysis assay
pt 1158 m, n, o ('E7, 8, 9')	AML-M4	CR after 2nd SCT and 4th DLI	0	MNCs, CD3+, CD4+	LAA analyses, Fluorolysis assay

Dgn. diagnosis; Pers persistant disease; Rel. relapse; CR complete remission; MDS myelodysplastic syndrome; RAEB Refractory Anaemia with Excess Blasts; AML-M3 acute myeloid leukemia FAB M3; CMML Chronic Myelomonocytic Leukemia; LAA leukemia associated antigen, overexpression analyses compared to healthy controls, detected by the RQ= $2^{-\Delta\Delta ct}$ method in MNCs; E effector cells prepared at different time points in the course of the disease or after stimulation with CD4 depleted LAA/HA1 protein loaded APCs; P passage after x stimulations with CD4 depleted LAA/HA1 protein loaded APCs, nd not done;

Table 3.1.3-1 Patients´ characteristics II:

A flow chart with the methodological strategy to create LAA-presenting nonCD4+ cells or mini-LCL is given in figure 3.1.3-1. We could demonstrate that in general it is possible to stimulate untouched CD4+ T-cells using the CD4 depleted fraction or 'mini-LCL' as stimulator fraction. Furthermore we enriched LAA-stimulated cells from these stimulation settings by either enrichment of proliferating, CD40L+CD4+ T-cells or by T-cell single-cell cloning after repeated stimulation with LAA- and HA-1- loaded stimulator cells. Resulting cells were characterized by proliferation, CD40L upregulation and cellular expansion.

I.

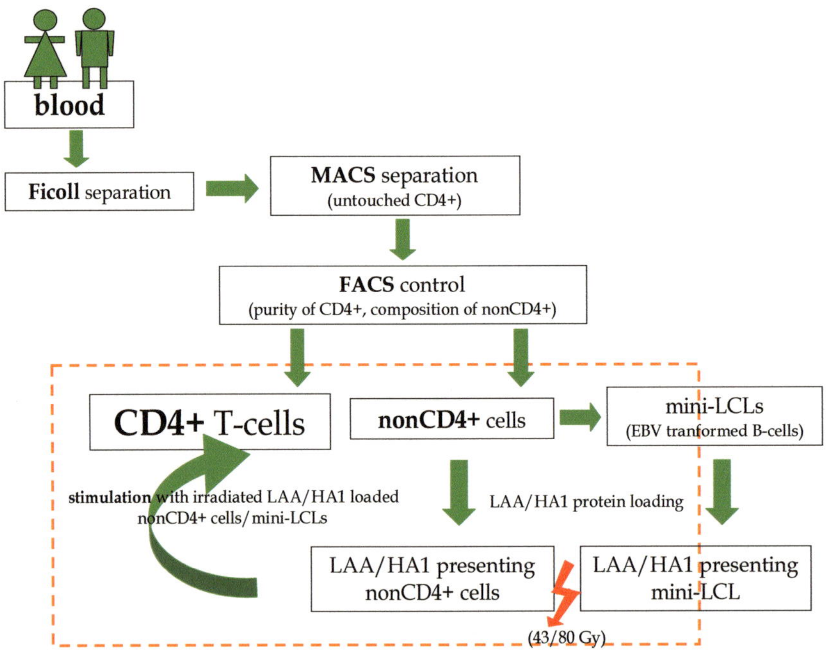

I. Untouched CD4+ T-cells prepared from AML patients' PBMNC after SCT were stimulated (6-16 stimulations with 20 U IL2 twice a month) with irradiated (43 Gy, 80 Gy by mini-LCLs), LAA/HA1 protein loaded nonCD4 cells (as APCs) for 24h. Alternative stimulation of the nonCD4+ cell fraction with EBV transformed B-cells as APCs.

II.

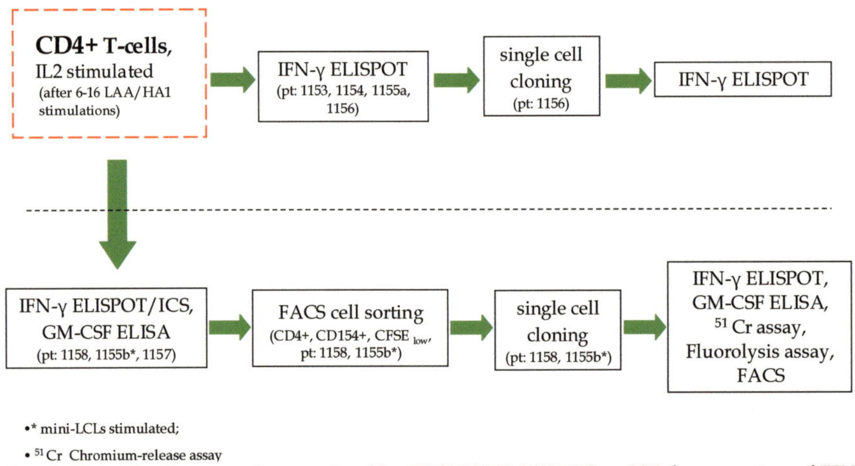

II. Stimulated CD4+ T-cells were characterized by ELISPOTS, ELISAS or ICS for secretion of IFN-γ or GM-CSF before or after single cell cloning or sorting of CD40L+ $CFSE_{low}$ cells. Moreover some cases were tested for antileukemic activity in a chromium-release or fluorolysis assay.

Fig. 3.1.3-1. Preparation, stimulation (I) and characterisation (II) of (untouched) CD4+T-cell

Patient 1158	T-cell source (stage of disease)	Cell source of the effectors	Proportion of blasts lysed by different effector cells (%)	Patient	T-cell source (stage of disease)	Cell source of the effectors	Proportion of blasts lysed by different effector cells (%)
pt 1158 a ('E1')	CR before SCT	MNCs	0	**pt 1158 f ('E6')**	CR after 1st SCT and 3 weeks after 1st DLI	CD4+ (after P17 and FACS Sort P14)	87
pt 1158 b ('E2')	CR after 1st SCT and 3 weeks after 1st DLI	CD4+	51	**pt 1158 e ('E11')**	CR after 1st SCT and 3 weeks after 1st DLI	CD4+ (after P16)	0
pt 1158 c ('E3')	CR after 1st SCT and 5 weeks after 1st DLI	CD4+	42	**pt 1158 d ('E12')**		CD4+ (after P5)	9
pt 1158 j ('E4')	Rel. after 1st SCT and 2nd DLI	CD4+	61	**pt 1158 g ('E13')**		CD4+ T-cell clones (after FACS Sort P14)	0
pt 1158 k ('E5')	CR after 2nd SCT and 4th DLI	CD4+	60	**pt 1158 h ('E14')**			16
pt 1158 m ('E7')	CR after 2nd SCT and 4th DLI	MNCs	22	**pt 1158 i ('E15')**			0
pt 1158 n ('E8')		CD3+	19 (38, 4h)	**pt 1158 l ('E16')**	CR after 2nd SCT and 4th DLI	CD4+	86
pt 1158 o ('E9')		CD4+	9	**pt 1158 q**	† Myocardial Chloroma		

E effector cells prepared at different time points in the course of the disease or after stimulation with CD4 depleted LAA/HA1 protein loaded APCs (table 3.1.3-1); P passage after x stimulations with CD4 depleted LAA/HA1 protein loaded APCs; † Exitus letalis

Table 3.1.3-2 Antileukemic functionality (demonstrated by fluorolysis assay) of different CD4+ effector cells prepared before or after specific stimulation, enrichment or sorting

In summary this means that the cytokine release of IFN-γ or GM-CSF is no reliable tool to detect specificity in T-cell fractions. We could demonstrate however, that CD4+ T-cells in general can mediate cytotoxic reactions and that certain enriched CD4+ subtypes might be even more promising candidates for GvL- without GvH reactions.

3.2 T-cells addressing unknown leukemia-specific antigens

3.2.1 Establishment and maintenance of protective immunity by dendritic cells derived from leukemic blasts (Groupleaders: Borkhardt, Buhmann, Schmetzer; Scientists: Fischbacher, Freudenreich, Grabrucker, Liepert, Merle, Reuther, Schick, Schuster, Vogt)

Although technically possible the preparation of LAA-specific T-cells is cumbersome, in case of MHC-Multimer technology HLA-A2 restricted and moreover restricted to defined LAAs. Therefore we wanted to study the antileukemic activity of T-cells stimulated with DC derived from leukemic blasts (DC_{leu}), as those DC bear the advantage of potentially presenting the whole (including as yet unknown) and patient-typical antigen pool of the leukemic cells. We have already established the methods to generate sufficient amounts of DC and especially of DC_{leu} in every given case with AML: in a minimalized assay we cultured DC in three different media and choose the DC generation method with the quantitatively highest DC/DC_{leu} counts (Kremser et al., 2010; Schmetzer et al., 2007). In a next step we study the functional profiles of DC/DC_{leu}-stimulated T-cells. We could demonstrate, that isolated *unstimulated* T-cells were able to lyse blasts in 47% of cases, whereas only 26% of those T-cells showed antileukemic activity after a 10 days culture with blasts - pointing to an establishment of a T-cell inhibitory microenvironment in the presence of blasts. However, stimulation of T-cells with blasts after their conversion to DC_{leu} resulted in a blast lytic activity in 58% of all cases. These data suggest, that the inhibitory (blast-induced) atmosphere could be abolished after blast-conversion to DC_{leu}, although not completely: even a DC/DC_{leu} stimulation of T-cells was not effective to induce antileukemic T-cells in every case (Schmetzer et al., 2011; Grabrucker et al., 2010). Therefore we analysed potential conditions and factors being responsible for these impaired immune reactions. We could demonstrate that the *quality of DC* – especially with respect to proportions of mature DC and DC_{leu} - is predictive for their activation capacity for antileukemic T-cells. Those DC/DC_{leu} induce an ‚antileukemically effective' T-cell composition of DC-stimulated T-cells, characterized by higher proportions of CD4+ and non-naive T-cells (Grabrucker et al., 2010; Liepert et al., 2010). A detailed analysis of T-cells' compositions in cases *with* compared to those *without* antileukemic activity revealed significantly higher proportions of naïve (T_{naive}) and central memory T-cells (T_{cm}) and lower proportions of effector memory regulatory ($T_{eff/em\ reg}$) as well as CD8+ regulatory T-cells (CD8+ T_{reg}) (Schick et al., 2011; Vogt et al., 2011). Moreover we could define soluble factors that are predictive for the mediation of antileukemic activity of DC/DC_{leu}-stimulated T-cells: A higher release of 'inflammatory' chemokines (CXCL8, CCL2) in DC culture supernatants or of 'T-cell-promoting' cytokines (IFN-γ, IL-6) in mixed lymphocyte culture (MLC) supernatants of T-cells with DC clearly correlated with antileukemic activity of DC-stimulated T-cells (Fischbacher et al., 2011; Merle et al., 2011; Schmetzer et al., 2011).

Detailed studies of T-cell subsets via spectratyping analysis could verify that especially after DC-stimulation CD4+ as well as CD8+ T-cells were characterized by a highly restricted Vß T-cell-receptor (TCR) repertoire (Schuster et al., 2008). Interestingly, in one patient studied comprehensively *in vitro* stimulation with DC/DC_{leu} resulted into an identical TCR β chain restriction pattern which could be identified *in vivo* in the patient's T-cells 3 months after allo-SCT (Reuther et al., 2011).

In summary, DC_{leu} are promising candidates to stimulate and enrich antileukemic T-cells without knowledge of defined antigen targets which is attended with the creation of an 'antileukemic cellular microenvironment' and could contribute to develop strategies to overcome immunological resistances. With our experimental *in vitro* models combining culture methods and functional flow cytometry with spectratyping we can moreover provide explanations for clinical observations and might provide predictive information about T-cellular response patterns *in vivo*. Further studies of selected T-cells (e.g. by their Vß type) for their phenotype and function will allow to understand clinical responses and to prepare T-cells for treatment.

3.2.2 Antileukemic T-cell profiles to predict antileukemic reactions of DC/DC_{leu} stimulated T-cells and prognosis of patients (Groupleaders: Schmetzer; Scientists: Fischbacher, Freudenreich, Grabrucker, Liepert, Merle, Schick, Vogt)

Since not every AML patient responds to immunotherapy (SCT, DLI) *in vivo* and since not every *ex vivo* T-cell stimulation with DC_{leu} results in antileukemia effector T-cells we wanted to elucidate responsible cells or soluble factors. We could evaluate 'cut-off values' for DC- or T-cell subtypes in the cellular settings and in addition amounts of soluble factors that allow a *prediction of the antileukemic function of T-cells* in this cellular or microenvironmental context or a *prediction of the clinical response to immunotherapy*. We could demonstrate, that T-cells stimulated with DC_{leu} in a 'favorable cellular and soluble chemokine/cytokine context' (with >45% mature DC and >65% DC_{leu} with a release of >200pg CXCL8, >100pg CCL2, >10pg IFN-γ or >15pg IL-6, resulting in >65% CD4+, $T_{non\text{-}naive}$ and <60% CD8+ T-cells and especially >3% T_{naive} , >11% T_{cm} and low proportions of CD8+ T_{reg} after the DC stimulation) had a more than 75% chance to gain antileukemic *ex vivo* activity. In addition we could demonstrate, that clinical responders to immunotherapy were characterized by a higher *ex vivo* generability of DC_{leu} and mature DC, a better *ex vivo* T-cell proliferation and CD4:CD8 and $T_{non\text{-}naive}$: T_{naive} ratios>1 and in addition by a high release of CCL2, IFN-γ and IL-6 (Liepert et al., 2010; Schmetzer et al., 2011; Fischbacher et al., 2011; Merle et al., 2011; Grabrucker et al., 2010).

That means, that we can not only associate cellular subtypes of T-cells and DC or cytokine/chemokine release patterns with antileukemic functions of T-cells in *ex vivo* settings and in the context of a clinical response to immunotherapies, but in addition define predictive 'cut-off values' , that means proportions of cells with certain cellular subtypes or concentrations of soluble factors, that allow a correlation with cellular responses or the clinical course of the disease after immunotherapy.

4. Clinical key results

4.1 Clinical use of donor T-cells for prevention and treatment of AML relapse after allogeneic SCT (Schmetzer, Schmid)

Donor lymphocyte infusion (DLI) for treatment of leukemic relapse after allogeneic hematopoietic stem cell transplantation (SCT) has been introduced in the early nineties (Kolb et al., 1990). Being extremely effective in chronic myeloid leukemia, the procedure was less successful in AML, although remissions were observed in selected cases (Kolb et al., 1995; Collins et al., 1997). Therefore, on behalf of the Acute leukemia Working Party of the European Group of Blood and Marrow Transplantation (EBMT) our group performed a retrospective analysis of patients who had been transplanted for AML in complete remission, and had suffered from leukemia relapse post SCT (Schmid et al., 2007). The analysis was based on the EBMT transplant registry, and included 399 adult patients who had received (n=171) or not (n=228) DLI as part of their treatment. With a median follow up of 27 and 40 months in the both groups, overall survival (OS) at two years was 21±3% for patients receiving, and 9±2% for patients not receiving DLI. After adjustment for differences between the groups, better outcome was associated with younger age (p= 0.008), remission duration >5 months after SCT (p<0.0001), and use of DLI (p=0.04). Among DLI recipients, a lower tumour burden at relapse (<35% of bone marrow blasts; p=0.006), female gender (p=0.02), favorable cytogenetics (p=0.004) and remission at time of DLI (p<0.0001) were predictive for survival in a multivariate analysis. Two-year survival was 56±10%, if DLI was performed in remission or with favourable karyotype, and 15±3% if DLI was given in aplasia or with active disease. Therefore, an algorithm for the clinical use of DLI in the treatment of relapsed AML after allogeneic SCT was developed, comprising the sequence of cytoreductive chemotherapy for disease control or induction of complete remission, followed by DLI for long term control of the leukemia based on cellular immune effects (Schmid et al., 2011).

In an approach to increase the antileukemic efficacy of donor T-cells against myeloid leukemias, systemic application of GM-CSF was studied after DLI for relapse of AML or MDS after SCT (Schmid et al., 2004). GM-CSF was chosen due to its capacity to contribute *in vitro* to the generation of antigen-presenting cells (APC) from leukemic blasts (Woiciechowsky et al., 2001; Kufner et al., 2005; Kremser et al., 2010; Dreyssig et al., 2011). As described above, blasts from myeloid leukemias should have the full genetic repertoire for effective antigen presentation, but might be ineffective stimulators or even induce specific anergy, due to inferior or aberrant expression of co-stimulatory molecules, such as CD80 or CD86. GM-CSF has been shown to induce up-regulation of these molecules on the surface of leukemia blasts and to improve cytotoxic efficacy of autologous and allogeneic T-cells. In a clinical pilot trial for AML relapse after SCT, mild chemotherapy with low dose AraC, infusion of donor T-cells together with stem cells for reconstitution of haematopoiesis, and s.c. or i.v. application of GM-CSF, was studied. Overall response rate was 67% among evaluable patients', overall survival at 2 years was 29%. Long term survival was associated with longer remission post transplant, disease control by low dose AraC and development of chronic GvHD. These results confirm the proposed strategy of initial cytoreduction by chemotherapy and induction of a GvL reaction for long term disease control. Systemic application of GM-CSF was safe in this setting, however, its clinical efficacy remains to be evaluated in randomised studies. Nevertheless, in accompanying *ex vivo* experiments we could demonstrate, that cases in which DC could be generated *ex vivo* using GM-CSF-based protocols showed a more favourable outcome after *in vivo* immunotherapy (Freudenreich et al., 2011).

Since overall, the outcome of patients with AML who relapse after allogeneic SCT is poor, strategies to prevent occurrence of overt haematological relapse are of increasing interest. Intensive monitoring of minimal residual disease and donor chimerism in different cellular compartments (CD34+, CD3+) of bone marrow or peripherial blood has gained in importance by allowing early interventions (chemotherapy, DLI, second SCT) *before* haematological relapse has occurred (Bornhauser et al., 2009). Our group has developed a protocol for the use of prophylactic or preemptive DLI (pDLI) for patients with high-risk AML. Starting from day +120 after SCT, patients in haematologic remission, free of immunosuppression for at least 30 days without clinically evident GvHD, and free of infections receive up to 3 courses of DLI in 4 weeks' intervals, using an escalating cell dose schedule. Patients receiving prophylactic DLI have been compared to a control group of high-risk AML patients, who were treated according to the same transplant protocol, would have fulfilled the criteria for pDLI, but did not receive the cells since their transplant centres did not take part at this part of the study (Schmid et al., 2005; Schleuning et al., unpublished results). Hence, patients receiving pDLI showed a significantly lower incidence of relapse and achieved a longer overall survival as compared to controls without pDLI. The treatment was also safe and induction of severe GvHD was a rare event in this setting.

In summary, although less effective as in CML, the clinical use of DLI for AML patients after SCT is an effective therapeutic tool for prevention or as part of treatment of relapses after SCT in AML.

5. Perspectives for future therapies: Adoptive transfer or *in vivo* activation of antileukemic T-cells? (Schmetzer, Schmid)

We could clearly demonstrate, that it is possible to *detect and monitor* leukemia specific T-cells by LAA-peptid specific (HLA-A2 restricted) MHC-Multimer analyses or LAA-protein specific CD4+ T-cells, especially if combined with spectratyping and cellular subtype analyses. Concerning 'known' (leukemia-) specific antigens we can conclude from our data, that LAA-peptide specific CD8+ T-cells can be prepared by MHC multimer-technology, LAA-protein specific CD4+ T-cells by preparation of (enriched) proliferating CD40L+ or cloned CD4+ cells and both cell types could be used for adoptive therapies. Moreover we could work out *in vitro* as well as in a dog model, that male specific antigens might be promising candidate antigens for immunotherapies. In addition we could show, that (enriched) T-cells addressing known as well as unknown leukemia-associated antigens, as demonstrated after DC/DC_{leu}-stimulation, can mediate cytotoxic reactions. Those cells could be promising candidates for *adoptive immunotherapies* in selected patients.

Since the manipulation and selection of antigen-specific T-cells is not only an oncological challenge, but has to be approved by special committees before a clinical application another strategy circumventing T-cell manipulations could be more promising: *applicating immune-modulators* and cytokines like GM-CSF or IFN-α *in vivo* could possibly induce the conversion of (residual) blasts in patients to DC_{leu}. In a small patients' cohort we could already show, that patients receiving GM-CSF in the context of a DLI-relapse therapy had a better chance to respond to this relapse therapy compared to patients without additionally applicated GM-CSF (Freudenreich et al., 2011). Moreover we could show, that the convertibility of blasts to DC_{leu} in *ex vivo* settings correlated with the clinical response and outcome to immunotherapies, what can be interpretated by an '*ex vivo* simulation' of the DC-generating potential out of blasts. We could even demonstrate, that cases, in that higher proportions of DC_{leu} could be generated *ex vivo* had a longer overall survival compared to cases with lower DC_{leu} proportions.

Our *ex vivo* focus in the future will therefore be to thoroughly investigate and optimize *in vivo* strategies with allo SCT applying different donor transplants or *ex vivo* 'manipulated' grafts. We further want to develop and test different 'immune modulating cocktails' (Ansprenger et al., 2011; Deen et al., 2011) that can be applicated to patients with the aim to induce leukemia-derived DC *in vivo* and in consequence to stimulate the generation of leukemia-specific T-cells *in vivo*. In parallel we want to further enlighten the role of different (enriched, selected) effector cells - e.g. CD4+, CD8+, NK, NK-T- cells) in the mediation of antileukemic reactions in order to find promising candidates for adoptive T-cell transfer.

6. Acknowledgements

This book chapter was written by Helga Schmetzer (experimental parts) and Christoph Schmid (clinical part) in cooperation with groupleaders and scientists as listed below. The chapter summarizes, among others, the main results worked out in the course of SFB TR project 36 (supported by the DFG; applicants Prof. Borkhardt, Prof. Busch, Prof. Kolb), Deutsche Jose Carreras Stiftung, DLR-grant 01GU 0516 (Prof. Kolb) and TRANSNET project MRTN-CT-2004-512253 (EC Marie Curie grant, Prof. Kolb).
All authors thank technicians, nurses and physicians for skilful work and data contribution. Many of the data presented were worked out in the course of theses (MD, PhD).

6.1 Groupleaders and responsibilities

Adamski Jerzi (Prof., PhD, Helmholtz Research Center for Environmental Health, responsible for male specific antigens)
Borkhardt Arndt (Prof., MD, Director of the Department for Pediatric Oncology and Hematology, University of Duesseldorf, responsible for spectratyping)
Buhmann Raymund (MD, Helmholtz Research Center for Environmental Health, responsible for LAA analyses, CD4 characterization)
Busch Dirk (Prof., MD, Director of the Department for Immunology, Technical University of Munich, responsible for MHC-Multimer analyses)
Kolb Hans-Jochem (Prof., MD, former head of the Stem Cell Transplantation Unit of the Med Dept III, University of Munich, responsible for clinical development and application of stem cell transplantation protocols , initiation and discussions of T-cell projects)
Milosevic Slavoljub (PhD, Helmholtz Research Center for Environmental Health, responsible for the LAA-specific CD4+ T-cells)
Schmetzer Helga (PhD, PD, Prof., Med Dept 3, dept for stem cell transplantations, University of Munich, responsible for DC-projects, CD4 project, author of this book chapter)
Schmid Christoph (PD, MD, head of the Stem Cell Transplantation unit at Klinikum Augsburg, responsible for clinical development and application of stem cell transplantation protocols, main coauthor of this book chaper)

6.2 Scientists and responsibilities

Ansprenger Christian (MD student; thesis student, Med Dept 3, working group Schmetzer, responsible for characterization of immunomodulatory DC culture additives)
Bund Dagmar (PhD student, Helmholtz Research Center for Environmental Health, working group Buhmann/Kolb; responsible for male specific antigens/T-cell reactions)
Deen Diana (MD student, thesis student, Med Dept 3, working group Schmetzer, responsible for characterization of immunomodulatory DC culture additives)
Doessinger Georg (PhD student, Department for Immunology, Technical University of Munich, working group Busch, responsible for MHC-Multimer analyses)
Fischbacher Dorothea (MD, thesis student, Med Dept 3, working group Schmetzer, responsible for cytokine analyses in DC project)
Freudenreich Markus (MD, thesis student, Med Dept 3, working group Schmetzer, responsible for *in vivo* correlations in DC project)
Gallo Antonio (PhD, formerly Helmholtz Research Center for Environmental Health, working group Kolb/Adamsky, responsible for male specific antigens/T-cell reactions)

Grabrucker Christine (MD, Med Dept 3, working group Schmetzer, responsible for DC- or blast-stimulations of T-cells in DC project)
Liepert Anja (MD, thesis student, Med Dept 3, working group Schmetzer, responsible for characterization of DC-stimulated T-cells in DC project)
Merle Marion (MD, thesis student, Med Dept 3, working group Schmetzer, responsible for chemokine analyses in DC project)
Reuther Susanne (PhD student, Department for Pediatric Oncology and Hematology, University of Duesseldorf, working group Borkhardt, responsible for spectratyping)
Schick Julia (MD student, thesis student, Med Dept 3, working group Schmetzer, responsible for characterization of regulatory T-cells in DC project)
Schuster Friedhelm (MD, Department for Pediatric Oncology and Hematology, University of Duesseldorf, working group Borkhardt, responsible for spectratyping)
Steger Brigitte (PhD student, Helmholtz Research Center for Environmental Health, working group Kolb/ Buhmann/ Schmetzer, responsible for LAA analyses and LAA-specific CD4+ T-cell project)
Vogt Valentin (MD student, thesis student, Med Dept 3, working group Schmetzer, responsible for characterization of T-cell differentiation subtypes)

7. References

Adhikary, D.; Behrends, U.; Feederle, R.; Delecluse, H.J.; Mautner, J. (2008). Standardized and highly efficient expansion of Epstein-Barr virus-specific CD4+ T-cells by using virus-like particles. *Journal of virology.*,Vol.82(8), pp. 3903-3911.

Ansprenger, C.; Grabrucker, C.; Kroell, T.; Schmetzer, H. (2011). Modulation of blasts using paraimmunity- inducing factors (PINDS) to improve leukemic antigen presenting function. Manuscript in preparation.

Barrett, AJ & Le Blanc, K. (2010). Immunotherapy prospects for acute myeloid leukemia. *Clin. Exp. Immunol.*;161(2), pp.223-32.

Blanco-Arias, P.; Sargent, C.A. & Affara N.A. (2002). The human-specific Yp11.2/Xq21.3 homology block encodes a potentially functional testis-specific TGIF-like retroposon. *Mamm Genome* 13, pp. 463-468.

Bornhauser,M., Oelschlaegel,U., Platzbecker,U., Bug,G.; Lutterbeck,K.; Kiehl,M.G.; Schetelig,J.; Kiani,A.; Illmer,T.; Schaich,M.; Theuser,C.; Mohr,B.; Brendel,C.; Fauser,A.A.; Klein,S.; Martin,H.; Ehninger,G.; Thiede,C. (2009). Monitoring of donor chimerism in sorted CD34+ peripheral blood cells allows the sensitive detection of imminent relapse after allogeneic stem cell transplantation. *Haematologica*, 94, pp. 1613-1617.

Buechner, T.; Hiddemann, W.; Berdel, W.E.; Wörmann, B.; Schoch, C.; Fonatsch, C.; Löffler, H.; Haferlach, T.; Ludwig, W.D.; Maschenmeyer, G.; Staib, P.; Aul, C.; Gruneisen, A.; Lengfelder, E.; Frickhofen, N.; Kern, W.; Serve, H.L.; Mesters, R.M.; Sauerland, M.C.; Heinecke, A.(2003). 6-Thioguanine, cytarabine and daunorubicin (TAD) and high-dose cytarabine and mitoxantrone (HAM) for induction, TAD for consolidation and either prolonged maintenance by reduced monthly TAD or TAD-HAM-TAD and one course of intensive consolidation by sequential HAM in adult patients at all ages with de novo acute myeloid leukemia (AML): a randomized trial of the German AML Cooperative Group. *J Clin Oncol* 21; pp. 4496-504.

Bund, D.; Schmetzer, H.; Buhmann, R.; Zorn, J.; Gökmen, F.; Kolb, H.J. (2011). UTY encoded minor-histocompatibility antigens as targets for a graft-versus-haematopoiesis (GvHm) and graft versus leukemia (GvL) in the canine model. Submitted for publication.

Collins, R.H., Jr.; Shpilberg, O.; Drobyski, W.R.; Porter, D.L.; Giralt, S.; Champlin, R.; Goodman, S.A.; Wolff, S.N.; Hu, W.; Verfaillie, C.; List, A.; Dalton, W.; Ognoskie, N.; Chetrit, A.; Antin, J.H.; Nemunaitis, J. (1997). Donor leukocyte infusions in 140 patients with relapsed malignancy after allogeneic bone marrow transplantation. *J Clin Oncol.* 15, pp. 433-444.

Deen, D.; Ansprenger, C.; Grabrucker, C.; Kroell, T.; Schmetzer, H. (2011). Modulation of blasts by a combination of immunomodulatory factors and cytokines to stimulate the generation of DC *in vivo* . Manuscript in preparation

Dreyßig, J.; Kremser, A.; Liepert, A.; Grabrucker, C.; Freudenreich, M.; Schmid, C.; Kroell, T.; Scholl, N.; Tischer, J.; Kufner, S.; Salih, H.; Kolb, H.J.; Schmetzer, H. (2011). Various 'Dendritic Cell Antigens' (DCA) are already expressed on uncultured blasts in acute myeloid Leukemia (AML) and myelodysplastic Syndromes (MDS). *J. Immunotherapy* 3(9), pp. 1113-1124.

Elbaz, O. & Shaltout, A. (2001). Implication of granulocyte-macrophage colony stimulating factor (GM-CSF) and interleukin-3 (IL-3) in children with acute myeloid leukaemia (AML). *Hematology* 5(5), pp. 383-388.

Fischbacher, D.; Merle, M.; Liepert, A.; Grabrucker, C.; Kremser, A.; Loibl, J.; Schmid, C.; Kroell, T.; Kolb, H.J.; Schmetzer, H. (2011). CXCL8 and CCL2 secretion by dendritic cells generated from AML-blasts (DCleu) as well as IFN-γ or IL-6 release in mixed lymphocyte cultures (MLC) from T-cells after DC-priming are predictive for antileukemic T-cell reactions. Manuscript in preparation

Freudenreich, M.; Schmid, C.; Kremser, A.; Dreyßig, J., Kroell, T.; Tischer, J.; Kolb, H.J.; Schmetzer, H. (2011). Clinical relevance of in vitro generation of Dendritic cells in patients with AML or MDS. Manuscript in preparation.

Gallo, A.; Bund, D.; Moeller, G.; Schmetzer, H.; Mindnich, R.; Buhmann, R.;, Kolb, H.J., Adamski, J. (2011) Characterization of TGIF2LY, a human Y chromosome encoded gene which is up-regulated in acute myelomonocytic leukemia. Submitted

Grabrucker, C.; Liepert, A.; Dreyssig, J.; Kremser, A.; Kroell, T.; Freudenreich, M.; Schmid, C.; Schweiger, C.; Tischer, J.; Kolb, H.J, Schmetzer, H. (2010). The quality and quantity of leukaemia-derived dendritic cells (DC) from patirnts with acute myeloid leukaemia and myelodysplastic syndrome are a predictive factor for the lytic potential of DC-primed leukemia-specific T-cells. *J. Immunotherapy* 33,5, pp. 523-537.

Greiner, J.; Döhner, H.; Schmitt, M. (2006). Cancer vaccines for patients with acute myeloid leukaemia-definition of leukaemia-associated antigens and current clinical protocols targeting these antigens. *Haematologica* 91(12), pp. 1653-61.

Hemmati, P.G.; Terwey, T.H.; le Coutre, P.; Vuong, L.G.; Massenkeil, G.; Dörken, B.; Arnold, R. (2011). A modified EBMT risk score predicts the outcome of patients with acute myeloid leukaemia receiving allogeneic stem cell transplants. *Eur.J. Haematol.* ;86(4), pp. 305-16.

Jordanova, E.S.; Gorter, A.; Ayachi, O.; Prins, F.; Durrant, L.G.; Kenter, G.G.; van der Burg, S.H.; Fleuren, G.J. (2008). Human leukocyte antigen class I, MHC class I chain-related molecule A, and CD8+/ regulatory T-cell ratio: which variable determines survival of cer vical cancer patients?. *Clin. Cancer Res.* 14(7), pp. 2028-2035.

Kahl, C.; Storer, B.E.; Sandmaier, B.M.; Mielcarek, M.; Maris, M.B.; Blume, K.G.; Niederwieser, D.; Chauncey, T.R.; Forman, S.J.; Agura, E.; Leis, J.F.; Bruno, B.; Langston, A.; Pulsipher, M.A.; McSweeney, P.A.; Wade, J.C.; Epner, E.; Bo Petersen, F.; Bethge, W.A.; Maloney, D.G.; Storb R. (2007). Relapse risk in patients with malignant diseases given allogeneic hematopoetic cell transplantation after nonmyeloablative conditioning. *Blood.* 110(7) pp. 2744-8.

Knabel, M.; Franz, T.J.; Schiemann, M.; Wulf, A.; Villmow, B.; Schmidt, B.; Bernhard, H.; Wagner, H.; Busch, D.H.(2002). Reversible MHC multimer staining for functional isolation of T-cell populations and effective adoptive transfer. *Nat. Med.* 8(6), pp. 631-7.

Kolb, H.J.; Mittermüller, J.; Clemm, C.; Holler, E.; Ledderose, G.; Brehm, G.; Heim, M.; Wilmanns, W.(1990). Donor leukocyte transfusions for treatment of recurrent chronic myelogenous leukemia in marrow transplant patients. *Blood,* 76(12), pp. 2462-2465.

Kolb, H.J.; Schattenberg, A.; Goldman, .JM.; Hertenstein, B.; Jacobsen, N.; Arcese, W.; Ljungman, P.; Ferrant, A.; Niederwieser, D.; van Rhee, F.; Mittermüller, J.; de Witte, T.; Holler, E.; Ansari, H. Graft-versus-leukemia effect of donor lymphocyte transfusions in marrow grafted patients. European Group for Blood and Marrow Transplantation Working Party Chronic Leukemia. *Blood.* ; 86, pp. 2041-2050.

Kolb, H.J.; Simoes, B.; Schmid, C. (2004). Cellular immunotherapy after allogeneic stem cell transplantation in hematologic malignancies. *Current Opinions in Oncology* 16, pp. 167-173.

Kremser, A.; Kufner, S.; Konheuser, E.; Kroell, T.; Doehner, C.; Schoch, C.; Kolb, H.J.; Zitzelsberger, H.; Schmetzer, H. (2011). Combined immunophenotyping and FISH ('FISH-IPA') with chromosome-specific DNA-probes allows the quantification and differentiation of *ex vivo* generated dendritic cells (DC), leukaemia-derived DC and clonal leukemic cells in patients with AML. Manuscript in preparation

Kufner, S.; Kroell, T.; Pelka-Fleischer, R.; Schmid, C.; Zitzelsberger, H.; de Valle, F; Zirpel, I; Schmetzer, H.(2005). Serum-free generation and quantification of functionally active leukemia-derived dendritic cells is possible from malignant blasts in acute myeloid leukemia (AML) and myelodysplastic syndromes (MDS). *Cancer Immunol. Immunother.* 54, pp. 953-970.

Liepert, A.; Grabrucker, C.; Kremser, A.; Dreyssig, J.; Ansprenger, C.; Freudenreich, M.; Kroell, T.; Reibke, R.; Tischer, J.; Schweiger, C.; Schmid. C.; Kolb, H.J; Schmetzer, H. (2010). Quality of T-cells after stimulation with leukaemia-derived dendritic cells (DC) from patients with acute myeloid leukaemia (AML) or myeloid dysplastic syndrome (MDS) is predictive for their leukaemia cytotoxic potential. *Cellular Immunity* 265, pp. 23-30.

Liu, D.W., Chen, S.T. & Liu, H.P. (2005). Choice of endogenours control for gene expression in nonsmall cell lung cancer. *Eur Respir J,.* 26, pp. 1002-1008.

Merle, M.; Fischbacher, D.; Liepert, A.; Kroell, T.; Grabrucker, C.; Kremser, A.; Dreyssig, J.; Freudenreich, M.; Schmid, C.; Kolb, H.J; Schmetzer, H. (2011). Chemokines secreted by dendritic cells (DC) from patients with acute myeloid leukaemia (AML)

are a predictive factor for the lytic potential of DC-stimulated leukaemia specific T-cells. Manuscript in preparation

Miryara, M. ; Yoshioka, Y. ; Kitoh, A. ;Shima, T. ; Wing, K. ; Niwa, A. ; Parizot, C. ; Taflin, C. ; Heike, T. ; Valeyre, D. ; Mathian, A. ; Nakahata, T. ; Yamaguchi, T. ; Nomura, T. ; Ono, M. ; Amoura Z. ; Gorochov, G. ; Sakaguchi, S. (2009). Functional delineation and differentiation dynamics of human CD4+ T-cells expressing the FoxP3 transcription factor. *Immunity* 30, pp. 899-911.

Moosmann , A.; Khan, N.; Cobbold, M.; Zentz, C.; Delecluse, H.J.; Hollweck, G.; Hislop, A.D.; Blake, N.W.; Croom-Carter, D.; Wollenberg, B.; Moss, P.A.; Zeidler, R.; Rickinson, A.B.; Hammerschmidt, W. (2002). B cells immortalized by a mini-Epstein-Barr virus encoding a foreign antigen efficiently reactivate specific cytotoxic T cells.*Blood,* 100(5), pp.1755-1764.

Moosmann,A.; Bigalke,I.; Tischer,J.; Schirrmann,L.; Kasten,J.; Tippmer,S.; Leeping,M.; Prevalsek,D.; Jaeger,G., Ledderose,G.; Mautner,J.; Hammerschmidt,W.; Schendel,D.J.; Kolb,H.J. (2010) Effective and long-term control of EBV PTLD after transfer of peptide-selected T cells. *Blood,* 115, pp. 2960-2970.

Mutis, T. & Goulmy, E. (2002). Hematopoietic system specific antigens as targets for cellular immunotherapy of hematological malignancies. *Semin Hematol.* ;39, pp. 23–31.

Neudorfer, J.; Schmidt, B.; Huster, K.M; Anderl, F.; Schiemann, M.; Holzapfel, G.; Schmidt, T.; Germeroth, L.; Wagner, H.; Peschel, C.; Busch, D.H; Bernhard, H. (2007). Reversible HLA multimers (Streptamers) for the isolation of human cytotoxic T lymphocytes functionally active against tumor-and virus-derived antigens. *J. of Immunol. Meth.* 320, pp. 119-131.

Nijman, H.W.; Houbiers, J.G.; Vierboom MP.; van der Burg, S.H.; Drijfhout, J.W.; D'Amaro, J.; Kenemans, P.; Melief, C.J.; Kast, W.M.(1993). Identification of peptide sequences that potentially trigger HLA- A2.1-restricted cytotoxic T lymphocytes. *Eur J Immunol.* 23, pp. 1215–1219.

Ofran, Y.; Kim, H.T.; Brusic, V.; Blake, L.; Mandrell, M.; Wu, C.J.; Sarantopoulos, S.; Bellucci, R.; Keskin, D.B.; Soiffer, R.J.; Antin, J.H.; Ritz, J. (2010). Diverse patterns of T-cell response against multiple newly identified human Y chromosome-encoded minor histocompatibility epitopes. *Clin Cancer Res.*16, pp. 1642–1651.

Parmar, S.; Fernandez-Vina, M; de Lima, M. (2011). Novel transplant strategies for generating graft-versus-leukemia effect in acute myeloid leukaemia. *Curr. Opin. Hematol.* 18(2), pp. 98-104.

Rammensee, H.G.; Friede, Stevanovic, S. (1995). MHC ligands and peptide motifs: 1st listing. *Immunogenetics.* 41, pp. 178-228.

Reuther, S.; Schuster, F.; Hubner, B.; Grabrucker, C.; Liepert, A.; Reibke, R.; Lichtner, P.; Yang, T.; Kroell, T.; Kolb, H.J.; Borkhardt, A.; Schmetzer, H.; Buhmann, R.(2011). *In vitro* induced response patterns of antileukemic T cells- characterization by spectratyping and immunophenotyping. Submitted for publication

Saller, R.D.; Howell, D.N.; Cresswell, P.(1985). Genes regulating HLA class I antigen expression in T-B lymphoblast hybrid. *Immunogenetics.* 1985;21, pp. 235–46.

Sallusto, F.; Geginat, J.; Lanzavecchia, A.(2004). Central memory and effector memory T cell subsets: function, generation and maintenance. *Annu. Rev. Immunol.* 22, pp. 745-763.

Schick, J.; Vogt, V.; Zerwes, M.; Kroell, T.; Schweiger, C.; Köhne H, Kolb H.J, Buhmann R, Schmetzer H. (2011). Antileukemic T-cell responses after dendritic cell (DC)

stimulation can be predicted by compositions of regulatory T-cell subpopulations, especially with respect to regulatory central memory and regulatory CD8 cells. *Bone Marrow Transplantation* 46, supp 1, abstr. 1048.

Schleuning, M.; Schmid, C.; Koenecke, C.; Hertenstein, B.; Baurmann, H.; Schwerdtfeger, R.; Kolb, H.J.(2008). Long term follow-up and matched pair analysis of adjuvant donor lymphocyte transfusions following allogeneic stem cell transplantation after reduced intensity conditioning for high-risk AML. ASH 2008; *Blood* supplement, abstract No 2142.

Schmetzer, H.; Kremser, A.; Loibl, J.; Kröll, T.; Kolb, H.J.(2007). Quantification of *ex vivo* generated dendritic cells (DC) and leukaemia-derived DC ('DCleu') contributes to estimate the quality of DC, to detect optimal DC-generating methods or to optimize DC-mediated T-cell-activation-procedures *ex vivo* or *in vivo*. *Leukemia* 21, pp. 1338-41.

Schmetzer, H. (2011). Antileukemic T-cell-mediated immune reactions: limitations and perspectives for future therapies. Editorial, *Immunotherapy* 3(7), pp. 809-811.

Schmid,C.; Schleuning,M; Aschan,J.; Ringden,O.; Hahn,J.; Holler,E.; Hegenbart,U.; Niederwieser,D.; Dugas,M.; Ledderose,G.; Kolb,H.J. (2004) Low-dose ARAC, donor cells, and GM-CSF for treatment of recurrent acute myeloid leukemia after allogeneic stem cell transplantation. *Leukemia*, 18, pp. 1430-1433.

Schmid, C.; Schleuning, M.; Ledderose, G.; Tischer, J.; Kolb, H.J.(2005). Sequential regimen of chemotherapy, reduced-intensity conditioning for allogeneic stem-cell transplantation, and prophylactic donor lymphocyte transfusion in high-risk acute myeloid leukemia and myelodysplastic syndrome. *J Clin Oncol.* 23, pp. 5675-5687.

Schmid, C.; Labopin,M.; Nagler,A.; Bornhauser,M.; Finke,J.; Fassas,A.; Volin,L.; Gurman,G.; Maertens,J.; Bordigoni,P.; Holler E.; Ehninger,G.; Polge,E.; Gorin,N.; Kolb,H.-J.; Rocha,V. (2007) Donor lymphocyte infusion in the treatment of first hematological relapse after allogeneic stem cell transplantation in adults with acute myeloid leukaemia: A retrospective risk factors analysis and comparison with other strategies by the EBMT Acute Leukemia Working Party Journal of Clinical Oncology, 2007, 25 (31); 4938-4945. *J Clin Oncol.*, 25, pp. 4938-4945.

Schmid, C., Labopin, M., Nagler, A., Niederwieser, D., Castagna, L., Tabrizi, R., Stadler, M., Kuball, J., Cornelissen, J., Vorlicek, J., Socié, G., Falda, M., Vindeløv, L., Ljungman, P., Jackson, G., Kröger, N., Rank, A., Polge, E., Rocha, V., Mohty, M. (2011) Treatment, risk factors, and outcome of adults with relapsed AML after reduced intensity conditioning for allogeneic stem cell transplantation. *Blood.* 2011 Dec 13. [Epub ahead of print]

Schmittel, A.; Keilholz, U.; Thiel, E.; Scheibenbogen, C. (2000). Quantification of tumor-specific T lymphocytes with the ELISPOT assay. *Journ. of Immunother*,23(3), pp.289-295.

Schuster, F.; Buhmann, R.; Reuther, S.; Hubner, B.; Grabrucker, C.; Liepert, A.; Reibke, R.; Lichtner, P.; Yang, T.; Kroell, T.; Kolb, H.J.; Borkhardt, A.; Schmetzer, H.(2008). Improved effector functions of leukaemia-specific T-lymphocyte clones trained with AML-derived dendritic cells. *Cancer Genomics Proteomics* 5, pp. 275-286.

Skaletsky,H; Kuroda-Kawaguchi, T; Minx, P.J.; Cordum, H.S.; Hillier, L.; Brown, L.G.; Repping, S.; Pyntikova, T.; Ali, J.; Bieri, T.; Chinwalla, A.; Delehaunty, A.; Delehaunty, K.; Du, H.; Fewell, G., Fulton, L.; Fulton, R.; Graves, T., Hou, S.F.;

Latrielle, P.; Leonard, S.; Mardis, E., Maupin, R., McPherson, J., Miner, T., Nash, W.; Nguyen, C.; Ozersky, P.; Pepin, K.; Rock, S.; Rohlfing, T.; Scott, K.; Schultz, B.; Strong, C.; Tin-Wollam, A.; Yang, S.P.; Waterston, R.H.; Wilson, R.K.; Rozen, S., Page, D.C.(2003). The male-specific region of the human Y chromosome is a mosaic of discrete sequence classes. *Nature*. 423, pp. 825–837.

Smits, EL.; Lee, C.; Hardwick, N.; Brooks, S.; Van Tendeloo, V.F.; Orchard, K.; Guinn, B.A. (2011). Clinical evalution of cellular immunotherapy in acute myeloid leukaemia. *Cancer Immunol. Immunother*. 60(6), pp.757-69.

Steger, B.; Schmetzer, H.; Floro, L.; Kroell, T.; Tischer, J.; Kolb, H.J.; Buhmann, R. (2011). Clinical relevance of mRNA overexpressions of the leukaemia associated antigens (LAA) WT1, PRAME and PR1 in patients (pts) with Acute Myeloid Leukaemia (AML): an analysis of (co)expressions in different cellular compartments, subtypes or stages. Submitted for publication.

Steger, B.; Milosevic, S.; Doessinger, G.; Reuther, S.; Liepert, A.; Zerwes, M.; Schick, J.; Vogt, V.; Schuster, F.; Kroell, T.; Busch D.; Borkhardt, A.; Kolb, H.J. ; Tischer, J.; Buhmann, R.; Schmetzer, H. (2012). $CD4^+$ as well as $CD8^+$ T-cells addressing leukaemia-associated antigens (LAA: WT1, PRAME, PR1) or minor histocompatibility antigens (mHags: HA1) can be identified and mobilized to recognize and kill leukemic cells expressing from patients (pts) with acute myeloid leukaemia (AML). Submitted for publication

Stern, M.; Loredana, R.; Mancusi, A.; Bernardo, M.E.; de Angelis, C.; Bucher, C.; Locatelli, F.; Aversa, F.; Velardi, A. (2008). Survival after T cell-depleted haploidentical stem cell transplantation is improved using the mother as donor. *Blood*, Vol 112(7), pp. 2990-2995.

van den Brink,M.R.; Porter,D.L.; Giralt,S.; Lu,S.X.; Jenq,R.R.; Hanash,A.; Bishop,M.R. (2010) Relapse after allogeneic hematopoietic cell therapy. *Biol.Blood Marrow Transplant.*, 16, pp. S138-S145.

Vignali, D.A.; Collison, L.W.; Workman, C.J.(2008). How regulatory T cells work. *Nat. Rev. Immunol*. 8(7), pp. 523-32.

Vogt, V.; Schick, J.; Zerwes, M.; Kroell, T.; Schweiger, C.; Koehne, C.; Buhmann, R.; Kolb, H.J.; Schmetzer, H. (2011). Antileukemic T cell responses after DC-stimulation can be predicted by composition of regulatory T-cell subpopulations, especially with respect to regulatory central memory and regulatory CD8 cells. Bone Marrow Transplantation 46, supp 1, abstr. 1047

Warren, E.H.; Gavin, M.A.; Simpson, E.; Chandler, P.; Page, D.C.; Disteche, C.; Stankey, K.A.; Greenberg, P.D.; Riddell, S.R. (2000). The human UTY gene encodes a novel HLA-B8-restricted H-Y antigen. J. Immunol.164(5), pp. 2807-2814

Woiciechowsky, A.; Regn, S.; Kolb, H.J.; Roskrow, M.(2001). Leukemic dendritic cells generated in the presence of FLT3 ligand have the capacity to stimulate an autologous leukemia-specific cytotoxicT cell response from patients with acute myeloid leukemia.*Leukemia*. 15, pp. 246–255.

Determination of Th1/Th2/Th17 Cytokines in Patients Undergoing Allogeneic Hematopoietic Stem Cell Transplantation

Adriana Gutiérrez-Hoya[1], Rubén López-Santiago[1], Jorge Vela-Ojeda[2], Laura Montiel-Cervantes[2], Octavio Rodríguez-Cortes[1] and Martha Moreno-Lafont[1]
[1]Departamento de Inmunología, Escuela Nacional de Ciencias Biológicas – Instituto Politécnico Nacional
[2]Unidad Médica de Alta Especialidad, Centro Médico Nacional La Raza, Instituto Mexicano del Seguro Social
México

1. Introduction

Allogeneic hematopoietic cell transplantation emerges as a therapy option to treat the sequels of exposure to radiation, a great concern at the beginning of the atomic age and cold war (Welniak et al., 2007). Hematopoietic cell transplantation emerged as a rescue strategy since there were already antecedents, like the study of Lorenz et al. in 1952 (as cited by Welniak et al., 2007), who showed that infusion of the bone marrow after lethal irradiation healed radiation disease in mice. This lay the foundations for the current consideration of allogeneic hematopoietic cell transplantation as the first-line therapy for many life-threatening oncological and hematological diseases. Today, it is primarily used to treat patients with hereditary anemias or immunological deficiencies through replacement of the hematopoietic system with cells from a healthy individual. It also allows cancer patients to be treated with myeloablative radiation and/or chemotherapy (known as myeloablative conditioning) in an attempt to eliminate tumoral cells, and although this strategy brings loss of bone marrow function, the latter can be recovered with infusion of normal hematopoietic cells (Jenq & Van den Brink, 2010).

2. Hematopoietic stem cell transplantation

Hematopoietic stem cell transplantation is today a simple procedure involving infusion of these cells intravenously. Once in the bloodstream, stem cells are able to migrate to the bone marrow in order to thus restore hematopoiesis during the first two weeks following transplantation (Léger & Nevill, 2004).

Stem cells giving rise to hematopoietic cells are known as hematopoietic stem cells due to their capacity for self-renewal, division and differentiation into a variety of specialized hematopoietic cells. This is the principle underlying hematopoietic stem cell transplantation (Hows, 2005).

One of the major difficulties in using hematopoietic stem cells has been their identification, since they are morphologically very similar to lymphocytes. This problem was solved with the use of biomarkers such as CD34, a transmembrane glycoprotein expressed by hematopoietic stem cells which is currently the main biomarker used to identify these cells.
Hematopoietic cell transplantations are classified as isotransplantation, allotransplantation or autotransplantation. These are not the only types of transplantation, but they are the most commonly used in medical practice. It calls allotransplantation when donor and recipient are the same species but are not identical twins. The advantages of this type of transplantation compared to the allogeneic type are that the cells infused are normal cells and therefore the incidence of relapse is lower, as well as the fact that the graft cell is infused with immunocompetent cells that are able to induce a graft-versus-leukemia effect. The major concern with this type of transplantation is development of graft-versus-host disease (GVHD) or infections caused by opportunistic microorganisms, since patients are treated with immunosuppressant drugs (Léger & Neville, 2004; Vela-Ojeda et al., 2005).

2.1 Hematopoietic stem cell mobilization by G-CSF

At present peripheral blood hematopoietic stem cells are preferably used since grafting (particularly of blood platelets) is faster. This procedure is not excessively invasive - as is bone marrow procurement - and better results are obtained when mobilized peripheral blood is used as the hematopoietic stem cells source (Jaime et al., 2004). Normally, peripheral blood contains only a small amount of hematopoietic stem cells (<0.1% of nucleated cells). Different methods are therefore used to induce their egress from the bone marrow into the bloodstream in order to be able to collect them by apheresis for subsequent infusion in the patient. Hematopoietic stem cells mobilization was an innovative development in the 1990s, in particular after it was seen that the number of stem cells obtained from mobilized peripheral blood contained 1-log more lymphocytes than the number obtained from bone marrow (Champlin, 2000). The established, widely-used method of mobilization involves the use of G-CSF, which induces mobilization by initiating a stress process through neutrophil and osteoclast activation. This results in dissociation of the cell membrane unions between stem cells and the stroma cells as well as stem cell proliferation and activation and/or adhesion molecule degradation. Hematopoietic stem cells mobilization is also seen when chemotherapy is exclusively used (Devetten & Armitage, 2007).
The mechanism through which G-CSF mobilizes $CD34^+$ hematopoietic stem cells from the bone marrow into the peripheral blood involves a series of steps. First of all, there is increased hematopoietic stem cell proliferation followed by exit of these cells from the bone marrow. Increased proliferation has been shown to occur with cytokines such as GM-CSF that temporarily increase the cell adhesion of $CD34^+$ hematopoietic stem cells to the bone marrow stroma, a process that in turn increases cell proliferation. The mobilization of hematopoietic stem cells from the bone marrow to peripheral blood comprises several mechanisms. One hypothesis is modification of the cellular interactions occurring between hematopoietic stem cells and the bone marrow stroma. Analyses of peripheral blood mononuclear cells mobilized by G-CSF reveal a decrease in the expression of VLA-4 (*Very late antigen 4* [CD49d/CD29]) integrin which normally binds firmly to its ligand VCAM-1 (*Vascular cell adhesion molecule-1*) as well as to an extracellular-matrix fibronectine fragment. Other molecules in which a marked decrease occurs are LFA-1 (*Leucocyte functional antigen 1*

[CD11aCD18]) and c-kit. These molecules are expressed in most hematopoietic stem cells and are involved in binding of these cells to bone marrow stroma, a process that is expressed by the ligands of cells of the latter VCAM-1, ICAM-1 (*Intercellular adhesion molecule-1*) and ICAM-2 for LFA-1 and c-kitL. G-CSF can also initiate mobilization through neutrophils, by secretion of gelatinase B, breaking extracellular matrix molecules and

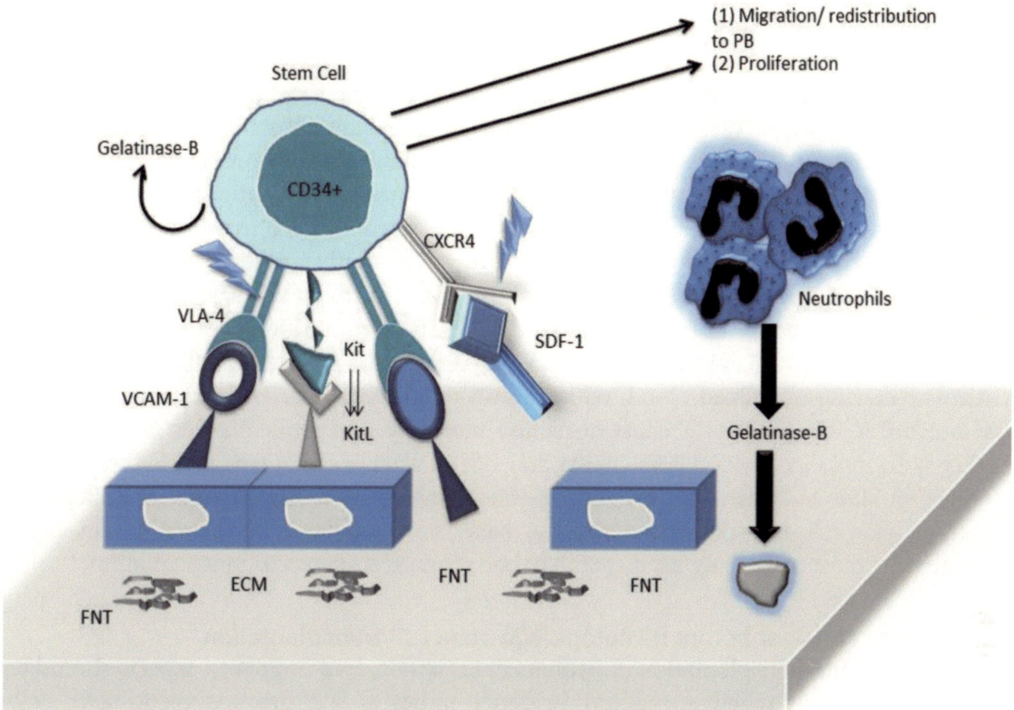

Fig. 1. Mechanism of hematopoietic stem cell mobilization from bone marrow to peripheral blood by stimulation with G-CSF. This process involves several steps, including modification of interactions occurring between stem cells and the stroma. Affected interactions include decreased expression of VLA-4 integrin, which normally binds to its ligand VCAM-1 and to FNT. Another molecule in which expression is decreased is c-kit; this process reduces the bond between this molecule and its ligand kit-L-. Decreased expression of these molecules reduces the bond between stem cells and the stromal cells. Another means by which G-CSF promotes stem cell mobilization is through neutrophils, by releasing gelatinase B, breaking extracellular matrix molecules and weakening adhesive interactions between stem cells and stroma cells. Yet another mechanism is stem cell secretion of gelatinase B, promoting faster migration of these cells to peripheral blood. A further mechanism involves indirect interaction of G-CSF with ligands, receptors/factors and stimulation of stem cell proliferation. PB: peripheral blood. G-CSF: granulocytic-colony stimulating factor. VCAM-1: vascular cell adhesion molecule-1, VLA-4: very late antigen 4 [CD49d/CD29] FNT: fibronectina (Modified from Gyger et al., 2000)

weakening adhesive interactions between stem cells and stroma cells. Stem cells have also been shown to secrete gelatinase B, a mechanism which may improve migration of these cells to peripheral blood (Figure 1). Finally, experimental evidence indicates that G-CSF may interact indirectly with a stem cell ligand to stimulate proliferation of these cells (Cashen et al., 2007; Cottler-Fox et al., 2003; Gyger et al., 2000).

In recent years, it has been demonstrated that G-CSF is also able to break one of the most important interactions between stem cells and stromal cells, which is formed by CXCL12 (formerly called SDF-1) and CXCR4 (Cashen et al., 2007).

2.1.1 Which stem cell source is best: bone marrow or mobilized peripheral blood?

This is one of the most commonly asked questions, since using mobilized peripheral blood rather than bone marrow in allogeneic transplantation is increasingly frequent. A retrospective study by Champlin et al. in 2000 examined the evolution of 288 allogeneic transplants in which mobilized peripheral blood was used and 536 in which bone marrow was used. They were followed-up during one year. The study found that in patients who had received mobilized peripheral blood, neutrophil engraftment was faster (14 days vs. 19) as was also platelet engraftment (19 days vs. 25). No significant differences were observed in development of acute GVHD. Chronic GVHD development was significantly higher in patients receiving mobilized blood, with a mean of 65% vs. 53%. The incidence of relapse did not differ significantly. Treatment-related mortality and leukemia-free survival were higher with mobilized blood transplants and hospitalization time was shorter. Additional studies similar to this one examining the benefits and drawbacks of allogeneic hematopoietic cell transplantation set the basis for mobilized peripheral blood being currently the most commonly used source in hematopoietic stem cell transplantation.

2.2 Points to consider before hematopoietic stem cell transplantation

When considering transplantation in patients in remission, two important aspects should be taken into account: whether medical evidence indicates that hematopoietic stem cells transplantation is more likely to heal the disease than other therapy forms and whether a suitable donor is available as a stem cell source. Although these are the major aspects to consider they are not the only ones. Other factors requiring consideration include biological characteristics at diagnosis, the specific disease that is being treated, presence of comorbidities that may complicate transplantation, and patient age (Deeg, 2010; Léger & Nevill, 2004).

2.3 Immunological typing of human leukocyte antigens

One reason for the progress that has taken place in hematopoietic cell transplantation is human leukocyte antigen (HLA) immunotyping. This is one of the major points to consider in allogeneic hematopoietic cell transplantation since, as already stated, it is important to have a suitable donor.

HLA proteins were first identified in 1950 by Jean Dausset upon observing that in many individuals, particularly those who previously received multiple blood transfusions or who were multiparous women, blood serum contained antibodies that reacted against a new kind of glycoprotein present on the outer surface of leukocytes of other members of the population; these glycoproteins were named human leukocyte antigens (HLAs). The latter behave as immunogenic markers making a person's cells distinct and are the major barrier to histocompatibility; they are therefore also called major histocompatibility complex (MHC) molecules. The importance of HLA molecules is not limited to the histocompatibility barrier, they are also essential in T-cell activation since HLA-molecules bind peptides to be presented to T cells. HLA class I molecules present peptides primarily to CD8+ T cells while CD4+ T cells recognize mainly peptides presented by class II molecules (Appelbaum, 2001; Bleakley & Riddell, 2004).

MHC genes are encoded on the short arm of chromosome 6 at locus p21. There are three different groups of genes called HLA-A, HLA-B and HLA-C, which individually code for the α chain of MHC class I. Similarly, there are three loci for the genes of class II MHC molecules known as HLA-DP, HLA-DQ and HLA-DR. Each of these includes genes coding for the α polypeptide chain and at least one β polypeptide chain. A person normally inherits two copies of the locus of each gene, one from each parent. Statistical data provided by the European Bioinformatics Institute (EBI) and the International Immunogenetics Organization Database (IMGT) suggest that the number of HLA class I and class II alleles discovered is on the increase. These data indicate that in human population there are 4,946 different class I alleles and 1,457 class II alleles, of which 1,601 are known alleles for HLA-A, 2,125 for HLA-B, 1,102 for HLA-C and 1,027 for HLA-DRβ of which 928 are HLA-DRβ1 alleles (these are the ones most commonly used to determine histocompatibility due to their high polymorphism) (Table 1).

Numbers of HLA Alleles

HLA Class I Alleles	4,946
HLA Class II Alleles	1,457
HLA Alleles	6,403

HLA Class I

Gene	**A**	**B**	**C**	**E**	**F**	**G**
Alleles	1,601	2,125	1,102	10	22	47
Proteins	1,176	1,641	808	3	4	15

HLA Class II

Gene	**DRA**	**DRB**	**DQA1**	**DQB1**	**DPA1**	**DPB1**	**DMA**	**DMB**	**DOA**	**DOB**
Alleles	7	1,027	44	153	32	149	7	13	12	13
Proteins	2	774	27	106	16	129	4	7	3	5

HLA Class II- DRB Alleles

Gene	**DRB1**	**DRB2**	**DRB3**	**DRB4**	**DRB5**	**DRB6**	**DRB7**	**DRB8**	**DRB9**
Alleles	928	1	57	15	19	3	2	1	1
Proteins	704	0	46	8	16	0	0	0	0

(Modified from http://www.ebi.ac.uk/imgt/hla/stats.html)

Table 1. Statistical data from the European Bioinformatics Institute (EBI) and the International Immunogenetics Organization database (IMGT) showing the number of each of the HLA alleles.

Historically, HLA immunotyping was performed by serological methods but now, with the advent of polymerase chain reaction (PCR), molecular immunotyping of the donor and recipient is possible. A study by Petersdorf et al. in 2001 examined patients who had previously undergone transplantation and were compatible by serological methods. When reexamined by molecular immunotyping, about 30%of these individuals were found to be incompatible in one or more alleles. These differences were correlated with increased GVHD and poor survival, indicating that a compatible donor and reliable HLA immunotyping are extremely important. The number of class I and class II HLA antigens is relatively large and therefore the probability of HLA matching between the recipient and an unrelated donor is extremely small.

2.4 Graft-versus-host disease and Th1/Th2/Th17 cytokines

Graft-versus-host disease (GVHD) may develop after hematopoietic stem cell transplantation. It is a reaction of immune cells from the donor against tissues of the host. Damage induced on epithelial cells of the host by activated T cells occurs after an inflammatory cascade that is unleashed by the conditioning regimen. Approximately 35-50%of allogenic hematopoietic cell transplantation recipients develop GVHD. The risk of developing the disease depends on several factors, primarily the stem cell source and donor cytokines, patient age, existing conditions and GVHD prophylaxis. GVHD involves mainly the skin, liver and gastrointestinal tract. Despite GVHD-related morbility and mortality, its development is often desirable since it has been found to be associated with a lower recurrence of malignant disease, in other words, it is important for establishment of the graft-versus-tumor effect (Ferrara & Levine, 2008; Léger & Nevill, 2004; Saliba et al., 2007; Weisdorf 2007).

The physiopathology of acute GVHD described by Ferrara & Levine (2008) is a three-stage phenomenon. The initial stage involves damage to tissues of the host due to inflammation derived from chemo- and/or radiotherapy during the recipient conditioning. In the second stage, antigen-presenting cells (APC) of both donor and recipient as well as inflammatory cytokines unleash the activation of donor-derived T cells, with expansion and differentiation of the latter into effector cells. Antigens (Ag) of the minor histocompatibility complex have a central role in this activation. The pathway of T-cell activation results in activation of genes coding for cytokines such as IL-2 and interferon gamma (IFNγ). Cells that produce these cytokines are considered to be Th1 profile, as opposed to cells producing predominantly IL-4, IL-5, IL-10 and IL-13 which are considered to be Th2 phenotype and are assumed to be the ones that modulate GVHD. During the third stage, also known as the effector stage, donor-derived activated T cells mediate cytotoxicity against target cells of the recipient through FasL-Fas, perforin and granzyme B interactions as well as additional production of tumor necrosis factor α (TNFα). This cytokine is produced by monocytes and macrophages, and secondarily by T lymphocytes and natural killer (NK) cells (Figure 2). (Ferrara & Levine, 2008; Jacobsohn & Vogelsang, 2007; Socie & Blazar 2009).

TNFα is firmly involved in GVHD physiopathology at several steps of the process including induction of apoptosis in target tissues through the TNFα receptor. It also induces the activation of macrophages, neutrophils, eosinophils, and B and T cells; stimulates production of inflammatory cytokines such as IL-1, IL-6, IL-12 and TNFα itself; increases the expression of HLA molecules; and promotes lysis by T lymphocytes. High levels of TNFα are associated with a higher incidence of GVHD in bone marrow transplant recipients. This allogeneic dysregulation, in addition to dysregulation of cytokines, leads to the acute tissue damage produced by GVHD.

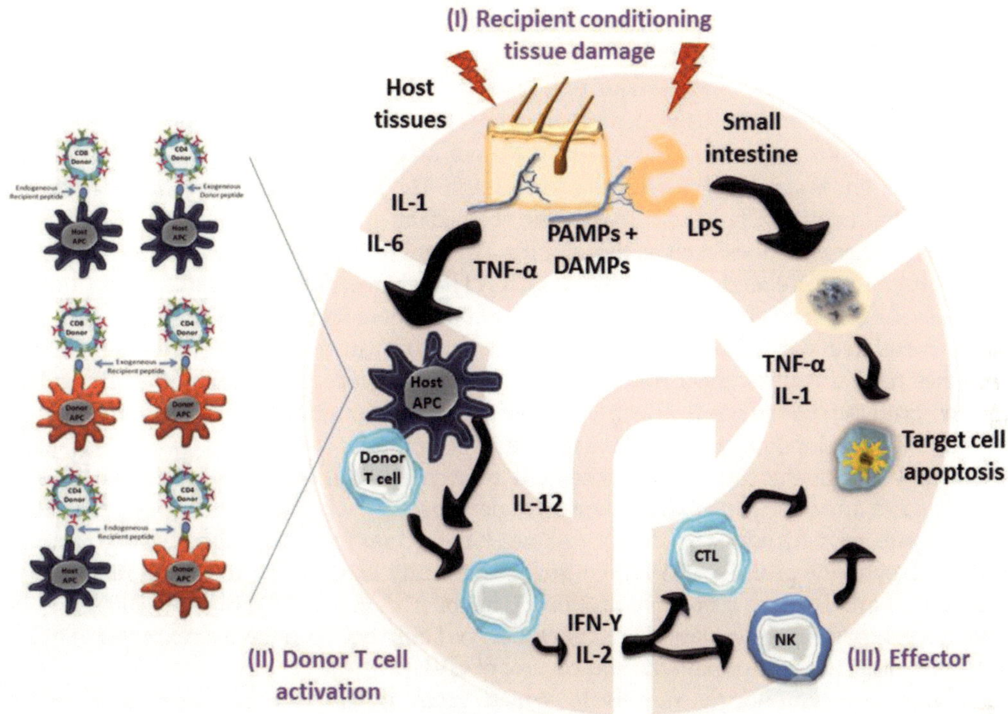

Fig. 2. Physiopathology of graft-versus-host disease. (I) Damage to host tissues by inflammation derived from the chemotherapy and/or radiotherapy conditioning regimen. (II) Antigen-presenting cells (APC) of both donor and recipient as well as inflammatory cytokines unleash the activation of donor-derived T cells, with expansion and differentiation of the latter into effector cells. The pathway of T-cell activation results in activation of the genes coding for cytokines such as IL-2 and IFNγ. (III) Effector stage: donor-derived activated T cells mediate cytotoxicity against target cells of the recipient through FasL-Fas, perforin and granzyme B interactions as well as additional production of tumor necrosis factor α (TNFα) (Modified from Ferrara & Levine, 2008).

In our laboratory, we have studied patients who underwent allogenic hematopoietic cell transplantation and developed GVHD, finding a correlation between these patients and an increased of CD14+ TNFα+ cells. Up to 32% of CD14+ cells secreting TNFα were found in a patient with stage II GVHD, increasing to 47% when the patient progressed to stage III, while in patients who did not develop GVHD this behavior was not observed. In patients with GVHD development, TNFα may promote increased expression of adhesion molecules such as VCAM-1, ICAM-1 and E-selectin in endothelium, therefore also promoting diapedesis of leukocytes in affected areas and degranulation of these cells with subsequent damage to tissues in which they are infiltrated (Aggarwal et al., 2000). On the other hand, activation of endothelium, release of nitric oxide, vasodilation and increased vascular permeability, and blood platelet activation are also promoted and, most importantly, increased expression of MHC-I and MHC-II molecules as well as co-stimulatory molecules such as CD40, CD80 and CD86 in dendritic cells, thus also activating T lymphocytes and giving rise to more effective antigen presentation, and along with this, more effective allorecognition (Steinman et al., 1998).

In addition to this, TNFα has a major role in the promotion of apoptosis (De Freitas et al., 2004) and due to the previously mentioned properties, TNFα overexpression contributes to GVHD emergence and severity (Figure 3). Recent studies reveal an increase not only in TNFα levels but also in TNF receptors (TNFR1 and TNFR2), and the latter are more stable and easier to quantify (Choi et al., 2009; Kitko et al., 2008).

The role of other cytokines such as those of the Th1 profile is worth noting. Diverse research groups initially correlated this profile with GVHD emergence, since large quantities of cytokines such as IFNγ, IL-2 and IL-12 were found in patients with GVHD development and this is correlated with GVHD severity (Das et al., 2001; Ju et al., 2005). However, these cytokines have a controversial role as they are necessary for development of both GVHD and the graft-versus-leukemia effect. We have found that mononuclear cells from patients with GVHD development in co-culture secrete large amounts of IFNγ and IL-2, but we have also observed that the capacity of these cells to secrete these cytokines is correlated with graft success unaccompanied by relapse or development of infections by opportunistic microorganisms, which tells us these cytokines have a dual role. Regarding the significance of the Th2 cytokine profile in GVHD control, some study teams point to the overexpression of IL-4, IL-5 and IL-10 as a positive prognostic factor (Das et al., 2001; Ju et al., 2005). In our laboratory, however, a correlation has been observed only between the overexpression of IL-10 by mononuclear cells of patients and control of GVHD.

There are new lymphocyte subsets to which great significance has been ascribed in inflammatory processes as well as in many pathologies previously thought to be associated with the Th1 profile. The subpopulation of Th17 cells discovered in 2005 is now known to have a controversial role as they are implicated in rejection of solid organ grafts (Kappel et al., 2008; Carlson 2009; Coghill et al., 2010).

In murine models, differences in GVHD development have been found between mice that were transferred CD4+ IL-17$^{-/-}$ T cells and mice that were transferred normal CD4+ cells. In the former, GVHD development took longer. However, no significant differences were noted between these groups in relation to mortality due to GVHD or in graft-versus-tumor activity. Another major finding was the fact that mice that were transferred CD4+ IL17$^{-/-}$ cells had fewer Th1 cells during early stages of GVHD. Also, a reduction occurred in the number of IFNγ-secreting macrophages and granulocytes as well as a decrease in the amount of pro-inflammatory cytokines. IL-17 is therefore believed to be essential for GVHD development and graft-versus-leukemia activity as it promotes pro-inflammatory cytokine production – all this in murine models (Kappel et al., 2008). These data and others showing the importance of the Th17 profile in inflammatory processes made several researchers think that this profile might be involved in GVHD development and severity in humans (Coghill, 2011). Our study team recently found that the Th17 profile is not relevant for GVHD development. We conducted a pilot study on the importance of this profile in six patients who underwent allogenic hematopoietic cell transplantation, following them for six months. This group was divided into patients with GVHD development and patients without GVHD. Peripheral blood and mononuclear cell cultures were analyzed at 30, 60, 100 and 180

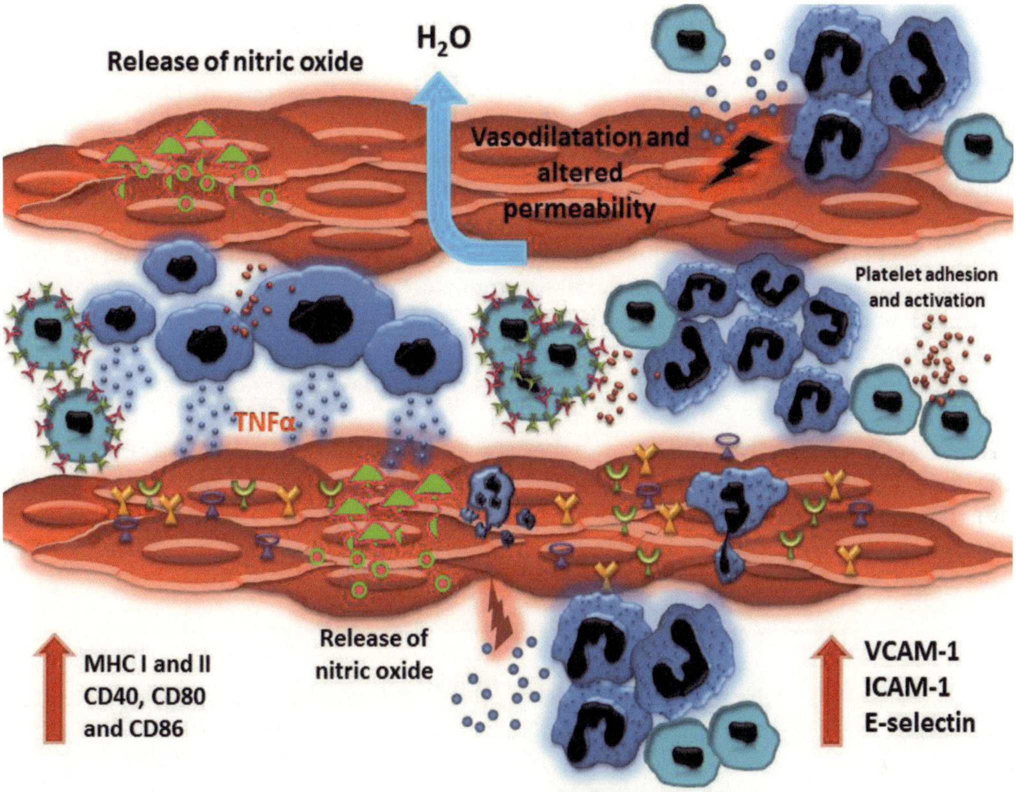

Fig. 3. The role of TNFα in GVHD emergence and exacerbation.

days after transplantation. Overproduction of Th17 profile was not observed neither in patients with GVHD nor in individuals without GVHD, as opposed to its production in healthy volunteers and in a patient who received a syngeneic transplant. Few months after concluding our study, Broady et al. (2010) published a similar study on patients who underwent allogenic hematopoietic cell transplantation, in which they evaluated Th1 and Th17 cells in tissue and peripheral blood in a cohort of 34 patients, of which 20 developed acute GVHD and 14 did not develop GVHD. The authors did not find an increase in the number of Th17 cells in patients with acute cutaneous GVHD as compared to healthy donors, but they did detect increased production of IFNγ-secreting cells (Broady, 2010).

GVHD diagnosis is suspected when the recipient develops all or some of the signs and symptoms that are characteristic of this disease such as dermatitis (rash), epidermal blistering, stomach cramps, abdominal pain with or without diarrhea which may be accompanied by passage of blood, nausea, persistent vomiting, and hepatitis (elevated bilirubins and/or liver enzymes). Typically, these signs and symptoms occur 100 days after allogenic hematopoietic cell transplantation but may also appear later. Because many of them are not specific of this complication, diagnosis must be supported with suitable biopsies, particularly if the symptoms are atypical or involve only the liver or gastrointestinal tract, since histological confirmation is extremely useful. A further reason

for biopsy-taking is to help differentiate GVHD from other diseases with similar symptoms, such as viral infections or reactions to pharmaceutical agents (Ferrara & Levine 2008; Jacobsohn & Vogelsang, 2007).
In GVHD, mature T cells of the donor that accompany the graft attack host tissues, particularly the skin, liver and gastrointestinal tract. This explains the signs and symptoms that characterize this disease. To prevent GVHD development, all patients receive some type of prophylaxis with immunosuppressors and in some cases T cell depletion , being one of the most commonly used methods although its effectiveness has not been fully proven since T-cell elimination contributes to absence of GVHD development but patients die from relapse because lack of GVT effect. Another prophylactic method involves pharmacological treatment with agents that affect T-cell function. These types of prophylaxis elicit adverse effects since mature T cells of the donor have a major role in mediating the reconstitution of the adaptive immune system, particularly in adults with low thymus function (Jacobsohn & Vogelsang, 2007).
GVHD is classified according to the number and extent of the organs involved (Table 2). In the current classification system, established in 1994, GVHD is divided into four groups (I - IV). Skin damage is evaluated by the percentage of the body surface area involved, liver damage by elevated bilirubins, and gastrointestinal tract damage by the amount of diarrhea (Jacobsohn & Vogelsang, 2007; Vela-Ojeda et al., 2008).

Stage	Skin	Liver (bilirubin)	Gut (stool output/day)
0	No GVHD rash	< 2 mg/dl	< 500 ml/day or persistent nausea
1	Maculopapular rash < 25% BSA	2-3 mg/dl	500-999 ml/day
2	Maculopapular rash < 25 - 50% BSA	3.1-6 mg/dl	1000-1500 ml/day
3	Maculopapular rash > 50% BSA	6.1-15 mg/dl	Adult: >1500 ml/day
4	Generalized erythroderma plus bullous formation	>15 mg/dl	Severe abdominal pain with or without ileus
Grade			
I	Stage 1-2	None	None
II	Stage 3 or	Stage 1 or	Stage 1
III	-	Stage 2-3 or	Stage 2-4
IV	Stage 4 or	Stage 4	-

(Modified from Jacobsohn & Volgelsang 2007).

Table 2. Classification of acute graft-versus-host disease.

Chronic GVHD (cGVHD) is one of the most common and significant problems affecting recipients of allogenic hematopoietic cell transplantation at long term. It appears approximately 100 days after allogenic hematopoietic cell transplantation, which is the critical time at which close to 50% of patients develop some degree of cGVHD. Increased use of hematopoietic stem cells obtained from peripheral blood rather than bone marrow, increased age of recipients, and the use of busulfan in the conditioning regimen have led to a higher incidence of cGVHD. Other clinical risk factors for cGVHD development include previous acute graft-versus-host disease (aGVHD) and a second transplant. cGVHD most commonly affects the skin, liver, eyes and mouth, although other sites may also be affected. Death from severe cGVHD is generally a result of infectious complications. The standard treatment for severe cGVHD is a combination of cyclosporine and prednisone. An alternating daily regimen of these two agents prolongs survival and reduces drug-related adverse effects. Topical therapy in affected areas is recommended for patients with grade 1-2 cutaneous disease. Survival at 10 years in patients who develop light aGVHD is approximately 80%, but it is drastically reduced in patients with severe cGVHD, in whom this rate is reported to be 5% (Horwitz & Sullivan 2006; Lee 2005; Vela-Ojeda et al., 2008).

Alloreactivity is the basis for cGVHD pathogenesis. However, the exact phenotype and the origin of alloreactive cells remain somewhat ambiguous. Donor-derived alloreactive T-cells transplanted with hematopoietic stem cells play a key role in acute and chronic GVHD. Current animal models of cGVHD implicate Th2 cells as the first cell type to induce damage. However, in humans with cGVHD, Th1/Th2-polarized $CD4^+$ cells have alloreactive properties. The formation of antibodies has been observed in experimental models and clinical studies of cGVHD. This suggests that B cells are implicated in the physiopathology of cGVHD as shown by antibody production in allogenic hematopoietic cell transplantation patients with donor of different sex, since antibodies against minor histocompatibility antigens are encoded in the Y chromosome. The presence of anti-nuclear, anti-double strand DNA and anti-smooth muscle antibodies in a frequency range of 11-62% has also been detected in patients with cGVHD as well as the presence of anti-cytoskeletal and anti-nucleolar antibodies. However, despite these findings, the role of antibodies in cGVHD remains unclear (Horwitz & Sullivan 2006; Lee 2005; Vela-Ojeda et al., 2008).

cGVHD can be classified according to type of clinical manifestations or extent of the disease. Most patients with cGVHD have previously had aGVHD. cGVHD may also be observed after achieving control of aGVHD. Similarly, patients may develop cGVHD without a previous history of aGVHD (*de novo* cGVHD). Classification according to the type and extent of clinical manifestations is shown in Table 3.

The International Center for Research on Bone Marrow and Blood Cell Transplantation estimates that 50,000 - 60,000 hematopoietic cell transplantations are performed each year throughout the world. Bone marrow is the main source of grafts for transplantation in children, although peripheral and umbilical cord blood are being increasingly used. Between 2004 and 2008 peripheral blood accounted for 27% and umbilical cord blood for 32% of transplants in patients fewer than 20 years in age. In patients over 20 years the most common source for allogenic hematopoietic cell transplantation is peripheral blood. Currently, very few adults receive grafts of umbilical cord blood but its use, although infrequent, increased 2-4% between 2004 and 2008. Mobilized peripheral blood is the main source for autologous transplantation, representing 91% of autotransplantations in children and 98% in adults. In recent years the number of hematopoietic cell transplantations, both allogeneic and autologous, in patients over 50 years old has increased. Approximately 40% of allogeneic transplants are unrelated donor transplantations. There has been a change lately: before 2002 the most commonly used source of hematopoietic stem cells was bone marrow but its use has declined since 2003 and peripheral and umbilical cord blood have been increasingly used.

Classification of GVHD	
Limited chronic GVHD	
Either or both:	
1	Localized skin involvement
2	Hepatic dysfunction due to chronic GVHD
Extensive chronic GVHD	
Either:	
1	Generalized skin involvement, or
2	Localized skin involvement and/or hepatic dysfunction due to chronic GVHD
Plus:	
3a	Liver histology showing chronic aggressive hepatitis, bridging necrosis, or cirrhosis, or
b	Involvement of eye (Schirmer test with <5-mm wetting), or
c	Involvement of minor salivary glands or oral mucosa demonstrated on labial biopsy, or
d	Involvement of any other target organ

Table 3. Classification of chronic graft-versus-host disease (Horwitz & Sullivan, 2006)

The most common cause of death after allogenic hematopoietic cell transplantation is relapse, although in unrelated donor transplants it is followed by GVHD and in HLA-matched sibling transplants by death from infections. Another major cause of death is interstitial pneumonitis (Pasquini & Wang, 2007).

2.5 Graft-versus-tumor effect

Existence of the graft-versus-tumor effect was first suggested in 1956 by Barnes et al., upon noting eradication of leukemia in irradiated mice receiving allogeneic bone marrow transplants, but not in those that after irradiation received syngeneic bone marrow transplants. The first evidence of this also occurring in humans came from studies reporting that the incidence of relapse was markedly lower in patients who developed GVHD than in those who did not (as cited in Appelbaum, 2001), and that just as in murine models, human allogenic hematopoietic cell transplantation recipients were at lower risk of relapse than recipients of syngeneic stem cell transplantations or allogenic hematopoietic cell transplantations in which T cells are previously depleted. Male patients receiving allogenic hematopoietic cell transplantation grafts from female donors are also seen to be a special group in that donor-derived T cells specific for receptors of minor histocompatibility antigens may contribute to development of the graft-versus-tumor or the graft-versus-leukemia effect as well as GVHD since the graft-versus-leukemia effect is associated with presence of GVHD. The potential impact of the graft-versus-leukemia effect can be observed when the patient is administered infusions of donor-derived lymphocytes. Graft versus tumor effect formed the basis of the so-called non-myeloablative transplants in which intensive chemotherapy regimen are substituted by the antitumoral effect of lymphocytes (Jacobsohn & Vogelsang, 2007).

Factors inducing GVHD or the graft-versus-leukemia effect are not well defined and may to some extent be considered speculative. Minor histocompatibility antigens have been suggested to play a key role since some of them have limited expression in hematopoietic cells, including leukemic cells. These antigens are perhaps targeted by the selective effect of the graft-versus-leukemia effect while minor histocompatibility antigens that are expressed in general may be targeted by GVHD, but this is still not firmly established (Jacobsohn & Vogelsang 2007; Kolb, 2008; Riddell & Appelbaum, 2007).

Various minor histocompatibility antigens are encoded in Y-chromosome genes that exhibit significant polymorphism with their homologues genes of the X chromosome. The former genes are responsible for the immune reactions occurring between men and women. Differential expression of the genes encoding for minor histocompatibility antigens in tissues has been proposed for potential use as a basis to separate the graft-versus-leukemia effect from GVHD. Cells that recognize minor histocompatibility antigens which are expressed only in hematopoietic cell receptors, including leukemic cells, may have a major role in the elimination of the latter without GVHD development. Several minor histocompatibility antigens such as HA-1, HA-2, HB-1 and BCL2A1 are expressed by hematopoietic cells and are therefore being examined as potential targets in order to promote the graft-versus-leukemia effect (Randolph et al., 2004; Kolb, 2008).

2.6 Clinical relevance of regulation and induction of tolerance after allogenic hematopoietic cell transplantation

The potential benefits of allogenic hematopoietic cell transplantation are offset by the moderate survival rate of the graft at long term. This is to a large extent due to immunosuppressant agents that unspecifically inhibit immune response in order to prevent rejection, but bring multiple adverse effects which are responsible for chronic rejection. Therefore, one of the major goals of allogenic hematopoietic cell transplantation is to achieve absence of immune response in the face of donor-derived alloantigens without requiring prolonged administration of immunosuppressants and to promote the graft-versus-leukemia effect. To this end, research has focused on the study of regulatory T cells (Treg), particularly those of the CD4+CD25highFoxP3+ phenotype (natural Treg, nTreg) as they have been shown to be able to control immune response in the face of donor-derived alloantigens and therefore possess great potential to establish in vivo tolerance to the transplant. These cells have been studied the most but are not the only ones described as having a regulatory role since regulatory functions have also been observed in other subpopulations of CD4+ cells, in T CD8+ cells, Tγδ cells, NK cells and CD3+CD4-CD8-CD16+56+ cells, also known as NKT cells. The latter have been shown to have a powerful anti-tumoral effect and are important in the maintenance of tolerance. This is why many studies are focusing on transfer of these cells. The strategy that is being sought for use in the near future is ex vivo stimulation of Treg purified with alloantigens of the donor or even FoxP3-mediated transfection of alloreactive $CD4+CD25^{-}$ cells (Bryceson & Ljunggren, 2007; Fehérvari & Sakaguchi, 2005).

Natural regulatory T cells (nTreg) express from the time of their differentiation in the thymus, $CD4^+$, $CD25^+$ and FoxP3 (*forkhead box P3 transcription factor*) expression patterns. They were identified by Sakaguchi et al. (1995) as a natural subset of $CD4^+$ T lymphocytes (approximately 5-10% of the T lymphocytes present in peripheral blood) that constitutively express the CD25 molecule and suppress the response of effector T lymphocytes ($CD4^+$ and $CD8^+$) *in vivo*. Another lymphocyte subpopulation in peripheral blood are $CD4^+CD25^-$ cells, which through the action of TGF-β and IL-2 may come to express CD25 and FoxP3, and are known as induced Treg (iTreg). Treg lymphocytes are also associated with low expression of CD127 which is positively correlated with regulatory function acquisition and negatively correlated with expression of FoxP3 (Curiel et al., 2004, Horwitz et al., 2008; Korn 2009; Liu et al., 2006; Seddiki et al., 2006).

CD4+CD25+FoxP3+ Treg cells have a key role in the maintenance of peripheral tolerance, and Treg deficiencies give rise to progressive autoimmune disorders. Similarly, improved Treg function can prevent graft rejection and suppress tumor immunity. In other words, an adequate balance between Treg and effector T cells is essential for maintenance of tolerance. In the context of allogenic hematopoietic cell transplantation, Treg have also been shown to have a major role in the establishment of tolerance between tissues of the recipient and donor-derived immunity. This was initially shown to occur in murine models where Treg depletion in the stem cell graft produced increased in GVHD and increased Treg numbers resulted in suppression of GVHD after transplantation (Lee 2005). In humans, patients with active cGVHD are reported to have a lower frequency of Treg than patients without cGVHD. These findings suggest that robust reconstitution of Treg post-allogenic hematopoietic cell transplantation is required to establish immune balance in order to be able to maintain adequate levels of peripheral tolerance. However, the mechanisms responsible for Treg reconstitution post-allogenic hematopoietic cell transplantation have not been adequately characterized and the factors contributing to inadequate recovery of Treg in patients who develop cGVHD are unknown (Matsuoka 2010).

Based on data obtained by Matsuoka et al. (2010) in patients examined during the first year after allogenic hematopoietic cell transplantation, thymus-dependent generation of Treg was considerably affected, but this subpopulation maintained high levels of cell proliferation in comparison to constitutive T cells. Such *in vivo* proliferation is apparently driven mainly by lymphopenia of CD4 cells. Among other findings, this study team has also shown that high levels of Treg proliferation were offset by higher susceptibility to apoptosis. Depletion of Treg in periphery in these patients was associated with development of extensive cGVHD.

On the other hand and in reference to lymphocyte populations that may take part in development of tolerance to alloantigens, the subpopulation of CD8+ Treg has been recently characterized. These lymphocytes, despite a long history in the field of immunology as described by Gershon & Kondo in 1970, have been difficult to characterize, and this factor combined with the discovery of CD4 Treg by Hall et al. (1990) and Sakaguchi et al. (1995) have considerably limited this area. The importance of CD8 Treg lies in the regulatory role observed in experimental autoimmune encephalitis (Lu & Cantor, 2008; Wang & Alexander 2009; Zheng et al., 2004).

The latter subpopulation has been said to be increased in peripheral blood analyses and infiltrated tissue of patients with colorectal cancer. Increased expression of TGF-β has also been found in these lymphocytes as compared to samples from healthy donors. These cells were able to suppress CD4+CD25- cell proliferation and cytokine Th1 production. The paper mentions the significance of immunological vigilance and the fact that CD8 Treg may promote tumoral growth. It also describes this cell subpopulation in patients with multiple sclerosis in whom lower numbers of these cells were correlated with relapse, a fact that may evidence their immunosuppressant role in the control of autoimmune diseases (Giovanni Frisullo, 2010; Chaput, 2009; Kiniwa, 2007).

Recently described populations of CD8+ Treg include CD8+ IL-10+ cells present in ovarian carcinoma, which are induced by plasmocytoid dendritic cells infiltrating the tumor. This differentiation towards CD8+ Treg was shown to be independent of $CD4^+$ Treg. All these antecedents combined with the generation of CD8+ Treg through continuous antigen stimulation may indicate the importance of the latter in GVHD control.

3. Conclusions

TNFα levels have a major role in development and severity of graft-versus-host disease and can be used as a negative prognostic factor.

Th1 response is essential and required in post-transplant patients to prevent relapse, graft loss and appearance of infections caused by opportunistic microorganisms.

The Th17 profile has no essential role in development or severity of GVHD and is therefore not a target for therapy in these patients.

4. Acknowledgments

The present study was financially supported through grants from Consejo Nacional de Ciencia y Tecnología (CONACyT), number 82559; and Secretaría de Investigación y Posgrado, Instituto Politécnico Nacional, numbers SIP-20100710 and SIP-20110504.

5. References

[1] Welniak LA, Blazar BR & Murphy WJ. (2007). Immunobiology of allogeneic hematopoietic stem cell transplantation. *Annu Rev Immunol.* Vol. 25, pp. (139-17), ISSN:0732-0582

[2] Jenq RR & van den Brink MR. (2010). Allogeneic haematopoietic stem cell transplantation: individualized stem cell and immune therapy of cancer. *Nat Rev Cancer.* Vol. 10, No. 3, (March 2010), pp. (213-221), ISSN 1474-1768

[3] Léger CS & Nevill TJ. (2004). Hematopoietic stem cell transplantation: a primer for the primary care physician. *CMAJ.* Vol. 170, No. 10, (May 2004), pp. (1569-1577), ISSN 1488-2329

[4] Hows J. (2005). Adult stem cell therapy beyond haematopoietic stem cell transplantation? An update. *Transpl Immunol.* Vol. 14, No. 3-4, (August 2005), pp. (221-223), ISSN 0966-3274

[5] Vela-Ojeda J & Ruiz-Esparza MA. (2005). Hematopoietic stem cell transplantation in multiple myeloma. *Rev Invest Clin.* Vol. 57, No. 2, (March-April 2005), pp. (305-313), ISSN 0034-9887

[6] Jaime Fagundo JC, Dorticós Balea E, Pavón Morán V & Cortina Rosales L. (2004). Trasplante de células progenitoras hematopoyéticas: tipos, fuentes e indicaciones. *Rev Cubana Hematol Inmunol Hemoter.* Vol. 20, No. 2. ISSN 0864-0289

[7] Champlin RE, Schmitz N, Horowitz MM, Chapuis B, Chopra R, Cornelissen JJ, Gale RP, Goldman JM, Loberiza FR Jr, Hertenstein B, Klein JP, Montserrat E, Zhang MJ, Ringdén O, Tomany SC, Rowlings PA, Van Hoef ME & Gratwohl A. (2000). Blood stem cells compared with bone marrow as a source of hematopoietic cells for allogeneic transplantation. IBMTR Histocompatibility and Stem Cell Sources Working Committee and the European Group for Blood and Marrow Transplantation (EBMT). *Blood.* Vol. 95, No. 12, (June 2000), pp. (3702-3709), ISSN 1528-0020

[8] Devetten M & Armitage JO. (2007). Hematopoietic cell transplantation: progress and obstacles. *Ann Oncol.* Vol. 18, No. 9, (September 2007), pp. (1450-1456), ISSN 1569-8041

[9] Cashen AF, Lazarus HM & Devine SM. (2007). Mobilizing stem cells from normal donors: is it possible to improve upon G-CSF? *Bone Marrow Transplant.* Vol. 39, No. 10, (May 2007), pp. (577-588), ISSN 0006-4971

[10] Cottler-Fox MH, Lapidot T, Petit I, Kollet O, DiPersio J & Link. (2003). Stem cell mobilization. *Ame Soc Hematol.* Vol. 1, pp. (419-437), ISSN 0001-2001

[11] Gyger M, Stuart RK & Perreault C. (2000). Immunobiology of allogeneic peripheral blood mononuclear cells mobilized with granulocyte-colony stimulating factor. *Bone Marrow Transplant.* Vol. 26, No. 1, (July 2000), pp. (1-16), ISSN 0268-3369

[12] Champlin RE, Schmitz N, Horowitz MM, Chapuis B, Chopra R, Cornelissen JJ, Gale RP, Goldman JM, Loberiza FR Jr, Hertenstein B, Klein JP, Montserrat E, Zhang MJ, Ringdén O, Tomany SC, Rowlings PA, Van Hoef ME & Gratwohl A. (2000). Blood stem cells compared with bone marrow as a source of hematopoietic cells for allogeneic transplantation. IBMTR Histocompatibility and Stem Cell Sources Working Committee and the European Group for Blood and Marrow Transplantation (EBMT). *Blood,* Vol. 15, No. 95, (June 2000), pp. (3702-3709), ISSN 1528-0020

[13] Deeg HJ & Sandmaier BM. (2010). Who is fit for allogeneic transplantation? *Blood.* Vol. 116, No. 23, (Dec 2010), pp. (4762-4770), ISSN 1528-0020

[14] Appelbaum FR. (2001). Haematopoietic cell transplantation as immunotherapy. *Nature.* Vol. 411, No. 6835, (May 2001), pp. (385-389), ISSN 0028-0836

[15] Bleakley M & Riddell SR. (2004). Molecules and mechanisms of the graft-versus-leukaemia effect. *Nat Rev Cancer.* Vol. 4, No. 5, (May 2004), pp. (371-380), ISSN 1474-1768

[16] Petersdorf EW, Hansen JA, Martin PJ, Woolfrey A, Malkki M, Gooley T, Storer B, Mickelson E, Smith A & Anasetti C. (2001). Major-histocompatibility-complex class I alleles and antigens in hematopoietic-cell transplantation. *N Engl J Med.* Vol. 345, No. 25, (December 2001), pp. (1794-1800), ISSN 0028-4793

[17] Ferrara J & Levine J. (2008). A new approach to therapy for acute GVHD. *Cell Ther Transplant*. Vol. 1, No. 1, (May 2008), pp. (25-27), ISSN 1866-8836

[18] Saliba RM, de Lima M, Giralt S, Andersson B, Khouri IF, Hosing C, Ghosh S, Neumann J, Hsu Y, De Jesus J, Qazilbash MH, Champlin RE & Couriel DR. (2007). Hyperacute GVHD: risk factors, outcomes, and clinical implications. *Blood*. Vol. 109, No. 7, (Apr 2007), pp. (2751-2758), ISSN 1528-0020

[19] Weisdorf D. (2007). GVHD- The nuts and bolts. *Hematology Am Soc Hematol Educ Program*, Vol. 1, No. 1, (January 2007), pp. (62-67), ISSN 1520-4391

[20] Jacobsohn DA & Vogelsang GB. (2007). Acute graft versus host disease. *Orphanet J Rare Dis*. Vol. 2, No. 35, (Sep 2007), ISSN 0212-1611

[21] Socié G & Blazar BR. (2009). Acute graft-versus-host disease: from the bench to the bedside. *Blood*. Vol. 114, No. 20, (Nov 2009), pp. (4327-4336), ISSN 1528-0020

[22] Aggarwal BB, Samanta A & Feldmann M. (2000). TNFα. *Cytokine ref*. DOI:1006/RWEY/150001.200. pp. (414-434).

[23] Steinman RM. (1988). Cytokines amplify the function of accessory cells. *Immunol Lett*. Vol. 17, No. 3, (March 1988), pp. (197-202), ISSN 0165-2478

[24] De Freitas I, Fernández-Somoza M, Essenfeld-Sekler E & Cardier JE. (2004). Serum levels of the apoptosis-associated molecules, tumor necrosis factor-alpha/tumor necrosis factor type-I receptor and Fas/FasL, in sepsis. *Chest*. Vol. 125, No. 6, (June 2004), pp. (2238-2246), ISSN 1931-3543

[25] Choi SW, Kitko CL, Braun T, Paczesny S, Yanik G, Mineishi S, Krijanovski O, Jones D, Whitfield J, Cooke K, Hutchinson RJ, Ferrara JL& Levine JE. (2008). Change in plasma tumor necrosis factor receptor 1 levels in the first week after myeloablative allogeneic transplantation correlates with severity and incidence of GVHD and survival. *Blood*. Vol. 112, No. 4, (August 2008), pp. (1539-1542), ISSN 1528-0020

[26] Kitko CL, Paczesny S, Yanik G, Braun T, Jones D, Whitfield J, Choi SW, Hutchinson RJ, Ferrara JL & Levine JE. (2008). Plasma elevations of tumor necrosis factor-receptor-1 at day 7 postallogeneic transplant correlate with graft-versus-host disease severity and overall survival in pediatric patients. *Biol Blood Marrow Transplant*. Vol. 14, No. 7, (July 2008), pp. (759-765), ISSN 1083-8791

[27] Das H, Imoto S, Murayama T, Mizuno I, Sugimoto T, Taniguchi R, Toda K, Isobe T, Nakagawa T, Nishimura R & Koizumi T. (2001). Kinetic analysis of cytokine gene expression in patients with GVHD after donor lymphocyte infusion. *Bone Marrow Transplant*. Vol. 27, No. 4, (February 2001), pp. (373-380), ISSN 0006-4971

[28] Ju XP, Xu B, Xiao ZP, Li JY, Chen L, Lu SQ & Huang ZX. (2005). Cytokine expression during acute graft-versus-host disease after allogeneic peripheral stem cell transplantation. *Bone Marrow Transplant*. Vol. 35, No. 12, (June 2005), pp. (1179-1186), ISSN 0006-4971

[29] Kappel LW, Goldberg GL, King CG, Suh DY, Smith OM, Ligh C, Holland AM, Grubin J, Mark NM, Liu C, Iwakura Y, Heller G & van den Brink MR. (2009). IL-17 contributes to CD4-mediated graft-versus-host disease. *Blood*. Vol. 113, No. 4, (January 2009), pp. (945-952), ISSN 1528-0020

[30] Carlson MJ, West ML, Coghill JM, Panoskaltsis-Mortari A, Blazar BR & Serody JS. (2009). *In vitro*-differentiated TH17 cells mediate lethal acute graft-versus-host disease with severe cutaneous and pulmonary pathologic manifestations. *Blood.* Vol. 113, No. 6, (February 2009), pp. (1365-1374), ISSN 1528-0020

[31] Coghill JM, Sarantopoulos S, Moran TP, Murphy WJ, Blazar BR & Serody JS. (2011). Effector CD4+ T cells, the cytokines they generate, and GVHD: something old and something new. *Blood.* Vol. 117, No. 12, (March 2011), pp. (3268-3276), ISSN 1528-0020

[32] Broady R, Yu J, Chow V, Tantiworawit A, Kang C, Berg K, Martinka M, Ghoreishi M, Dutz J & Levings MK. (2010). Cutaneous GVHD is associated with the expansion of tissue-localized Th1 and not Th17 cells. *Blood.* Vol. 116, No. 25, (December 2010), pp. (5748-5751), ISSN 1528-0020

[33] Vela-Ojeda J, Ruiz-Esparza MA & Borbolla-Escoboza JR. (2008). *Trasplante de células progenitoras hematopoyéticas.* 1a. Ed, Ed. Prado, pp. (73-123, 485-515). México.

[34] Horwitz ME & Sullivan KM. (2006). Chronic graft-versus-host disease. *Blood Rev.* Vol. 20, No. 1, (January 2006), pp. (15-27), ISSN 1528-0020

[35] Lee SJ. (2005). New approaches for preventing and treating chronic graft-versus-host disease. *Blood.* Vol. 105, No. 11, (June 2005), pp. (4200-4206), ISSN 1528-0020

[36] Pasquini M & Wang Z. (2007). CIBMTR Newsletter. *Medical College of Wisconsin.* Vol. 14, pp. (1-16).

[37] Kolb HJ. (2008). Graft-versus-leukemia effects of transplantation and donor lymphocytes. *Blood.* Vol. 112, No. 12, (December 2008), pp. (4371-4383), ISSN 1528-0020

[38] Riddell SR & Appelbaum FR. (2007). Graft-versus-host disease: a surge of developments. *PLoS Med.* Vol. 4, No. 7, (July 2007), pp. (e198), ISSN 1549-1676

[39] Randolph SS, Gooley TA, Warren EH, Appelbaum FR & Riddell SR. (2004). Female donors contribute to a selective graft-versus-leukemia effect in male recipients of HLA-matched, related hematopoietic stem cell transplants. *Blood.* Vol. 103, No. 1, (January 2004), pp. (347-352), ISSN 1528-0020

[40] Bryceson YT & Ljunggren HG. (2007). Lymphocyte effector functions: armed for destruction? *Curr Opin Immunol.* Vol. 19, No. 3, (June 2007), pp. (337-338), ISSN 0952-7915

[41] Fehérvari Z & Sakaguchi S. (2005). CD4+ regulatory cells as a potential immunotherapy. *Philos Trans R Soc London B Biol Sci.* Vol. 360, No. 1461, (September 2005), pp. (1647-1661), ISSN 1471-2970

[42] Sakaguchi S, Sakaguchi N, Asano M, Itoh M & Toda M. (1995). Immunologic self-tolerance maintained by activated T cells expressing IL-2 receptor alpha-chains (CD25). Breakdown of a single mechanism of self-tolerance causes various autoimmune diseases. *J Immunol.* Vol. 155, No. 3, (August 1995), pp. (1151-1164), ISSN 1550-6606

[43] Curiel TJ, Coukos G, Zou L, Alvarez X, Cheng P, Mottram P, Evdemon-Hogan M, Conejo-Garcia JR, Zhang L, Burow M, Zhu Y, Wei S, Kryczek I, Daniel B, Gordon A, Myers L, Lackner A, Disis ML, Knutson KL, Chen L & Zou W. (2004). Specific recruitment of regulatory T cells in ovarian carcinoma fosters immune privilege

and predicts reduced survival. Nat Med. Vol. 10, No. 9 (September 2004), pp. (942-949), ISSN 1078-8956

[44] Horwitz DA, Zheng SG & Gray JD. (2003). The role of the combination of IL-2 and TGF-beta or IL-10 in the generation and function of CD4+ CD25+ and CD8+ regulatory T cell subsets. *J Leukoc Biol.* Vol. 74, No. 4, (October 2003), pp. (471-478), ISSN 1938-3673

[45] Korn T, Bettelli E, Oukka M & Kuchroo VK. (2009). IL-17 and Th17 cells. *Annu Rev Immunol.* Vol. 27, pp. (485-517), ISSN 0732-0582

[46] Liu W, Putnam AL, Xu-Yu Z, Szot GL, Lee MR, Zhu S, Gottlieb PA, Kapranov P, Gingeras TR, Fazekas de St Groth B, Clayberger C, Soper DM, Ziegler SF & Bluestone JA. (2006). CD127 expression inversely correlates with FoxP3 and suppressive function of human CD4+ T reg cells. *J Exp Med.* Vol. 203, No. 7, (July 2006), pp. (1701-1711), ISSN 1940-5901

[47] Seddiki N, Santner-Nanan B, Martinson J, Zaunders J, Sasson S, Landay A, Solomon M, Selby W, Alexander SI, Nanan R, Kelleher A & Fazekas de St Groth B. (2006). Expression of interleukin (IL)-2 and IL-7 receptors discriminates between human regulatory and activated T cells. *J Exp Med.* Vol. 203, No. 7, (July 2006), pp. (1693-1700), ISSN 1940-5901

[48] Matsuoka K, Kim HT, McDonough S, Bascug G, Warshauer B, Koreth J, Cutler C, Ho VT, Alyea EP, Antin JH, Soiffer RJ & Ritz J. (2010). Altered regulatory T cell homeostasis in patients with CD4+ lymphopenia following allogeneic hematopoietic stem cell transplantation. *J Clin Invest.* Vol. 120, No. 5, (May 2010), pp. (1479-1493), ISSN 0021-9738

[49] Gershon RK & Kondo K. (1970). Cell interactions in the induction of tolerance: The role of thymic lymphocytes. *Immunology.* Vol. 18, No. 5 (May 1970), pp. (723-737), ISSN 0019-2805

[50] Hall BM, Pearce NW, Gurley KE & Dorsch SE. (1990). Specific unresponsiveness in rats with prolonged cardiac allograft survival after treatment with cyclosporine. III. Further characterization of the CD4+ suppressor cell and its mechanisms of action. *J Exp Med.* Vol. 171, No. 1, (January 1990), pp. (141-157), ISSN 1940-5901

[51] Lu L & Cantor H. (2008). Generation and regulation of CD8(+) regulatory T cells. *Cell Mol Immunol.* Vol. 5, No. 6, (December 2008), pp. (401-406), ISSN 1672-7681

[52] Wang YM & Alexander SI. (2009). CD8 regulatory T cells: what's old is now new. *Immunol Cell Biol,* Vol.87, No.3, (March 2009) pp. (192-193), ISSN 0818-9641

[53] Zheng SG, Wang JH, Koss MN, Quismorio F Jr, Gray JD & Horwitz DA. (2004). CD4+ and CD8+ regulatory T cells generated ex vivo with IL-2 and TGF-beta suppress a stimulatory graft-versus-host disease with a lupus-like syndrome. *J Immunol.* Vol. 172, No. 3, (February 2004), pp. (1531-1539), ISSN 1550-6606

[54] Frisullo G, Nociti V, Iorio R, Plantone D, Patanella AK, Tonali PA & Batocchi AP. (2010). CD8(+)Foxp3(+) T cells in peripheral blood of relapsing-remitting multiple sclerosis patients. *Hum Immunol.* Vol. 71, No. 5, (May 2010), pp. (437-441), ISSN 0198-8859

[55] Chaput N, Louafi S, Bardier A, Charlotte F, Vaillant JC, Ménégaux F, Rosenzwajg M, Lemoine F, Klatzmann D & Taieb J. (2009). Identification of CD8+CD25+Foxp3+ suppressive T cells in colorectal cancer tissue. *Gut*. Vol. 58, No. 4, (April 2009), pp. (520-529), ISSN 1468-3288

[56] Kiniwa Y, Miyahara Y, Wang HY, Peng W, Peng G, Wheeler TM, Thompson TC, Old LJ & Wang RF. (2007). CD8+ Foxp3+ regulatory T cells mediate immunosuppression in prostate cancer. *Clin Cancer Res*. Vol. 13, No. 23, (December 2007), pp. (6947-6958), ISSN 1557-3265

6

Licensed to Kill: Towards Natural Killer Cell Immunotherapy

Diana N. Eissens, Arnold van der Meer and Irma Joosten
Radboud University Nijmegen Medical Centre
The Netherlands

1. Introduction

Allogeneic stem cell transplantation (SCT) is often the final treatment modality for patients with leukemia or other hematological malignancies (Schaap et al., 1997; Thomas & Blume, 1999). However, relapse of the underlying malignancy is still a major complication post SCT. To prevent the occurrence of relapse post SCT, patients are treated with pre-emptive donor lymphocyte infusions (DLI) consisting of donor-derived T cells from the same donor used for allogeneic SCT in order to boost the donor-derived immune system to terminally eradicate residual tumour cells (Barge et al., 2003; Peggs et al., 2004; Schaap et al., 2001). Unfortunately, DLI treatment using donor-derived T cells does not only provoke graft-versus-leukemia (GVL) reactivity, but also increases the risk for the development of graft-versus-host disease (GVHD). Thus, post allogeneic SCT, long-term remission is still greatly dependent on effective GVL reactivity, while strictly controlling GVHD. Therefore, the development of treatment strategies augmenting GVL reactivity while reducing GVHD is of clinical importance.

In allogeneic SCT, natural killer (NK) cells have shown to play an important role in GVL reactivity within the first months after transplantation (Cook et al., 2004; Giebel et al., 2006; Ruggeri et al., 2002; van der Meer et al., 2008). Ruggeri *et al.* showed that alloreactive donor NK cells were able to lyse recipient tumour cells *in vitro*, implying that these NK cells may be able to provide immune reactivity by targeting residual tumour cells still present in the recipient (Ruggeri et al., 1999). Moreover, fast recovery of NK cells and predicted GVL reactivity towards host tumour cells has been associated with reduced GVHD, decreased relapse rates, and better overall survival of the patient (Kim et al., 2005a; Kim et al., 2006; Ruggeri et al., 2007; Savani et al., 2007). Altogether, this makes NK cells important candidates for immunotherapeutic use in the treatment of leukemia and other malignancies.

2. Natural Killer cells

NK cells are members of the innate immune system. They are important in the initial phase of defense against infections and play an important role in tumour surveillance. Their name was based on their ability to kill target cells without prior sensitization (Kiessling et al., 1975a; Kiessling et al., 1975b). As they are morphologically recognized as relatively large lymphocytes containing azurophilic granules, they have also been known as large granular lymphocytes (LGL) (Herberman, 1986). NK cells comprise approximately 5-15% of the peripheral blood lymphocyte (PBL) population and are also found in lymph nodes, spleen, bone marrow, lung, liver, intestine, omentum and placenta (Vivier, 2006). NK cells are believed to originate from the same common lymphoid progenitor lineage as T and B cells in bone marrow (Spits et al., 1995). However, they do not rearrange T cell receptor genes or immunoglobulin (Ig) like, respectively, T and B cells do.

Resting NK cells can be recognized based on their expression of CD56 (neural cell adhesion molecule; NCAM). Since CD56 is also expressed by other immune cells, NK cells are identified by the expression of CD56 combined with the lack of CD3, which are both present on NKT cells. NK cells can be further characterized based on their expression of CD56 in combination with CD16, a low affinity Fc receptor (Figure 1) (Cooper et al., 2001).

In peripheral blood, approximately 90% of NK cells shows low expression of CD56 and high expression of CD16 on the cell surface. These cells are collectively referred to as the $CD56^{dim}CD16^{+}$ or $CD56^{dim}$ NK cell subset. The other 10% of NK cells shows a high level of CD56 expression and almost no expression of CD16 on the cell surface. Together, these cells are known as the $CD56^{bright}CD16^{+/-}$ or $CD56^{bright}$ NK cell subset. The $CD56^{dim}$ NK cell subset is characterized by a highly cytolytic behaviour towards target cells, whereas $CD56^{bright}$ NK cells abundantly produce cytokines, such as IFN-γ and TNF-α, upon activation. The production of cytokines by NK cells influences the T_H1/T_H2 bias of the adaptive immune response by activating T_H1 cells. Thereby, NK cells form a bridge between innate and adaptive immunity (Seaman, 2000).

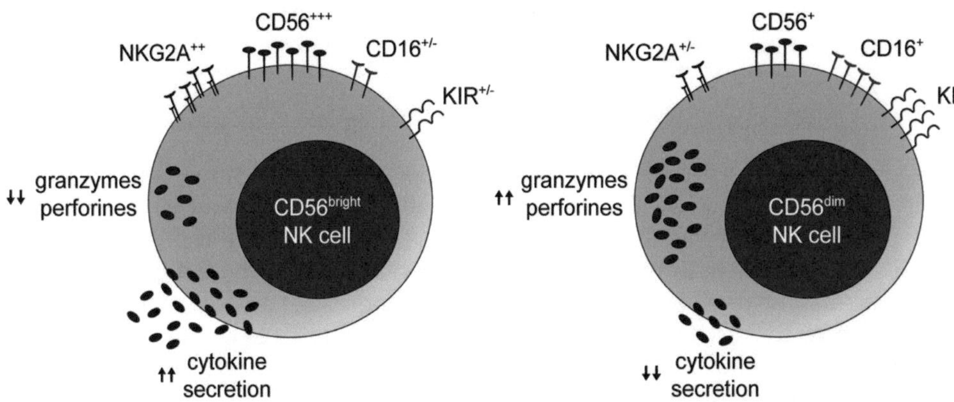

Fig. 1. Natural Killer cell subsets. NK cells can be divided in two major subsets based on their expression of CD56 and CD16. $CD56^{bright}$ NK cells have a high CD56 and low CD16 expression profile, and are specialized in cytokine secretion (e.g. TNF-α, IFN-γ). In addition they highly express inhibitory receptor NKG2A. $CD56^{dim}$ NK cells have a low CD56 and high CD16 expression profile, and are highly cytolytic. Their function is predominantly inhibited through KIR.

NK cells lyse susceptible target cells (e.g. virus infected cells, malignant transformed cells) by one of two mechanisms: "natural killing" (no prior sensitization) or antibody dependent cellular cytotoxicity (ADCC). Natural killing is initiated by activating signals from a variety of stimulatory receptors that can be inhibited by a variety of inhibitory receptors. In ADCC, the activating receptor FcRγIII (CD16) binds to the Fc piece of antibodies bound to target cells. In both mechanisms, target cells are lysed by the release of cytolytic proteins (i.e. granzymes and perforines) or by the induction of apoptosis (Bossi & Griffiths, 1999).

2.1 NK cell receptors

The NK cell receptor repertoire forms the basis for NK cell immune surveillance and NK cell activity. NK cell immune surveillance is regulated through the recognition of HLA class I molecules by inhibitory receptors. Subsequent activation of NK cells is triggered through the recognition of activating ligands by stimulatory receptors.

2.1.1 Inhibitory receptors

NK cells survey potential target cells for the absence or loss of expression of classical HLA class I molecules or non-classical HLA class I specific signals through inhibitory killer cell immunoglobulin-like receptors (KIRs) and lectin-like receptors (Farag & Caligiuri, 2006; Papamichail et al., 2004). The cytoplasmic domains of all inhibitory NK cell receptors contain an immunoreceptor tyrosine-based inhibitory motif (ITIM) (Burshtyn et al., 1996; Fry et al., 1996; Vivier & Daeron, 1997). These domains recruit intracellular tyrosine phosphatases SHP-1 or SHP-2 that mediate the inhibition of cytotoxicity and cytokine release (Burshtyn et al., 1996; Fry et al., 1996; Le et al., 1998; Olcese et al., 1996).

The lectin-like receptor complex CD94:NKG2A forms an inhibitory receptor that recognizes non-classical HLA-E molecules (Braud et al., 1998). As the expression of HLA-E is promoted by the binding of signal sequence-derived peptides from HLA class I molecules, it is thought that HLA-E expression serves as a barometer of classical HLA class I expression (Braud et al., 1997). The purpose of the inhibitory CD94:NKG2A receptor complex may, therefore, be to monitor the overall HLA class I expression. KIRs, on the other hand, allow for a more subtle immune surveillance as these receptors scan the presence of specific classical HLA class I molecules.

KIRs are encoded by a family of polymorphic and highly homologous genes, and recognize polymorphic epitopes present on HLA-A, -B, or -C molecules; HLA-A3 and -A11 are recognized by KIR3DL2, HLA-Bw4 is recognized by KIR3DL1 and receptors KIR2DL1, KIR2DL2, and KIR2DL3 are able to distinguish HLA-C into HLA group C1 and HLA group C2 molecules (Parham, 2005). The various KIRs are classified by the number of immunoglobulin-like (Ig) extracellular domains as 2-domain (2D) or 3-domain (3D). They are further subdivided on the basis of the length of their cytoplasmic tail L (long) or S (short) (Vilches & Parham, 2002). Different KIRs sharing the same number of Ig domains and length of the cytoplasmic tail are distinguished by number at the end of their name, e.g. KIR2DL2 or KIR2DL3. KIRs are clonally distributed among NK cells within each individual, which creates a complex combinatorial repertoire of NK cell specificities for HLA class I molecules (Vilches & Parham, 2002).

2.1.2 Stimulatory receptors

Some members of the CD94:NKG2 receptor and KIR family have stimulatory properties. NKG2C is a stimulatory member of the CD94:NKG2 family and competes with CD94:NKG2A for the recognition of HLA-E (Braud et al., 1998; Houchins et al., 1997). KIR2DS and KIR3DS are stimulatory members of the KIR family (Biassoni et al., 1996; Bottino et al., 1996; Moretta et al., 1995). Instead of ITIM, these stimulatory receptors contain a positively charged amino acid residue in their transmembrane domain that associates with the negatively charged DAP-12 molecule. DAP-12 contains an immunoreceptor tyrosine-based activating motif (ITAM) (Lanier et al., 1998). There is evidence that the stimulatory KIRs bind self-HLA class I molecules with lower affinity as compared with the inhibitory receptors (Vales-Gomez et al., 1998a; Vales-Gomez et al., 1998b). Thus autoimmunity can be prevented by a balance towards negative NK cell regulation. Similar to the inhibitory receptors, the HLA class I-specific stimulatory receptors are expressed in a variegated and predominantly stochastic fashion by NK cells (Raulet et al., 2001).

Besides stimulatory members of the CD94:NKG2 receptor and KIR family, NK cells also express a variety of other stimulatory receptors. The biological roles of many of these receptors are not well understood, primarily because the ligands for these receptors have not been fully identified. The main triggering receptors for NK cell activity are the natural cytotoxicity receptors (NCR) and NKG2D (Arnon et al., 2001; Bauer et al., 1999; Mandelboim et al., 2001; Sivori et al., 1997). Their stimulation causes direct killing of target cells and their stimulatory signals can even override the inhibition of NK cells.

NCR consist of three members; NKp30, NKp44, and NKp46. NCR belong to the immunoglobulin superfamily and molecular cloning of NCR confirmed that they are structurally distinct (Bottino et al., 2000; Pende et al., 1999). The ligands for NCR remain controversial. Some groups have proposed viral antigens as being the ligands for NCR based on their role in the lysis of virus infected cells (Arnon et al., 2001; Arnon et al., 2005; Mandelboim et al., 2001). NCR have also been shown to mediate lysis of tumour cells and that NK cells with a NCR^{dull} phenotype are unable to kill tumour cells, suggesting that their ligands may be upregulated or induced upon malignant transformation of cells (Bottino et al., 2005; Fauriat et al., 2005; Fauriat et al., 2007). NKp30 and NKp46 are both uniquely expressed on resting and activated NK cells, whereas NKp44 is only present on IL-2 activated NK cells (Cantoni et al., 1999).

Unlike NKG2A, NKG2D is not associated with CD94, but is a homodimer that needs association to the adaptor molecule DAP10 for stable cell surface expression (Wu et al., 1999). NKG2D recognizes HLA class I-like molecules, such as MIC A and MIC B. It has been shown that the expression of NKG2D ligands, i.e. MIC A/B, ULPB1, ULPB2, and ULPB3, are upregulated by cells in times of stress, virus infection, and malignant transformation (Bauer et al., 1999; Cosman et al., 2001; Farrell et al., 2000). NKG2D is constitutively expressed on all human NK cells and can be upregulated through stimulation by IL-15, IL-12 and IFN-α (Diefenbach et al., 2000; Sutherland et al., 2006). Stimulation of NKG2D complements NCR activation in mediating NK cell lysis of tumour cells (Pende et al., 2001). Similarly, cooperation between NKG2D and stimulatory KIRs has been shown for both cytolytic activity and IFN-γ secretion (Wu et al., 2000). Therefore, it is possible that NKG2D may serve both as a primary stimulatory receptor, whose engagement triggers cytotoxicity, and also as a co-stimulatory receptor, which cooperates with other activating receptors (e.g. activating KIR or NCR) for cytokine secretion. A similar phenomenon is seen on cytomegalovirus-specific T cells, where NKG2D acts as a co-stimulatory receptor for TCR-dependent signals (Das et al., 2001; Groh et al., 2001; Ugolini & Vivier, 2001).

Other stimulatory receptors that are involved in NK cell activation are co-stimulatory receptors NKp80 and 2B4 (CD244) (Biassoni et al., 2001). NKp80 and 2B4 both function synergistically with NCR (Sivori et al., 2000; Vitale et al., 2001). In addition, CD16, CD69 and DNAM-1 have been shown to trigger NK cell-mediated lysis in redirected cytotoxicity assays (Lanier et al., 1988; Moretta et al., 1991; Shibuya et al., 1996).

2.2 NK cell allorecognition

2.2.1 The "missing self" hypothesis

In 1976, Snell *et al.* observed a correlation between the susceptibility of target cells to NK cell lysis and the absence or loss of expression of HLA class I molecules on the target cells (Snell, 1976). Absent or low expression of HLA class I molecules is common in virus infected and malignant transformed cells, which are the usual targets for NK cell lysis. Therefore, they proposed that NK cell receptors may not only interact with HLA class I molecules, but that these receptors are also able to detect a decrease in HLA class I expression.

It was not until 10 years later that Kärre and Ljunggren demonstrated the regulation of NK cell activity (Kärre et al., 1986a). They showed that murine lymphoma cells with low, or absent, MHC class I expression were less malignant than wild-type cells after low dose inoculation in syngeneic mice, and that the rejection of these cells was regulated through innate immunity, preferably through NK cell-mediated lysis (Kärre et al., 1986b). Resistance to NK cell-mediated lysis of tumours with low MHC class I expression could be restored by reintroduction of MHC class I molecules (Franksson et al., 1993; Ljunggren et al., 1990). Based on these data, they proposed the "missing self" hypothesis, which is nowadays still appreciated as the basic model for NK cell activation (Figure 2) (Ljunggren & Kärre, 1990).

2.2.2 Licensed to kill

As the KIR repertoire of NK cells is encoded by a set of highly polymorphic genes and segregates independently from HLA class I genes during NK cell development, it is essential that the KIR repertoire of NK cells properly corresponds with the HLA environment to provide self-tolerance and prevent autoimmunity. There have been several hypotheses on the acquisition of self-tolerance by NK cells. Raulet *et al.* proposed that an individual NK cell can simultaneously express multiple inhibitory KIRs in a stochastic fashion (Raulet et al., 2001). The only rule appears to be that every NK cell has at least one inhibitory KIR specific for self-HLA class I in order to avoid autoreactivity. This is referred to as the "at least one receptor" model (Raulet et al., 2001). Others have suggested a "receptor calibration" model in which the acquisition of the KIR repertoire may be related to changes in the HLA class I environment and is dependent on the HLA class I haplotype (Salcedo et al., 1997; Sentman et al., 1995; Valiante et al., 1997). However, these studies involved *in vitro* cultures that could alter the intrinsic features of NK cells that may be different from the *in vivo* situation.

Recently, the acquisition of self-tolerance was demonstrated in an *in vivo* murine study. Kim *et al.* showed that NK cells from MHC-deficient mice were functionally immature as they were defective in cytokine secretion upon *ex vivo* stimulation as compared with wild type NK cells, indicating that MHC-specific receptors are involved in the acquisition of functional competence (Kim et al., 2005b). They also found a correlation between the expression of an inhibitory receptor for self-MHC class I and the capacity of an individual naive NK cell to be activated to produce cytokines and lyse susceptible target cells. Based on their findings they proposed the "licensing" model, in which NK cells acquire functional competence through "licensing" by self-HLA class I molecules, resulting in two types of self-tolerant NK cells: licensed or unlicensed. This model was confirmed by others, both in mice and human, demonstrating that NK cells without expression of known self-receptors were found to be hyporesponsive (Cooley et al., 2007; Fernandez et al., 2005). Thus, in order for NK cells to get their "license to kill", they need to fulfil the requirement of HLA class I-specific receptor engagement by self-HLA. There appears to be one exception to this rule: Orr *et al.* showed in a mouse model that, contrary to the licensing hypothesis, unlicensed NK cells were the main mediators of NK cell-mediated control of mouse cytomegalovirus infection (MCMV) *in vivo* (Orr et al., 2010). It would be highly relevant to check whether such a cell population harbours increased allo- or tumour reactivity.

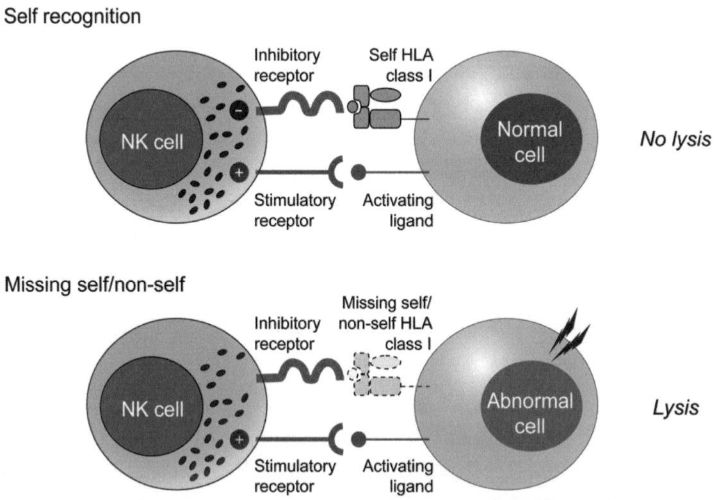

Fig. 2. "Missing self" hypothesis. NK cell activity depends on a balance between inhibitory (i.e. KIR, CD94/NKG2A) and stimulatory (e.g. NCR, NKG2D) signals. In steady state, NK cells are inhibited from activation by the recognition of self-HLA class I molecules, which overrules potential stimulatory signals (self recognition). In case of virus infection or malignant transformation, cells may downregulate self-HLA class I molecules, while upregulating activating ligands that trigger NK cells to respond resulting in lysis of the infected/transformed cells (missing self). After allogeneic SCT, donor NK cells may be triggered by host leukemic cells due to reduced HLA-matched class I molecules (HLA-matched SCT), or the presence of non-self HLA class I molecules (HLA-mismatched SCT), combined with strong stimulation by upregulated activating ligands.

3. NK cells and their therapeutic role in SCT

3.1 Evidence for GVL

Deficient HLA class I expression has been described for leukemic cells making them susceptible targets for NK cell-mediated lysis. However, this phenomenon was not ubiquitously observed in the autologous setting for patients with different forms of leukemia. In CML, NK cell numbers and NK cell function have been shown to decrease progressively during the spontaneous course of the disease, but could be recovered upon IFN-α treatment (Pawelec et al., 1995; Pierson & Miller, 1996). Moreover, activated autologous NK cells were shown to suppress the growth of primitive CML progenitors in long-term *in vitro* cultures (Cervantes et al., 1996). In AML, however, autologous NK cells were demonstrated to be impaired in their cytolytic function, which correlated with a low NCR cell surface density (NCR^{dull}) (Fauriat et al., 2007). Moreover, these NK cells were impaired in regulating DC physiology (killing the surplus of immature DCs), which could lead to specific T cell tolerization by expanded immature DCs expressing leukemia-derived antigens (Fauriat et al., 2005). Allogeneic SCT may overcome the impairment of NK cell-mediated lysis.

In different allogeneic SCT settings, NK cells have been shown to play an important role in the anti-tumour response within the first months after transplantation (Cook et al., 2004; Giebel et al., 2006; Ruggeri et al., 2002; van der Meer et al., 2008). In haploidentical SCT, Ruggeri *et al.* showed that donor alloreactive NK cells isolated from peripheral blood of the recipient were able to lyse tumour cells derived from the recipient, implying that within one month after SCT, NK cells may be able to provide some degree of immune reactivity by targeting residual tumour cells still present in the recipient (Ruggeri et al., 1999). Moreover, fast recovery of NK cells and predicted GVL reactivity towards host tumour cells has been associated with decreased relapse rates and better overall survival of the patient (Kim et al., 2006; Ruggeri et al., 2007; Savani et al., 2007). Altogether these data suggest that NK cells play an important role in the control and clearance of leukemic cells after allogeneic SCT.

3.2 Evidence for the prevention of GVHD

The haploidentical transplantations performed by Ruggeri *et al.* additionally suggested that NK cells may prevent the development of GVHD (Ruggeri et al., 2002). For patients, the prevalence of GVHD was significantly lower using grafts with potential NK cell alloreactivity in the GVL direction as compared with grafts without potential NK cell alloreactivity. In a murine model, they demonstrated that mice transplanted with non-T cell depleted grafts could be rescued from GVHD upon infusion of alloreactive NK cells. Mice infused with non-alloreactive NK cells died as they were not protected from GVHD. They also demonstrated that alloreactive NK cells are able to lyse recipient antigen presenting cells (APCs), thereby preventing interaction with donor T cells which otherwise would initiate GVHD. Recently, a novel mechanism for NK cell-mediated GVHD reduction was demonstrated, whereby alloreactive donor NK cells were able to inhibit and lyse alloreactive donor T cells during the initiation of GVHD (Olson et al., 2010). Overall, these studies demonstrate that alloreactive NK cells may, directly or indirectly, reduce or prevent the occurrence of GVHD while retaining GVL reactivity.

4. Exploitation of NK cell alloreactivity

4.1 Infusion of mature donor NK cells as part of the graft

Immediately after allogeneic SCT, alloreactive NK cells have been shown to be beneficial not only for boosting the anti-tumour response, but also for the prevention of GVHD as well as infections. In these cases, optimal functional activity of NK cells already in the early phase after SCT is essential, and therefore the presence of NK cells in the graft appears to be beneficial for transplant outcome (Bethge et al., 2006; Kim et al., 2005a).

As part of a prospective randomized phase III study, we directly compared the alloreactive potential of allogeneic donor NK cells between patients having either received a $CD3^+/CD19^+$ cell depleted graft (containing substantial NK cell numbers) or a conventional $CD34^+$ selected graft (devoid of NK cells) in the setting of HLA-matched SCT (Eissens et al., 2010a). Results demonstrate that patients having received a $CD3^+/CD19^+$ cell depleted graft, exhibited a faster recovery of NK cells and a functional NK cell receptor repertoire of inhibitory and stimulatory receptors as compared with patients having received a conventional $CD34^+$ graft. Furthermore, transplantation with a $CD3^+/CD19^+$ cell depleted graft resulted in the development of a functionally different NK cell population that was more prone to activation via the CD94:NKG2C receptor complex and less sensitive to inhibition via the CD94:NKG2A receptor complex. Although it was demonstrated that human cytomegalovirus (CMV) infection may result in increased CD94:NKG2C expression levels and subsequent loss of CD94:NKG2A expression (Guma et al., 2004; Guma et al., 2006; van Stijn et al., 2008), this phenomenon remained present in the CD3/19 depletion group after exclusion of CMV positive patients from analysis. Unfortunately, later interim analysis on 25 patients per group showed that the primary objectives of this clinical study could not be reached resulting in early termination of the study. Thus, the alternative reconstitution of the NK cell receptor repertoire using $CD3^+/19^+$ depleted grafts, characterized by the change in balance of CD94:NKG2A$^+$ NK cells to more CD94:NKG2C$^+$ NK cells, and its impact on clinical outcomes after HLA-matched SCT remains a subject for further study.

Recently, the reconstitution of allogeneic donor NK cells was evaluated in haploidentical SCT after reduced intensity conditioning (RIC) using $CD3^+/19^+$ depleted grafts (Federmann et al., 2011). Data showed similar results as compared with our study, including fast recovery of NK cells and fast immune reconstitution of the NK cell receptor repertoire. In addition, a similar decrease of NKG2A$^+$ NK cells was seen post SCT. However, the expression of NKG2C was not evaluated. Nevertheless, this study confirms our findings that different graft manipulation methods may trigger differential NK cell reconstitution, which may be beneficial for transplant outcome. Previously, Gentilini *et al.* even showed a significant faster and sustained recovery of NK cells in a group of patients after RIC allogeneic SCT with $CD3^+/CD19^+$ depleted grafts in comparison with patients with myeloablative allogeneic SCT with $CD34^+$ selected grafts combined with adoptive NK cell infusion two days post SCT (Gentilini et al., 2007).

Overall, these studies suggest that the use of NK cell rich grafts is favourable for the facilitation of fast and sustained NK cell recovery and differential reconstitution of the NK cell receptor repertoire, which may lead to improved donor NK cell alloreactivity in the GVL direction (Eissens et al., 2010a). Further prospective comparisons of the different graft manipulation methods for allogeneic SCT in the HLA-matched or haploidentical setting are warranted for more detailed analysis of the impact of graft composition on immune reconstitution. Subsequently, the impact of adoptive NK cell infusions after allogeneic SCT for boosting GVL reactivity needs to be studied in further detail.

4.2 Skewing donor NK cell alloreactivity before SCT

In allogeneic SCT, donor NK cell alloreactivity can be facilitated by allowing mismatches for specific HLA molecules (e.g. HLA-C) between donor and recipient. This is referred to as the "ligand-ligand" model. The introduction of certain HLA mismatches has been shown to induce NK cell-mediated GVL reactivity, without inducing severe GVHD, and to contribute to decreased relapse, better engraftment and improved overall survival (Ruggeri et al., 1999; Ruggeri et al., 2004; Ruggeri et al., 2007). However, others state that the induction of NK cell alloreactivity is not dependent on HLA mismatching, but is rather induced by the presence of an inhibitory KIR in the donor's genotype with the absence of the corresponding KIR-ligand in the recipient's HLA repertoire ("receptor-ligand" model) (Hsu et al., 2005; Leung et al., 2004; Leung et al., 2005). This makes the exploitation of NK cell alloreactivity not only feasible for HLA-mismatched settings, but may also be promising for HLA-matched settings.

For further exploitation, however, the "licensing/education" model needs to be considered as well. Upon maturation, NK cells obtain their "license to kill" through interactions of inhibitory KIRs with self-HLA class I molecules (Parham, 2006; Raulet & Vance, 2006; Vivier et al., 2008). NK cells that fail to interact with self-HLA class I molecules remain functionally immature and will reside in a hyporesponsive state. Recently, it was shown that the strength of response by an individual NK cell is even quantitatively controlled by the extent of inhibitory signals that are received from HLA class I molecules during NK cell education (Brodin et al., 2009). Concerning adoptive transfer of mature NK cells for immunotherapeutic purposes, this suggests that the presence of inhibitory KIR on donor NK cells in absence of its cognate ligand in the recipient ("receptor-ligand" model) as well as the HLA-background of the donor NK cells ("licensing/education" model) are two key factors that need to be taken into account for the successful exploitation of alloreactive donor NK cell responses.

4.3 Interference by immunosuppressive drugs

For optimal NK cell-mediated GVL reactivity, NK cells need to be fully functional in the early phase after allogeneic SCT, despite that at this stage a high level of immunosuppressive treatment is given. Among the various immunosuppressive drugs (ISDs), cyclosporin A (CsA), rapamycin (Rapa) and mycophenolate mofetil (MMF) have successfully been applied for the prevention of GVHD (Cutler et al., 2007; Haentzschel et al., 2008; Neumann et al., 2005; Schleuning et al., 2008; Vogelsang & Arai, 2001). Therefore, we studied the influence of CsA, Rapa and mycophenolic acid (MPA; the active metabolite of MMF) on NK cell phenotype and function in an *in vitro* cytokine-based culture system (Eissens et al., 2010b). Results showed that the modulation of the NK cell receptor repertoire during culture was arrested by Rapa and MPA treatment. This was reflected in the cytolytic activity, as MPA- and Rapa-treated NK cells, in contrast to CsA-treated NK cells, lost their cytotoxicity against leukemic target cells. In contrast, IFN-γ production was not only impaired by MPA and Rapa, but also by CsA upon target encounter. A recent study, however, suggested that IFN-γ production upon target encounter may be limited to the $CD56^{dim}$ NK cell subset, whereas the $CD56^{bright}$ NK cell subset produces IFN-γ upon cytokine-stimulation (Fauriat et al., 2010). Thus, as $CD56^{bright}$ NK cells were still abundantly present in the CsA-treated cultures, in contrast to MPA- and Rapa-treated cultures, the IFN-γ production upon cytokine stimulation may largely be preserved after CsA treatment. This was confirmed in a study showing sustained IFN-γ production by CsA-treated NK cell cultures upon IL-12 and IL-18 stimulation (Wang et al., 2007), suggesting that IFN-γ-mediated GVL reactivity after allogeneic SCT should remain intact when using CsA as GVHD prophylaxis.

Our findings on the effect of CsA and MPA on the cytolytic response by *in vitro* cytokine-stimulated NK cells are in concert with previous findings on this subject (Ohata et al., 2011). Besides CsA and MPA, they also evaluated the effect of tacrolimus (TAC) and methotrexate (MTX), which are also successfully used as GVHD prophylaxis after allogeneic SCT (Alyea et al., 2008; Cutler et al., 2007; Ho et al., 2009). Both ISDs did not interfere with NK cell-mediated cytolytic activity against different leukemic cell lines. However, a dose range of each ISD is lacking in this study, which would be more appropriate when studying the effect of ISDs on NK cell functionality.

Overall, these *in vitro* studies clearly suggest that the choice of immunosuppressive treatment might affect the outcome of NK cell immunotherapy *in vivo* after transplantation. Additional studies on NK cell phenotype and function of patients after allogeneic SCT using different immunosuppressive strategies are warranted to survey the *in vivo* effect of the different immunosuppressive regimens in more detail.

4.4 Clinical grade NK cell products for adoptive cancer immunotherapy

4.4.1 Development of clinical grade NK cell products

The facilitation of donor NK cell alloreactivity is not restricted to HLA-matched/mismatched allogeneic SCT, but may also be exploited for adoptive immunotherapy in non-transplantation settings. Previously, several clinical studies have examined the feasibility of allogeneic NK cells for adoptive immunotherapy using allogeneic NK cells selected from leukapheresis products by immunomagnetic beads selection protocols (Iyengar et al., 2003; Klingemann & Martinson, 2004; Koehl et al., 2005; McKenna, Jr. et al., 2007; Meyer-Monard et al., 2009; Passweg et al., 2004). In all these studies, the adoptive transfer of allogeneic NK cells proved to be safe and well tolerated by patients. Nevertheless, for optimal exploitation of NK cell adoptive immunotherapy, the development of innovative strategies producing allogeneic NK cell products with high cell numbers, high purity and functionality are needed. In this respect, recent studies have developed culture systems for large scale *ex vivo* expansion of allogeneic NK cells using either hematopoietic progenitor cells from bone marrow or UCB (Carayol et al., 1998; Kao et al., 2007; Miller et al., 1992). However, most of these culture systems are unsuitable for clinical application due to the use of animal sera, animal-derived proteins, and/or supportive feeder cells.

Previously, in our centre a cytokine-based method for high log-scale *ex vivo* expansion of functional allogeneic NK cells from hematopoietic stem and progenitor cells from umbilical cord blood (UCB) using a novel clinical grade medium was developed (Spanholtz et al., 2010). The *ex vivo* generated NK cell products are of high purity and contain developmentally mature NK cell populations expressing inhibitory NKG2A and KIRs, and a variety of stimulatory receptors. Furthermore, these NK cell products show the ability to efficiently kill myeloid leukemia and melanoma tumour cell lines. The findings in this study provide an important advance for the clinical application of *ex vivo* generated NK cell products to be exploited for adoptive immunotherapy either following allogeneic SCT for boosting NK cell-mediated GVL reactivity or in the non-transplant setting following lymphodepleting immunosuppressive regimens. In addition, human *in vitro* studies and *in vivo* evidence in mice suggest that NK cell-based immunotherapy may also be beneficial for patients with melanoma or renal cell carcinoma when applied in the setting of the "receptor-ligand" model (Burke et al., 2010; Igarashi et al., 2004; Lakshmikanth et al., 2009). Overall, further exploitation of this culture system may provide a broad clinical application for NK cell-based immunotherapy against hematological and non-hematological malignancies.

4.4.2 Clinical feasibility of *ex vivo* generated NK cells: a phase I trial

Recently, the NK cell generation protocol described by Spanholtz et al. was transferred to clinical applicable conditions (Spanholtz et al., 2011) and medical ethical approval was given to study the feasibility of adoptive transfer of *ex vivo* generated NK cell products in elderly patients diagnosed with poor prognosis AML (Spanholtz et al., 2010). In the second quartile of 2011, a phase I dose-escalating study has started for a group of 12 AML patients (age >65 years), not eligible for allogeneic SCT, who have achieved clinical remission after standard remission-induction chemotherapy and who have completed consolidation chemotherapy. Patients will receive allogeneic NK cells generated *ex vivo* from $CD34^+$ UCB cells in a single escalating dose up to $10x10^7$ donor NK cells/kg body weight after completing standard chemotherapy and preparative immunosuppressive conditioning consisting of fludarabine and cyclophosphamide in order to prevent rejection. The primary goal is to evaluate the safety and dose-limiting toxicity of adoptive transfer of the allogeneic *ex vivo* generated NK cells. Secondly, the *in vivo* lifespan of the adoptively transferred allogeneic NK cells will be evaluated together with an assessment of NK cell-mediated GVL reactivity in study participants.

5. Towards NK cell immunotherapy

Within the last decade, different NK cell-based immunotherapy strategies have successfully been developed. They have either already been proved safe in clinical phase I/II trials or are currently under clinical evaluation. Today, applications for NK cell-based immunotherapy can generally be divided into three main categories: (1) the use of enriched NK cell grafts for allogeneic SCT; (2) administering NK-DLI after allogeneic SCT and; (3) adoptive NK cell transfer in non-transplantation settings (Figure 3).

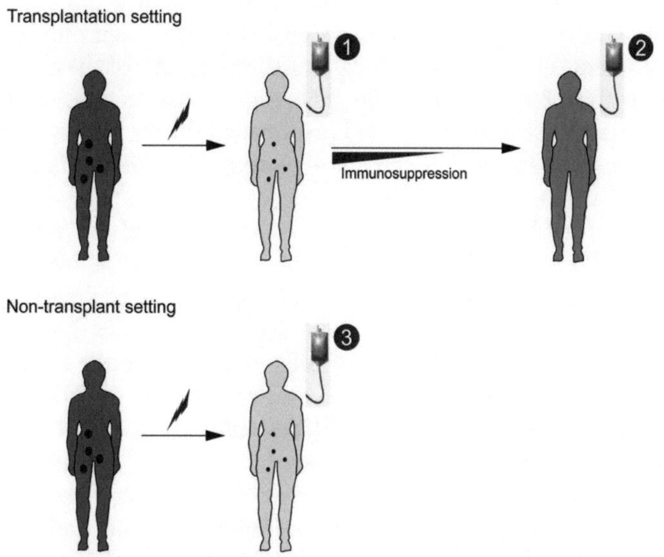

Fig. 3. Applications for NK cell-based immunotherapy. In the allogeneic SCT setting, patients are first treated with a conditioning regimen to eradicate healthy hematopoietic recipient cells and to minimize the tumour burden within the recipient. Subsequently, enriched NK cell grafts (1) and/or NK-DLI (2) may be applied in the further course of allogeneic SCT in order to facilitate alloreactive donor NK cell-mediated GVL reactivity. In the non-transplant setting (e.g. patients no longer eligible for allogeneic SCT or with solid tumours), adoptive NK cell transfer can be used as a NK cell-based immunotherapeutic strategy in patients after completing standard chemotherapy and preparative immunosuppressive conditioning in order to prevent rejection.

5.1 Allogeneic SCT setting

The use of enriched NK cell grafts for allogeneic SCT has shown to be beneficial for transplant outcome and provides enhanced GVL reactivity directly after transplantation as compared with conventional grafts that lack NK cells (Bethge et al., 2006; Eissens et al., 2010a; Kim et al., 2006). In addition, the head start in GVL reactivity may lead to better outcomes in terms of infectious events and overall survival after allogeneic SCT (Kim et al., 2005a). Altogether, this indicates the immunotherapeutic value of the exploitation of donor NK cells in the allogeneic SCT setting. In order to further increase the beneficial effects of such NK cell-based immunotherapy strategies in the setting of allogeneic SCT, donor NK cells can also be administered as part of the conditioning regimen prior to transplantation instead of, or in combination with, the use of enriched NK cell grafts. This can have three potential beneficial effects. First, NK cell-mediated GVL reactivity could provide anti-tumour activity prior to allogeneic SCT (Lundqvist et al., 2007; Ruggeri et al., 2002). Second, NK cell-mediated depletion of host dendritic cells before transplantation could prevent the development of acute GVHD allowing for a less stringent depletion of T cells in the graft (Lundqvist et al., 2007; Ruggeri et al., 2002). Third, NK cells may facilitate better engraftment through eradication of host T cells, thereby reducing the need for toxic myeloablative regimens and shortening the neutropenic period (Ruggeri et al., 2002). However, to be able to implement the combinatorial strategy of NK cell infusions as part of the conditioning regimen together with the use of enriched NK cell grafts in the clinic, the necessary amount of NK cells used for infusion prior to allogeneic SCT to provoke beneficial immunotherapeutic effects still needs to be established. Insufficient numbers may cause the necessity to choose between the use of NK cell infusions as part of the conditioning regimen and the use of enriched NK cell grafts. In this respect, also in case of the combinatorial strategy, it is important that more research is performed on the effects of conditioning and immunosuppressive regimens on NK cells in exploiting NK cell-mediated (GVL) reactivity in the allogeneic SCT setting.

The use donor T cells for DLI after allogeneic SCT has developed into an effective treatment of recurrent hematological malignancy as well as prophylactic treatment in high-risk leukemia and lymphoma (Kolb et al., 2004). Still, the main risk of T-DLI is the induction of life-threatening GVHD. To minimize the risk of GVHD, studies have been initiated to modify conventional DLI by using donor NK cells instead of donor T cells (Passweg et al., 2004; Passweg et al., 2005). Additionally, the administration of NK-DLI may facilitate engraftment and induce NK cell-mediated GVL reactivity (Passweg et al., 2004; Passweg et al., 2005). Although no firm conclusions can be drawn on the clinical efficacy of NK-DLI after allogeneic SCT at this point, data indicate that NK-DLI is safe and well tolerated, and can generate GVL reactivity and long-term remission in some patients after leukemia relapse. As non-malignant tissues generally do not overexpress ligands for activating NK cell receptors, NK-DLI should not cause GVHD (Bottino et al., 2005; Ruggeri et al., 2004; Ruggeri et al., 2006). Until now, NK-DLI has been tested in the haploidentical SCT setting. Thus, further research on the efficacy of NK-DLI in the HLA-matched SCT setting is warranted.

5.2 Non-transplant settings

In non-transplant settings, donor NK cells can be exploited for immunotherapeutic strategies for the treatment of hematological and non-hematological malignancies.

Previously, Miller et al. showed that haploidentical donor NK cell infusions after high-dose cyclophosphamide and fludarabine treatment resulted in long-term survival and *in vivo* expansion of donor NK cells in patients with metastatic melanoma (n=10), metastatic renal cell carcinoma (n=13), refractory non-Hodgkin's disease (n=1), and poor-prognosis AML (n=19) (Miller et al., 2005). The *in vivo* NK cell expansion was associated with increased levels of endogenous IL-15, which were possibly responsible for driving the survival and proliferation of donor NK cells. In general, the donor NK cell infusions were well tolerated without evidence for the induction of GVHD. Furthermore, 5 out of 19 patients with poor-prognosis AML achieved complete remission. Only 4 of the 19 AML patients were KIR-ligand (HLA) mismatched in the graft-versus-host direction. Interestingly, out of these 4 patients, 3 achieved complete remission. These findings indicate that haploidentical donor NK cells can persist and expand *in vivo* and may have a role in the treatment of (non-)hematological malignancies in non-transplant settings or in combination with allogeneic SCT. In addition, when using haploidentical donors the choice of a KIR-ligand mismatched donor, based on the "ligand-ligand" model, may be needed to obtain successful results in future clinical trials (Miller et al., 2005). In case of HLA-matched donors, the choice of a "receptor-ligand" mismatched donor is preferred.

In parallel, adoptive transfers are currently being performed with the NK cell line NK-92. This cell line can be cultured under good manufacturing practice (GMP) conditions and shows significant cytotoxicity against several tumour cell lines (Tam et al., 2003). Infusions of NK-92 cells have been administered to more than 20 patients with advanced renal-cell carcinoma and malignant melanoma. This proved to be safe and generated anti-tumor effects in some cases (Klingemann, 2005). Furthermore, NK-92 cells can easily be obtained in high numbers during GMP culture providing sufficient amounts of cells for adoptive immunotherapeutic strategies. However, to prevent that the success of NK cell adoptive immunotherapy is solely based on the use of one NK cell line and the use of donor NK cells are still preferred, recent studies have developed culture systems for large scale *ex vivo* expansion of allogeneic NK cells using either hematopoietic progenitor cells from bone marrow or UCB (Carayol et al., 1998; Kao et al., 2007; Miller et al., 1992; Spanholtz et al., 2010). After transferring these culture systems to GMP conditions, these allogeneic donor NK cells may also be proven safe for use in NK cell adoptive immunotherapy.

6. Conclusion

Several issues remain crucial for the development and implementation of successful NK cell-based immunotherapy in the future. In the non-transplant setting , these include issues relating to the type of NK cell preparation to be used (activation, degree of enrichment and possible selection or skewing of specific subpopulations), criteria for donor selection ("ligand-ligand" versus "receptor-ligand" model, KIR genotyping and phenotyping, and size of the alloreactive subset), conditioning of patients prior to immunotherapy, clinical context of therapy, criteria for patient selection, and strategies for the identification of susceptible tumours within patient groups. Besides these issues, the effects of immunosuppressive regimens given after allogeneic SCT are also important when implementing NK cell-based immunotherapy in the setting of allogeneic SCT.

7. References

Alyea, E.P.; Li, S.; Kim, H.T.; Cutler, C.; Ho, V.; Soiffer, R.J. & Antin, J.H. (2008). Sirolimus, Tacrolimus, and Low-Dose Methotrexate as Graft-versus-Host Disease Prophylaxis in Related and Unrelated Donor Reduced-Intensity Conditioning Allogeneic Peripheral Blood Stem Cell Transplantation. *Biology of Blood and Marrow Transplantation,* 14(8), 920-926.

Arnon, T.I.; Achdout, H.; Levi, O.; Markel, G.; Saleh, N.; Katz, G.; Gazit, R.; Gonen-Gross, T.; Hanna, J.; Nahari, E.; Porgador, A.; Honigman, A.; Plachter, B.; Mevorach, D.; Wolf, D.G. & Mandelboim, O. (2005). Inhibition of the NKp30 activating receptor by pp65 of human cytomegalovirus. *Nature Immunology,* 6(5), 515-523.

Arnon, T.I.; Lev, M.; Katz, G.; Chernobrov, Y.; Porgador, A. & Mandelboim, O. (2001). Recognition of viral hemagglutinins by NKp44 but not by NKp30. *European Journal of Immunology,* 31(9), 2680-2689.

Barge, R.M.; Osanto, S.; Marijt, W.A.; Starrenburg, C.W.; Fibbe, W.E.; Nortier, J.W.; Falkenburg, J.H. & Willemze, R. (2003). Minimal GVHD following in-vitro T cell-depleted allogeneic stem cell transplantation with reduced-intensity conditioning allowing subsequent infusions of donor lymphocytes in patients with hematological malignancies and solid tumors. *Experimental Hematology,* 31(10), 865-872.

Bauer, S.; Groh, V.; Wu, J.; Steinle, A.; Phillips, J.H.; Lanier, L.L. & Spies, T. (1999). Activation of NK cells and T cells by NKG2D, a receptor for stress-inducible MICA. *Science,* 285(5428), 727-729.

Bethge, W.A.; Haegele, M.; Faul, C.; Lang, P.; Schumm, M.; Bornhauser, M.; Handgretinger, R. & Kanz, L. (2006). Haploidentical allogeneic hematopoietic cell transplantation in adults with reduced-intensity conditioning and CD3/CD19 depletion: fast engraftment and low toxicity. *Experimental Hematology,* 34(12), 1746-1752.

Biassoni, R.; Cantoni, C.; Falco, M.; Verdiani, S.; Bottino, C.; Vitale, M.; Conte, R.; Poggi, A.; Moretta, A. & Moretta, L. (1996). The human leukocyte antigen (HLA)-C-specific "activatory" or "inhibitory" natural killer cell receptors display highly homologous extracellular domains but differ in their transmembrane and intracytoplasmic portions. *Journal of Experimental Medicine,* 183(2), 645-650.

Biassoni, R.; Cantoni, C.; Pende, D.; Sivori, S.; Parolini, S.; Vitale, M.; Bottino, C. & Moretta, A. (2001). Human natural killer cell receptors and co-receptors. *Immunological Reviews,* 181, 203-214.

Bossi, G. & Griffiths, G.M. (1999). Degranulation plays an essential part in regulating cell surface expression of Fas ligand in T cells and natural killer cells. *Nature Medicine,* 5(1), 90-96.

Bottino, C.; Biassoni, R.; Millo, R.; Moretta, L. & Moretta, A. (2000). The human natural cytotoxicity receptors (NCR) that induce HLA class I-independent NK cell triggering. *Human Immunology,* 61(1), 1-6.

Bottino, C.; Castriconi, R.; Moretta, L. & Moretta, A. (2005). Cellular ligands of activating NK receptors. *Trends in Immunology,* 26(4), 221-226.

Bottino, C.; Sivori, S.; Vitale, M.; Cantoni, C.; Falco, M.; Pende, D.; Morelli, L.; Augugliaro, R.; Semenzato, G.; Biassoni, R.; Moretta, L. & Moretta, A. (1996). A novel surface molecule homologous to the p58/p50 family of receptors is selectively expressed on a subset of human natural killer cells and induces both triggering of cell functions and proliferation. *European Journal of Immunology*, 26(8), 1816-1824.

Braud, V.M.; Allan, D.S.; O'Callaghan, C.A.; Soderstrom, K.; D'Andrea, A.; Ogg, G.S.; Lazetic, S.; Young, N.T.; Bell, J.I.; Phillips, J.H.; Lanier, L.L. & McMichael, A.J. (1998). HLA-E binds to natural killer cell receptors CD94/NKG2A, B and C. *Nature*, 391(6669), 795-799.

Braud, V.M.; Jones, E.Y. & McMichael, A. (1997). The human major histocompatibility complex class Ib molecule HLA- E binds signal sequence-derived peptides with primary anchor residues at positions 2 and 9. *European Journal of Immunology*, 27(5), 1164-1169.

Brodin, P.; Lakshmikanth, T.; Johansson, S.; Karre, K. & Hoglund, P. (2009). The strength of inhibitory input during education quantitatively tunes the functional responsiveness of individual natural killer cells. *Blood*, 113(11), 2434-2441.

Burke, S.; Lakshmikanth, T.; Colucci, F. & Carbone, E. (2010). New views on natural killer cell-based immunotherapy for melanoma treatment. *Trends in Immunology*, 31(9), 339-345.

Burshtyn, D.N.; Scharenberg, A.M.; Wagtmann, N.; Rajagopalan, S.; Berrada, K.; Yi, T.; Kinet, J.P. & Long, E.O. (1996). Recruitment of tyrosine phosphatase HCP by the killer cell inhibitor receptor. *Immunity*, 4(1), 77-85.

Cantoni, C.; Bottino, C.; Vitale, M.; Pessino, A.; Augugliaro, R.; Malaspina, A.; Parolini, S.; Moretta, L.; Moretta, A. & Biassoni, R. (1999). NKp44, a triggering receptor involved in tumor cell lysis by activated human natural killer cells, is a novel member of the immunoglobulin superfamily. *Journal of Experimental Medicine*, 189(5), 787-796.

Carayol, G.; Robin, C.; Bourhis, J.H.; naceur-Griscelli, A.; Chouaib, S.; Coulombel, L. & Caignard, A. (1998). NK cells differentiated from bone marrow, cord blood and peripheral blood stem cells exhibit similar phenotype and functions. *European Journal of Immunology*, 28(6), 1991-2002.

Cervantes, F.; Pierson, B.A.; McGlave, P.B.; Verfaillie, C.M. & Miller, J.S. (1996). Autologous activated natural killer cells suppress primitive chronic myelogenous leukemia progenitors in long-term culture. *Blood*, 87(6), 2476-2485.

Cook, M.A.; Milligan, D.W.; Fegan, C.D.; Darbyshire, P.J.; Mahendra, P.; Craddock, C.F.; Moss, P.A. & Briggs, D.C. (2004). The impact of donor KIR and patient HLA-C genotypes on outcome following HLA-identical sibling hematopoietic stem cell transplantation for myeloid leukemia. *Blood*, 103(4), 1521-1526.

Cooley, S.; Xiao, F.; Pitt, M.; Gleason, M.; McCullar, V.; Bergemann, T.; McQueen, K.L.; Guethlein, L.A.; Parham, P. & Miller, J.S. (2007). A subpopulation of human peripheral blood NK cells that lacks inhibitory receptors for self MHC is developmentally immature. *Blood*, 110(2), 578-586.

Cooper, M.A.; Fehniger, T.A. & Caligiuri, M.A. (2001). The biology of human natural killer-cell subsets. *Trends in Immunology*, 22(11), 633-640.

Cosman, D.; Mullberg, J.; Sutherland, C.L.; Chin, W.; Armitage, R.; Fanslow, W.; Kubin, M. & Chalupny, N.J. (2001). ULBPs, novel MHC class I-related molecules, bind to CMV glycoprotein UL16 and stimulate NK cytotoxicity through the NKG2D receptor. *Immunity*, 14(2), 123-133.

Cutler, C.; Li, S.; Ho, V.T.; Koreth, J.; Alyea, E.; Soiffer, R.J. & Antin, J.H. (2007). Extended follow-up of methotrexate-free immunosuppression using sirolimus and tacrolimus in related and unrelated donor peripheral blood stem cell transplantation. *Blood,* 109(7), 3108-3114.

Das, H.; Groh, V.; Kuijl, C.; Sugita, M.; Morita, C.T.; Spies, T. & Bukowski, J.F. (2001). MICA engagement by human Vgamma2Vdelta2 T cells enhances their antigen-dependent effector function. *Immunity,* 15(1), 83-93.

Diefenbach, A.; Jamieson, A.M.; Liu, S.D.; Shastri, N. & Raulet, D.H. (2000). Ligands for the murine NKG2D receptor: expression by tumor cells and activation of NK cells and macrophages. *Nature Immunology,* 1(2), 119-126.

Eissens, D.N.; Schaap, N.P.; Preijers, F.W.; Dolstra, H.; van, C.B.; Schattenberg, A.V.; Joosten, I. & van der, M.A. (2010a). CD3(+)/CD19(+)-depleted grafts in HLA-matched allogeneic peripheral blood stem cell transplantation lead to early NK cell cytolytic responses and reduced inhibitory activity of NKG2A. *Leukemia,* 24(3), 583-591.

Eissens, D.N.; van der, M.A.; van, C.B.; Preijers, F.W. & Joosten, I. (2010b). Rapamycin and MPA, but not CsA, impair human NK cell cytotoxicity due to differential effects on NK cell phenotype. *American Journal of Transplantation,* 10(9), 1981-1990.

Farag, S.S. & Caligiuri, M.A. (2006). Human natural killer cell development and biology. *Blood Reviews,* 20(3), 123-137.

Farrell, H.; gli-Esposti, M.; Densley, E.; Cretney, E.; Smyth, M. & vis-Poynter, N. (2000). Cytomegalovirus MHC class I homologues and natural killer cells: an overview. *Microbes and Infection,* 2(5), 521-532.

Fauriat, C.; Just-Landi, S.; Mallet, F.; Arnoulet, C.; Sainty, D.; Olive, D. & Costello, R.T. (2007). Deficient expression of NCR in NK cells from acute myeloid leukemia: evolution during leukemia treatment and impact of leukemic cells in NCRdull phenotype induction. *Blood,* 109(1), 323-330.

Fauriat, C.; Moretta, A.; Olive, D. & Costello, R.T. (2005). Defective killing of dendritic cells by autologous natural killer cells from acute myeloid leukemia patients. *Blood,* 106(6), 2186-2188.

Fauriat, C.; Long, E.O.; Ljunggren, H.G. & Bryceson, Y.T. (2010). Regulation of human NK cell cytokine and chemokine production by target cell recognition. *Blood,* 115(11), 2167-2176.

Federmann, B.; Hagele, M.; Pfeiffer, M.; Wirths, S.; Schumm, M.; Faul, C.; Vogel, W.; Handgretinger, R.; Kanz, L. & Bethge, W.A. (2011). Immune reconstitution after haploidentical hematopoietic cell transplantation: impact of reduced intensity conditioning and CD3/CD19 depleted grafts. *Leukemia,* 25(1), 121-129.

Fernandez, N.C.; Treiner, E.; Vance, R.E.; Jamieson, A.M.; Lemieux, S. & Raulet, D.H. (2005). A subset of natural killer cells achieves self-tolerance without expressing inhibitory receptors specific for self-MHC molecules. *Blood,* 105(11), 4416-4423.

Franksson, L.; George, E.; Powis, S.; Butcher, G.; Howard, J. & Karre, K. (1993). Tumorigenicity conferred to lymphoma mutant by major histocompatibility complex-encoded transporter gene. *Journal of Experimental Medicine,* 177(1), 201-205.

Fry, A.M.; Lanier, L.L. & Weiss, A. (1996). Phosphotyrosines in the killer cell inhibitory receptor motif of NKB1 are required for negative signaling and for association with protein tyrosine phosphatase 1C. *Journal of Experimental Medicine,* 184(1), 295-300.

Gentilini, C.; Hägele, M.; Meussig, A.; Nogai, A.; Kliem, C.; Bartsch, K.; Bazara, N.; Vogel, W.; Faul, C.; Kanz, L.; Thiel, E.; Niederwieser, D.W.; Uharek, L. & Bethge, W.A. (2007). NK cell recovery and immune reconstitution after haploidentical hematopoietic cell transplantation using either CD34 selected grafts and adoptive NK cell transfer or CD3/CD19 depleted grafts: comparison of two strategies for NK cell based immunotherapy. *Blood (ASH Annual Meeting Abstracts),* 110, 2988.

Giebel, S.; Nowak, I.; Wojnar, J.; Markiewicz, M.; Dziaczkowska, J.; Wylezol, I.; Krawczyk-Kulis, M.; Bloch, R.; Kusnierczyk, P. & Holowiecki, J. (2006). Impact of activating killer immunoglobulin-like receptor genotype on outcome of unrelated donor-hematopoietic cell transplantation. *Transplantation Proceedings,* 38(1), 287-291.

Groh, V.; Rhinehart, R.; Randolph-Habecker, J.; Topp, M.S.; Riddell, S.R. & Spies, T. (2001). Costimulation of CD8alphabeta T cells by NKG2D via engagement by MIC induced on virus-infected cells. *Nature Immunology,* 2(3), 255-260.

Guma, M.; Angulo, A.; Vilches, C.; Gomez-Lozano, N.; Malats, N. & Lopez-Botet, M. (2004). Imprint of human cytomegalovirus infection on the NK cell receptor repertoire. *Blood,* 104(12), 3664-3671.

Guma, M.; Budt, M.; Saez, A.; Brckalo, T.; Hengel, H.; Angulo, A. & Lopez-Botet, M. (2006). Expansion of CD94/NKG2C+ NK cells in response to human cytomegalovirus-infected fibroblasts. *Blood,* 107(9), 3624-3631.

Haentzschel, I.; Freiberg-Richter, J.; Platzbecker, U.; Kiani, A.; Schetelig, J.; Illmer, T.; Ehninger, G.; Schleyer, E. & Bornhauser, M. (2008). Targeting mycophenolate mofetil for graft-versus-host disease prophylaxis after allogeneic blood stem cell transplantation. *Bone Marrow Transplantation,* 42(2), 113-120.

Herberman, R.B. (1986). Natural killer cells. *Annual Review of Medicine,* 37, 347-352.

Ho, V.T.; Aldridge, J.; Kim, H.T.; Cutler, C.; Koreth, J.; Armand, P.; Antin, J.H.; Soiffer, R.J. & Alyea, E.P. (2009). Comparison of Tacrolimus and Sirolimus (Tac/Sir) versus Tacrolimus, Sirolimus, and Mini-Methotrexate (Tac/Sir/MTX) as Acute Graft-versus-Host Disease Prophylaxis after Reduced-Intensity Conditioning Allogeneic Peripheral Blood Stem Cell Transplantation. *Biology of Blood and Marrow Transplantation,* 15(7), 844-850.

Houchins, J.P.; Lanier, L.L.; Niemi, E.C.; Phillips, J.H. & Ryan, J.C. (1997). Natural killer cell cytolytic activity is inhibited by NKG2-A and activated by NKG2-C. *Journal of Immunology,* 158(8), 3603-3609.

Hsu, K.C.; Keever-Taylor, C.A.; Wilton, A.; Pinto, C.; Heller, G.; Arkun, K.; O'Reilly, R.J.; Horowitz, M.M. & Dupont, B. (2005). Improved outcome in HLA-identical sibling hematopoietic stem-cell transplantation for acute myelogenous leukemia predicted by KIR and HLA genotypes. *Blood,* 105(12), 4878-4884.

Igarashi, T.; Wynberg, J.; Srinivasan, R.; Becknell, B.; McCoy, J.P.; Takahashi, Y.; Suffredini, D.A.; Linehan, W.M.; Caligiuri, M.A. & Childs, R.W. (2004). Enhanced cytotoxicity of allogeneic NK cells with killer immunoglobulin -like receptor (KIR) ligand incompatibility against melanoma and renal cell carcinoma cells. *Blood,* 104(1), 170-177.

Iyengar, R.; Handgretinger, R.; Babarin-Dorner, A.; Leimig, T.; Otto, M.; Geiger, T.L.; Holladay, M.S.; Houston, J. & Leung, W. (2003). Purification of human natural killer cells using a clinical-scale immunomagnetic method. *Cytotherapy,* 5(6), 479-484.

Kao, I.T.; Yao, C.L.; Kong, Z.L.; Wu, M.L.; Chuang, T.L. & Hwang, S.M. (2007). Generation of natural killer cells from serum-free, expanded human umbilical cord blood CD34+ cells. *Stem Cells and Development,* 16(6), 1043-1051.

Kärre, K.; Ljunggren, H.G.; Piontek, G. & Kiessling, R. (1986a). Selective rejection of H-2-deficient lymphoma variants suggests alternative immune defence strategy. *Nature,* 319(6055), 675-678.

Kärre, K.; Ljunggren, H.G.; Piontek, G. & Kiessling, R. (1986b). Selective rejection of H-2-deficient lymphoma variants suggests alternative immune defence strategy. *Nature,* 319(6055), 675-678.

Kiessling, R.; Klein, E.; Pross, H. & Wigzell, H. (1975a). "Natural" killer cells in the mouse. II. Cytotoxic cells with specificity for mouse Moloney leukemia cells. Characteristics of the killer cell. *European Journal of Immunology,* 5(2), 117-121.

Kiessling, R.; Klein, E. & Wigzell, H. (1975b). "Natural" killer cells in the mouse. I. Cytotoxic cells with specificity for mouse Moloney leukemia cells. Specificity and distribution according to genotype. *European Journal of Immunology,* 5(2), 112-117.

Kim, D.H.; Sohn, S.K.; Lee, N.Y.; Baek, J.H.; Kim, J.G.; Won, D.I.; Suh, J.S.; Lee, K.B. & Shin, I.H. (2005a). Transplantation with higher dose of natural killer cells associated with better outcomes in terms of non-relapse mortality and infectious events after allogeneic peripheral blood stem cell transplantation from HLA-matched sibling donors. *European Journal of Hematology,* 75(4), 299-308.

Kim, S.; Poursine-Laurent, J.; Truscott, S.M.; Lybarger, L.; Song, Y.J.; Yang, L.; French, A.R.; Sunwoo, J.B.; Lemieux, S.; Hansen, T.H. & Yokoyama, W.M. (2005b). Licensing of natural killer cells by host major histocompatibility complex class I molecules. *Nature,* 436(7051), 709-713.

Kim, D.H.; Won, D.I.; Lee, N.Y.; Sohn, S.K.; Suh, J.S. & Lee, K.B. (2006). Non-CD34(+) Cells, Especially CD8(+) Cytotoxic T Cells and CD56(+) Natural Killer Cells, Rather Than CD34 Cells, Predict Early Engraftment and Better Transplantation Outcomes in Patients with Hematologic Malignancies after Allogeneic Peripheral Stem Cell Transplantation. *Biology of Blood and Marrow Transplantation,* 12(7), 719-728.

Klingemann, H.G. (2005). Natural killer cell-based immunotherapeutic strategies. *Cytotherapy,* 7(1), 16-22.

Klingemann, H.G. & Martinson, J. (2004). Ex vivo expansion of natural killer cells for clinical applications. *Cytotherapy,* 6(1), 15-22.

Koehl, U.; Esser, R.; Zimmermann, S.; Tonn, T.; Kotchetkov, R.; Bartling, T.; Sorensen, J.; Gruttner, H.P.; Bader, P.; Seifried, E.; Martin, H.; Lang, P.; Passweg, J.R.; Klingebiel, T. & Schwabe, D. (2005). Ex vivo expansion of highly purified NK cells for immunotherapy after haploidentical stem cell transplantation in children. *Klinische Pädiatrie,* 217(6), 345-350.

Kolb, H.J.; Simoes, B. & Schmid, C. (2004). Cellular immunotherapy after allogeneic stem cell transplantation in hematologic malignancies. *Current Opinion in Oncology,* 16(2), 167-173.

Lakshmikanth, T.; Burke, S.; Ali, T.H.; Kimpfler, S.; Ursini, F.; Ruggeri, L.; Capanni, M.; Umansky, V.; Paschen, A.; Sucker, A.; Pende, D.; Groh, V.; Biassoni, R.; Hoglund, P.; Kato, M.; Shibuya, K.; Schadendorf, D.; Anichini, A.; Ferrone, S.; Velardi, A.; Karre, K.; Shibuya, A.; Carbone, E. & Colucci, F. (2009). NCRs and DNAM-1 mediate NK cell recognition and lysis of human and mouse melanoma cell lines in vitro and in vivo. *Journal of Clinical Investigation,* 119(5), 1251-1263.

Lanier, L.L.; Corliss, B.C.; Wu, J.; Leong, C. & Phillips, J.H. (1998). Immunoreceptor DAP12 bearing a tyrosine-based activation motif is involved in activating NK cells. *Nature,* 391(6668), 703-707.

Lanier, L.L.; Ruitenberg, J.J. & Phillips, J.H. (1988). Functional and biochemical analysis of CD16 antigen on natural killer cells and granulocytes. *Journal of Immunology,* 141(10), 3478-3485.

Le, D.E.; Vely, F.; Olcese, L.; Cambiaggi, A.; Guia, S.; Krystal, G.; Gervois, N.; Moretta, A.; Jotereau, F. & Vivier, E. (1998). Inhibition of antigen-induced T cell response and antibody-induced NK cell cytotoxicity by NKG2A: association of NKG2A with SHP-1 and SHP-2 protein-tyrosine phosphatases. *European Journal of Immunology,* 28(1), 264-276.

Leung, W.; Iyengar, R.; Triplett, B.; Turner, V.; Behm, F.G.; Holladay, M.S.; Houston, J. & Handgretinger, R. (2005). Comparison of killer Ig-like receptor genotyping and phenotyping for selection of allogeneic blood stem cell donors. *Journal of Immunology,* 174(10), 6540-6545.

Leung, W.; Iyengar, R.; Turner, V.; Lang, P.; Bader, P.; Conn, P.; Niethammer, D. & Handgretinger, R. (2004). Determinants of antileukemia effects of allogeneic NK cells. *Journal of Immunology,* 172(1), 644-650.

Ljunggren, H.G. & Kärre, K. (1990). In search of the 'missing self': MHC molecules and NK cell recognition. *Immunology Today,* 11(7), 237-244.

Ljunggren, H.G.; Sturmhofel, K.; Wolpert, E.; Hammerling, G.J. & Kärre, K. (1990). Transfection of beta 2-microglobulin restores IFN-mediated protection from natural killer cell lysis in YAC-1 lymphoma variants. *Journal of Immunology,* 145(1), 380-386.

Lundqvist, A.; McCoy, J.P.; Samsel, L. & Childs, R. (2007). Reduction of GVHD and enhanced antitumor effects after adoptive infusion of alloreactive Ly49-mismatched NK cells from MHC-matched donors. *Blood,* 109(8), 3603-3606.

Mandelboim, O.; Lieberman, N.; Lev, M.; Paul, L.; Arnon, T.I.; Bushkin, Y.; Davis, D.M.; Strominger, J.L.; Yewdell, J.W. & Porgador, A. (2001). Recognition of haemagglutinins on virus-infected cells by NKp46 activates lysis by human NK cells. *Nature,* 409(6823), 1055-1060.

McKenna, D.H., Jr.; Sumstad, D.; Bostrom, N.; Kadidlo, D.M.; Fautsch, S.; McNearney, S.; Dewaard, R.; McGlave, P.B.; Weisdorf, D.J.; Wagner, J.E.; McCullough, J. & Miller, J.S. (2007). Good manufacturing practices production of natural killer cells for immunotherapy: a six-year single-institution experience. *Transfusion,* 47(3), 520-528.

Meyer-Monard, S.; Passweg, J.; Siegler, U.; Kalberer, C.; Koehl, U.; Rovo, A.; Halter, J.; Stern, M.; Heim, D.; ois Gratwohl, J.R. & Tichelli, A. (2009). Clinical-grade purification of natural killer cells in haploidentical hematopoietic stem cell transplantation. *Transfusion,* 49(2), 362-371.

Miller, J.S.; Soignier, Y.; Panoskaltsis-Mortari, A.; McNearney, S.A.; Yun, G.H.; Fautsch, S.K.; McKenna, D.; Le, C.; Defor, T.E.; Burns, L.J.; Orchard, P.J.; Blazar, B.R.; Wagner, J.E.; Slungaard, A.; Weisdorf, D.J.; Okazaki, I.J. & McGlave, P.B. (2005). Successful adoptive transfer and in vivo expansion of human haploidentical NK cells in patients with cancer. *Blood,* 105(8), 3051-3057.

Miller, J.S.; Verfaillie, C. & McGlave, P. (1992). The generation of human natural killer cells from CD34+/DR- primitive progenitors in long-term bone marrow culture. *Blood,* 80(9), 2182-2187.

Moretta, A.; Poggi, A.; Pende, D.; Tripodi, G.; Orengo, A.M.; Pella, N.; Augugliaro, R.; Bottino, C.; Ciccone, E. & Moretta, L. (1991). CD69-mediated pathway of lymphocyte activation: anti-CD69 monoclonal antibodies trigger the cytolytic activity of different lymphoid effector cells with the exception of cytolytic T lymphocytes expressing T cell receptor alpha/beta. *Journal of Experimental Medicine,* 174(6), 1393-1398.

Moretta, A.; Sivori, S.; Vitale, M.; Pende, D.; Morelli, L.; Augugliaro, R.; Bottino, C. & Moretta, L. (1995). Existence of both inhibitory (p58) and activatory (p50) receptors for HLA-C molecules in human natural killer cells. *Journal of Experimental Medicine,* 182(3), 875-884.

Neumann, F.; Graef, T.; Tapprich, C.; Vaupel, M.; Steidl, U.; Germing, U.; Fenk, R.; Hinke, A.; Haas, R. & Kobbe, G. (2005). Cyclosporine A and Mycophenolate Mofetil vs Cyclosporine A and Methotrexate for graft-versus-host disease prophylaxis after stem cell transplantation from HLA-identical siblings. *Bone Marrow Transplantation,* 35(11), 1089-1093.

Ohata, K.; Espinoza, J.L.; Lu, X.; Kondo, Y. & Nakao, S. (2011). Mycophenolic Acid inhibits natural killer cell proliferation and cytotoxic function: a possible disadvantage of including mycophenolate mofetil in the graft-versus-host disease prophylaxis regimen. *Biology of Blood and Marrow Transplantation,* 17(2), 205-213.

Olcese, L.; Lang, P.; Vely, F.; Cambiaggi, A.; Marguet, D.; Blery, M.; Hippen, K.L.; Biassoni, R.; Moretta, A.; Moretta, L.; Cambier, J.C. & Vivier, E. (1996). Human and mouse killer-cell inhibitory receptors recruit PTP1C and PTP1D protein tyrosine phosphatases. *Journal of Immunology,* 156(12), 4531-4534.

Olson, J.A.; Leveson-Gower, D.B.; Gill, S.; Baker, J.; Beilhack, A. & Negrin, R.S. (2010). NK cells mediate reduction of GVHD by inhibiting activated, alloreactive T cells while retaining GVT effects. *Blood,* 115(21), 4293-4301.

Orr, M.T.; Murphy, W.J. & Lanier, L.L. (2010). 'Unlicensed' natural killer cells dominate the response to cytomegalovirus infection. *Nature Immunoly,* 11(4), 321-327.

Papamichail, M.; Perez, S.A.; Gritzapis, A.D. & Baxevanis, C.N. (2004). Natural killer lymphocytes: biology, development, and function. *Cancer Immunology, Immunotherapy,* 53(3), 176-186.

Parham, P. (2005). MHC class I molecules and KIRs in human history, health and survival. *Nature Reviews Immunology,* 5(3), 201-214.

Parham, P. (2006). Taking license with natural killer cell maturation and repertoire development. *Immunological Reviews,* 214, 155-160.

Passweg, J.R.; Stern, M.; Koehl, U.; Uharek, L. & Tichelli, A. (2005). Use of natural killer cells in hematopoetic stem cell transplantation. *Bone Marrow Transplantation,* 35(7), 637-643.

Passweg, J.R.; Tichelli, A.; Meyer-Monard, S.; Heim, D.; Stern, M.; Kuhne, T.; Favre, G. & Gratwohl, A. (2004). Purified donor NK-lymphocyte infusion to consolidate engraftment after haploidentical stem cell transplantation. *Leukemia,* 18(11), 1835-1838.

Pawelec, G.; Da, S.P.; Max, H.; Kalbacher, H.; Schmidt, H.; Bruserud, O.; Zugel, U.; Baier, W.; Rehbein, A. & Pohla, H. (1995). Relative roles of natural killer- and T cell-mediated anti-leukemia effects in chronic myelogenous leukemia patients treated with interferon-alpha. *Leukemia and Lymphoma,* 18(5-6), 471-478.

Peggs, K.S.; Thomson, K.; Hart, D.P.; Geary, J.; Morris, E.C.; Yong, K.; Goldstone, A.H.; Linch, D.C. & Mackinnon, S. (2004). Dose-escalated donor lymphocyte infusions following reduced intensity transplantation: toxicity, chimerism, and disease responses. *Blood,* 103(4), 1548-1556.

Pende, D.; Cantoni, C.; Rivera, P.; Vitale, M.; Castriconi, R.; Marcenaro, S.; Nanni, M.; Biassoni, R.; Bottino, C.; Moretta, A. & Moretta, L. (2001). Role of NKG2D in tumor cell lysis mediated by human NK cells: cooperation with natural cytotoxicity receptors and capability of recognizing tumors of nonepithelial origin. *European Journal of Immunology.*

Pende, D.; Parolini, S.; Pessino, A.; Sivori, S.; Augugliaro, R.; Morelli, L.; Marcenaro, E.; Accame, L.; Malaspina, A.; Biassoni, R.; Bottino, C.; Moretta, L. & Moretta, A. (1999). Identification and molecular characterization of NKp30, a novel triggering receptor involved in natural cytotoxicity mediated by human natural killer cells. *Journal of Experimental Medicine,* 190(10), 1505-1516.

Pierson, B.A. & Miller, J.S. (1996). CD56+bright and CD56+dim natural killer cells in patients with chronic myelogenous leukemia progressively decrease in number, respond less to stimuli that recruit clonogenic natural killer cells, and exhibit decreased proliferation on a per cell basis. *Blood,* 88(6), 2279-2287.

Raulet, D.H. & Vance, R.E. (2006). Self-tolerance of natural killer cells. *Nature Reviews Immunology,* 6(7), 520-531.

Raulet, D.H.; Vance, R.E. & McMahon, C.W. (2001). Regulation of the natural killer cell receptor repertoire. *Annual Review of Immunology,* 19, 291-330.

Ruggeri, L.; Aversa, F.; Martelli, M.F. & Velardi, A. (2006). Allogeneic hematopoietic transplantation and natural killer cell recognition of missing self. *Immunological Reviews,* 214, 202-218.

Ruggeri, L.; Capanni, M.; Casucci, M.; Volpi, I.; Tosti, A.; Perruccio, K.; Urbani, E.; Negrin, R.S.; Martelli, M.F. & Velardi, A. (1999). Role of natural killer cell alloreactivity in HLA-mismatched hematopoietic stem cell transplantation. *Blood,* 94(1), 333-339.

Ruggeri, L.; Capanni, M.; Mancusi, A.; Urbani, E.; Perruccio, K.; Burchielli, E.; Tosti, A.; Topini, F.; Aversa, F.; Martelli, M.F. & Velardi, A. (2004). Alloreactive natural killer cells in mismatched hematopoietic stem cell transplantation. *Blood Cells, Molecules and Diseases,* 33(3), 216-221.

Ruggeri, L.; Capanni, M.; Urbani, E.; Perruccio, K.; Shlomchik, W.D.; Tosti, A.; Posati, S.; Rogaia, D.; Frassoni, F.; Aversa, F.; Martelli, M.F. & Velardi, A. (2002). Effectiveness of donor natural killer cell alloreactivity in mismatched hematopoietic transplants. *Science,* 295(5562), 2097-2100.

Ruggeri, L.; Mancusi, A.; Capanni, M.; Urbani, E.; Carotti, A.; Aloisi, T.; Stern, M.; Pende, D.; Perruccio, K.; Burchielli, E.; Topini, F.; Bianchi, E.; Aversa, F.; Martelli, M.F. & Velardi, A. (2007). Donor natural killer cell allorecognition of missing self in haploidentical hematopoietic transplantation for acute myeloid leukemia: challenging its predictive value. *Blood,* 110(1), 433-440.

Salcedo, M.; Diehl, A.D.; Olsson-Alheim, M.Y.; Sundback, J.; Van, K.L.; Karre, K. & Ljunggren, H.G. (1997). Altered expression of Ly49 inhibitory receptors on natural killer cells from MHC class I-deficient mice. *Journal of Immunology*, 158(7), 3174-3180.

Savani, B.N.; Mielke, S.; Adams, S.; Uribe, M.; Rezvani, K.; Yong, A.S.; Zeilah, J.; Kurlander, R.; Srinivasan, R.; Childs, R.; Hensel, N. & Barrett, A.J. (2007). Rapid natural killer cell recovery determines outcome after T-cell-depleted HLA-identical stem cell transplantation in patients with myeloid leukemias but not with acute lymphoblastic leukemia. *Leukemia*, 10, 2145-2152.

Schaap, N.; Schattenberg, A.; Bar, B.; Preijers, F.; Geurts van, K.A.; van der, M.R.; de, B.T. & de, W.T. (1997). Outcome of transplantation for standard-risk leukaemia with grafts depleted of lymphocytes after conditioning with an intensified regimen. *British Journal of Haematology*, 98(3), 750-759.

Schaap, N.; Schattenberg, A.; Bar, B.; Preijers, F.; van de Wiel van Kemenade & de Witte, T. (2001). Induction of graft-versus-leukemia to prevent relapse after partially lymphocyte-depleted allogeneic bone marrow transplantation by pre-emptive donor leukocyte infusions. *Leukemia*, 15(9), 1339-1346.

Schleuning, M.; Judith, D.; Jedlickova, Z.; Stubig, T.; Heshmat, M.; Baurmann, H. & Schwerdtfeger, R. (2008). Calcineurin inhibitor-free GVHD prophylaxis with sirolimus, mycophenolate mofetil and ATG in Allo-SCT for leukemia patients with high relapse risk: an observational cohort study. *Bone Marrow Transplantation*, 43(9), 717-723.

Seaman, W.E. (2000). Natural killer cells and natural killer T cells. *Arthritis & Rheumatism*, 43(6), 1204-1217.

Sentman, C.L.; Olsson, M.Y. & Karre, K. (1995). Missing self recognition by natural killer cells in MHC class I transgenic mice. A 'receptor calibration' model for how effector cells adapt to self. *Seminars in Immunology*, 7(2), 109-119.

Shibuya, A.; Campbell, D.; Hannum, C.; Yssel, H.; Franz-Bacon, K.; McClanahan, T.; Kitamura, T.; Nicholl, J.; Sutherland, G.R.; Lanier, L.L. & Phillips, J.H. (1996). DNAM-1, a novel adhesion molecule involved in the cytolytic function of T lymphocytes. *Immunity*, 4(6), 573-581.

Sivori, S.; Parolini, S.; Falco, M.; Marcenaro, E.; Biassoni, R.; Bottino, C.; Moretta, L. & Moretta, A. (2000). 2B4 functions as a co-receptor in human NK cell activation. *European Journal of Immunology*, 30(3), 787-793.

Sivori, S.; Vitale, M.; Morelli, L.; Sanseverino, L.; Augugliaro, R.; Bottino, C.; Moretta, L. & Moretta, A. (1997). p46, a novel natural killer cell-specific surface molecule that mediates cell activation. *Journal of Experimental Medicine*, 186(7), 1129-1136.

Snell, G.D. (1976). Recognition structures determined by the H-2 complex. *Transplantation Proceedings*, 8(2), 147-156.

Spanholtz, J.; Preijers, F.; Tordoir, M.; Trilsbeek, C.; Paardekooper, J.; de, W.T.; Schaap, N. & Dolstra, H. (2011). Clinical-grade generation of active NK cells from cord blood hematopoietic progenitor cells for immunotherapy using a closed-system culture process. *PLoS ONE*, 6(6), e20740.

Spanholtz, J.; Tordoir, M.; Eissens, D.; Preijers, F.; van der, M.A.; Joosten, I.; Schaap, N.; de Witte, T.M. & Dolstra, H. (2010). High Log-Scale Expansion of Functional Human

Natural Killer Cells from Umbilical Cord Blood CD34-Positive Cells for Adoptive Cancer Immunotherapy. *PLoS ONE*, 5(2), e9221.

Spits, H.; Lanier, L.L. & Phillips, J.H. (1995). Development of human T and natural killer cells. *Blood*, 85(10), 2654-2670.

Sutherland, C.L.; Rabinovich, B.; Chalupny, N.J.; Brawand, P.; Miller, R. & Cosman, D. (2006). ULBPs, human ligands of the NKG2D receptor, stimulate tumor immunity with enhancement by IL-15. *Blood*, 108(4), 1313-1319.

Tam, Y.K.; Martinson, J.A.; Doligosa, K. & Klingemann, H.G. (2003). Ex vivo expansion of the highly cytotoxic human natural killer-92 cell-line under current good manufacturing practice conditions for clinical adoptive cellular immunotherapy. *Cytotherapy*, 5(3), 259-272.

Thomas, E.D. & Blume, K.G. (1999). Historical markers in the development of allogeneic hematopoietic cell transplantation. *Biology of Blood and Marrow Transplantation*, 5(6), 341-346.

Ugolini, S. & Vivier, E. (2001). Multifaceted roles of MHC class I and MHC class I-like molecules in T cell activation. *Nature Immunology*, 2(3), 198-200.

Vales-Gomez, M.; Reyburn, H.T.; Erskine, R.A. & Strominger, J. (1998a). Differential binding to HLA-C of p50-activating and p58-inhibitory natural killer cell receptors. *Proceedings of the National Academy of Sciences*, 95(24), 14326-14331.

Vales-Gomez, M.; Reyburn, H.T.; Mandelboim, M. & Strominger, J.L. (1998b). Kinetics of interaction of HLA-C ligands with natural killer cell inhibitory receptors. *Immunity*, 9(3), 337-344.

Valiante, N.M.; Uhrberg, M.; Shilling, H.G.; Lienert, W.K.; Arnett, K.L.; D'Andrea, A.; Phillips, J.H.; Lanier, L.L. & Parham, P. (1997). Functionally and structurally distinct NK cell receptor repertoires in the peripheral blood of two human donors. *Immunity*, 7(6), 739-751.

van der Meer, A.; Schaap, N.P.M.; Schattenberg, A.V.M.B.; van Cranenbroek, B.; Tijssen, H.J. & Joosten, I. (2008). KIR2DS5 is associated with leukemia free survival after HLA identical stem cell transplantation in chronic myeloid leukemia patients. *Molecular Immunology*, 45(13), 3631-3638.

van Stijn, A.; Rowshani, A.T.; Yong, S.L.; Baas, F.; Roosnek, E.; ten Berge, I.J.M. & van Lier, R.A.W. (2008). Human Cytomegalovirus Infection Induces a Rapid and Sustained Change in the Expression of NK Cell Receptors on CD8+ T Cells. *The Journal of Immunology*, 180(7), 4550-4560.

Vilches, C. & Parham, P. (2002). KIR: diverse, rapidly evolving receptors of innate and adaptive immunity. *Annual Review of Immunology*, 20, 217-251.

Vitale, M.; Falco, M.; Castriconi, R.; Parolini, S.; Zambello, R.; Semenzato, G.; Biassoni, R.; Bottino, C.; Moretta, L. & Moretta, A. (2001). Identification of NKp80, a novel triggering molecule expressed by human NK cells. *European Journal of Immunology*, 31(1), 233-242.

Vivier, E. (2006). What is natural in natural killer cells? *Immunology Letters*, 107(1), 1-7.

Vivier, E. & Daeron, M. (1997). Immunoreceptor tyrosine-based inhibition motifs. *Immunology Today*, 18(6), 286-291.

Vivier, E.; Tomasello, E.; Baratin, M.; Walzer, T. & Ugolini, S. (2008). Functions of natural killer cells. *Nature Immunology*, 9(5), 503-510.

Vogelsang, G.B. & Arai, S. (2001). Mycophenolate mofetil for the prevention and treatment of graft-versus-host disease following stem cell transplantation: preliminary findings. *Bone Marrow Transplantation,* 27(12), 1255-1262.

Wang, H.; Grzywacz, B.; Sukovich, D.; McCullar, V.; Cao, Q.; Lee, A.B.; Blazar, B.R.; Cornfield, D.N.; Miller, J.S. & Verneris, M.R. (2007). The unexpected effect of cyclosporin A on CD56+CD16- and CD56+CD16+ natural killer cell subpopulations. *Blood,* 110(5), 1530-1539.

Wu, J.; Cherwinski, H.; Spies, T.; Phillips, J.H. & Lanier, L.L. (2000). DAP10 and DAP12 form distinct, but functionally cooperative, receptor complexes in natural killer cells. *Journal of Experimental Medicine,* 192(7), 1059-1068.

Wu, J.; Song, Y.; Bakker, A.B.; Bauer, S.; Spies, T.; Lanier, L.L. & Phillips, J.H. (1999). An activating immunoreceptor complex formed by NKG2D and DAP10. *Science,* 285(5428), 730-732.

7

Dendritic Cells in Hematopoietic Stem Cell Transplantation

Yannick Willemen[1,*], Khadija Guerti[1,2,3,*], Herman Goossens[2], Zwi Berneman[1], Viggo Van Tendeloo[1] and Evelien Smits[1]

[1]*Laboratory of Experimental Hematology, Vaccine and Infectious Disease Institute*

[2]*Laboratory of Medical Microbiology, Vaccine and Infectious Disease Institute*

[3]*Laboratory of Immunology, Antwerp University Hospital*

University of Antwerp

Belgium

1. Introduction

Dendritic cells (DC) are highly specialized antigen-presenting cells (APC) that are pivotal in regulating the balance between immune tolerance and protective immunity. This functional versatility is highlighted in the context of allogeneic hematopoietic stem cell transplantation (allo-HSCT), where DC are crucial for the induction and modulation of graft-versus-host reactions. Furthermore, in the process of immune restoration after allo-HSCT, DC play a central role in generating protective immunity against pathogens. The importance of DC in directing the immune system during the complex immunological situation after allo-HSCT warrants further research, aimed at uncovering the therapeutic potential they hold in this setting.

2. Role of dendritic cells in the development of acute graft-versus-host disease following allogeneic hematopoietic stem cell transplantation

Allo-HSCT is a well-established and valuable therapeutic option for a variety of life-threatening malignant and non-malignant diseases (Gratwohl et al., 2010). In cancer, allo-HSCT has been mainly applied to treat leukemia and lymphoma patients (Gratwohl et al., 2010). Immunologic graft-versus-leukemia (GVL) effects mediated by allogeneic lymphocytes present in the graft are major contributors to its success. A number of distinct donor cell subsets have been identified that may play a role in the GVL responses after allo-HSCT. These include natural killer cells (Gill et al., 2009; Ruggeri et al., 2007), T cells reactive to tumor-specific or tumor-associated antigens (TAA; Molldrem et al., 2002; K. Rezvani & Barrett, 2008), and T cells reactive to host minor histocompatibility (miHC) antigens (Falkenburg et al., 2002, 2003; Riddell et al., 2002, 2003).

---

* Both authors contributed equally

2.1 Development of acute graft-versus-host disease following allogeneic hematopoietic stem cell transplantation

A major obstacle that substantially limits the therapeutic potential of allo-HSCT is the occurrence of graft-versus-host reactions against healthy host tissues, resulting in graft-versus-host disease (GVHD). GVHD is a major cause of morbidity and mortality following allo-HSCT. The overall incidence lies between 30% and 60% with a mortality rate of approximately 50% (Barton-Burke et al., 2008). It is a complex multi-step process, involving innate and adaptive immunity and affecting many organs, including skin, liver and the gastrointestinal tract (Ball & Egeler, 2008; Ferrara et al., 2009).

Billingham was the first to describe GVHD (Billingham, 1966). According to the Billingham criteria, three conditions must exist in order for GVHD to occur after allogeneic transplantation: (1) the donor graft must contain viable and immunologically functional effector cells, (2) the donor and recipient must be histoincompatible, and (3) the recipient must be immunocompromised.

The series of events that contribute to the development of acute GVHD (as described by Ferrara & Reddy, 2006; Goker et al., 2001) can be divided in three phases (Goker et al., 2001). The first phase - conditioning phase - starts before the engraftment. This phase involves tissue damage caused by pre-transplantation myeloablative radiation/chemotherapy regimens, followed by release of lipopolysaccharide and secretion of proinflammatory cytokines, upregulation of adhesion molecules and enhanced expression of major histocompatibility complex (MHC) molecules on recipient tissues. The proinflammatory environment will also activate APC. The second phase - induction and expansion phase - starts with the recognition of the histoincompatible host tissue antigens by donor T cells. This phase involves T cell activation, stimulation, proliferation and differentiation. Activated host APC play a key role in the second phase of the graft-versus-host reaction by presenting mismatched recipient antigens to donor T cells. The first two phases constitute the afferent phase of GVHD. Finally, the third phase - effector phase - represents the actual clinical phase of acute GVHD and involves direct and indirect damage to host cells contributing to aggravation of GVHD.

From these models, it is clear that donor T cells play a crucial role in evoking GVHD after allo-HSCT. Simultaneously, donor T cells represent major mediators of GVL effects. Therefore, research efforts are aimed at separating GVL reactions from GVHD (Li et al., 2009a; Mackinnon et al., 1995; A.R. Rezvani & Storb, 2008). A key question is whether GVL activity and GVHD are fundamentally different mechanisms, or whether they are both clinical manifestations of similar graft-versus-host reactions.

Preclinical model systems and clinical trials designed to investigate the possibility of selectively activating graft-versus-host reactions that result in GVL effects without GVHD, have led to new insights in the pathophysiology of GVL responses and GVHD after allo-HSCT (Li et al., 2009a; A.R. Rezvani & Storb, 2008). In a more complex model of human GVHD and GVL pathophysiology (Li et al., 2009a), differentiation of activated T cells into the distinct subsets T helper (Th)1/cytotoxic T cell (Tc)1, Th2/Tc2, Th17 or regulatory T (Treg) cells is taken into account. These T cell subsets differ both in cytokine profiles and in their graft-versus-host activities. Activated Th1/Tc1 cells can directly attack host tissue and initiate specific inflammatory immune responses that lead to both GVL responses and acute GVHD. Th2 cells on the other hand, evoke antigen-specific cellular and humoral immune responses resulting in GVL responses, but also in chronic GVHD. Notably, Th2 cytokines may inhibit the development of acute GVHD. Activated Th17 cells potentiate inflammation and lead to acute GVHD, whereas the Th1 cytokine interferon (IFN)-γ can suppress Th17 responses to decrease GVHD. Donor Treg cells suppress GVHD, but the effect of Treg cells on GVL responses remains to be further elucidated. The T cell subsets that are most likely associated with shifting the balance away from GVHD towards GVL responses are Th1/Tc1, γδ T and Treg cells (Li et al., 2009a).

2.2 Dendritic cells in the development of acute graft-versus-host disease

Early models have mainly focused on the central role of T cell activation and cytokine release in the pathophysiologic process of GVHD. The 1966 Billingham criteria clearly accounted for the presence of viable and immunologically functional effector cells as a prerequisite for the development of GVHD (Billingham, 1966). More recent models of GVHD (Choi et al., 2010; Ferrara & Reddy, 2006; Goker et al., 2001; Li et al., 2009a) also take into account the key role of antigen presentation in its development by stating that activation of APC precedes activation and clonal expansion of T cells in the immune cascade. Host APC play a crucial role in the graft-versus-host reaction by presenting mismatched recipient antigens to donor T cells (Goker et al., 2001). In allo-HSCT with a histocompatible donor, the relevant antigens are miHC antigens (Falkenburg et al., 2002, 2003; Ridell et al., 2002, 2003). APC digest miHC antigens into short peptides that are linked to MHC molecules and presented on the surface of APC as allopeptide-MHC complexes. Physical interaction between the allopeptide-MHC complexes and antigen-specific T cell receptors (TCR) then leads to recognition and activation of antigen-specific T cells (Clark & Chakraverty, 2002; Goker et al., 2001).

Following allo-HSCT, a unique situation is created in which both host- and donor-derived APC co-exist within the host. Thus, foreign miHC antigens can be presented by either host-derived or donor-derived APC. The latter case implies effective cross-presentation of recipient miHC antigens by donor-derived APC (Shlomchik, 2003). The roles of host- and donor-derived APC in the development of GVHD have been examined in experimental mouse studies. In a murine allogeneic bone marrow transplantation (BMT) model, Shlomchik and colleagues showed that host-derived APC were necessary and sufficient to initiate GVHD (Shlomchik et al., 1999). Donor APC on the other hand, while redundant for the onset of GVHD, were required to maximize the GVHD (Matte et al., 2004). A model focusing on the role of host-derived APC in the effector phase of GVHD demonstrated that tissue-resident APC control migration of alloreactive donor T cells into the tissues and subsequent local development of GVHD (Zhang et al., 2002).

APC represent a heterogeneous population of cells with varying antigen-presenting capacities. As the most specialized and professional APC of the immune system, DC are highly efficient in processing and presenting antigens (Mellman & Steinman, 2001). The role of DC in GVHD has been investigated and confirmed in various experimental settings (Mohty, 2007; Mohty & Gaugler, 2008; Xu et al., 2008).

Allo-HSCT can change the origin (host- versus donor-derived), number, lineage and activation level of DC in the host (Clark & Chakraverty, 2002). Several studies have examined the role of DC counts and subsets in the development and severity of GVHD. Based on their immunophenotype and functional properties, DC can be classified into myeloid conventional DC (cDC) and plasmacytoid DC (pDC) (Liu, 2001). A murine BMT model demonstrated that host-derived DC are necessary and sufficient for priming donor T cells to cause acute GVHD (Duffner et al., 2004). In humans, peripheral blood DC chimerism experiments have been performed following allo-HSCT to analyze the contribution of different DC subsets to GVHD (Boeck et al., 2006; Chan et al., 2003; Pihusch et al., 2005). Findings of Chan et al (Chan et al., 2003) confirmed the importance of host DC, because persistence of the DC at day 100 after allo-HSCT was correlated with GVHD. On the other hand, graft-versus-host reactions were also detected in patients that had DC exclusively of donor origin (Boeck et al., 2006). Lower counts of cDC and pDC in patients were associated with an increased risk for acute GVHD (Horváth et al., 2009; Lau et al., 2007; Rajasekar et al., 2008; Reddy et al., 2004; Vakkila et al., 2005). In addition, higher numbers of donor pDC following allo-HSCT decreased the risk of developing chronic GVHD, but also increased the risk of relapse, possibly due to interference with GVL reactions (Waller et al., 2001). In contrast to these data, higher pDC numbers in the graft or in the recipient after allo-HSCT have also been found to correlate with the development of chronic GVHD (Clark et al., 2003; Rossi et al., 2002). Next to absolute numbers, also activation status can be predictive of GVHD, with activated cDC being highly correlated with acute GVHD (Lau et al., 2007). Taken together, experimental data suggest that different DC subsets have different effects on GVHD and GVL reactions, but further research is required to unravel the exact role of each subset.

3. Dendritic cell-based therapy and allogeneic hematopoietic stem cell transplantation

Over the past decade several approaches of DC-based therapy in allo-HSCT settings have been scrutinized, yielding some promising results with regard to decreasing GVHD, optimizing GVL reactions and restoring protective immunity against pathogens.

3.1 Dendritic cell-based therapy to reduce graft-versus-host-disease and enhance graft-versus-leukemia effects

Allogeneic T cells have the capacity to kill residual malignant cells in the host, but also to destruct normal host tissue contributing to GVHD, which can be life-threatening and limits the use of allo-HSCT. While T cell depletion of the graft is a very effective way of reducing the risk of GVHD, it also diminishes the GVL effect, thereby increasing the risk of relapse. Hence, a more refined approach is needed to balance graft-versus-host reactions after allo-HSCT. Given the inherent key regulatory function of DC, DC-based therapy is considered an attractive approach to shift the balance in favor of GVL reactions.

3.1.1 Dendritic cell-based therapy to reduce graft-versus-host-disease

The finding in murine BMT models that host APC are necessary for GVHD to develop (Matte et al., 2004; Shlomchik et al., 1999), led the authors to suggest that depletion of host APC before the conditioning regimen should prevent GVHD without the need for prolonged immunosuppressive treatment.

Antibody-mediated depletion of DC was investigated in a chimeric human/mouse model of GVHD, in which severe combined immunodeficient (SCID) mice received a xenogeneic transplantation with human peripheral blood mononuclear cells (PBMC) (Wilson et al., 2009). Antibodies against the DC activation marker CD83 were injected in host mice 3 hours before injection of human PBMC. This therapeutic intervention almost completely prevented lethal GVHD, whereas negative control mice all developed severe GVHD.

Moreover, mice treated with anti-CD83 antibodies required no further immunosuppressive therapy and possessed functional T cell immunity *in vitro* (Wilson et al., 2009).

These data support further investigation of *in vivo* depletion of host and/or donor APC as a way of preventing GVHD in allo-HSCT recipients. This strategy makes redundant both T cell depletion, thereby preserving the memory T cell pool, and T cell-targeted immunosuppression, which greatly hampers GVL responses and protective immunity. However, the effect of DC depletion on GVL responses still needs to be investigated in animal allo-HSCT models including *in vivo* leukemic challenge. Some concern can be raised about potential interference of DC depletion with the GVL effect, because in mice studies antigen-presentation by host APC has been shown to be important in mediating GVL responses following donor lymphocyte infusions (DLI) (Chakraverty et al., 2006; Mapara et al., 2002). Furthermore, DC depletion might result in a delayed restorage of immunity against pathogens (Clark & Chakraverty, 2002).

More thorough elucidation of the role of distinct DC subsets in allo-antigen responses after allo-HSCT will pave the way to depletion of undesirable or expansion of desirable DC subsets. In this context, a study of Li et al. (Li et al., 2009b) has shown that manipulating the content of donor APC subsets in allo-HSCT grafts can enhance the GVL effect without increasing GVHD. In their study, leukemia-bearing mice that received hematopoietic stem cells (HSC) and CD11b-negative donor APC had substantially enhanced survival compared to recipients of HSC alone, HSC and T cells, or HSC and CD11b-positive APC.

Another promising strategy to modulate allo-antigen responses following allo-HSCT involves DC engineered to boost their tolerogenic or regulatory capacities.

In a study of Reichardt et al (Reichardt et al., 2008), DC were isolated directly from mice bone marrow and spleen cells using positive magnetic cell selection and exposed to rapamycin for 24 hours. Adoptive transfer of rapamycin-treated DC of host origin, but not donor origin, administered together with the bone marrow transplant, reduced GVHD severity and led to improved survival of recipient mice in a dose-dependent way. The reduced expansion of alloreactive T cells could account for the beneficial effects on GVHD and survival, but carries the risk of reducing the GVL effect.

In two other studies with similar methodology (Chorny et al., 2006; Sato et al., 2003), DC were generated from murine bone marrow cells using granulocyte macrophage colony-stimulating factor (GM-CSF) and either interleukin (IL)-10 and transforming growth factor (TGF)-β1 or vasoactive intestinal peptide (VIP) for 6 days. Then, lipopolysaccharide (LPS) was added for 2 days to induce activation, followed by injection of the DC 2 days after BMT. Results of both studies demonstrated that host-matched DC, but not host-mismatched DC, prevented the onset of severe GVHD in recipient mice in a dose-dependent way. In order to study the effect of DC therapy on GVL responses, mice were challenged with P815 or A20 malignant cells. BMT recipient mice that received host-matched DC were not only protected from lethal GVHD, but also maintained a strong GVL effect and survived significantly longer than control animals (Chorny et al., 2006; Sato et al., 2003).

In conclusion, the administration of specifically engineered DC appears to be a favorable means of modulating alloreactivity after allo-HSCT, because they are able to reduce the risk of severe GVHD, while maintaining the benefits of the GVL effect. Clinical trials will have to show if these beneficial effects can also be seen in humans. Considering that only host-matched DC were able to protect the recipient from severe GVHD and conserve a strong GVL effect in these murine models, it seems likely that the DC will have to be tailored to every individual patient. Although this will be costly and labor-intensive, it can be cost-effective if proven beneficial in allo-HSCT.

3.1.2 Dendritic cell-based therapy to enhance the graft-versus-leukemia effect without aggravating graft-versus-host-disease

Donor alloreactive T cells responsible for the GVL effect target a broad range of allogeneic antigens and may thereby lead to GVHD. Hence, there is much interest in developing strategies that can direct the immune reaction towards specific antigens only or primarily expressed on malignant cells, so-called TAA.

As key regulators of the immune system, DC are inherently capable of inducing tumor-specific immune responses (Steinman & Banchereau, 2007). Various clinical studies have already explored the use of DC loaded with TAA as cellular cancer vaccines for hematological malignancies (Smits et al., 2011; Van de Velde et al., 2008). Thus far, results are often modest, but there is proof of principle that a DC vaccine can lead to eradication of malignant cells in an antigen-specific manner. Promisingly, in a phase I/II study by our group, vaccination with autologous monocyte-derived DC loaded with Wilms' tumor 1 (WT1) protein-encoding mRNA was able to convert partial remission into complete molecular remission in two patients in the absence of any other therapy (Van Tendeloo et al., 2010). These clinical responses were correlated with vaccine-associated increases in WT1-specific CD8+ T cell frequencies.

While DC vaccines are thoroughly being investigated in clinical trials for their capacity to induce tumor-specific immune responses, only few trials addressed their use in the setting of HSCT. In the context of autologous hematopoietic stem cell transplantation (auto-HSCT) for multiple myeloma (MM), a clinical trial in 27 patients suggested a benefit in overall survival of vaccination with autologous idiotype-pulsed APC, given at 4 time points after auto-HSCT, compared to historical controls (Lacy et al., 2009).

In allo-HSCT, the dynamic immunological situation that follows transplantation due to the scollision of donor and host immune system adds complexity to the development of DC-based therapy. Hitherto, it is unclear whether donor- or host-derived DC would be best suited for use in immunotherapy aimed at increasing GVL responses. In this regard, murine models demonstrated that host APC are crucial for GVL reactions and that donor APC, although not strictly necessary, can contribute to the GVL effect (Matte et al., 2004; Reddy et al., 2005). This is similar to what is observed in GVHD, which is not surprising given that both are manifestations of graft-versus-host immunity. Therefore, avoiding aggravation of GVHD is an important concern when developing DC-based strategies aiming to augment GVL immunity after allo-HSCT.

Next to GVHD, another concern regarding DC vaccination to boost GVL responses is its effectiveness when given shortly after allo-HSCT, considering the immunosuppressive state of patients at that time. Murine vaccination studies have shown, however, that tumor lysate-pulsed bone marrow-derived DC administered early after auto- or allo-HSCT can elicit effective anti-tumor immunity (Asavaroengchai et al., 2002; Moyer et al., 2006). Furthermore, DC vaccination around the time of HSCT could have some benefits, such as lower tumor burden, donor T cells that are not tolerant to host antigens and low numbers of host Treg cells (Hashimoto et al., 2011).

A total of 6 patients have been involved in three clinical reports of DC vaccines after allo-HSCT. Donor monocyte-derived DC were used for vaccination, pulsed with recipient tumor cells (Fujii et al., 2001; Tatsugami et al., 2004) or with WT1 peptide (Kitawaki et al., 2008). Only one trial reported clinical, but transient responses in 4 relapsed patients with hematological malignancies in the absence of GVHD (Fujii et al., 2001). There were no detectable responses nor GVHD in the other two cases (renal cell carcinoma and acute myeloid leukemia). In the trial reporting clinical responses, patients were infused both with donor monocyte-derived DC pulsed with irradiated patient tumor cells and with donor T cells primed by these DC, which might both have contributed to the observed responses.

A fourth study involving 20 MM patients investigated DLI and/or host-derived DC vaccination (Levenga et al., 2010). The authors concluded that partial T cell-depleted allo-HSCT can be combined with pre-emptive DLI and recipient monocyte-derived DC vaccination to increase graft-versus-myeloma effects with limited GVHD.

In conclusion, early results of clinical DC vaccination in the context of HSCT are promising with no or limited GVHD, but because of the small study populations and lack of controls, further research is required.

Instead of engineering DC *ex vivo* and then transferring them to patients, another approach is to directly target them *in vivo*. To our knowledge, this approach has not been tested yet in the allo-HSCT setting.

However, in 8 patients with Hodgkin disease, non-Hodgkin lymphoma or advanced-stage breast cancer, auto-HSCT was followed by immunotherapy with fms-like tyrosine kinase receptor-3-ligand (Flt3-L) for 6 weeks (Chen et al., 2005). Flt3-L is a hematopoietic growth factor, essential for the development of DC from progenitor cells. This phase I study demonstrated that vaccination with Flt3-L was safe and well-tolerated, resulting in increased frequencies and absolute numbers of circulating immature DC and their precursors in patients' blood without affecting other mature cell lineages. The expanded DC were mostly pDC and were shown here to enhance T cell activation and NK cell cytotoxicity against tumor cells *in vitro* after Toll-like receptor 9-ligand administration, but are also known to play a role in antiviral immunity and in preventing GVHD (Arpinati et al., 2003). In correspondence with these data, others have also suggested that mobilization of specific DC subsets through Flt3-L administration might be a feasible way to target DC *in vivo* (Eto et al., 2002; Teshima et al., 2002), but more research is needed to unravel the functional diversity of these mobilized DC.

3.2 Dendritic cell-based therapy to restore protective immunity against pathogens

Viral and fungal infections are an important cause of morbidity and mortality in patients following HSCT (Gratwohl et al., 2005). These patients have increased susceptibility for primary infection, reinfection and also reactivation of latent viruses due to hampering of their immune system by two main factors (Smits & Berneman, 2010). Firstly, there is the immunosuppressed state accompanying HSCT, often further increased by medication given to prevent GVHD. Secondly, the intense pre-transplantation chemotherapy conditioning regimen, intended to destroy a large part of blood cells, is believed to eliminate memory T cells. Furthermore, early after HSCT dysfunctional DC lead to severely impaired development of antigen-specific T cells (Safdar, 2006). Considering their central role in innate and adaptive immunity, DC seem the ideal candidate for immunotherapy aimed at bringing about the swift restoration of immunity against pathogens in this particular setting.

With regard to DC-based therapy for antifungal immunity after allo-HSCT, much knowledge was obtained from research by the group of Romani. They showed in murine

models of allo-HSCT that DC discriminate between different fungal morphotypes or their corresponding RNA with regard to maturation, cytokine production and Th1 cell priming both *in vitro* and *in vivo* (Bacci et al., 2002; Bozza et al., 2003; d'Ostiani et al., 2000). Similarly, also human monocyte-derived DC were found to react differently in terms of cytokine production and activation of IFN-γ-producing T cells.
Subcutaneous vaccination of mice with DC pulsed with Candida yeasts or Aspergillus conidia (or transfected with the corresponding RNA) on days 1 and 7 after T cell-depleted allo-BMT dramatically increased the recovery of antifungal resistance to subsequent fungal challenge (Bacci et al., 2002; Bozza et al., 2003).
They also demonstrated that Flt3-L-expanded and thymosin α1-treated IL-4-expanded monocyte-derived DC were capable of inducing antifungal immunity as well as allogeneic transplant tolerance (Romani et al., 2006). Overall, the findings of the group of Romani suggest a role for active DC vaccination very shortly after allo-HSCT to restore antifungal immunity and show that expansion of distinct DC might allow more specific regulation of post-transplantation immunity (Montagnoli et al., 2008; Perruccio et al., 2004).
Over the last 10 years, DC have established a firm foothold in immune-based strategies aimed at restoring antiviral (and especially anti-CMV) immunity following allo-HSCT. Monocyte-derived DC from CMV-seropositive HSCT donors pulsed with CMV peptide/lysate or transfected with an adenoviral vector encoding CMV-peptide, have been used with great success to expand CMV-specific cytotoxic T lymphocytes (CTL) *ex vivo* (Micklethwaite et al., 2008; Peggs et al., 2001; Szmania et al., 2001). Clinical trials examining adoptive transfer of these DC-expanded CMV-specific CTL to allo-HSCT recipients demonstrated that this is a safe method capable of restoring functional anti-CMV immunity early after transplantation (Micklethwaite et al., 2007, 2008; Peggs et al., 2009). Although a minority of the patients developed GVHD after adoptive transfer of CMV-specific CTL, this was most likely not related to the infusion itself.
Another study showed that DC transfected with CMV pp65-encoding RNA can successfully expand autologous CMV-specific CTL *in vitro* from both seropositive and -negative patients after allo-HSCT, suggesting that CMV-loaded DC vaccination could provide a valid clinical alternative to adoptive CTL transfer (Heine et al., 2006).
Also for measles virus (MV), DC vaccination could be a favorable approach as results of an *in vitro* study with MV-loaded DC from HSCT patients showed that these DC significantly induced autologous MV-specific T cells from the naïve repertoire (Nashida et al., 2006). Clinical trials are needed, however, to validate whether viral antigen-loaded DC vaccination can indeed live up to the promising results obtained with adoptive virus-specific CTL transfer.

4. Conclusion

DC have been the subject of intensive investigation in mouse models to reduce the occurrence of GVHD and enhance GVL reactions following allo-HSCT. Also in humans, it is clear that DC play an important role in initiating and balancing graft-versus-host reactions. Further clarification of differences between DC subsets in their capacity to shift the balance away from GVHD towards GVL and anti-microbial reactions will help to translate the promising mouse data into clinical success. Questions to be solved are which would be the best time frame and strategy of immunotherapy to use in allotransplant patients. DC-based approaches to be further investigated include DC vaccines, adoptive transfer of *in vitro* primed T cells and *in vivo* targeting of DC.

5. Acknowledgment

E.S. is postdoctoral researcher of the Research Foundation Flanders (FWO-Vlaanderen). This work was supported in part by research grants of the FWO-Vlaanderen (G.0082.08), the Belgian Foundation against Cancer, the Vlaamse Liga tegen Kanker, the National Cancer Plan Action 29, the Agency for Innovation by Science and Technology (IWT-TBM) and the Methusalem program of the Flemish Government.

6. References

Arpinati, M.; Chirumbolo, G.; Urbini, B.; Perrone, G.; Rondelli, D. & Anasetti, C. (2003). Role of plasmacytoid dendritic cells in immunity and tolerance after allogeneic hematopoietic stem cell transplantation. *Transplant Immunology,* Vol.11, No.3-4, (July-September 2003), pp. 345-356, ISSN 0966-3274

Asavaroengchai, W.; Kotera, Y. & Mulé, J.J. (2002). Tumor lysate-pulsed dendritic cells can elicit an effective antitumor immune response during early lymphoid recovery. *Proceedings of the National Academy of Sciences of the United States of America,* Vol.99, No.2, (January 2002), pp. 931-936, ISSN 0027-8424

Bacci, A.; Montagnoli, C.; Perruccio, K.; Bozza, S.; Gaziano, R.; Pitzurra, L.; Velardi, A.; d'Ostiani, CF.; Cutler, J.E. & Romani, L. (2002). Dendritic cells pulsed with fungal RNA induce protective immunity to Candida albicans in hematopoietic transplantation. *Journal of Immunology,* Vol.168, No.6, (March 2002), pp. 2904-2913, ISSN 0022-1767

Ball, L.M. & Egeler, R.M. (2008). Acute GvHD: pathogenesis and classification. *Bone Marrow Transplantation,* Vol.41, Suppl.2, (June 2008), pp. 58-64, ISSN 0268-3369

Barton-Burke, M.; Dwinell, D.M.; Kafkas, L.; Lavalley, C.; Sands, H.; Proctor, C. & Johnson E. (2008). Graft-versus-host disease: a complex long-term side effect of hematopoietic stem cell transplant. *Oncology Williston Park NY,* Vol.22, No.11, (October 2008), pp. 31-45, ISSN 0890-9091

Billingham, R.E. (1966). The biology of graft-versus-host reactions. *Harvey Lectures 1966-67,* Vol.62, pp. 21-78, ISSN 0073-0874

Boeck, S.; Hamann, M.; Pihusch, V.; Heller, T.; Diem, H.; Rolf, B.; Pihusch, R.; Kolb, H.J. & Pihusch, M. (2006). Kinetics of dendritic cell chimerism and T cell chimerism in allogeneic stem cell recipients. *Bone Marrow Transplantation,* Vol.37, No.1, (January 2006), pp. 57-64, ISSN 0268-3369

Bozza, S.; Perruccio, K.; Montagnoli, C.; Gaziano, R.; Bellocchio, S.; Burchielli, E.; Nkwanyuo, G.; Pitzurra, L.; Velardi, A. & Romani, L. (2003). A dendritic cell vaccine against invasive aspergillosis in allogeneic hematopoietic transplantation. *Blood,* Vol.102, No.10, (November 2003), pp. 3807-3814, ISSN 0006-4971

Chakraverty, R.; Eom, H.S.; Sachs, J.; Buchli, J.; Cotter, P.; Hsu, R.; Zhao, G. & Sykes, M. (2006). Host MHC class II+ antigen-presenting cells and CD4 cells are required for CD8-mediated graft-versus-leukemia responses following delayed donor leukocyte infusions. *Blood,* Vol.108, No.6, (September 2006), pp. 2106-2113, ISSN 0006-4971

Chan, G.W.; Gorgun, G.; Miller, K.B. & Foss, F.M. (2003). Persistence of host dendritic cells after transplantation is associated with graft-versus-host disease. *Biology of Blood and Marrow Transplantation*, Vol.9, No.3, (March 2003), pp. 170-176, ISSN 1083-8791

Chen, W.; Chan, A.S.; Dawson, A.J.; Liang, X.; Blazar, B.R. & Miller, J.S. (2005). FLT3 ligand administration after hematopoietic cell transplantation increases circulating dendritic cell precursors that can be activated by CpG oligodeoxynucleotides to enhance T-cell and natural killer cell function. *Biology of Blood and Marrow Transplantation,* Vol.11, No.1, (January 2005), pp. 23-34, ISSN 1083-8791

Choi, S.W.; Levine, J.E. & Ferrara, J.L. (2010). Pathogenesis and management of graft-versus-host disease. *Immunology and Allergy Clinics of North America,* Vol.30, No.1, (February 2010), pp. 75-101, ISSN 0889-8561

Chorny, A.; Gonzalez-Rey, E.; Fernandez-Martin, A.; Ganea, D. & Delgado, M. (2006). Vasoactive intestinal peptide induces regulatory dendritic cells that prevent acute graft-versus-host disease while maintaining the graft-versus-tumor response. *Blood,* Vol.107, No.9, (May 2006), pp. 3787-3794, ISSN 0006-4971

Clark, F.J. & Chakraverty R. (2002). Role of dendritic cells in graft-versus-host disease. *Journal of Hematotherapy & Stem Cell Research,* Vol.11, No.4, (August 2002), pp. 601-616, ISSN 1061-6128

Clark, F.J.; Freeman, L.; Dzionek, A.; Schmitz, J.; McMullan, D.; Simpson, P.; Mason, J.; Mahedra, P.; Craddock, C.; Griffiths, M.; Moss, P.A. & Chakraverty, R. (2003). Origin and subset distribution of peripheral blood dendritic cells in patients with chronisch graft-versus-host disease. *Transplantation,* Vol.75, No.2, (January 2003), pp. 221-225, ISSN 0041-1337

d'Ostiani, C.F.; Del Sero, G.; Bacci, A.; Montagnoli, C.; Spreca, A.; Mencacci, A.; Ricciardi-Castagnoli, P. & Romani, L. (2000). Dendritic cells discriminate between yeasts and hyphae of the fungus Candida albicans. Implications for initiation of T helper cell immunity in vitro and in vivo. *The Journal of Experimental Medicine,* Vol.191, No.10, (May 2000), pp. 1661-1674, ISSN 0022-1007

Duffner, U.A.; Maeda, Y.; Cooke, K.R.; Reddy, P.; Ordemann, R.; Liu, C.; Ferrara, J.L. & Teshima, T. (2004). Host dendritic cells alone are sufficient to initiate acute graft-versus-host disease. *The Journal of Immunology,* Vol.172, No.12, (June 2004), pp. 7393-7398, ISSN 0022-1767

Eto, M.; Hackstein, H.; Kaneko, K.; Nomoto, K. & Thomson, A.W. (2002). Promotion of skin graft tolerance across MHC barriers by mobilization of dendritic cells in donor hemopoietic cell infusions. *The Journal of Immunology,* Vol.169, No.5, (September 2002), pp. 2390-2396, ISSN 0022-1767

Falkenburg, J.H.; Marijt, W.A.; Heemskerk, M.H. & Willemze, R. (2002). Minor histocompatibility antigens as targets of graft-versus-leukemia reactions. *Current Opinion in Hematology,* Vol.9, No.6, (November 2002), pp. 497-502, ISSN 1065-6251

Falkenburg, J.H.; van de Corput, L.; Marijt, E.W. & Willemze, R. (2003). Minor histocompatibility antigens in human stem cell transplantation. *Experimental Hematology,* Vol.31, No.9, (September 2003), pp. 743-751, ISSN 0301-472X

Ferrara, J.L. & Reddy, P. (2006). Pathophysiology of graft-versus-host disease. *Seminars in Hematology,* Vol.43, No.1, (January 2006), pp. 3-10, ISSN 0037-1963

Ferrara, J.L.; Levine, J.E.; Reddy, P. & Holler E. (2009). Graft-versus-host disease. *The Lancet,* Vol.373, No.9674, (May 2009), pp. 1550-1561, ISSN 0140-6736

Fujii, S.; Shimizu, K.; Fujimoto, K.; Kiyokawa, T.; Tsukamoto, A.; Sanada, I. & Kawano, F. (2001). Treatment of post-transplanted, relapsed patients with hematological malignancies by infusion of HLA-matched, allogeneic-dendritic cells (DCs) pulsed with irradiated tumor cells and primed T cells. *Leukemia & Lymphoma,* Vol.42, No.3, (July 2001), pp. 357-369, ISSN 1026-8022

Gill, S.; Olson, J.A. & Negrin, R.S. (2009). Natural killer cells in allogeneic transplantation: effect on engrafment, graft-versus-tumor, and graft-versus-host responses. *Biology of Blood and Marrow Transplantation,* Vol.15, No.7, (July 2009), pp. 765-776, ISSN 1083-8791

Goker, H.; Haznedaroglu, I.C. & Chao, N.J. (2001). Acute graft-vs-host disease: pathobiology and management. *Experimental Hematology,* Vol.29, No.3, (March 2001), pp. 259-277, ISSN 0301-472X

Gratwohl, A.; Brand, R.; Frassoni, F.; Rocha, V.; Niederwieser, D.; Reusser, P.; Einsele, H. & Cordonnier, C. (2005). Cause of death after allogeneic haematopoietic stem cell transplantation (HSCT) in early leukaemias: an EBMT analysis of lethal infectious complications and changes over calendar time. *Bone Marrow Transplantation,* Vol.36, No.9, (November 2005), pp. 757-769, ISSN 0268-3369

Gratwohl, A.; Baldomero, H.; Aljurf, M.; Pasquini, M.C.; Bouzas, L.F.; Yoshimi, A.; Szer, J.; Lipton, J.; Schwendener, A.; Gratwohl, M.; Frauendorfer, K.; Niederwieser, D.; Horowitz, M. & Kodera, Y. (2010). Hematopoietic stem cell transplantation: a global perspective. *Journal of the American Medical Association,* Vol.303, No.16, (April 2010), pp. 1617-1624, ISSN 0098-7484

Hashimoto, D. & Merad, M. (2011). Harnessing dendritic cells to improve allogeneic hematopoietic cell transplantation outcome. *Seminars in Immunology,* Vol.23, No.1, (February 2011), pp. 50-57, ISSN 1044-5323

Heine, A.; Grünebach, F.; Holderried, T.; Appel, S.; Weck, M.M.; Dörfel, D.; Sinzger, C. & Brossart, P. (2006). Transfection of dendritic cells with in vitro-transcribed CMV RNA induces polyclonal CD8+- and CD4+-mediated CMV-specific T cell responses. *Molecular Therapy,* Vol.13, No.2, (February 2006), pp. 280-288, ISSN 1525-0016

Horváth, R.; Budinský, V.; Kayserová, J.; Kalina, T.; Formánková, R.; Starý, J.; Bartůňková, J.; Sedláček, P. & Špíšek, R. (2009). Kinetics of dendritic cells reconstitution and costimulatory molecules expression after myeloablative allogeneic haematopoetic stem cell transplantation: implications for the development of acute graft-versus-host disease. *Clinical Immunology,* Vol.131, No.1, (April 2009), pp. 60-69, ISSN 1521-6616

Kitawaki, T.; Kadowaki, N.; Kondo, T.; Ishikawa, T.; Ichinohe, T.; Teramukai, S.; Fukushima, M.; Kasai, Y.; Maekawa, T. & Uchiyama, T. (2008). Potential of dendritic-cell immunotherapy for relapse after allogeneic hematopoietic stem cell transplantation, shown by WT1 peptide- and keyhole-limpet-hemocyanin-pulsed, donor-derived dendritic-cell vaccine for acute myeloid leukemia. *American Journal of Hematology,* Vol.83, No.4, (April 2008), pp. 315-317, ISSN 0361-8609

Lacy, M.Q.; Mandrekar, S.; Dispenzieri, A.; Hayman, S.; Kumar, S.; Buadi, F.; Dingli, D.; Litzow, M.; Wettstein, P.; Padley, D.; Kabat, B.; Gastineau, D.; Rajkumar, S.V. & Gertz, M.A. (2009). Idiotype-pulsed antigen-presenting cells following autologous transplantation for multiple myeloma may be associated with prolonged survival.

American Journal of Hematology, Vol.84, No.12, (December 2009), pp. 799-802, ISSN 0361-8609

Lau, J.; Sartor, M.; Bradstock, K.F.; Vuckovic S.; Munster, D.J. & Hart, D.N. (2007). Activated circulating dendritic cells after hematopoietic stem cell transplantation predict acute graft-verus-host disease. *Transplantation*, Vol.83, No.7, (April 2007), pp. 839-846, ISSN 0041-1337

Levenga, H.; Schaap, N.; Maas, F.; Esendam, B.; Fredrix, H.; Greupink-Draaisma, A.; de Witte, T.; Dolstra, H. & Raymakers, R. (2010). Partial T cell-depleted allogeneic stem cell transplantation following reduced-intensity conditioning creates a platform for immunotherapy with donor lymphocyte infusion and recipient dendritic cell vaccination in multiple myeloma. *Biology of Blood and Marrow Transplantation*, Vol.16, No.3, (March 2010), pp. 320-332, ISSN 1083-8791

[a]Li, J.M.; Giver, C.R.; Lu, Y.; Hossain, M.S.; Akhtari, M. & Waller, E.K. (2009). Separating graft-versus-leukemia from graft-versus-host disease in allogeneic hematopoietic stem cell transplantation. *Immunotherapy*, Vol.1, No.4, (July 2009), pp. 599-621, ISSN 1750-743X

[b]Li, J.M.; Southerland, L.T.; Lu, Y.; Darlak, K.A.; Giver, C.R.; McMillin, D.W.; Harris, W.A.; Jaye, D.L. & Waller, E.K. (2009). Activation, immune polarization, and graft-versus-leukemia activity of donor T cells are regulated by specific subsets of donor bone marrow antigen-presenting cells in allogeneic hemopoietic stem cell transplantation. *Journal of Immunology*, Vol.183, No.12, (December 2009), pp. 7799-7809, ISSN 0022-1767

Liu, Y.J. (2001). Dendritic cell subsets and lineages, and their functions in innate and adaptive immunity. *Cell*, Vol.106, No.3, (August 2001), pp. 259-262, ISSN 0092-8674

Mackinnon, S.; Papadopoulos, E.B.; Carabasi, M.H.; Reich, L.; Collins, N.H.; Boulad, F.; Castro-Malaspina, H.; Childs, B.H.; Gillio, A.P.; Kernan, N.A.; Small, T.N.; Young, J.W. & O'Reilly, R.J. (1995). Adoptive immunotherapy evaluating escalating doses of donor leukocytes for relapse of chronic myeloid leukemia after bone marrow transplantation: separation of graft-versus-leukemia responses from graft-versus-host disease. *Blood*, Vol.86, No.4, (August 1995), pp. 1261-8, ISSN 0006-4971

Mapara, M.Y.; Kim, Y.M.; Wang, S.P.; Bronson, R.; Sachs, D.H. & Sykes, M. (2002). Donor lymphocyte infusions mediate superior graft-versus-leukemia effects in mixed compared to fully allogeneic chimeras: a critical role for host antigen-presenting cells. *Blood*, Vol.100, No.5, (September 2002), pp. 1903-1909, ISSN 0006-4971

Matte, C.C.; Liu, J.; Cormier, J.; Anderson, B.E.; Athanasiadis, I.; Jain, D.; McNiff, J. & Shlomchik, W.D. (2004). Donor APCs are required for maximal GVHD but not for GVL. *Nature Medicine*, Vol.10, No.9, (September 2004), pp. 987-992, ISSN 1078-8956

Mellman, I. & Steinman, R.M. (2001). Dendritic cells: specialized and regulated antigen processing mahines. *Cell*, Vol.106, No.3, (August 2001), pp. 255-258, ISSN 0092-8674

Micklethwaite, K.; Hansen, A.; Foster, A.; Snape, E.; Antonenas, V.; Sartor, M.; Shaw, P.; Bradstock, K. & Gottlieb, D. (2007). Ex vivo expansion and prophylactic infusion of CMV-pp65 peptide-specific cytotoxic T-lymphocytes following allogeneic hematopoietic stem cell transplantation. *Biology of Blood and Marrow Transplantation*, Vol.13, No.6, (June 2007), pp. 707-714, ISSN 1083-8791

Micklethwaite, K.P.; Clancy, L.; Sandher, U.; Hansen, A.M.; Blyth, E.; Antonenas, V.; Sartor, M.M.; Bradstock, K.F. & Gottlieb, D.J. (2008). Prophylactic infusion of cytomegalovirus-specific cytotoxic T lymphocytes stimulated with Ad5f35pp65 gene-modified dendritic cells after allogeneic hemopoietic stem cell transplantation. *Blood,* Vol.112, No.10, (November 2008), pp. 3974-3981, ISSN 0006-4971

Mohty, M. (2007). Dendritic cells and acute graft-versus-host disease after allogeneic stem cell transplantation. *Leukemia & Lymphoma,* Vol.48, No.9, (September 2007), pp. 1696-1701, ISSN 1042-8194

Mohty, M. & Gaugler, B. (2008). Inflammatory cytokines and dendritic cells in acute graft-versus-host disease after allogeneic stem cell transplantation. *Cytokine & Growth Factor Reviews,* Vol.19, No.1, (February 2008), pp. 53-63, ISSN 1359-6101

Molldrem, J.J.; Komanduri, K. & Wieder, E. (2002). Overexpressed differentiation antigens as targets of graft-versus-leukemia reactions. *Current Opinion in Haematology,* Vol.9, No.6, (November 2002), pp. 503-508, ISSN 1065-6251

Montagnoli, C.; Perruccio, K.; Bozza, S.; Bonifazi, P.; Zelante, T.; De Luca, A.; Moretti, S.; D'Angelo, C.; Bistoni, F.; Martelli, M.; Aversa, F.; Velardi, A. & Romani, L. (2008). Provision of antifungal immunity and concomitant alloantigen tolerization by conditioned dendritic cells in experimental hematopoietic transplantation. *Blood cells, Molecules & Diseases,* Vol.40, No.1, (January-February 2008), pp. 55-62, ISSN 1079-9796

Moyer, J.S.; Maine, G. & Mulé, J.J. (2006). Early vaccination with tumor-lysate-pulsed dendritic cells after allogeneic bone marrow transplantation has antitumor effects. *Biology of Blood and Marrow Transplantation,* Vol.12, No.10, (October 2006), pp. 1010-1019, ISSN 1083-8791

Nashida, Y.; Kumamoto, T.; Azuma, E.; Hirayama, M.; Araki, M.; Yamada, H.; Dida, F.; Iwamoto, S.; Tamaki, S.; Ido, M.; Ihara, T. & Komada, Y. (2006). Development of a dendritic cell vaccine against measles for patients following hematopoietic cell transplantation. *Transplantation,* Vol.82, No.8, (October 2006), pp. 1104-1107, ISSN 0041-1337

Peggs, K.; Verfuerth, S. & Mackinnon, S. (2001). Induction of cytomegalovirus (CMV)-specific T-cell responses using dendritic cells pulsed with CMV antigen: a novel culture system free of live CMV virions. *Blood,* Vol.97, No.4, (February 2001), pp. 994-1000, ISSN 0006-4971

Peggs, K.S.; Verfuerth, S.; Pizzey, A.; Chow, S.L.; Thomson, K. & Mackinnon, S. (2009). Cytomegalovirus-specific T cell immunotherapy promotes restoration of durable functional antiviral immunity following allogeneic stem cell transplantation. *Clinical Infectious Diseases,* Vol.49, No.12, (December 2009), pp. 1851-1860, ISSN 1058-4838

Perruccio, K.; Bozza, S.; Montagnoli, C.; Bellocchio, S.; Aversa, F.; Martelli, M.; Bistoni, F.; Velardi, A. & Romani, L. (2004). Prospects for dendritic cell vaccination against fungal infections in hematopoietic transplantation. *Blood cells, Molecules & Diseases,* Vol.33, No.3, (November-December 2004), pp. 248-255, ISSN 1079-9796

Pihusch, M.; Boeck, S.; Hamann, M.; Pihusch, V.; Heller, T.; Diem, H.; Rolf, B.; Pihusch, R.; Andreesen, R.; Holler, E. & Kolb, H.J. (2005). Peripheral dendritic cell chimerism in allogeneic stem cell recipients. *Transplantation,* Vol.80, No.6, (September 2005), pp. 843-849, ISSN 0041-1337

Rajasekar, R.; Mathews, V.; Lakshmi, K.M.; Sellathamby, S.; George, B.; Viswabandya, A.; Daniel, D.; Chandy, M. & Srivastava, A. (2008). Plasmacytoid dendritic cell count on day 28 in HLA-matched related allogeneic peripheral blood stem cell transplant predicts the incidence of acute and chronic GVHD. *Biology of Blood and Marrow Transplantation*, Vol.14, No.3, (March 2008), pp. 344-350, ISSN 1083-8791

Reddy, V.; Iturraspe, J.A.; Tzolas, A.C.; Meier-Kriesche, H.U.; Schold, J. & Wingard, J.R. (2004). Low dendritic cell count after allogeneic hematopoietic stem cell transplantation predicts relapse, death, and acute graft-versus-host disease. *Blood*, Vol.103, No.11, (June 2004), pp. 4330-4335, ISSN 0006-4971

Reddy, P.; Maeda, Y.; Liu, C.; Krijanovski, O.I.; Korngold, R. & Ferrara, J.L. (2005). A crucial role for antigen-presenting cells and alloantigen expression in graft-versus-leukemia responses. *Nature Medicine*, Vol.11, No.11, (November 2005), pp. 1244-1249, ISSN 1078-8956

Reichardt, W.; Dürr, C.; von Elverfeldt, D.; Jüttner, E.; Gerlach, U.V.; Yamada, M.; Smith, B.; Negrin, R.S. & Zeiser, R. (2008). Impact of mammalian target of rapamycin inhibition on lymphoid homing and tolerogenic function of nanoparticle-labeled dendritic cells following allogeneic hematopoietic cell transplantation. *Journal of Immunology*, Vol.181, No.7, (October 2008), pp. 4770-4779, ISSN 0022-1767

Rezvani, A.R. & Storb, R.F. (2008). Separation of graft-vs.-tumor effects from graft-vs.-host disease in allogeneic hematopoietic cell transplantation. *Journal of Autoimmunity*, Vol.30, No.3, (May 2008), pp. 172-179, ISSN 0896-8411

Rezvani, K. & Barrett, A.J. (2008). Characterizing and optimizing immune responses to leukaemia antigens after allogeneic stem cell transplantation. *Best Practice and Research Clinical Haematology*, Vol.21, No.3, (September 2008), pp. 437-453, ISSN 1521-6926

Riddell, S.R.; Murata, M.; Bryant, S. & Warren, E.H. (2002). Minor histocompatibility antigens - targets of graft versus leukemia responses. *International Journal of Hematology*, Vol.76, Suppl.2, (August 2002), pp. 155-161, ISSN 0925-5710

Riddell, S.R.; Berger, C.; Murata, M.; Randolph, S. & Warren, E.H. (2003). The graft versus leukemia response after allogeneic hematopoietic stem cell transplantation. *Blood Reviews*, Vol.17, No.3, (September 2003), pp. 153-162, ISSN 0268-960X

Romani, L.; Bistoni, F.; Perruccio, K.; Montagnoli, C.; Gaziano, R.; Bozza, S.; Bonifazi, P.; Bistoni, G.; Rasi, G.; Velardi, A.; Fallarino, F.; Garaci, E. & Puccetti, P. (2006). Thymosin alpha1 activates dendritic cell tryptophan catabolism and establishes a regulatory environment for balance of inflammation and tolerance. *Blood*, Vol.108, No.7, (October 2006), pp. 2265-2274, ISSN 0006-4971

Rossi, M.; Arpinati, M.; Rondelli, D. & Anasetti, C. (2002). Plasmacytoid dendritic cells: do they have a role in immune responses after hematopoietic cell transplantation? *Human immunology*, Vol.63, No.12, (December 2002), pp. 1194-1200, ISSN 0198-8859

Ruggeri, L.; Mancusi, A.; Burchielli, E.; Aversa, F.; Martelli, M.F. & Velardi, A. (2007). Natural killer cell alloreactivity in allogeneic hematopoietic transplantation. *Current Opinion in Oncology*, Vol.19, No.2, (March 2007), pp. 142-147, ISSN 1040-8746

Safdar, A. (2006). Strategies to enhance immune function in hematopoietic transplantation recipients who have fungal infections. *Bone Marrow Transplantation*, Vol.38, No.5, (September 2006), pp. 327-337, ISSN 0268-3369

Sato, K.; Yamashita, N.; Yamashita, N.; Baba, M. & Matsuyama, T. (2003). Regulatory dendritic cells protect mice from murine acute graft-versus-host disease and leukemia relapse. *Immunity,* Vol.18, No.3, (March 2003), pp. 367-379, ISSN 1074-7613

Shlomchik, W.D.; Couzens, M.S.; Tang, C.B.; McNiff, J.; Robert, M.E.; Liu, J.; Shlomchik M.J. & Emerson, S.G. (1999). Prevention of graft versus host disease by inactivation of host antigen-presenting cells. *Science,* Vol.285, No.5426, (July 1999), pp. 412-415, ISSN 0036-8075

Shlomchik, W.D. (2003). Antigen presentation in graft-vs-host disease. *Experimental Hematology,* Vol.31, No.12, (December 2003), pp. 1187-1197, ISSN 0301-472X

Smits, E.L. & Berneman, Z.N. (2010). Viral infections following allogeneic stem cell transplantation: how to cure the cure? *Leukemia & Lymphoma,* Vol.51, No.6, (June 2010), pp. 965-966, ISSN 1026-8022

Smits, E.L.; Lee, C.; Hardwick, N.; Brooks, S.; Van Tendeloo, V.F.; Orchard, K.; Guinn, B.A. (2011) Clinical evaluation of cellular immunotherapy in acute myeloid leukaemia. *Cancer Immunology Immunotherapy,* Vol.60, No.6, (June 2011), pp. 757-769, ISSN 1432-0851

Steinman, R.M. & Banchereau, J. (2007). Taking dendritic cells into medicine. *Nature,* Vol.449, No.7161, (September 2007), pp. 419-426, ISSN 0028-0836

Szmania, S.; Galloway, A.; Bruorton, M.; Musk, P.; Aubert, G.; Arthur, A.; Pyle, H.; Hensel, N.; Ta, N.; Lamb, L.Jr.; Dodi, T.; Madrigal, A.; Barrett, J.; Henslee-Downey, J. & van Rhee, F. (2001). Isolation and expansion of cytomegalovirus-specific cytotoxic T lymphocytes to clinical scale from a single blood draw using dendritic cells and HLA-tetramers. *Blood,* Vol.98, No.3, (August 2001), pp. 505-512, ISSN 0006-4971

Tatsugami, K.; Eto, M.; Harano, M.; Nagafuji, K.; Omoto, K.; Katano, M.; Harada, M. & Naito, S. (2004). Dendritic-cell therapy after non-myeloablative stem-cell transplantation for renal-cell carcinoma. *The Lancet Oncology,* Vol.5, No.12, (December 2004), pp. 750-752, ISSN 1470-2045

Teshima, T.; Reddy, P.; Lowler, K.P.; KuKuruga, M.A.; Liu, C.; Cooke, K.R. & Ferrara, J.L. (2002). Flt3 ligand therapy for recipients of allogeneic bone marrow transplants expands host CD8 alpha(+) dendritic cells and reduces experimental acute graft-versus-host disease. *Blood,* Vol.99, No.5, (March 2002), pp. 1825-1832, ISSN 0006-4971

Vakkila, J.; Thomson, A.W.; Hovi, L.; Vettenranta, K. & Saarinen-Pihkala, U.M. (2005). Circulating dendritic cell subset levels after allogeneic stem cell transplantation in children correlate with time post transplant and severity of acute graft-versus-host disease. *Bone Marrow Transplantation,* Vol.35, No.5, (March 2005), pp. 501-507, ISSN 0268-3369

Van de Velde, A.; Berneman, Z.N.; Van Tendeloo, V.F. (2008) Immunotherapy of hematological malignancies using dendritic cells. *Bulletin du Cancer,* Vol.95, No.3, (March 2008), pp. 320-326, ISSN 0007-4551

Van Tendeloo, V.F.; Van de Velde, A.; Van Driessche, A.; Cools, N.; Anguille, S.; Ladell, K.; Gostick, E.; Vermeulen, K.; Pieters, K.; Nijs, G.; Stein, B.; Smits, E.L.; Schroyens, W.A.; Gadisseur, A.P.; Vrelust, I.; Jorens, P.G.; Goossens, H.; de Vries, I.J.; Price, D.A.; Oji, Y.; Oka, Y.; Sugiyama, H. & Berneman, Z.N. (2010). Induction of complete and molecular remissions in acute myeloid leukemia by Wilms' tumor 1 antigen-targeted dendritic cell vaccination. *Proceedings of the National Academy of Sciences of the United States of America,* Vol.107, No.31, (August 2010), pp. 13824-13829, ISSN 0027-8424

Waller, E.K.; Rosenthal, H.; Jones, T.W.; Peel, J.; Lonial, S.; Langston, A.; Redei, I.; Jurickova, I.; Boyer, M.W. (2001). Larger numbers of CD4(bright) dendritic cells in donor bone marrow are associated with increased relapse after allogeneic bone marrow transplantation. *Blood,* Vol.97, No.10, (May 2001), pp. 2948-2956, ISSN 0006-4971

Wilson, J.; Cullup, H.; Lourie, R.; Sheng, Y.; Palkova, A.; Radford, K.J.; Dickinson, A.M.; Rice, A.M.; Hart, D.N. & Munster, D.J. (2009). Antibody to the dendritic cell surface activation antigen CD83 prevents acute graft-versus-host disease. *The Journal of Experimental Medicine,* Vol.206, No.2, (February 2009), pp. 387-398, ISSN 0022-1007

Xu, J.; Zhou, T. & Zhang, Y. (2008). Role of dendritic cells and chemokines in acute graft-versus-host disease. *Frontiers in Bioscience,* Vol.13, (January 2008), pp. 2065-2074, ISSN 1093-9946

Zhang, Y.; Shlomchik, W.D.; Joe, G.; Louboutin, J.P.; Zhu, J.; Rivera, A.; Giannola, D. & Emerson, S.G. (2002). APCs in the liver and spleen recruit activated allogeneic CD8+ T cells to elicit hepatic graft-versus-host disease. *The Journal of Immunology,* Vol.169, No.12, (December 2002), pp. 7111-7118, ISSN 0022-1767

8

Mesenchymal Stem Cells as Immunomodulators in Transplantation

Nadia Zghoul[1], Mahmoud Aljurf[2] and Said Dermime[1]
[1]*Immunology & Innovative Cell Therapy Unit, Department of Biomedical Research, Dasman Diabetes Institute*
[2]*Adult Haematology/Oncology, King Faisal Specialist Hospital and Research Centre,*
[1]*Kuwait*
[2]*Kingdom of Saudi Arabia*

1. Introduction

In recent years it has become evident that mesenchymal stem cells (MSCs), also termed mesenchymal stromal cells, have potent immunomodulatory effects in addition to their known ability of organ regeneration and recruitment to sites of injured or inflamed tissue (Caplan 1991; Garin, Chu et al. 2007; Nauta and Fibbe 2007; Bianco, Robey et al. 2008; Shi, Hu et al. 2010). While the ability of MSCs to mediate tissue and organ repair replacing damaged tissue has initially been attributed to their multilineage differentiation potential, it is now widely attributed to their ability to home to site of injury secreting cytokines and growth factors that mediate the repair process inducing proliferation and differentiation of progenitor cells (Karp and Leng Teo 2009; Hoogduijn, Popp et al. 2010; Sordi, Melzi et al. 2010). Whether MSCs migrate from the bone marrow in case of injury or if the stem cell niche available in the diseased organ replaces dying cells remains to be elucidated.

MSCs have been demonstrated to exhibit a profound immunomodulatory effect on T cells, B cells, and Natural Killer (NK) cells. This effect has been recently shown to be mediated via soluble factors, and this can be enhanced further if direct cell–cell contact between the MSCs and immune cells is allowed (Ren, Zhao et al. 2010). Via these factors, MSCs inhibit T cell proliferation (Di Nicola, Carlo-Stella et al. 2002; Aggarwal and Pittenger 2005), maturation and differentiation of B cells (Corcione, Benvenuto et al. 2006; Tabera, Perez-Simon et al. 2008; Asari, Itakura et al. 2009), maturation of dendritic cells (DCs) (Nauta, Kruisselbrink et al. 2006), generation of cytotoxic T cells (Angoulvant, Clerc et al. 2004) and proliferation and cytotoxic activity of NK cells (Rasmusson, Ringden et al. 2003; Sotiropoulou, Perez et al. 2006; Spaggiari, Capobianco et al. 2008), while inducing regulatory T cells (Tregs) (Prevosto, Zancolli et al. 2007; Di Ianni, Del Papa et al. 2008; Crop, Baan et al. 2010).

MSCs have been shown to induce immunologic peripheral tolerance, suggesting their potential application in a therapeutic approach for immune mediated disorders. In this context, MSCs can be used to support the function of standard pharmacological immunosuppressants to reduce their dosage or even replace such toxic immunosuppressants promoting long-term survival of the transplanted organ (Crop, Baan et al. 2009).

Limited information is available about the molecular mechanisms responsible for the immunomodulation By MSCs and there is no single mechanism responsible for their observed tolerogenic effect. However, MSCs have been considered to potentially work through multiple mechanisms and have the ability to affect immunological, inflammatory and regenerative pathways supporting or replacing current pharmacological agents.

Considering their immunosuppressive properties in addition to their low inherent immunogenicity (Rasmusson, Uhlin et al. 2007) makes MSCs an attractive treatment option in cell and organ transplantation potentially improving the graft outcome and eliminating a long immunosuppressive treatment regimen (Ryan, Barry et al. 2005; Eggenhofer, Steinmann et al. 2011). Both the immunosuppressive effects of MSCs and their regenerative potential participate to facilitate grafting of a transplanted organ as well as repair and regeneration of the organ after transplantation.

In both allogeneic hematopoietic stem cell transplantation (AHSCT) and organ transplantation setting, major problems exist due to the lack of suitable donors. High histo-incompatability between donor and recipient is often associated with an increased risk of graft rejection or graft versus host disease which MSCs might ameliorate if infusion of MSCs along with the organ transplant increases organ engraftment making this immunoprivilege useful for transplantation. Indeed, in one of the first in vivo studies showing the advantageous immunosuppressive effect of MSCs, allogeneic MSCs were demonstrated to prolong (MHC)-mismatched skin allograft survival in baboons (Bartholomew, Sturgeon et al. 2002).

Furthermore, the immunosuppressive properties of MSCs, in addition to their low immunogenicity features, have prompted researchers to investigate co-transplantation of these cells in AHSCT setting to promote hematopoietic stem cells (HSCs) engraftment and to prevent graft versus host reactions as well as host versus graft reactivity. Additionally, MSCs provide support for the growth and development of HSCs further promoting engraftment. In this, MSCs were shown to interact with HSCs through production of growth factors that influence HSCs-homing and –differentiation (Krampera, Cosmi et al. 2006). This effect was attributed via either cell-cell contact or production of soluble factors by MSCs such as the CXCL12 chemokine which may attract HSCs through its interaction with the CXCR4 ligand (Krampera, Cosmi et al. 2006) hence improving HSCs engraftment.

First clinical trials have been undertaken to assess the safety of MSCs administration as well as a potential treatment option for graft versus host disease (GvHD). Encouraging results have been obtained in patients with steroid resistant GvHD and in the management of chronic GvHD after AHSCT. Interestingly, MSCs treatment improved the overall outcome and successfully attenuated GvHD in these patients (Le Blanc, Rasmusson et al. 2004; Ringden and Le Blanc 2011).

MSCs were shown to attenuate graft rejection and in combination with immunosuppressive therapy were able to prolong cardiac allograft survival when co-administered with immunosuppressive therapy. Indeed, they promote donor-specific graft tolerance and ameliorate the alloimmune response where the use of low dose therapy alone was not sufficient to maintain the graft; only combination therapy with MSCs maintained the cardio graft (Ge, Jiang et al. 2009).

MSCs have also been suggested as a promising cell immunotherapy tool to promote tolerance for organ transplants and to control allograft rejection in post-transplant therapy settings facilitating both transplant acceptance and physiologic functions (Crop, Baan et al. 2009; Dahlke, Hoogduijn et al. 2009). MSCs have also been recently shown to have an ameliorating effect in a model of acute lung injury were transplantation of MSCs resulted in a significant increase in the level of protective/immunomodulatory Tregs (Sun, Han et al. 2011). Finally, clinical trials are ongoing to determine the efficacy of MSCs as an effective tool for the treatment of autoimmune diseases, such as multiple sclerosis, inflammatory bowel diseases, rheumatoid arthritis, and type 1 diabetes.

2. The biology and originality of mesenchymal stem cells and their generation and expansion in vitro for therapeutic use

MSCs are a self-renewing heterogeneous population of multipotent cells, originally isolated from bone marrow as shown in pioneer experiments by Friedenstein and colleagues and first referred to as colony-forming unit fibroblasts (Friedenstein, Chailakhyan et al. 1974). Since then, MSCs have been isolated from many other adult tissues such a umbilical cord blood, placenta, amniotic fluid, peripheral blood, adipose tissue and various other somatic tissues sharing the regenerative as well as the immunomodulatory properties of MSCs (In 't Anker, Scherjon et al. 2004; Zuk 2010; Bianco 2011) but have mainly been characterized after isolation from bone marrow.

MSCs can be expanded in vitro as plastic adherent cells with a fibroblast-like morphology and can be differentiated into cells of mesodermal lineage (osteocytes, chondrocytes and adipocytes) (Pittenger, Mackay et al. 1999) as well as cells from other embryonic lineages (Sanchez-Ramos 2006; Schwartz, Brick et al. 2008). At present, MSCs lack a definitive marker and no single marker has been identified distinguishing MSCs however the International Society for Stem Cell Research has outlined minimal criteria to characterize human MSCs, these cells have been reported to be positive for CD73, CD90, CD105 and major histocompatibility complex class I (HLA-ABC). They are devoid of the hematopoietic markers such as CD14, CD 19, CD34, CD45 and for major histocompatibility complex class II, (HLA-DR) (Dominici, Le Blanc et al. 2006).

An attractive characteristic of MSCs is that they can be easily expanded maintaining a relatively stable phenotype and karyotype along with their potential to differentiate into multiple mesodermal tissues. The possibility that MSCs might undergo malignant transformation does exist, however this might be directly linked to the origin of the tissue. Interestingly, it has been shown that human bone marrow derived MSCs could be expanded in vitro and despite decreased proliferative capacity upon prolonged expansion and eventual cell senescence, no chromosomal abnormalities were detected rendering these cells suitable for cell therapeutic approaches (Bernardo, Zaffaroni et al. 2007).

However, there is a general consensus that MSCs should be used at low passages when applied in cell therapy as chromosomal modifications and loss of function can occur after prolonged in vitro expansion.

3. The Immunomodulatory properties of mesenchymal stem cells and their role in transplantation

Understanding the mechanisms by which MSCs exert their immunomodulatory effects will have profound therapeutic implications in designing new therapies that can lead to more efficient use of these cells in novel treatment regimens. Currently, there is no single clear mechanism which clearly clarifies the immunomodulatory effect of MSCs and several mechanisms which sometimes seem paradoxical, have been suggested. Several in vitro experimental studies have shown that the immunosuppressive effect of MSCs is sustained in transwell experiments suggesting that soluble factors are responsible for such inhibition (Di Nicola, Carlo-Stella et al. 2002; Meisel, Zibert et al. 2004; Aggarwal and Pittenger 2005; Gao, Wu et al. 2008; Ren, Zhang et al. 2008; Selmani, Naji et al. 2008), while other studies have claimed a required cell-cell contact which may be due to the use of different systems and cells by the individual research groups (Quaedackers, Baan et al. 2009; Ren, Zhao et al. 2010).

MSCs have been shown to modulate the immune response mainly by inhibiting the proliferation of effector immune cells preventing further damage to injured tissue allowing repair after injury (Uccelli, Moretta et al. 2008). Moreover, MSCs can stimulate the activation and proliferation of Tregs which in turn have a beneficial immunosuppressive effect (Crop, Baan et al. 2010).

When MSCs home to site of tissue inflammation or injury, they release various growth factors which enhance repair at site of defect including : fibroblast growth factor, epidermal growth factor, platelet-derived growth factor, transforming growth factor-β, vascular endothelial growth factor and insulin-like growth factor (Ng, Boucher et al. 2008). MSCs mediated inhibition of effector T cell proliferation seems to be dependent on the microenvironment. In this respect, the presence of pro-inflammatory cytokines such as IFN-γ activates MSCs and this was more effective for the treatment of GvHD (Polchert, Sobinsky et al. 2008). Indeed, it has been shown, that the immunosuppressive capacity of MSCs was enhanced strongly under inflammatory conditions while their differentiation capacity was preserved and suggested that in vitro preconditioning provides MSCs with improved properties for immediate clinical immune therapy (Crop, Baan et al. 2010).

MSCs have been shown to affect almost all cell types of the immune system. It has been demonstrated that MSCs can alter the cytokine secretion profile of immune cells such as decreasing TNF-α and IFN-γ secretion and increasing the secretion of suppressive cytokines (IL-4 and IL-10) and this shift from a pro-inflammatory to a beneficial anti-inflammatory response is of therapeutic advantage for the management of GvHD (Dahlke, Hoogduijn et al. 2009; Hoogduijn, Popp et al. 2010). Moreover, several studies have demonstrated that the MSCs immunoregulatory properties are partially mediated by transforming growth factor (TGF-β), prostaglandin E2 (PGE2), hepatocyte growth factor (HGF) and Indoleamine 2, 3-dioxgenase (IDO) secreted by MSCs in addition to their induction and activation of Tregs all of which can lead to amplification of an effective immunosuppressive response (English, Ryan et al. 2009; Kim, Wee et al. 2011).

3.1 MSC–soluble factors and immunosuppression induction

Different studies have attributed the immunosuppressive effect of MSCs to several immunosuppressive factors leading to different mechanisms of immune cell inhibition. These include, IDO (Meisel, Zibert et al. 2004; DelaRosa, Lombardo et al. 2009), PGE2 (Aggarwal and Pittenger 2005), TGF-β and HGF (Di Nicola, Carlo-Stella et al. 2002), HLA-G (Selmani, Naji et al. 2008), nitric oxide (Ren, Zhang et al. 2008), interleukin (IL)-10 (Gao, Wu et al. 2008) and haeme oxygenase-1 (Chabannes, Hill et al. 2007). One important mechanism is that MSCs suppression is mediated by IDO, a tryptophan catabolising enzyme (Meisel, Zibert et al. 2004). In this, activated T cells or NK cells produce elevated levels of IFN-γ which in turn stimulates MSCs to produce IDO. IDO metabolizes tryptophan to kynurenine leading to essential tryptophan depletion and accumulation of metabolites in the medium which in turn inhibits proliferation of activated T cells and NK cells (Meisel, Zibert et al. 2004). Competitive inhibition of IDO activity did not completely abrogate MSC mediated immunosuppression and inhibition of IFN-γ was required for complete abrogation (Krampera, Cosmi et al. 2006; DelaRosa, Lombardo et al. 2009). This suggests that co-administration of MSCs with graft T cells-derived IFN-γ activates the immunomodulatory properties of MSCs. Furthermore, there is a species variation in the immunosuppressive mechanisms mediated by MSCs. It has been shown that while human IDO is a major effector molecule for MSCs immunosuppression, mouse MSCs mediate their inhibitory effect of immune responses via nitric oxide playing a central role in such immunosuppression (Sato, Ozaki et al. 2007; Ren, Zhang et al. 2008).

3.2 MSC-immune cell interaction and immunosuppression induction

MSCs have been demonstrated to exhibit a profound immunomodulatory effect on Tregs, cytotoxic and T helper cells, NK cells, B cells and dendritic cells.

3.2.1 Regulatory T cells

Tregs are a subset of T cells that regulate the immune response by suppressing the proliferation and cytokine production of effector T cells. Tregs are thus important for protecting our body by suppressing auto-reactive T cells (Thornton and Shevach 1998; Ng, Duggan et al. 2001). Tregs were shown to be upregulated in the presence of MSCs suggesting that MSCs constitute a suitable niche for Tregs (Crop, Baan et al. 2010). MSCs have been suggested to play a role in Tregs recruitment, regulating and maintaining the T regulatory phenotype and function over time (Prevosto, Zancolli et al. 2007; Di Ianni, Del Papa et al. 2008). Tregs induction has been suggested to be mediated by PGE2, synthesized by cyclooxygenase enzymes (COX) which are expressed by MSCs (Le Blanc and Ringden 2007). Furthermore, Treg induction has been shown to be mediated by direct cell-cell contact between MSCs and $CD4^{+}$ T cells and the presence of soluble MSC derived factors such as TGF-β1 and PGE2 (English, Ryan et al. 2009). In a recent interesting study by (Sundin, D'Arcy et al. 2011) it was demonstrated for the first time that MSCs share features with regulatory T cells , such as the expression of the Tregs specific transcription factor FOXP3 (Yagi, Nomura et al. 2004) at variable levels. However, the MSC immunosuppressive function is not as tightly linked to FOXP3 expression as is the case for Tregs (Sundin, D'Arcy et al. 2011).

3.2.2 Cytotoxic T cells and T helper cells

MSCs were shown to inhibit the proliferation of $CD4^{+}$ T helper cells (Di Nicola, Carlo-Stella et al. 2002; Aggarwal and Pittenger 2005). In addition to indirect inhibitory factors produced by MCSs such as TGF-β1, HGF, IL-10, IFN-γ and TNF-α, there is evidence that cell-membrane interaction between MCSs and T helper cells (Quaedackers, Baan et al. 2009) via the intracellular adhesion molecule (ICAM-1) or vascular cell adhesion molecule (VCAM-1) play a crucial role in such immunosuppression (Ren, Zhao et al. 2010). MSCs have also been shown to suppress the induction of cytotoxic T cell response to allo-antigens (Angoulvant, Clerc et al. 2004). However, once cytotoxic T cells are activated, MSCs show no inhibitory effect (Rasmusson, Ringden et al. 2003; Le Blanc and Ringden 2007). On the other hand, it is not clear how helper and cytotoxic T cells affect MSCs development and function. It has been shown that MSCs have a low immunophenotype as they express low levels of HLA class I, no HLA class II, no co-stimulatory molecules such as CD80 and CD86, and therefore do not induce immune responses (Beggs, Lyubimov et al. 2006; Rasmusson, Uhlin et al. 2007). This should make MSCs transplantable across HLA barriers (Le Blanc, Tammik et al. 2003). However, in contrast to these data, there is evidence that MSCs are immunogenic and can induce memory T-cell responses both in animals (Badillo, Beggs et al. 2007) and human studies (Nauta, Westerhuis et al. 2006). Furthermore, it has recently been reported that MSCs are susceptible for lysis by $CD8^+$ cytotoxic T cells (Crop, Korevaar et al. 2011). Designing tools to escape allogeneic MSCs destruction by cytotoxic T cells should render MSCs a promising therapeutic option for transplantation across MHC barriers.

3.2.3 Natural Killer cells

MSCs have been shown to inhibit NK cell proliferation, and cytotoxicity (Sotiropoulou, Perez et al. 2006). It has been demonstrated that MSCs can mediate this inhibitory effect trough inhibition of cytokine production in addition to the central role played by IDO and PGE2 (Spaggiari, Capobianco et al. 2008). Inversely, it has been shown that MSCs are susceptible for lysis by NK cells (Spaggiari, Capobianco et al. 2006; Crop, Korevaar et al. 2011) as MSCs express the activating NK cell-receptor ligands NKG2D and UL16 (Poggi, Prevosto et al. 2007). Moreover, intravenously administered MSCs have been demonstrated to disappear within days after infusion in immunocompetent mice (Popp, Eggenhofer et al. 2008). It is possible that lysis by cytotoxic T cells (and not NK cells) is responsible for the disappearance of the infused MSCs (Eliopoulos, Stagg et al. 2005). The demonstration of tumour engraftment after administration of autologous MSCs in immunodeficient mice (NK cells intact) further strengthens this possibility.

3.2.4 B cells

MSCs have been shown to suppress B cell terminal differentiation (Asari, Itakura et al. 2009) and modulate their function (Corcione, Benvenuto et al. 2006; Tabera, Perez-Simon et al. 2008). Human MSCs were shown to inhibit antibody production induced in vitro by allo-stimulation (Comoli, Ginevri et al. 2008; English, French et al. 2010). They have the ability to regulate Immunoglobulin production by B cells through soluble factors affecting B cells directly or through an indirect MSCs effect by altering the amount of free alloantigen due to the overall suppression of graft damage (Ge, Jiang et al. 2009).

3.2.5 Dendritic cells

MSCs inhibit monocyte maturation into dendritic cells (Jiang, Zhang et al. 2005), the most potent antigen presenting cells, inhibiting their migration to lymph nodes and thereby reducing their ability to activate allo-reactive T cells (Aggarwal and Pittenger 2005). In this, MSCs were shown to reduce secretion of pro-inflammatory cytokines such as IFN-γ, IL-12 and TNF-α by DCs while IL-10, a suppressive cytokine, was increased leading to the inhibition of DCs maturation and the inability to activate allo-reactive T cells resulting in a state of an immunologic tolerance. The inhibitory effect of MSCs on DC differentiation is mainly mediated through cell-cell contact involving activation of the Notch signalling pathway (Li, Paczesny et al. 2008) as well via soluble factors (Nauta, Kruisselbrink et al. 2006). Moreover, PGE2 produced by MSCs following TNF-α or IFN-γ stimulation blocks differentiation of monocytes into DCs and stimulates macrophages to produce Il-10.

3.3 MSCs-Galectins and immunosuppression induction

Recently, Sioud et al (Sioud, Mobergslien et al. 2010; Sioud 2011)and Gieseke et al (Gieseke, Bohringer et al. 2010) have described another mechanism of MSCs immunosuppression that involves a family of beta galactosidase-binding proteins named Galectins which are involved in immune tolerance (Garin, Chu et al. 2007). It has been demonstrated that Galectin-1, Galectin-3, Galectin-8 and Galectin-9 are constitutively expressed by human bone marrow MSCs with Galectin-1 and Galectin -3 being further secreted and expressed on the outer plasma membrane (Wada and Makino 2001). New evidence has shown the involvement of Galectin-3 as a regulator of MSCs immunosuppression function inhibiting allogeneic T cell proliferation by MSCs (Sioud 2011). This group demonstrated that gene knockdown of Galectin-3 resulted in less immunosuppressive effect on T cell proliferation (Sioud, Mobergslien et al. 2010). Galectins were also shown to regulate the secretion of pro-inflammatory cytokines and promote expansion of IL-10 producing peripheral Tregs (Sioud 2011). Gieseke et al have also reported that galectin-1 is expressed by MSCs and has immunosuppressive activity on both $CD4^+$ and $CD8^+$ T cells (Gieseke, Bohringer et al. 2010).

4. Clinical prospects of mesenchymal stem cells

Currently suggested clinical uses of MSCs include; differentiation and repairing of damaged tissue e.g. in osteogenesis imperfecta, promoting hematopoietic cell engraftment in AHSCT e.g. in leukaemia and their immunosuppressive abilities in autoimmune-induced inflammatory bowel disease, GvHD induced upon allogeneic cell or organ transplantation as well as autoimmune diseases. Treating autoimmunity with MSCs was first investigated for experimental autoimmune encephalomyelitis as a model for multiple sclerosis in preclinical studies followed by collagen-induced arthritis, autoimmune type 1 diabetes, experimental colitis and lupus nephritis (Uccelli and Prockop 2010).

Still, very little is known about the mechanisms underlying the MSCs immunomodulatory effect in vivo and survival of MSCs upon injection and whether they remain present after systemic injection or local transplantation remains unresolved. However, the clinical studies performed so far look encouraging but still the mechanisms as to how the MSCs regulate immune cells in vivo are still missing. Furthermore, direct methods to assess MSCs mobilization and homing in response to injury or inflammation is essential to eventually understand the underlying mechanism (Karp and Leng Teo 2009).

MSCs are immune-privileged and were shown in most studies to 'escape' the immune system and be tolerated when transplanted across MHC barriers engrafting and failing to induce an immune response. It is important to mention that most of these clinical studies have used MSCs from HLA-identical or near identical donors (Le Blanc and Ringden 2007).

It has been reported that AHSCT recipients have specific tolerance, immune-unresponsiveness, directed towards MSCs but not to other cells from the MSC donor (Sundin, Barrett et al. 2009). This suggests that totally HLA mismatched MSCs from one donor can be expanded ex vivo, cryopreserved and used for the treatment of multiple patients. Furthermore, the inability of MSCs to induce donor specific tolerance suggests that

MSCs co-transplantation with a solid organ most probably facilitates engraftment through the immunosuppressive action of MSCs and not by inducing specific tolerance to the transplant (Sundin, D'Arcy et al. 2011).

4.1 Graft versus Host Disease

GvHD occurs in donated organ recipients and patho-physiologically presents itself as damage in skin, mucosa, gastrointestinal tract, the liver, connective tissue and exocrine glands. It is considered chronic when it persists more than 100 days after transplantation and it is distinguished from acute GvHD which often cannot be ameliorated because of a coexistent immunosuppression. It results from the attack of donor own T cells in the graft on recipient tissues after activation by recipient MHC molecules as well as activation of cytotoxic T cells, antigen presenting cells and NK cells. In GvHD the observation of an increased secretion of pro-inflammatory cytokines such as TNF-α, IFN-γ, IL-1, IL-2 and IL-12 suggests that therapeutic approaches to inhibit these cytokines should lead to a decreased severity of GvHD (Aggarwal and Pittenger 2005).

The immunosuppressive effects of MSCs have been evidenced in a successful clinical trial reported by Ringden et al (Ringden and Le Blanc 2011) in patients presented with acute GvHD after AHSCT and resistant to conventional immunosuppressive therapy. This study showed a dramatic improvement after MSCs infusion with acute GvHD disappearing completely in patients with steroid resistant GvHD suggesting that MSCs could be considered as a promising alternative immunosuppressive therapy with improved long term outcome for the treatment of GvHD (Le Blanc, Rasmusson et al. 2004; English, French et al. 2010; Ringden and Le Blanc 2011; Tolar, Villeneuve et al. 2011).

4.2 Type 1 diabetes

MSCs have been suggested as a prospective cell therapy of autoimmune type 1 diabetes known to be associated with an inflammation in the pancreas and degeneration of β cells. In a preclinical study, it has been shown that co-transplantation of bone marrow cells and syngeneic or allogeneic MSCs initiated endogenous pancreatic regeneration and improved blood glucose levels in streptozotocin induced diabetic mice (Ulicna, Danisovic et al. 2009). The success in the treatment was suggested to be due to MSCs aiding in the regeneration of recipient derived pancreatic beta cells, as well as maintaining the β cell reserve through MSC inhibition of T cell-mediated immune responses against the newly formed β cells preventing further cell degeneration (Voltarelli, Couri et al. 2007; Couri and Voltarelli 2009; Ulicna, Danisovic et al. 2009; Sordi, Melzi et al. 2010). Hence, despite of the persisting hostile autoimmune response in the pancreas, when MSCs are applied at an early stage of the disease they potentially protect and allow the endogenous regeneration of the remaining intact β cell reserve.

4.3 Solid organ transplantation

In solid organ transplantation (e.g. liver transplant), rejection of a transplanted organ is typically caused by induction of T cell proliferation and presentation of allo-antigens to naive and memory T cells by antigen presenting cells (both host and donor) leading to activation and differentiation into effector T cells (Popp, Renner et al. 2009). Both the regenerative and immunomodulatory properties of MSCs are of medical importance in organ transplantation studies. MSCs have been shown to home to site of allograft and help

the body accommodate the new organ through immunosuppression or preventing rejection and acquisition of a state of tolerance (English, French et al. 2010). Furthermore, the regenerative properties of MSCs possibly maintain the organ to be transplanted stretching its lifespan in case of delayed transplantation (Hematti 2008; Hoogduijn, Popp et al. 2010).
For MSCs to be used as a sole therapy or as a combination therapy replacing immunosuppressive drugs, reducing the burden on the patient and the risk for long term side effects, further investigation will be needed to fully explain the immunomodulatory properties of MSCs. However, MSCs were shown to increase immunosuppression when co-administered complementing the therapy (Sundin, D'Arcy et al. 2011). In this regard, an interesting animal study of allogeneic heart graft transplantation that MSCs combined with the immunosuppressive drug mycophenolate mofetil (MMF) promoted the elimination of activated T cells in secondary lymphatic organs, delayed antigen presenting cells activation, and protected the graft from cellular infiltration by modulating the endothelium (Eggenhofer, Steinmann et al. 2011).

5. Quality considerations and regulations in MSC based therapy

Clinical use of MSCs for cellular therapeutic approaches will require that the bio-safety of these cells has to be adequately optimised. This requires the absence of chromosomal, functional, and phenotypical alterations in ex vitro expanded MSCs before considering their injection in patients. The safety of these cells has to be guaranteed upon their administration as immunocompromised patients might have an increased risk of tumour induction or the potential risk of stimulating the growth of a previously undetected cancer. The risk of tumorigenicity and genomic instability remains an obstacle for any given stem-cell based therapy product and will have to be adequately assessed prior to approval for human use (Dittmar, Simann et al. 2010). In this respect, the first clinical trials have been undertaken to assess the safety of MSC administration and potential treatment options for GvHD and it has been reported that autologous and allogeneic MSCs are safe to be injected in patients with life threatening acute GvHD not responding to conventional immunosuppressive therapy without acute toxicity and no signs of ectopic tissue formation (Lazarus, Koc et al. 2005).
MSCs administered for therapy might carry viruses (e.g. herpes simplex virus, cytomegalo virus (CMV), Epstein-Barr Virus (EBV)) transmitted from the donor to the host tissue especially when the patient is immunocompromised. Interestingly, MSCs were shown to exert differential effects on alloantigen and virus-specific T cell responses (Karlsson, Samarasinghe et al. 2008). In this regard, it has been reported that despite MSC infusion as a cellular immunotherapy for GvHD, effector functions of virus specific T cells were retained with very little effect on T cell responses to pathogenic CMV and EBV contrasting the strong immunosuppressive effect on allo-reactive T cells suggesting that MSCs to be a promising cellular immunotherapy (Karlsson, Samarasinghe et al. 2008). This is an important advantageous aspect of MSCs therapy especially since after allogeneic organ transplantation; infections are a major cause of morbidity and mortality in immunocompromised patients. Therefore, MSCs can be safely administered without exacerbating their susceptibility to infectious pathogens.
Currently large clinical trials are being carried out in using MSCs for the treatment of autoimmune diseases including diabetes as well as to treat or ameliorate symptoms of GvHD. Further, clinical trials are also testing the MSCs ability to facilitate the prevention of the rejection upon organ transplantation thereby reducing or completely eliminating the

need for immunosuppressive therapies. Reviewing the clinical trials registered by the U.S. National Institutes of Health (NIH) in the United States and around the world (U.S. National Institutes of Health, http://clinicaltrials.gov/), 168 registered studies (with a start date between September 2004 and March 2011) were found upon entering the search criteria mesenchymal + stem + cells including clinical trials in regenerating organs (e.g. bone fractures, diabetic foot and foot ulcer), liver failure, the administration of MSCs for the treatment of type 1 Diabetes Mellitus as well as Type 2, acute GvHD including studies with patients who have failed to respond to steroid therapy (or with steroid resistance) and poor graft function. In addition to these trials, ongoing clinical studies investigating the safety and efficacy of MSCs promoting engraftment of allogeneic hematopoietic stem cells are in hand (Le Blanc, Rasmusson et al. 2004).

The translation into cellular therapies satisfying safety and efficacy criteria by the regulatory authorities will have to ensure the identity, purity, potency, and lack of tumorigenicity. Questions regarding the expansion of MSCs in vitro and the passage at which they are used as well as the culture conditions containing foetal calf serum, optimal dosing, timing and HLA matching still remain to be answered (Le Blanc, Samuelsson et al. 2007; Sundin, Ringden et al. 2007).

Scientists and clinicians should adhere to local, national and international guidelines and regulations that govern transfer of cells into patients. The clinical trial and the eventual cell therapeutic product is assessed and approved by a national regulatory agency, such as the European Medicines Agency (EMA) or the U.S. Food and Drug Administration (FDA).The International Society for Stem Cell Research (ISSCR) has published guidelines for the clinical translation of stem cells emphasising on the scientific, clinical and ethical issues that should be addressed for a responsible translation of basic stem cell research into suitable clinical applications. The guidelines give attention to the main areas of clinical translational stem cell research namely, cell processing and manufacture, the necessity of preclinical studies and clinical research promoting maximum safety and quality of the cells to be used (Hyun, Lindvall et al. 2008).

6. Conclusion

Understanding the immunomodulatory properties of MSCs and possibly identifying genes that regulate MSC inhibitory function, and genes regulating inflammation which play a major role in transplant rejection and inflammatory processes should help in developing applicable MSC based cellular therapies for solid organ transplantation, GvHD and autoimmune diseases.

The currently available data, in vitro and in vivo, suggest that MSCs can be applied in a wide range of clinical approaches, ranging from tissue repair and regeneration, drug or gene delivery to injured tissue, treatment and prevention of GvHD and AHSCT engraftment offering a promising option for treating autoimmune mediated disorders as well as organ transplantation. Administration of MSCs provides novel modalities for the treatment of patients with allograft rejection with fewer side effects than existing immunosuppressive therapies following organ transplantation.

There is an enormous amount of excitement and the scope of possible stem cell based therapies has expanded in the recent years due to rapid advances in stem cell research. But today the range of diseases where stem cell treatments have been shown to be beneficial in responsibly conducted clinical trials is still extremely restricted.

The best defined and most extensively used is hematopoietic stem cell transplantation in blood malignancies and aplastic anaemia. To design safe and effective cellular therapies, the long term effects of MSCs injection will still need to be shown in relevant clinical trials and further studies are needed to optimize cell-dosing, time of injection and the combination with immunosuppressive drugs to confirm both efficacy and safety of this cell therapy, are still needed within the careful regulations of the EMA and FDA.

7. References

Aggarwal, S. and M. F. Pittenger (2005). "Human mesenchymal stem cells modulate allogeneic immune cell responses." Blood 105(4): 1815-1822.

Angoulvant, D., A. Clerc, et al. (2004). "Human mesenchymal stem cells suppress induction of cytotoxic response to alloantigens." Biorheology 41(3-4): 469-476.

Asari, S., S. Itakura, et al. (2009). "Mesenchymal stem cells suppress B-cell terminal differentiation." Exp Hematol 37(5): 604-615.

Badillo, A. T., K. J. Beggs, et al. (2007). "Murine bone marrow stromal progenitor cells elicit an in vivo cellular and humoral alloimmune response." Biol Blood Marrow Transplant 13(4): 412-422.

Bartholomew, A., C. Sturgeon, et al. (2002). "Mesenchymal stem cells suppress lymphocyte proliferation in vitro and prolong skin graft survival in vivo." Exp Hematol 30(1): 42-48.

Beggs, K. J., A. Lyubimov, et al. (2006). "Immunologic consequences of multiple, high-dose administration of allogeneic mesenchymal stem cells to baboons." Cell Transplant 15(8-9): 711-721.

Bernardo, M. E., N. Zaffaroni, et al. (2007). "Human bone marrow derived mesenchymal stem cells do not undergo transformation after long-term in vitro culture and do not exhibit telomere maintenance mechanisms." Cancer Res 67(19): 9142-9149.

Bianco, P. (2011). "Back to the future: Moving beyond "mesenchymal stem cells"." J Cell Biochem 112(7): 1713-1721.

Bianco, P., P. G. Robey, et al. (2008). "Mesenchymal stem cells: revisiting history, concepts, and assays." Cell Stem Cell 2(4): 313-319.

Caplan, A. I. (1991). "Mesenchymal stem cells." J Orthop Res 9(5): 641-650.

Chabannes, D., M. Hill, et al. (2007). "A role for heme oxygenase-1 in the immunosuppressive effect of adult rat and human mesenchymal stem cells." Blood 110(10): 3691-3694.

Comoli, P., F. Ginevri, et al. (2008). "Human mesenchymal stem cells inhibit antibody production induced in vitro by allostimulation." Nephrol Dial Transplant 23(4): 1196-1202.

Corcione, A., F. Benvenuto, et al. (2006). "Human mesenchymal stem cells modulate B-cell functions." Blood 107(1): 367-372.

Couri, C. E. and J. C. Voltarelli (2009). "Stem cell therapy for type 1 diabetes mellitus: a review of recent clinical trials." Diabetol Metab Syndr 1(1): 19.

Crop, M. J., C. C. Baan, et al. (2009). "Donor-derived mesenchymal stem cells suppress alloreactivity of kidney transplant patients." Transplantation 87(6): 896-906.

Crop, M. J., C. C. Baan, et al. (2010). "Inflammatory conditions affect gene expression and function of human adipose tissue-derived mesenchymal stem cells." Clin Exp Immunol 162(3): 474-486.

Crop, M. J., C. C. Baan, et al. (2010). "Human adipose tissue-derived mesenchymal stem cells induce explosive T-cell proliferation." Stem Cells Dev 19(12): 1843-1853.

Crop, M. J., S. S. Korevaar, et al. (2011). "Human mesenchymal stem cells are susceptible to lysis by CD8+ T-cells and NK cells." Cell Transplant.

Dahlke, M. H., M. Hoogduijn, et al. (2009). "Toward MSC in solid organ transplantation: 2008 position paper of the MISOT study group." Transplantation 88(5): 614-619.

DelaRosa, O., E. Lombardo, et al. (2009). "Requirement of IFN-gamma-mediated indoleamine 2,3-dioxygenase expression in the modulation of lymphocyte proliferation by human adipose-derived stem cells." Tissue Eng Part A 15(10): 2795-2806.

Di Ianni, M., B. Del Papa, et al. (2008). "Mesenchymal cells recruit and regulate T regulatory cells." Exp Hematol 36(3): 309-318.

Di Nicola, M., C. Carlo-Stella, et al. (2002). "Human bone marrow stromal cells suppress T-lymphocyte proliferation induced by cellular or nonspecific mitogenic stimuli." Blood 99(10): 3838-3843.

Dittmar, K. E., M. Simann, et al. (2010). "Quality of Cell Products: Authenticity, Identity, Genomic Stability and Status of Differentiation." Transfus Med Hemother 37(2): 57-64.

Dominici, M., K. Le Blanc, et al. (2006). "Minimal criteria for defining multipotent mesenchymal stromal cells. The International Society for Cellular Therapy position statement." Cytotherapy 8(4): 315-317.

Eggenhofer, E., J. F. Steinmann, et al. (2011). "Mesenchymal stem cells together with mycophenolate mofetil inhibit antigen presenting cell and T cell infiltration into allogeneic heart grafts." Transpl Immunol 24(3): 157-163.

Eliopoulos, N., J. Stagg, et al. (2005). "Allogeneic marrow stromal cells are immune rejected by MHC class I- and class II-mismatched recipient mice." Blood 106(13): 4057-4065.

English, K., A. French, et al. (2010). "Mesenchymal stromal cells: facilitators of successful transplantation?" Cell Stem Cell 7(4): 431-442.

English, K., J. M. Ryan, et al. (2009). "Cell contact, prostaglandin E(2) and transforming growth factor beta 1 play non-redundant roles in human mesenchymal stem cell induction of CD4+CD25(High) forkhead box P3+ regulatory T cells." Clin Exp Immunol 156(1): 149-160.

Friedenstein, A. J., R. K. Chailakhyan, et al. (1974). "Stromal cells responsible for transferring the microenvironment of the hemopoietic tissues. Cloning in vitro and retransplantation in vivo." Transplantation 17(4): 331-340.

Gao, F., D. Q. Wu, et al. (2008). "In vitro cultivation of islet-like cell clusters from human umbilical cord blood-derived mesenchymal stem cells." Transl Res 151(6): 293-302.

Garin, M. I., C. C. Chu, et al. (2007). "Galectin-1: a key effector of regulation mediated by CD4+CD25+ T cells." Blood 109(5): 2058-2065.

Ge, W., J. Jiang, et al. (2009). "Infusion of mesenchymal stem cells and rapamycin synergize to attenuate alloimmune responses and promote cardiac allograft tolerance." Am J Transplant 9(8): 1760-1772.

Gieseke, F., J. Bohringer, et al. (2010). "Human multipotent mesenchymal stromal cells use galectin-1 to inhibit immune effector cells." Blood 116(19): 3770-3779.

Hematti, P. (2008). "Role of mesenchymal stromal cells in solid organ transplantation." Transplant Rev (Orlando) 22(4): 262-273.

Hoogduijn, M. J., F. Popp, et al. (2010). "The immunomodulatory properties of mesenchymal stem cells and their use for immunotherapy." Int Immunopharmacol 10(12): 1496-1500.

Hoogduijn, M. J., F. C. Popp, et al. (2010). "Advancement of mesenchymal stem cell therapy in solid organ transplantation (MISOT)." Transplantation 90(2): 124-126.

Hyun, I., O. Lindvall, et al. (2008). "New ISSCR guidelines underscore major principles for responsible translational stem cell research." Cell Stem Cell 3(6): 607-609.

In 't Anker, P. S., S. A. Scherjon, et al. (2004). "Isolation of mesenchymal stem cells of fetal or maternal origin from human placenta." Stem Cells 22(7): 1338-1345.

Jiang, X. X., Y. Zhang, et al. (2005). "Human mesenchymal stem cells inhibit differentiation and function of monocyte-derived dendritic cells." Blood 105(10): 4120-4126.

Karlsson, H., S. Samarasinghe, et al. (2008). "Mesenchymal stem cells exert differential effects on alloantigen and virus-specific T-cell responses." Blood 112(3): 532-541.

Karp, J. M. and G. S. Leng Teo (2009). "Mesenchymal stem cell homing: the devil is in the details." Cell Stem Cell 4(3): 206-216.

Kim, Y. H., Y. M. Wee, et al. (2011). "IL-10 induced by CD11b+cells and IL-10 activated regulatory T cells play a role in immune modulation of mesenchymal stem cells in rat islet allograft." Mol Med.

Krampera, M., L. Cosmi, et al. (2006). "Role for interferon-gamma in the immunomodulatory activity of human bone marrow mesenchymal stem cells." Stem Cells 24(2): 386-398.

Lazarus, H. M., O. N. Koc, et al. (2005). "Cotransplantation of HLA-identical sibling culture-expanded mesenchymal stem cells and hematopoietic stem cells in hematologic malignancy patients." Biol Blood Marrow Transplant 11(5): 389-398.

Le Blanc, K., I. Rasmusson, et al. (2004). "Treatment of severe acute graft-versus-host disease with third party haploidentical mesenchymal stem cells." Lancet 363(9419): 1439-1441.

Le Blanc, K. and O. Ringden (2007). "Immunomodulation by mesenchymal stem cells and clinical experience." J Intern Med 262(5): 509-525.

Le Blanc, K., H. Samuelsson, et al. (2007). "Generation of immunosuppressive mesenchymal stem cells in allogeneic human serum." Transplantation 84(8): 1055-1059.

Le Blanc, K., C. Tammik, et al. (2003). "HLA expression and immunologic properties of differentiated and undifferentiated mesenchymal stem cells." Exp Hematol 31(10): 890-896.

Li, Y. P., S. Paczesny, et al. (2008). "Human mesenchymal stem cells license adult CD34+ hemopoietic progenitor cells to differentiate into regulatory dendritic cells through activation of the Notch pathway." J Immunol 180(3): 1598-1608.

Meisel, R., A. Zibert, et al. (2004). "Human bone marrow stromal cells inhibit allogeneic T-cell responses by indoleamine 2,3-dioxygenase-mediated tryptophan degradation." Blood 103(12): 4619-4621.

Nauta, A. J. and W. E. Fibbe (2007). "Immunomodulatory properties of mesenchymal stromal cells." Blood 110(10): 3499-3506.

Nauta, A. J., A. B. Kruisselbrink, et al. (2006). "Mesenchymal stem cells inhibit generation and function of both CD34+-derived and monocyte-derived dendritic cells." J Immunol 177(4): 2080-2087.

Nauta, A. J., G. Westerhuis, et al. (2006). "Donor-derived mesenchymal stem cells are immunogenic in an allogeneic host and stimulate donor graft rejection in a nonmyeloablative setting." Blood 108(6): 2114-2120.

Ng, F., S. Boucher, et al. (2008). "PDGF, TGF-beta, and FGF signaling is important for differentiation and growth of mesenchymal stem cells (MSCs): transcriptional profiling can identify markers and signaling pathways important in differentiation of MSCs into adipogenic, chondrogenic, and osteogenic lineages." Blood 112(2): 295-307.

Ng, W. F., P. J. Duggan, et al. (2001). "Human CD4(+)CD25(+) cells: a naturally occurring population of regulatory T cells." Blood 98(9): 2736-2744.

Pittenger, M. F., A. M. Mackay, et al. (1999). "Multilineage potential of adult human mesenchymal stem cells." Science 284(5411): 143-147.

Poggi, A., C. Prevosto, et al. (2007). "NKG2D and natural cytotoxicity receptors are involved in natural killer cell interaction with self-antigen presenting cells and stromal cells." Ann N Y Acad Sci 1109: 47-57.

Polchert, D., J. Sobinsky, et al. (2008). "IFN-gamma activation of mesenchymal stem cells for treatment and prevention of graft versus host disease." Eur J Immunol 38(6): 1745-1755.

Popp, F. C., E. Eggenhofer, et al. (2008). "Mesenchymal stem cells can induce long-term acceptance of solid organ allografts in synergy with low-dose mycophenolate." Transpl Immunol 20(1-2): 55-60.

Popp, F. C., P. Renner, et al. (2009). "Mesenchymal stem cells as immunomodulators after liver transplantation." Liver Transpl 15(10): 1192-1198.

Prevosto, C., M. Zancolli, et al. (2007). "Generation of CD4+ or CD8+ regulatory T cells upon mesenchymal stem cell-lymphocyte interaction." Haematologica 92(7): 881-888.

Quaedackers, M. E., C. C. Baan, et al. (2009). "Cell contact interaction between adipose-derived stromal cells and allo-activated T lymphocytes." Eur J Immunol 39(12): 3436-3446.

Rasmusson, I., O. Ringden, et al. (2003). "Mesenchymal stem cells inhibit the formation of cytotoxic T lymphocytes, but not activated cytotoxic T lymphocytes or natural killer cells." Transplantation 76(8): 1208-1213.

Rasmusson, I., M. Uhlin, et al. (2007). "Mesenchymal stem cells fail to trigger effector functions of cytotoxic T lymphocytes." J Leukoc Biol 82(4): 887-893.

Ren, G., L. Zhang, et al. (2008). "Mesenchymal stem cell-mediated immunosuppression occurs via concerted action of chemokines and nitric oxide." Cell Stem Cell 2(2): 141-150.

Ren, G., X. Zhao, et al. (2010). "Inflammatory cytokine-induced intercellular adhesion molecule-1 and vascular cell adhesion molecule-1 in mesenchymal stem cells are critical for immunosuppression." J Immunol 184(5): 2321-2328.

Ringden, O. and K. Le Blanc (2011). "Mesenchymal stem cells for treatment of acute and chronic graft-versus-host disease, tissue toxicity and hemorrhages." Best Pract Res Clin Haematol 24(1): 65-72.

Ryan, J. M., F. P. Barry, et al. (2005). "Mesenchymal stem cells avoid allogeneic rejection." J Inflamm (Lond) 2: 8.

Sanchez-Ramos, J. (2006). "Stem cells from umbilical cord blood." Semin Reprod Med 24(5): 358-369.

Sato, K., K. Ozaki, et al. (2007). "Nitric oxide plays a critical role in suppression of T-cell proliferation by mesenchymal stem cells." Blood 109(1): 228-234.

Schwartz, P. H., D. J. Brick, et al. (2008). "Differentiation of neural lineage cells from human pluripotent stem cells." Methods 45(2): 142-158.

Selmani, Z., A. Naji, et al. (2008). "Human leukocyte antigen-G5 secretion by human mesenchymal stem cells is required to suppress T lymphocyte and natural killer function and to induce CD4+CD25highFOXP3+ regulatory T cells." Stem Cells 26(1): 212-222.

Shi, Y., G. Hu, et al. (2010). "Mesenchymal stem cells: a new strategy for immunosuppression and tissue repair." Cell Res 20(5): 510-518.

Sioud, M. (2011). "New insights into mesenchymal stromal cell-mediated T-cell suppression through galectins." Scand J Immunol 73(2): 79-84.

Sioud, M., A. Mobergslien, et al. (2010). "Evidence for the involvement of galectin-3 in mesenchymal stem cell suppression of allogeneic T-cell proliferation." Scand J Immunol 71(4): 267-274.

Sordi, V., R. Melzi, et al. (2010). "Mesenchymal cells appearing in pancreatic tissue culture are bone marrow-derived stem cells with the capacity to improve transplanted islet function." Stem Cells 28(1): 140-151.

Sotiropoulou, P. A., S. A. Perez, et al. (2006). "Interactions between human mesenchymal stem cells and natural killer cells." Stem Cells 24(1): 74-85.

Spaggiari, G. M., A. Capobianco, et al. (2008). "Mesenchymal stem cells inhibit natural killer-cell proliferation, cytotoxicity, and cytokine production: role of indoleamine 2,3-dioxygenase and prostaglandin E2." Blood 111(3): 1327-1333.

Spaggiari, G. M., A. Capobianco, et al. (2006). "Mesenchymal stem cell-natural killer cell interactions: evidence that activated NK cells are capable of killing MSCs, whereas MSCs can inhibit IL-2-induced NK-cell proliferation." Blood 107(4): 1484-1490.

Sun, J., Z. B. Han, et al. (2011). "Intrapulmonary Delivery of Human Umbilical Cord Mesenchymal Stem Cells Attenuates Acute Lung Injury by Expanding CD4CD25(+) Forkhead Boxp3 (FOXP3) (+) Regulatory T Cells and Balancing Anti- and Pro-inflammatory Factors." Cell Physiol Biochem 27(5): 587-596.

Sundin, M., A. J. Barrett, et al. (2009). "HSCT recipients have specific tolerance to MSC but not to the MSC donor." J Immunother 32(7): 755-764.

Sundin, M., P. D'Arcy, et al. (2011). "Multipotent mesenchymal stromal cells express FoxP3: a marker for the immunosuppressive capacity?" J Immunother 34(4): 336-342.

Sundin, M., O. Ringden, et al. (2007). "No alloantibodies against mesenchymal stromal cells, but presence of anti-fetal calf serum antibodies, after transplantation in allogeneic hematopoietic stem cell recipients." Haematologica 92(9): 1208-1215.

Tabera, S., J. A. Perez-Simon, et al. (2008). "The effect of mesenchymal stem cells on the viability, proliferation and differentiation of B-lymphocytes." Haematologica 93(9): 1301-1309.

Thornton, A. M. and E. M. Shevach (1998). "CD4+CD25+ immunoregulatory T cells suppress polyclonal T cell activation in vitro by inhibiting interleukin 2 production." J Exp Med 188(2): 287-296.

Tolar, J., P. Villeneuve, et al. (2011). "Mesenchymal stromal cells for graft-versus-host disease." Hum Gene Ther 22(3): 257-262.

Uccelli, A., L. Moretta, et al. (2008). "Mesenchymal stem cells in health and disease." Nat Rev Immunol 8(9): 726-736.
Uccelli, A. and D. J. Prockop (2010). "Why should mesenchymal stem cells (MSCs) cure autoimmune diseases?" Curr Opin Immunol 22(6): 768-774.
Ulicna, M., L. Danisovic, et al. (2009). "Diabetes--adult stem cells as an future alternative therapy?" Bratisl Lek Listy 110(12): 773-776.
Voltarelli, J. C., C. E. Couri, et al. (2007). "Autologous nonmyeloablative hematopoietic stem cell transplantation in newly diagnosed type 1 diabetes mellitus." JAMA 297(14): 1568-1576.
Wada, J. and H. Makino (2001). "Galectins, galactoside-binding mammalian lectins: clinical application of multi-functional proteins." Acta Med Okayama 55(1): 11-17.
Yagi, H., T. Nomura, et al. (2004). "Crucial role of FOXP3 in the development and function of human CD25+CD4+ regulatory T cells." Int Immunol 16(11): 1643-1656.
Zuk, P. A. (2010). "The adipose-derived stem cell: looking back and looking ahead." Mol Biol Cell 21(11): 1783-1787.

9

Endovascular Methods for Stem Cell Transplantation

Johan Lundberg and Staffan Holmin
Department of Clinical Neuroscience, Karolinska Institutet and Department of Neuroradiology, Karolinska University Hospital, Stockholm, Sweden

1. Introduction

Results from cell transplantation research have received interest and attention both from a clinical, a scientific and a public point of view. This chapter discusses new endovascular transplantation methods for different cell systems. First different cell based therapies are presented, followed by an overview of pathological conditions wherein several cell based strategies are implemented. Thereafter the delivery of cells in broad terms, and then specifically by endovascular technique compared to surgical technique, is presented. A general description of the active process of diapedesis is provided, as it is understood for immunological cells, since this is most probably a fundamental process for endovascular transplantation of cells as well.

1.1 Cell based therapies

Cell based strategies are sought after as a way of repairing or facilitating self renewal in pathological organ systems that have little or no intrinsic regenerative capacity. The plethora of different diseases in organ systems that might have a regenerative capacity, but is limited through physiological processes, is almost boundless (Bajada et al. 2008). Cell based therapies have been successfully used in the clinical practice for distinct pathological conditions during a relatively long period of modern medicine. One of the broadest success stories are the transplantation of cells to patients suffering from hematological diseases (Buckner et al. 1974; Thomas et al. 1975; Thomas et al. 1975; Slavin et al. 1998). Hematological stem cell transplantation has also been expanded to comprise autologous transplantations following chemotherapy of solid tumor forms (Childs et al. 2000). Following the isolation of stem cell lines from human blastocysts (Thomson et al. 1998) and other adult sources such as the central nervous system (CNS) (Johansson et al. 1999), bone marrow (Bruder et al. 1997), multi-lineage mesenchymal (Pittenger et al. 1999), adipose tissue (Zuk et al. 2001) among others, new cell based approaches to disease treatment can be envisioned. The potential for *in vivo* expansion of these cells followed by transplantation, re-implantation and/or tissue engineering becomes possible (Vacanti et al. 1999). These findings open up possibilities for strategies aimed at ameliorating disease burden, or in the long run, obtaining curative goals through cell therapies in a clinical setting. Proposed treatments can broadly be divided into stimulation of an endogenous population or transplantation of cells (Lindvall et al. 2006), be them homologous, from a donor or across the xeno-barrier. The wider implications of the prospect of cell based treatments are summarized in a review article where the coining of the effort and/or subject of Regenerative Medicine is presented (Daar et al. 2007).

The CNS has attracted attention since the potential of restored function could be very valuable for patients. Particularly since pathological conditions in the CNS can be severely disabling. Cell based therapies are in clinical trials in *e.g.* Parkinsons Disease (Freed et al. 2001; Gordon et al. 2004), ischemic stroke (Kondziolka et al. 2000; Bang et al. 2005) and spinal cord lesions (Sykova et al. 2006). Outside the CNS, other clinical trials with cell based therapies aimed at *e.g.* muscle dystrophy (Gussoni et al. 1997; Miller et al. 1997), ischemic heart disease (Stamm et al. 2003), graft versus host disease (Le Blanc et al. 2004; Ringden et al. 2006) and type I diabetes mellitus (Scharp et al. 1991; Shapiro et al. 2000; Korsgren et al. 2008) have yielded promising results. So far, many of the cell therapies are still in trials since both safety and effects must be thoroughly evaluated.

In many areas of pre-clinical, and to some extent clinical research, cell based therapies deliver positive results. However, as with the definition of stem cells/progenitor cells (Potten et al. 1990) the field of cell based therapies are a very heterogeneous one (Bajada et al. 2008). One feasible way of applying taxonomy to this field is by discussing different basic components in the treated diseases. The first example would be in pathological conditions with a definable population of cells being defect. Examples of diseases with certain cell types being depleted are Morbus Parkinson - dopamine producing cells, (Lindvall et al.), muscle dystrophy -satellite cells of the muscles (Gussoni et al. 1997) and type I diabetes - insulin producing cells (White et al. 2001). Such specialized cells might be easier to replace than the second general idea of cell transplantation wherein attempts of transplantation is aimed at more intricately functioning physiological systems. The complexity increases steeply when transplanted cells must differentiate into subpopulations of cell and/or interact within networks (*e.g.* the CNS). An example of that would be the transplantation of neural progenitor cells with the aim of full neural integration (Nikolic et al. 2009). A third strategy is to exert effects on existing cells/organs through transplantation but without functional integration. This strategy might explain some of the results from studies aimed at integration of cells, but without engraftment in the target organ, albeit with positive functional results observed (Borlongan et al. 2004). One explanation for that phenomenon is that in more complex situations, such as following CNS insults, some of the reported beneficial effects might be associated with immune modulation or local secretion of growth factors, thereby rescuing cells from apoptosis. Examples of immune modulation could be Fas-ligand expressing cells (Ghio et al. 1999; Nagata 1999; Lee et al. 2008) whereas secretion of growth factors could be exemplified by an over-expression of IGF-1 from transplanted mesenchymal cells (Haider et al. 2008). Modulation of the immunological response by cell transplantation has also been shown to favorably treat graft versus host reactions in clinical practice (Le Blanc et al. 2004; Ringden et al. 2006). The concept of transplantation of cells serving as self renewing, local, biologically active, pharmacological factories are attractive for many parts of regenerative medicine (Amar et al. 2003). Local, self-sustaining, treatments that are only affecting niche parts of organs have many benefits that might include, but are not limited to, higher local concentration, less risk of adverse events and customization to different pathological conditions.

1.2 Delivery of cells

For cell transplantation, different percutaneous techniques assisted by modern imaging are viable through minimal invasive methods (Bale et al. 2007) and most parts and locales of the human body can be reached with that approach. On the other hand, for organs with less accessible anatomical location, parenchymal access can be associated with significant surgical risks (Villiger et al. 2005; Ben-Haim et al. 2009). In situations where engraftment rate after intravenous or intra-arterial cell administration is low and when a high anatomical specificity is required, such as the scenario when replacing a distinct cell type, direct puncture of the parenchyma might be preferable. For CNS applications this can be done with stereotactic needle puncture or in a combination with open surgery (Hagell et al. 2002; Wennersten et al. 2004).

Direct parenchymal access can also be achieved by endovascular technique. An example of this is a system that adds a possibility to, via large veins, administrate cells to the heart parenchyma (Thompson et al. 2003). The design of that system, requiring a large diameter catheter and without a closure device for the penetration site makes it usable only in large vessels on the venous side, more specifically, in the coronary sinus of the heart (Thompson et al. 2003; Siminiak et al. 2005). Other organs that might be difficult to reach, such as the CNS and the pancreas, are not reachable by the transvenous technique due to the design of that system requiring a large catheter diameter. Furthermore, venous navigation to most parts of the CNS and the pancreas, and to certain parts of the heart, is very difficult due to the more unpredictable venous anatomy and the venous valves.

Insulin producing cells are today transplanted by a hybrid method with percutaneous access to the portal vein and then intra-luminal cell release in the bloodstream. The concept of the intra portal transplantation is considered superior to open surgical techniques, due to the un-acceptable risk of adverse events (Kandaswamy et al. 1999; Humar et al. 2000). The risk-analysis naturally differs substantially for different surgical procedures and transplantations, both with respect to organs and cells. Risks with portal vein transplantation include portal vein thrombosis, hemorrhages, and transient increase of transaminase values (Shapiro et al. 1995; Ryan et al. 2001). A continuous work of reaching a balance between the risk of bleeding after the procedure and portal vein thrombosis is of utmost importance for portal vein transplantations and results are improving steadily in clinical trials. Further adding to the risk side of the comparison is the need for immunosuppressant treatment after transplantation. The potential benefit of the procedure must also, as in the case with diabetes and insulin producing cells, be compared to the standard treatment of insulin injection and or pumps. Patients eligible for portal vein transplantation of insulin producing cells are thus: patients already subject to immunosuppressive therapy due to previous transplantations, patients with unstable glycemia, unawareness hypoglycemia, or patients with progressive chronic complications despite intensive insulin treatment (Bertuzzi et al. 2006). It has been suggested that it would be of great benefit if insulin producing cells could be transplanted directly to the parenchyma of the pancreas. Advantages with pancreas as the target locale are *e.g.* the possibility of mimicking the physiological release of insulin and the more hospital micro-environment for the insulin producing cells; the pancreas has a higher oxygen tension compared to the liver (Merani et al. 2008).

Different strategies of cell based therapies are currently being evaluated in both pre-clinical and clinical trials but the cell delivery methods per se have received limited interest. One of the larger obstacles that have been observed, is that the lung acts as a kind of clearance filter for the intravenous cell infusion, resulting in pulmonary trapping (Barbash et al. 2003; Fischer et al. 2009). An intra-arterial selective approach would possibly result in higher transplantation efficiency in certain conditions. The versatility of cell suspensions must not be underestimated and limit the way of thinking when considering treatments with cell based approaches (Nikolic et al. 2009). The cell suspensions can easily be handled and administrated through tubing and catheters, thus providing the possibility to by-pass the lung and selectively reach designated target vessels/parenchyma with a first passage effect. Those possibilities of cell handling forms the basis for catheter based strategies for cell transplantation.

Endovascular treatments are continuously providing a third option to open surgical or percutaneous approaches. From the establishment of the Seldinger technique (Seldinger 1953) and the first use of digital subtraction angiography (DSA) (Meaney et al. 1980) to the modern interventional lab with 3D road maps (Soderman et al. 2005) and CT like capacities of the C-arm (Soderman et al. 2008) the path has been long but rapidly progressing. The driving force up until today, that has made the leap ward style of improvements possible, is both rapid developments in computational power and material sciences. The arteries and veins can today be regarded as "internal routes" for navigation, diagnosis and intervention. The shift from open surgical options is for example illustrated by the patients that used to undergo thoracotomy and that now are being referred for percutaneous coronary intervention or the established coiling of intracranial aneurysms instead of open neurosurgical operation.

1.3 Scaling from bench to bedside

As this chapter hopes to illustrate; the rapid development of endovascular technique has implications on cell transplantation methods as well. To illustrate these implications as opposed to organ transplantation, one can visualize a liver transplantation. The wound in the abdominal wall must at least be big enough for the liver to go into the patient. This severely limits the possibility of minimal invasive transplantation of organs. On the other hand, the versatility of cell suspensions could make intra-luminal techniques the natural way of access. In experimental trials, open surgical options are used in e.g. rodent models for transplantation with positive results. This presents a limitation of scalability for clinical translation. For instance, when evaluating pre clinical CNS transplantation schemes in rodents one or two burr holes are established and cells are transplanted. One or two injections in the rat brain covers a relatively large volume but when scaling that to the human brain following *e.g.* a middle cerebral artery ischemic event, a large number of percutaneous trajectories would be required to cover a human brain volume corresponding to the experimental situation. The migratory capacity for transplanted mesenchymal cells after stereo-tactical transplantation has been shown to be around two millimeter over 14 days (Chen et al. 2001). Furthermore, the mechanical neuronal injury and the risk for intracerebral hemorrhage would increase with each injection trajectory. In brain stimulation procedures, the literature is somewhat divergent; the risk for hemorrhage could be as high as 5% per injection (Ben-Haim et al. 2009). The easier clinical scalability of endovascular access comes in the terms of cells delivered to a larger volume of tissue by taking advantage of the already existing vascular system. If an ischemic stroke occurs due to a vessel occlusion at some point, it would be very tempting to intraluminally disperse cells from the same point via the vascular system in order to reach the affected parenchyma. A normal cell dose for a human adult would probably at least be hundredfold higher than in the rodent and needs to spread out over a vastly larger volume thus requiring many injection trajectories. The average human brain weight is quoted at around 1400 grams as opposed to the adult rat at 2 grams to give some sense to the scale proportions.

Endovascular intervention is not without risks either, exemplified by the risk of adverse events reported at 0 to 4.0%, commonly reported at 0.5%, in different cerebral interventions (Raymond et al. 1997; Cognard et al. 1998; Ng et al. 2002; Murayama et al. 2003; Gonzalez et al. 2004; Cronqvist et al. 2005). Added to the risk of the procedure *per se* is the risk of cell transplantations. The risk of intra-arterial transplantation has already been documented in humans for up to 24 months without severe complications, albeit in a small material (Sykova et al. 2006). Other pre-clinical studies with intra-arterial coronary injections performed in healthy dogs revealed micro-infarction of the heart parenchyma (Vulliet et al. 2004) whereas that has not been observed in clinical studies (Stamm et al. 2003). Factors that might be limiting are the size of the cells injected versus the size of the capillary system, the proportion of shunts in the microcirculation, the stickyness of the cells and the amount of cells administrated. Many of the cells have a much larger diameter (10 up to 70 μm) as opposed to the capillaries 5 to 8 μm (Chien et al. 1975). The shunting zones in the microcirculation could potentially lead also large cells to the post-capillary venules where diapedesis usually occurs (Tuma 2008). The main reason for leukocyte adhesion/diapedesis through venules is the usually restricted expression of adhesion molecules on venular but not arteriolar or capillary endothelium (Tuma 2008). In intravital microscopy studies, adipose mesenchymal cells have been shown to act as embolic material (Furlani et al. 2009). This risk could be speculated on to be lower than the comparable trauma of the surgical methods although a final risk assessment requires a randomized clinical trial.

1.4 Leaving the bloodstream - diapedesis

One limiting factor to specific intra-arterial and intravenous transplantations is that an intra-parenchymal approach yields a higher efficacy. It has been shown that when performing transplantations to a rodent stroke model and comparing intravenous, intra-ventricular and intra-striatal injections, the highest efficacy in sheer number of cells were obtained with the intra-striatal route (Jin et al. 2005). Further limiting the selective intra-luminal approach is the speculation that some cell systems, such as insulin producing cells, appear incapable of leaving the bloodstream (Hirshberg et al. 2002).

In all implementations aimed at intra-luminal administration (intra-luminal encompassing both intravenous and selective intra-arterial administration) the ability of the cells to leave the bloodstream, or perform diapedesis (Fulton et al. 1957) is fundamental. The diapedesis function has previously been studied predominantly in immunological cells (Fulton 1957), it is in fact the active process whereby cells leave the bloodstream. The diapedesis of immunologically active cells has been thoroughly studied since the discovery of the significant multistep, ordered, cross-talk procedure of leukocyte-endothelial cell interaction both *in vitro* and *in vivo* (Butcher 1991; Springer 1994).

The barricades limiting the cells from haphazardly leaving the bloodstream are many. All blood vessels contain an endothelium and several organized barriers. In the CNS for instance, tight junctions are located between endothelial cells, predominantly to permit the conservation of the water fraction of the blood stream. The liquid pressure gradient composed of the blood pressure can be broken down to one force directed with the axis of the laminar flow of blood and one force aimed perpendicular to the flow on to the wall (Glagov et al. 1992; Fay 1994), thereby providing an evolutionary rationale for sealing the blood stream tightly. Situated underneath the endothelial cells and forming their structural base is the basal layer; a specific protein structure composed of extra cellular matrix (ECM) proteins abundant with expression of elastin, laminin and collagen type I, III and IV (Mayne 1986). Collagen and elastin are the major structural components of blood vessels of all sizes throughout the mammalian body. The cross-banded fibrils in the tunica media and tunica adventitia, formed by type I and type III collagen, provide the tensile strength and comprise probably 80 to 90% of the total collagen present. The other major structural protein component in elastic arteries is elastin; the protein that provides the elastic component of the blood vessels (Mayne 1986). Around most capillaries, pericytes are situated which further seal the blood stream. The pericytes are less abundant in the post capillary venules where most of the diapedesis occurs. As a general note, the post capillary venules are the most "leaky" part of the vascular tree. It should be noted that blood vessels are not merely the plumbing in the mammal body, they are in all respects a vividly living part of the organism, reorganized as a response to stress and demand (Gibbons et al. 1994) and among other things have a self-generating electro potential towards the bloodstream that is important for the clotting cascade (Danon et al. 1976).

As previously mentioned, in a diverging literature, it has been speculated that some cells totally lack the function of exiting the blood stream, thereby limiting intra-luminal based techniques. Even though insulin producing cells appears incapable of leaving the bloodstream, they are still transplanted through the portal vein (Hirshberg et al. 2002). The mechanism of action is believed to be microembolization of the cells to the liver parenchyma in the low pressure system that the portal vein constitutes (Lehmann et al. 2007). It has been speculated that arterial blood from the hepatic artery pushes back into the portal vein during the mixing of the two flows which would stop the cells and create advantageous conditions for engraftment. The cells found functioning in the livers of transplanted patients are, however, situated as plaques in the hepatic artery tree thus adding yet another hurdle to the understanding of portal vein transplantation. The hypothesis is that the cells cannot perform diapedesis but are instead primarily forming a mural thrombus after, with low probability, being displaced against the blood flow into the arterial tree and then encapsulated by endothelial cells. The plaque formation can thereby provide capillary ingrowth, an absolute requirement for endocrine function (Korsgren et al. 2008). Such a hypothesis would also shed further light on the low efficacy of portal transplantations methods. Nevertheless, the portal vein approach to transplantation is today the golden standard since the existing open surgical options of total pancreatic transplantation carries a mortality risk of 10% during the first year of follow-up (Kandaswamy et al. 1999; Humar et al. 2000).

The process of diapedesis is basically divided into tethering, rolling and stopping of leukocytes prior to diapedesis into inflamed tissue. It is thought of as a multistep procedure involving complex crosstalk between cells in the bloodstream and the endothelial cells. On the endothelial side intra-cellular adhesion molecule -1 (ICAM-1) (Dustin et al. 1986; Rothlein et al. 1986), vascular cell adhesion molecule -1 (VCAM-1) (Elices et al. 1990; Pulido et al. 1991) and junctional adhesion molecule –A (JAM-A) have all been implicated to play crucial roles. Interestingly, all of these receptors are part of the Ig-super family. Leukocytes capacity for homing and diapedesis is pivotal for the development of the inflammatory response to injury and starts as a tightly controlled up-regulation of endothelial E- and P-selectins that stimulate the leukocytes (Vestweber et al. 1999). These leukocytes then respond by activation of G protein-coupled receptors that increase the affinity for endothelial VCAM-1 and ICAM-1 (Muller 2003). Interaction with VCAM-1 is maintained through the heterodimer CD29 and CD49d forming the cell surface antigen Very Late Antigen - 4 (VLA-4) (Elices et al. 1990) and the CD11aCD18 heterodimer interacting with ICAM-1 (Meerschaert et al. 1995). VCAM-1 and ICAM-1 have previously been shown to be up-regulated on the endothelium as a response to inflammation (Dustin et al. 1986; Pulido et al. 1991). This interaction starts the diapedesis itself wherein the leukocytes "crawl" through, either a para-cellular or a trans-cellular pathway interacting with both PECAM-1 and/or CD99 or members of the JAM family of proteins (Petri et al. 2006). Thus significant crosstalk is needed to initiate and execute the active process of diapedesis.

VCAM-1 is also implicated both as a model of treatment for immunological modulation of the inflammatory responses following CNS insults and as surface antigen for cell treatment of CNS insults. In one study, fluorescence activated cell sorting (FACS) was performed for the expression of CD49d and identifying it as one of the important factors for directing diapedesis (Guzman et al. 2008). This finding has received a lot of interest from different fields, among other it has been shown that VCAM-1 expression has a critical role in transplantation of cells in dystrophic muscle (Gavina et al. 2006). Another utilization of VCAM-1 is the selective blocking by the monoclonal antibody drug Natalizumab used in multiple sclerosis, thereby inhibiting diapedesis for immunologically active cells (Stuve et al. 2008). VCAM-1 blockade has also been tested for neuroprotective effects following ischemic events in pre-clinical trials with disappointing results (Justicia et al. 2006).

2. Results and discussion with emphasis on the central nervous system

In this chapter three different types of endovascular methods of cell transplantation are discussed. The first two are the selective intra-arterial and the intravenous methods, these are compared for efficacy. The third method is the trans-vessel wall technique by using the Extroducer; an endovascular catheter system developed within our group for penetrating the vessel wall from the inside to out, thereby creating a working channel to extravascular tissue.

For certain cell types, the selective intra-arterial method is superior to the intravenous one after TBI in the rat. This was measured by the level of engraftment of hMSC at one and five days following TBI, without thrombo-embolic complications (Lundberg et al. 2009). Selective intra-arterial transplantation method for rNPC after TBI in the rat is also superior compared to intravenous method. We were, however, not able to engraft hNPC, *ceteris paribus*, thus indicating that diapedesis and engraftment is an active process and that different cell systems have different capabilities to engraft following intraluminal delivery (Lundberg et al. 2011). We show, by indirect methods, that CD29CD49dVCAM-1 interactions might be one of the factors with impact on engraftment in an intra-luminal transplantation setting (Lundberg et al. 2011). Further, we show that it is possible to perform minimally invasive parenchymal injections by trans-vessel wall technique by the development of the Extroducer (Lundberg et al. 2010). The Extroducer have no adverse long term effects on the blood vessels up to 3 months following interventions (Lundberg et al. 2011) and it is feasible to use the system as a novel approach for transplantation of e.g. insulin producing cells to the CNS, to the pancreas or other cell types to organs that are difficult or risky to reach by traditional methods. The trans-vessel wall technique thereby adds a new possibility when transplanting cell populations without the necessary properties for performing diapedesis. The establishment of a working channel to the extravascular tissue, by endovascular method, also opens up several other possible applications, such as different methods for sampling.

2.1 Selective intra-arterial method versus intravenous method

We established a model for selective intra-arterial transplantation (Lundberg et al. 2009) that we applied to a model of TBI in the rat (Feeney et al. 1981). We compared the selective intra-arterial method to intra-venous methods for different cell systems in the same TBI setting. Several different variables were tested such as cell concentration, days post injury for transplantation, time for infusion, level of immunosuppressant drugs etc. The majority of transplantation experiments were not successful from the start and many different variables were tested before robust engraftments could be reached. Too few cells had a dire impact on the success of transplantation; 200.000 (unpublished results) and 500.000 hMSCs did only result in very low engraftment. The failure of low cell numbers can obviously be interpreted as an indication that the efficacy of the intra arterial method in this setting is quite low although it is still superior to intravenous alternatives.

Results of engraftment levels were obtained through IHC methods by counting engrafted cells in sectioned brains (Fig 1). We found that engraftment levels of rNPCs were more than five-fold higher than in the control group ($p=0.034$) and hMSC were more than fifteen-fold higher than the in control group in absolute values ($p=0.007$), with a large spread within the intra-arterial groups (Fig 2). Few studies compare selective intra-arterial and intravenous methods but recent clinical data suggest an advantage for the selective intra-arterial route (Sykova et al. 2006). A known problem with all intravenous methods is the fact that the lung acts as a kind of clearance filter during the first passage (Barbash et al. 2003; Fischer et al. 2009). Following intravenous cell infusion, the blood transports the cells after venous passage to the right ventricle and then through the lung where up to 80% of cells are trapped during the first passage (Fischer et al. 2009). Thus as low as 20% of the cells transplanted might be ejected for the first total body distribution through the aorta. Of all the blood leaving the left ventricle of the heart, only somewhere between 1.8 to 8.5 percent of the blood actually reaches the brain of the rat (Pannier et al. 1973) leaving only 0.01 to 1.7 percent of the initially transplanted cells to reach the brain in the first passage. All cells not distributed to the brain are then again re-transported for another passage through the lung, acting as a filter. This phenomenon has been studied *e.g.* by PET for mesenchymal stem cells

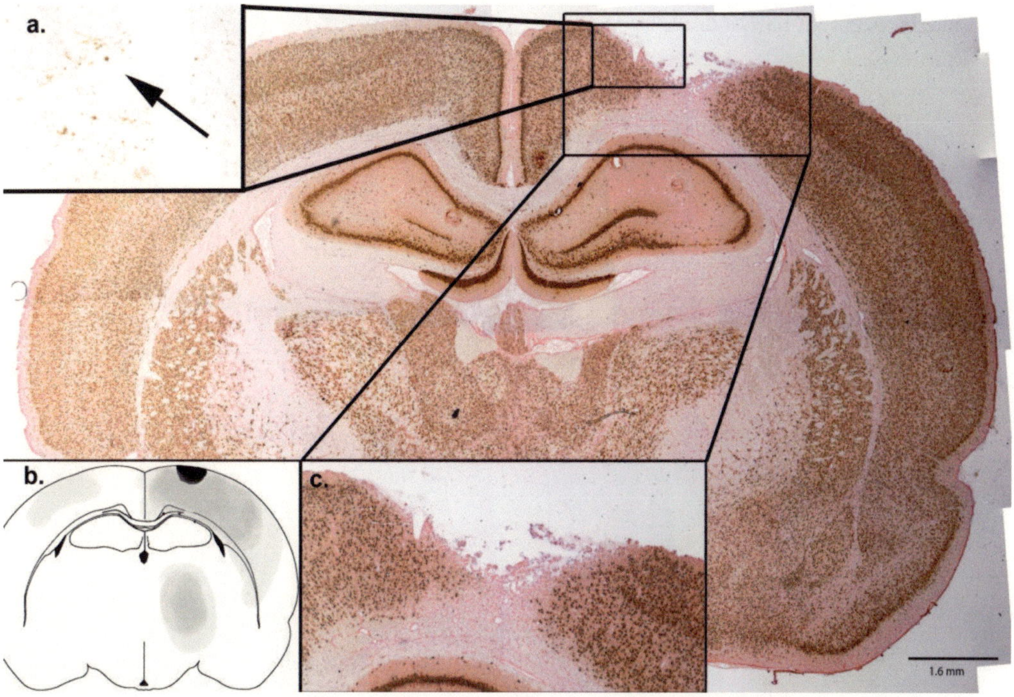

Fig. 1. Representation of intra-arterial transplantation in the TBI model
In the background is a reconstructed image showing a coronal section of the rat brain 2.5 mm posterior to bregma stained with GFAP (red) and NeuN (Brown) five days after traumatic brain injury. In blow-up a. a low magnification single staining with HuN (human nuclear antigen, MAB 1281) without any counterstaining, along the peri-lesional zone is shown. Brown dots represent HuN positive, transplanted cells (arrow indicates an example of a positive cell). In blow-up b. the black area represents the contusion zone and the grey areas represent primary localization of engrafted mesenchymal stem cells. In blow-up c. a magnification of the injury area itself is presented.

(Ma et al. 2005). This situation can readily be changed by placing a micro-catheter in the arteries supplying the organ of interest. An interesting finding in the present study is that our hMSCs predominantly were found in the spleen with few transplanted cells in the lung at 24 hours post injection, a bio-distribution phenomenon, from the lung to the spleen, previously described in rat following intravenous hMSC transplantation (Detante et al. 2009). To increase the understanding of the role of cell line properties for engraftment, we conducted a study with transplantation of different cell lines through either intra-arterial or intravenous routes. This analysis showed that there were dramatic differences between the different cell lines; hNPCs did not engraft at all after intra-luminal delivery whereas there was a significant difference between the hMSCs and the rNPCs using the same transplantation method. Noteworthy is the large variability in the engraftment levels of the latter two cell lines. One important factor that might contribute to this is that neither hMSC nor rNPCs are defined homologous cell lines, succinctly there can be important variations within the cell systems transplanted. The remarkable finding that no engraftment was obtained following hNPC transplantations is even more noteworthy since the hNPCs have previously been robustly transplanted by open surgical technique (Wennersten et al. 2004; Akesson et al. 2007).

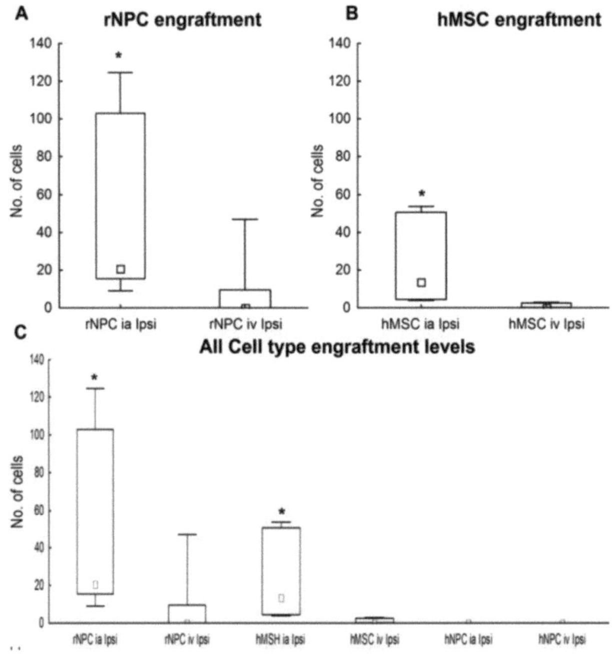

Fig. 2. Graph illustrating engraftment levels
Engrafted cells were counted per section and is reported with median (marker), quartiles (box) and max - min (whiskers). In A a significant difference between engraftment levels per section following selective intra-arterial and intravenous transplantation in the ipsilateral hemisphere of the rNPC group. In B a significant difference between engraftment levels per section following selective intra-arterial and intravenous transplantation in the ipsilateral hemisphere of the hMSC group and finally in C a panel of all engraftment levels per section in the ipsilateral hemispheres of all groups. * marks $p < 0.05$

2.2 Gene expression profiling of cell systems

After discovering that there were differences in engraftment capability between the different cell lines, the opportunity to investigate the bio-molecular basis of said differences presented itself. We started with characterizing and confirming earlier results (Clausen et al. 2007) that our TBI model leads to up-regulation of VCAM-1 expression in the endothelium. That result may suggest that the injured CNS parenchyma could provide cues for diapedesis and migration of engrafted cells in similar ways as immunological cells respond to inflammatory cues (Butcher 1991; Springer 1994). As a screening method, we started by performing microarray on the human cells. rNPCs were not included in the microarray analysis due to problems with cross-species comparisons in the microarray chips used. hMSCs showed a broad expression of integrins, commonly expressed by immune cells, that are important for diapedesis through the vessel endothelium and subsequent migration into the parenchyma. Specifically, analysis of the heterodimers forming receptors for ICAM-1 and VCAM-1 were analyzed, based on previous work indicating CD49d expression as important for successful intra-vasal transplantations (Guzman et al. 2008). Thus, probably the most interesting finding was the CD49d signal of 68 in hMSC as opposed to 0.4 in hNPC (p=0.0047). This was then confirmed with RT-qPCR data from all cell lines with average CD49d mRNA levels that were highest in the hMSC (0.98) followed by the rNPC (0.0057) and finally a dwindling finding in hNPC (0.0012). CD29 and CD49d forms a heterodimer named very late antigen -4 (VLA-4) which was expressed in falling order in hMSC, rNPC and hNPC (Fig 3). The difference in mRNA levels between rNPC and hNPC is not large but might reflect larger differences in protein translation. Further studies at the protein and functional level are required to elucidate the importance of CD49d for diapedesis of these cell systems. CD11a mRNA was detected in hNPC (0.0029) albeit with low CD18 mRNA (0.00025), suggesting that CD11a-CD18-ICAM-1 interaction may be dispensable for engraftment. In contrast, hMSC displayed high CD18 mRNA levels (0.43) but no detectable CD11a expression, suggesting neglectable ICAM-1 dependent engraftment in the hMSCs.

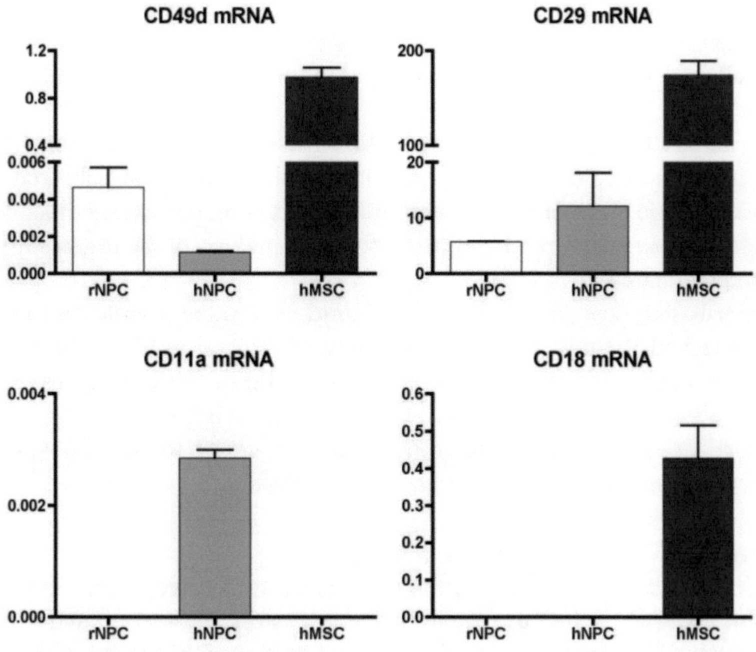

Fig. 3. RT-qPCR results from Cell systems
Bars represent relative levels of integrin CD49d, CD29, CD11a and CD18 mRNA expression in rNPC, hNPC and hMSC compared to respective endogenous TBP mRNA levels. Error bars represent the distribution between the biological replicates.

Finally, both our own findings regarding hNPCs and other previously known cells incapable of diapedesis, such as insulin producing cells (Hirshberg et al. 2002), shows the apparent need for surgical techniques. In some organs that are hard to reach and/or when surgical technique comes with a high risk of adverse events for the patients, the need for an alternative strategy becomes apparent. Thus, we also initiated the development of the trans-vessel wall approach.

2.3 Trans-vessel wall transplantation

An endovascularly based system that could penetrate the vessel wall would, in instances where the target parenchyma is either hard to reach or carries a significant surgical risk, be a method with both the merits of accurate placement and reduction of patient risk. Further, it would solve the problem for certain cells to leave the bloodstream. A proposed solution to this problem is a endovascular catheter system that we have named the Extroducer (Lundberg et al. 2010).

2.3.1 Extroducer in vivo testing - small animals

After extensive computer simulations and *ex vivo* testing, *in vivo* short term testing were performed in rat by creating arterial access from the medial tail artery and performing the Extroducer trans-vessel wall technique passage in either the subclavian or carotid artery. Two different stages of the procedure was tested; first the trans-vessel wall technique passage per se with surgical microscope monitoring of hemorrhage or other adverse events, and thereafter the deployment of the distal penetrating tip through the vascular wall and retracting the proximal part of the system. No cases of intra-operative hemorrhage or intra-luminal thrombosis occurred. Thus, the vascular penetration procedure was uneventful and the vessel wall completely sealed around the Extroducer, thereby preventing leakage of blood.

The second group with deposited Extroducer tips also showed absolute hemostasis during the primary intervention. Fourteen days post intervention, this group showed no signs of pain or discomfort. No signs of dissection of the vessels or impairment of blood-flow distal to intervention sites were observed and macroscopical analysis of the organ supplied by the vessel, showed no infarcts.

With computer-based flow simulations, we found that there should be no blood-flow through the detached Extroducer interior lumen at physiological blood pressure. This was also tested *in vivo* by cannulating the deployed distal tip of the prototypes with a nitinol mandrel. This was done to reassure that even when removing possible clotting inside the prototype, it still prevented bleeding from inside the vessel to the extravascular space. Furthermore, no signs of delayed hemorrhage were detected.

2.3.2 Extroducer in vivo testing - large animals

After successful trials in small animals, the Extroducer system was tested in large animals. An adaptation towards clinical use was that these prototypes were manufactured from a longer nitinol tube, 1700 mm *vis-à-vis* 300 mm that was used in the rat. We evaluated the prototypes in the rabbit together with standard clinical catheters and angiographical equipment.

The Extroducer prototypes within the microcatheters were visible at high magnification fluoroscopy and thereby maneuvered into the subclavian artery (SCA) (Fig 4). The rabbit SCA was chosen since it is close to the intended target vessel size of 0.5 to 3 millimeters, the SCA is fairly easy to access in order to perform simultaneous open surgical monitoring and it had been used in the rat. A slight amount of pressure was required on the protecting plastic catheter to advance the system through the microcatheter to the desired vessel wall, thereafter the Extroducer was gently advanced out through the vessel wall to the extra-vascular space. Hemorrhages were neither observed by simultaneous direct observation through a surgical microscope, nor by high resolution angiographical series (DSA), during and after the intervention (Fig 4). Further, no thromboembolic complications or vascular dissections were observed using high resolution DSA. No navigational problems were encountered with respect to Extroducer prototype integration with clinical catheters.

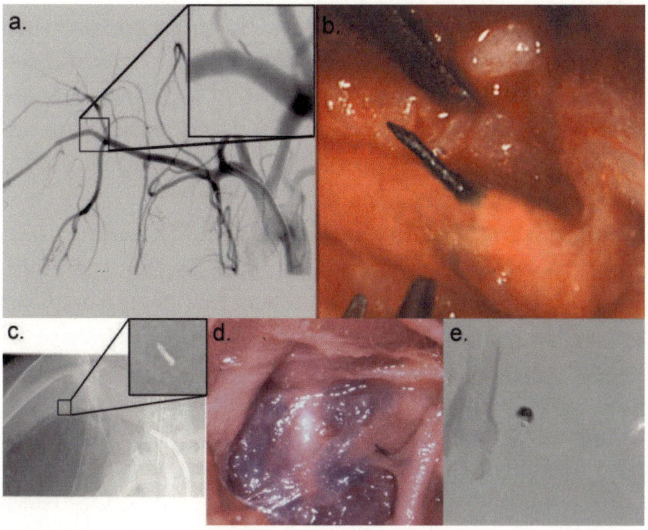

Fig. 4. Trans-vessel wall interventions
For full control over the procedure in the large animal trials, both a surgical microscope and high resolution angiographical series was used. In a. digital subtraction angiogram showing a detached Extroducer tip without hemorrhage, dissection or thromboembolic complications. In b. photograph showing the microsurgical view of the detached Extroducer tip. In c. x-ray image showing the detached Extroducer tip with guide catheter. In d. photograph from post-operative dissection showing the detached Extroducer tip with methylene blue injected in the surrounding tissue. In e. digital subtraction angiogram showing an extra vascular injection of 25 μl contrast agent through the Extroducer system.

Finally, electrolysis detachment of the distal Extroducer distal tip was tested in rabbit. We chose electrolysis since it was the easiest way of performing detachment in our hands. Our design was based on the work of the first detachable coils (Guglielmi et al. 1991). An important difference compared to the detachment zone in coils was, however, that we needed a hollow detachment zone which required additional development. After navigation to the designated intervention site and after methylene blue or contrast agent had been deposited in the extravascular space, a tension of 8V was applied and the distal tip was then detached after, on average, five minutes (range three to nine minutes). This was also uneventful without observation of hemorrhage from around the body of the distal tip or through the inner lumen. The procedure was successfully performed both with simultaneous microscopical monitoring via surgical access, and with fluoroscopical/angiographical guidance solely. In a previous work describing a method for penetrating large veins (Thompson et al. 2003), that system design required a much larger catheter and also lacked a method for sealing the vessel wall, thus making penetration through the arteries impossible. This severely limits the use of that system to their testing vessel, i.e. the sinus coronarius of the heart, whereas the Extroducer is applicable in both arteries and veins of any sizes down to approximately 0.5 mm in diameter. Another system, in which vessel perforations are performed, is the trans-jugular intrahepatic portacaval stent shunt (TIPS) technique (Richter et al. 1990). That system also does not have the requirement of sealing the vessel wall when finishing the procedure, since a patent blood flow through the stent is the preferred result. On the contrary, thrombosis of the stent might be considered the main problem (Merli et al. 1998) which requires rigorous follow-up.

Thus, the Extroducer system is unique in the ways that it permits safe exit of both arteries and veins and that it is usable in vessels with large dimensions as well as in the microvasculature with inner lumen diameters down to approximately 0.5 mm.

2.2.3 Extroducer testing with long time follow up

In a long term follow up, the end points five days, one month and three months after the deployment of the device was selected. No stenosis or late hemorrhagic complications were noted in any animals. No alterations in behavior or other measures of discomfort were noted either (Lundberg et al. 2011).

The distribution of followed up animals were as follows; two at the five days end point , five at the 30 days end point and six at the 80 days end point with histological analysis of one resulting in a total number of 19 detached Extroducer tips.

In the follow up DSA we also found that four of 19 (21 %) of the detached tips were no longer placed through the vessel wall but had instead been "pushed" or "migrated" through the endothelium to the extravascular space immediately adjacent to the penetration site. No vascular stenosis or other adverse reactions were observed around those tips. Also for the rest of the tips, that were located through the vessel wall, no adverse reactions were detected. An important endpoint was biocompatibility of the detached tips. The nitinol alloy used was selected for its many advantageous properties. Nitinol is a nickel/titanium alloy with both memory and super elastic properties (Adler et al. 1990). These special properties are the foundation for the use of nitinol in stent fabrication in clinical practice. Thus, the excellent biocompatibility of nitinol (Castleman et al. 1976) has been extensively studied, especially with respect to vessel wall interactions in the use of stents (Stoeckel et al. 2004). However, the compatibility of nitinol when placed through arterial walls has, to our knowledge, not been studied. The parylene, used as coating in our device, is also FDA approved and CE marked when used in pacemaker electrodes. To evaluate interaction between the deposited tips and the endothelium, we performed histological analysis with the prototypes *in situ* by a specialized grind-cutting technique and then consulted an external, independent, evaluator with expertise in the field of titanium implants. This evaluation showed full biocompatibility with a very small fibrotic capsule (< 1 µm) formed around the detached distal tips (fig 5). No ongoing inflammation was observable around any of the distal tips. Around one (of the total number 14 left in place) of the day five animal implants, three macrophages were indentified in the area of the detached distal tip. Apart from those three macrophages, no other signs of inflammation were observed. The endothelia showed no signs of alterations adjacent to the deposited tips.

In conclusion, the biocompatibility of the distal tips was comparable to titanium implants. The interactions of nitinol with the interior of the vascular wall and the extra vascular space has not been as extensively studied since stents most often are positioned inside the vascular lumen with direct contact only with the bloodstream and the endothelium. Therefore, the present histological analysis in this new application and new position of nitinol, adds important knowledge about biocompatibility for possible future applications.

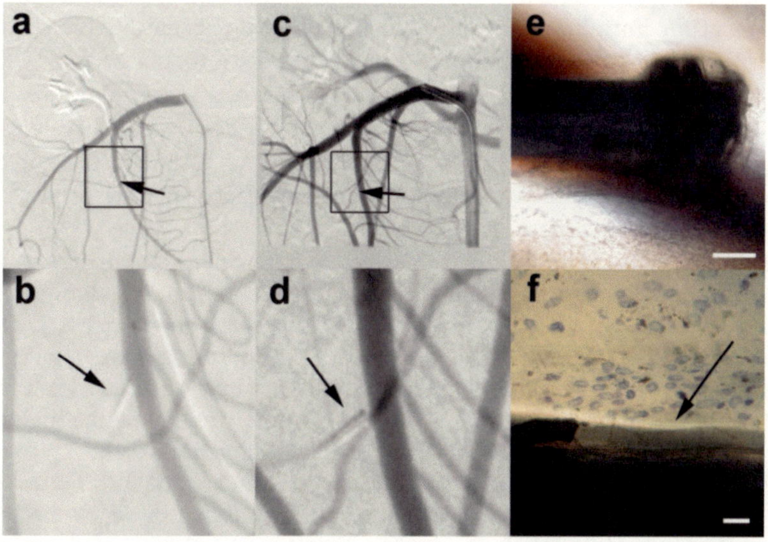

Fig. 5. Long term follow-up of the trans-vessel wall intervention
In a. the initial follow up angiogram directly following detachment in the Superior Mesenteric Artery (SMA) is shown with a square marking the blow-up in b. Arrows indicate the detached distal tip. In c. an SMA angiogram, performed 80 days after the intervention in the same animal, is shown with a square indicating the blow-up in d. Arrows indicate the detached distal tip. In e. a microphotograph of a histological van Geeson and toulene blue staining prepared by grind-cutting with the detached tip in situ, is shown. Scale bar = 100 µm. The blow-up in f shows the parylene coating surrounding the detached tip which is marked by an arrow. Note that no fibrous response or inflammation is observable. Scale bar = 4 µm.

The prototypes excluded due to failed detachment were analyzed by sweep electron microscopy (SEM), since monitoring of electric current gave limited prognostic information about the failure to detach. We found that, in failed detachments, large amounts of chloride were observed by surface spectroscopy (6-10 wt %) indicating titanium chloride ion formations in a passive layer, thus providing current transmission without electrolysis. Titanium chloride molecules can provide a surface area that lets electron pass through but without formation of soluble titanium or nickel ions. This problem is probably due to the simple technique used for creating the insulation defect in our hands. However, numerous available solutions for detachment are available on the market that can easily be integrated with the trans-vessel wall technique in an industrialized manufacturing process.

3. General discussion

The rationale for our previous studies within the field of endovascular cell transplantation is the path to translating cell based regenerative medicine to patients. In a translational perspective the actual route of transplantation will be important. For operative techniques there is a problem with scalability and for intravenous techniques there is a problem with efficacy. We show that by selective intra-arterial methods it is possible to increase the level of cerebral engraftment with certain cell types with six to fifteen fold yields. Further, not all cells are optimal for intra-luminal transplantation. It is, for example, known that insulin producing cells lack the capability to perform diapedesis (Hirshberg et al. 2002) and we show that hNPC, a cell system previously transplanted by open surgical means (Wennersten et al. 2004) also lacks the capability to perform diapedesis. For these cell systems, and other applications, we have developed the Extroducer as a tool to establish a direct, minimal invasive working channel with parenchymal access in organs that are difficult or risky to reach with traditional techniques.

The first passage of cells delivered through selective intra-arterial approaches, compared to systemic intravenous delivery, results in a higher local concentration, shorter blood stream exposure and less mechanical stress factors before cells reach the target site. These factors could be of importance for successful engraftment. Supporting that hypothesis are the present results showing significantly higher total cerebral uptake of cells after intra-arterial compared to intra-venous administration. The next supporting fact is the higher uptake of cells in the ipsilateral hemisphere after intra-arterial administration and by the absence of difference between the hemispheres after intravenous transplantation. Future studies are needed for elucidating molecules responsible for diapedesis, e.g by direct methods such as knock-ins, of for example CD49d, in non-functioning cell systems and knock down or blocking in other cell systems.

The absence of adverse effects in transplanted animals suggests that in the short term, the selective intra-arterial transplantation method is safe for delivering even high concentrations of cells. It has been reported ischemic events following intra-arterial approaches to the heart in dogs (Vulliet et al. 2004) thus indicating the need for thorough studies prior to translation into clinical practice. However, in clinical studies on intracoronary infusions (Stamm et al. 2003) and in spinal cord artery infusions (Sykova et al. 2006) no embolic events have been recorded. Connected to the need for safety studies, it could be argued that the higher engraftment rates in our intra-arterial groups would be a consequence of microembolization of cells. In that scenario we would, however, have detected ischemic histological changes and localization of the transplanted cells within arterioles and capillaries, which we did not.

For all intra-luminal approaches, this thesis shows that a thorough investigation must be performed to clarify if the cells actually can perform diapedesis prior to choosing transplantation strategy. hNPCs has previously been shown to have an impact on neurological outcome when transplanted with open surgical techniques, but in their present form they seem unsuitable for intra-luminal transplantation. Therefore, based on the findings in this thesis, considerations should be made that CD29CD49dVCAM-1 interaction is one of the best studied crosstalk mechanisms on how immunological cells leave the bloodstream to perform homing to disease ridden tissue where inflammation occurs (Elices et al. 1990). This line of reasoning is supported by other studies implicating VCAM-1 interaction in both ischemic stroke and muscle dystrophy pre-clinical studies (Gavina et al. 2006; Guzman et al. 2008). An interesting comparison can be made regarding diapedesis in cell systems aimed for transplantation compared to immunological cells. In such an exercise it could be considered blatantly ignorant to immediately dismiss the fylogenetically conserved mechanism of the immunological system and suggest a hitherto unknown method of diapedesis for these cells. Applying Occam's razor to the hypothesis of diapedesis for cell transplantation, two conclusions can potentially be made; i) the process of diapedesis is most probably an active process for cell systems transplanted, since there are no known inactive ways of diapedesis and ii) the most likely system for diapedesis crosstalk should be found within the same systems that are used by the immunological system, meaning that proteins are highly likely to come from the Ig-super family such as VCAM-1. The other, more complex, explanation would be that a hitherto unknown system for diapedesis exists. That it also unlikely since mutations in such a system would lead to genetical diseases that should be known, but without reasonable explanations. A research program dedicated on endovascular transplantation of different cell systems in different diseases should include a variety of diseases and cell systems to increase the understanding of both cell-endothelium interactions and the effects on target niches.

When designing both pre-clinical and clinical cell transplantation studies, consideration should be taken to how the cells are hypothesized to reach their designated targets. The first option might be open surgical techniques for small niche locations in a target parenchyma. Downsides to open surgical/percutaneous techniques are *e.g.* impracticalities of accruing the desired extent of target tissue volume, especially if the volume is relatively large, such as following a major CNS insult. Further, the patient risk for adverse events might be unacceptable in relation to a potential benefit, in particular for "difficult to reach" organs such as the CNS, the pancreas and/or the heart. If the disadvantages of open surgical/percutaneous techniques are unacceptable, intra-luminal options could be explored. For intra-luminal approaches, the concept of how transplanted cells are presumed to leave the blood stream becomes an issue with several potential solutions. For a cell system without the necessary features for diapedesis, one could use knock-in methods to provide the necessary adhesion molecule set up, otherwise intra-parenchymal injections must be considered. For knock-in methods, an intra-arterial approach would probably still have benefits in efficacy over intravenous methods through the first passage effect and by avoiding pulmonary trapping. As previously discussed, an interesting purely academical calculation when performing intravenous cell transplantations could result in such a low cell dose to the brain as 0.01 to 1.7% after intravenous injection, assuming that 80% of cells are trapped in the first passage through the lung (Barbash et al. 2003; Fischer et al. 2009). This can readily be changed by placing a micro-catheter in vessels supplying the target parenchyma with blood thereby providing a chance for all cells to perfuse the target parenchyma.

For more discrete functioning cells, an intra-parenchymal injection might be even more attractive for reasons such as shielding the cells from the exposure to the bloodstream and accurate anatomical placement. In difficult to reach organs, minimally invasive direct parenchymal transplantation could be performed by the trans-vessel wall technique described within this thesis. That technique might, however, not be suitable for treating a large ischemic lesion in humans. On the contrary, in discrete lesions where only a niche cell needs to be replaced, such as in type I diabetes, where the cells do not possess properties for diapedesis, the Extroducer could potentially really show its worth. The next natural step is transplantation of insulin producing cells to swine before planning for clinical trials.

The Extroducer is not limited to cell transplantation. Its main design is to provide a working channel by endovascular technique to the parenchyma in various, otherwise, inaccessible organs. Through that working channel, other procedures such as local chemotherapy-, irradiation-, growth factor administration, tissue sampling, electrophysiological diagnostics and thermo-therapy becomes possible. Further, combined with optical spectroscopic analysis, it might even be possible to perform infra-light histological analysis of tissue via the Extroducer.

3.1 Conclusions and future research

We have discussed the rationale for using endovascular methods for cell transplantation and described findings from our group and other groups, showing that selective intra-arterial administration is a safe way, with a short follow up time, to increase engraftment levels compared to intravenous delivery. However, not all cell systems are optimal for intra-luminal transplantation. These factors might be dependent on integrin expression and endothelium interactions. For cells that lack the capacity to perform diapedesis, and especially for more specific niche cell systems in organ systems that are difficult to reach, we have also developed a system for trans-vessel wall parenchymal access. The Extroducer system has been evaluated both for long term effects and feasibility for pancreas access. *Ergo,* endovascular intervention should provide a number of methods for efficient and safe cell transplantation in current and future clinical practice. The transplantation method must be decided on a disease to disease, cell to cell and patient to patient manner. Future research should investigate the possibilities of providing cells meant for intravasal transplantation with the necessary properties for performing diapedesis.

4. References

Adler, P. H., W. Yu, et al. (1990). "On the tensile and torsional properties of pseudoelastic Ni-Ti." *Scripta Metallurgica et Materila* 24: 943-947

Akesson, E., J. H. Piao, et al. (2007). "Long-term culture and neuronal survival after intraspinal transplantation of human spinal cord-derived neurospheres." *Physiol Behav* 92(1-2): 60-6

Amar, A. P., B. V. Zlokovic, et al. (2003). "Endovascular restorative neurosurgery: a novel concept for molecular and cellular therapy of the nervous system." *Neurosurgery* 52(2): 402-12; discussion 412-3

Bajada, S., I. Mazakova, et al. (2008). "Updates on stem cells and their applications in regenerative medicine." *J Tissue Eng Regen Med* 2(4): 169-83

Bale, R. and G. Widmann (2007). "Navigated CT-guided interventions." *Minim Invasive Ther Allied Technol* 16(4): 196-204

Bang, O. Y., J. S. Lee, et al. (2005). "Autologous mesenchymal stem cell transplantation in stroke patients." *Ann Neurol* 57(6): 874-82

Barbash, I. M., P. Chouraqui, et al. (2003). "Systemic delivery of bone marrow-derived mesenchymal stem cells to the infarcted myocardium: feasibility, cell migration, and body distribution." *Circulation* 108(7): 863-8

Ben-Haim, S., W. F. Asaad, et al. (2009). "Risk factors for hemorrhage during microelectrode-guided deep brain stimulation and the introduction of an improved microelectrode design." *Neurosurgery* 64(4): 754-62; discussion 762-3

Bertuzzi, F., S. Marzorati, et al. (2006). "Islet cell transplantation." *Curr Mol Med* 6(4): 369-74

Borlongan, C. V., M. Hadman, et al. (2004). "Central nervous system entry of peripherally injected umbilical cord blood cells is not required for neuroprotection in stroke." *Stroke* 35(10): 2385-9

Bruder, S. P., N. Jaiswal, et al. (1997). "Growth kinetics, self-renewal, and the osteogenic potential of purified human mesenchymal stem cells during extensive subcultivation and following cryopreservation." *J Cell Biochem* 64(2): 278-94

Buckner, C. D., R. A. Clift, et al. (1974). "Marrow transplantation for the treatment of acute leukemia using HL-A-identical siblings." *Transplant Proc* 6(4): 365-6

Butcher, E. C. (1991). "Leukocyte-endothelial cell recognition: three (or more) steps to specificity and diversity." *Cell* 67(6): 1033-6

Castleman, L. S., S. M. Motzkin, et al. (1976). "Biocompatibility of nitinol alloy as an implant material." *J Biomed Mater Res* 10(5): 695-731

Chen, J., Y. Li, et al. (2001). "Therapeutic benefit of intracerebral transplantation of bone marrow stromal cells after cerebral ischemia in rats." *J Neurol Sci* 189(1-2): 49-57

Chien, S., R. G. King, et al. (1975). "Viscoelastic properties of human blood and red cell suspensions." *Biorheology* 12(6): 341-6

Childs, R., A. Chernoff, et al. (2000). "Regression of metastatic renal-cell carcinoma after nonmyeloablative allogeneic peripheral-blood stem-cell transplantation." *N Engl J Med* 343(11): 750-8

Clausen, F., T. Lorant, et al. (2007). "T lymphocyte trafficking: a novel target for neuroprotection in traumatic brain injury." *J Neurotrauma* 24(8): 1295-307

Cognard, C., A. Weill, et al. (1998). "Intracranial berry aneurysms: angiographic and clinical results after endovascular treatment." *Radiology* 206(2): 499-510

Cronqvist, M., R. Wirestam, et al. (2005). "Diffusion and perfusion MRI in patients with ruptured and unruptured intracranial aneurysms treated by endovascular coiling: complications, procedural results, MR findings and clinical outcome." *Neuroradiology* 47(11): 855-73

Daar, A. S. and H. L. Greenwood (2007). "A proposed definition of regenerative medicine." *J Tissue Eng Regen Med* 1(3): 179-84

Danon, D. and E. Skutelsky (1976). "Endothelial surface charge and its possible relationship to thrombogenesis." *Ann N Y Acad Sci* 275: 47-63

Detante, O., A. Moisan, et al. (2009). "Intravenous administration of 99mTc-HMPAO-labeled human mesenchymal stem cells after stroke: in vivo imaging and biodistribution." *Cell Transplant* 18(12): 1369-79

Dustin, M. L., R. Rothlein, et al. (1986). "Induction by IL 1 and interferon-gamma: tissue distribution, biochemistry, and function of a natural adherence molecule (ICAM-1)." *J Immunol* 137(1): 245-54

Elices, M. J., L. Osborn, et al. (1990). "VCAM-1 on activated endothelium interacts with the leukocyte integrin VLA-4 at a site distinct from the VLA-4/fibronectin binding site." *Cell* 60(4): 577-84

Fay, J. A. (1994). Introduction to Fluid Mechanics. Boston, MIT press.

Feeney, D. M., M. G. Boyeson, et al. (1981). "Responses to cortical injury: I. Methodology and local effects of contusions in the rat." *Brain Res* 211(1): 67-77

Fischer, U. M., M. T. Harting, et al. (2009). "Pulmonary passage is a major obstacle for intravenous stem cell delivery: the pulmonary first-pass effect." *Stem Cells Dev* 18(5): 683-92

Freed, C. R., P. E. Greene, et al. (2001). "Transplantation of embryonic dopamine neurons for severe Parkinson's disease." *N Engl J Med* 344(10): 710-9

Fulton, G. P. (1957). "Microcirculatory terminology." *Angiology* 8(1): 102-4

Fulton, G. P. and B. R. Lutz (1957). "The use of the hamster cheek pouch and cinephotomicrography for research on the microcirculation and tumor growth, and for teaching purposes." *Bmq* 8(1): 13-9

Furlani, D., M. Ugurlucan, et al. (2009). "Is the intravascular administration of mesenchymal stem cells safe? Mesenchymal stem cells and intravital microscopy." *Microvasc Res* 77(3): 370-6

Gavina, M., M. Belicchi, et al. (2006). "VCAM-1 expression on dystrophic muscle vessels has a critical role in the recruitment of human blood-derived CD133+ stem cells after intra-arterial transplantation." *Blood* 108(8): 2857-66

Ghio, M., P. Contini, et al. (1999). "Soluble HLA class I, HLA class II, and Fas ligand in blood components: a possible key to explain the immunomodulatory effects of allogeneic blood transfusions." *Blood* 93(5): 1770-7

Gibbons, G. H. and V. J. Dzau (1994). "The emerging concept of vascular remodeling." *N Engl J Med* 330(20): 1431-8

Glagov, S., R. Vito, et al. (1992). "Micro-architecture and composition of artery walls: relationship to location, diameter and the distribution of mechanical stress." *J Hypertens Suppl* 10(6): S101-4

Gonzalez, N., Y. Murayama, et al. (2004). "Treatment of unruptured aneurysms with GDCs: clinical experience with 247 aneurysms." *AJNR Am J Neuroradiol* 25(4): 577-83

Gordon, P. H., Q. Yu, et al. (2004). "Reaction time and movement time after embryonic cell implantation in Parkinson disease." *Arch Neurol* 61(6): 858-61

Guglielmi, G., F. Vinuela, et al. (1991). "Electrothrombosis of saccular aneurysms via endovascular approach. Part 1: Electrochemical basis, technique, and experimental results." *J Neurosurg* 75(1): 1-7

Gussoni, E., H. M. Blau, et al. (1997). "The fate of individual myoblasts after transplantation into muscles of DMD patients." *Nat Med* 3(9): 970-7

Guzman, R., A. De Los Angeles, et al. (2008). "Intracarotid injection of fluorescence activated cell-sorted CD49d-positive neural stem cells improves targeted cell delivery and behavior after stroke in a mouse stroke model." *Stroke* 39(4): 1300-6

Hagell, P., P. Piccini, et al. (2002). "Dyskinesias following neural transplantation in Parkinson's disease." *Nat Neurosci* 5(7): 627-8

Haider, H., S. Jiang, et al. (2008). "IGF-1-overexpressing mesenchymal stem cells accelerate bone marrow stem cell mobilization via paracrine activation of SDF-1alpha/CXCR4 signaling to promote myocardial repair." *Circ Res* 103(11): 1300-8

Hirshberg, B., S. Montgomery, et al. (2002). "Pancreatic islet transplantation using the nonhuman primate (rhesus) model predicts that the portal vein is superior to the celiac artery as the islet infusion site." *Diabetes* 51(7): 2135-40

Humar, A., R. Kandaswamy, et al. (2000). "Decreased surgical risks of pancreas transplantation in the modern era." *Ann Surg* 231(2): 269-75

Jin, K., Y. Sun, et al. (2005). "Comparison of ischemia-directed migration of neural precursor cells after intrastriatal, intraventricular, or intravenous transplantation in the rat." *Neurobiol Dis* 18(2): 366-74

Johansson, C. B., M. Svensson, et al. (1999). "Neural stem cells in the adult human brain." *Exp Cell Res* 253(2): 733-6

Justicia, C., A. Martin, et al. (2006). "Anti-VCAM-1 antibodies did not protect against ischemic damage either in rats or in mice." *J Cereb Blood Flow Metab* 26(3): 421-32

Kandaswamy, R., A. Humar, et al. (1999). "Vascular graft thrombosis after pancreas transplantation: comparison of the FK 506 and cyclosporine eras." *Transplant Proc* 31(1-2): 602-3

Kondziolka, D., L. Wechsler, et al. (2000). "Transplantation of cultured human neuronal cells for patients with stroke." *Neurology* 55(4): 565-9

Korsgren, O., T. Lundgren, et al. (2008). "Optimising islet engraftment is critical for successful clinical islet transplantation." *Diabetologia* 51(2): 227-32

Le Blanc, K., I. Rasmusson, et al. (2004). "Treatment of severe acute graft-versus-host disease with third party haploidentical mesenchymal stem cells." *Lancet* 363(9419): 1439-41

Lee, S. T., K. Chu, et al. (2008). "Anti-inflammatory mechanism of intravascular neural stem cell transplantation in haemorrhagic stroke." *Brain* 131(Pt 3): 616-29

Lehmann, R., R. A. Zuellig, et al. (2007). "Superiority of small islets in human islet transplantation." *Diabetes* 56(3): 594-603

Lindvall, O. and Z. Kokaia "Stem cells in human neurodegenerative disorders--time for clinical translation?" *J Clin Invest* 120(1): 29-40

Lindvall, O. and Z. Kokaia (2006). "Stem cells for the treatment of neurological disorders." *Nature* 441(7097): 1094-6

Lundberg, J., S. Jonsson, et al. (2010). "New endovascular method for transvascular exit of arteries and veins: developed in simulator, in rat and in rabbit with full clinical integration." *PLoS One* 5(5): e10449

Lundberg, J., S. Jonsson, et al. (2011). "Long Term Follow-up of the Endovascular Trans-Vessel Wall Technique for Parenchymal Access in Rabbit with Full Clinical Integration." *PLoS One In press*

Lundberg, J., K. Le Blanc, et al. (2009). "Endovascular transplantation of stem cells to the injured rat CNS." *Neuroradiology*

Lundberg, J., E. Sodersten, et al. (2011). "Targeted Intra-arterial Transplantation of Stem Cells to the Injured CNS is More Effective than Intravenous Administration - Engraftment is Dependent on Cell Type and Adhesion Molecule Expression." *Cell Transplant*

Ma, B., K. D. Hankenson, et al. (2005). "A simple method for stem cell labeling with fluorine 18." *Nucl Med Biol* 32(7): 701-5

Mayne, R. (1986). "Collagenous proteins of blood vessels." *Arteriosclerosis* 6(6): 585-93

Meaney, T. F., M. A. Weinstein, et al. (1980). "Digital subtraction angiography of the human cardiovascular system." *AJR Am J Roentgenol* 135(6): 1153-60

Meerschaert, J. and M. B. Furie (1995). "The adhesion molecules used by monocytes for migration across endothelium include CD11a/CD18, CD11b/CD18, and VLA-4 on monocytes and ICAM-1, VCAM-1, and other ligands on endothelium." *J Immunol* 154(8): 4099-112

Merani, S., C. Toso, et al. (2008). "Optimal implantation site for pancreatic islet transplantation." *Br J Surg* 95(12): 1449-61

Merli, M., F. Salerno, et al. (1998). "Transjugular intrahepatic portosystemic shunt versus endoscopic sclerotherapy for the prevention of variceal bleeding in cirrhosis: a randomized multicenter trial. Gruppo Italiano Studio TIPS (G.I.S.T.)." *Hepatology* 27(1): 48-53

Miller, R. G., K. R. Sharma, et al. (1997). "Myoblast implantation in Duchenne muscular dystrophy: the San Francisco study." *Muscle Nerve* 20(4): 469-78

Muller, W. A. (2003). "Leukocyte-endothelial-cell interactions in leukocyte transmigration and the inflammatory response." *Trends Immunol* 24(6): 327-34

Murayama, Y., Y. L. Nien, et al. (2003). "Guglielmi detachable coil embolization of cerebral aneurysms: 11 years' experience." *J Neurosurg* 98(5): 959-66

Nagata, S. (1999). "Fas ligand-induced apoptosis." *Annu Rev Genet* 33: 29-55

Ng, P., M. S. Khangure, et al. (2002). "Endovascular treatment of intracranial aneurysms with Guglielmi detachable coils: analysis of midterm angiographic and clinical outcomes." *Stroke* 33(1): 210-7

Nikolic, B., S. Faintuch, et al. (2009). "Stem cell therapy: a primer for interventionalists and imagers." *J Vasc Interv Radiol* 20(8): 999-1012

Pannier, J. L. and I. Leusen (1973). "Circulation to the brain of the rat during acute and prolonged respiratory changes in the acid-base balance." *Pflugers Arch* 338(4): 347-59

Petri, B. and M. G. Bixel (2006). "Molecular events during leukocyte diapedesis." *Febs J* 273(19): 4399-407

Pittenger, M. F., A. M. Mackay, et al. (1999). "Multilineage potential of adult human mesenchymal stem cells." *Science* 284(5411): 143-7

Potten, C. S. and M. Loeffler (1990). "Stem cells: attributes, cycles, spirals, pitfalls and uncertainties. Lessons for and from the crypt." *Development* 110(4): 1001-20

Pulido, R., M. J. Elices, et al. (1991). "Functional evidence for three distinct and independently inhibitable adhesion activities mediated by the human integrin VLA-4. Correlation with distinct alpha 4 epitopes." *J Biol Chem* 266(16): 10241-5

Raymond, J. and D. Roy (1997). "Safety and efficacy of endovascular treatment of acutely ruptured aneurysms." *Neurosurgery* 41(6): 1235-45; discussion 1245-6

Richter, G. M., G. Noeldge, et al. (1990). "Transjugular intrahepatic portacaval stent shunt: preliminary clinical results." *Radiology* 174(3 Pt 2): 1027-30

Ringden, O., M. Uzunel, et al. (2006). "Mesenchymal stem cells for treatment of therapy-resistant graft-versus-host disease." *Transplantation* 81(10): 1390-7

Rothlein, R., M. L. Dustin, et al. (1986). "A human intercellular adhesion molecule (ICAM-1) distinct from LFA-1." *J Immunol* 137(4): 1270-4

Ryan, E. A., J. R. Lakey, et al. (2001). "Clinical outcomes and insulin secretion after islet transplantation with the Edmonton protocol." *Diabetes* 50(4): 710-9

Scharp, D. W., P. E. Lacy, et al. (1991). "Results of our first nine intraportal islet allografts in type 1, insulin-dependent diabetic patients." *Transplantation* 51(1): 76-85

Seldinger, S. I. (1953). "Catheter replacement of the needle in percutaneous arteriography; a new technique." *Acta radiol* 39(5): 368-76

Shapiro, A. M., J. R. Lakey, et al. (1995). "Portal vein thrombosis after transplantation of partially purified pancreatic islets in a combined human liver/islet allograft." *Transplantation* 59(7): 1060-3

Shapiro, A. M., J. R. Lakey, et al. (2000). "Islet transplantation in seven patients with type 1 diabetes mellitus using a glucocorticoid-free immunosuppressive regimen." *N Engl J Med* 343(4): 230-8

Siminiak, T., D. Fiszer, et al. (2005). "Percutaneous trans-coronary-venous transplantation of autologous skeletal myoblasts in the treatment of post-infarction myocardial contractility impairment: the POZNAN trial." *Eur Heart J* 26(12): 1188-95

Slavin, S., A. Nagler, et al. (1998). "Nonmyeloablative stem cell transplantation and cell therapy as an alternative to conventional bone marrow transplantation with lethal cytoreduction for the treatment of malignant and nonmalignant hematologic diseases." *Blood* 91(3): 756-63

Soderman, M., D. Babic, et al. (2008). "Brain imaging with a flat detector C-arm : Technique and clinical interest of XperCT." *Neuroradiology* 50(10): 863-8

Soderman, M., D. Babic, et al. (2005). "3D roadmap in neuroangiography: technique and clinical interest." *Neuroradiology* 47(10): 735-40

Springer, T. A. (1994). "Traffic signals for lymphocyte recirculation and leukocyte emigration: the multistep paradigm." *Cell* 76(2): 301-14

Stamm, C., B. Westphal, et al. (2003). "Autologous bone-marrow stem-cell transplantation for myocardial regeneration." *Lancet* 361(9351): 45-6

Stoeckel, D., A. Pelton, et al. (2004). "Self-expanding nitinol stents: material and design considerations." *Eur Radiol* 14(2): 292-301

Stuve, O., R. Gold, et al. (2008). "alpha4-Integrin antagonism with natalizumab: effects and adverse effects." *J Neurol* 255 Suppl 6: 58-65

Sykova, E., A. Homola, et al. (2006). "Autologous bone marrow transplantation in patients with subacute and chronic spinal cord injury." *Cell Transplant* 15(8-9): 675-87

Thomas, E., R. Storb, et al. (1975). "Bone-marrow transplantation (first of two parts)." *N Engl J Med* 292(16): 832-43

Thomas, E. D., R. Storb, et al. (1975). "Bone-marrow transplantation (second of two parts)." *N Engl J Med* 292(17): 895-902

Thompson, C. A., B. A. Nasseri, et al. (2003). "Percutaneous transvenous cellular cardiomyoplasty. A novel nonsurgical approach for myocardial cell transplantation." *J Am Coll Cardiol* 41(11): 1964-71

Thomson, J. A., J. Itskovitz-Eldor, et al. (1998). "Embryonic stem cell lines derived from human blastocysts." *Science* 282(5391): 1145-7

Tuma, R. F. (2008). Microcirculation. Amsterdam, Academic Press.

Vacanti, J. P. and R. Langer (1999). "Tissue engineering: the design and fabrication of living replacement devices for surgical reconstruction and transplantation." *Lancet* 354 Suppl 1: SI32-4

Wennersten, A., X. Meier, et al. (2004). "Proliferation, migration, and differentiation of human neural stem/progenitor cells after transplantation into a rat model of traumatic brain injury." *J Neurosurg* 100(1): 88-96

Vestweber, D. and J. E. Blanks (1999). "Mechanisms that regulate the function of the selectins and their ligands." *Physiol Rev* 79(1): 181-213

White, S. A., R. F. James, et al. (2001). "Human islet cell transplantation--future prospects." *Diabet Med* 18(2): 78-103

Villiger, P., E. A. Ryan, et al. (2005). "Prevention of bleeding after islet transplantation: lessons learned from a multivariate analysis of 132 cases at a single institution." *Am J Transplant* 5(12): 2992-8

Vulliet, P. R., M. Greeley, et al. (2004). "Intra-coronary arterial injection of mesenchymal stromal cells and microinfarction in dogs." *Lancet* 363(9411): 783-4

Zuk, P. A., M. Zhu, et al. (2001). "Multilineage cells from human adipose tissue: implications for cell-based therapies." *Tissue Eng* 7(2): 211-28

10

Dynamic Relationships of Collagen Extracellular Matrices on Cardiac Differentiation of Human Mesenchymal Stem Cells

Pearly Yong, Ling Qian,
YingYing Chung and Winston Shim
Research and Development Unit, National Heart Centre
Singapore

1. Introduction

Myocardial infarction (MI) results in necrosis, inflammation and scar formation in the myocardium. Such pathological insults place increasing mechanical demands on surviving cardiomyocytes (Boudoulas & Hatzopoulos, 2009). As cardiomyocytes have limited regenerative potential, loss of functional healthy tissue and subsequent left ventricular (LV) remodelling, eventually leads to pathological hypertrophic cardiomyopathy. Hypertrophy of the LV has been documented as a chronic response to MI and invariably progresses to heart failure (Hannigan et al., 2007). Chronic heart failure is a major health problem with patients experiencing a debilitating quality of life.

Cardiac remodelling after MI is characterised by progressive and pathological interstitial fibrosis. During acute phase of cardiac repair, degradation of myocardial extracellular matrix (ECM) coupled with an influx of inflammatory cells and cytokines permits deposition of granulation tissue in the infarct region. At the site of tissue injury, granulation tissue composes of macrophages, myofibroblasts and neovascularisation. Activated myofibroblasts synthesise collagen and other ECM proteins to form dense scar tissue in the infarct in response to inflammatory mediators such as angiotensin II (Ang II) and transforming growth factor-β1 (TGF-β1). Macrophages drive the production of TGF-β1, an essential growth factor for fibroblast production, collagen synthesis and inhibition of collagen degradation (O'Kane & Ferguson, 1997; Sun & Weber, 2000). At the site of MI, increased expression of adhesion molecules (inter-cellular adhesion molecule-1, ICAM-1) and chemoattractant cytokines (monocyte chemotactic protein-1, MCP-1) facilitate migration of inflammatory cells (e.g. macrophages) enabling scavenging of necrotic tissues (Lu et al., 2004). This couples with elevated expression of matrix metalloproteinase-1 (MMP-1) results in remodelling of myocardial ECM by degradation of existing collagen I and III in the injured myocardium (Lu et al., 2004). Furthermore, MMP-9 has been implicated in tissue remodelling by cleaving collagen V at the amino-terminus (Niyibizi et al., 1994). Consequently, this process compromises structural integrity of the ventricles, resulting in myocyte slippage, wall thinning and rupture

(Cleutjens et al., 1995b). Derangements in cardiomyocyte-ECM interactions cause the loss of cellular tensegrity and initiates anoikis in neighbouring healthy tissue (Michel, 2003). It is now well recognised that structural changes in the myocardial ECM can alter collagen-integrin-cytoskeletal-myofibril relations, thus affecting overall geometry and function of the heart (Spinale, 2007).

In non-cartilaginous tissues like the heart, collagen I, III and V are the predominant subtypes of the ECM (Breuls et al., 2009; Linehan et al., 2001). Collagen I is primarily a structural element of the myocardial ECM while collagen V represents a minor, but important component sequestered within collagen I fibres. However, collagen V levels increase in inflammation and scar tissue. The relative resistance of collagen V to mammalian collagenases makes it transiently available during tissue remodelling. The temporal availability of collagen V during active extracellular remodelling implies that it may play an important role in ECM remodelling and tissue stiffness (Breuls et al., 2009; Ruggiero et al., 1994). In fact, collagen V plays a deterministic role in collagenous fibril structure, matrix organisation and stiffness (Fichard et al., 1995).

Binding of ECM to integrins provides a linkage between the ECM and cellular cytoskeleton. Integrins are heterodimeric receptors composed of non-covalently bound α and β subunits. (Brancaccio et al., 2006). Dynamic integrin-ECM interactions result in bidirectional signalling and determines cell morphology, gene expression, migration, proliferation, differentiation and death. Perkins et al. (2010) showed that integrin-mediated adhesion is mandatory for maintenance of the sarcomeric architecture. They proposed that disintegration of the Z-line and progressive muscle degeneration can occur once the adhesion complex comprising of integrins, talin or integrin linked kinase (ILK) is not replenished. In the myocardium, integrins can function as mechanotransducers that transmit mechanical ECM cues to the myocyte, resulting in changes to myocyte biology and function (Ross & Borg, 2001). Integrins $\alpha_2\beta_1$, $\alpha_1\beta_1$, $\alpha_3\beta_1$, $\alpha_v\beta_3$, $\alpha_{IIb}\beta_3$ are collagen binding heterodimers and adhesion to collagen V has been reported to be primarily mediated by integrin $\alpha_2\beta_1$ and $\alpha_1\beta_1$ (Ruggiero et al., 1994). Integrins $\alpha_2\beta_1$ and $\alpha_1\beta_1$ may play a significant role in remodelling of the heart where there is increased collagen synthesis and collagen V expression, although we have previously shown $\alpha_v\beta_3$, but not $\alpha_2\beta_1$, in a collagen V associated cardiac differentiation of human mesenchymal stem cells (hMSCs) (Tan et al., 2010).

Increased ejection fraction (EF) and fractional shortening (FS) parameters, coupled with a reduction in the amount of fibrotic scar tissue have been highlighted following cellular therapy (Chacko et al., 2009). Our previous study showed that cardiomyocyte-like cells (CLCs) that were differentiated from MSCs, improved systolic performance without compromising end-diastolic pressure of the infarcted myocardium when compared to MSCs. CLCs may facilitate hemodynamic recovery by preserving tissue elasticity in the collagen V-expressing peri-infarct borders. This unique cell/matrix relationship may be more conducive to a functionally adaptive remodelling response in maintaining contractile efficiency of post-infarcted myocardium (Tan et al., 2010).

Experimental data show that MSC transplantation inhibits LV remodelling and improves heart function in animals with MI (Xu et al., 2005). Despite the ability of angiogenic mechanisms to reduce infarct mass, only partial restoration of ventricular contraction occurs as myocytes are not regenerated (Gaudette & Cohen, 2006). In addition, cardiac differentiation and retention of surviving transplanted MSCs in-vivo is limited (Feygin et al.,

2007). Influence of ECM proteins and integrin interactions on MSC differentiation have been widely investigated for chrondrogenic and osteogenic differentiation (Djouad et al., 2007; Gronthos et al., 2001). Conversely, studies investigating ECM role in cardiac differentiation of MSCs is limited. Unravelling of integrin roles in cardiac differentiation of MSCs would aid in understanding of mechanisms leading to retention and integration of stem cells in myocardium.

We have previously reported in-vitro differentiation of human MSCs towards CLCs and shown that collagen V promoted adhesion and cardiac gene expression in CLCs (Shim et al., 2004; Tan et al., 2010). In the present study, we further examine the role of individual integrins in cardiac differentiation of CLCs.

2. Materials and methods

2.1 Isolation and culture of bone marrow derived MSCs

Bone marrow was isolated from the sternum of patients undergoing open-heart surgery. They were collected in 17 IU/ml heparin using a 23-gauge needle. Bone marrow aspirates were topped up to 15 ml with Dulbecco's modified Eagle's medium-low glucose (DMEM-LG, GIBCO) supplemented with 10% fetal bovine serum (FBS, Hyclone) and 1% penicillin-streptomycin (Gibco, Invitrogen). To deplete bone marrow asiprates of mature blood lineages, 15 ml of bone marrow blood mixture was overlaid onto 15 ml of Histopaque®-1077 (Sigma-Aldrich) and centrifuged for 1500 rpm (Kubota Centrifuge) for 30 minutes at 4°C. The enriched cell fraction was collected from the interphase, washed once with 5 ml of media and centrifuged at 1200 rpm (Kubota Centrifuge) for 10 minutes. Resuspended cells were then transferred into tissue culture flasks with basal normal growth medium (NGM) comprising DMEM-LG supplemented with 10% FBS for 9 - 11 days to yield plastic adherent MSCs. Subconfluent cells were harvested using 1X Trypsin-EDTA solution for endothelial cell culture (Sigma-Aldrich), 14 – 21 days after initial plating and maintained as MSCs in basal NMG or differentiated towards CLCs in a myogenic differentiation medium (MDM) as previously described (Shim et al., 2004).

Type V collagen (Sigma-Aldrich) and Type I collagen (BD™) were coated on 6-well plates or tissue culture flasks at 10μg/cm² for 3 hours at room temperature. Plates and flasks were washed twice with phosphate buffered saline (PBS) and kept at 4°C until required.

2.2 Fluorescence microscopy

Frozen tissue sections of the explanted ventricular rat hearts were fixed in 4% paraformaldehyde (PFA), permeabilised with 0.1% Triton X-100, and further blocked in 5% bovine serum albumin (BSA). This was followed by overnight incubation at 4°C with primary antibodies, including collagen I (Southern Biotech), collagen III (Affinity Bioregent) collagen V (Biotrend) and anti-α-sarcomeric actinin (Sigma-Aldrich) diluted in 1% BSA. Sections were incubated with Alexa Fluor® 488/555/660 - conjugated secondary antibodies (Molecular Probes) in 0.1% BSA at room temperature for 3 hours before staining the nuclei with DAPI. Immunofluorescence microscopy was performed with Zeiss Axiovert 200 M fluorescence microscope, using the Metamorph software (version 6.2, Molecular Devices) or Leica MZ 16 FA Fluorescence Steromicroscope, using the Leica Application Suite software (Version 3.3.0, Leica).

2.3 Flow cytometry

Sternum-derived bone marrow MSCs were differentiated into CLCs and characterised by flow cytometry after 14 days in a MDM. CLCs cultured on uncoated, collagen I or V coated tissue culture flasks were stained with antibodies directed towards integrin subunits α_1 (Abcam), α_2 (Santa Cruz), α_v (Fitzgerald), β_1 (Chemicon) and β_3 (Cell Signaling). Cells were treated with Fix & Perm® Cell Permeabilisation Kit (Invitrogen) and subsequently blocked in PBS containing 5% BSA, 1% FBS and 5 mM ethylenediaminetetracetic acid (EDTA) for 30 minutes at 4°C on a roller. CLCs were then incubated with directly conjugated antibodies for 30 minutes at 4°C. Indirectly conjugated antibodies were incubated for 2 hours at 4°C and subsequently stained with their respective Alexa Fluor® 555 conjugated secondary antibodies (Invitrogen) for 2 hours at 4°C. Isotype controls were stained in parallel with the test samples. Samples were washed in PBS containing 2% BSA, 2% FBS and 5 mM EDTA after each antibody staining and fixation step. All samples were fixed in PBS containing 4% PFA/PBS, washed and resuspended in PBS containing 2% FBS and 0.09% sodium azide (NaZ). Data analysis was performed using FACSDiva software (version 6.1.2, BD™), FlowJo software (version 6.4, Tree Star, Inc.). Histogram overlays were performed and the change in median fluorescence intensity and overton subtraction percentages were computed.

2.4 Integrin neutralisation assays

Integrin neutralisation assays were performed on CLCs using neutralising antibodies against the integrin α_1 (Millipore) subunit and $\alpha_v\beta_3$ (Millipore) heterodimer, at 1μg/ml and 10 μg/ml respectively. CLCs treated with 1μg/ml or 10 μg/ml isotype IgG (Abcam/Dako) antibodies and untreated CLCs served as controls. After trypsin digestion, CLCs were incubated with neutralising and isotype control antibodies for 2 hours at 4°C. 50,000 untreated and treated CLCs were seeded on collagen V pre-coated 6-well plates. Plated CLCs were harvested after 72 hours of culture at 37°C, 5% CO_2. Total RNA was extracted using the RNeasy Mini Kit (Qiagen) and treated with RNAse free DNase solution (Qiagen). DNAse treated RNA samples were stored at -80°C until required.

2.5 Real-time reverse transcriptase polymerase chain reaction for quantitation of cardiac gene expression

First strand cDNA was synthesised from total RNA using the SuperScript™ III First-Strand Synthesis System (Invitrogen) and equal concentrations of cDNA were loaded into tubes containing QuantiFast SYBR Green PCR mastermix (Qiagen). Real-time reverse transcriptase polymerase chain reaction (RT-PCR) was performed on the Rotor-Gene Q thermocycler (Qiagen) using standard cycling parameters and relative gene expression of the following cardiac transcripts was quantitated using the $\Delta\Delta C_T$ method. These transcripts include β actin (BA), cardiac α-actin (CAA), skeletal muscle α-actin (SKAA), troponin T (Trop T), troponin C (Trop C), Nkx2.5 and GATA4 (Sigma-Aldrich). Target gene expression values were normalised relative to the untreated CLCs. BA served as a housekeeping gene for the real time RT-PCR experiments. No template controls were concurrently processed with test samples to rule out the presence of contaminated reagents and nucleic acids.

Gene	Acession Number	Primer Sequence	Product Size (bp)
CAA	NM_005159	5'-CTTCTAAGATGCCTTCTCTCTCCA-3' 5'-TAT TAG AAG CAC AAA CAA ATT GCA-3'	147
SKAA	NM_001100.3	5'-CGAGACCACCTACAACAGCA-3' 5'-GCGGTGATCTCTTTCTGCAT-3'	132
Trop C	NM_003280.2	5'-CTACAAGGCTGCGGTAGAGC-3' 5'-CAGCACGAAGATGTCGAAGG-3'	76
Trop T	NM_000364.2	5'-ATCCCCGATGGAGAGAGAGT-3' 5'-ACGAGCTCCTCCTCCTCTTT-3'	128
Nkx2.5	NM_004387	5'-GATTCCGCAGAGCAACTCG-3' 5'-GGAGCTGTTGAGGTGGGATCG-3'	105
GATA4	NM_002052.3	5'-TCCAAACCAGAAAACGGAAG-3' 5'-AAGGCTCTCACTGCCTGAAG-3'	77
BA	NM_001101	5'-TCCCTGGAGAAGAGCTACGA-3' 5'-AGCACTGTGTTGGCGTACAG-3'	194

Table 1. Primer sequences for real time reverse transcriptase polymerase chain reaction (RT-PCR). Transcripts obtained from RT-PCR assays were all less than 200 bp. CAA, cardiac α-actin; SKAA, skeletal muscle α-actin; Trop T, troponin T; Trop C, troponin C; BA, β actin.

2.6 Cell labelling

CLCs were labelled with 1 mmol/L Vybrant CellTracker chloromethyldialkylcarbocyanie (CM-Dil; Molecular Probes) overnight at 37°C and rinsed 3 times before trypsin disgestion and transplantation. MSCs were labelled with 10mmol/L Vybrant carboxy fluorescein diacetate succinimidyl ester (CFDA-SE; Molecular Probes). Cells were resuspended in a final concentration of 1×10^6/0.1 ml to 5×10^6/0.2 ml.

2.7 Rat myocardial infarction model

MI was created in n=20 female Wistar rats per group. Each rat weighed approximately 350 - 400g in body weight. The animals were subjected to left thoracotomy and the left anterior descending artery (LAD) was exposed and ligated. After which rats were allowed a week for recovery before given treatment of either injection with labelled cells or placebo to the area of infarction. Cyclosporin A was administered at a dose of 5mg/kg body weight at 3 days before and daily following treatment for 6 weeks until end point.

2.8 Echocardiography

Baseline echocardiography was performed on each rat before MI and 6 weeks after treatment. Echocardiography images were acquired using Vivid 7 ultrasound machine (General Electric VingMed) equipped with i13L linear probe operated at 14MHz. Rats were anaesthetised using 1% - 2% isofluorane with 1L/hr oxygen and then fixed in the supine position on a heated platform. Rats were then shaved at the chest and abdominal areas before electrocardiography (ECG) electrodes were placed onto the left and right leg as well as the left upper extremity. All analysis was performed offline with EchoPAC workstation (General Electric Healthcare).

2.9 Statistical analysis

One-way analysis of variance (ANOVA) was used to determine statistical significance between different treatment groups. Tukey Honestly Significant Difference (HSD) post-hoc analyses were used to determine statistical significance between treatment groups using SPSS 13 software (SPSS Inc.). $p < 0.05$ was considered statistically significant. All data are presented as mean ± standard deviation (SD).

3. Results

3.1 Integrin expression and cardiac differentiation

Flow cytometric analysis showed that α_v and β_1 were the predominant subunits of integrins in CLCs, independent of substrate surface (Table 2). In comparison to collagen V matrix, CLCs cultured on collagen I showed a higher expression of integrin α_1 (59.4 ± 13.7% vs. 78.0 ± 0.9%) and β_3 (44.7 ± 10.6% vs. 56.0 ± 21.8%) subunits. Furthermore, with the exception of α_1 subunit, α_2, α_v, β_1 and β_3 integrins in CLCs cultured on either collagen matrices showed a reduction of expression in comparison to CLCs cultured on polystyrene tissue culture surface.

	CLC		
Integrin Subunit	Uncoated	Collagen I	Collagen V
$\alpha 1$	58.5 ± 1.8	78.0 ± 0.9	59.4 ± 13.7
$\alpha 2$	59.4 ± 6.1	39.9 ± 11.5	44.3 ± 12.8
αV	91.6 ± 0.5	83.8 ± 5.7	81.8 ± 4.6
$\beta 1$	93.4 ± 0.8	78.5 ± 8.6	79.1 ± 8.6
$\beta 3$	57.6 ± 2.4	56.0 ± 21.8	44.7 ± 10.6

Table 2. Flow cytometric analysis showed that integrin α_v and β_1 were the predominant subunits in CLCs. CLCs cultured on collagen I showed increased levels of α_1 and β_3. Data were derived from 3 independent experiments and the overton percentage positive results are expressed as mean ± SD. MSCs, Mesenchymal stem cells; CLCs, Cardiomyocyte-like cells.

3.2 CLCs enhance cardiac gene expression via integrin α_1 and $\alpha_v\beta_3$ on collagen V matrices

We previously reported that collagen V matrix enhanced cardiac gene expression when compared to CLCs seeded on collagen I matrix. Collagen V selectively upregulated expression of cardiac transcription factors (GATA4, Nkx2.5), calcium handling transporter (RyR2) and sarcomeric myofilament proteins (Trop T, Trop C, SKAA) in CLCs (Tan et al., 2010). Neutralisation of $\alpha_v\beta_3$ integrin or α_1 subunit in this study did not affect CAA and SKAA gene expression in CLCs that were cultured on collagen V matrix. Furthermore, no significant changes in Nkx2.5 or GATA4 expression was observed in α_1 subunit neutralised CLCs. However, Nkx2.5 down regulation was observed in CLCs neutralised with $\alpha_v\beta_3$, although similar down regulation was also evident in the isotype control experiment. Gene expression of Trop C reduced significantly after $\alpha_v\beta_3$ integrin neutralisation. In contrary, α_1

subunit neutralisation upregulated Trop C expression. Furthermore, there was a concomitant upregulation of Trop T following α_1 integrin neutralisation.

Integrin Neutralisation Antibody	CAA		SKAA		Trop C		Nkx2.5		Trop T		GATA4	
	Isotype	Test	Isotype	Test	Isotype	Test	Isotype	Test	Isotype	Test	Isotype	Test
$\alpha_v \beta_3$	0.87 ± 0.00	1.00 ± 0.10	1.32 ± 0.13	1.15 ± 0.00	0.94 ± 0.10	0.12 ± 0.02	0.23 ± 0.21	0.22 ± 0.16	N.D.	N.D.	N.D.	N.D.
α_1	0.52 ± 0.05	0.91 ± 0.02	0.99 ± 0.12	1.02 ± 0.10	1.17 ± 0.11	1.45 ± 0.40	1.02 ± 0.22	0.90 ± 0.22	1.04 ± 0.05	1.63 ± 0.16	0.87 ± 0.00	0.63 ± 0.45

Table 3. CLCs cultured on collagen V were treated with integrin α_1 (1 μg/ml) neutralising antibodies. Untreated and isotype IgG (1 μg/ml) treated CLCs served as controls for this experiment. Optimal concentrations of test and control antibodies were predetermined in a series of titration experiments. CLCs cultured on collagen V were treated with integrin $\alpha_v\beta_3$ (10 μg/ml) neutralising antibodies. Untreated and isotype IgG (10 μg/ml) treated CLCs served as controls for this experiment. Results are expressed as mean ± SD. CAA, cardiac α-actin; SKAA, skeletal muscle α-actin; Trop T, troponin T; Trop C, troponin C, N.D., not done; MSCs, Mesenchymal stem cells; CLCs, Cardiomyocyte-like cells.

3.3 CLCs integrate into collagen V-rich cardiac syncytium

Consistent with our previous report (Tan et al., 2010), collagen I as the main constituent of cardiac ECM in intact rat myocardium, was found to co-localise with collagen III matrix in the epicardium and perimysial space between major muscle bundles dispersed throughout the myocardium (Fig. 1A). On the other hand, collagen V was predominantly observed in the endomysial space surrounding healthy cardiomyocytes and in the perivascular structures within the myocardium. Following MI by ligating the LAD artery, significant wall thinning was observed in the anterior wall of the LV (Fig.1B). Accumulation of collagen matrices was evident in the infarcted and non-infarcted zones 7 weeks post infarction. Spatial remodelling and redistribution of collagen matrices were observed whereby perimysial collagen I diminished significantly and upregulation of collagen I and III were observed in the pericardium and epicardium of infarcted as well as non-infarcted zones. Furthermore, fibrosis consisted mainly of collagen I matrix was prominently found in the endocardium of infarcted zone while it co-localised with collagen III matrix in the pericardium/epicardium of the non-infarcted and infarcted zones (Fig. 1Bii & 1Biii). In contrast, collagen V fibrils were sparsely detected in the epicardium of infarct, but were prominently found in the peri-vascular structures within the infarct (Fig. 1Biii). In contrast to redistribution of collagen I matrix, collagen V remained in the endomysial matrix of individual muscle fibres in the non-infarcted borders (Fig.1Bii) and also surrounding isolated, but viable cardiac fibres in the infarct.

Myocardial transplanted CLCs were closely associated with collagen V matrix in the endomysial space in the peri-infarct border of the myocardium (Fig. 2A). In contrast, similarly transplanted MSCs were only found in collagen I-rich infarct despite the presence of isolated, collagen V-expressing, myofibres at the infarct borders (Fig. 2B). Furthermore, CLCs were often intimately engrafted among α-actinin stained native cardiomyocytes that were surrounded by collagen V, but not collagen I, matrices (Fig. 2C and 2D). On the contrary, transplanted MSCs were sequestered in the infarct that was dominated with collagen I matrix and isolated from viable and α-actinin stained myocardium that expressed collagen V matrix (Fig. 2E).

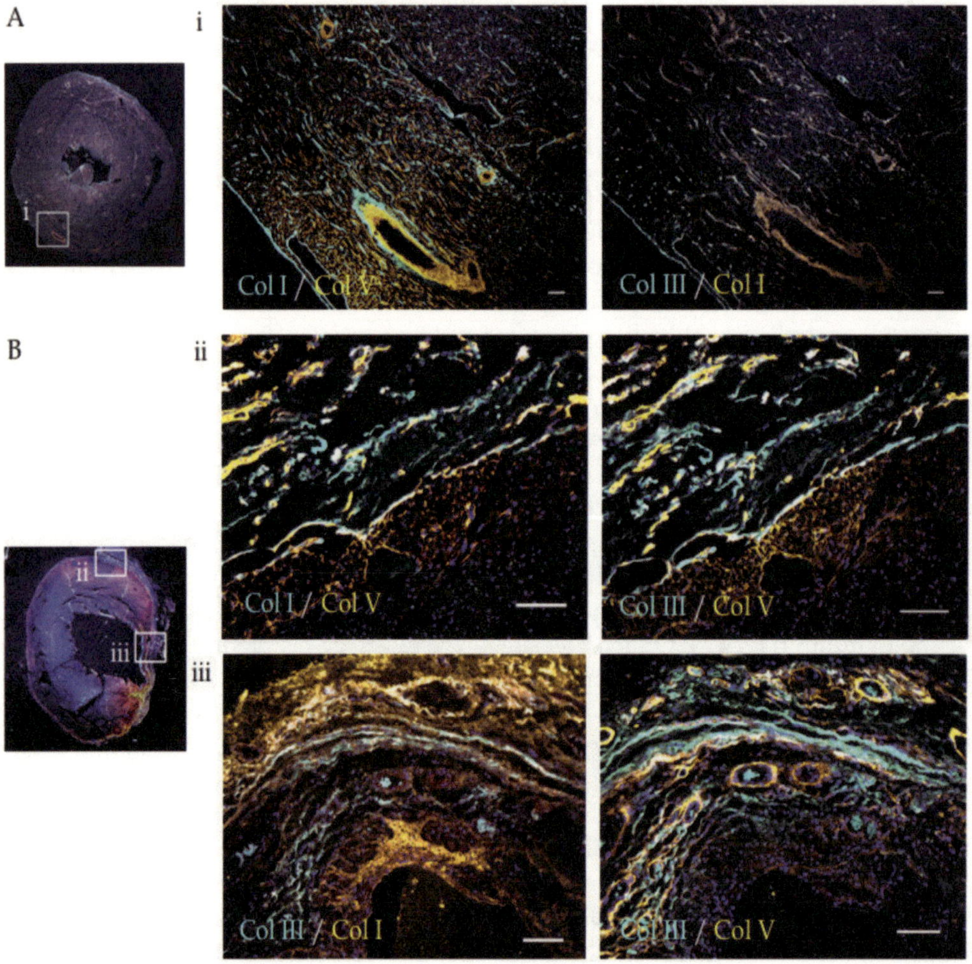

Fig. 1. (A) Collagen distribution in an intact myocardium. (Ai) Higher magnification of the boxed area showing collagen I and III distribution in the perimysium while collagen V was expressed in the endomysial space. (B) Collagen distribution in an infarcted myocardium. (Bii) Higher magnification of the myocardium, epicardium and pericardium at the boxed area. Collagen V was predominantly expressed at the peri-infarct border surrounding viable myocytes and vasculature structures in the infarct region. Collagen I and III were positively stained in the infarcted epicardium and pericardium. (Biii) Higher magnification demonstrating severe thinning of the LV anterior wall. Co-localisation of collagen I and III extended from the pericardium into the infarcted myocardium whereas collagen I was primarily localised in the endocardium. Collagen V was expressed in the vessels and sparsely in the infarct. Scale bar: 200μm.

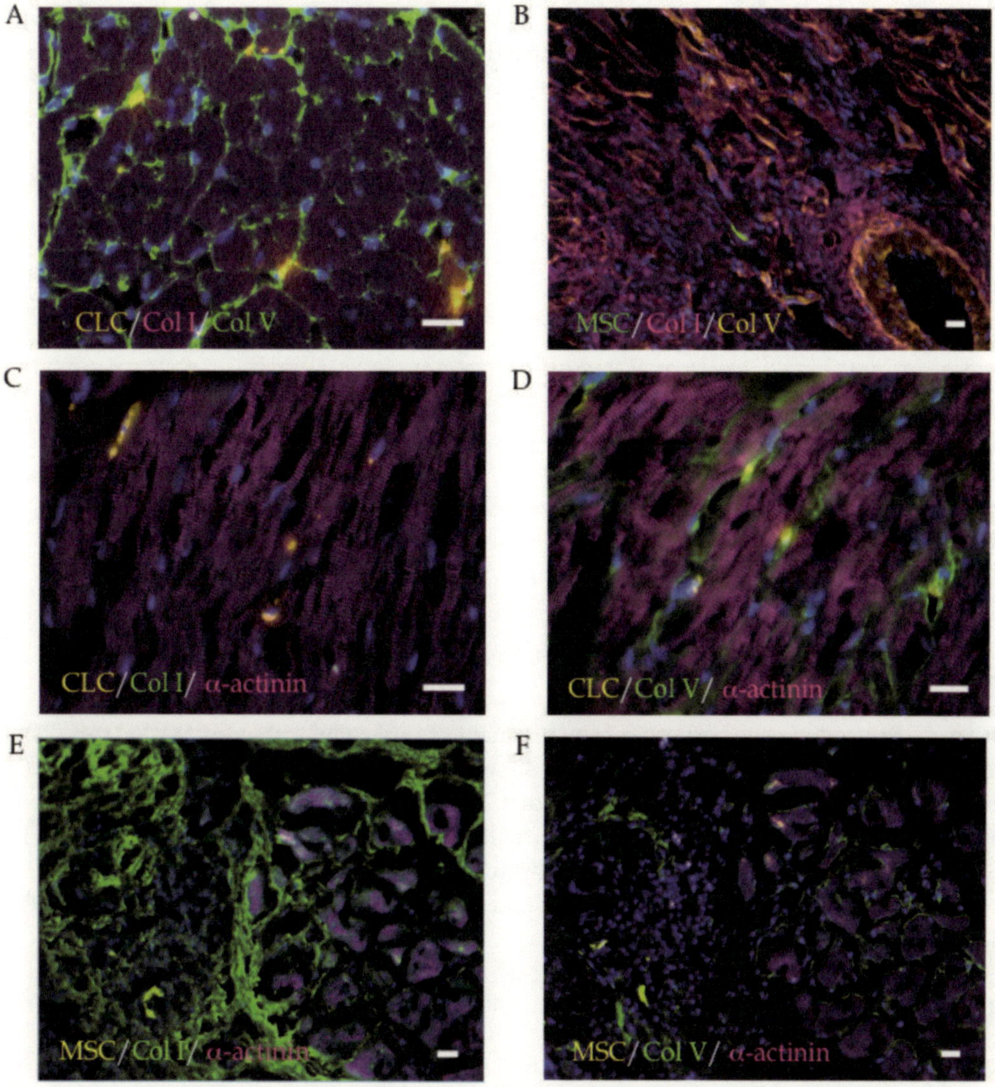

Fig. 2. (A) Transplanted CLCs preferentially home to the collagen V-rich myocardial ECM. (B) Transplanted MSCs localised in the collagen I enriched infarct zone away from the collagen V peri-infarct region. (C) Engraftment of CLCs in the α-actinin stained myocardium (D) showing an affinity towards collagen V matrix in the absence of collagen I staining. (E) MSCs were embedded in the collagen I-rich infarct zone and were isolated from α-actinin expressing cardiomyocytes. (F) Collagen V was sparsely distributed in the infarcted region, but mainly surrounded viable myocytes at the peri-infarct border. Scale bar: 20μm. MSCs: Mesenchymal stem cells; CLCs: Cardiomyocyte-like cells; Col I: Collagen I; Col V: Collagen V.

3.4 CLC therapy at high doses improve cardiac hemodynamics

Consistent with their muscular engraftment, LV echocardiography confirmed a better cardiac performance of transplanted CLCs, 6 weeks post cell transplant (Table 4).

Transplanted CLCs (2.2 ± 0.3 mm, $p<0.05$), but not MSCs (2.1 ± 0.3 mm), improved LV anterior wall thickness as compared to control infarcted animal (1.8 ± 0.4 mm). Nevertheless, other cardiac parameters indicated that CLCs and MSCs contributed comparably to functional improvements by reducing chamber dilatation and moderating negative LV remodelling.

M-Mode	SF (n=15)	MSC (n=19)	CLC (n=17)	Statistical Significance
LVIDed (mm)	7.7 ± 0.9	7.3 ± 0.6	7.1 ± 0.7	NS
LVIDes (mm)	5.0 ± 1.1	4.2 ± 0.6^{+}	4.1 ± 0.9*	$^{+}p<0.05$ vs. SF $^{*}p<0.01$ vs. SF
IVSed (mm)	1.3 ± 0.2	1.3 ± 0.1	1.4 ± 0.1	NS
IVSes (mm)	1.8 ± 0.4	2.1 ± 0.3	2.2 ± 0.3*	$^{*}p<0.05$ vs. SF
AWT (%)	49.8 ± 16.3	53.6 ± 14.7^{+}	54.0 ± 15.6*	$^{+}p<0.05$ vs. SF $^{*}p<0.05$ vs. SF
FS (%)	35.0 ± 5.6	42.4 ± 5.3^{+}	43.1 ± 6.6*	$^{+}p<0.05$ vs. SF $^{*}p<0.005$ vs. SF
EF (%)	60.4 ± 7.9	68.7 ± 6.3^{+}	69.8 ± 9.5*	$^{+}p<0.05$ vs. SF $^{*}p<0.005$ vs. SF

Table 4. Ultrasound echocardiography assessment of post cellular therapy treated rats. 2D ultrasound echocardiography assessments showed significant improvements in cell transplanted animals. SF: Serum free control; CLC: Cardiomyocyte-like-cells; MSCs: Mesenchymal stem cells; LVIDed: Left ventricular internal dimension at end diastolic; LVID: Left ventricular internal dimension at end systolic; IVSed: Interventricular septum at end diastolic, IVSes: Interventricular septum at end systolic; AWT: Anterior wall thickening; FS: Fractional shortening; EF: Ejection fraction.

4. Discussion

Integrins and ECM are important modulators of stem cell behaviours. To date, cardiac cell therapy supported only modest benefits, likely due to low engraftment of transplanted cells in the infarcted myocardium. Exploration of specific integrin/ECM interaction may improve engraftment and survival of transplanted cells and ultimately, mechanical function of the heart. Our current study examines integrin/ECM interactions on cardiac gene expression of CLCs and distribution of transplanted CLCs in infarcted myocardium.

The distribution and quantity of type I and III collagens in the heart play an important role in maintaining cardiac function. Alterations of collagen population and distribution in the myocardium affect size and shape of the heart chambers as well as myocardial diastolic and systolic function (Cleutjens et al., 1995a; Janicki & Brower, 2002). However, it is unclear if such alterations could affect stem cell migration and differentiation in the myocardium.

We have previously demonstrated that CLCs showed preferential adhesion to collagen V over collagen I matrix by interacting with subsets of integrins (Shim et al., 2004; Tan et al., 2010). van Laake et al. (2010) reported that pre and post transplanted human embryonic cardiomyocytes (hESC-CM) express integrins matching ECM types they encountered in their environment. Therefore, the integrin modulating role of collagen V may aid in the observed retention of the myocardial transplanted CLCs. Furthermore, intimate engraftment of the transplanted CLCs with collagen V-expressing, α-actinin positive, native cardiomyocytes supports an unique role of collagen V in the myocardium. Moreover, differential expression of α_1 and β_3 integrin between collagen I and V cultured CLCs coupled with the preferential homing demonstrated between transplanted MSCs and CLCs suggested a key role of collagen V matrix, not only in cellular retention, but cardiac differentiation of the transplanted stem cells. This is consistent with modulation of cardiac gene expression of CLCs demonstrated in relation to α_1 and $\alpha_v\beta_3$ neutralisation in vitro, although such relationship was not examined in vivo. Nevertheless, the comparable cardiac outcomes achieved in spite of selective homing of the transplanted cells, indicate that different reparative mechanisms may be initiated by MSCs and CLCs. Despite a positive trend of systolic improvement by CLCs, further mechanistic studies are warranted to discern their specific contribution to systolic and diastolic components of cardiac performance.

Integrin α_1 is known to transduce ECM signals to the cytoskeleton that activate downstream mitogen activated protein kinase (MAPK) and extracellular signal-regulated kinase 1 (ERK1) signalling pathways that phosphorylate and activate GATA4 (Akazawa & Komuro, 2003). However, GATA4 expression was unaffected by integrin α_1 neutralisation despite the upregulated Trop C and Trop T belonging to downstream genes known to be activated by GATA4 (Liang et al., 2001; Tidyman et al., 2003). Similarly, neutralisation of $\alpha_v\beta_3$ integrin attenuated Trop C expression despite GATA4 was previously shown to be unaffected by neutralisation of $\alpha_v\beta_3$ (Tan et al., 2010). It is unclear if the modulation of myofilamental gene expression demonstrated was secondary to other nuclear transcription factors. However, integrins are known to mechanotransduce signals to activate Raf-MEK-ERK-1/2 cascade that has been shown to elicit cardiomyocyte growth, increased fetal-gene expression and cytoskeletal reorganisation in neonatal cardiomyocytes (Lorenz et al., 2009). Nevertheless, it is unclear if reduced expression of integrin demonstrated on either collagen surface as compared to CLCs cultured on uncoated polystrene surface was associated with enhanced proliferation of CLCs as previously reported (Tan et al 2010). However, contrary to our previous data, SKAA was not down regulated by integrin $\alpha_v\beta_3$ neutralisation in the current study. This could be due to donor variations. Indeed, donor variation in integrin expression has been documented from different bone marrow isolates and passage numbers, resulting in different growth and proliferation potential (ter Brugge et al., 2002).

Despite beneficial effects of collagen V on cardiac gene expression and stem cell distribution, it should be noted that collagen distribution in the infarcted rat hearts may be different from humans during MI. Furthermore, a 3D structure like the heart may transmit different environmental cues to integrins as compared to 2D environments provided in tissue culture experiments. It remains to be determined whether inhibitory antibodies may transactivate other integrin receptors during epitope occupancy. In addition, the promiscuity of integrins renders it technically challenging to identify whether a single integrin or interplay of synergistic interactions between a few integrins is required for regulation of cardiac gene expression. Future studies employing siRNA techniques that selectively silence α_1 or $\alpha_v\beta_3$ integrin may provide additional information regarding the regulation of cardiac gene expression of CLCs on collagen V matrix ex vivo or in the transplanted milieu of infarcted myocardium.

5. Conclusion

In conclusion, our study indicates that α_1 and $\alpha_v\beta_3$ integrins drive cardiac gene expression of CLCs. Integrin families and ECM are important regulators of cardiac differentiation and myocardial distribution of adult MSCs and CLCs. Specific modulation of interaction between subclasses of collagen and integrin subunits in the post-infarct myocardial ECM could potentially offer a unique opportunity in cardiac regenerative medicine.

6. Acknowledgements

This study was supported by grants from National Medical Research Council, Biomedical Research Council and National Research Foundation of Singapore to W.S.

7. References

Akazawa, H., & Komuro, I. (2003). Roles of cardiac transcription factors in cardiac hypertrophy. *Circ Res, 92*(10), 1079-1088.

Boudoulas, K. D., & Hatzopoulos, A. K. (2009). Cardiac repair and regeneration: The rubik's cube of cell therapy for heart disease. *Dis Model Mech, 2*(7-8), 344-358.

Brancaccio, M., Hirsch, E., Notte, A., Selvetella, G., Lembo, G., & Tarone, G. (2006). Integrin signalling: The tug-of-war in heart hypertrophy. *Cardiovasc Res, 70*(3), 422-433.

Breuls, R. G., Klumpers, D. D., Everts, V., & Smit, T. H. (2009). Collagen type v modulates fibroblast behavior dependent on substrate stiffness. *Biochem Biophys Res Commun, 380*(2), 425-429.

Chacko, S. M., Khan, M., Kuppusamy, M. L., Pandian, R. P., Varadharaj, S., Selvendiran, K., et al. (2009). Myocardial oxygenation and functional recovery in infarct rat hearts transplanted with mesenchymal stem cells. *Am J Physiol Heart Circ Physiol, 296*(5), H1263-1273.

Cleutjens, J. P., Kandala, J. C., Guarda, E., Guntaka, R. V., & Weber, K. T. (1995b). Regulation of collagen degradation in the rat myocardium after infarction. *J Mol Cell Cardiol, 27*(6), 1281-1292.

Cleutjens, J. P., Verluyten, M. J., Smiths, J. F., & Daemen, M. J. (1995a). Collagen remodeling after myocardial infarction in the rat heart. *Am J Pathol, 147*(2), 325-338.

Djouad, F., Delorme, B., Maurice, M., Bony, C., Apparailly, F., Louis-Plence, P., et al. (2007). Microenvironmental changes during differentiation of mesenchymal stem cells towards chondrocytes. *Arthritis Res Ther, 9*(2), R33.

Feygin, J., Mansoor, A., Eckman, P., Swingen, C., & Zhang, J. (2007). Functional and bioenergetic modulations in the infarct border zone following autologous mesenchymal stem cell transplantation. *Am J Physiol Heart Circ Physiol, 293*(3), H1772-1780.

Fichard, A., Kleman, J. P., & Ruggiero, F. (1995). Another look at collagen v and xi molecules. *Matrix Biol, 14*(7), 515-531.

Gaudette, G. R., & Cohen, I. S. (2006). Cardiac regeneration: Materials can improve the passive properties of myocardium, but cell therapy must do more. *Circulation, 114*(24), 2575-2577.

Gronthos, S., Simmons, P. J., Graves, S. E., & Robey, P. G. (2001). Integrin-mediated interactions between human bone marrow stromal precursor cells and the extracellular matrix. *Bone, 28*(2), 174-181.

Hannigan, G. E., Coles, J. G., & Dedhar, S. (2007). Integrin-linked kinase at the heart of cardiac contractility, repair, and disease. *Circ Res, 100*(10), 1408-1414.

Janicki, J. S., & Brower, G. L. (2002). The role of myocardial fibrillar collagen in ventricular remodeling and function. *J Card Fail, 8*(6 Suppl), S319-325.

Liang, Q., De Windt, L. J., Witt, S. A., Kimball, T. R., Markham, B. E., & Molkentin, J. D. (2001). The transcription factors gata4 and gata6 regulate cardiomyocyte hypertrophy in vitro and in vivo. *J Biol Chem, 276*(32), 30245-30253.

Linehan, K. A., Seymour, A. M., & Williams, P. E. (2001). Semiquantitative analysis of collagen types in the hypertrophied left ventricle. *J Anat, 198*(Pt 1), 83-92.

Lorenz, K., Schmitt, J. P., Vidal, M., & Lohse, M. J. (2009). Cardiac hypertrophy: Targeting raf/mek/erk1/2-signaling. *Int J Biochem Cell Biol, 41*(12), 2351-2355.

Lu, L., Zhang, J. Q., Ramires, F. J., & Sun, Y. (2004). Molecular and cellular events at the site of myocardial infarction: From the perspective of rebuilding myocardial tissue. *Biochem Biophys Res Commun, 320*(3), 907-913.

Michel, J. B. (2003). Anoikis in the cardiovascular system: Known and unknown extracellular mediators. *Arterioscler Thromb Vasc Biol, 23*(12), 2146-2154.

Niyibizi, C., Chan, R., Wu, J. J., & Eyre, D. (1994). A 92 kda gelatinase (mmp-9) cleavage site in native type v collagen. *Biochem Biophys Res Commun, 202*(1), 328-333.

O'Kane, S., & Ferguson, M. W. (1997). Transforming growth factor beta s and wound healing. *Int J Biochem Cell Biol, 29*(1), 63-78.

Perkins, A. D., Ellis, S. J., Asghari, P., Shamsian, A., Moore, E. D., & Tanentzapf, G. (2010). Integrin-mediated adhesion maintains sarcomeric integrity. *Dev Biol, 338*(1), 15-27.

Ross, R. S., & Borg, T. K. (2001). Integrins and the myocardium. *Circ Res, 88*(11), 1112-1119.

Ruggiero, F., Champliaud, M. F., Garrone, R., & Aumailley, M. (1994). Interactions between cells and collagen v molecules or single chains involve distinct mechanisms. *Exp Cell Res, 210*(2), 215-223.

Shim, W. S., Jiang, S., Wong, P., Tan, J., Chua, Y. L., Tan, Y. S., et al. (2004). Ex vivo differentiation of human adult bone marrow stem cells into cardiomyocyte-like cells. *Biochem Biophys Res Commun, 324*(2), 481-488.

Spinale, F. G. (2007). Myocardial matrix remodeling and the matrix metalloproteinases: Influence on cardiac form and function. *Physiol Rev, 87*(4), 1285-1342.

Sun, Y., & Weber, K. T. (2000). Infarct scar: A dynamic tissue. *Cardiovasc Res, 46*(2), 250-256.

Tan, G., Shim, W., Gu, Y., Qian, L., Chung, Y. Y., Lim, S. Y., et al. (2010). Differential effect of myocardial matrix and integrins on cardiac differentiation of human mesenchymal stem cells. *Differentiation, 79*(4-5), 260-271.

ter Brugge, P. J., Torensma, R., De Ruijter, J. E., Figdor, C. G., & Jansen, J. A. (2002). Modulation of integrin expression on rat bone marrow cells by substrates with different surface characteristics. *Tissue Eng, 8*(4), 615-626.

Tidyman, W. E., Sehnert, A. J., Huq, A., Agard, J., Deegan, F., Stainier, D. Y., et al. (2003). In vivo regulation of the chicken cardiac troponin t gene promoter in zebrafish embryos. *Dev Dyn, 227*(4), 484-496.

van Laake, L. W., van Donselaar, E. G., Monshouwer-Kloots, J., Schreurs, C., Passier, R., Humbel, B. M., et al. (2010). Extracellular matrix formation after transplantation of human embryonic stem cell-derived cardiomyocytes. *Cell Mol Life Sci, 67*(2), 277-290.

Xu, X., Xu, Z., Xu, Y., & Cui, G. (2005). Effects of mesenchymal stem cell transplantation on extracellular matrix after myocardial infarction in rats. *Coron Artery Dis, 16*(4), 245-255.

Part 2

Clinical Aspects of Stem Cell Transplantation

11

Sources of Hematopoietic Stem Cells

Piotr Rzepecki, Sylwia Oborska and Krzysztof Gawroński
BMT Unit, Military Institute of Medicine, Warsaw
Poland

1. Introduction

Hematopoietic stem cell transplantation (HSCT) has the potential to cure a variety of malignant and non- malignant diseases. Sources of hematopoietic stem cells for transplantation have expanded progressively since the beginning of the modern era of transplantation in the late 1960s. Although bone marrow was the main source of stem cells in the early years of transplantation, in the past 10 to 15 years peripheral blood has assumed increasing importance. The initial impetus for the use of PBSCs for transplantation was to be able to offer transplantation to patients who were not candidates for the use of bone marrow cells (tumor contamination of the marrow or those with hypocellular marrows). Subsequent studies demonstrated that PBSCs could be mobilized from the bone marrow with either hematopoietic growth factors (GM-CSF, G-CSF) or a combination of chemotherapy and growth factors, which increased the number of hematopoietic progenitors collected from the blood by 10- to 1000- fold compared with steady-state conditions. Umbilical cord blood represents the newest source of stem cells for transplantation. At now peripheral blood is the main source in the autologous setting. Within the allogeneic setting, multiple sources of stem cells are possible and include those derived from individuals related or unrelated to the patient.

Hematopoietic progenitor cell (HPC) products contain hematopoietic stem and lineage-committed progenitor cells capable of providing hematopoietic and immune reconstitution after myeloablative or reduced-intensity preparative regimens. HPCs administered intravenously migrate to the marrow, where they adhere, expand, selfrenew (stem cells only), and differentiate. The differentiated cells are released into the blood, restoring blood counts and immunity. The time from administration of HPCs to recovery of adequate or normal blood counts is variable. Recipients of peripheral blood stem cells recover counts faster than recipients of bone marrow. Cord blood tends to be the slowest to engraft.

The minimum number of HPCs necessary for engraftment in a myeloablated recipient has not been established. Different products have widely different numbers of progenitors and stem cells. However, eligibility criteria for some protocols usually dictate a minimum number of cells to be collected and infused.

Several methods are used to measure the number of cells in an HPC collection. Simple cell count may be adequate for many marrow collections. Most centers use flow cytometric enumeration of CD34+ cells for the majority of cellular products. The discovery of the CD34 antigen in the early 1980s revolutionized our understanding of hematopoiesis. Cells expressing CD34 are capable of reconstituting hematopoiesis in lethally irradiated animals and humans, indicating that the putative hematopoietic stem cell expresses CD34.

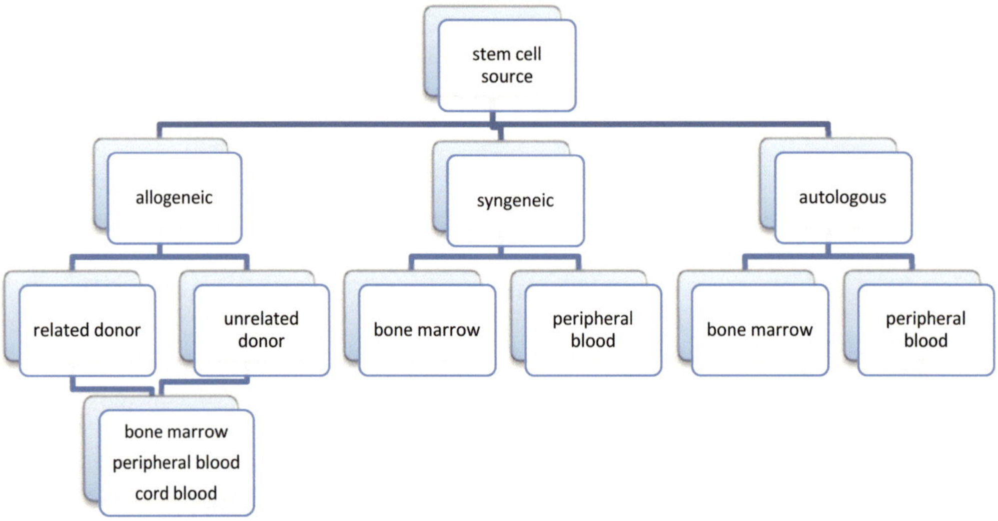

Fig. 1. Sources of haematopoietic stem cells for different types of transplantation

This type I transmembrane glycoprotein is expressed on:
1- 3% of bone marrow mononuclear cells
0,01- 0,1% of peripheral blood mononuclear cells
0,1- 0,4% of umbilical cord blood cells

2. Marrow as a source of stem cells

Marrow is collected in the day surgery suite using either general or regional anesthesia. With proper fluid and blood replacement, overnight hospitalization should not be required. For the healthy donor, the risks of serious complications from either general or regional anesthesia are about the same. Spinal or epidural anesthesia avoid the nausea that may occur with general anesthesia, especially for young women, but hypotension from loss of vascular tone in the lower extremities often occurs as the volume of marrow is collected. General anesthesia is preferable for the donor with comorbid disorders such as cardiovascular or cerebral vascular disease because of the better control of donor airway and lower risk of hypotension during the harvest procedures. Local anesthesia is acceptable only if a very limited harvest is being performed, as large quantities of lidocaine are cardiotoxic and local anesthesia does not achieve anesthesia of the marrow space. The technique of bone marrow harvest is straightforward and involves repeated aspirations of small volumes (10ml) of marrow. The marrow is removed sterilely from both posterior iliac crests by two operators simultaneously to minimize anesthesia time. Occasionally marrow is obtained from the anterior iliac crest or sternum.Typically we do two puncture through the skin and multiple bone punctures. Do not take more than 25 ml of bone marrow per kilogram of donor - this is the upper limit on the volume of collected bone marrow. The marrow is placed in a sterile container with an electrolyte solution and an appropriate anticoagulant.

The cell suspension is passed through sterile filters to remove fat, bone particles, and cellular debris [4,5]. For patients with a history of radiation or tumor involvement of one pelvic crest, adequate cells can be harvested from the anterior and posterior crests of the other side [6-8].

2.1 Toxicity and adverse events associated with bone marrow collection

Anesthesia complications present the major health risk to the donor. Marrow aspiration is generally well tolerated. Major complications occur in approximately 0.27% of healthy allogeneic donors and up to 0.97% of autologous transplant patients **[9-10].** Complications include hemorrhage and infections at skin puncture sites. Severe hematomas and neuralgias rarely occur, but attention to pelvic anatomy is required to decrease the risk of damage to vessels and nerves lying under or adjacent to the iliac crest harvest sites. Irritation of the sacral nerves may result from needle penetration through the pelvic bone or from blood tracking into the nerve roots and requires several months of convalescence. Localized pain is common, may last for several days, and may require a brief period of medication with opioid/acetaminophen combinations **[4].** In a survey of almost 500 donors for unrelated marrow transplantation, the average time for recovery was 15.8 days, although 10% of donors required more than 30 days for self-reported complete recovery **[11].** Most donors are able to return to routine activities 1 to 2 days after harvesting. In a study of related donors, an equivalent level of pain was reported by donors undergoing bone marrow harvesting and those receiving filgrastim for mobilization of PBSC **[12].** Minor complications occur in 6-20% of marrow donations **[11].** These include such events as hypotension, syncope, severe post-spinal headache, excess pain, unexpected hospitalization and minor

Symptom	Women %	Men %
Tired	85	76
Collection site pain	78	75
Back pain	67	68
Nausea	63	40
Sore throat	62	57
Pain sitting	62	57
Lightheadedness	53	42
Headache	40	32
Vomiting	39	17
Intravenous site pain	37	23
Fever	22	22
Bandage pain	19	26
Bleeding at site	10	8
Fainting	7	5

Table 1. Symptoms reported by National Marrow Donor Program (NMDP) bone marrow donors, 1987-2000 (n=9601) **[8]**

infections. In National Marrow Donor Program (NMDP) observation the frequency of serious adverse events following marrow donation is estimated at 0,1-0,3%. Reactions to the anesthetic agents administrated may be adverse reactions, hypersesitivity or idiosyncratic reactions. One NMDP donor experienced laryngospasm following extubation. Additionally, several NMDP donors have experienced profound bradycardia during anesthesia, including regional anesthesia (spinal and epidural), that required emergency treatment. Death has occurred among normal marrow donors. A recent revive of 7857 marrow collections reported two deaths **[13].**

2.2 Quantity of bone marrow cells for transplantation

Generally, 10-20ml marrow/kg of donor weight is harvested to achieve a minimum mononuclear cell (MNC) count of 2 x 10 to 8 MNC/kg of recipient body weight, although ideally up to 4,0 x 10 to 8 MNC/kg is preferred to compensate for cell loss during processing and to ensure adequate engraftment. The only setting in which higher numbers of MNC/kg definitely have been shown to be of benefit is aplastic anemia, in which low cell counts have been associated with an increased risk of rejection. Marrow contains mature red cells, white cells, platelets, mast cells, fat cells, plasma cells, committed progenitors of all lineages and hematopoietic stem cells. The most common modifications of allogeneic marrow are to decrease the volume of ABO-incompatible red cells, remove ABO- incompatible plasma, isolate CD34+ cells and remove donor T lymphocytes. The most common modification of autologous marrow is to reduce the volume by removing plasma and red cells before cryopreservation **[4]**.

2.3 "Rich" bone marrow

Pretreatment of the marrow donor with filgrastim (granulocyte- colony-stimulating factor G-CSF) may increase the number of myeloid progenitor cells harvested and decrease the period of posttransplant aplasia to that achievable by PBSC transplantation **[14].** Hematologic recovery in patients who are treated with autologous stem cells taken from bone marrow after G-CSF stimulation (rich bone marrow - RBM) is faster than in patients without G-CSF. In the Polish study **[15]** were estimated engraftment outcomes of patients who received bone marrow unstimulated or stimulated with G-CSF. The median and range for neutrophil engraftment times in this study when they used stimulated bone marrow were comparable with those published by Lemoli et al. **[16].** It seems that a better method of obtaining stem cells from bone marrow is the RBM. Using of stimulated bone marrow can faster engraftment comparing to non-stimulated bone marrow and can help patients, who fail to collect adequate number of stem cells from their peripheral blood. It is generally accepted that RBM engraft more rapidly than unstimulated bone marrow and that RBM appear to have similar engraftment times to peripheral blood stem cell transplantations suggesting that it is prior G-CSF exposure, not the anatomic site, which influences engraftment **[14,17,18].** CD34+ cell dose now is being correlated with transplant outcomes, with more rapid engraftment kinetics, possibly lower transplant-related mortality, and better overall survival, for example, in recipients of allogeneic products containing higher quantities of CD34+ cells. Otherwise patients receiving filgrastim- primed bone marrow had significantly less steroid-refractory acute graft-vs-host disease (GvHD), less chronic GvHD and fewer days of immunosuppressive therapy. Download stimulated bone marrow may favor the acquisition of more hematopoietic cells and facilitate reconstitution after myeloablative therapy with no significant increase in risk of GvHD **[19]**.

	PLT> 20	ANC> 0.5 G/L	Length of hospitalization
Median time to recovery (RBM)	12,6 days	13,0 days	17,3
Median time to recovery (bone marrow without stimulation of G-CSF)	18,8 days	17,8 days	23,1

Table 2. Time to recovery of platelets (PLT) and neutrophils (ANC) and length of hospitalization when we used bone marrow after stimulation G-CSF and without stimulation of G-CSF **[15]**

3. Peripheral blood stem cells (PBSC)

Stem cells were detected in the peripheral blood of mice in 1962 and in human in 1971 **[20]**. However, interest in peripheral blood stem cell transplantation did not develop until mid-1980s. During 1985-1986 several centers from different parts of the world reported encouraging results of the autologous transplants using hematopoietic progenitor cells collected from peripheral blood **[21,22]**. The first allogeneic transplant with peripheral blood progenitor cells was reported in 1989 **[23]**. Use of PBSC became more common after 1995 with the publication reports of successful allogeneic transplants with PBSC **[24-26]**. By stimulating the donor with either hematopoietic growth factors, or chemotherapy and growth factors, a sufficient number of circulating stem cells for marrow rescue can be collected in one to three apheresis procedures. Mobilized PBSC products are routinely used as an alternative source of HSCs for transplantation **[27,28]**. A PBSC collection is performed with a cell separation device originally developed for therapeutic leukapheresis, plasmapheresis and platelet donation procedures. This apheresis device uses a centrifuge to separate and collect mononuclear cells, including HSCs, from the blood. In order to achieve an adequate HSCs for transplantation it is necessary to process 12 to 25 liters of blood or 2.5 to 6.0 times the patient's/ healthy donor's calculated blood volume. Investigators have reported that the yield of CD34+ cells increases continuously as more blood volumes are processed. Although up to six times the donor's blood volume can be safety processed, some donors are not able to tolerate 4 or 6 hours being connected to an apheresis machine. Therefore, these donors require repeated collections on sequential days once the peripheral CD34+ count has increased to acceptable levels (>10 CD34+ cells/μl) for collection. There are currently three different commercially available apheresis instruments. In each case, instrument settings such as inlet flow rate, centrifuge speed, collect pump flow rate and anticoagulant: whole blood ratio vary, depending on the target cell type to be collected. The three instruments operate differently. The Amicus (Baxter) and the COM.TEC (Fresenius) are more automated, computer-controlled instruments; however, the Spectra is more widely used **[1,29]**. Because an apheresis procedure is used to collect PBSC products, they contain very few erythrocytes or granulocytes, compared to marrow, and are primarily composed of mononuclear cells (MNCs). PBSC products also contain larger numbers of HPCs than either marrow or cord blood, and therefore facilitate faster engraftment and shorter hospital stays.

Patients who have been heavily pretreated with multiple rounds of chemotherapy or radiation therapy often mobilize poorly and require multiple collection episodes **[30].**

Although the peripheral blood of healthy individuals contains fewer than 0,1% HSCs, this number increases dramatically during recovery from cytotoxic therapy and even more so more so then recombinant CSFs such as G-CSF are administered. Various mobilizing techniques are used by centers and generally consist of growth factor administration alone or in combination with various types of chemotherapy. G-CSF and high-dose cyclophosphamide (CY) are the most commonly used agents **[28, 31-33].** Currently, almost all of autologous and a majority of allogeneic transplants are performed with PBSC. Advantages of PBSC over bone marrow include **[21,34]**:

- Elimination of the need of general anesthesia, pain and other side effects of bone marrow aspiration.
- Patients with bone marrow metastases could be transplanted with autologous PBSC as there is a potential for tumor cell free collection.
- The hematological recovery with PBSC was faster than bone marrow significantly reducing the time to transfusion independence.

PBSC products contain larger numbers of T cells than marrow collections, and thus present a greater risk of causing graft-versus-host disease (GvHD) in the allogeneic setting, although the rate of acute GvHD is less than originally feared **[35-39].** In a prospective randomized study comparing PBSC donation with bone marrow, PBSC products contained double the number of CD34+ cells, eight fold more of T and NK cells than bone marrow collection. The advantages to the recipient were faster recovery of both neutrophil and platelets with PBSC compared to marrow without increasing risk of graft-vs-host disease. Another multi-center study also reported faster neutrophil and platelet recovery but significantly more frequent acute and chronic GvHD in PBSC recipients than recipients of bone marrow cells. There were no significant differences in transplant related mortality, relapse rate and overall survival were found **[40-42].**

3.1 Mobilization regimens

Recent advances in stem cell mobilization techniques have exploited the interactions between stem cells and the bone marrow microenvironment. Composed of stromal cells, endothelial cells, osteoblasts and other matrix components (collagens, fibronectins, proteoglycans), the bone marrow microenvironment anchors hematopoietic stem cells though a wide range of adhesive interactions **[43].** Hematopoietic stem cells express a broad array of cell surface receptors, namely the adhesion molecules lymphocyte function-associated Ag-1, very late Ag-4, and Mac-1; the chemokine receptors CXCR4 and CXCR2; the cell surface glycoproteins CD44 and CD62L; and the tyrosine kinase receptor c-kit **[43-45].** The bone marrow stroma contains stromal cell-derived factor-1 (SDF-1), CXC chemokine GRO-β, vascular cell adhesion molecule-1, kit-ligand, P-selectin glycoprotein ligand-1 and hyaluronic acid, all of which are cognate ligands for the stem cell adhesion molecules **[43-45].** Data from a number of preclinical models showed that inhibition of these receptor-ligand interactions resulted in enhanced progenitor cell mobilization **[46-48].**

Several exogenous hematopoietic cytokines that can mobilize hematopoietic progenitor cells into circulation are now available. Granulocyte colony stimulating factor (G-CSF) (filgrastim, lenograstim, pegfilgrastim) and granulocyte macrophage colony stimulating factor (GM-CSF) (sargramostim) are most commonly used for mobilization. They are used

	G-CSF	GM-CSF	Erythropoietin	Plerixafor
Adhesion factors	↓	Additive effects with G-CSF. May inhibit osteoclast activation.	Exact mechanism unknown, but additive when used in combination with G-CSF or GM-CSF	
CXCR4	↓ or ↑			No effect or ↓
SDF-1	↓			
MMPs	↑			
Neutrophil elastase	↑			

MMP- matrix metalloproteinase; SDF-1: stromal-derived factor 1α

Table 3. Factors using for mobilization- mechanism of action **[31]**

alone to mobilize hematopoietic stem cells into circulation in healthy donors for allogeneic transplantation and in patients who are in complete remission or in heavily pretreated ones for autologous transplantation. Approximately 4 days after the daily administration of G-CSF at a dose of 8-20 mcg/kg body weight, there is an increase in the number of CD34+ cells in the blood of patients and healthy stem cell donors. With this method, the progenitor cell peak is reached between days 5 and 6 **[49].** The extent of mobilization is determined by the age of the patient, the diagnosis, the earlier cytotoxic therapy, the dose of G-CSF and the sequence of G-CSF administration **[50].** Kroger et al **[51]** showed that mobilization was more effective if the total dose of G-CSF was divided into two equal fractions and given in the morning and in the evening rather than as a single dose once daily. G-CSF administered every 12 h at doses of 5 μg/kg provides better CD34+ cell yield than 10 μg/kg once a day in normal donors which may translate into a decrease in the number of apheresis required to obtain enough numbers of CD34+ cells for allogeneic PBSC transplant.

3.2 Which growth factor should we choose?

Filgrastim is commonly used to mobilize peripheral blood stem cells. Pegylation of filgrastim (pegfilgrastim) leads to prolongation of its half- life without loss of activity. Attachment of the polyethylene glycol (PEG) moiety reduces renal excretion and masks proteolytic cleavage sites resulting in elevated G-CSF serum levels for up to 14 days after a single injection. While filgrastim is also cleared by the kidneys, pegfilgrastim is mainly eliminated via a neutrophil-mediated clearance mechanism **[52-54].** The initial results of using pegfilgrastim obtained in limited numbers of patients with solid tumours showed that pegylated G-CSF was principally capable of mobilizing haematopoietic progenitor cells **[54].** In 2001 and 2003, two studies were designed for the treatment of myeloma patients. The treatment regimens were comparable and employed DT-PACE **[55,56]** as mobilization chemotherapy. In the first study patients received twice a day filgrastim, until completion of stem cell collection; in the second study two doses of pegfilgrastim were administered after DT-PACE. After a cycle of DT-PACE, pegfilgrastim 6 mg was given subcutaneously on days

+6 and +13. If the WBC count > 100 G/l by day +13 the second dose of pegfilgrastim was not administered. Investigators found some advantages when pegfilgrastim was used:

- A higher percentage of patients collected 15 x 10^6/kg in the first three days ($p < 0.001$)
- The median number of CD34 cells/kg collected on day 1 was higher ($p = 0.004$)
- The median number of growth factor injections was 2 versus 26 ($p < 0.0001$)
- Post- transplantation neutrophil recovery was faster after first and second transplant ($p < 0.001$)
- Platelet recovery was faster after first transplant (when less stem cells were infused) ($p = 0.01$)
- Authors concluded, that pegfilgrastim may be considered the standard of care for stem cell mobilization.

Kobbe et al. **[54]** found that in patients with multiple myeloma, a single dose of 6 mg pegfilgrastim after cyclophosphamide (4g/m^2) is equally effective in terms of the mobilization of haematopoietic progenitor cells as daily administration of conventional G-CSF. There is no increase in this effect if the dose is doubled to 12 mg. This finding is consistent with the results of a recently published, blinded, placebo- controlled multicenter study conducted in patients with malignant lymphomas **[57]**. These results were also confirmed by other groups of investigators **[58,59]**. There are no studies on the use of pegfilgrastim to mobilize haematopoietic stem cells in patients with acute leukemia. Pegylated G-CSF has not also been studied in the so-called "poor mobilizers".

Two formulations of recombinant human (rh) G-CSF, one glycosylated form and one non-glycosylated, are available. The glycosylated form, lenograstim, possesses at least 25% greater bioactivity *in vitro*. Some comparative studies into the preparation's potential to mobilize haematopoietic stem cells were performed to assess the potential greater efficacy of lenograstim *in vivo*. Ataergin S et al. **[60]** investigated whether a 25% reduced dose of lenograstim at 7.5 μg/kg/day is equivalent to 10 μg/kg/day filgrastim for autologous peripheral blood stem cell mobilization and transplantation. The two evaluated patients' cohorts were similar in regard to disease, sex, body, weight, body surface area, conditioning regimens, previous chemotherapy cycles and radiotherapy. Each growth factor was administered for 4 consecutive days. The first PBSC apheresis was done on the 5th day. In the posttransplant period, the same G-CSF was given at 5 μg/kg/day until leukocyte engraftment. No significant difference was seen in the median number of CD34+ cells mobilized, as well as the median number of apheresis, median volume of apheresis, percentage of CD34+ cells, and CD34+ cell number. Leukocyte and platelet engraftments, the number of days requiring G-CSF and parenteral antibiotics, the number of transfusions were similar in both groups in the posttransplant period. In conclusion, lenograstim 7.5 μg/kg/day is as efficious as filgrastim 10 μg/kg/day for autologous PBSC mobilization and transplantation.

Another study **[61]** explored the efficacy of the IGEV regimen (ifosfamide, gemcytabine, vinorebline and prednisone) combined with a fixed dose of lenograstim (263 μg/day) to mobilize PBSCs in 90 Hodgkin's lymphoma patients. An adequate number of CD34+ cells (> 3 x 10^6/kg) were collected in 98,7% mobilized patients. Hematological and non-hematological side effects were acceptable, and no toxic deaths occurred. These results confirm that the IGEV regimen with lenograstim support can be used successfully and safely to mobilize PBSCs.

Kopf B et al.**[62]**, in 2006 conducted a prospective randomized clinical trial to assess the mobilizing efficacy of filgrastim, lenograstim and molgramostim (GM-CSF) following a disease-specific chemotherapy regimen. In conclusion, all three growth factors were efficacious in mobilizing peripheral blood progenitor cells with no statistically significant difference between CD34+ cell yield and the different regimens, and the time to apheresis is likely confounded by the different mobilization regimens. An advantage on platelet recovery with molgramostim, suggested by other authors **[63,64]** was not confirmed by results of this study.

3.3 G-CSF in conjunction with chemotherapy

Growth factor combined with chemotherapy are given for patients, who are candidates to autologous transplantation and need additional treatment as debulking therapy or to achieve complete/ partial remission before transplantation **[65]**. A variety of chemotherapeutic agents are used with G-CSF or GM-CSF to mobilize hematopoietic stem cells for autologous transplantation. Administration of CY with G-CSF is widespread, as this regimen mobilizes hematopoietic stem cells effectively and is highly active against tumor cells **[65]**. Successful chemomobilization regimens used in combination with G-CSF are often disease oriented. Some of the most frequently used chemotherapeutic regimens in lymphoma patients include IEV (ifosfamide, epirubicin and etoposide), DHAP and ESHAP (etoposide, Ara-C, methylprednisolone and cisplatin) **[66,67]**. Patients with other hematologic malignancies are frequently treated with ICE (ifosfamide, carboplatin and etoposide) plus G-CSF for mobilization **[68]**.

Chemomobilization is widely used in clinical practice because the addition of a myelosuppressive chemotherapy agent to a cytokine mobilization regimen results in higher CD34+ cell yields, which may promise better outcomes for patients. In particular, mobilization with CY and G-CSF rather than with G-CSF alone improves CD34+ cell collection significantly in patients with either MM **[69,70]** or NHL **[71,72]**. However, it has been noted that the use of CY plus G-CSF severely depletes T cells and spares regulatory T cells, which could negatively affect immune reconstitution **[73]**. The benefits of adding chemotherapeutic agents to a G-CSF mobilization regimen may be offset by the increased risk of complications to the patients. Compared with mobilization regimens using G-CSF alone, chemomobilization is associated with increased morbidity, greater risk of infection, more hospital admissions, transfusions, antibiotic therapy and considerably greater cost overall. **[28]**. Although treatment-related mortality is rare, significant morbidity related to neutropenia that can often require hospitalization has been described, and many reports point to greater resource utilization with chemomobilization than with cytokine-alone mobilization **[28]**.

Koumakis et al [74] compared various time schedules of granulocyte colony-stimulating factor (G-CSF) treatment in a clinical model of patients who received high-dose cyclophosphamide (4.5 g/m^2). They found, that:

- G-CSF administration after high-dose cyclophosphamide has a similar effect upon the incidence and duration of severe leukopenia and thrombocytopenia
- Severe leukopenia is shorter when G-CSF starts up to 72 hours after high-dose of cyclophosphamide
- The length of G-CSF administration and its cost is also in favor of early initiation of treatment as well as the number of febrile days and antibiotic use

- Delayed (> 72 hours) or supportive treatment indicate more febrile days, antibiotic use and higher cost when compared to the early groups

For these reasons, preemptive rather than therapeutic administration of G-CSF is indicated in patients who receive high-dose cyclophosphamide as treatment for mobilization CD34+ cells into peripheral blood.

3.4 Toxicity of growth factors

Cytokine-mobilized PBSC collected from healthy sibling or unrelated donors are increasingly used as stem-cell source in the allogeneic setting. PBSCs have been shown to be superior to BM-derived stem cells during the early posttransplant course. The high CD34+ cell yield in PBSC grafts significantly shortens the posttransplant aplasia and the need for blood component support, especially platelets **[40,41]**. Both cytokine administration and harvest procedure cause unphysiological conditions in healthy individuals. The effects of rhG-CSF on peripheral blood count and leukocyte function, cytokine release and response, coagulation parameters and metabolic changes have been reviewed by Anderlini et al **[75,76]**. Limited data exist on the late effects of cytokine mobilization and PBSC collection. Until now, there is no convincing evidence that rhG-CSF administration leads to an increased risk to develop hematological malignancies in healthy individuals **[77]**. Leitner GC et al **[78]** investigated the actual quality of life (QoL) and health status of the donors as well as the need for medical treatment since PBSC donation by a questionnaire (151 donors were evaluated). The questionnaire was sent to donors at a median of 4 (range: 0.2-11) years after donation. rhG-CSF mobilization as well as subsequent PBSC collection is shown to be well tolerated in the short- and long-term profiles in these sibling donors. It had no negative influence on health status and QoL in the majority of them. Investigators observed no increased risk for hematological or oncological disorders. However, to acquire profound knowledge about rhG-CSF- and donation-related long term risks, consecutive monitoring of more donors for at least 10 years has to be performed. Hasenclever and Sextro **[79]** stated that to exclude or to show a 10-fold increase in the 10-year cancer incidence, a long-term prospective follow-up of several thousand donors for at least 10 years would be necessary. In 2009 the report from European Group for Blood and Marrow Transplantation Group was published **[80]**. Three hundred and thirty-eight allogeneic transplant teams from 35 European countries were asked to report numbers of fatalities, severe adverse events and hematologic malignancies occurring among their hematopoietic stem cell donors. 51024 first allogeneic hematopoietic stem cell transplantations were evaluated, of which 27770 were bone marrow and 23254 peripheral blood. They observed five donor fatalities, one after a bone marrow donation and four after peripheral blood donation (incidence 0.98 per 10000 donations; 95% CI 0.32-2.29), 37 severe adverse events (7.25/10000; 95% CI 5.11-9.99), of which 12 in bone marrow donors (4,32/10000; 95% CI 2.24-7.75) and 25 in peripheral blood donors (10.76/10000; 95% CI 6.97-15.85; $p< 0.05$) and 20 hematologic ,malignancies (3.92/10000; 95% CI 2.39-6.05), of which 8 after donating bone marrow and 12 after donating peripheral blood stem cells. The observed incidence rate of hematologic malignancies did not exceed the expected incidence in an age- and sex-adjusted general population. Authors concluded, that hematopoietic stem cell donation is associated with a small but define risk of fatalities and serious adverse events. True incidences might be higher, due to potential underreporting by study design. A continuous, standardized donor follow-up is needed to define donor risk groups and to monitor intermediate and long-term sequelae.

Healthy donors who were mobilized using lenograstim and who were undergoing peripheral hematopoietic cell collection with apheresis were enrolled in a surveillance protocol. The study was conducted by Martino et al **[81]**. The median dose of lenograstim was 10 μg/kg (range 5-15). 184 healthy donors have been assessed with a median follow-up of 62 months (range 2-155). The short-term adverse events:

- Bone pain 71,2%
- Headache 27,7%
- Insomnia 22,3%
- Fatigue 19,0%
- Nausea 12,0%
- Fever 5,4%
- Increased spleen size 4,3%

No vascular disorders, no cardiac disease

Long-term follow-up included monitoring of adverse events, neoplastic disease or other pathologies:

Type of adverse event	Number of donors suffered from	Time of occurrence after donation [months]
Transit ischaemic attack	1	39
Ankylosing spondylitis	1	28
Secondary polyglobulia	1	50
Lung cancer	1	19
No haematological disease was observed		

During hematopoietic stem cell mobilization in healthy donors, slight thrombocytopenia is common and is attributed to the leukapheresis procedure or to splenomegaly. The platelet depletion is a recognized effect of continuous flow leukapheresis, particularly large-volume leukapheresis **[75]**.

About one-third of neutropenic patients chronically treated with rh-G-CSF develop palpable splenomegaly and there have been reports of spontaneous spleen rupture in rhG-CSF or cyclophosphamide plus rh G-CSF- mobilized individuals or even in patients treated with rhG-CSF or rhGM-CSF after chemotherapy for acute leukemia or lymphoma. Few data are available on changes in spleen size as a result of a brief course of rhG-CSF **[82]**. Picardi et al. evaluated spleen size, comparing palpation with ultrasound (US)-evaluated longitudinal diameter and volume, in 13 healthy donors and 22 patients with a hematological malignancy who were undergoing PBSC mobilization with rhG-CSF-including regimens. When evaluated by sensitive methods, rhG-CSF caused spleen enlargement in almost all individuals treated. US-calculated volume proved to be an excellent method, much better than longitudinal diameter, for detecting non-palpable splenomegaly induced by rhG-CSF **[83]**.

G-CSF has been also discussed as the causative agent for the occurrence of **[84-94]:**

- Sweet syndrome
- Leukocytoclastic vasculitis
- Interstitial pneumonitis
- Adult respiratory distress syndrome (ARDS)
- Pyoderma gangraenosum

- Capillary leakage
- Stroke
- Acute gouty arthritis
- Iritis
- Severe anaphylactoid reactions
- Non-traumatic rupture of the spleen
- Severe pyogenic infection (painful perianal abscess and apical abscess)

3.5 How can we predict the optimal time for adequate collection of peripheral blood progenitor cells after chemotherapy ?

The absolute number of circulating Cd34+ cells was found to correlate closely with the CD34+ cell yield of the corresponding leucapheresis product. The number of CD34+ cells circulating in the peripheral blood reliably predicts both:

- The number of CD34+ cells
- The number of CFU-GM harvested

By monitoring the level of circulating CD34+ cells during the mobilization period, generally starting at about day 9 or 10, the day to perform the leucapheresis can be planned. The optimal time for collecting PBSC is when a peripheral blood sample contains at least 20 x 10^3 circulating CD34 cells/ml. This value provided at least 2 x 120^6 CD34+ cells/kg in a single leucapheresis, harvested the following day, in 94% of the collections regardless of the patient's diagnosis or mobilization regimen **[95]**. In Hill's study **[96]** a total of 168 adult patients with haematological malignancy were primed using low-moderate dose cyclophosphamide (1.5-3 g/m^2) with G-CSF 5-10 µg/kg per day. Harvesting was booked and peripheral blood counts first checked between 6 and 10 days post-priming. The peripheral blood CD34+ cell count correlated significantly with harvest yield ($r = 0.8448$, $p < 0.0001$). A peripheral blood CD34+ count ≥ 10/ µl predicted a collection of ≥ 2 x 10^6/kg (positive-predictive value of 61%, negative-predictive value 100%). There was no benefit to checking the peripheral blood CD34+ count or booking apheresis before day 9 post-cyclophosphamide.

3.6 What is the minimal CD34+ cells threshold collected from peripheral blood for sufficient engraftment?

What is the optimal CD34+ cells number for transplantation?

Gandhi et al.[97] advocated a minimum CD34+ threshold of > 1.0 x 10^6/kg in patients without extensive prior chemoradiotherapy, and ≥ 2.0 x 10^6/kg in all other patients. In this study all patients infused with grafts containing CD34+ cell doses between 1.0 and 2.0 x 10^6/kg engrafted by day 51. The only variable associated with slow platelet recovery was exposure to stem cell toxins (BCNU, melphalan, CCNU and mustine). The majority of patients with CD34+ > 1.0 x 10^6/kg achieved rapid and sustained engraftment and the only predictive factor of delayed recovery is prior exposure to stem cell toxins.

Villalon et al. **[98]** analyzed the factors affecting mobilization and engraftment in autologous peripheral blood progenitor cell transplantation according to the number of CD34+ re-infused (< 2.0 x 10^6/kg CD34+ vs > 2.0 x 10^6/kg CD34+). They found that,:

- Neutrophil and platelet engraftment was significantly longer with < 2.0 x 10^6/kg (12 vs 10 days, $p = 0.014$ and 16 vs 13 days, $p = 0.0001$ respectively)

- Platelet recovery was affected by exposure to alkylating agents ($p = 0.04$), refractory disease ($p= 0.02$), and AML ($p= 0.0001$), but only the last two variables remained significant in Cox regression ($p < 0.01$)
- Granulocyte engraftment was longer in CML (univariate, $p = 0.04$) and in refractory disease (multivariate, $p= 0.02$)
- In patients re-infused with > 2.0 x 10^6/CD34+/kg, the Cox model did not identify prognostic factors for haematopoietic recovery

They concluded, that although mobilization schedules and disease status influenced not only the yield of progenitor cells, but also the engraftment kinetics, the number of CD34+ re-infused was the main predictor of hematopoietic recovery. While engraftment succeeded in most of the cases, the re-infusion of > 2.0 x 10^6/kg resulted in significantly shorter recovery times. Thus, for autologous stem cell transplantation, it is common practice to infuse at least 2.0 x 10^6/CD34+/kg to ensure rapid engraftment. But investigators **[99-102]** found, that when comparing patients receiving at least 5 x 10^6/kg and 2-5 x 10^6/kg CD34+ cells there are:

- A significant reduction in the median number of days with fever
- Incidence of fever
- Duration of antibiotic treatment
- Faster neutrophil recovery

There was no significant difference in the number of platelet or red cell transfusions

Thus, transplantations with stem cell dose of at least 5.0 x 10^6/kg reduce infectious complications and should thereby increase the safety of this type of therapy while reducing duration (and cost) of antibiotic therapy.

In Bolwell study **[103]** investigators checked whether patients collecting high numbers of CD34+ cells ("super mobilizers") have a better outcome than other patients. Super mobilizers were defined as collecting a minimum of 8.0 x 10^6/CD34+/kg. In this study super mobilizers were younger and more likely to have received two or fewer prior chemotherapy regimens. Median CD34+ cell dose for the super mobilizing group was 13.7 x 10^6 CD34+/kg versus 4.4 x 10^6 CD34+/kg in the standard collecting group. The super mobilizer group had a superior overall survival ($p = 0.006$). In multivariable analysis, favorable disease status and younger age at transplant, and super mobilization were associated with improved survival.

It is reasonable to believe that the CD34+ cell dose has a positive influence on engraftment and survival. But, when compared with BM, PBSC grafts contain significantly more nucleated cells, more CD34+ hematopoietic stem cells, and more CD3+ lymphocytes **[104,105]**. It has been shown that granulocyte colony-stimulating factor (G-CSF)-mobilized CD34+ hematopoietic stem cells not only participate in engraftment, but also have an immunogenic role **[106-108]**. In the setting of T-cell-depleted allogeneic transplants using CD34+ positive selection as T-cell-depletion method, Urbano- Ispizua et al **[109]** could show that a high CD34+ cell dose not only does not improve the clinical results, but also actually may be associated with a poorer outcome. There is the suggestion, that transplantation with a higher CD34+ cell dose was detrimental in terms of chronic GvHD in allogeneic CD34+ cell-selected peripheral blood stem cell transplantation. Moreover, this association may also exist in the context of allogeneic unmanipulated PBSCT where a higher T-cell content or extremely high dose of CD34+ cells can be involved. It should be noted that CD34+ cells not only participate in engraftment, but also have an immunogenic role. However, although cGvHD is a leading cause of late mortality in allogeneic settings, it also plays a positive role in preventing relapse, especially in advanced hematological malignancies with a high risk of relapse [110]. Sohn et al [111] investigated the impact of the CD34+ cell dose on chronic graft-versus-host disease and the clinical outcome in adult patients submitted to allogeneic peripheral blood stem cell transplantation from HLA-identical siblings. The patients were classified into "low" or "high" CD34+ cell dose groups based on whether they received less or more than a median CD34+ cell dose of 10.5 x 10^6/kg, respectively. There was a significant difference in the incidence of extensive cGvHD and relapse between the two groups. With a median follow up of 335 days, the 3-year survival estimate for whole population was 47.9%, while that for the low and high groups was 29.9 and 67.8% respectively (p = 0.0434). An inverse relation was noted between the relapse rate and the incidence of extensive cGvHD (p = 0.043). Authors concluded, that is would appear reasonable that the optimal dose of CD34+ cells should be determined based on the disease status or aggressiveness of the malignant cells in each patient. Thus, in the case of patients with a high risk of relapse, transplantation with a CD34+ cell dose > 10.5 x 10^6/kg would seem to be acceptable to minimize the risk of relapse. Mohty et al [110], Sohn et al [111] investigated whether there was a correlation between the composition of PBSC grafts (CD34+ and CD3+ cells) and hematological recovery, GvHD, relapse and relapse-free survival after myeloablative HLA-identical sibling PBSCT. Neither hematological recovery, acute or chronic GvHD, nor relapse, was significantly associated with CD3+ cell dose. Increasing CD34+ stem cells was associated with faster neutrophil and platelet recovery. The probability of extensive cGvHD at 4 years was 34% in patients receiving a "low" CD34+ cell dose (< 8.3 x 10^6/kg) as compared to 62% in patients receiving a "high" CD34+ cell dose (> 8.3 x 10^6/kg) (p = 0.01). At a median follow-up of 59 months, this has not translated into a difference in relapse. In patients evaluable for cGvHD relapse free survival was significantly higher in patients receiving "low" CD34+ cell dose as compared to those receiving a "high" CD34+ cell dose (p = 0.04). This difference was mainly because of a significantly higher cGvHD-associated mortality (p = 0.01).

3.7 Factors to predict the efficiency of blood progenitor cell mobilization

It is important to note that most patients mobilize adequate numbers of CD34+ cells using a regimen of G-CSF alone. Although the addition of chemotherapy improves CD34+ yield, this comes at the expense of increased short-term toxicity and, possibly, the increased risk of secondary myelodysplastic syndrome [112]. Even with chemotherapy-growth factor combination regimens, it may be difficult to achieve an adequate CD34+ cell yield in some patients. Several studies have identified predictors of poor PBSC yield.

- The most important factor is the amount of myelosuppressive therapy (chemotherapy +/- radiation therapy) received prior to mobilization
- Using stem cell toxic agents prior to mobilization: nitrogen mustard, procarbazine, melphalan, carmustine and > 7.5 g of cyclophosphamide
- The number of chemotherapeutic regimens > 6 and ≥ 11
- Duration of exposure to chemotherapy (> 12 months)
- Short time interval since last chemotherapy < 6 months and < 65 days
- Previous radiation therapy
- Hypocellular marrow
- Refractory disease

A recent scoring system based on previous therapy may be useful in predicting CD34-positive cell yield.

Treatment score was built by Drake et al **[113]**, and then improved by others **[114,115]**. Treatment score was evaluated (as given in detail by Drake et al) **[113]** by Clark **[114]**. There was one except in Clark study, in addition, arbitrarily allocated a score of 2 for ifosfamide per treatment cycle.

Briefly, chemotherapy drugs are assigned a toxicity factor as follows:

0. prednisolone, dexamethasone;
1. vincristine, vinblastine, bleomycin, alpha interferon;
2. cyclophosphamide, anthracyclines, cisplatin, etoposide, ifosfamide;
3. chlorambucil, procarbazine;
4. melphalan, carmustine, mechlorethamine, lomustine.

The number of courses of each drug received was multiplied by its toxicity factor, and the score for each drug administered was summed to yield an overall treatment score. An additional 2 points were added if mediastinal radiotherapy was administered. In this study was validated this scoring system on an independent group of 99 patients undergoing 103 harvesting episodes. In 61 patients mobilized with cyclophosphamide 1.5 g/m^2 and G-CSF, those with treatment scores less than 21 yielded significantly more CD34-positive cells than patients with scores greater than 63 (P = 0.0012). Previous treatment with melphalan or carmustine was associated with a significantly lower yield of CD34-positive cells (P = 0.0001). No relationship was seen between the time from previous chemoradiotherapy and harvest outcome. Patients with treatment scores less than 21 required a shorter duration of G-CSF therapy (P = 0.05). Similar findings were seen in 42 further mobilization cycles undertaken with alternative mobilization schedules. The data suggest that a score summarizing previous treatment can be used to predict CD34 yields, and could be of clinical use to identify poor PBPC mobilisers in advance. The next improvement in scoring system was done by Jantunen et al. [115]. Results are shown in the table.

0	Prednisolone, dexamethasone, METHYLPREDNISOLONE
1	Vincristine, vinblastine, bleomycin, METHOTREXATE, alpha-interferon, CYTOSINE ARABINOSIDE
2	MITOGUAZON, cyclophosphamide, IFOSFAMIDE, cisplatin, anthracyclines, MITOXANTRONE, etoposide
3	Chlorambucil[a], procarbazine, FLUDARABINE, DACARBAZINE
4	Melphalan, carmustine, metchlorethamine, lomustine

Changes to the original scoring system proposed by Drake et al [113] are shown in capital letters.

[a]oral continuous treatment (4–6 mg/day) for a month=1 cycle.

Table 4. An improved chemotherapy scoring system [115]

3.8 Strategies of remobilization/ second-line stem cell harvest of patients who fail to achieve minimal progenitor thresholds at the first attempt

After an initial mobilization attempt, if too few CD34+ cells are collected to ensure prompt engraftment, patients often undergo additional mobilization attempts, which increase the risks associated with treatment **[116]**. Several salvage regimens have been developed to improve mobilization in patients in whom a first mobilization attempt with G-CSF alone fails to result in collection of an adequate cell dose. In recent years, several investigational agents have been developed that may prove useful for amplifying yields of CD34+ cells without introducing additional toxicity. As the understanding of stem cell interactions with the BM microenvironment grows, new mobilizing agents will emerge.

STRATEGY
second PBSC mobilization using the same regimen
steady-state bone marrow
stimulated BM-"rich bone marrow"
Increase chemotherapy dose: 4-7 g/m^2 CTX apprears more effective than 2-4 g/m^2
Prolong duration between last chemotherapy and planned collection
Increase growth factor dose (up to 24 µg/kg G-CSF) and use twice daily schedule
Use investigational agents

Table 5. Strategies of remobilization/ second-line stem cell harvest of patients who fail to achieve minimal progenitor thresholds at the first attempt [27,28,117]

Mobilization agent	*Mechanism*
	Cytokines approved for mobilization
GM-CSF	Stimulates production of granulocytes and macrophages
G-CSF	Granulocyte expansion/activation, protease release and cleavage of adhesion molecules
	Chemotherapeutic agents commonly used for mobilization
CY **Paclitaxel** **Etoposide**	Expansion and activation of granulocytes after bone marrow suppression
	Investigational mobilization agents
Pegylated G-CSF	Granulocyte expansion/activation, protease release and cleavage of adhesion molecules
EPO	Stimulates erythropoiesis
Stem cell factor	G-CSF potentiation
Plerixafor	Disrupts CXCR4/SDF-1 α interactions
SB-251353	GRO-β analog involved in directing the movement of stem cells and other leukocytes
TPO	Regulates megakaryocyte development, G-CSF synergism
Parathyroid hormone	Activates osteoblasts, which produce HGFs in the stem cell niche

Abbreviations: CXCR4=chemokine receptor 4; GRO- β=a human CXC chemokine; SDF-1=stromal cell-derived-factor-1. TPO- thrombopoietin, HGFs- hematopoietic growth factors

Table 6. Mobilization agents currently used in auto-HSCT [27,28,117]

3.8.1 Erythropoietin- EPO

EPO, which is commonly used to preserve blood hemoglobin concentrations in patients undergoing chemotherapy, has also been shown to potentiate the mobilization effect of G-CSF or GM-CSF [**118**]. Although the mechanism of this cooperative effect is unknown, it is thought that expression of EPO receptors on CD34+ progenitor cells primed with G-CSF or GM-CSF may promote survival of these cells [**119**]. However, EPO use is generally regarded as inefficient and, as such, has not become a standard of care. Studies evaluating the mobilization efficacy of EPO plus G-CSF or G-CSF alone in various doses have generated mixed results **[120-122]**. Further investigation is required to ascertain the clinical benefits of EPO combined with G-CSF **[123].**

3.8.2 High-dose G-CSF

High-dose G-CSF was investigated as a primary mobilization regimen throughout the 1990s. **[49,124,125]**. Although seldom used today for primary mobilization, high-dose G-CSF regimens are occasionally used for remobilization **[126,127]**. Although there is no standard protocol for high-dose G-CSF administration, doses ranging from 16 to 32 µg/kg s.c. daily to 12–16 µg/kg s.c. twice daily have been considered high-dose regimens even though some of the available data apply to patients with aplastic anemia, solid tumors or hematological malignancies other than NHL **[124,126,128]**.

3.8.3 GM-CSF plus G-CSF

Significant synergism has been reported between GM-CSF and G-CSF in the formation of granulocytic colonies in vitro **[129]**. Mobilization regimens combining GM-CSF with G-CSF have consisted of sequential or concurrent administration of these agents at a range of doses (G-CSF, 5–10 µg/kg; GM-CSF, 5 µg/kg–250 µg/m2), with or without chemotherapeutic agents **[130-133]**. These combination regimens have not been shown to have substantial benefits over regimens that use G-CSF alone; therefore, GM-CSF and G-CSF are not commonly administered together for primary mobilization. However, the combination of G-CSF and GM-CSF is used as a salvage mobilization regimen when mobilization with G-CSF alone has been unsuccessful **[126,134,135]**.

3.8.4 SCF

The c-kit ligand SCF is produced in BM stromal cells and acts as a potent co-mitogen for many hematopoietic growth factors **[32]**. Recombinant methionyl human SCF (ancestim, Stemgen, Amgen Inc.) administered sc. in combination with G-CSF has been shown to enhance mobilization and may fasten recovery in transplant recipients **[32, 136].** The combination of SCF and G-CSF exerts a sustained mobilization effect that persists longer than does the effect of G-CSF alone, which persists for up to 7 days, as shown by an increase in the numbers of circulating CD34+ cells for up to 13 days in patients with breast cancer (BC) who received the combination treatment **[32]**. Despite the efficacy of SCF, its use is hindered by the infrequent occurrence of severe anaphylactoid reactions and the resultant need to closely monitor patients after SCF administration **[137]**. Although approved for use in Canada and New Zealand, ancestim is not currently available in the United States, and it is seldom used in Europe because of the relatively high risk of side effects.

3.8.5 Plerixafor

Plerixafor is a reversible bicyclam inhibitor of hematopoietic stem cells (HSC) binding to SDF-1α on marrow stromal cells via the chemokine receptor 4 (CXCR4) on HSC **[138-140]**. Plerixafor used in conjunction with G-CSF has been shown in a phase 2 studies to quickly and predictably enhance the numbers of CD34+ cells circulating in the peripheral blood **[141]**. In patients in whom mobilization with G-CSF either alone or in combination with chemotherapy has previously failed, CD34+ cell yields have been noted to increase by 5- to

100-fold in response to administration of plerixafor plus G-CSF [**141,142**]. Preliminary results of two phase 3 multicenter randomized placebo-controlled studies indicated that the addition of plerixafor to a G-CSF regimen resulted in greater efficacy than was seen with a regimen of G-CSF alone [**140, 143**]. In general, treatment with plerixafor and G-CSF was associated with side effects similar to those seen with treatment with G-CSF alone. Most treatment-related AEs appeared to be mild and transient. The most common AEs were gastrointestinal tract effects, such as diarrhea, nausea and vomiting, and injection-site reactions, such as erythema or edema[**140, 143**].

3.8.6 SB-251353

SB-251353 is another investigational mobilization agent currently in preclinical studies [**144,145**]. SB-251353 is an analog of GRO-β, a human CXC chemokine involved in directing the movement of stem cells and leukocytes. Although human data are lacking, this agent, when combined with G-CSF in rhesus monkeys, was shown to greatly increase mobilization of stem cells and progenitor cells in comparison with G-CSF alone. Further research is necessary to determine the efficacy and potential toxicities of this treatment in humans. [**144,145**].

3.8.7 Thrombopoietin- TPO

Endogenous TPO is the primary regulator of megakaryocyte development. Recombinant human TPO (rhTPO) has been shown to act synergistically with G-CSF to enhance stem cell mobilization [**146**]. This regimen has not been shown to be more efficacious or safer than existing mobilization regimens; however, a few studies document encouraging results [**147**]. AEs associated with the use of rhTPO plus G-CSF appear to be similar to those seen with the use of G-CSF alone [**146,147**], however, cytopenias owing to neutralizing antibodies to TPO have been reported in a small number of patients who were given rhTPO to treat chemotherapy-induced thrombocytopenia [**148**]. Currently, no TPOs have been approved by the FDA for mobilization [**149**].

3.8.8 Parathyroid hormone

Parathyroid hormone (PTH) activates osteoblasts, which produce hematopoietic growth factors in the stem cell niche, thereby increasing the numbers of circulating stem cells [**150,151**]. The efficacy and safety of PTH have yet to be established. In a recent phase 1 study, 20 patients with MM, NHL, HD or AML, in whom one or two previous mobilization attempts had failed, received escalating doses of 40, 60, 80 and 100 µg of PTH (s.c.) for 14 days; PTH doses were combined with G-CSF 10 µg/kg on the last 4 days of treatment [**151**]. Overall, 47% of patients in whom one previous mobilization attempt had failed reached the mobilization criterion of >5 CD34+ cells/µl in the peripheral blood, and 40% of patients who had previously experienced two failed mobilization attempts reached the mobilization criterion. No dose-limiting toxicity was evident, and PTH was well tolerated; AEs included headache, muscle pain, back pain, fatigue and hypothermia [**151**].

3.9 BM vs. PBSC for whom? [152]

BM for:

a. Good risk younger patients
   - Children

- Aplastic anemia, CML in first chronic phase (CP1)

b. PBSC for:
   - Advanced phase of diseases
   - If graft manipulation is needed
   - Reduced intensity transplants
   - Donor's preference

4. Umbilical cord blood

The use of allogeneic blood and bone marrow stem cell transplantations are limited by the availability of suitably HLA- matched donors. Only 30% of patients have HLA-identical sibling donors and through the National Marrow Donors Program and other registries worldwide nearly 75% of Caucasians, but far fewer racial minorities find suitably HLA-matched donors. In 1988 umbilical cord blood (UCB) hematopoietic stem cells from a related sibling were transplanted successfully into a 5-year-old child with Fanconi anemia. Fifteen years later, this patient is doing well with full donor hematopoietic and lymphoid reconstitution. UCB offers the advantages of easy procurement, no risk to donors, the reduced risk of transmitting infections, immediate availability of cryopreserved units and acceptable partial HLA mismatches. Nearly all patients can find at least one potential 4 of 6 HLA-matched UCB units through either Netcord or other banks. Limited cell dose has been the major limitation to the wide use of UCB for allogeneic transplantation.

The blood in the umbilical cord of newborn babies contains large numbers of stem cells, which have been shown to be capable of long-term engraftment in children and some adults after transplantation. Cord blood cells obtained from the umbilical cord at the time of delivery are used mainly for unrelated allogeneic stem cell transplantation. Patients without time to find an unrelated stem cell donor or who do not have a HLA 10/10 or 9/10 unrelated bone marrow graft donor should be considered for cord blood cells transplantation (CBCT). The cord blood cells may also be used for family- member transplantation, particularly in children for both malignant or nonmalignant diseases **[153-155]**.

Advantages of CBCT compared with bone marrow or mobilized peripheral blood transplantations **[156-161]** :

- significantly faster graft availability (patients receiving CBCT in a median of 3-5 weeks earlier than those receiving an unrelated bone marrow graft),
- lack of risk to the donor,
- higher frequency of rare haplotypes compared to bone marrow donor registries,
- CD34+/CD38 cord blood cells proliferate more rapidly and generate large number of progenitor cells **[162]**
- extension of the donor pool due to tolerance of 1-2 HLA mismatches out of 6,
- lower incidence and severity of acute graft-versus-host disease,
- lower risk of infections by latent viruses - such as Epstein-Barr virus or cytomegalovirus

Disadvantages of CBCT compared with bone marrow or mobilized peripheral blood transplantations:

- cell dose is based per kilogram of recipient weight, which can be limiting. Target is 1,7-3,5 x 10^7 total nucleated cells/kg,

- low number of hematopoietic progenitor cells- single cord blood unit transplantation (100ml) contains 0,3x10^6 CD34+ cells/kg (small recipient, below 40kg),
- in adults, single cord blood unit transplantation is associated with very delayed engraftment, combining two products seems to provide more rapid engraftment, albeit at the expense of higher rates of acute graft-versus-host disease,
- donor cannot be used for donor lymphocyte infusion,
- donor cannot be used for treatment of graft rejection or failure.

The cell dose infused is consistently an important marker for improved engraftment and survival. The lower dose of CD34+ cells translate into increased risk of graft failure, delayed hematopoietic engraftment **[160]** and delayed immune reconstitution **[162,163]**. Because of the delayed immune reconstitution infections are a serious problem in cord blood transplantation **[164]**.

Many tests have been performed to enhance collection of hematopoietic stem cells in cord blood units. Examples include: in vivo or ex vivo expansion of cord blood cells **[165,166]**, injecting cord blood cells directly into the bone marrow **[167]**, use of double unit CBCT **[168,169]**, use of reduced intensity conditioning regimen **[170,171]**, coinfusion with a haploidentical T cell depleted graft **[172,173]** or mesenchymal stem cells **[174]**.

5. Storage [1]

Hematopoietic progenitor cell products are stored using various methods depending on the required duration of storage. Products used fresh can be refrigerated for at least 24 hours before infusion. If there is a > 48-hour delay before infusion, most products are frozen to maintain viability. Most frozen products are stored in the vapor phase of liquid nitrogen (< -50 C). Products may be stored for up to 10 years although no longevity limit has yet been determined. Long-term storage is generally done in the liquid phase of liquid nitrogen. Hematopoietic progenitor cells have been frozen using cryoprotectant solutions. The most commonly used cryoprotectant is 10% DMSO and a protein additive such as human serum albumin.

6. References

[1] Leemhuis T, Sacher RA. Hematopoietic progenitor cells: biology and processing. In: Rossi`s principles of transfusion medicine, 4th edn. Simon TL, Snyder EL, Solheim BG, Stowell CP, Strauss RG and Petrides M (eds.). Blackwell Publishing, Singapore, 2009: 508- 520.

[2] Rashidi HH, Rowley SD. Hematopoietic progenitor cells: autologous transplantation. In: Rossi`s principles of transfusion medicine, 4th edn. Simon TL, Snyder EL, Solheim BG, Stowell CP, Strauss RG and Petrides M (eds.). Blackwell Publishing, Singapore, 2009: 521-541.

[3] Alousi AM, Giralt SA. Hematopoietic progenitor cells: allogeneic transplantation. In: Rossi`s principles of transfusion medicine, 4th edn. Simon TL, Snyder EL, Solheim BG, Stowell CP, Strauss RG and Petrides M (eds.). Blackwell Publishing, Singapore, 2009: 542-558.

[4] Scott D, Donato R, Donato M. Practical Aspects of stem cell collection. In: Hoffman R, Furie B, Benz, E Jr. Hematology: Basic Principles and Practice, Churchill Livingstone, Philadelphia, 2009: 1695-1711.

[5] Zikos P, Van Lint MT, Frassoni F, et al. Low Transplant Mortality in Allogeneic Bone Marrow Transplantation for Acute Myeloid Leukemia: A Randomized Study of Low-Dose Cyclosporin Versus Low-Dose Cyclosporin and Low-Dose Methotrexate. Blood 1998; 91: 3503-3508

[6] Thomas ED, Storb R. Technique for human marrow grafting. Blood 1970; 36: 507-515.

[7] Brandwein JM, Callum J, Rubinger M, et al: An evaluation of outpatient bone marrow harvesting. J Clin Oncol 1989; 7:648-650.

[8] Confer DL. Hematopoietic Cell Donors. In: Thomas' Hematopoietic Cell Transplantation Third Edition. Blume KG, Forman SJ, Appelbaum FR (red.). Blackwell Publishing; Massachusetts 2004; 538-549

[9] Buckner CD, Clift RA, Sanders JE, et al: Marrow harvesting from normal donors. Blood 1984; 64: 630-634.

[10] Jin NR, Hill RS, Petersen FB, et al: Marrow harvesting for autologous marrow transplantation. Exp Hematol 1985; 13:879-884.

[11] Stroncek DF, Holland PV, Bartch G, et al: Experiences of the first 493 unrelated marrow donors in the National Marrow Donor Program. Blood 1993; 81:1940-1946.

[12] Rowley SD, Donaldson G, Lilleby K, et al: Experiences of donors enrolled in a randomized study of allogeneic bone marrow or peripheral blood stem cell transplantation. Blood 2001; 97:2541-2548.

[13] Anderlini P, Rizzo JD, Nugent ML, et al. Peripheral blood stem cell donation. An analysis from the Interrnational Bone Marrow Transplant Registry (IBMTR) and European Group for Blood and Marrow Transplant (EBMT) databases. Bone Marrow Transplant 2001; 27: 689-692.

[14] Weisdorf D, Jiller J, Verfaillie C, et al: *Cytokine-primed bone marrow stem cells vs. peripheral blood stem cells for autologous transplantation: A randomized comparison of GM-CSF vs. G-CSF.* Biol Blood Marrow Transplant 1997; 3:217-223.

[15] Gawronski K, Rzepecki P, Oborska S. Hematologic recovery in patients who are treated with autologous stem cells taken from bone marrow after G-CSF stimulation. EBMT 2011, Abstr. P1103.

[16] Lemoli RM, de Vivo A, Damiani D, et al. Autologous transplantation of granulocyte colony-stimulating factor-primed bone marrow is effective in supporting myeloablative chemotherapy in patients with hematologic malignancies and poor peripheral blood stem cell mobilization. *Blood* 2003; 102: 1595–1600.

[17] Damiani D, Fanin R, Silvestri F, et al. Randomized trial of autologous filgrastim-primed bone marrow transplantation versus filgrastim-mobilized peripheral blood stem cell transplantation in lymphoma patients. *Blood* 1997; 90: 36–42.

[18] al-Fiar F, Prince HM, Imrie K, et al. Bone marrow mononuclear cell count does not predict neutrophil and platelet recovery following autologous bone marrow transplant: value of the colony-forming unit granulocyte-macrophage (CFU-GM) assay. *Cell Transplant* 1997; 6: 491–495.

[19] Morton J., Hutchins C., Durrant S. Granulocyte-colony-stimulating factor (G-CSF)-primed allogeneic bone marrow: significantly less graft-versus-host disease and comparable engraftment to G-CSF-mobilized peripheral blood stem cells. Blood, 2001, 98, 3186-3191

[20] Reddy RL. Mobilization and collection of peripheral blood progenitor cells for transplantation. Transfus Apher Sci 2005; 32: 63- 72.

[21] Kessinger A, Armitage JO, Landmark JD, Wisenburger DD. Reconstitution of human hematopoietic function with autologous cryopreserved circulating stem cell. Exp Hematol 1986; 14: 192-196.
[22] Reiffers J, Bernard P, David B. Successful autologous transplantation with peripheral blood hematopoietic cells in a patient with acute leukemia. Exp Hematol 1986; 14: 312-315.
[23] Kessinger A, Smith DM, Strandford SW, et al. Allogeneic transplantation of blood-derived, T-cell-depleted hematopoietic stem cells after myeloablative treatment in a patient with acute lymphoblastic leukemia. Bone Marrow Transplant 1989; 4: 643-646.
[24] Bensinger WI, Weaver CH, Appelbaum FR, et al. Transplantation of allogeneic peripheral blood stem cells mobilized by recombinant human granulocyte colony-stimulated factor. Blood 1995; 85: 1655-1658.
[25] Korbling M, Przepiorka D, YO H, et al. Allogeneic blood cell transplantation for leukemia and lymphoma: potential advantage of blood over marrow allografts. Blood 1995; 85: 16590-1665.
[26] Schmitz N, Dreger P, Suttorp M. Primary transplantation of allogeneic peripheral blood progenitor cells mobilized by filgrastim (granulocyte colony-stimulating factor). Blood 1995; 85: 1666-1672.
[27] Reddy RL. Mobilization and collection of peripheral blood progenitor cells for transplantation. Transfus Apher Sci 2005; 32: 63- 72.
[28] Bensinger W, DiPersio JF, McCarty JM. Improving stem cell mobilization strategies: future directions. Bone Marrow Transplant 2009; 43: 181-195.
[29] Bojko P, Stellberg W, Kudde C, et al. Kinetic study of CD34 cells during peripheral blood stem cecollections. J Clin Apher 1999; 14: 18-25.
[30] Haas R, Mohle R, Fruhauf S, et al. Patient characteristics associated with successful mobilizing and autografting of peripheral blood progenitor cells in malignant lymphoma. Blood 1994; 83: 3787-3794.
[31] Montgomery M, Cottler-Fox M. Mobilization and collection of autologous hematopoietic progenitor/stem cells. Clin Adv Hematol Oncol 2007; 5: 127-136.
[32] Glaspy JA, Shpall EJ, LeMaistre CF, et al. Peripheral blood progenitor cell mobilization using stem cell factor in combination with filgrastim in breast cancer patients. Blood 1997; 90: 2939-2951.
[33] Bensinger W, Singer J, Appelbaum F, et al. Autologous transplantation with peripheral blood mononuclear cells collected after administration of recombinant granulocyte stimulating factor. Blood 1993; 81: 3158-3163.
[34] Beyer J, Schwella N, Zingsem J, et al. Hematopoietic rescue after high-dose chemotherapy using autologous peripheral-blood progenitor cells or bone marrow: a randomized comparison. J Clin Oncol 1995; 13: 1328-1335.
[35] Storek J, Gooley T, Siadak M, et al. Allogeneic peripheral blood stem cell transplantation may be associated with a high risk of chronic graft-versus-host disease. Blood 1997; 90: 4705-4709.
[36] Champlin R, Schitz N, Horowitz MM, et al. Blood stem cells compared with bone marrow as a source of hematopoietic cells for allogeneic transplantation. Blood 2000; 95: 3702-3709.

[37] Schrezenmeier H, Passweg JR, Marsh JCW, et al. Worse outcome and more chronic GVHD with peripheral blood progenitor cells than bone marrow in HLA- matched sibling donor transplants for young patients with severe aplastic anemia. Blood 2007; 110: 1397-1400.

[38] Przepiorka D, Smith TL, Folloder J, et al. Risk factors for acute graft-versus-host disease after allogeneic blood stem cell transplantation. Blood 1999; 94: 1465-1470.

[39] Bensinger WI, Clift R, Martin P, et al. Allogeneic peripheral blood stem cell transplantation in patients with advanced hematologic malignancies: A retrospective comparison with marrow transplantation. Blood 1996; 88: 2794-2800.

[40] Favre G, Beksac M, Bacigalupo A, et al. Differences between graft product and donor side effects following bone marrow or stem cell donation. Bone Marrow Transplant 2003; 32: 873-880.

[41] Bensinger WI, Martin PJ, Storer B, et al. Transplantation of bone marrow as compared with peripheral- blood cells from HLA-identical relatives in patients with hematologic cancers. New Engl J Med 2001; 344: 175-181.

[42] Schmitz N, Beksac M, Hasenclever D, et al. Transplantation of mobilized peripheral blood cells to HLA-identical siblings with standard-risk leukemia. Blood 2002; 100: 761- 767.

[43] Nervi B, Link DC, DiPersio JF. Cytokines and hematopoietic stem cell mobilization. J Cell Biochem 2006; 99: 690-705.

[44] Winkler IG, Levesque JP. Mechanisms of hematopoietic stem cell mobilization: when innate immunity assails the cells that make blood and bone. Exp Hematol 2006; 34: 996-1009.

[45] Cashen AF, Link D, Devine S, DiPersio J. Cytokines and stem cell mobilization for autologous and allogeneic transplantation. Curr Hematol Rep 2004; 3: 406-412.

[46] Nakamura Y, Tajima F, Ishiga K, et al. Soluble c-kit receptor mobilizes hematopoietic stem cells to peripheral blood in mice. Exp Hematol 2004; 32: 390-396.

[47] Papayannopoulou T, Priestley GV, Nakamato B, et al. Synergistic mobilization of hematopoietic progenitor cells using concurrent beta1 and beta2 integrin blockade or beta2-deficient mice. Blood 2001; 97: 1282-1288.

[48] Craddock CF, Nakamoto B, Andrews RG, et al. Antibodies to VLA4 integrin mobilize long-term repopulating cells and augment cytokine-induced mobilization in primates and mice. Blood 1997; 90: 4779-4788.

[49] Kobbe G, Sohngen D, Bauser U, et al. Factors influencing G-CSF-mediated mobilization of hematopoietic progenitor cells during steady-state hematopoiesis in patients with malignant lymphoma and multiple myeloma. Ann Hematol 1999; 78: 456-462.

[50] Jacoub JF, Suryadevara U, Pereyra V, et al. Mobilization strategies for the collection of peripheral blood progenitor cells: Results from a pilot study of delayed addition G-CSF following chemotherapy and review of the literature. Exp Hematol 2006; 34: 1443-1450.

[51] Kroger N, Renges H, Kruger W, et al. A randomized comparison of once versus twice daily recombinant human granulocyte colony-stimulating factor (filgrastim) for stem cell mobilization in healthy donors for allogeneic transplantation. Br J Haematol 2000; 111: 761-765.

[52] Lord BI, Woolford LB, Molineux G. Kinetics of neutrophil production in normal and neutropenic animals during the response to filgrastim (r-metHu G-SCF) or filgrastim SD/01 (PEG-r-metHu G-CSF). Clin Cancer Res 2001; 7: 2085-2090.
[53] Molineux G. The design and development of pegfilgrastim (PEG-rmetHu G-CSF, Neulasta) Current Pharmaceut Design 2004; 10: 1235- 1244.
[54] Kobbe G, Bruns I, Fenk R, et al. Pegfilgrastim for PBSC mobilization and autologous haematopoietic SCT. Bone Marrow Transplant 2009; 43: 669-677.
[55] Lee CK, Barlogie B, Munshi N, et al. DT-PACE: an effective, novel combination chemotherapy with thalidomide for previously treated patients with myeloma. J Clin Oncol 2003; 21: 2732-2739.
[56] Tricot G, Barlogie B, Zangari M, et al. Mobilization of peripheral blood stem cells in myeloma with either pegfilgrastim or filgrastim following chemotherapy. Haematologica 2008; 93: 1739-1742.
[57] Russell N, Mesters R, Schubert J, et al. A phase 2 pilot study of pegfilgrastim and filgrastim for mobilizing peripheral blood progenitor cells in patients with non-Hodgkin's lymphoma receiving chemotherapy. Haematologica 2008; 93: 405-412.
[58] Isidori A, Tani M, Bonifazi F, et al. Phase II study of a single pegfilgrastim injection as an adjuct to chemotherapy to mobilize stem cells into the peripheral blood of pretreated lymphoma patients. Haematologica 2005; 90: 225-231.
[59] Nosari A, Cairoli R, Ciapanna D, et al. Efficacy of single dose pegfilgrastim in enhancing the mobilization of CD34+ peripheral blood stem cells in aggressive lymphoma patients treated with cisplatin-aracytin-containing regimens. Bone Marrow Transplant 2006; 38: 413-416.
[60] Ataergin S, Arpaci F, Turan M, et al. Reduced dose of lenograstim is as efficacious as standard dose of filgrastim for peripheral blood stem cell mobilization and transplantation: A randomized study in patients undergoing autologous peripheral stem cell transplantation. Am J Hematol 2008; 83: 644-648.
[61] Magagnoli M, Spina M, Balzarotti M, et al. IGEV regimen and a fixed dose of lenograstim: an effective mobilization regimen in pretreated Hodgkin's lymphoma patients. Bone Marrow Transplant 2007; 40: 1019-1025.
[62] Kopf B, DeGiorgi U, Vertogen B, et al. A randomized study comparing filgrastim versus lenograstim versus molgramostim plus chemotherapy for peripheral blood progenitor cell mobilization. Bone Marrow Transplant 2006; 38: 407-412.
[63] Gianni AM, Siena S, Bregni M, et al. Granulocyte-macrophage colony stimulating factor to harvest circulating hemopoietic stem cells for autotransplantation. Lancet 1989; 2: 580-585.
[64] Hoeglund M, Smedmyr B, Bengtsson M, et al. Mobilization of CD34+ cells by glycosylated and nonglycosylated G-CSF in healthy volunteers- a comparative study. Eur J Haematol 1997; 59: 177-183.
[65] Fu S, Liesveld J. Mobilization of hematopoietic stem cells. Blood Rev 2000; 14: 205–218.
[66] Pavone V, Gaudio F, Guarini A, et al. Mobilization of peripheral blood stem cells with high-dose cyclophosphamide or the DHAP regimen plus G-CSF in non-Hodgkin's lymphoma. Bone Marrow Transplant 2002; 29: 285–290.
[67] Watts MJ, Ings SJ, Leverett D, et al. ESHAP and G-CSF is a superior blood stem cell mobilizing regimen compared to cyclophosphamide 1.5g m-2 and G-CSF for pre-

treated lymphoma patients: a matched pairs analysis of 78 patients. Br J Cancer 2000; 82: 278–282.
[68] Moskowitz CH, Bertino JR, Glassman JR, et al. Ifosfamide, carboplatin, and etoposide: a highly effective cytoreduction and peripheral-blood progenitor-cell mobilization regimen for transplant-eligible patients with non-Hodgkin's lymphoma. J Clin Oncol 1999; 17: 3776–3785.
[69] Alegre A, Tomas JF, Martinez-Chamorro C, et al. Comparison of peripheral blood progenitor cell mobilization in patients with multiple myeloma: high-dose cyclophosphamide plus GM-CSF vs G-CSF alone. Bone Marrow Transplant 1997; 20: 211–217.
[70] Desikan KR, Barlogie B, Jagannath S, et al. Comparable engraftment kinetics following peripheral-blood stem-cell infusion mobilized with granulocyte colony-stimulating factor with or without cyclophosphamide in multiple myeloma. J Clin Oncol 1998; 16: 1547–1553.
[71] Narayanasami U, Kanteti R, Morelli J, et al. Randomized trial of filgrastim versus chemotherapy and filgrastim mobilization of hematopoietic progenitor cells for rescue in autologous transplantation. Blood 2001; 98: 2059–2064.
[72] Dazzi C, Cariello A, Rosti G, et al. Is there any difference in PBPC mobilization between cyclophosphamide plus G-CSF and G-CSF alone in patients with non-Hodgkin's lymphoma? Leuk Lymphoma 2000; 39: 301–310.
[73] Condomines M, Quittet P, Lu ZY, Nadal L, Latry P, Lopez E et al. Functional regulatory T cells are collected in stem cell autografts by mobilization with high-dose cyclophosphamide and granulocyte colony-stimulating factor. J Immunol 2006; 176: 6631–6639.
[74] Koumakis G, Vassilomanolakis M, Barbounis V, et al. Optimal timing (preemptive versus supportive) of granulocyte colony-stimulating factor administration following high-dose cyclophosphamide. Oncology 1999; 56: 28- 35.
[75] Anderlini P, Przepiorka D, Champlin R, Korbling M. Biologic and clinical effects of granulocyte colony-stimulating factor in normal individuals. Blood 1996; 88: 2819-2825.
[76] Anderlini P, Champlin RE. Biologic and molecular effects of granulocyte colony-stimulating factor in healthy individuals: recent findings and current challenges. Blood 2008; 111: 1767-1772.
[77] Pamphilon D, Nacheva E, Navarrete C, et al. The use of granulocyte-colony-stimulating factor in volunteer untrelated hemopoietic stem cell donors. Transfusion 2008; 48: 1495-1501.
[78] Leitner GC, Baumgartner K, Kalhs P, et al. Regeneration, health status and quality of life after rhG-CSF-stimulated stem cell collection in healthy donors: a cross-sectional study.
[79] Hasenclever D, Sextro M. Safety of alloPBPCT donors: biometrical considerations on monitoring long term risks. Bone Marrow Transplant 1996; 17 (suppl 2): S28-S30.
[80] Halter J, Kodera Y, Ispizua UI, et al. Severe events in donors after allogeneic hematopoietic stem cell donation. Haematologica 2009; 94: 94-101.
[81] Martino M, Console G, Dattola A, et al. Short and long-term safety of lenogastim administration in healthy peripheral haematopoietic progenitor cell donors: a single centre experience. Bone Marrow Transplant 2009; 44: 163-168.

[82] Kuendgen A, Fenk R, Bruns I, et al. Splenic rupture following administration of pegfilgrastim in a patient with multiple myeloma undergoing autologous peripheral blood stem cell transplantation. Bone Marrow Transplant 2006; 38: 69-70.
[83] Picardi M, De Rosa G, Selleri C, et al. Spleen enlargement following recombinant human granulocyte colony-stimulating factor administration for peripheral blood stem cell mobilization. Haematologica 2003; 88: 794-800.
[84] Anderlini P, Donato M, Chan KW, et al. Allogeneic blood progenitor cell collection in normal donors after mobilization with filgrastim: the MD Anderson Cancer Center experience. Transfusion 1999; 39: 555-560.
[85] Spiekermann K, Roesler J, Emmendoerffer A, et al. Functional features of neutrophils induced by G-CSF and GM-CSF treatment: differential effects and clinical implications. Leukemia 1997; 11: 466-478.
[86] Ross HJ, Moy LA, Kaplan R, Figlin RA. Bullous pyoderma gangrenosum after granulocyte colony-stimulating factor treatment. Cancer 1991; 68: 441-443.
[87] Johnson ML, Grimwood RE. Leukocyte colony-stimulating factors. A review of associated neutrophilic dermatoses and vasculitides. Arch Dermatol 1994; 130: 77-81.
[88] Oeda E, Shinohara K, Kamei S, et al. Capillary leak syndrome likely the result of granulocyte-stimulating-factor after high-dose chemotherapy. Int Med 1994; 33: 115-19.
[89] Spitzer T, McAfee S, Poliquin C, Colby C. Acute gouty arthritis following recombinant human granulocyte colony-stimulating factor in an allogeneic blood stem cell donor (letter). Bone Marrow Transplant 1998; 21: 966-967.
[90] Parkkali T, Volin L, Siren MK, Ruutu T. Acute iritis induced by granulocyte colony-stimulating factor used for mobilization in a volunteer unrelated peripheral blood progenitor cell donor. Bone Marrow Transplant 1996; 17: 433-434.
[91] Becker PS, Wagle M, Matous S, et al. Spontaneous splenic rupture following administration of granulocyte colony stimulating factor (G-CSF): occurrence in an allogeneic donor of peripheral blood stem cells. Biol Blood Marrow Transplant 1997; 3: 45-49.
[92] Adkins DR. Anaphylactoid reaction in a normal donor given granulocyte colony-stimulating factor (letter). J Clin Oncol 1998; 16: 812.
[93] Falzetti F, Aversa F, Minelli O, Tabilio A. Spontaneous rupture of spleen during peripheral blood stem cell mobilization in healthy donor. Lancet 1999; 353: 555.
[94] Hilbe W, Nussbaumer W, Bonatti H, et al. Unusual adverse events following peripheral blood stem cell (PBSC) mobilization using granulocyte colony stimulating factor (G-CSF) in healthy donors. Bone Marrow Transplant 2000; 26; 811-813.
[95] Armitage A, Hargreaves R, Samson D, et al. CD34 counts to predict the adequate collection of peripheral blood progenitor cells. Bone Marrow Transplant 1997; 20: 587-591.
[96] Hill QA, Buxton D, Pearce R, et al. An analysis of the optimal timing of peripheral blood stem cell harvesting following priming with cyclophosphamide and G-CSF. Bone Marrow Transplant 2007; 40: 925-930.

[97] Gandhi MK, Jestice K, Scott MA, et al. The minimum CD34 threshold depends on prior chemotherapy in autologous peripheral blood stem cell recipients. Bone Marrow Transplant 1999; 23: 9-13.
[98] Villalon L, Odriozola J, Larana JG, et al. Autologous peripheral blood progenitor cell transplantation with < 2.0 x 10^6/kg: an analysis of variables concerning mobilization and engraftment. The Hematology Journal 2000; 1: 374-381.
[99] Scheid C, Draube A, Reiser M, et al. Using at least 5 x 10^6/kg CD34+ cells for autologous stem cell transplantation significantly reduces febrile complications and use of antibiotics after transplantation. Bone Marrow Transplant 1999; 23: 1177-1181.
[100] Weaver CH, Hazelton B, Birch R, et al. An analysis of engraftment kinetics as a function of the CD34 content of peripheral blood progenitor cell collections in 692 patients after the administration of myeloablative chemotherapy. Blood 1995; 86: 3961-3969.
[101] Kiss JE, Rybka WB, Winckelstein A, et al. Relationship of CD34+ cell dose to early and late hematopoiesis following autologous peripheral blood stem cell transplantation. Bone Marrow Transplant 1997; 19: 303-310.
[102] Ketterer N, Salles G, Raba M, et al. High CD34+ cell counts decrease hematologic toxicity of autologous peripheral blood progenitor cell transplantation. Blood 1998; 91: 3148-3155.
[103] Bolwell BJ, Pohlman B, Rybicki L, et al. Patients mobilizing large numbers of CD34+ cells ("super mobilizers") have improved survival in autologous stem cell transplantation for lymphoid malignancies. Bone Marrow Transplant 2007; 40: 437-441.
[104] Korbling M, Andrelini P. Peripheral blood stem cell versus bone marrow allotransplantation: does the source of hematopoietic stem cells matter? Blood 2001; 98: 2900-2908.
[105] Bacigalupo A, Francesco F, Brinch L, Van Lint MT. Bone marrow or peripheral blood as a source of stem cells for allogeneic transplants. Curr Opin Hematol 2000; 7: 343-347.
[106] Ryncarz RE, Anasetti C. Expression of CD86 on human marrow CD34+ cells identifies immunocompetent committed precursors of macrophages and dendritic cells. Blood 1998; 91: 3892-3900.
[107] Rondelli D, Anasetti C, Fortuna A, et al. T cell alloreactivity induced by normal G-CSF-mobilized CD34+ blood cells. Bone Marrow Transplant 1998; 21: 1183-1191.
[108] Van Rhee F, Jiang YZ, Vigue F, et al. Human G-CSF-mobilized CD34-positive peripheral blood progenitor cells can stimulate allogeneic T-cell responses: implications for graft rejection in mismatched transplantation. Br J Haematol 1999; 105: 1014-1024.
[109] Urbano-Ispizua A, Carreras E, Marin P, et al. Allogeneic transplantation of CD34(+) selected cells from peripheral blood from human leukocyte antigen-identical siblings: detrimental effect of a high number of donor CD34(+) cells? Blood 2001; 98: 2352-2357.
[110] Mohty M, Bilger K, Jourdan E, et al. Higher doses of CD34+ peripheral blood stem cells are associated with increased mortality from chronic graft-versus-host disease after allogeneic HLA-identical sibling transplantation. Leukemia 2003; 17: 869-875.

[111] Sohn SK, Kim JG, Kim DH, et al. Impact of transplanted CD34+ cell dose in allogeneic unmanipulated peripheral blood stem cell transplantation. Bone Marrow Transplant 2003; 31: 967-972.

[112] Krishnan A, Bhatia S, Slovak M, et al. Predictors for therapy-related leukemia and myelodysplasia following autologous transplantation for lymphoma: an assessment of risk factors. Blood 2000; 95: 1588-1593.

[113] Drake M, Ranaghan L, Morris TCM et al. Analysis of the effect of prior chemotherapy on progenitor cell yield: use of a chemotherapy scoring system. Br J Haematol 1997; 98: 745-749.

[114] Clark RE and Brammer CG. Previous treatment predicts the efficiency of blood progenitor cell mobilisation: validation of a chemotherapy scoring system. Bone Marrow Transplant 1998; 22: 859-863.

[115] Jantunen E, Fruehauf S. Importance of blood graft characteristics in auto-SCT: implications for optimizing mobilization regimens. Bone Marrow Transplantation 2011; 46: 627–635.

[116] Gazitt Y, Freytes CO, Callender N, et al. Successful PBSC mobilization with high-dose G-CSF for patients failing a first round of mobilization. J Hematother 1999; 8: 173-183.

[117] Watts MJ, Ings SJ, Flynn M, et al. Remobilization of patients who fail to achieve minimal progenitor thresholds at the first attempt is clinically worthwhile. Br J Haematol 2000; 111: 287-291.

[118] Olivieri A, Offidani M, Cantori I, et al. Addition of erythropoietin to granulocyte colony-stimulating factor after priming chemotherapy enhances hematopoietic progenitor mobilization. Bone Marrow Transplant 1995; 16: 765–770.

[119] Kaushansky K. Lineage-specific hematopoietic growth factors. N Engl J Med 2006; 354: 2034–2045.

[120] Hart C, Grassinger J, Andreesen R and Hennemann B. EPO in combination with G-CSF improves mobilization effectiveness after chemotherapy with ifosfamide, epirubicin and etoposide and reduces costs during mobilization and transplantation of autologous hematopoietic progenitor cells. Bone Marrow Transplantation 2009 43, 197–206.

[121] Olivieri A, Scortechini I, Capelli D, et al. Combined administration of alpha-erythropoietin and filgrastim can improve the outcome and cost balance of autologous stem cell transplantation in patients with lymphoproliferative disorders Bone Marrow Transplantation 2004; 34: 693–702.

[122] Perillo A, Ferrandina G, Pierelli L, et al. Cytokines alone for PBPC collection in patients with advanced gynaecological malignancies: G-CSF vs G-CSF plus EPO. Bone Marrow Transplant 2004; 34: 743–744.

[123] Miller CB, Lazarus HM. Erythropoietin in stem cell transplantation. Bone Marrow Transplant 2001; 27: 1011–1016.

[124] Zeller W, Gutensohn K, Stockschlader M, et al. Increase of mobilized CD34-positive peripheral blood progenitor cells in patients with Hodgkin's disease, non-Hodgkin's lymphoma, and cancer of the testis. Bone Marrow Transplant 1996; 17: 709–713.

[125] Sheridan WP, Begley CG, To LB, et al. Phase II study of autologous filgrastim (G-CSF)-mobilized peripheral blood progenitor cells to restore hemopoiesis after high-dose

chemotherapy for lymphoid malignancies. Bone Marrow Transplant 1994; 14: 105–111.
[126] Boeve S, Strupeck J, Creech S, et al. Analysis of remobilization success in patients undergoing autologous stem cell transplants who fail an initial mobilization: risk factors, cytokine use and cost. Bone Marrow Transplant 2004; 33: 997–1003.
[127] Wang S, Nademanee A, Qian D, et al. Peripheral blood hematopoietic stem cell mobilization and collection efficacy is not an independent prognostic factor for autologous stem cell transplantation. Transfusion 2007; 47: 2207–2216.
[128] Gazitt Y. Homing and mobilization of hematopoietic stem cells and hematopoietic cancer cells are mirror image process, utilizing similar signaling pathways and occurring concurrently: circulating cancer cells constitute an ideal target for concurrent treatment with chemotherapy and anti- lineage- specific antibodies. Leukemia 2004; 18: 1-10.
[129] Bot FJ, van Eijk L, Schipper P, et al. Synergistic effects between GM-CSF and G-CSF or M-CSF on highly enriched human marrow progenitor cells. Leukemia 1990; 4: 325–328.
[130] Gazitt Y. Comparison between granulocyte colony-stimulating factor and granulocyte-macrophage colony-stimulating factor in the mobilization of peripheral blood stem cells. Curr Opin Hematol 2002; 9: 190–198.
[131] Weaver CH, Schulman KA, Wilson-Relyea B, et al. Randomized trial of filgrastim, sargramostim, or sequential sargramostim and filgrastim after myelosuppressive chemotherapy for the harvesting of peripheral-blood stem cells. J Clin Oncol 2000; 18: 43–53.
[132] Quittet P, Ceballos P, Lopez E, et al. Low doses of GM-CSF (molgramostim) and G-CSF (filgrastim) after cyclophosphamide (4 g/m2) enhance the peripheral blood progenitor cell harvest: results of two randomized studies including 120 patients. Bone Marrow Transplant 2006; 38: 275–284.
[133] Gazitt Y, Callander N, Freytes CO, et al. Peripheral blood stem cell mobilization with cyclophosphamide in combination with G-CSF, GM-CSF, or sequential GM-CSF/G-CSF in non-Hodgkin's lymphoma patients: a randomized prospective study. J Hematother Stem Cell Res 2000; 9: 737–748.
[134] Pusic I, Jiang SY, Landua S, et al. Impact of mobilization and remobilization strategies on achieving sufficient stem cell yields for autologous transplantation. Biol Blood Marrow Transplant 2008; 14: 1045–1056.
[135] Bashey A, Corringham S, Gilpin E, et al. Simultaneous administration of G-CSF and GM-CSF for re-mobilization in patients with inadequate initial progenitor cell collections for autologous transplantation. Cytotherapy 2000; 2: 195–200.
[136] Lapidot T, Petit I. Current understanding of stem cell mobilization: the roles of chemokines, proteolytic enzymes, adhesion molecules, cytokines, and stromal cells. Exp Hematol 2002; 30: 973–981.
[137] Costa JJ, Demetri GD, Harrist TJ, et al. Recombinant human stem cell factor (kit ligand) promotes human mast cell and melanocyte hyperplasia and functional activation in vivo. J Exp Med 1996; 183: 2681–2686.
[138] Stiff P, Micallef I, McCarthy P, et al. Treatment with plerixafor in non-Hodgkin's lymphoma and multiple myeloma patients to increase the number of peripheral

blood stem cells when given a mobilizing regimen of G-CSF: Implications for the heavily pretreated patient. Biol Blood Marrow transplant 2009; 15: 249-256.
[139] Fowkler CJ, Dunn A, Hayes-Lattin B, et al. Rescue from failed growth factor and/or chemotherapy HSC mobilization with G-CSF and plerixafor (AMD3100): an institutional experience. Bone Marrow Transplant 2009; 43: 909-917.
[140] DiPersio J, Stadtmauer EA, Nademanee AP, et al. A phase III, multicenter, randomized, double-blind, placebo controlled, comparative trial of AMD3100 (plerixafor)+G-CSF vs G-CSF+placebo for mobilization in multiple myeloma (MM) patients for autologous hematopoietic stem cell (aHSC) transplantation. Blood (ASH Annual Meeting Abstracts) 2007; 110: 445.
[141] Flomenberg N, Devine SM, Dipersio JF, et al. The use of AMD3100 plus G-CSF for autologous hematopoietic progenitor cell mobilization is superior to G-CSF alone. Blood 2005; 106: 1867–1874.
[142] Calandra G, McCarty J, McGuirk J, et al. AMD3100 plus G-CSF can successfully mobilize CD34+ cells from non-Hodgkin's lymphoma, Hodgkin's disease and multiple myeloma patients previously failing mobilization with chemotherapy and/or cytokine treatment: compassionate use data. Bone Marrow Transplant 2008; 41: 331–338.
[143] DiPersio JF, Micallef I, Stiff PJ, et al. A phase III, multicenter, randomized, double-blind, placebo controlled, comparative trial of AMD3100 (plerixafor)+G-CSF vs. placebo+G-CSF in non-Hodgkin's lymphoma (NHL) patients for autologous hematopoietic stem cell (aHSC) transplantation. Blood (ASH Annual Meeting Abstracts) 2007; 110: 601.
[144] King AG, Horowitz D, Dillon SB, et al. Rapid mobilization of murine hematopoietic stem cells with enhanced engraftment properties and evaluation of hematopoietic progenitor cell mobilization in rhesus monkeys by a single injection of SB-251353, a specific truncated form of the human CXC chemokine GRObeta. Blood 2001; 97: 1534–1542.
[145] Pelus LM, Fukuda S. Peripheral blood stem cell mobilization: the CXCR2 ligand GRObeta rapidly mobilizes hematopoietic stem cells with enhanced engraftment properties. Exp Hematol 2006; 34: 1010–1020.
[146] Somlo G, Sniecinski I, ter Veer A, et al: Recombinant human thrombopoietin in combination with granulocyte colony-stimulating factor enhances mobilization of peripheral blood progenitor cells, increases peripheral blood platelet concentration, and accelerates hematopoietic recovery following high-dose chemotherapy. Blood 1999; 93: 2798-2806.
[147] Linker C, Anderlini P, Herzig R, et al. Recombinant human thrombopoietin augments mobilization of peripheral blood progenitor cells for autologous transplantation. Biol Blood Marrow Transplant 2003; 9: 405–413.
[148] Li J, Yang C, Xia Y, Bertino A, Glaspy J, Roberts M et al. Thrombocytopenia caused by the development of antibodies to thrombopoietin. Blood 2001; 98: 3241–3248. Basser RL, O'Flaherty E, Green M, Edmonds M, Nichol J, Menchaca DM et al. Development of pancytopenia with neutralizing antibodies to thrombopoietin after multicycle chemotherapy supported by megakaryocyte growth and development factor. Blood 2002; 99: 2599–2602.

[149] Basser R. The impact of thrombopoietin on clinical practice. Curr Pharm Des 2002; 8: 369-377.
[150] Adams GB, Martin RP, Alley IR, et al. Therapeutic targeting of a stem cell niche. Nat Biotechnol 2007; 25: 238–243.
[151] Ballen KK, Shpall EJ, Avigan D, et al. Phase I trial of parathyroid hormone to facilitate stem cell mobilization. Biol Blood Marrow Transplant 2007; 13: 838–843.
[152] Ispizua Urbano A. Stem cell product for allo-SCT. Tenth ESH-EBMT training course on haematopoietic stem cell transplantation. Warsaw, Poland 2006.
[153] Wagner JE, Kernan NA, Steinbuch M, et al. Allogeneic sibling umbilical cord-blood transplantation in children with malignant and non-malignant disease. Lancet 1995; 346: 214-219.
[154] Rocha V, Kabbara N, Ionescu et al. Pediatric related and unrelated cord blood transplantation for malignant diseases. Bone Marrow Transplant 2009; 44: 653-659.
[155] Rocha V, Locatelli F. Searching for alternative hematopoietic stem cell donors for pediatric patients. Bone Marrow Transplant 2008; 41: 207–214.
[156] Rocha V, Cornish J, Sievers EL, et al. Comparison of outcomes of unrelated bone marrow and umbilical cord blood transplants in children with acute leukaemia. Blood 2001; 97: 2962–2971.
[157] Rocha V, Wagner JE, Jr., Sobocinski KA, et al. Graft-versus-host disease in children who have received a cord-blood or bone marrow transplant from an HLA-identical sibling. Eurocord and International Bone Marrow Transplant Registry Working Committee on Alternative Donor and Stem Cell Sources. N Engl J Med 2000; 342: 1846–1854.
[158] Rocha V, Labopin M, Sanz G, et al. Transplants of umbilical-cord blood or bone marrow from unrelated donors in adults with acute leukemia. N Engl J Med 2004; 351: 2276–2285.
[159] Eapen M, Rubinstein P, Zhang MJ, et al. Outcomes of transplantation of unrelated donor umbilical cord blood and bone marrow in children with acute leukaemia: a comparison study. Lancet 2007; 369: 1947–1954.
[160] Laughlin M, Eapen M, Rubinstein P, et al. Outcomes after transplantation of cord blood or bone marrow from unrelated donors in adults with leukaemia. N Engl J Med 2004; 351: 2265–2275.
[161] Niehues T, Rocha V, Filipovich AH, et al. Factors affecting lymphocyte subset reconstitution after either related or unrelated cord blood transplantation in children- a Eurocord analysis. Br J Haematol 2001; 114: 42–48.
[162] Hao QL, Shah AJ, Thiemann FT, et al. A functional comparison of CD34+ CD38- cells in cord blood and bone marrow. Blood 1995;86: 3745-3753.
[163] Komanduri KV, St John LS, De Lima M, et al. Delayed immune reconstitution after cord blood transplantation is characterized by impaired thymopoiesis and late memory T-cell skewing. Blood 2007; 110: 4543–4551.
[164] Saavedra S, Sanz GF, Jarque I, et al. Early infections in adult patients undergoing unrelated donor cordbloodtransplantation. Bone Marrow Transplant 2002; 30: 937-943.
[165] Delaney C, Varnum-Finney B, Aoyama K, et al. Dose-dependent effects of the Notch ligand Delta1 on ex vivo differentiation and in vivo marrow repopulating ability of cord blood cells. Blood. 2005; 106: 2693–2699.

[166] de Lima M, McMannis J, Gee A, et al. Transplantation of ex vivo expanded cord blood cells using the copper chelator tetraethylenepentamine: a phase I/II clinical trial. Bone Marrow Transplant 2008;41: 771-778.

[167] Frassoni F, Gualandi F, Podestà M, et al. Direct intrabone transplant of unrelated cord-blood cells in acute leukaemia: a phase I/II study. Lancet Oncol 2008; 9: 831–839.

[168] Barker JN, Weisdorf DJ, DeFor TE, et al. Transplantation of 2 partially HLA-matched umbilical cord blood units to enhance engraftment in adults with hematologic malignancy. Blood. 2005; 105: 1343–1347.

[169] Brunstein CG, Barker JN, Weisdorf DJ, et al. Umbilical cord blood transplantation after nonmyeloablative conditioning: impact on transplantation outcomes in 110 adults with hematologic disease. Blood. 2007; 110: 3064–3070.

[170] Barker JN, Weisdorf DJ, DeFor TE, et al. Rapid and complete donor chimerism in adult recipients of unrelated donor umbilical cord blood transplantation after reduced-intensity conditioning. Blood. 2003; 102: 1915-1919.

[171] Ballen KK, Spitzer TR, Yeap BY, et al. Double unrelated reduced-intensity umbilical cord blood transplantation in adults. Biol Blood Marrow Transplant 2007; 13: 82–89.

[172] Fernandes J, Rocha V, Robin M, et al. Second transplant with two unrelated cord blood units for early graft failure after haematopoietic stem cell transplantation. Br J Haematol 2007; 137: 248–251.

[173] Bautista G, Cabrera JR, Regidor C, et al. Cord blood transplants supported by co-infusion of mobilized hematopoietic stem cells from a third-party donor. Bone Marrow Transplant 2009; 43: 365–373.

[174] MacMillan ML, Blazar BR, DeFor TE, et al. Transplantation of ex-vivo culture-expanded parental haploidentical mesenchymal stem cells to promote engraftment in pediatric recipients of unrelated donor umbilical cord blood: results of a phase I-II clinical trial. Bone Marrow Transplant. 2009; 43: 447–454.

12

Cryopreservation of Hematopoietic and Non-Hematopoietic Stem Cells – A Review for the Clinician

David Berz[1] and Gerald Colvin[2]
[1]*City of Hope, NCI-designated Comprehensive Cancer Center, Duarte, California*
[2]*Brown University, Providence, Rhode Island*
USA

1. Introduction

The transplantation of stem cells can be performed in an autologous, where the recipient donates his own stem cells for later use, or in an allogeneic fashion, where the donor and the recipient are two different persons (Berz, McCormack, Winer, Colvin, & Peter J Quesenberry, 2007; A. Gratwohl et al., 2010). In some clinical settings, stem cells can be utilized within a 72 hour timeframe without the need for extensive storage (Fleming & A Hubel, 2006; Allison Hubel, Carlquist, Clay, & Jeff McCullough, 2004; Pettengell, Woll, O'Connor, Dexter, & Testa, 1994). The autologous transplantation of cellular products and the therapeutic use of umbilical cord stem cells rely heavily on the preservation of stem cells after initial collection to be utilized at a later point in time (Berz, McCormack, Winer, Colvin, & Peter J Quesenberry, 2007).

Hematopoietic stem cell transplantation has been in clinical use for decades to treat benign and malignant hematologic and non-hematologic conditions (C. J. Hunt, 2011). The principal sources for those therapeutic strategies are bone marrow, peripheral blood hematopoietic progenitor cells and stem cells derived from umbilical cord blood. Although clinicians have decades of experience with the use of hematopoietic stem cells(HSC), the interest in the clinical use of non-hematopoietic stem cells, such as embryonic(C. J. Hunt, 2011; Leeb et al., 2011), mesenchymal (Ding, Shyu, & Lin, 2011; Leclerc et al., 2011; Puglisi et al., 2011) and induced pluripotent stem cells (Leeb et al., 2011; Sohni & Verfaillie, 2011) is rapidly expanding. The therapeutic strategies utilizing such cellular products, depend heavily on the effective preservation of those cell products for clinical use at a later point in time. The need for ready availability of such products calls for storage procedures with favorable graft survival rates and a tolerable toxicity profile. While cryopreservation protocols for HSC are well developed, the field for non-hematopoietic stem cells still remains to be defined.

The cryopreservation of all stem cells follows certain principal steps; First, Cytoreduction and prefreezing processing of the freshly collected graft. Second, the cryopreservative medium is prepared and added. Third, the graft is assessed for viability and integrity. Microbial contamination is ruled out. Fourth, freezing of the stem cells is performed. Fifth, thawing of the cryopreserved graft is conducted. Last, the post thawing processing is performed, including in certain situations washing of the stem cell concentrate. A standard algorithm for the preservation of HSC populations is depicted in figure 1

No single, standardized cryopreservation protocol has been universally used and differences in techniques continue to exist in between different centers.

We at our institution utilize a standardized NIH protocol. We collect Hematopoietic Progenitor Stem Cells in a minimally manipulated fashion as defined by the Foundation for the Accreditation of Hematopoietic Cell Therapy (FAHCT) with a minimal cell dose of at least $1.5\text{-}3.0 \times 10^6\text{–}5.0 \times 10^6$ CD34(+) cells/kg body weight, depending on the clinical indication. The specimen is then centrifuged to develop the cell rich pellet. In autologous transplants donor plasma is used for re-suspension, if available. Human serum albumin solution is an alternative. Then, a solution of heparinized Plasmalyte and 10% DMSO (Dimethylsulfoxide) is added. This usually eventuates into a cellular concentration of 200×10^6 cells in the cryopreservate. We store the bone marrow or peripheral blood stem cell product at initially −4°C. Then the sample is frozen down to the target temperature of-135° to -156°Celsius to be placed into the vapor phase of the nitrogen tank.

2. Types of stem cells

a. Hematopoietic stem cells (HSC) are in therapeutic use for more than 50 years (THOMAS, LOCHTE, CANNON, SAHLER, & FERREBEE, 1959). In 2006 50,417 hematopoietic stem cell transplants were performed globally, 21 516 allogeneic (43%) and 28 901 autologous (57%) (A. Gratwohl et al., 2010). The principal sources for HSCs are the bone marrow (BM), mobilized peripheral blood stem cells (PBSC) and umbilical cord stem cells (UCB)(Berz, McCormack, Winer, Colvin, & Peter J Quesenberry, 2007). Allogeneic grafts are often used shortly after collection, but autologous grafts and cord blood units are generally cryopreserved. The amount of cryopreserved cord blood units is rising and the storage time in cord blood banks often exceeds a decade (Valeri & Pivacek, 1996).
b. Mesenchymal stem cells (MSC) were initially identified in 1976 as bone marrow stromal cells with the capability to form mesenchymal components such as fat, cartilage and bone(Friedenstein, Gorskaja, & Kulagina, 1976). Along with the large regenerative potential of damaged mesenchymal tissues, MSCs are powerful immune modulators with promising results in autoimmune diseases and GVHD(Dazzi & Krampera, 2011; McGuirk & Weiss, 2011).
c. Human embryonic stem cells(hESC) are primitive precursor cells with an unlimited potential for self- renewal and the capability to differentiate into any cell type derived from all three germ cell layers(Murdoch et al., 2011). This pluripotent property makes hESCs powerful candidates for regenerative cellular therapies.
d. Induced pluripotent stem cells (iPSC) are somatic cells which are transformed by genetic reprogramming into cells with pluripotent stem cell character. Initially described in 2007, such cells would be optimal candidates for autologous regenerative approaches without the ethical concerns about hESCs (Walia, Satija, Tripathi, & Gangenahalli, 2011). However, immunologic barriers have recently been identified even in autologous settings. Before a viable clinical use, such obstacles have first to be overcome (Zhao, Z.-N. Zhang, Rong, & Y. Xu, 2011).

Although most of our clinical expertise exists with HSC, efforts to develop appropriate cryopreservation protocols for non-hematopoietic stem cells are ongoing. This chapter will primarily focus on the recent developments in the field of cryopreservation of hematopoietic stem cells, but also outline some of the similarities and differences between the cryopreservation of HSCs and non-hematopoietic stem cells.

3. Processing prior to cryostorage

3.1 Liquid phase storage

Several centers, particularly in remote rural settings rely on the performance of autologous bone marrow transplantation without a local croypreservation expertise. Along with that, umbilical cord blood processing is only performed in highly specialized cell processing facilities, which are often geographically remote from the place of collection. In such clinical settings, the initial cell collections have to be transported to a center with the necessary expertise in a liquid form (Fleming & A Hubel, 2006; Rodrigues et al., 2008). Liquid storage, either for transport purposes or to bridge a short time span prior to definitive clinical use has been successfully used for different clinical indications (Corato et al., 2000; Pettengell, Woll, O'Connor, Dexter, & Testa, 1994; Pettengell, Woll, Thatcher, Dexter, & Testa, 1995).
Several studies have investigated the effects of temperature, total liquid phase duration and storage media with varying cell concentrations on functional outcomes. A short term storage temperature of 4° Celsius has been shown to be suitable for UB cord blood collections (Burger, A Hubel, & J McCullough, 1999; Allison Hubel, Carlquist, Clay, & Jeffrey McCullough, 2003) and non cord blood hematopoietic progenitor cells (Burger et al., 1999; Allison Hubel, Carlquist, Clay, & Jeff McCullough, 2004; Allison Hubel et al., 2003). The major rationale for the choice of 4° Celsius was the concern about crystal formation and corresponding cell death at lower temperatures (Matsumoto et al., 2002). However, the actual freezing/crystallization point of human plasma has been shown to be at -0.8° Celsius (K. B. Storey & J. M. Storey, 1990). Hence, different temperatures have been investigated for short term storage. A study from Japan established the feasibility of storage of HPC at -2 Celsius in University of Wisconcin medium without cryopreservative additive (CPA) for up to 72 hours. In this short term storage conditions, appropriate for transport purposes, the post-thaw nucleated cell count recoveries and functional assay outcomes were both above 90%. This was superior to a control, which was stored at -80° Celsius with DMSO for the same amount of time (Matsumoto et al., 2002). The results of this and similar studies suggest that storage around the freezing point is more appropriate in short term storage for periods of less than 72 hours than the usual deep freezing temperatures.

3.2 Centrifugation and resuspension

The pre-cryostorage processing entails the actual collection procedure, the removal of cell bulk, volume reduction with concentration of the stem cells and addition of the cryo-medium. The process is performed in a strictly sterile environment. It also includes the identification of the donor with labeling of the bag, general assessment of the collected specimen (such as weighing and cell enumeration) and ascertainment of a sterile specimen with microbiologic studies (S. M. Watt, Austin, & S. Armitage, 2007).
Subsequently the specimen undergoes centrifugation and resuspension of the pellet in order to achieve volume reduction and concentration of the target cell population(Laroche et al., 2005; Rebulla, 2002; Rowley, Bensinger, Gooley, & Buckner, 1994).
The volume reduction process is well documented to be associated with loss of active cells (Koliakos et al., 2007). This is of particular relevance in UCB specimens, in which a paucity

of donor cells is a concern (Koliakos et al., 2007; Laroche et al., 2005; J. C. Wang, Doedens, & Dick, 1997).
Hence, volume reduction during pre-freezing processing remains a field of ongoing research to limit the cell loss. One example is a study, by Koliakos et al, which achieved limited cell loss by a careful double processing in the presence of 2% HES(Koliakos et al., 2007).

3.3 Cell concentration

Triggered by concerns of toxicities of the cryopreservative additive to the cells, high stem cell concentrations in the cryopreserved unit were initially deemed to be detrimental. Also the osmotic shock during manipulation of the specimen could potentially be associated with untoward effects (Luciano et al., 2009). Hence the recommended cell concentrations in cryopreserved hematopoietic stem cells was suggested to be not above $2X10^7$/mL (Aird, Labopin, Gorin, & Antin, 1992; Gorin, 1986; Rowley, 1992; Rowley, Bensinger, Gooley, & Buckner, 1994; Silberstein & Jefferies, 1996). This would lead to a total storage space of 7 liters for every patient needing a routinely used cell dose (Rowley, Bensinger, Gooley, & Buckner, 1994). Early work by Law et al., established that stem cells have a high osmotic resistance (Law, Alsop, D C Dooley, & Meryman, 1983), what led to the investigation of higher cell concentrations with a smaller corresponding storage volumes.
After initial preclinical experiments, several studies established that higher cell concentrations were compatible with good functional outcomes and engraftment kinetics (Cabezudo et al., 2000; Yoshifumi Kawano et al., 2004; Rowley, Bensinger, Gooley, & Buckner, 1994; Villalón et al., 2002). Cell counts between 1 and $2X10^8$ were found to be safe and feasible (Alencar et al., 2010; Cabezudo et al., 2000; Yoshifumi Kawano et al., 2004; Rowley, Bensinger, Gooley, & Buckner, 1994; Villalón et al., 2002). A more recent study from Brazil (Alencar et al., 2010) compared the effect of PBSC concentrations on cell viability, functional assays and engraftment kinetics. No significant differences were observed between cell concentrations of $1X10^8$/mL and $2X10^8$mL. Today cell concentrations of $2X10^8$/mL and above are considered safe and are used in many cryopreservation centers.

3.4 Creation and addition of the cryopreservation medium

The cryopreseved unit for storage consists of the stem cell collection, the diluent medium and the cryopreservative additive(CPA). Often the cell concentrate is added to the cryopreservative medium (diluent medium+CPA) in a 1:1 volume ratio. The optimal consistency of the cryopreservation medium is still a matter of active research and is poorly standardized (S. M. Watt, Austin, & S. Armitage, 2007; Zeisberger et al., 2010).

3.4.1 The diluent

The diluent consists most frequently of cryoprecipitated autologous or allogeneic plasma. If unavailable human albumin solution is widely used (Berz, McCormack, Winer, Colvin, & Peter J Quesenberry, 2007; Rodrigues et al., 2008; S. M. Watt, Austin, & S. Armitage, 2007; Zeisberger et al., 2010). However, particularly in non-hematologic stem cell cryopreservation protocols, fetal bovine serum(FBS) is still used as a standard diluent (Ellerström et al., 2006; Holm et al., 2010; Unger, Skottman, Blomberg, Dilber, & Hovatta, 2008; Zeisberger et al., 2010). The use of FBS, a xenobiotic mix with variable consistency, is associated with certain tangible risks. Immunologic responses to animal serum components

(Mackensen, Dräger, Schlesier, Mertelsmann, & Lindemann, 2000) or stimulated by hapten formation(Martin, Muotri, Gage, & Varki, 2005) and the transmission of known or unknown animal pathogens to the recipient(Elsaadany et al., 2011; Will et al., 1996) have been raised as concerns. Hence, current research focuses on the development of standardized, xeno-free cryopreservation practices for human stem cells (Ellerström et al., 2006; Holm et al., 2010; Zeisberger et al., 2010). Recently, Zeisberger et al. presented excellent viabilities with a xeno-free, predefined cryomedium, suitable for hematopoietic and mesenchymal stem cells.

3.4.2 The cryopreservative additive

3.4.2.1 DMSO and its Toxicities

Dimethylsulfoxide is the most widely used and was already described as cryoprotectant in 1959(LOVELOCK & BISHOP, 1959). After having been initially synthesized by Alexander Mikhaylovich Zaitsev in 1867 (Lewis, 1994a; 1994b), DMSO has been used in the wood industry since the 19th century(Ruiz-Delgado et al., 2009). It also found medical application in a wide spectrum of musculoskeletal, autoimmune and metabolic diseases, including gonarthrosis, interstitial cystitis and amyloidosis (Albanell & Baselga, 2000; Eberhardt, Zwingers, & Hofmann, 1995; Iwasaki, Hamano, Aizawa, Kobayashi, & Kakishita, 1994; McCammon, Lentzner, Moriarty, & Schellhammer, 1998; Morassi, Massa, Mesesnel, Magris, & D'Agnolo, 1989). As a small amphiphatic molecule, DMSO penetrates also into stem cells and acts as a strong hydrogen bond disrupter and hence exerts colligative effects (Ruiz-Delgado et al., 2009; N. C. Santos, Prieto, Morna-Gomes, Betbeder, & Castanho, 1997).
The pulmonary excretion of DMSO accounts for the typical garlic like smell, noticed during stem cell infusion (Jacob & Herschler, 1983). Secondary to the wide pharmacokinetic distribution volume, including good blood brain barrier penetration, DMSO affects multiple organ systems with a wide spectrum of toxicities. Those adverse effects include CNS-, respiratory-, hemolytic-, gastrointestinal-, hepatic-, dermatologic-, cardiovascular- and renal toxicities.
The overall most frequently reported side effects are of gastrointestinal and cardiovascular nature. Nausea and abdominal cramping are observed with incidences of up to 70%. Early studies by Davis et al. (J. M. Davis, Rowley, Braine, Piantadosi, & G. W. Santos, 1990; J. Davis, Rowley, & G. W. Santos, 1990) demonstrated that the incidence of side effects rose with the volume of DMSO and amount of cell lysis products infused.
Vasovagal reactions with hypotension and bradycardia are observed with a high incidence. In a multinational survey study performed by Windrum et al., data from 97 EBMT transplant centers were included. DMSO related toxicities other than nausea and vomiting were observed in about one out of 50 transplants with a mean incidence of 2.2% in all administered units. Cardiovascular side effects were most frequently observed, witnessed in 27% of the participating centers (Windrum, Morris, Drake, D Niederwieser, & Ruutu, 2005). The vagolytic effects of DMSO have been documented in physiologic experiments and the hypotension can be attributed to DMSO induced histamine release (J. M. Davis, Rowley, Braine, Piantadosi, & G. W. Santos, 1990; KLIGMAN, 1965). However, confounding is the fact, that stem cell concentrates are usually infused after thawing to a slushy state at a temperature of 4°-8°C. Hence, a substantial amount of the frequently observed bradycardia and hypotension could be calorically induced. Along with this, recipients of stem cell concentrates are usually premedicated with IV glucocorticosteroids, hydration, mannitol and anti-histamines (J. M. Davis, Rowley, Braine, Piantadosi, & G. W. Santos, 1990; J. Davis,

Rowley, & G. W. Santos, 1990). Several of the cardiovascular side effects, including the frequently observed hypertensive episodes, could hence be multifactorial, like the small vessel smooth muscle constriction by DMSO, hydration in combination with mannitol and the influence of glucocorticosteroids in the presence of antihistamine premedication. Hypotensive attacks are more frequently observed in the absence of antihistamine premedication (English et al., 1989; O'Donnell, Burnett, Sheehan, Tansey, & G. A. McDonald, 1981). Other cardiovascular toxicities, which are potentially related to DMSO are electrocardiographic abnormalities (J. M. Davis, Rowley, Braine, Piantadosi, & G. W. Santos, 1990), pulmonary edema and rarely observed case fatalities, such as cardiac arrest(Baum, Weissman, Tsukamoto, Buckle, & Peault, 1992; PEGG & KEMP, 1960; Ruiz-Delgado et al., 2009).

Respiratory side effects are also frequently observed with stem cell infusions and can often be attributed to DMSO toxicity. Mild bronchospasm and subclinical reductions of pulmonary capacity are frequently observed, but severe respiratory depressions with the need for pulmonary resuscitation have been reported (Benekli et al., 2000; Miniero, Vai, Giacchino, Giubellino, & Madon, n d).

Other, less frequently observed toxicities are anaphylaxis, renal failure, seizures, acute hepatotoxicity and hemolysis.

3.4.2.2 Alternatives to standard DMSO

The above described toxicities led to efforts to reduce the patients' exposure to DMSO. Principally five different approaches are possible. First, the reduction of DMSO concentration in the stem cell concentrate, second creating a stem cell product with a higher cell concentration and corresponding smaller volumes and a lower cumulative exposure to DMSO, third a prolonged infusion with a less intense exposure to DMSO, fourth DMSO depletion in post-thaw processing and last, using alternative CPAs, either alone or in combination with DMSO.

A DMSO concentration of 10% in the stem cell concentrate is still considered standard in most centers around the world. However, several investigators examined the effects of lower DMSO concentrations on recovered cell counts, viability and colony forming capacity. DMSO concentration as low as 2% were used(Bakken, O Bruserud, & J F Abrahamsen, 2003; Balint et al., 1999; Galmés et al., 1999; Halle et al., 2001; Syme et al., 2004; Zeisberger et al., 2010). Overall, 5% DMSO concentrations delivered comparable results to the standard 10% concentration, whilst DMSO concentrations of less or equal to 2% revealed inferior cell integrity, at least in some reports (Zeisberger et al., 2010). In a recent retrospective report from Norway, 103 consecutive patients underwent autologous PBSC transplants after high dose chemotherapy for lymphoma or myeloma (Akkök et al., 2008). The stem cell concentrates were preserved with 10% DMSO in the initial period and with 5% DMSO in the later period. Clinical outcomes, such as transfusion requirements and neutrophil/platelet recovery were essentially the same.

In recent years, the disaccharides trehalose and sucrose have been evaluated as cryopreservative additives (Buchanan et al., 2004; Rodrigues et al., 2008; Scheinkönig, Kappicht, Kolb, & Schleuning, 2004; E J Woods et al., 2000; X. B. Zhang et al., 2003). The exact mode of action of those small molecules remains elusive, but it has been well demonstrated that the integrity of membrane layers and proteins during cryopreservation remains well preserved in the presence of those molecules (J H Crowe et al., 2001; John H Crowe, 2007). Along with a favorable toxicity profile of those compounds, certain pharmacokinetic properties may be of benefit. Trehalose is not able to permeate into the interior of the cells. This facilitates removal during post-thaw washing (Rodrigues et al., 2008). Sucrose/trehalose/DMSO combinations with DMSO concentrations as low as 2.5% have been shown to be comparable with standard 10% DMSO as cryo-additive. Cell count recovery, viability and clonogenicity were similar in several studies (Rodrigues et al., 2008). Similarly encouraging results have been noticed in the presence of catalase type natural bio-oxidants in the cryopreservation medium (Motta, Gomes, L F Bouzas, Paraguassú-Braga, & Porto, 2010; Sasnoor, Kale, & Limaye, 2003), although the exact role of the natural bio-oxidant is still not known.

Another group of substances which are excellent candidates as cryo-preservative additives are hydrophilic macromolecules. They also have the pharmacokinetic benefit of being restricted to the extracellular space and some follow a first order elimination kinetic. Albumin, modified gelatin, hydroxyethyl starch(HES), polyvinylpyrrolidone and polyethylene oxide are members of this group of substances. HES has been most widely studied, particularly to reduce the DMSO concentration (Clapisson et al., 2004; Jeffrey McCullough et al., 2010). A combination of 5% DMSO and 6%HES has been shown to be associated with successful long term storage of PBSC(Jeffrey McCullough et al., 2010). In a blinded randomized phase III, patients underwent high dose chemotherapy with autologous PBSC support, either preserved with a standard solution containing 10% dimethylsulfoxide (DMSO, v/v) or 5% DMSO in combination with 6% hydroxyethylstarch (HES, w/v). One hundred and forty eight patients received PBSC frozen with 10% DMSO and 146 received cells frozen in 5% DMSO/6% HES. Whilst platelet recovery and the median amount of blood products transfused did not differ in between the two groups, the patients obtaining cell concentrates, cryopreserved with the DMSO/HES combination obtained neutrophil recovery in average one day earlier and needed one day less antibiotic administration($p=0.04$)(Rowley et al., 2003).

Unfortunately hematopoietic and other stem cells, particularly human embryonic stem cells, undergo apoptotic transformation during the cryopreservation process (Sangeetha, Kale, & Limaye, 2010; Stroh et al., 2002). This led to the study of caspase inhibitors as cryopreservatives. The initial encouraging results on cell cultures (Heng, Clement, & T. Cao, 2007; Stroh et al., 2002), were recently supported by in vivo experiments(Sangeetha, Kale, & Limaye, 2010).

Other cryo-preservative additives, such as α tocopherol(Neunert et al., 2009; E J Woods et al., 2000), are under investigation and may be of future interest.

3.5 The freezing process

3.5.1 Freezing rate

The optimal method of freezing the cell concentrate to the target storage temperature still remains a matter of debate. Only few high quality studies are available to guide the clinician and laboratorian in the choice of the optimal technique. The controlled rate freezing(CRF) procedure still remains the defined standard in many countries (S. M. Watt, Austin, & S. Armitage, 2007). The principal rationale for CRF as choice is the limited cell damage during the freezing process (Donaldson et al., 1996; Douay, 1985; Yang et al., 2001), particularly at the eutectic transition point. At the eutectic transition point the liquid phase transits into a solid phase and fusion heat is released. Prolonged cell exposure at this point is regarded detrimental to the survival of cells (Douay, 1985). The principal time trajectory during controlled rate freezing involves initially slow freezing, at a rate of -1° to -2° Celsius per minute, then very rapidly around the eutectic point, to be then further cooled at a steady, preset rate to a target temperature to be finally placed into nitrogen for durable storage. Several modifications of this technique have been described in the literature (Berz, McCormack, Winer, Colvin, & Peter J Quesenberry, 2007; Donaldson et al., 1996; Gorin, 1986; Perez-Oteyza et al., 1998; Valeri & Pivacek, 1996; S. M. Watt, Austin, & S. Armitage, 2007). Briefly, one typical algorithm for controlled rate freezing using the KRYO10 freezer encompasses, pre-cooling of the specimen to 6° Celsius and placement into a cryo-cassette. Then the cooling process proceeds at a rate of-1° to -2° Celsius/minute to a temperature of -5° C, followed by rapid cooling around the eutectic point to avoid damage by fusion heat release. Then the cell concentrate continues to be cooled at a rate of -1 Celsius /minute to a temperature of -40 Celsius followed by a rate of -5 Celsius /minute to a target of -135 to -160 Celsius. Finally, the product is placed into permanent nitrogen storage.

Unfortunately, this technique requires sophisticated, costly equipment and a certain personal expertise, which are not available at every center (Almici et al., 2003; Perez-Oteyza et al., 1998).

Several single armed studies have reported excellent viability and hematopoietic recovery rates with uncontrolled freezing procedures in bone marrow (Clark, Pati, & D. McCarthy, 1991; Stiff, Murgo, Zaroulis, DeRisi, & Clarkson, 1983) and peripheral blood progenitor specimens (Almici et al., 2003; Cilloni et al., 1999; Halle et al., 2001).

Few studies directly compared outcomes of controlled rate to uncontrolled rate freezing processes.

A preclinical study, using murine bone marrow samples compared two controlled rate freezing protocols with one uncontrolled rate protocol. Nuclear cell count, viability and several functional assays (MRA, CFU-S and CFU-GM) were assessed after thawing and washing. Although comparable cell counts and viability assays were achieved, superior functional assays (particularly CFU-S and CFU-GM) were achieved with the controlled rate freezing procedure (Balint et al., 1999).

In a clinical study from Japan, two different freezing protocols were compared in the cryopreservation of peripheral blood stem cells (PBSC) (Y Takaue et al., 1994). The PBSC were cryopreserved by either controlled rate or un-controlled rate freezing methods. No differences were observed in the granulocyte/platelet engraftment times and transfusion requirements were similar in both groups (Y Takaue et al., 1994).

The interpretation of this study is somewhat complicated by the fact that the cells in the uncontrolled rate freezing arm also differed in the cryopreservative additive from the standard arm. In addition, only twelve patients obtained the cells, preserved by uncontrolled rate freezing and the study design was not randomized, using matched historical controls as reference.

In a prospective, randomized controlled multicenter study from Spain, apheresis products from 105 patients, who obtained a peripheral stem cell transplant for various malignancies, were split into two bags. One was processed with controlled rate, the other one with uncontrolled rate freezing. Nucleated cell counts, viability and a committed functional assay (CFU-GM) were assessed after thawing. No difference in the loss of cell counts and viability were observed, but again the CFU-GM assay performed superior in the controlled rate freezing arm (Perez-Oteyza et al., 1998). Although the strength of the study was that each sample functioned as its own control, i.e. all other the outcome influencing parameters were kept identical in both arms, the study was not powered to detect differences in the patients' clinical outcomes, such as engraftment kinetics and survival. It remains hence unclear, if the differences in the functional assay outcome hold clinical relevance.

In conclusion, we feel that the controlled rate and the uncontrolled rate freezing procedure are both viable techniques for the preservation of stem cells. At our institution, we perform controlled rate freezing only for bone marrow. Future studies, sufficiently powered to examine relevant clinical outcomes will be needed to determine if the more elaborate controlled rate approach is truly superior.

3.5.2 Storage temperature

The minimal requirements for a long term storage temperature are technical feasibility and a successful clinical outcome. The storage temperature for hematopoietic stem cells varies between different centers. Temperatures range from -196°C to -80°Celsius (Aird, Labopin, Gorin, & Antin, 1992; Cilloni et al., 1999; Galmés et al., 1999; Halle et al., 2001; Rubinstein et al., 1995; Son, Heo, Park, H. H. Kim, & K. S. Lee, 2010; Valeri & Pivacek, 1996). The initially used storage temperatures of -196°C, reflecting the liquid phase nitrogen storage, have largely been replaced by temperatures of

-156°C to -135°C, reflecting the vapor phase storage (Berz, McCormack, Winer, Colvin, & Peter J Quesenberry, 2007) in BM-, PB- as well as umbilical cord derived stem cells(Aird, Labopin, Gorin, & Antin, 1992; Cilloni et al., 1999; Galmés et al., 1999; Halle et al., 2001; Rubinstein et al., 1995; Son, Heo, Park, H. H. Kim, & K. S. Lee, 2010; Valeri & Pivacek, 1996). This shift in culture was mainly induced by the observation that infectious pathogens can survive and be propagated in the liquid nitrogen phase (Bielanski & Vajta, 2009). This fact received a large deal of public attention, when in the mid-1990s six recipients of BM- and PBSC- transplants developed icteric hepatitis B. Follow up examinations revealed that leakage from one auto-donor's stem cell product induced viral contamination of five patients' units, which were stored in the same tank (Tedder et al., 1995).

A recent study by McCullough et al. compared 5 different protocols for five year PBSC storage. Along with other variables, liquid phase nitrogen storage (-196°C) was also compared to a mechanical freezer temperature at -135°C. No significant outcome differences were observed between both temperatures (Jeffrey McCullough et al., 2010).

For human embryonic stem cells, which are stored after vitrification, the storage temperature is a more crucial subject. Even slight temperature variation can lead to de-vitrification of the metastable cryopreservate with crystal formation and corresponding cell damage (Baicu, M. J. Taylor, Z. Chen, & Rabin, 2008; Wusteman, Robinson, & D. Pegg, 2004). Techniques have been developed to minimize those potentially detrimental temperature fluctuations (C. J. Hunt, 2011; Rowley & Byrne, 1992).

3.5.3 Infectious considerations

The contamination of stem cell products with infectious pathogens can occur at several points during the stem cell processing-The marrow harvest, cord blood collection or apheresis, the transport of the product to the cryopreservation facility, the pre-thaw processing, the thawing and washing process as well as the infusion of the final product are all processing points with the potential for microbial contamination. Infectious contaminations during those processes are observed with varying degrees, generally 0-4.5% for PBSC and up to 26% for bone marrow, even if strictly aseptic protocols are followed(Attarian, Bensinger, Buckner, D. L. McDonald, & Rowley, 1996; Espinosa, Fox, Creger, & Lazarus, 1996; Jestice et al., 1996; Kamble et al., 2005; Larrea et al., 2004; Lowder & Whelton, 2003; Majado et al., 2007; D. J. Padley et al., 2003; D. Padley, Koontz, Trigg, Gingrich, & Strauss, 1996; Schwella et al., 1994; Webb et al., 1996). Along with this, bacteria, fungi and viruses are able to survive in the liquid nitrogen storage phase and cross-contaminations between different units, stored in the same container have been observed (Bielanski, 2005; Bielanski, Bergeron, Lau, & Devenish, 2003; Bielanski & Vajta, 2009).

If microbial contaminations in other blood products, such as red blood cells, platelet concentrates or plasma products are detected, they are usually discarded. However, in stem cell concentrates this decision is complicated by the fact, that the total amount of stem cells is limited. Several reports described case fatalities when stem cell recipients received microbially contaminated products("Current good tissue practice for human cell, tissue, and cellular and tissue-based product establishments; inspection and enforcement. Final rule.," 2004; "Eligibility determination for donors of human cells, tissues, and cellular and tissue-based products. Final rule.," 2004; Kamble et al., 2005). The FDA recommends caution with the use of contaminated products and estimated that seven deaths per year could be prevented by the elimination of contaminated stem cell units.

The predominant cultured bacterial subspecies are part of the skin flora and other commensal organisms, with the remainder being mainly enteric organisms. However, other opportunistic and non-opportunistic organisms like Stenotrophomonas maltophilia and MRSA have been identified (Kamble et al., 2005; Larrea et al., 2004; Patah et al., 2007; Vanneaux et al., 2007).

The clinical relevance of microbial contamination of stem cell products has been questioned by several authors. Two recent studies examined cultures of stem cell collections at different time points of the cryopreservation process. All patients were routinely prophylaxed with fluoroquinolones, acyclovir and fluconazole. Further, organism specific antimicrobial therapy was added when positive cultures were detected prior to the infusion(Majado et al., 2007; Patah et al., 2007).

A study from the MD Anderson cancer center reported the experience with 3078 autologous or allogeneic hematopoietic stem cell infusions over a six year time span. Cultures were taken at apheresis or marrow harvest, infusion for fresh products (mainly allogeneic) and thawing/infusion of the cellular products. The overall rate of positive cultures in this study was relative low at 1.2% and coagulase negative Staphylococci were the most frequently detected organisms. None of the 21.6% of the patients who died in the post-transplant period died from a cause, ascribed to the contaminated stem cell product (Patah et al., 2007).

Another study from Spain summarized the experience with 152 patients receiving a total of 617 bags of autologous, cryopreserved PBSC. Cultures were taken pre-freezing and at the end of the infusion from the post-thaw specimen. Overall, 31 were found to be contaminated (5%) and skin commensals were again the most frequently identified species. No increased mortality or other severe clinical sequelae were observed in the patients obtaining the contaminated stem cell infusions. The length of hospitalization in the acute transplant period was longer in this group of patients. However, this could not be attributed to infections with the contaminant, suggesting other patient specific factors in this patient group(Majado et al., 2007).

Other studies with a similar methodology were performed and confirmed all the following results(Kamble et al., 2005; Larrea et al., 2004; Vanneaux et al., 2007): First, skin commensals are the most prevalent cause of microbial contaminations; Second, the relatively increased incidence of positive cultures after bone marrow harvest when compared to peripheral blood apheresis products seems to be decreasing, possibly reflecting the implementation of recent asepsis guidelines. Third, the rate of positive blood cultures is the highest directly after stem cell collection and lowest after cryopreservation. This eludes to a certain antimicrobial effect of the cryopreservation procedure and possibly the cryopreservative DMSO.

Cultured organism	Overall incidence of positive cultures[%]	Fraction of positive cultures[%]
Staph. epidermidis and other coagulase negative Staphylococcus (CNS)	3-11.7	53.1-87.2
Propionibacterium acni	0.6–2.2	0.1-27.2
Staphylococcus aureus	0–1.6	0-2.3
Bacillus cereus and other Bacillus spec.	0.06–0.35	0-0.8
Pseudomonas spec.(aeruginosa, putida and fluoresces)	0–0.32	0–0.8
Corynebacterium spec.	0–0.3	0-6.5
Stenotrohomonas maltophilia	0-0.3	0-5.9
Aspergillus fumigatus	0–0.3	0-7.6
Mixed cultures	0.1–15.6	0-0.82

Table 1. The table above describes the growth of different organisms when cultures were assessed in after thawing, before infusion of the unit. The studies, used to extract the table are referenced in the paragraph above. The middle column reflects the overall incidence of positive cultures by organisms over different studies, the right column lists the fraction those organisms represented. All values are expressed in percentages.

However, bacteria and other organisms can survive cryopreservation and have to be considered when antibiotic coverage after infusion is administered. Fourth, the clinical impact of infusion with contaminated stem cell products seems to be managable, when handled by expert hands. Table 1 summarizes the results of the referenced studies.

3.5.4 Durability

The durability of a stem cell graft, is defined as the timespan a stem cell graft can be preserved and still exert the desired clinical effect when utilized. Although the time span between collection and infusion into the recipient for most allogeneic and autologous hematologic transplants is relatively short, for cord blood it can be much longer than a decade. The maximal possible cryopreservation time span is still unknown.

Several of the above listed functional substitute assays have been used to determine the functional integrity of cryopreserved stem cell grafts. For bone marrow, peripheral blood as well as cord blood stem cells, it has been shown that clonogenic assays, such as the BFU-E and CFU-GM are compromised fairly early during the course of cryopreservation. In

contrast, the recovery of nucleated cells (NC) and CD34+ cells remains relatively well preserved(Attarian, Feng, Buckner, MacLeod, & Rowley, 1996; Hal E Broxmeyer et al., 2003; Buchanan et al., 2004; Cilloni et al., 1999; E J Woods et al., 2000; Xiao & Douglas C Dooley, 2003; Yang et al., 2001; X. B. Zhang et al., 2003). In addition, the engraftment capacity in the immunodeficient NOD/SCID mice remains relatively well preserved for long periods of time (Bock, Orlic, Dunbar, H E Broxmeyer, & D M Bodine, 1995; Vormoor et al., 1994). The durability of cord blood stem cells beyond 15 years of cryopreservation has been shown by several authors with different functional substitute assays (Hal E Broxmeyer et al., 2003; Kobylka, Ivanyi, & Breur-Vriesendorp, 1998; Mugishima et al., 1999). Kobylka et al. and Mugishima et al. proved the durability after up to 15 years of storage with flow cytometric and clonogenic assays, whilst Broxmeier et al. also demonstrated hematopoietic reconstitution of sublethally irradiated NOD/SCID mice (Hal E Broxmeyer et al., 2003).
The clinical experience with long term cryopreserved stem cell grafts remains somehow limited. Anecdotal reports confirmed successful trilineage engraftment with BM derived stem cell grafts, which were stored for 7 years (Walter et al., 1999). A systematic review evaluating the combined experience of the Brigham and Women's Hospital and the EBMT Group(Aird, Labopin, Gorin, & Antin, 1992) noticed that HSC can be effectively cryopreserved for up to 11 years. A retrospective study from Seattle revealed full trilineage recovery in patients receiving HSC, stored for up to 7.8 years without consistent detrimental effects (Attarian, Feng, Buckner, MacLeod, & Rowley, 1996). It also remains unclear if any of the currently used cryopreservation protocols reveals superior results when used for clinical applications in humans. A recent report by McCullough et al. compared four different protocols for cryopreservation of PBSC. After five years of storage the relative integrity of PBSCs was preserved, regardless which of the cryopreservation protocols was used (Jeffrey McCullough et al., 2010).

3.5.5 Storage containers

Containers for the long term storage of stem cell concentrates need to fulfill at least closure integrity, sample stability over long periods of time and easy accessibility as minimal requirements(Erik J Woods & Thirumala, 2011). With the increasing financial and environmental strain on health care systems, even in developed nations, and the increasing amount of umbilical cord stem cells banked, additional points of relevance are a low cost, potential for reuse, environmental sustainability and economic use of available long term storage space. Container devices prior to the actual long term cryopreservation need to allow for convenient collection of the stem cells and storage in the liquid phase at supra-freezing temperatures for up to 72 hours. In addition, the processing of the stem cell collection often entails centrifugation and hence the requirement for the container to be resistant to significant gravitational forces. To date, most institutions use several containers or bags prior to the actual cryopreservation bag. In addition, to prevent spillage and cross contamination of concentrates stored in the same storage nitrogen tank, the ISCT recommends over-rap bags on their web site. Historically most stem cell storage institutions arose from blood banks with long standing expertise with bags as storage containers (Khuu et al., 2002; Thirumala, Goebel, & Erik J Woods, 2009; Erik J Woods & Thirumala, 2011). Hence, freezing bags are the most frequently used long term storage containers for stem cells and ethylene vinyl acetate(EVA) based products represent the majority in the market(Thirumala, Goebel, & Erik J Woods, 2009) .

A 2002 publication reported breakage of EVA based bags, storing PBSC and lymphocyte concentrates, with a microbial contamination rate of 42% (Khuu et al., 2002). Along with that, a recent publication described a breakage rate of 3.5% over a 6.5 year storage period on umbilical cord concentrates (Thyagarajan, Michael Berger, Sumstad, & McKenna, 2008). Similar concerns were raised in the past with different plastic materials (Valeri & Pivacek, 1996).

The historical experience with catastrophic viral cross infection (Tedder et al., 1995) as well as such recent reports triggered increasing concerns about EVA based products. EVA experiences a glass transition below -15°C, what renders it brittle and potentially fragile below this temperature (Kempe MD, Jorgensen GJ, Terwilliger KM, McMahon TJ, Kennedy CE, n d). Those concerns, along with the desire to produce reusable materials, have prompted the exploration of other materials as base for cryocontainers (Eakins MN, n d).

Trade name	Material
• Cryocyte/Baxter® • CellFlex/Maco Pharma® • Cell Freeze™ Charter Medical • Pall Medical® Freezing Bag 791-05 • Cryostore EVA/Origen Biomedical Inc. • Thermogenesis® Corp./Freezing bag 80346-0 • Origen Cryostore® • EVA CryoMACS®	EVA* based
• KryoSure® • American Fluoroseal® • Fresenius Hemocare/Hemofreeze® • Origen Biomedical Inc./Permalife Bag	FEP# FEP/Teflon Teflon/Kaplon FEP/Polyimide

Table 2. The table above lists several commercially available bags for cryopreservation of stem cells. Please note, that some of them may not be currently available or FDA approved.

A recent publication by Woods et al described a cyclic olefin co-polymer based container system with favorable cryo-physical properties. The small size of those containers at 2mL and 5mL would be prohibitive for large scale HSC storage but was successfully assessed for suitability of dental pulp derived MSC(Erik J Woods & Thirumala, 2011).

Another report from the Czech republic demonstrated success cryopreservation in a reusable stainless steel container specifically designed for PBSC(Měricka et al., 1991). Table 2 describes some of the products used for cryopreservation of stem cells around the world.

3.6 Thawing and post-thaw processing

At our institution the thawing of the cryopreserved unit is performed by the technician at the bedside. The unit is then handed over to the treating physician or nurse for infusion into the patient. Depending on the regulatory environment, in certain countries the unit is retrieved from the storage tank by the technician to be handed over for transport to the patient and thawing by the physician or supporting clinical staff. Regardless of the logistic specifics, the unit is thawed to a slushy state and then slowly infused into the recipient under close clinical observation. Several elements of the post thaw process are still a matter of debate and we will here discuss relevant points on washing and the functional post-thaw assessment of stem cells.

3.6.1 Washing of stem cells

Since increased toxicities of DMSO have been observed with increasing amounts, attempts to develop protocols to deplete the DMSO content in the infused stem cell product have been undertaken. Several protocols to reduce the infused DMSO content have been published (Lemarie et al., 2005; Syme et al., 2004). Some of those procedures achieve 2 log reductions in the infused DMSO concentration. Theoretical concerns have been raised that extensive washing can negatively impact on engraftment kinetics. Unfortunately, only few prospective studies have examined the influence of post-thaw washing on toxicities and hematopoietic reconstitution (Akkök, Holte, Tangen, Ostenstad, & Oystein Bruserud, 2009; Lemarie et al., 2005; Syme et al., 2004).

In a recent study by Akkök et al., 53 patients obtained high dose chemotherapy with PBSC transplant for multiple myeloma, non Hodgkins lymphoma, amyloidosis or POEMS syndrome. The maximal liquid storage time prior to cryopreservation was less than 24 hours. The cryomedium contained AB allogeneic plasma and controlled rate freezing was routinely performed. The patients either received the graft either directly post-thaw (n=34) or cells, DMSO depleted cells by virtue of a one-step washing procedure(n=19). Only patients with a CD34+ count of at least $2.5X10^6$ as backup were allowed for depletion. The washing solution consisted of an ACD-saline mixture. Directly after thawing, the graft was mixed with the washing solution and centrifuged at 850Xg for 6 minutes at 22°C. The supernatant was discarded and the pellet re-suspended with ACD-saline solution to a total volume of maximal 150mL. The total ex-vivo handling time after removal from liquid nitrogen storage was 50-60 minutes longer for the depleted grafts(Akkök et al., 2009). A statistically significant 23.1% loss of CD34+ cells was observed. However, the extracellular DMSO concentration and number of neutrophils were both reduced in the washed stem cell product. Overall less infusion related adverse events were observed in patients receiving the washed autografts (16% vs. 36%, p=0.024). The neutrophil recovery and length of neutropenic fever episodes were comparable in both arms, but the platelet recovery was delayed by two days (14 vs 12 days) in the group obtaining a washed graft. This was associated with a statistically not significant increase in clinically meaningful hemorrhage events. The authors concluded that in certain patient populations, a simple one-step washing procedure can reduce DMSO related adverse infusion reactions with a tolerable compromise of platelet engraftment kinetics.

Although post-thaw washing is not a routinely performed technique at our center for PBSC and BM derived stem cell concentrates, it holds a higher relevance for UCB products. Umbilical cord blood transplantation is gaining increasing popularity for a variety of malignant and non-malignant diseases. To date, more than 20,000 UCB transplants have been performed on children and adults, and more than 400,000 UCB units are available in more than 50 public CB banks (Solves, Mirabet, & Roig, 2010). The way cord blood units are collected from the donor, renders them rich in hypertonic cryopreservative, red cells, plasma and cell debris. This demands additional processing, not routinely applied in PBSC and BM derived stem cells. This processing is partially performed before and partially after the cryopreservation. Initially, after collection, the collected unit contains a large amount of

red blood cell bulk and plasma. Volume reduction is essential to reduce the total amount of required storage space. The standard New York Blood Center protocol implies a two-step procedure. Here, 6% HES is added in a 1:5 volume ration to the collected, anticoagulated UCB. This increases the erythrocyte sedimentation rate. The mixture is then centrifuged at 50Xg for 5 minutes at 10°C. The leukocyte rich supernatant is expressed and again centrifuged at 400Xg for 10 minutes. The pellet is resuspensed in supernatant plasma to a total volume of 20mL, which then is used to create the finally cryopreserved unit. The post-thaw washing procedure entails the addition of an equal volume 2.5% albumin, 40 dextran solution and centrifugation at 400Xg for 10 minutes. The sedimented cells are then re-dispensed in fresh albumin/dextran solution prior to infusion (Rubinstein et al., 1995).
Although the washing UCB units has been shown to increase the viability of remaining cells(Rubinstein et al., 1995), concerns of cell loss in a product which intrinsically suffers from a paucity of cells have been raised(Laroche et al., 2005). Hence, alternative washing procedures, excluding the post-thaw centrifugation step have been proposed. Here, the deleterious osmotic shift is prevented by introduction of a dextran/albumin solution under laboratory conditions, but no centrifugation is subsequently performed. In a recent report from the Memorial Sloan Kettering Cancer Center (MSKCC), such a technique was applied in 54 consecutive patients, obtaining double cord blood transplants (Barker et al., 2009). The sustained donor engraftment rate was excellent(94%)and the amount of DMSO infused compared favorably with autologous transplants. The infusion reactions were manageable with the most prominent concern being the episodes of renal compromise, probably induced by residual cell debris.

3.6.2 Functional substitute assays

The ultimate functional stem cell assay is the engraftment with subsequent reconstitution of the desired physiologic function, such as cell count recovery in hematopoietic stem cells in the myeloablated human host.
However, for obvious reasons we are relying on functional substitute assays to estimate the functional integrity of the stem cell graft. Such substitute assays consists of cell counting experiments enumerating cell populations with a high stem cell potential, viability tests, quantifying certain biologic functions, clonogenic assays, reflecting the potential of cell population to give rise to other subspecialized cell types and direct engraftment experiments in previously myelo-ablated or immune-deficient mammals (see table 3).
The most frequently used cell counting assays are the total nucleated cell count and the count of CD34 positive cells (Kurtz, Seetharaman, N. Greco, & Moroff, 2007; Yang et al., 2001). Increasing numbers of both have been shown to correlate positively with the hematopoietic recovery potential in HSC transplantation (Kurtz, Seetharaman, N. Greco, & Moroff, 2007; Yang et al., 2001). Another flow-cytometry marker, which has recently been found to correlate positively with the enumeration of viable stem cell populations is aldehyde dehydrogenase (ALDH+). The particular benefit of this maker is, that ALDH positive cells show a very low or absent apoptotic potential (D. A. Hess et al., 2004; Kurtz, Seetharaman, N. Greco, & Moroff, 2007). Trypan blue-, 7-aminoactinomycinD (7-AAD) and propidium iodide-exclusion have been shown to be particularly strongly expressed in immature, viable cell populations, such as stem cells (Kurtz, Seetharaman, N. Greco, & Moroff, 2007; K. Liu et al., 2003; M. Solomon, Wofford, C. Johnson, Regan, & Creer, 2010; Ware, Nelson, & Blau, 2005; E J Woods et al., 2000; Xiao & Douglas C Dooley, 2003). They

function as true viability assays and are diminished or absent in apoptotic cell populations. The simple and fast feasibility of such tests qualifies them as excellent functional substitute assays, especially before freezing and after thawing of stem cell concentrates.
Clonogenic assays such as the CFU-Sd12, CFU-GM, CFU-GEMM, BFU-E, LTC-IC and direct engraftment assays performed on NOD/SCID mice are more sophisticated functional substitute assays. Figure two displays images of clonogenic experiments. However, such assays are expensive, time intense, require specific personal expertise and laboratory equipment and are poorly standardized. Hence, they are not routinely used in standard qualitative assessments of stem cell grafts.

Biologic function	Assay type	Corresponding references
Cell enumeration assays	Nucleated Cell counts	(Donnenberg et al., 2002; Shlebak et al., 1999; E J Woods et al., 2000)
	Flow cytometry for CD34+cells	(Donnenberg et al., 2002; Sasnoor et al., 2003)
Viability/apoptosis tests	Trypan blue	(K. Liu et al., 2003; Ware, Nelson, & Blau, 2005)
	7-AminoactinomycinD	(Xiao & Douglas C Dooley, 2003; Yang et al., 2001)
	Propidium Iodide	(K. W. Johnson et al., 2007)
	SYTO16 assay	(Sparrow & Tippett, 2005)
	Flow cytometry for ALDH+ reactivity	(D. A. Hess et al., 2004; Kurtz, Seetharaman, N. Greco, & Moroff, 2007)
Clonogenic assays	CFU-sd12, CFU-GM, CFU-GEMM	(Balint et al., 1999; Kobylka et al., 1998; Perseghin et al., 1997; Shlebak et al., 1999)
	BFU-E	(Balint et al., 1999; Kobylka et al., 1998; Perseghin et al., 1997)
	LTC-IC	(Barker & Wagner, 2003; Ito et al., 2010)
Direct engraftment experiments	Engraftment in NOD/SCID mice	(Halle et al., 2001; Matsumoto et al., 2002; Perez-Oteyza et al., 1998; Valeri & Pivacek, 1996)

Table 3. The table above outlines different methods to assess a stem cell graft. The right column displays the biologic function assessed, the middle column the actually used assay and the right column references for review of the corresponding subject. long-term culture-initiating cells (LTC-IC), CFU-GM-colony forming unit granulocyte/macrophage; -GEMM-colony forming unit granulocyte/erythrocyte/macrophage/megakaryocyte; BFU-burst forming unit-erythrocyte, CFU-Sd12-colony forming unit spleen on day 12.

No final consensus is reached as to the optimal assessment of a donor graft. Cell counting assays, either with or without the enumeration of CD34 positive cells are still the most frequently used tests. Several problems have recently been outlined with the use of those assays. First, a substantial inter-laboratory variability in CD34 counts can be observed (Dzik, Sniecinski, & Fischer, 1999; Moroff et al., 2006). Second, although the total NC did not significantly change in several studies, the number of CD34+ cells can be significantly reduced after thawing (Kurtz, Seetharaman, N. Greco, & Moroff, 2007). Third, a significant number of CD34 bright cells are not functional when tested on viability assays. In a recent study by Pranke only about 1.8% of the bright CD34+ cells were alive, whereas a small part (19.0%) was actively undergoing apoptosis and most of them (79.2%) were dead, when judged by the 7-AAD exclusion assay (Pranke et al., 2006). Last, when novel enzymatic, such as the ALDH (N. J. Greco, W. R. Lee, Kurtz, Seetharaman, & Moroff, 2003), or apoptotic, such as SYTO16 (Sparrow & Tippett, 2005), markers are used, a more sophisticated assessment of the clonogenic/regenerative potential of the collected cell population may be possible. In a recent study by Kurtz et al., three different methods to enumerate CD34+ cells were assessed on pre-freeze and post-thaw samples, along with the viability marker 7-AAD and the apoptotic marker SYTO16. Although only minor changes in NC and CD34+ counts were observed, the functional assays were significantly impacted by the freezing/thawing process (Kurtz, Seetharaman, N. Greco, & Moroff, 2007).
The above mentioned concerns notwithstanding, Yang et al demonstrated that viable CD34+ cell counts are reliable indictors for successful clinical hematopoietic recovery on pre-freeze and post-thaw samples (Yang et al., 2005). This observation reflects most likely the dogma of stem cell biology, that principally only one cell with stem cell potential is necessary to regenerate the entire hematopoietic system of a myeloablated host. Although, the theoretical definition of a stem cell is relatively clear, it still remains unclear how to define its immunophenotype and although the CD34+ cell population certainly harbors cells with stem cell potential, those represent certainly only a small fraction (K. W. Johnson, M. Dooner, & P J Quesenberry, 2007; Peter J Quesenberry, M. S. Dooner, & Aliotta, 2010). Table 3 depicts different methods to assess the functionality of a stem cell graft. At our center, we routinely perform NC-, CD34+- count as well as the trypan blue viability assay before freeze and after thaw with excellent observed engraftment correlations.

3.7 Cryopreservation of Non-Hematologic Stem Cells

The cryopreservation for the different hematologic stem cell populations are well defined with rather subtle evolving refinements. In contrast, the cryopreservative technology for non-hematologic stem cells is still evolving. The highest level of sophistication to date has been reached with the MSCs. A comprehensive discussion of cryopreservation for non-hematopoietic stem cells would be beyond the scope of this chapter. Detailed reviews have been recently published (Hanna & Allison Hubel, 2009; C. J. Hunt, 2011). We will briefly outline some of the parallels and differences between non-hematologic and hematologic stem cell cryopreservation in the following paragraph.

a. Mesenchymal stem cells(MSC). No consensus has been reached as to the optimal cryopreservation protocol of MSC, but most of the published efforts were guided by cryopreservation approaches for HSCs (G. Liu et al., 2008; Thirumala, Gimble, & Devireddy, 2010; Thirumala, Goebel, & Erik J Woods, 2009; Erik J Woods et al., 2009). Several particular concerns were identified. First, MSCs seem to lose their viability very rapidly post-thaw, what can most likely be attributed to the rapid development of apoptotic processes. When Pal et al. investigated the behavior of MSC post thawing, a rapid decrease in viability from >80% at 2 hours to <40% at 8 hours was identified, even when the cells were maintained at 4°C (Pal, Hanwate, & Totey, 2008). Protocols utilizing anti-apoptotic agents, such as ROCK-inhibitors, have shown promise in preclinical experiments (Heng, 2009). Second, the reliability of standard viability assays for MSCs has been questioned. Trypan blue, which is still in wide application in the clinical use of HSCs in the US, may not as accurately as 7-AAD, the preferred HSC-viability-assay in Germany, reflect the viability of MSCs (D. E. Pegg, 1989). Third, MSCs may play a role in the treatment of acute and chronic processes of the heart muscle. DMSO as cryopreservative additive(CPA) with the associated cardio-toxicity may not be the best choice for these clinical scenarios (Zenhäusern, Tobler, Leoncini, O. M. Hess, & P. Ferrari, 2000). Fourth, the alternative CPAs for MSCs are still poorly defined. Trehalose, glycerol and proline did have good efficacy in HSC but do not work in MSC (Grein et al., 2010; Y. Liu et al., n d). Last, most published protocols to date encompass the use of xenobiotic components, such as fetal calf serum(FCS). This is associated with the corresponding immunonologic as well as infectious concerns. Recent developments with xeno-free cryopreservation protocols seem to be viable options for MSCs (Zeisberger et al., 2010).

b. Human embroynic stem cells(hESC). Initial attempts to preserve hESC with protocols analogous to the slow controlled rate freezing HSC protocols were associated with recovery rates below 30% and high post-thaw differentiation (Ha et al., 2005; Reubinoff, Pera, Vajta, & Trounson, 2001; C.-quan Zhou, Q.-yun Mai, T. Li, & Zhuang, 2004). Hence, alternative methods for crypreservation of hESC were explored. The vitrification technique is for long in use for cryopreservation of embryos and ooctyes in veterinary and human application (Yan et al., 2011). Vitrification represents the direct transformation of a substance into a glass like state without the formation of ice crystals (M. Ojovan, n d). This can be reached by very high cooling rates and high concentrations of cryopreservative. InReubinoff's original open pulled straw (OPS) protocol 20% DMSO, 20% ethylene glycol and 0.5 mol/l sucrose were used as CPA(Reubinoff et al., 2001). The specimen consisting of a 20µL solution in an open straw, was then plunged directly into liquid nitrogen, achieving a cooling rate of approximately -75°C/second. This technique results in high post-thaw viabilities and low post-thaw differentiation rates. However, several practical problems exist with the OPS method. First, the volume preserved in one pulled open straw is with 20µL very small and hence logistically prohibitive for a large scale clinical use. Second, the open character of the system with storage in liquid nitrogen is associated with infectious concerns. Third, the glass like state achieved by this vitrification method is metastable and requires consistent temperatures below -135°Celsius, what is often a problem in conventional freezing tanks, which imply internal temperature gradients (C. J. Hunt, 2011; Rowley & Byrne, 1992). Fourth, most formulations use animal proteins in the solution, what corresponds to infectious and allergic concerns. Fifth, substitute assays, to measure the functionaltiy of hESC for extended periods of time are still poorly defined (C. J. Hunt, 2011). Last, the method requires technical and personal sophistication, which is not necessarily everywhere available.

Closed straw as well as potentially scalable techniques have been recently published(T. Li, Q. Mai, J. Gao, & C. Zhou, 2010; Richards, Fong, S. Tan, Chan, & Bongso, 2004). The reason for the difficulties with the preservation of hESC with traditional slow cooling protocols may be the cooperative growth behavior and the inter-cellular interaction via gap junctions. Hence, ice crystal niduses formed within a hESC colony can easily cause mechanical damage throughout the entire cell cluster. The awareness of this idiosyncrasy in the growth behavior of hESC has led to a series of publications, trying to modify the different cryo-biological parameters, such as cooling rate, ice nucleation, cryoprotectant concentration, osmotic effects associated with the introduction and removal of the cryoprotectant and matrix systems allowing for the seeding of cell clusters, maintaining continuous cell-to-cell contact in slow cooling protocols (Baran & Ware, 2007; J. Y. Lee et al., 2010; Yang Li, J.-C. Tan, & L.-S. Li, 2010; Valbuena et al., 2008). The recovery success in those experiments was promising.

c. Most of the principals applying to hESC also appear to apply to iPS and so far cryopreservation approaches similar to hESCs have been explored (C. J. Hunt, 2011; Nishigaki et al., 2011).

4. Summary and conclusions

4.1 Summary

The cryopreservation of stem cells is a crucial component of their therapeutic use in hematologic disorders and regenerative medicine. Although protocols for the preservation of HSC are well defined, the standardization still remains poor. Protocols for the preservation of non-HSCs have been developed and will experience further clinical validation if hESC, MSC and iPSC will find wider clinical application in the future. Liquid phase storage is safe and effective for short term intervals. It is probably superior to cryopreservation at -80° to -156°C for period less than 72hrs, because it avoids the unnecessary physical and mechanical damage by more elaborate procedures. Volume and cell bulk reduction, although associated with a certain loss of cells, is necessary to accommodate the increasing logistical demands in the cellular therapy sector. Cell concentrations of $2X10^8$/mL or even higher are safe, effective and necessary to limit the spatial needs, particularly considering the rising amount of cryopreserved UCB units. DMSO at concentrations of 5-10% is widely used and alternative cryoprevatives have been developed. Future studies are needed to define the optimal CPA with the most favorable side effect profile. For most stem cell products, controlled rate freezing is still considered standard in most countries, but uncontrolled rate freezing approaches are safe and associated with good clinical outcomes. The optimal long term storage temperature remains to be defined but the vapor phase of nitrogen tanks is suitable for most cellular products. In conclusion, future clinical and translational studies will be needed to define and standardize the optimal cryopreservation techniques with optimal clinical outcomes and minimal clinical, environmental and financial adverse effects.

4.2 Future directives

Efforts in the future should focus on clinical and basic science studies to further the understanding of the cryobiology of target cell population, to better explain the damage caused to the cells by the cryopreservation process. This will help with many issues, including the identification of optimal cryobiologic parameters, such as cell concentration, pre-freezing cell processing, freezing rate, storage temperature and others. In addition with a widening use of stem cells in regenerative medicine, less toxic cryopreservatives are in demand. Standardization and the implementation of good manufacturing practices remain a global issue. The existing and increasing financial strain on public health systems around the world calls clearly for a global standardization with protocols, feasible also in less affluent parts of the world. Finally, an educational exchange across borders to ascertain optimal skill sets in laboratorians and clinicians, dealing with stem cell cryopreservation and utilization, has to be encouraged.

4.3 Appendix

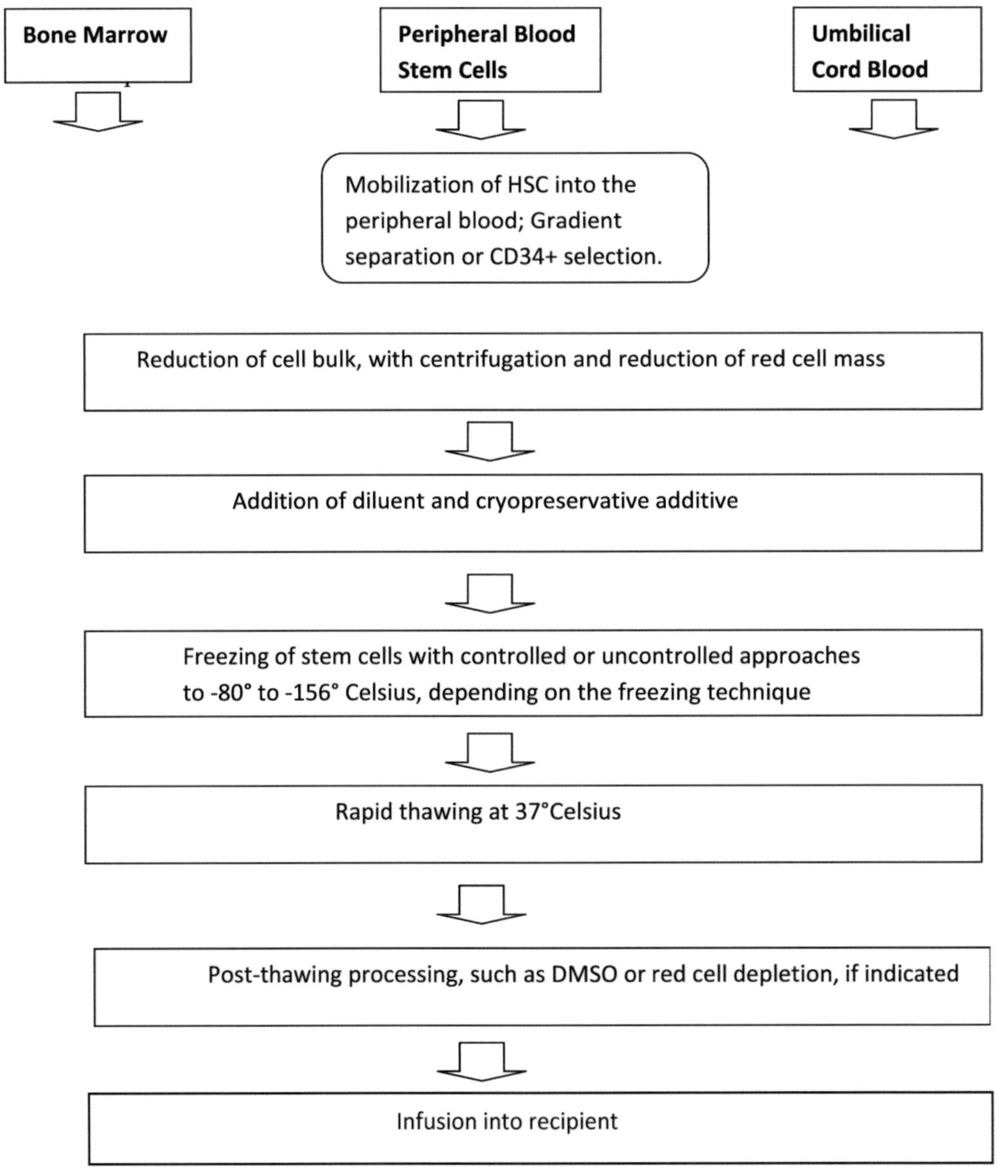

Graphic 1.

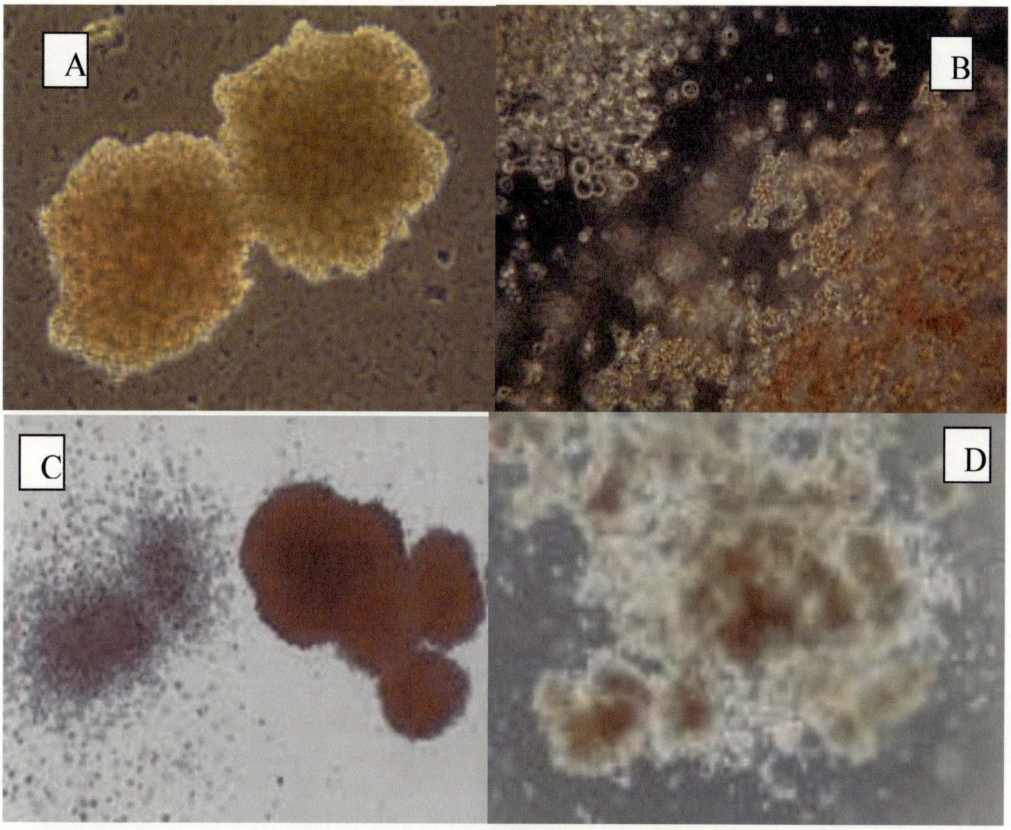

The images above display: A-CFU-Sd12 (colony forming unit spleen on day 12);
B- CFU-GM(colony forming unit granulocyte/macrophage); C- LTC-IC(long-term culture-initiating cells);
D- GEMM-(colony forming unit granulocyte/erythrocyte/macrophage/megakaryocyte)

Graphic 2.

5. References

Aird, W., Labopin, M., Gorin, N. C., & Antin, J. H. (1992). Long-term cryopreservation of human stem cells. *Bone marrow transplantation, 9*(6), 487-90. Retrieved June 14, 2011, from http://www.ncbi.nlm.nih.gov/pubmed/1628134.

Akkök, C. A., Holte, M. R., Tangen, J. M., Ostenstad, B., & Bruserud, Oystein. (2009). Hematopoietic engraftment of dimethyl sulfoxide-depleted autologous peripheral blood progenitor cells. *Transfusion, 49*(2), 354-61. doi: 10.1111/j.1537-2995.2008.01949.x.

Akkök, C. A., Liseth, K., Nesthus, I., Løkeland, T., Tefre, K., Bruserud, Oystein, et al. (2008). Autologous peripheral blood progenitor cells cryopreserved with 5 and 10 percent dimethyl sulfoxide alone give comparable hematopoietic reconstitution after transplantation. *Transfusion, 48*(5), 877-83. doi: 10.1111/j.1537-2995.2008.01648.x.

Albanell, J., & Baselga, J. (2000). Systemic therapy emergencies. *Seminars in oncology, 27*(3), 347-61. Retrieved June 16, 2011, from http://www.ncbi.nlm.nih.gov/pubmed/10864222.

Alencar, S., Garnica, M., Luiz, R. R., Nogueira, C. M., Borojevic, R., Maiolino, A., et al. (2010). Cryopreservation of peripheral blood stem cell: the influence of cell concentration on cellular and hematopoietic recovery. *Transfusion, 50*(11), 2402-12. doi: 10.1111/j.1537-2995.2010.02743.x.

Almici, C., Ferremi, P., Lanfranchi, A., Ferrari, E., Verardi, R., Marini, M., et al. (2003). Uncontrolled-rate freezing of peripheral blood progenitor cells allows successful engraftment by sparing primitive and committed hematopoietic progenitors. *Haematologica, 88*(12), 1390-5. Retrieved June 30, 2011, from http://www.ncbi.nlm.nih.gov/pubmed/14687993.

Attarian, H., Bensinger, W. I., Buckner, C. D., McDonald, D. L., & Rowley, S. D. (1996). Microbial contamination of peripheral blood stem cell collections. *Bone marrow transplantation, 17*(5), 699-702. Retrieved July 1, 2011, from http://www.ncbi.nlm.nih.gov/pubmed/8733684.

Attarian, H., Feng, Z., Buckner, C. D., MacLeod, B., & Rowley, S. D. (1996). Long-term cryopreservation of bone marrow for autologous transplantation. *Bone marrow transplantation, 17*(3), 425-30. Retrieved June 30, 2011, from http://www.ncbi.nlm.nih.gov/pubmed/8704699.

Baicu, S., Taylor, M. J., Chen, Z., & Rabin, Y. (2008). Cryopreservation of carotid artery segments via vitrification subject to marginal thermal conditions: correlation of freezing visualization with functional recovery. *Cryobiology, 57*(1), 1-8. doi: 10.1016/j.cryobiol.2008.03.002.

Bakken, A. M., Bruserud, O, & Abrahamsen, J F. (2003). No differences in colony formation of peripheral blood stem cells frozen with 5% or 10% dimethyl sulfoxide. *Journal of hematotherapy & stem cell research, 12*(3), 351-8. doi: 10.1089/152581603322023089.

Balint, B., Ivanović, Z., Petakov, M., Taseski, J., Jovcić, G., Stojanović, N., et al. (1999). The cryopreservation protocol optimal for progenitor recovery is not optimal for preservation of marrow repopulating ability. *Bone marrow transplantation, 23*(6), 613-9. doi: 10.1038/sj.bmt.1701623.

Baran, S. W., & Ware, C. B. (2007). Cryopreservation of rhesus macaque embryonic stem cells. *Stem cells and development, 16*(2), 339-44. doi: 10.1089/scd.2007.900-de.

Barker, J. N., Abboud, M., Rice, R. D., Hawke, R., Schaible, A., Heller, G., et al. (2009). A "no-wash" albumin-dextran dilution strategy for cord blood unit thaw: high rate of engraftment and a low incidence of serious infusion reactions. *Biology of blood and marrow transplantation : journal of the American Society for Blood and Marrow Transplantation, 15*(12), 1596-602. doi: 10.1016/j.bbmt.2009.08.009.

Barker, J. N., & Wagner, J. E. (2003). Umbilical-cord blood transplantation for the treatment of cancer. *Nature reviews. Cancer, 3*(7), 526-32. doi: 10.1038/nrc1125.

Baum, C. M., Weissman, I. L., Tsukamoto, A. S., Buckle, A. M., & Peault, B. (1992). Isolation of a candidate human hematopoietic stem-cell population. *Proceedings of the National Academy of Sciences of the United States of America, 89*(7), 2804-8. Retrieved June 16, 2011, from http://www.pubmedcentral.nih.gov/articlerender.fcgi?artid=48751&tool=pmcentrez&rendertype=abstract.

Benekli, M., Anderson, B., Wentling, D., Bernstein, S., Czuczman, M., & McCarthy, P. (2000). Severe respiratory depression after dimethylsulphoxide-containing autologous stem cell infusion in a patient with AL amyloidosis. *Bone marrow transplantation, 25*(12), 1299-301. doi: 10.1038/sj.bmt.1702452.

Berz, D., McCormack, E. M., Winer, E. S., Colvin, G. A., & Quesenberry, Peter J. (2007). Cryopreservation of hematopoietic stem cells. *American journal of hematology, 82*(6), 463-72. doi: 10.1002/ajh.20707.

Bielanski, A. (2005). Non-transmission of bacterial and viral microbes to embryos and semen stored in the vapour phase of liquid nitrogen in dry shippers. *Cryobiology, 50*(2), 206-10. doi: 10.1016/j.cryobiol.2004.12.004.

Bielanski, A., Bergeron, H., Lau, P. C. K., & Devenish, J. (2003). Microbial contamination of embryos and semen during long term banking in liquid nitrogen. *Cryobiology, 46*(2), 146-52. Retrieved December 2, 2010, from http://www.ncbi.nlm.nih.gov/pubmed/12686204.

Bielanski, A., & Vajta, G. (2009). Risk of contamination of germplasm during cryopreservation and cryobanking in IVF units. *Human reproduction (Oxford, England), 24*(10), 2457-67. doi: 10.1093/humrep/dep117.

Bock, T. A., Orlic, D., Dunbar, C. E., Broxmeyer, H E, & Bodine, D M. (1995). Improved engraftment of human hematopoietic cells in severe combined immunodeficient (SCID) mice carrying human cytokine transgenes. *The Journal of experimental medicine, 182*(6), 2037-43. Retrieved June 30, 2011, from http://www.pubmedcentral.nih.gov/articlerender.fcgi?artid=2192238&tool=pmcentrez&rendertype=abstract.

Broxmeyer, Hal E, Srour, E. F., Hangoc, G., Cooper, S., Anderson, S. A., & Bodine, David M. (2003). High-efficiency recovery of functional hematopoietic progenitor and stem cells from human cord blood cryopreserved for 15 years. *Proceedings of the National Academy of Sciences of the United States of America, 100*(2), 645-50. doi: 10.1073/pnas.0237086100.

Buchanan, S. S., Gross, S. A., Acker, Jason P, Toner, M., Carpenter, J. F., & Pyatt, D. W. (2004). Cryopreservation of stem cells using trehalose: evaluation of the method using a human hematopoietic cell line. *Stem cells and development, 13*(3), 295-305. doi: 10.1089/154732804323099226.

Burger, S. R., Hubel, A, & McCullough, J. (1999). Development of an infusible-grade solution for non-cryopreserved hematopoietic cell storage. *Cytotherapy, 1*(2), 123-33. Retrieved June 14, 2011, from http://www.ncbi.nlm.nih.gov/pubmed/19746589.

Cabezudo, E., Dalmases, C., Ruz, M., Sanchez, J. A., Torrico, C., Sola, C., et al. (2000). Leukapheresis components may be cryopreserved at high cell concentrations without additional loss of HPC function. *Transfusion, 40*(10), 1223-7. Retrieved June 14, 2011, from http://www.ncbi.nlm.nih.gov/pubmed/11061859.

Cilloni, D., Garau, D., Regazzi, E., Sammarelli, G., Savoldo, B., Caramatti, C., et al. (1999). Primitive hematopoietic progenitors within mobilized blood are spared by uncontrolled rate freezing. *Bone marrow transplantation, 23*(5), 497-503. doi: 10.1038/sj.bmt.1701601.

Clapisson, G., Salinas, C., Malacher, P., Michallet, M., Philip, I., & Philip, T. (2004). Cryopreservation with hydroxyethylstarch (HES) + dimethylsulfoxide (DMSO) gives better results than DMSO alone. *Bulletin du cancer, 91*(4), E97-102. Retrieved June 30, 2011, from http://www.ncbi.nlm.nih.gov/pubmed/15562559.

Clark, J., Pati, A., & McCarthy, D. (1991). Successful cryopreservation of human bone marrow does not require a controlled-rate freezer. *Bone marrow transplantation, 7*(2), 121-5. Retrieved June 30, 2011, from http://www.ncbi.nlm.nih.gov/pubmed/2049555.

Corato, A., Ambrosetti, A., Rossi, B., Vincenzi, C., Lambiase, A., Perona, G., et al. (2000). Transplantation potential of peripheral whole blood primed by VACOP-B chemotherapy plus filgrastim (r-metHuG-CSF) in patients with aggressive non-Hodgkin's lumphoma. *Journal of hematotherapy & stem cell research, 9*(5), 673-82. doi: 10.1089/15258160050196713.

Crowe, J H, Crowe, L. M., Oliver, A. E., Tsvetkova, N., Wolkers, W., & Tablin, F. (2001). The trehalose myth revisited: introduction to a symposium on stabilization of cells in the dry state. *Cryobiology, 43*(2), 89-105. doi: 10.1006/cryo.2001.2353.

Crowe, John H. (2007). Trehalose as a "chemical chaperone": fact and fantasy. *Advances in experimental medicine and biology, 594*, 143-58. doi: 10.1007/978-0-387-39975-1_13.

Current good tissue practice for human cell, tissue, and cellular and tissue-based product establishments; inspection and enforcement. Final rule. (2004). *Federal register, 69*(226), 68611-88. Retrieved July 1, 2011, from http://www.ncbi.nlm.nih.gov/pubmed/15562555.

Davis, J. M., Rowley, S. D., Braine, H. G., Piantadosi, S., & Santos, G. W. (1990). Clinical toxicity of cryopreserved bone marrow graft infusion. *Blood, 75*(3), 781-6. Retrieved June 16, 2011, from http://www.ncbi.nlm.nih.gov/pubmed/2297578.

Davis, J., Rowley, S. D., & Santos, G. W. (1990). Toxicity of autologous bone marrow graft infusion. *Progress in clinical and biological research, 333*, 531-40. Retrieved June 16, 2011, from http://www.ncbi.nlm.nih.gov/pubmed/2308999.

Dazzi, F., & Krampera, M. (2011). Mesenchymal stem cells and autoimmune diseases. *Best practice & research. Clinical haematology, 24*(1), 49-57. doi: 10.1016/j.beha.2011.01.002.

Ding, D.-C., Shyu, W.-C., & Lin, S.-Z. (2011). Mesenchymal stem cells. *Cell transplantation, 20*(1), 5-14. doi: 10.3727/096368910X.

Donaldson, C., Armitage, W. J., Denning-Kendall, P. A., Nicol, A. J., Bradley, B. A., & Hows, J. M. (1996). Optimal cryopreservation of human umbilical cord blood. *Bone marrow transplantation, 18*(4), 725-31. Retrieved June 30, 2011, from http://www.ncbi.nlm.nih.gov/pubmed/8899187.

Donnenberg, A. D., Koch, E. K., Griffin, D. L., Stanczak, H. M., Kiss, J. E., Carlos, T. M., et al. (2002). Viability of cryopreserved BM progenitor cells stored for more than a decade. *Cytotherapy, 4*(2), 157-63. doi: 10.1080/146532402317381866.

Douay, L. (1985). Hematopoietic stem cells controls for autologous bone marrow transplantation monitoring. *Revue française de transfusion et immuno-hématologie, 28*(5), 397-409. Retrieved June 30, 2011, from http://www.ncbi.nlm.nih.gov/pubmed/3912919.

Dzik, W., Sniecinski, I., & Fischer, J. (1999). Toward standardization of CD34+ cell enumeration: an international study. Biomedial Excellence for Safer Transfusion Working Party. *Transfusion, 39*(8), 856-63. Retrieved June 30, 2011, from http://www.ncbi.nlm.nih.gov/pubmed/10504122.

Eakins MN. (n.d.). New Plastic for old vials. Retrieved July 1, 2011, from file:///C:/Users/PC1/AppData/Local/Mendeley Ltd/Mendeley Desktop/Downloaded/Unknown - Unknown - 0306ar07_77872a.pdf (applicationpdf-objekt).html.

Eberhardt, R., Zwingers, T., & Hofmann, R. (1995). [DMSO in patients with active gonarthrosis. A double-blind placebo controlled phase III study]. *Fortschritte der Medizin, 113*(31), 446-50. Retrieved June 16, 2011, from http://www.ncbi.nlm.nih.gov/pubmed/8529985.

Eligibility determination for donors of human cells, tissues, and cellular and tissue-based products. Final rule. (2004). *Federal register, 69*(101), 29785-834. Retrieved July 1, 2011, from http://www.ncbi.nlm.nih.gov/pubmed/15160713.

Ellerström, C., Strehl, R., Moya, K., Andersson, K., Bergh, C., Lundin, K., et al. (2006). Derivation of a xeno-free human embryonic stem cell line. *Stem cells (Dayton, Ohio), 24*(10), 2170-6. doi: 10.1634/stemcells.2006-0130.

Elsaadany, S., Tyshenko, M. G., Oraby, T., Malcolm, A. J., Senior, P., Shapiro, A. M. J., et al. (2011). The risk to human islet cell transplant recipients of acquiring variant creutzfeldt-jakob disease: a provisional quantitative risk assessment. *Transplantation, 92*(1), e2-4. doi: 10.1097/TP.0b013e31821eabab.

English, D., Lamberson, R., Graves, V., Akard, L. P., McCarthy, L. J., & Jansen, J. (1989). Semiautomated processing of bone marrow grafts for transplantation. *Transfusion, 29*(1), 12-6. Retrieved June 16, 2011, from http://www.ncbi.nlm.nih.gov/pubmed/2643209.

Espinosa, M. T., Fox, R., Creger, R. J., & Lazarus, H. M. (1996). Microbiologic contamination of peripheral blood progenitor cells collected for hematopoietic cell transplantation. *Transfusion, 36*(9), 789-93. Retrieved July 1, 2011, from http://www.ncbi.nlm.nih.gov/pubmed/8823451.

Fleming, K. K., & Hubel, A. (2006). Cryopreservation of hematopoietic and non-hematopoietic stem cells. *Transfusion and apheresis science : official journal of the World Apheresis Association : official journal of the European Society for Haemapheresis, 34*(3), 309-15. doi: 10.1016/j.transci.2005.11.012.

Friedenstein, A. J., Gorskaja, J. F., & Kulagina, N. N. (1976). Fibroblast precursors in normal and irradiated mouse hematopoietic organs. *Experimental hematology, 4*(5), 267-74. Retrieved March 14, 2011, from http://www.ncbi.nlm.nih.gov/pubmed/976387.

Galmés, A., Besalduch, J., Bargay, J., Novo, A., Morey, M., Guerra, J. M., et al. (1999). Long-term storage at -80 degrees C of hematopoietic progenitor cells with 5-percent dimethyl sulfoxide as the sole cryoprotectant. *Transfusion, 39*(1), 70-3. Retrieved June 30, 2011, from http://www.ncbi.nlm.nih.gov/pubmed/9920169.

Gorin, N. C. (1986). Collection, manipulation and freezing of haemopoietic stem cells. *Clinics in haematology, 15*(1), 19-48. Retrieved June 14, 2011, from http://www.ncbi.nlm.nih.gov/pubmed/2870831.

Gratwohl, A., Baldomero, H., Aljurf, M., Pasquini, M. C., Bouzas, Luis Fernando, Yoshimi, A., et al. (2010). Hematopoietic stem cell transplantation: a global perspective. *JAMA : the journal of the American Medical Association, 303*(16), 1617-24. doi: 10.1001/jama.2010.491.

Greco, N. J., Lee, W. R., Kurtz, J., Seetharaman, S., & Moroff, G. (2003). Characterization of multiple CD34+ cell populations in cord blood. *Journal of hematotherapy & stem cell research, 12*(2), 199-213. doi: 10.1089/152581603321628340.

Grein, T. A., Freimark, D., Weber, C., Hudel, K., Wallrapp, C., & Czermak, P. (2010). Alternatives to dimethylsulfoxide for serum-free cryopreservation of human mesenchymal stem cells. *The International journal of artificial organs, 33*(6), 370-80. Retrieved June 30, 2011, from http://www.ncbi.nlm.nih.gov/pubmed/20669142.

Ha, S. Y., Jee, B. C., Suh, C. S., Kim, H. S., Oh, S. K., Kim, S. H., et al. (2005). Cryopreservation of human embryonic stem cells without the use of a programmable freezer. *Human reproduction (Oxford, England), 20*(7), 1779-85. doi: 10.1093/humrep/deh854.

Halle, P., Tournilhac, O., Knopinska-Posluszny, W., Kanold, J., Gembara, P., Boiret, N., et al. (2001). Uncontrolled-rate freezing and storage at -80 degrees C, with only 3.5-percent DMSO in cryoprotective solution for 109 autologous peripheral blood progenitor cell transplantations. *Transfusion, 41*(5), 667-73. Retrieved June 30, 2011, from http://www.ncbi.nlm.nih.gov/pubmed/11346704.

Hanna, J., & Hubel, Allison. (2009). Preservation of stem cells. *Organogenesis, 5*(3), 134-7. Retrieved June 30, 2011, from http://www.pubmedcentral.nih.gov/articlerender.fcgi?artid=2781093&tool=pmcentrez&rendertype=abstract.

Heng, B. C. (2009). Effect of Rho-associated kinase (ROCK) inhibitor Y-27632 on the post-thaw viability of cryopreserved human bone marrow-derived mesenchymal stem cells. *Tissue & cell, 41*(5), 376-80. doi: 10.1016/j.tice.2009.01.004.

Heng, B. C., Clement, M. V., & Cao, T. (2007). Caspase inhibitor Z-VAD-FMK enhances the freeze-thaw survival rate of human embryonic stem cells. *Bioscience reports, 27*(4-5), 257-64. doi: 10.1007/s10540-007-9051-2.

Hess, D. A., Meyerrose, T. E., Wirthlin, L., Craft, T. P., Herrbrich, P. E., Creer, M. H., et al. (2004). Functional characterization of highly purified human hematopoietic repopulating cells isolated according to aldehyde dehydrogenase activity. *Blood, 104*(6), 1648-55. doi: 10.1182/blood-2004-02-0448.

Holm, F., Ström, S., Inzunza, J., Baker, D., Strömberg, A.-M., Rozell, B., et al. (2010). An effective serum- and xeno-free chemically defined freezing procedure for human embryonic and induced pluripotent stem cells. *Human reproduction (Oxford, England), 25*(5), 1271-9. doi: 10.1093/humrep/deq040.

Hubel, Allison, Carlquist, D., Clay, M., & McCullough, Jeff. (2004). Liquid storage, shipment, and cryopreservation of cord blood. *Transfusion, 44*(4), 518-25. doi: 10.1111/j.1537-2995.2004.03238.x.

Hubel, Allison, Carlquist, D., Clay, M., & McCullough, Jeffrey. (2003). Short-term liquid storage of umbilical cord blood. *Transfusion, 43*(5), 626-32. Retrieved June 14, 2011, from http://www.ncbi.nlm.nih.gov/pubmed/12702185.

Hunt, C. J. (2011). Cryopreservation of Human Stem Cells for Clinical Application: A Review. *Transfusion medicine and hemotherapy : offizielles Organ der Deutschen Gesellschaft fur Transfusionsmedizin und Immunhamatologie, 38*(2), 107-123. doi: 10.1159/000326623.

Ito, C. Y., Kirouac, D. C., Madlambayan, G. J., Yu, M., Rogers, I., & Zandstra, P. W. (2010). The AC133+CD38-, but not the rhodamine-low, phenotype tracks LTC-IC and SRC function in human cord blood ex vivo expansion cultures. *Blood, 115*(2), 257-60. doi: 10.1182/blood-2009-07-228106.

Iwasaki, T., Hamano, T., Aizawa, K., Kobayashi, K., & Kakishita, E. (1994). A case of pulmonary amyloidosis associated with multiple myeloma successfully treated with dimethyl sulfoxide. *Acta haematologica, 91*(2), 91-4. Retrieved June 16, 2011, from http://www.ncbi.nlm.nih.gov/pubmed/8023651.

Jacob, S. W., & Herschler, R. (1983). Dimethyl sulfoxide after twenty years. *Annals of the New York Academy of Sciences, 411*, xiii-xvii. Retrieved June 16, 2011, from http://www.ncbi.nlm.nih.gov/pubmed/6349491.

Jestice, H. K., Farrington, M., Hunt, C, Matthews, I., Scott, M. A., Foreman, J., et al. (1996). Bacterial contamination of peripheral blood progenitor cells for transplantation. *Transfusion medicine (Oxford, England), 6*(2), 103-10. Retrieved July 1, 2011, from http://www.ncbi.nlm.nih.gov/pubmed/8809956.

Johnson, K. W., Dooner, M., & Quesenberry, P J. (2007). Fluorescence activated cell sorting: a window on the stem cell. *Current pharmaceutical biotechnology, 8*(3), 133-9. Retrieved July 1, 2011, from http://www.ncbi.nlm.nih.gov/pubmed/17584086.

Kamble, R., Pant, S., Selby, G. B., Kharfan-Dabaja, M. A., Sethi, S., Kratochvil, K., et al. (2005). Microbial contamination of hematopoietic progenitor cell grafts-incidence, clinical outcome, and cost-effectiveness: an analysis of 735 grafts. *Transfusion, 45*(6), 874-8. doi: 10.1111/j.1537-2995.2005.04178.x.

Kawano, Yoshifumi, Lee, C. L., Watanabe, Tsutomu, Abe, Takanori, Suzuya, H., Okamoto, Y., et al. (2004). Cryopreservation of mobilized blood stem cells at a higher cell concentration without the use of a programmed freezer. *Annals of hematology, 83*(1), 50-4. doi: 10.1007/s00277-003-0817-8.

Kempe MD, Jorgensen GJ, Terwilliger KM, McMahon TJ, Kennedy CE, B. T. (n.d.). ScienceDirect - Solar Energy Materials and Solar Cells : Acetic acid production and glass transition concerns with ethylene-vinyl acetate used in photovoltaic devices. *Acetic acid production and glass transition concerns with ethylenevinyl acetate used in photovoltaic devices.* Retrieved June 30, 2011, from http://www.sciencedirect.com/science/article/pii/S0927024806004107.

Khuu, H. M., Cowley, H., David-Ocampo, V., Carter, C. S., Kasten-Sportes, C., Wayne, A. S., et al. (2002). Catastrophic failures of freezing bags for cellular therapy products: description, cause, and consequences. *Cytotherapy, 4*(6), 539-49. doi: 10.1080/146532402761624700.

Kligman, A. M. (1965). Topical Pharmacology and Toxicology of Dimethyl Sulfoxide. 1. *JAMA : the journal of the American Medical Association, 193*, 796-804. Retrieved June 16, 2011, from http://www.ncbi.nlm.nih.gov/pubmed/14329978.

Kobylka, P., Ivanyi, P., & Breur-Vriesendorp, B. S. (1998). Preservation of immunological and colony-forming capacities of long-term (15 years) cryopreserved cord blood cells. *Transplantation, 65*(9), 1275-8. Retrieved June 30, 2011, from http://www.ncbi.nlm.nih.gov/pubmed/9603182.

Koliakos, G., Alamdari, D. H., Tsagias, N., Kouzi-Koliakos, K., Michaloudi, E., & Karagiannis, V. (2007). A novel high-yield volume-reduction method for the cryopreservation of UC blood units. *Cytotherapy, 9*(7), 654-9. doi: 10.1080/14653240701508445.

Kurtz, J., Seetharaman, S., Greco, N., & Moroff, G. (2007). Assessment of cord blood hematopoietic cell parameters before and after cryopreservation. *Transfusion, 47*(9), 1578-87. doi: 10.1111/j.1537-2995.2007.01327.x.

Laroche, V., McKenna, D. H., Moroff, G., Schierman, T., Kadidlo, D., & McCullough, Jeffrey. (2005). Cell loss and recovery in umbilical cord blood processing: a comparison of postthaw and postwash samples. *Transfusion, 45*(12), 1909-16. doi: 10.1111/j.1537-2995.2005.00638.x.

Larrea, L., Rubia, J. de la, Soler, M. A., Ribas, P., Fernández, J. M., Picón, I., et al. (2004). Quality control of bacterial contamination in autologous peripheral blood stem cells for transplantation. *Haematologica, 89*(10), 1232-7. Retrieved July 1, 2011, from http://www.ncbi.nlm.nih.gov/pubmed/15477209.

Law, P., Alsop, P., Dooley, D C, & Meryman, H. T. (1983). Studies of cell separation: a comparison of the osmotic response of human lymphocytes and granulocyte-monocyte progenitor cells. *Cryobiology, 20*(6), 644-51. Retrieved June 14, 2011, from http://www.ncbi.nlm.nih.gov/pubmed/6661913.

Leclerc, T., Thepenier, C., Jault, P., Bey, E., Peltzer, J., Trouillas, M., et al. (2011). Cell therapy of burns. *Cell proliferation, 44 Suppl 1*, 48-54. doi: 10.1111/j.1365-2184.2010.00727.x.

Lee, J. Y., Lee, J. E., Kim, D. K., Yoon, T. K., Chung, H. M., & Lee, D. R. (2010). High concentration of synthetic serum, stepwise equilibration and slow cooling as an efficient technique for large-scale cryopreservation of human embryonic stem cells. *Fertility and sterility, 93*(3), 976-85. doi: 10.1016/j.fertnstert.2008.10.017.

Leeb, C., Jurga, M., McGuckin, C., Forraz, N., Thallinger, C., Moriggl, R., et al. (2011). New perspectives in stem cell research: beyond embryonic stem cells. *Cell Proliferation, 44*, 9-14. doi: 10.1111/j.1365-2184.2010.00725.x.

Lemarie, C., Calmels, B., Malenfant, C., Arneodo, V., Blaise, D., Viret, F., et al. (2005). Clinical experience with the delivery of thawed and washed autologous blood cells, with an automated closed fluid management device: CytoMate. *Transfusion, 45*(5), 737-42. doi: 10.1111/j.1537-2995.2005.04126.x.

Lewis, D. E. (1994a). The University of Kazan-Provincial Cradle of Russian Organic Chemistry: Part I. Nikolai Zinin and the Butlerov School. *Journal of Chemical Education, 71*(1), 39. American Chemical Society. doi: 10.1021/ed071p39.

Lewis, D. E. (1994b). The University of Kazan-Provincial Cradle of Russian Organic Chemistry: Part II: Aleksandr Zeitsev and his Students. *Journal of Chemical Education, 71*(2), 93. American Chemical Society. doi: 10.1021/ed071p93.

Li, T., Mai, Q., Gao, J., & Zhou, C. (2010). Cryopreservation of human embryonic stem cells with a new bulk vitrification method. *Biology of reproduction, 82*(5), 848-53. doi: 10.1095/biolreprod.109.080713.

Li, Yang, Tan, J.-C., & Li, L.-S. (2010). Comparison of three methods for cryopreservation of human embryonic stem cells. *Fertility and sterility, 93*(3), 999-1005. doi: 10.1016/j.fertnstert.2008.10.052.

Liu, G., Zhou, H., Li, Yulin, Li, G., Cui, L., Liu, W., et al. (2008). Evaluation of the viability and osteogenic differentiation of cryopreserved human adipose-derived stem cells. *Cryobiology, 57*(1), 18-24. doi: 10.1016/j.cryobiol.2008.04.002.

Liu, K., Gao, Z., Jiang, Y., Dong, W., Li, D., Wang, Y., et al. (2003). [Collection, processing and cryopreservation of placental cord blood hematopoietic stem cells]. *Beijing da xue xue bao. Yi xue ban = Journal of Peking University. Health sciences, 35*(2), 119-22. Retrieved June 30, 2011, from http://www.ncbi.nlm.nih.gov/pubmed/12920823.

Liu, Y., Xu, X., Ma, X., Martin-Rendon, E., Watt, S., & Cui, Z. (n.d.). Cryopreservation of human bone marrow-derived mesenchymal stem cells with reduced dimethylsulfoxide and well-defined freezing solutions. *Biotechnology progress, 26*(6), 1635-43. doi: 10.1002/btpr.464.

Lovelock, J. E., & Bishop, M. W. (1959). Prevention of freezing damage to living cells by dimethyl sulphoxide. *Nature, 183*(4672), 1394-5. Retrieved June 16, 2011, from http://www.ncbi.nlm.nih.gov/pubmed/13657132.

Lowder, J. N., & Whelton, P. (2003). Microbial contamination of cellular products for hematolymphoid transplantation therapy: assessment of the problem and strategies to minimize the clinical impact. *Cytotherapy, 5*(5), 377-90. doi: 10.1080/14653240310003044.

Luciano, A. M., Chigioni, S., Lodde, V., Franciosi, F., Luvoni, G. C., & Modina, S. C. (2009). Effect of different cryopreservation protocols on cytoskeleton and gap junction mediated communication integrity in feline germinal vesicle stage oocytes. *Cryobiology, 59*(1), 90-5. doi: 10.1016/j.cryobiol.2009.05.002.

M. Ojovan, W. L. (n.d.). ScienceDirect - Journal of Non-Crystalline Solids : Connectivity and glass transition in disordered oxide systems. *356, 2534-2540.* Retrieved June 30, 2011, from http://www.sciencedirect.com/science/article/pii/S0022309310002176.

Mackensen, A., Dräger, R., Schlesier, M., Mertelsmann, R., & Lindemann, A. (2000). Presence of IgE antibodies to bovine serum albumin in a patient developing anaphylaxis after vaccination with human peptide-pulsed dendritic cells. *Cancer immunology, immunotherapy : CII, 49*(3), 152-6. Retrieved June 30, 2011, from http://www.ncbi.nlm.nih.gov/pubmed/10881694.

Majado, M. J., García-Hernández, A., Morales, A., González, C., Martínez-Sánchez, V., Menasalvas, A., et al. (2007). Influence of harvest bacterial contamination on autologous peripheral blood progenitor cells post-transplant. *Bone marrow transplantation, 39*(2), 121-5. doi: 10.1038/sj.bmt.1705549.

Martin, M. J., Muotri, A., Gage, F., & Varki, A. (2005). Human embryonic stem cells express an immunogenic nonhuman sialic acid. *Nature medicine, 11*(2), 228-32. doi: 10.1038/nm1181.

Matsumoto, N., Yoshizawa, H., Kagamu, H., Abe, T, Fujita, N., Watanabe, S., et al. (2002). Successful liquid storage of peripheral blood stem cells at subzero non-freezing temperature. *Bone marrow transplantation, 30*(11), 777-84. doi: 10.1038/sj.bmt.1703692.

McCammon, K. A., Lentzner, A. N., Moriarty, R. P., & Schellhammer, P. F. (1998). Intravesical dimethyl sulfoxide for primary amyloidosis of the bladder. *Urology, 52*(6), 1136-8. Retrieved June 16, 2011, from http://www.ncbi.nlm.nih.gov/pubmed/9836571.

McCullough, Jeffrey, Haley, R., Clay, M., Hubel, Allison, Lindgren, B., & Moroff, G. (2010). Long-term storage of peripheral blood stem cells frozen and stored with a conventional liquid nitrogen technique compared with cells frozen and stored in a mechanical freezer. *Transfusion, 50*(4), 808-19. doi: 10.1111/j.1537-2995.2009.02482.x.

McGuirk, J. P., & Weiss, M. L. (2011). Promising cellular therapeutics for prevention or management of graft-versus-host disease (a review). *Placenta*. doi: 10.1016/ j.placenta.2011.04.013.

Miniero, R., Vai, S., Giacchino, M., Giubellino, C., & Madon, E. (n.d.). Severe respiratory depression after autologous bone marrow infusion. *Haematologica, 77*(1), 98-9. Retrieved June 16, 2011, from http://www.ncbi.nlm.nih.gov/pubmed/1383102.

Morassi, P., Massa, F., Mesesnel, E., Magris, D., & D'Agnolo, B. (1989). [Treatment of amyloidosis with dimethyl sulfoxide (DMSO)]. *Minerva medica, 80*(1), 65-70. Retrieved June 16, 2011, from http://www.ncbi.nlm.nih.gov/pubmed/2915815.

Moroff, G., Eichler, H., Brand, A., Kekomäki, R., Kurtz, J., Letowska, M., et al. (2006). Multiple-laboratory comparison of in vitro assays utilized to characterize hematopoietic cells in cord blood. *Transfusion, 46*(4), 507-15. doi: 10.1111/j.1537-2995.2006.00758.x.

Motta, J. P. R., Gomes, B. E., Bouzas, L F, Paraguassú-Braga, F. H., & Porto, L. C. (2010). Evaluations of bioantioxidants in cryopreservation of umbilical cord blood using natural cryoprotectants and low concentrations of dimethylsulfoxide. *Cryobiology, 60*(3), 301-7. doi: 10.1016/j.cryobiol.2010.02.002.

Mugishima, H., Harada, K., Chin, M., Suzuki, T., Takagi, K., Hayakawa, S., et al. (1999). Effects of long-term cryopreservation on hematopoietic progenitor cells in umbilical cord blood. *Bone marrow transplantation, 23*(4), 395-6. doi: 10.1038/sj.bmt.1701580.

Murdoch, A., Braude, P., Courtney, A., Brison, D., Hunt, Charles, Lawford-Davies, J., et al. (2011). The Procurement of Cells for the Derivation of Human Embryonic Stem Cell Lines for Therapeutic Use: Recommendations for Good Practice. *Stem cell reviews*. doi: 10.1007/s12015-011-9288-9.

Měricka, P., Schustr, P., Vins, M., Dudek, A., Vávra, L., Cervinka, M., et al. (1991). Containers for freezing and storage of bone marrow stem cells. *Sborník vědeckých prací Lékařské fakulty Karlovy university v Hradci Králové, 34*(4), 367-87. Retrieved June 30, 2011, from http://www.ncbi.nlm.nih.gov/pubmed/1815317.

Neunert, G., Polewski, P., Walejko, P., Markiewicz, M., Witkowski, S., & Polewski, K. (2009). Glycosidic moiety changes the spectroscopic properties of dl-alpha-tocopherol in DMSO/water solution and in organic solvents. *Spectrochimica acta. Part A, Molecular and biomolecular spectroscopy, 73*(2), 301-8. doi: 10.1016/j.saa.2009.02.032.

Nishigaki, T., Teramura, Y., Nasu, A., Takada, K., Toguchida, J., & Iwata, H. (2011). Highly efficient cryopreservation of human induced pluripotent stem cells using a dimethyl sulfoxide-free solution. *The International journal of developmental biology, 55*(3), 305-11. doi: 10.1387/ijdb.103145tn.

O'Donnell, J. R., Burnett, A. K., Sheehan, T., Tansey, P., & McDonald, G. A. (1981). Safety of dimethylsulphoxide. *Lancet, 1*(8218), 498. Retrieved June 16, 2011, from http://www.ncbi.nlm.nih.gov/pubmed/6110113.

Padley, D. J., Greiner, C. W., Heddlesten, T. L., Hopkins, M. K., Maas, M. L., & Gastineau, D. A. (2003). Endogenous microbial contamination of cultured autologous preparations in trials of cancer immunotherapy. *Cytotherapy, 5*(2), 147-52. Retrieved July 1, 2011, from http://www.ncbi.nlm.nih.gov/pubmed/12769074.

Padley, D., Koontz, F., Trigg, M. E., Gingrich, R., & Strauss, R. G. (1996). Bacterial contamination rates following processing of bone marrow and peripheral blood progenitor cell preparations. *Transfusion, 36*(1), 53-6. Retrieved July 1, 2011, from http://www.ncbi.nlm.nih.gov/pubmed/8607154.

Pal, R., Hanwate, M., & Totey, S. M. (2008). Effect of holding time, temperature and different parenteral solutions on viability and functionality of adult bone marrow-derived mesenchymal stem cells before transplantation. *Journal of tissue engineering and regenerative medicine, 2*(7), 436-44. doi: 10.1002/term.109.

Patah, P. A., Parmar, S., McMannis, J., Sadeghi, T., Karandish, S., Rondon, G., et al. (2007). Microbial contamination of hematopoietic progenitor cell products: clinical outcome. *Bone marrow transplantation, 40*(4), 365-8. doi: 10.1038/sj.bmt.1705731.

Pegg, D. E. (1989). Viability assays for preserved cells, tissues, and organs. *Cryobiology, 26*(3), 212-31. Retrieved June 30, 2011, from http://www.ncbi.nlm.nih.gov/pubmed/2743785.

Pegg, D. E., & Kemp, N. H. (1960). Collection, storage, and administration of autologous human bone-marrow. *Lancet, 2*(7166), 1426-9. Retrieved June 16, 2011, from http://www.ncbi.nlm.nih.gov/pubmed/13733911.

Perez-Oteyza, J., Bornstein, R., Corral, M., Hermosa, V., Alegre, A., Torrabadella, M., et al. (1998). Controlled-rate versus uncontrolled-rate cryopreservation of peripheral blood progenitor cells: a prospective multicenter study. Group for Cryobiology and Biology of Bone Marrow Transplantation (CBTMO), Spain. *Haematologica, 83*(11), 1001-5. Retrieved June 30, 2011, from http://www.ncbi.nlm.nih.gov/pubmed/9864921.

Perseghin, P., Epis, R., Vigano, M., Malacrida, A., Pastorini, A., & Camerone, G. (1997). Satisfactory recovery and viability of stem cells cryopreserved at high cell concentration. *Transfusion science, 18*(3), 399-403. doi: 10.1016/S0955-3886(97)00037-4.

Pettengell, R., Woll, P. J., O'Connor, D. A., Dexter, T. M., & Testa, N. G. (1994). Viability of haemopoietic progenitors from whole blood, bone marrow and leukapheresis product: effects of storage media, temperature and time. *Bone marrow transplantation, 14*(5), 703-9. Retrieved August 11, 2010, from http://www.ncbi.nlm.nih.gov/pubmed/7889003.

Pettengell, R., Woll, P. J., Thatcher, N., Dexter, T. M., & Testa, N. G. (1995). Multicyclic, dose-intensive chemotherapy supported by sequential reinfusion of hematopoietic progenitors in whole blood. *Journal of clinical oncology : official journal of the American Society of Clinical Oncology, 13*(1), 148-56. Retrieved June 14, 2011, from http://www.ncbi.nlm.nih.gov/pubmed/7528271.

Pranke, P., Hendrikx, J., Alespeiti, G., Nardi, N., Rubinstein, P., & Visser, J. (2006). Comparative quantification of umbilical cord blood CD34+ and CD34+ bright cells using the ProCount-BD and ISHAGE protocols. *Brazilian journal of medical and biological research = Revista brasileira de pesquisas médicas e biológicas / Sociedade Brasileira de Biofísica [et al.], 39*(7), 901-6. Retrieved June 30, 2011, from http://www.ncbi.nlm.nih.gov/pubmed/16862281.

Puglisi, M. A., Saulnier, N., Piscaglia, A. C., Tondi, P., Agnes, S., & Gasbarrini, A. (2011). Adipose tissue-derived mesenchymal stem cells and hepatic differentiation: old concepts and future perspectives. *European review for medical and pharmacological sciences, 15*(4), 355-64. Retrieved June 30, 2011, from http://www.ncbi.nlm.nih.gov/pubmed/21608430.

Quesenberry, Peter J, Dooner, M. S., & Aliotta, J. M. (2010). Stem cell plasticity revisited: the continuum marrow model and phenotypic changes mediated by microvesicles. *Experimental hematology, 38*(7), 581-92. doi: 10.1016/j.exphem.2010.03.021.

Rebulla, P. (2002). Cord blood banking 2002: 112,010 of 7,914,773 chances. *Transfusion, 42*(10), 1246-8. Retrieved June 16, 2011, from http://www.ncbi.nlm.nih.gov/pubmed/12423505.

Reubinoff, B. E., Pera, M. F., Vajta, G., & Trounson, A. O. (2001). Effective cryopreservation of human embryonic stem cells by the open pulled straw vitrification method. *Human reproduction (Oxford, England), 16*(10), 2187-94. Retrieved June 30, 2011, from http://www.ncbi.nlm.nih.gov/pubmed/11574514.

Richards, M., Fong, C.-Y., Tan, S., Chan, W.-K., & Bongso, A. (2004). An efficient and safe xeno-free cryopreservation method for the storage of human embryonic stem cells. *Stem cells (Dayton, Ohio), 22*(5), 779-89. doi: 10.1634/stemcells.22-5-779.

Rodrigues, J. P., Paraguassú-Braga, F. H., Carvalho, L., Abdelhay, E., Bouzas, L F, & Porto, L. C. (2008). Evaluation of trehalose and sucrose as cryoprotectants for hematopoietic stem cells of umbilical cord blood. *Cryobiology, 56*(2), 144-51. doi: 10.1016/j.cryobiol.2008.01.003.

Rowley, S. D. (1992). Hematopoietic stem cell cryopreservation: a review of current techniques. *Journal of hematotherapy, 1*(3), 233-50. Retrieved June 14, 2011, from http://www.ncbi.nlm.nih.gov/pubmed/1365030.

Rowley, S. D., Bensinger, W. I., Gooley, T. A., & Buckner, C. D. (1994). Effect of cell concentration on bone marrow and peripheral blood stem cell cryopreservation. *Blood, 83*(9), 2731-6. Retrieved June 14, 2011, from http://www.ncbi.nlm.nih.gov/pubmed/7513212.

Rowley, S. D., & Byrne, D. V. (1992). Low-temperature storage of bone marrow in nitrogen vapor-phase refrigerators: decreased temperature gradients with an aluminum racking system. *Transfusion, 32*(8), 750-4. Retrieved June 30, 2011, from http://www.ncbi.nlm.nih.gov/pubmed/1412683.

Rowley, S. D., Feng, Z., Chen, L., Holmberg, L., Heimfeld, S., MacLeod, B., et al. (2003). A randomized phase III clinical trial of autologous blood stem cell transplantation comparing cryopreservation using dimethylsulfoxide vs dimethylsulfoxide with hydroxyethylstarch. *Bone marrow transplantation, 31*(11), 1043-51. doi: 10.1038/sj.bmt.1704030.

Rubinstein, P., Dobrila, L., Rosenfield, R. E., Adamson, J. W., Migliaccio, G., Migliaccio, A. R., et al. (1995). Processing and cryopreservation of placental/umbilical cord blood for unrelated bone marrow reconstitution. *Proceedings of the National Academy of Sciences of the United States of America, 92*(22), 10119-22. Retrieved June 16, 2011, from http://www.pubmedcentral.nih.gov/articlerender.fcgi?artid=40747&tool=pmcentrez&rendertype=abstract.

Ruiz-Delgado, G. J., Mancías-Guerra, C., Tamez-Gómez, E. L., Rodríguez-Romo, L. N., López-Otero, A., Hernández-Arizpe, A., et al. (2009). Dimethyl sulfoxide-induced toxicity in cord blood stem cell transplantation: report of three cases and review of the literature. *Acta haematologica, 122*(1), 1-5. doi: 10.1159/000227267.

Sangeetha, V. M., Kale, V. P., & Limaye, L. S. (2010). Expansion of cord blood CD34 cells in presence of zVADfmk and zLLYfmk improved their in vitro functionality and in vivo engraftment in NOD/SCID mouse. *PloS one, 5*(8), e12221. doi: 10.1371/journal.pone.0012221.

Santos, N. C., Prieto, M. J. E., Morna-Gomes, A., Betbeder, D., & Castanho, M. A. R. B. (1997). Structural characterization (shape and dimensions) and stability of polysaccharide/lipid nanoparticles. *Biopolymers, 41*(5), 511-520. doi: 10.1002/(SICI)1097-0282(19970415)41:5<511::AID-BIP3>3.0.CO;2-P.

Sasnoor, L. M., Kale, V. P., & Limaye, L. S. (2003). Supplementation of conventional freezing medium with a combination of catalase and trehalose results in better protection of surface molecules and functionality of hematopoietic cells. *Journal of hematotherapy & stem cell research, 12*(5), 553-64. doi: 10.1089/152581603322448268.

Scheinkönig, C., Kappicht, S., Kolb, H.-J., & Schleuning, M. (2004). Adoption of long-term cultures to evaluate the cryoprotective potential of trehalose for freezing hematopoietic stem cells. *Bone marrow transplantation, 34*(6), 531-6. doi: 10.1038/sj.bmt.1704631.

Schwella, N., Zimmermann, R., Heuft, H. G., Blasczyk, R., Beyer, J., Rick, O., et al. (1994). Microbiologic contamination of peripheral blood stem cell autografts. *Vox sanguinis, 67*(1), 32-5. Retrieved July 1, 2011, from http://www.ncbi.nlm.nih.gov/pubmed/7975449.

Shlebak, A. A., Marley, S. B., Roberts, I. A., Davidson, R. J., Goldman, J. M., & Gordon, M. Y. (1999). Optimal timing for processing and cryopreservation of umbilical cord haematopoietic stem cells for clinical transplantation. *Bone marrow transplantation, 23*(2), 131-6. doi: 10.1038/sj.bmt.1701551.

Silberstein, L. E., & Jefferies, L. C. (1996). Placental-blood banking--a new frontier in transfusion medicine. *The New England journal of medicine, 335*(3), 199-201. doi: 10.1056/NEJM199607183350310.

Sohni, A., & Verfaillie, C. M. (2011). Multipotent adult progenitor cells. *Best practice & research. Clinical haematology, 24*(1), 3-11. doi: 10.1016/j.beha.2011.01.006.

Solomon, M., Wofford, J., Johnson, C., Regan, D., & Creer, M. H. (2010). Factors influencing cord blood viability assessment before cryopreservation. *Transfusion, 50*(4), 820-30. doi: 10.1111/j.1537-2995.2009.02491.x.

Solves, P., Mirabet, V., & Roig, R. (2010). Volume reduction in routine cord blood banking. *Current stem cell research & therapy, 5*(4), 362-6. Retrieved June 16, 2011, from http://www.ncbi.nlm.nih.gov/pubmed/20528760.

Son, J. H., Heo, Y. J., Park, M. Y., Kim, H. H., & Lee, K. S. (2010). Optimization of cryopreservation condition for hematopoietic stem cells from umbilical cord blood. *Cryobiology, 60*(3), 287-92. doi: 10.1016/j.cryobiol.2010.01.007.

Sparrow, R. L., & Tippett, E. (2005). Discrimination of live and early apoptotic mononuclear cells by the fluorescent SYTO 16 vital dye. *Journal of immunological methods, 305*(2), 173-87. doi: 10.1016/j.jim.2005.07.017.

Stiff, P. J., Murgo, A. J., Zaroulis, C. G., DeRisi, M. F., & Clarkson, B. D. (1983). Unfractionated human marrow cell cryopreservation using dimethylsulfoxide and hydroxyethyl starch. *Cryobiology, 20*(1), 17-24. Retrieved June 30, 2011, from http://www.ncbi.nlm.nih.gov/pubmed/6187518.

Storey, K. B., & Storey, J. M. (1990). Frozen and alive. *Scientific American, 263*(6), 92-7. Retrieved June 14, 2011, from http://www.ncbi.nlm.nih.gov/pubmed/2270462.

Stroh, C., Cassens, U., Samraj, A. K., Sibrowski, W., Schulze-Osthoff, K., & Los, M. (2002). The role of caspases in cryoinjury: caspase inhibition strongly improves the recovery of cryopreserved hematopoietic and other cells. *The FASEB journal : official publication of the Federation of American Societies for Experimental Biology, 16*(12), 1651-3. doi: 10.1096/fj.02-0034fje.

Syme, R., Bewick, M., Stewart, D., Porter, K., Chadderton, T., & Glück, S. (2004). The role of depletion of dimethyl sulfoxide before autografting: on hematologic recovery, side effects, and toxicity. *Biology of blood and marrow transplantation : journal of the American Society for Blood and Marrow Transplantation, 10*(2), 135-41. doi: 10.1016/j.bbmt.2003.09.016.

Takaue, Y, Abe, T, Kawano, Y, Suzue, T., Saito, S., Hirao, A., et al. (1994). Comparative analysis of engraftment after cryopreservation of peripheral blood stem cell autografts by controlled- versus uncontrolled-rate methods. *Bone marrow transplantation, 13*(6), 801-4. Retrieved June 30, 2011, from http://www.ncbi.nlm.nih.gov/pubmed/7920317.

Tedder, R. S., Zuckerman, M. A., Goldstone, A. H., Hawkins, A. E., Fielding, A., Briggs, E. M., et al. (1995). Hepatitis B transmission from contaminated cryopreservation tank. *Lancet, 346*(8968), 137-40. Retrieved June 30, 2011, from http://www.ncbi.nlm.nih.gov/pubmed/7603227.

Thirumala, S., Gimble, J. M., & Devireddy, R. V. (2010). Cryopreservation of stromal vascular fraction of adipose tissue in a serum-free freezing medium. *Journal of tissue engineering and regenerative medicine, 4*(3), 224-32. doi: 10.1002/term.232.

Thirumala, S., Goebel, W. S., & Woods, Erik J. (2009). Clinical grade adult stem cell banking. *Organogenesis, 5*(3), 143-54. Retrieved June 30, 2011, from http://www.pubmedcentral.nih.gov/articlerender.fcgi?artid=2781095&tool=pmcentrez&rendertype=abstract.

Thomas, E. D., Lochte, H. L., Cannon, J. H., Sahler, O. D., & Ferrebee, J. W. (1959). Supralethal whole body irradiation and isologous marrow transplantation in man. *The Journal of clinical investigation, 38*, 1709-16. doi: 10.1172/JCI103949.

Thyagarajan, B., Berger, Michael, Sumstad, D., & McKenna, D. H. (2008). Loss of integrity of umbilical cord blood unit freezing bags: description and consequences. *Transfusion, 48*(6), 1138-42. doi: 10.1111/j.1537-2995.2008.01688.x.

Unger, C., Skottman, H., Blomberg, P., Dilber, M. S., & Hovatta, O. (2008). Good manufacturing practice and clinical-grade human embryonic stem cell lines. *Human molecular genetics, 17*(R1), R48-53. doi: 10.1093/hmg/ddn079.

Valbuena, D., Sánchez-Luengo, S., Galán, A., Sánchez, E., Gómez, E., Poo, M. E., et al. (2008). Efficient method for slow cryopreservation of human embryonic stem cells in xeno-free conditions. *Reproductive biomedicine online, 17*(1), 127-35. Retrieved June 30, 2011, from http://www.ncbi.nlm.nih.gov/pubmed/18616900.

Valeri, C. R., & Pivacek, L. E. (1996). Effects of the temperature, the duration of frozen storage, and the freezing container on in vitro measurements in human peripheral blood mononuclear cells. *Transfusion, 36*(4), 303-8. Retrieved June 30, 2011, from http://www.ncbi.nlm.nih.gov/pubmed/8623128.

Vanneaux, V., Foïs, E., Robin, M., Rea, D., Latour, R. P. de, Biscay, N., et al. (2007). Microbial contamination of BM products before and after processing: a report of incidence and immediate adverse events in 257 grafts. *Cytotherapy, 9*(5), 508-13. doi: 10.1080/14653240701420427.

Villalón, L., Odriozola, J., Ramos, P., Ramos, M. L., Herrera, P., & Oteyza, J. P. de. (2002). Cryopreserving with increased cellular concentrations of peripheral blood progenitor cells: clinical results. *Haematologica, 87*(2), ELT06. Retrieved June 14, 2011, from http://www.ncbi.nlm.nih.gov/pubmed/11836183.

Vormoor, J., Lapidot, T., Pflumio, F., Risdon, G., Patterson, B., Broxmeyer, H E, et al. (1994). Immature human cord blood progenitors engraft and proliferate to high levels in severe combined immunodeficient mice. *Blood, 83*(9), 2489-97. Retrieved June 30, 2011, from http://www.ncbi.nlm.nih.gov/pubmed/7513200.

Walia, B., Satija, N., Tripathi, R. P., & Gangenahalli, G. U. (2011). Induced Pluripotent Stem Cells: Fundamentals and Applications of the Reprogramming Process and its Ramifications on Regenerative Medicine. *Stem cell reviews*. doi: 10.1007/s12015-011-9279-x.

Walter, Z., Szostek, M., Weglarska, D., Raguszewska, D., Jabłoński, M., Lorenz, F., et al. (1999). [Methods for freezing, thawing and viability estimation of hemopoietic stem cells]. *Przegląd lekarski, 56 Suppl 1*, 34-9. Retrieved June 30, 2011, from http://www.ncbi.nlm.nih.gov/pubmed/10494180.

Wang, J. C., Doedens, M., & Dick, J. E. (1997). Primitive human hematopoietic cells are enriched in cord blood compared with adult bone marrow or mobilized peripheral blood as measured by the quantitative in vivo SCID-repopulating cell assay. *Blood, 89*(11), 3919-24. Retrieved June 16, 2011, from http://www.ncbi.nlm.nih.gov/pubmed/9166828.

Ware, C. B., Nelson, A. M., & Blau, C. A. (2005). Controlled-rate freezing of human ES cells. *BioTechniques, 38*(6), 879-80, 882-3. Retrieved June 30, 2011, from http://www.ncbi.nlm.nih.gov/pubmed/16018548.

Watt, S. M., Austin, E., & Armitage, S. (2007). Cryopreservation of hematopoietic stem/progenitor cells for therapeutic use. *Methods in molecular biology (Clifton, N.J.), 368*, 237-59. Retrieved June 16, 2011, from http://www.ncbi.nlm.nih.gov/pubmed/18080475.

Webb, I. J., Coral, F. S., Andersen, J. W., Elias, A. D., Finberg, R. W., Nadler, L. M., et al. (1996). Sources and sequelae of bacterial contamination of hematopoietic stem cell components: implications for the safety of hematotherapy and graft engineering. *Transfusion, 36*(9), 782-8. Retrieved July 1, 2011, from http://www.ncbi.nlm.nih.gov/pubmed/8823450.

Will, R. G., Ironside, J. W., Zeidler, M., Cousens, S. N., Estibeiro, K., Alperovitch, A., et al. (1996). A new variant of Creutzfeldt-Jakob disease in the UK. *Lancet, 347*(9006), 921-5. Retrieved December 1, 2010, from http://www.ncbi.nlm.nih.gov/pubmed/8598754.

Windrum, P., Morris, T. C. M., Drake, M. B., Niederwieser, D, & Ruutu, T. (2005). Variation in dimethyl sulfoxide use in stem cell transplantation: a survey of EBMT centres. *Bone marrow transplantation, 36*(7), 601-3. doi: 10.1038/sj.bmt.1705100.

Woods, E J, Liu, J., Derrow, C. W., Smith, F. O., Williams, D. A., & Critser, J. K. (2000). Osmometric and permeability characteristics of human placental/umbilical cord blood CD34+ cells and their application to cryopreservation. *Journal of hematotherapy & stem cell research, 9*(2), 161-73. Retrieved June 30, 2011, from http://www.ncbi.nlm.nih.gov/pubmed/10813529.

Woods, Erik J, Perry, B. C., Hockema, J. J., Larson, L., Zhou, D., & Goebel, W. S. (2009). Optimized cryopreservation method for human dental pulp-derived stem cells and their tissues of origin for banking and clinical use. *Cryobiology, 59*(2), 150-7. doi: 10.1016/j.cryobiol.2009.06.005.

Woods, Erik J, & Thirumala, S. (2011). Packaging Considerations for Biopreservation. *Transfusion medicine and hemotherapy : offizielles Organ der Deutschen Gesellschaft fur Transfusionsmedizin und Immunhamatologie*, *38*(2), 149-156. doi: 10.1159/000326083.

Wusteman, M., Robinson, M., & Pegg, D. (2004). Vitrification of large tissues with dielectric warming: biological problems and some approaches to their solution. *Cryobiology*, *48*(2), 179-89. doi: 10.1016/j.cryobiol.2004.01.002.

Xiao, M., & Dooley, Douglas C. (2003). Assessment of cell viability and apoptosis in human umbilical cord blood following storage. *Journal of hematotherapy & stem cell research*, *12*(1), 115-22. doi: 10.1089/152581603321210190.

Yan, J., Suzuki, J., Yu, X.-M., Qiao, J., Kan, F. W. K., & Chian, R.-C. (2011). Effects of duration of cryo-storage of mouse oocytes on cryo-survival, fertilization and embryonic development following vitrification. *Journal of assisted reproduction and genetics*. doi: 10.1007/s10815-011-9563-3.

Yang, H., Acker, J P, Cabuhat, M., Letcher, B., Larratt, L., & McGann, L. E. (2005). Association of post-thaw viable CD34+ cells and CFU-GM with time to hematopoietic engraftment. *Bone marrow transplantation*, *35*(9), 881-7. doi: 10.1038/sj.bmt.1704926.

Yang, H., Acker, J P, Hannon, J., Miszta-Lane, H., Akabutu, J. J., & McGann, L. E. (2001). Damage and protection of UC blood cells during cryopreservation. *Cytotherapy*, *3*(5), 377-86. doi: 10.1080/146532401753277193.

Zeisberger, S. M., Schulz, J. C., Mairhofer, M., Ponsaerts, P., Wouters, G., Doerr, D., et al. (2010). Biological and physicochemical characterization of a serum- and xeno-free chemically defined cryopreservation procedure for adult human progenitor cells. *Cell transplantation*. doi: 10.3727/096368910X547426.

Zenhäusern, R., Tobler, A., Leoncini, L., Hess, O. M., & Ferrari, P. (2000). Fatal cardiac arrhythmia after infusion of dimethyl sulfoxide-cryopreserved hematopoietic stem cells in a patient with severe primary cardiac amyloidosis and end-stage renal failure. *Annals of hematology*, *79*(9), 523-6. Retrieved June 30, 2011, from http://www.ncbi.nlm.nih.gov/pubmed/11043425.

Zhang, X. B., Li, K., Yau, K. H., Tsang, K. S., Fok, T. F., Li, C. K., et al. (2003). Trehalose ameliorates the cryopreservation of cord blood in a preclinical system and increases the recovery of CFUs, long-term culture-initiating cells, and nonobese diabetic-SCID repopulating cells. *Transfusion*, *43*(2), 265-72. Retrieved June 30, 2011, from http://www.ncbi.nlm.nih.gov/pubmed/12559024.

Zhao, T., Zhang, Z.-N., Rong, Z., & Xu, Y. (2011). Immunogenicity of induced pluripotent stem cells. *Nature*, *474*(7350), 212-5. doi: 10.1038/nature10135.

Zhou, C.-quan, Mai, Q.-yun, Li, T., & Zhuang, G.-lun. (2004). Cryopreservation of human embryonic stem cells by vitrification. *Chinese medical journal*, *117*(7), 1050-5. Retrieved June 30, 2011, from http://www.ncbi.nlm.nih.gov/pubmed/15265381.

13

Hematopoietic Stem Cell Transplantation for Adult Acute Lymphoblastic Leukaemia

Pier Paolo Piccaluga[1,4], Stefania Paolini[1], Francesca Bonifazi[1], Giuseppe Bandini[1], Giuseppe Visani[2] and Sebastian Giebel[3,4]
[1]*Department of Hematology and Oncological Sciences "L. and A. Seràgnoli", Hematopathology Unit, S. Orsola-Malpighi Hospital, University of Bologna*
[2]*Hematology and Hematopoietic Stem Cell Transplant Centre, San Salvatore Hospital, Pesaro*
[3]*Department of Bone Marrow Transplantation, Maria Sklodowska-Curie Memorial Cancer Center and Institute of Oncology, Gliwice Branch; Gliwice,*
[4]*European Leukaemia Net- Working Party on Acute Lymphoblastic Leukaemia (ELN-EWALL)*
[1,2]*Italy*
[3]*Poland*

1. Introduction

The most recent clinical trials on adult acute lymphoid leukaemia (ALL) have shown complete remission and disease-free survival (DFS) rates of 80-85% and 30-40%, respectively (Annino, *et al*, Durrant, *et al*, Kantarjian, *et al*, Larson, *et al*, Ribera, *et al*, Rowe). Intensified consolidation, particularly with high-dose methotrexate and high-dose cytarabine, may be one of the reasons for the improved outcome in recent series (Bassan and Hoelzer, Hoelzer and Gokbuget, Kebriaei and Larson). In addition, risk-adapted and subtype-oriented therapy may have contributed to this better outcome. However, the long term outcome of adult patients is still dismal, with approximately one third of the cases only being cured. At present, therapeutic options include conventional chemotherapy (CHT), high dose therapy with autologous and, especially, allogeneic stem cells transplantation (SCT) and, for certain subsets, such as *BCR-ABL1*$^{+}$ ALL, specific targeted therapy (Piccaluga, *et al*).

Although SCT has been used in adult ALL for more than 20 years, its role remains controversial as demonstrated by conflicting results in various studies. Previous case-controlled studies did not show that allogeneic SCT (alloSCT) provided any advantage over CHT (Horowitz, *et al*, Zhang, *et al*) while in some studies there was an advantage, but restricted to young adults (Oh, *et al*). The number of controlled published or ongoing trials is remarkably small and some of them did not include both standard-risk and high-risk patients. Thus, it is difficult to draw definitive conclusions from their results. In fact, while some authors did not report any differences between alloSCT and chemotherapy or autologous SCT (ASCT)(Gupta, *et al*, Labar, *et al*), others only found differences favouring allogeneic SCT in standard risk (Goldstone, *et al*) or high-risk ALL patients (Sebban, *et al*, Thiebaut, *et al*, Thomas, *et al*).

In this article, the Authors reviewed data concerning alloSCT in adult ALL and discuss current controversial and possible perspectives.

2. Rationale for allogeneic SCT: the graft versus leukaemia effect in ALL

Initial studies were conduced in patients with advanced stage disease, aiming to overcome drug resistance by intensifying the dose of CHT and rescuing patients with syngeneic or allogeneic haemopoietic stem cells (SC) (Fefer, *et al*, Thomas, *et al*). In the first study, 16 patients with refractory ALL received high dose therapy followed by infusion of syngeneic bone marrow (Fefer, *et al*). Six out of these patients achieved a durable remission, demonstrating the efficacy and the potential curative effect of high dose therapy in ALL. On the other hand, the role of donor-derived immunologic anti-leukaemic effect - so called "graft versus leukaemia" (GVL) effect – was first established in animal models (Chester, *et al*). The first evidence of GVL effect in humans was actually demonstrated in ALL patients treated with alloSCT (Mathe, *et al*, Thomas, *et al*). In this series, patients who developed grade ≥2 graft versus host disease (GVHD) had a significantly lower relapse rate after transplant if compared with patients with grade 1 or no GVHD or with those who received syngeneic bone marrow ($p<0.01$). Subsequent studies further underlined this phenomenon (Doney, *et al*, Horowitz, *et al*, Lee, *et al*), although its relevance is less convincing than in myeloproliferative diseases such as chronic myeloid leukaemia, or chronic lymphoproliferative disease and possibly lymphomas.

3. Designation of risk factors in ALL

There is a considerable variation in the relapse risk which can be predicted by various risk factors (Table 1) (Hoelzer D 2002, Moorman, *et al*), including unfavourable cytogenetics,

	Good	*Adverse*
Age	15-35* 35-50	>50-60
WBC at diagnosis	<30x10⁹/L (B-ALL) <100x10⁹/L (T-ALL)	>30x10⁹/L (B-ALL) >100x10⁹/L (T-ALL)

Cytogenetics	del(9p)
	high hyperdiploidy
		t(9;22);
		t(4;11);
		t(8;14)
		hypodiploidy/almost triploidy
		complex karyotype

Time to CR	< 4 weeks	> 4 weeks
Immunophenotype	Pre-B Common Thymic	Pro-B Pro-T Pre-T Mature T
MRD	Negative Stable during follow up	Persistent (≥10⁻³) after induction Increasing during treatment/follow up

* Special benefit appear to be related in this group to the application of paediatric (i.e. intensified) regimens
** intermediate prognosis
WBC = white blood cells

Table 1. Prognostic factors in adult ALL

poor/slow response to induction CHT, leukocyte count at diagnosis and phenotype. Accordingly, patients can be stratified as standard-risk (no risk factors) or high risk (one or more risk factors) with 45-60% vs less than 30% DFS, respectively (Hoelzer D). In particular, detection of minimal residual disease (MRD) after treatment was shown to be associated with impending relapse, independent of any other risk factor (Bassan, *et al*, Mortuza, *et al*)(Gazzola A et al, submitted). According to some investigators the status of MRD may be sufficient for stratification and the role conventional clinical risk factors appears questionable. [29]

4. Preparative regiments

4.1 Ablative regimens

Conditioning regimens used for alloSCT must ideally: 1) achieve adequate immunosuppression of the recipient to prevent rejection of the donor SC, and 2) destroy residual malignant cells while causing minimal toxicity. Most preparative regimens for ALL use total body irradiation (TBI) and cyclophosphamide (Cy) while some centres use etoposide and cytosine arabinoside (Ara-C) in addition to or instead of Cy. Busulphan-cyclophosphamide (BU/Cy) has also been used to avoid radiation but an IBMTR study showed an inferior outcome with this regimen with a 3-year probability of DFS of 35% versus 50% when TBI and Cy were used (Davies, *et al*). Because most chemo-radiation regimens are at the limits of toxicity, any further dose escalation in an attempt to reduce the risk of relapse would probably increase the regimen-related toxicity to unacceptable levels, particularly in older or heavily pretreated patients. The addition of biological agents like monoclonal antibodies or radioconjugates to conventional conditioning regimens can potentially provide an augmented anti-leukaemic effect without increasing treatment related toxicity.

4.2 Non-myeloablative regimens

High-dose CHT and alloSCT carry substantial treatment related morbidity and mortality in older patients (>50 years), those with compromised organ function (e.g., congestive heart failure), coexisting infections, or those who were heavily pre-treated prior to SCT. In all of these patients, treatment-related mortality can exceed 50%, making them ineligible for SCT. More recently, new strategies for allografting have explored an approach of less intensive conditioning therapy with the aim of allowing partial engraftment of donor immune and haematopoietic systems with eventual replacement of the host's own haematopoiesis and immunity. A variety of regimens based on low-dose TBI or fludarabine are increasingly used (Giralt, *et al*, Khouri, *et al*, Slavin, *et al*, Storb, *et al*). Such reduced intensity conditioning (RIC) regimens may be more suitable for patients with indolent malignancies where there is sufficient time for a graft versus malignancy effect to operate. Patient with acute leukaemia who have active disease transplanted with non-myeloablative regimens have a high relapse rate. Another major problem with submyeloablative regimens is an increased rate of graft failure ranging from 5% to 30% versus 1% to 5% in patients undergoing full myeloablation prior to alloSCT. The use of lymphodepleting antibodies or a combination of monoclonal antibodies in addition to cytotoxic or immunosuppressive drugs could potentially decrease rejection rates (Brenner, *et al*, Kottaridis, *et al*).

Examples of preparative regimens used in ALL are summarized in Table 2.

Author	Regimen
Thomas et al 1982	Cy 120 mg/Kg + TBI (SD 10 Gy)
Vitale et al. 1983	Cy 120 mg/Kg + fractioned TBI (2 Gy x6)
Tutschka et al. 1987	BU 16 mg/Kg + Cy 120 mg/Kg
Blume et al. 1993	Etoposide/TBI vs. BU/Cy
Rowe et al. 2001	Etoposide/TBI
Bieri et al. 2001	TBI at different doses*

BU = busulfan; Cy = cyclophosphamide; SD = single dose

* retrospective comparison

Table 2. Examples of preparative regimens adopted in ALL

5. Allogeneic stem cell transplantation for Ph-negative ALL

Many factors need to be considered before recommending alloSCT including:

1. Cure rate with conventional CHT;
2. Transplant related mortality (TRM) and morbidity;
3. Cure rate with alloSCT.

As TRM and cure rates strongly depend on the disease phase at the time of transplant, different phases will be considered separately in our description.

5.1 Acute lymphoid leukaemia (ALL) in first remission

As the risk of transplant related mortality needs to be balanced against potential cure with standard CHT alone, alloSCT as post-consolidation therapy is favoured in patients who are at a high risk of relapse after standard CHT. While in children the likelihood of cure with modern intensive CHT range is 70–80% (Pui and Evans), in adults with ALL the cure rate with CHT is only 25–40% and, despite the intensification of CHT regimens, the majority of patients will relapse and die of their disease (Bassan and Hoelzer). In particular, the presence of any risk factor (Table 1) reduces the chance of long-term DFS with CHT to between 15% and 25%, making the patient a candidate for allografting in first remission (CR1). In fact, alloSCT in CR1 was suggested to provide a better anti-leukaemic effect and to improve patients' outcome compared with a conventional post-remission CHT regimen.
Retrospective studies evaluating the role of alloSCT in CR1 were encouraging with DFS ranging between 40 and 63% and relapse rates of 10-40% (Barrett, *et al*, Blume, *et al*, Blume, *et al*, Chao, *et al*, De Witte, *et al*, Thomas, *et al*, Vernant, *et al*, Vey, *et al*). However, most of the studies were small (with less than 50 patients each) and were limited by selection bias (see Table 3 for details). In addition, few studies retrospectively compared alloSCT vs. autologous SCT (ASCT) or CHT in ALL performed in CR1. None of these studies showed an improved survival in transplanted patients (Table 4). In particular, Horowitz and Colleagues, on behalf of the IBMTR, reported the outcome of adult ALL patients (aged 15-45 years) treated with intensive CHT (IC, N=484) vs. allogeneic SCT (N=251) in CR1 (Horowitz, *et al*).
Allografted patients had significantly reduced risk of relapse (26% vs. 59%). However, the DFS was not superior after SCT, reflecting the higher mortality rate observed after SCT (38% vs. 4%) (Horowitz, *et al*).

Author	Number of patients/ characteristics	Age	GVHD grade 2-4 (%)	TRM (%)	RR (%)	5-y DFS (%)
Blume et al. 1987	38	16-41	31	26	16	63
Vernant et al. 1988	27	15-36	44	30	11	59
Barrett et al. 1989	243	16-48	40	37	30	39
Chao et al. 1991	53/high risk	1-45	11	28	10	61
Vey et al. 1994	29	16-41	27	29	10	62
De Witte et al. 1994	22/high risk	15-51	NA	23	23	58

GVHD=graft versus host disease
TRM=transplant related mortality
RR=relapse risk
5-y DFS=disease free survival at 5 years

Table 3. Retrospective non-comparative studies considering alloSCT in ALL

Only few studies prospectively compared alloSCT to other post-remission strategies in ALL (Table 4) (Attal, *et al*, Fielding, *et al*, Fiere, *et al*, Goldstone, *et al*, Hunault, *et al*, Rowe, Sebban, *et al*, Thiebaut, *et al*)*,* (Labar, *et al*). The BGMT group conduced a prospective trial evaluated alloSCT vs. ASCT in CR1. According to *genetic randomization*, patients with (n = 43) or without an HLA-identical sibling (n = 77) were assigned to receive allogeneic or autologous SCT, respectively. The 3-year post-CR probability of DFS was significantly higher in the HLA-identical sibling group than in the non-HLA-identical sibling group (68% v 26%; P<0.001) (Attal, *et al*). Recently, the 10-year follow-up results confirmed the marked superiority of alloSCT to CHT in terms of survival (44 vs 11%, P=0.009)(Marks, *et al*).
The prospective randomized LALA87 trial compared alloSCT vs. ASCT/CHT. Patients aged 15-40 years who achieved CR and had a matched related donor were assigned to allograft, whereas those without a donor were randomized to receive ASCT vs CHT. Seventy-six percent (436/572 evaluable patients) achieved CR. Intention to treat analysis, comparing patients assigned to alloSCT (N=116) vs. ASCT/CHT, showed increased survival in allografted patients (10 years overall survival - OS- 46% vs. 31%, p=0.04). In high risk patients, the advantage was more evident (44% vs. 115, p=0.009). On the contrary, standard risk patients did not significantly benefit from SCT (p=0.06) (Thiebaut, *et al*).
The GOELAL02 trial evaluated the impact in high risk ALL patients of early alloSCT or delayed unpurged ASCT for patients who had no HLA-matched sibling donor or who were older than 50 years. Among 198 patients, the median age was 33 years. The CR rate was 80% with standard induction therapy. AlloSCT was performed after 2 consolidation courses while ASCT was delayed after 1 additional reinduction. Intensified conditioning regimen before transplantation included etoposide, cyclophosphamide, and total body irradiation (TBI). Median follow-up was 5.1 years. The median overall survival (OS) was 29 months, with a 6-year OS of 41%. On an intent-to-treat analysis for patients younger than 50 years, alloSCT significantly improved the 6-year OS (75% versus 40% after ASCT; $P = .0027$). On the other hand, randomized interferon-α maintenance had no effect on relapse or survival after ASCT (Hunault, *et al*).

In the EORTC ALL-3 study, Labar and collegues evaluated the role of alloSCT based on a genetic randomization.(Labar, *et al*). Patients achieving CR and having an HLA-identical sibling were intended for alloSCT after completion of consolidation while the remaining ones were randomized to receive either ASCT or maintenance. Among 68 patients in the donor arm the transplantation was performed in 47 cases. According to the intention-to-treat analysis the incidence of relapse at 6 years was decreased (38% vs. 56%) while the incidence non-relapse mortality increased (24% vs. 7%) in the "donor" compared with "no donor" arm. The probability of DFS was superimposable (38% vs. 37%, respectively). Among 116 patients lacking a donor only 45 were randomized making the vs. maintenance comparison statistically underpowered.

Recently, Ribera and Colleagues reported on 222 high risk ALL patients enrolled in the PETHEMA ALL-93 trial (Ribera, *et al*). All received a standard five-drug/five-week induction course. Patients in CR with an HLA identical family donor were assigned to alloSCT (n=84) and the remaining were randomized to ASCT (n=50) or to delayed intensification followed by maintenance CHT up to 2 years in complete remission (n=48). Overall, 183 patients achieved complete remission (82%). With a median follow-up of 70 months, the median DFS and OS were 17 and 23 months, respectively. The 5-year DFS and OS were 35% (95% CI, 30%- 41%) and 34% (95% CI, 28%-39%), respectively. Patients allocated to the CHT, allogeneic and autologous SCT were comparable in the main pre-treatment ALL characteristics and the rate of response to therapy. Intention-to-treat analysis showed no differences between patients according to whether they had or did not have a donor in DFS (39%, 95% CI 30-48% vs. 33%, 95% CI 23-41%) and OS (44%, 95% CI 35- 52% vs. 35%, 95% CI 25-44%), as well as for autologous SCT vs. CHT comparisons (DFS: 40%, 95% CI 28-52% vs. 51%, 95% CI 37-67%; overall survival: 43%, 95% CI 29-58% vs. 52%, 95% CI 39-65%). No differences were observed when the analysis was made on the basis of the treatment actually performed (Ribera, *et al*).

More recently, the final results of the MRC UKALL XII/ECOG 2993 protocol were available, regarding a large prospective randomized study aiming to establish the best post-remission therapy for adult ALL (Avivi and Goldstone, Rowe, *et al*, Rowe). In this study, patients (N=1,646) received 2 phases of induction and, if in remission, were assigned to allogeneic transplantation if they had a compatible sibling donor. Other patients were randomized to chemotherapy for 2.5 years versus an ASCT. A donor versus no-donor analysis showed that Ph^- patients with a donor had a 5-year improved OS (53% versus 45%; $p = .01$), and the relapse rate was significantly lower ($p \leq .001$). The survival difference was significant in standard-risk patients, but not in high-risk patients due to a high treatment related mortality (TRM) in the high-risk donor group. Interestingly, patients randomized to chemotherapy had a higher 5-year OS (46%) than those randomized to autologous transplantation (37%; $p = .03$) (Goldstone, *et al*). Importantly, a dedicated analysis was carried on Ph+ cases (N=267) (Fielding, *et al*). In this setting, in the comparison of the outcome after any alloSCT with the outcome after chemotherapy alone, OS ($p = .001$), EFS ($p < .001$), and RFS ($P < .001$) were all significantly superior for patients receiving any alloSCT over those receiving chemotherapy alone. Of note, there was a marked difference in the cause of death between alloSCT and chemotherapy recipients. Whereas the leading cause of death in chemotherapy treated patients was relapse, the leading cause of death after transplantation was TRM, which was 27% after sibling-SCT and 39% after VUD-SCT (Fielding, *et al*).

Author	Post-remission therapy	Number of patients	Age	TRM (%)	RR (%)	5-y DFS
Blaise et al. 1990	Allo vs. Auto	25 vs. 22	4-36 vs. 7-47	20 vs. 9	9 vs. 57	71 vs. 40
Horowitz et al. 1991	Allo vs. CHT	234 vs. 484	15-45 vs. 15-45	39 vs. 4	26 vs. 59	44 vs. 38
Attal et al. 1995	Allo vs. Auto	43 vs. 77	15-55 vs. 15-55	12 vs. 2	12 vs. 62	68 vs. 26 (3-y)
Oh et al. 1998	Allo vs. CHT	127 vs. 38	<30	32 vs. 3	22 vs. 69	30 vs. 30
	Allo vs. CHT	87 vs. 38	>30	NA	37 vs. 70	30 vs. 26
Fiere et al. 1993	Allo vs. Auto vs. CHT	116 vs. 95 vs. 96	15-40 vs. 15-50	18	37	46 vs. 39 vs. 32
Fiere et al.1998	Allo vs. Auto vs. CHT	116 vs. 95 vs. 96	15-40 vs. 15-50	NA	NA	46 vs. 34 vs. 30
Rowe et al. 2001	Allo vs. Auto/CHT	190 vs. 253	15-50 vs. 15-50	NA	24 vs. 60	52 vs. 36
Hunault 2004	Allo vs. Auto*	41 vs. 115	15-52 vs. 15-57	15 vs. 3	10 vs. 49	75 vs. 38
Labar et al. 2004	Allo vs. (Auto or CHT)	68 vs. 116	<50	24 v.7	38 vs. 56	38 vs. 37
Ribera et al. 2006	Allo vs. Auto vs. CHT	84 vs. 50 vs. 48	16-49 vs. 15-50 vs. 15-50	8 vs. 3 vs. 0	62 vs. 57 vs. 46	44 vs. 54 vs. 45
Goldstone et al. 2009	Allo vs. CHT/ASCT	443 vs. 588	<54 vs. < 64	36-20% vs. 14-7%	24% vs. 49%	53% vs. 45%§

* early alloSCT vs. delayed ASCT
TRM=transplant related mortality
RR=relapse risk
5-y DFS=disease free survival at 5 years
§ Overall survival

Table 4. Studies comparing alloSCT vs. CHT/ASCT in ALL

5.2 ALL in second remission

More than 50% of adult ALL patients will relapse after initial CR. Half of these patients will actually obtain a second CR (CR2) following re-induction therapy. However, CR2 is not durable and only less of 10% of the patients will be cured with CHT or ASCT. By contrast, DFS after alloSCT performed in CR2 approaches 25-38% (Avivi and Goldstone, Herzig, *et al*, Thomas, *et al*). Indeed, alloSCT appears to be the best option for patients in CR2, irrespectively of the initial risk factors. However, unfortunately, no randomized trials compared alloSCT and CHT in this setting. Noteworthy, the outcome of alloSCT in CR2 is also affected by CR1 duration, being definitely better when CR1 was longer than 24 months (Forman, Kersey, *et al*, Thomas, *et al*).

5.3 ALL after second relapse or primary induction failure

Once patients are beyond second remission, the results of all allografting procedures worsen considerably, with only 10–15% of patients becoming long-term disease-free survivors (Goldman, *et al*, Storb, *et al*). Similarly, the outcome with transplant is poor with only 10–

15% DFS for patients who fail primary induction therapy. However, though these results are not brilliant, SCT still offers the only prospect of cure (Biggs, *et al*). Biggs et al reported 23% DFS in 38 ALL patients treated with alloSCT for primary induction failure disease; the 3 years probability of persistent/recurrent disease was 59% (Biggs, *et al*). The result is, however, encouraging with almost one quarter of patients rescued after initial treatment failure. On the other hand, it should be considered that 26% of the cases in this study were aged under 19 years (and 8 less under 10) and 39% achieved a partial remission before transplant. TRM was significantly higher in patients older than 30 years or with poor performance status (PS). Thus, it appears that 10-20% of patients who fail to achieve CR after induction can benefit from alloSCT; however, patients to be referred to this procedure should be carefully selected basing on age and PS.

6. Stem cell transplantation for Ph+ ALL

The presence of Ph chromosome is one of the most relevant adverse prognostic factors in ALL. To date, alloSCT is the unique approach which can be considered as potentially curative, thus being the treatment of choice in adult Ph+ ALL (Avivi and Goldstone, Barrett, *et al*, Chao, *et al*, Dombret, *et al*, Fielding, *et al*, Ottmann and Pfeifer, Piccaluga, *et al*, Piccaluga, *et al*). Twenty-seven to 65% long-term survival has been reported for patients undergoing alloSCT in first complete remission (CR1), (Avivi and Goldstone, Fielding, *et al*). Unfortunately, relapse occurs in approximately 30% of patients, representing the primary cause of treatment failure together with treatment-related mortality, which increases with age and advanced disease (Avivi and Goldstone, Martin and Gajewski). Achievement and maintenance of a CR prior to SCT is an important prerequisite for a favourable outcome after SCT (Wassmann, *et al*). In fact, when performed further than first remission, only a small minority of patients can be cured by alloSCT; nevertheless, in such cases, although

Author	N	TRM	REL	Outcome
Barret et al, 1992	33 22	42% 40%	34% 32%	38% at 2 years (DFS) 41%
Chao et al, 1995	38	NA	NA	46% (CR1) at 2 years (DFS) 28% (>CR1)
Snyder et al, 1999	23	30%	12%	65% at 3 years (DFS)
Goldstone et al, 2001	72	37% (Sib) vs 43% (VUD)	32%	42% at 5 years (DFS)
Dombret et al, 2002	56	25%	37%	37% at 3 years (OS)
Fielding et al, 2009	267	27% (Sib.) vs. 39% (VUD)	32%	41%-36%* at 5 years (EFS) 44%-36%* at 5 years (OS)

N= number of patients
TRM= transplant related mortality
REL= relapse rate
DFS= disease free survival
CR1= first complete remission
* Sibling donor vs. VUD

Table 5. Results of allogeneic stem cells transplantation in adult Ph+ ALL

the probability of success is limited, alloSCT may be still curative in a subset of patients and it remains the treatment of choice (Garcia-Manero and Thomas). The probability of DFS (DFS) at 2 years after alloSCT in second or third remission or as salvage therapy for refractory disease has been reported to be 17% and 5%, respectively (Cornelissen, *et al*). In patients failing allogeneic SCT, further treatment is rarely successful. With this regard, the availability of novel tyrosine-kinase inhibitors (TKI), including imatinib, dasatinib and nilotinib, has became of great interest (Piccaluga, *et al*). In fact, TKI either can be added to CHT potentially increasing the number of patients to be referred to alloSCT in CR1 (Wassmann, *et al*), and constitutes a possible salvage treatment in case of relapse after alloSCT (Olavarria, *et al*, Wassmann, *et al*).
Results with allogeneic stem cells transplantation are summarized in Table 5 (Avivi and Goldstone, Barrett, *et al*, Chao, *et al*, Fielding, *et al*, Snyder, *et al*).

7. Source of stem cell for allogeneic transplantation

7.1 Bone marrow versus mobilized peripheral blood

Cytokine-mobilized allogeneic peripheral blood stem cell (PBSC) harvest has become an alternative to bone marrow as a source of stem cells for matched-sibling transplants. Early phase II studies showed that this source of stem cells resulted in faster engraftment, no increase in acute GVHD (perhaps due to a G-CSF-mediated shift to Th2 helper cells) but an increased incidence of chronic GVHD (Bensinger, *et al*, Korbling, *et al*). A prospective randomized study of allogeneic PBSC compared to marrow showed a 2-year actuarial overall survival of 54% in patients receiving marrow compared with 66% in those receiving PBSCs (Bensinger, *et al*). Differences in survival were significant for patients with unfavourable-risk diseases but not for those with favourable-risk diseases (Bensinger, *et al*). Faster engraftment, similar GVHD and improvement in overall survival was also reported in a Canadian and New Zealand study (Couban, *et al*). In a retrospective multivariate analysis from the IBMTR comparing the results of 288 HLA-identical sibling blood stem cell transplants with the results of 536 HLA-identical sibling bone marrow transplants, the relapse incidence between the two transplant groups did not differ significantly (Champlin, *et al*). However, treatment-related mortality rates were lower and leukaemia-free survival rates were higher with use of blood stem cell transplants in patients with advanced versus early leukaemia (Champlin, *et al*). While the results of more studies should become available over the next few years, the current experience suggests that peripheral blood should be the preferred source of stem cells for patients with high-risk disease and/or advanced phase. For patients with low-risk disease, the increased risk of chronic GVHD needs to be balanced against the risk of relapse.

7.2 Alternative donor sources

Most patients do not have a suitable histocompatible sibling donor for an allograft and the use of alternative donors such as voluntary unrelated donor (VUD), haploidentical donor and umbilical cord blood (UCB) grafts. As regards VUD, the results are actually comparable to those obtained with matched sibling donors and many groups routinely refer to this source for patients in CR1. Conversely, to date, the other two options have been mainly reserved for patients in ≥CR2. Few large-scale studies compare these forms of alternative donor transplantation and the choice of stem cell source will also be influenced by clinical

urgency and the time taken to procure haematopoietic stem cells. Single centre results must be analysed carefully when deciding what plan to recommend in the individual patient.
If a matched donor is not available, some data indicate that a single antigen mismatched family donor allograft may result in the same outcome as using a matched UD, but a perspective randomized comparison has never been performed. In general, if there is no molecularly matched UD, then a single allele mismatched UD, an UCB donor or a haploidentical stem cell donor could be considered as possible alternative (Marks, *et al*).
In this regard, few small studies focused on the use of UCB in ALL, and basing on these data it is difficult to advise investigators how to approach potential patients. Small patient numbers, heterogeneity of remission status, variability of matching and incomplete information about UCB cell dose make data summary difficult. In general, these results are promising, but relapse remains a significant issue as well as does engraftment failure. Somewhere between 20 and 50% of patients in CR1 and an occasional CR2 patient can become long-term survivors. Double UCB transplants or intra-bone SC infusion (Castello, *et al*) may represent an advance, particularly addressing the issue of engraftment.
Haploidentical transplant have the major advantage of donor availability and results have been possibly improved. In particular, the high rejection rate and incidence of severe GVHD are reduced by combining high-intensity conditioning and an infusion of a large dose of purified SC in adult ALL patients transplanted in CR1 or CR2. These data, however, have not been corroborated in multi-centre trials, and a variety of opportunistic infections continue to plague this approach that does not appear useful in the setting of advanced disease. Further avenues of improvement include the exploration of killer cell immunoglobulin-like receptor (KIR) mismatching between donor grafts and recipients.
Data on alternative donor source in ALL have been recently reviewed by Marks et al (Marks, *et al*) and are summarized in Tables 6-7.

8. Post transplant strategies to improve clinical outcome

8.1 GVHD prophylaxis

GVHD results from alloreactivity between donor and recipient. The two major prophylactic regimens employed to prevent this complication are pharmacological (administration of immunosuppressive drugs), and immunological (in vitro T-cell depletion). The standard pharmacologic prophylaxis has been cyclosporine (CsA) and short-course methotrexate, but recent studies suggest that the incidence of GVHD may be lower if FK506 is substituted for CsA (Nash, *et al*). MMF also shows promise in animal models, and its combination with CsA is being evaluated in clinical trials. Ex vivo T-cell depletion reduces the risk of both acute and chronic GVHD and may allow higher tolerance of mismatching but may also increase the risk of rejection and delay immune reconstitution. A confounding feature for interpreting the value of T-cell depletion is that a variety of methodologies are employed to remove T cells, including physical methods, poly- and mono-clonal antibodies. Some techniques produce a pan-T depletion, whereas others use antibodies with more restricted T-subset specificities. An IBMTR study showed a better outcome when antibodies with narrow specificities are used (Champlin, *et al*).

8.2 Reducing the risk of relapse

The major cause of failure after transplant for ALL is relapse. The outcome of patients who relapse after allogeneic SCT is very poor. Remissions are possible with standard CHT but are not durable. Those who relapse more than 1-year post-transplant are more likely to achieve a further remission. Donor leukocyte infusions have been used but their success rate in patients with ALL is much lower than in patients with myeloid malignancies (Porter and Antin). Strategies to reduce the risk of relapse include intensifying conditioning regimens, altering the timing of transplant and augmenting the graft versus leukaemia effect. The risk of relapse may also be reduced by more precisely defining the biological risk factors that justify transplant in first remission, thus circumventing the possibility of selecting for leukaemia resistance during prolonged CHT. Patients who receive an alloSCT for ALL and develop graft versus host disease (GVHD) have a lower probability of relapse than patients who do not suffer this complication. A recent study showed that in patients with high levels of minimal residual disease (MRD) pre-transplant, the presence of acute or chronic GVHD may be needed to prevent relapse(Uzunel, *et al*). The likely explanation for this observations is that the alloreactive T cells in the donor graft are able to destroy residual host leukaemia cells. In support of this contention, administering lower doses of CsA reduces the relapse risk and improves DFS in children undergoing SCT for leukaemia (Locatelli, *et al*). Adoptive immunotherapy with donor mononuclear cells is less successful in ALL than in the myeloid leukaemias, although there is some evidence that use of immunostimulatory cytokines such as IL2 may amplify graft versus leukaemia mechanisms and induce remissions in patients who have failed to respond to donor lymphocyte infusions. Immune modulation post-transplant may therefore be a therapeutic alternative to reduce the risk of relapse. One means of reducing the risk of GVHD is to administer antigen-specific cytotoxic T-cell lines (CTL) lines when a specific antigen is known. Potential targets include minor antigens differentially expressed on haemopoietic cells (Mutis, *et al*) or lineage specific antigens, such as WTI or proteinase 3 (Ohminami, *et al*). Such lines could potentially mediate cytotoxic activity directed at recipient haemopoiesis (and leukaemia) but not donor haemopoiesis. An alternative approach when a tumour antigen is not defined is to increase the immunogenicity of the tumour and allow selection of the tumour antigen by responding immune system effector cells. This approach has shown efficacy in a number of animal models using molecules that modify antigen presentation such as Class I MHC molecules or GMCSF, co-stimulatory molecules such as B7 or T-cell activation factors such as IL2. For example pre B ALL cells lack B7-1(CD80) and induce allo-specific T-cell anergy. If ALL cells are transduced with B7, their antigen presentation capacity is improved and they are able to generate autologous leukaemia-specific CTL lines from marrow from the majority of patients (Cardoso, *et al*). Murine data also suggests that combination of molecules acting on different phases of the immune response may produce increased anti-tumour activity (Dilloo, *et al*) and this approach is currently being tested in a clinical trial in patients with ALL (Rousseau, *et al*).

9. Autologous stem cell transplantation

The role of in adults with ALL has not been clearly defined. Prospective studies comparing with alloSCT in high-risk patients demonstrated increased risk of relapse and reduced probability of survival in the autologous setting (Attal, *et al*, Hunault, *et al*). Trials comparing with maintenance chemotherapy did not show survival benefit from high-dose treatment, however, a pooled analysis of three subsequent trials by the French group (LALA-85/87/94) suggested that ASCT may contribute to decreased risk of relapse (66% vs. 78%, p=0.05) (Dhedin, *et al*, Labar, *et al*, Ribera, *et al*, Thiebaut, *et al*). In the MRC UKALLXII/ECOG 2993 study investigators tested if ASCT may be administered instead of consolidation + maintenance for patients lacking an HLA-identical sibling, irrespective of the presence of risk factors. OS was significantly worse in the ASCT compared to chemotherapy arm (46% vs. 37% at 5 years, p=0.03) leading to conclusion that early ASCT cannot substitute consolidation.

Disappointing results of prospective trials evaluating the role of ASCT led to a general tendency to abandon this treatment option in adults with ALL. It must be stressed, however, that in view of current knowledge design of the studies could have been suboptimal. First of all, the trials have been conducted before the era of routine MRD assessment. Low tumor burden is prerequisite of successful ASCT as in other conditions the transplant material may be contaminated by residual leukemic blasts leading to their re-transplantation and early relapse. Evaluation of MRD status in bone marrow reflecting the tumor load may therefore be crucial for selection of patients who could benefit from ASCT. Indeed, in a retrospective analysis by the European Working Group for Adult ALL, the MRD level was demonstrated the most important prognostic factor with DFS of 58% for patients with MRD<0.1% vs. 19% for those with MRD ≥0.1% in bone marrow (p=0.0002) (Giebel, *et al*). In a setting of 123 ASCT recipients with HR ALL, the incidence of non-relapse mortality was 2.4%. In the analysis of a selected group of 50 patients in whom the MRD was evaluated with sufficient sensitivity to define the MRD status as negative at the level of 0.01%, the 5-years probability of DFS was 69%.

In so far conducted prospective trials ASCT was administered instead of either maintenance or consolidation + maintenance and was considered the last stage of the treatment. There is, however, another option, which has not been prospectively tested i.e. incorporating ASCT in consolidation, followed by maintenance. Interim analysis of a retrospective study by the Central and Eastern Leukemia Group indicate significantly increased probability of the DFS if maintenance based on mercaptopurine + methotrexate was administered after ASCT (70% vs. 38% at 5 years, p=0.03). The effect was independent on the MRD status (HR=2.6, p=0.007) (unpublished data).

Finally, ASCT followed by interferon alpha administration has been shown to be quite effective for Ph+ patients (Piccaluga, *et al*, Visani, *et al*).

Taken together, ASCT is associated with low risk of transplant-related mortality and morbidity. Its efficacy should still be prospectively evaluated in patients with low MRD level, with post-transplant maintenance as an option.

10. Conclusion

Data from clinical trials trying to compare alloSCT and other post-remission therapies for ALL patients in CR1 are sometimes difficult to interpret because of varying proportions of patients with different risk factors in each study (some of them are actually limited to high risk patients as many centres do not include alloSCT in the therapeutic program of standard risk patients in CR1). Prospective randomized studies have been carried out by various cooperative groups. These studies have compared patients with HLA identical siblings assigned to the allogeneic SCT group to those without HLA identical sibs assigned to either a standard chemotherapy or ASCT group. However, as up to 70% of adults with ALL cannot be allocated to SCT because of a lack of a matched related sibling, comorbidities, or severe infections, an objective and unbiased comparison among treatments is difficult (Avivi and Goldstone, Martin and Gajewski, Popat, *et al*). A major limitation of these studies is that the proportion of patients in the allogeneic SCT group actually undergoing transplant varies from study to study. Despite this limitation, a donor versus no donor comparison is the best way to answer the question about role of allogeneic SCT in first remission, even though the so called "genetic randomization" (based on the availability of HLA-matched donor) is not a proper randomization. In addition, in some studies, patients receiving ASCT/CHT included cases over 50 years not suitable for alloSCT. Notable features of all these studies include a high treatment related mortality and a lower risk of relapse in allogeneic transplant arm. On the other hand, the most recent study from the MRC/ECOG group (MRC UKALL XII/ECOG E2993) (Fielding, *et al*, Goldstone, *et al*) represented the largest series in which the role of alloSCT in ALL was prospectively evaluated. Indeed, in this study alloSCT was shown to provide the best outcome specially in standard risk patients, thus offering a strong indication to this procedure in the majority of ALL cases.

On the other hand, the randomized studies so far available did not compare alloSCT to the most recent risk-adapted, intensified consolidation schedules, which provided significant benefits in terms of both relapse free and overall survival in standard risk ALL cases (Bassan and Hoelzer). In addition, not all the patients can benefit from it, at present. Thus, optimizing SCT strategies (see above) as well as improving post SCT therapy are necessary. In this light, the possible interventions regards either cellular therapies and pharmacological approaches. For example, the development of grade ≥2 GVHD is correlated with reduced relapse rate but relevant morbidity and mortality are related to GVHD. Thus, future strategies should consider graft manipulation aiming to maintain GVL effect though reducing GVHD and TRM.

Finally, the presence of minimal residual disease after SCT is associated with impending relapse. Thus, careful MRD monitoring after SCT should be performed and treatment of molecular relapse should be considered. In this setting, targeted therapy, such as TKI and interferon alpha (Piccaluga, *et al*) or novel TKI for Ph+ ALL, FLT3 inhibitors for MLL^+ cases, nelarabine/forodesine for T-ALL and, more in general, monoclonal antibodies (anti-CD19, anti-CD20, anti-CD22 and anti-CD33) surely warrants further evaluation in clinical trials.

Accordingly, based on the currently available data, allogeneic SCT is a reasonable treatment option for adults with high-risk and possibly standard risk ALL in CR1. Patients in CR2 with matched donor should be probably always referred to alloSCT unless major clinical contraindications are recorded. Finally, refractory patients as well as cases beyond CR2 should be carefully evaluated and selected for alloSCT basing on age, PS and amount of previous CHT.

11. Acknowledgements

Centro Interdipartimentale per la Ricerca sul Cancro "G. Prodi", BolognAIL, AIRC (IG10519; 5xMille 10007), RFO (Prof. Pileri, Dr. Piccaluga), Fondazione Cassa di Risparmio in Bologna, Fondazione della Banca del Monte e Ravenna, Progetto Strategico di Ateneo 2006 (Prof. Pileri and Dr. Piccaluga).

12. References

Annino, L., Vegna, M.L., Camera, A., Specchia, G., Visani, G., Fioritoni, G., Ferrara, F., Peta, A., Ciolli, S., Deplano, W., Fabbiano, F., Sica, S., Di Raimondo, F., Cascavilla, N., Tabilio, A., Leoni, P., Invernizzi, R., Baccarani, M., Rotoli, B., Amadori, S. &

Mandelli, F. (2002) Treatment of adult acute lymphoblastic leukemia (ALL): long-term follow-up of the GIMEMA ALL 0288 randomized study. *Blood,* 99, 863-871.

Attal, M., Blaise, D., Marit, G., Payen, C., Michallet, M., Vernant, J.P., Sauvage, C., Troussard, X., Nedellec, G., Pico, J. & et al. (1995) Consolidation treatment of adult acute lymphoblastic leukemia: a prospective, randomized trial comparing allogeneic versus autologous bone marrow transplantation and testing the impact of recombinant interleukin-2 after autologous bone marrow transplantation. BGMT Group. *Blood,* 86, 1619-1628.

Avivi, I. & Goldstone, A.H. (2003) Bone marrow transplant in Ph+ ALL patients. *Bone Marrow Transplant,* 31, 623-632.

Barrett, A.J., Horowitz, M.M., Ash, R.C., Atkinson, K., Gale, R.P., Goldman, J.M., Henslee-Downey, P.J., Herzig, R.H., Speck, B., Zwaan, F.E. & et al. (1992) Bone marrow transplantation for Philadelphia chromosome-positive acute lymphoblastic leukemia. *Blood,* 79, 3067-3070.

Barrett, A.J., Horowitz, M.M., Gale, R.P., Biggs, J.C., Camitta, B.M., Dicke, K.A., Gluckman, E., Good, R.A., Herzig, R.H., Lee, M.B. & et al. (1989) Marrow transplantation for acute lymphoblastic leukemia: factors affecting relapse and survival. *Blood,* 74, 862-871.

Bassan, R. & Hoelzer, D. (2011) Modern therapy of acute lymphoblastic leukemia. *J Clin Oncol,* 29, 532-543.

Bassan, R., Spinelli, O., Oldani, E., Intermesoli, T., Tosi, M., Peruta, B., Rossi, G., Borlenghi, E., Pogliani, E.M., Terruzzi, E., Fabris, P., Cassibba, V., Lambertenghi-Deliliers, G., Cortelezzi, A., Bosi, A., Gianfaldoni, G., Ciceri, F., Bernardi, M., Gallamini, A., Mattei, D., Di Bona, E., Romani, C., Scattolin, A.M., Barbui, T. & Rambaldi, A. (2009) Improved risk classification for risk-specific therapy based on the molecular study of minimal residual disease (MRD) in adult acute lymphoblastic leukemia (ALL). *Blood,* 113, 4153-4162.

Bensinger, W.I., Martin, P.J., Storer, B., Clift, R., Forman, S.J., Negrin, R., Kashyap, A., Flowers, M.E., Lilleby, K., Chauncey, T.R., Storb, R. & Appelbaum, F.R. (2001) Transplantation of bone marrow as compared with peripheral-blood cells from HLA-identical relatives in patients with hematologic cancers. *N Engl J Med,* 344, 175-181.

Bensinger, W.I., Weaver, C.H., Appelbaum, F.R., Rowley, S., Demirer, T., Sanders, J., Storb, R. & Buckner, C.D. (1995) Transplantation of allogeneic peripheral blood stem cells mobilized by recombinant human granulocyte colony-stimulating factor. *Blood,* 85, 1655-1658.

Biggs, J.C., Horowitz, M.M., Gale, R.P., Ash, R.C., Atkinson, K., Helbig, W., Jacobsen, N., Phillips, G.L., Rimm, A.A., Ringden, O. & et al. (1992) Bone marrow transplants may cure patients with acute leukemia never achieving remission with chemotherapy. *Blood,* 80, 1090-1093.

Blume, K.G., Forman, S.J., Snyder, D.S., Nademanee, A.P., O'Donnell, M.R., Fahey, J.L., Krance, R.A., Sniecinski, I.J., Stock, A.D., Findley, D.O. & et al. (1987) Allogeneic bone marrow transplantation for acute lymphoblastic leukemia during first complete remission. *Transplantation,* 43, 389-392.

Blume, K.G., Kopecky, K.J., Henslee-Downey, J.P., Forman, S.J., Stiff, P.J., LeMaistre, C.F. & Appelbaum, F.R. (1993) A prospective randomized comparison of total body irradiation-etoposide versus busulfan-cyclophosphamide as preparatory regimens for bone marrow transplantation in patients with leukemia who were not in first remission: a Southwest Oncology Group study. *Blood,* 81, 2187-2193.

Brenner, M.K., Wulf, G.G., Rill, D.R., Luo, K.L., Goodell, M.A., Mei, Z., Kuehnle, I., Brown, M.P., Pule, M., Heslop, H.E. & Krance, R.A. (2003) Complement-fixing CD45 monoclonal antibodies to facilitate stem cell transplantation in mouse and man. *Ann N Y Acad Sci,* 996, 80-88.

Cardoso, A.A., Seamon, M.J., Afonso, H.M., Ghia, P., Boussiotis, V.A., Freeman, G.J., Gribben, J.G., Sallan, S.E. & Nadler, L.M. (1997) Ex vivo generation of human anti-pre-B leukemia-specific autologous cytolytic T cells. *Blood,* 90, 549-561.

Castello, S., Podesta, M., Menditto, V.G., Ibatici, A., Pitto, A., Figari, O., Scarpati, D., Magrassi, L., Bacigalupo, A., Piaggio, G. & Frassoni, F. (2004) Intra-bone marrow injection of bone marrow and cord blood cells: an alternative way of transplantation associated with a higher seeding efficiency. *Exp Hematol,* 32, 782-787.

Champlin, R.E., Passweg, J.R., Zhang, M.J., Rowlings, P.A., Pelz, C.J., Atkinson, K.A., Barrett, A.J., Cahn, J.Y., Drobyski, W.R., Gale, R.P., Goldman, J.M., Gratwohl, A., Gordon-Smith, E.C., Henslee-Downey, P.J., Herzig, R.H., Klein, J.P., Marmont, A.M., O'Reilly, R.J., Ringden, O., Slavin, S., Sobocinski, K.A., Speck, B., Weiner, R.S. & Horowitz, M.M. (2000a) T-cell depletion of bone marrow transplants for leukemia from donors other than HLA-identical siblings: advantage of T-cell antibodies with narrow specificities. *Blood,* 95, 3996-4003.

Champlin, R.E., Schmitz, N., Horowitz, M.M., Chapuis, B., Chopra, R., Cornelissen, J.J., Gale, R.P., Goldman, J.M., Loberiza, F.R., Jr., Hertenstein, B., Klein, J.P., Montserrat, E., Zhang, M.J., Ringden, O., Tomany, S.C., Rowlings, P.A., Van Hoef, M.E. & Gratwohl, A. (2000b) Blood stem cells compared with bone marrow as a source of hematopoietic cells for allogeneic transplantation. IBMTR Histocompatibility and Stem Cell Sources Working Committee and the European Group for Blood and Marrow Transplantation (EBMT). *Blood,* 95, 3702-3709.

Chao, N.J., Blume, K.G., Forman, S.J. & Snyder, D.S. (1995) Long-term follow-up of allogeneic bone marrow recipients for Philadelphia chromosome-positive acute lymphoblastic leukemia. *Blood,* 85, 3353-3354.

Chao, N.J., Forman, S.J., Schmidt, G.M., Snyder, D.S., Amylon, M.D., Konrad, P.N., Nademanee, A.P., O'Donnell, M.R., Parker, P.M., Stein, A.S. & et al. (1991) Allogeneic bone marrow transplantation for high-risk acute lymphoblastic leukemia during first complete remission. *Blood,* 78, 1923-1927.

Chester, S.J., Esparza, A.R., Flinton, L.J., Simon, J.D., Kelley, R.J. & Albala, M.M. (1977) Further development of a successful protocol of graft versus leukemia without fatal graft-versus-host disease in AKR mice. *Cancer Res,* 37, 3494-3496.

Cornelissen, J.J., Carston, M., Kollman, C., King, R., Dekker, A.W., Lowenberg, B. & Anasetti, C. (2001) Unrelated marrow transplantation for adult patients with poor-risk acute lymphoblastic leukemia: strong graft-versus-leukemia effect and risk factors determining outcome. *Blood,* 97, 1572-1577.

Couban, S., Simpson, D.R., Barnett, M.J., Bredeson, C., Hubesch, L., Howson-Jan, K., Shore, T.B., Walker, I.R., Browett, P., Messner, H.A., Panzarella, T. & Lipton, J.H. (2002) A randomized multicenter comparison of bone marrow and peripheral blood in recipients of matched sibling allogeneic transplants for myeloid malignancies. *Blood,* 100, 1525-1531.

Davies, S.M., Ramsay, N.K., Klein, J.P., Weisdorf, D.J., Bolwell, B., Cahn, J.Y., Camitta, B.M., Gale, R.P., Giralt, S., Heilmann, C., Henslee-Downey, P.J., Herzig, R.H., Hutchinson, R., Keating, A., Lazarus, H.M., Milone, G.A., Neudorf, S., Perez, W.S., Powles, R.L., Prentice, H.G., Schiller, G., Socie, G., Vowels, M., Wiley, J., Yeager, A. & Horowitz, M.M. (2000) Comparison of preparative regimens in transplants for children with acute lymphoblastic leukemia. *J Clin Oncol,* 18, 340-347.

De Witte, T., Awwad, B., Boezeman, J., Schattenberg, A., Muus, P., Raemaekers, J., Preijers, F., Strijckmans, P. & Haanen, C. (1994) Role of allogenic bone marrow transplantation in adolescent or adult patients with acute lymphoblastic leukaemia or lymphoblastic lymphoma in first remission. *Bone Marrow Transplant,* 14, 767-774.

Dhedin, N., Dombret, H., Thomas, X., Lheritier, V., Boiron, J.M., Rigal-Huguet, F., Vey, N., Kuentz, M., Reman, O., Witz, F., Delannoy, A., Kovacsovics, T., Bradstock, K., Charrin, C., Boucheix, C., Gabert, J., Blaise, D., Fiere, D. & Vernant, J.P. (2006) Autologous stem cell transplantation in adults with acute lymphoblastic leukemia in first complete remission: analysis of the LALA-85, -87 and -94 trials. *Leukemia,* 20, 336-344.

Dilloo, D., Bacon, K., Holden, W., Zhong, W., Burdach, S., Zlotnik, A. & Brenner, M. (1996) Combined chemokine and cytokine gene transfer enhances antitumor immunity. *Nat Med,* 2, 1090-1095.

Dombret, H., Gabert, J., Boiron, J.M., Rigal-Huguet, F., Blaise, D., Thomas, X., Delannoy, A., Buzyn, A., Bilhou-Nabera, C., Cayuela, J.M., Fenaux, P., Bourhis, J.H., Fegueux, N., Charrin, C., Boucheix, C., Lheritier, V., Esperou, H., MacIntyre, E., Vernant, J.P. & Fiere, D. (2002) Outcome of treatment in adults with Philadelphia chromosome-positive acute lymphoblastic leukemia--results of the prospective multicenter LALA-94 trial. *Blood,* 100, 2357-2366.

Doney, K., Fisher, L.D., Appelbaum, F.R., Buckner, C.D., Storb, R., Singer, J., Fefer, A., Anasetti, C., Beatty, P., Bensinger, W. & et al. (1991) Treatment of adult acute lymphoblastic leukemia with allogeneic bone marrow transplantation. Multivariate analysis of factors affecting acute graft-versus-host disease, relapse, and relapse-free survival. *Bone Marrow Transplant,* 7, 453-459.

Durrant, I.J., Prentice, H.G. & Richards, S.M. (1997) Intensification of treatment for adults with acute lymphoblastic leukaemia: results of U.K. Medical Research Council randomized trial UKALL XA. Medical Research Council Working Party on Leukaemia in Adults. *Br J Haematol,* 99, 84-92.

Fefer, A., Einstein, A.B., Thomas, E.D., Buckner, C.D., Clift, R.A., Glucksberg, H., Neiman, P.E. & Storb, R. (1974) Bone-marrow transplantation for hematologic neoplasia in 16 patients with identical twins. *N Engl J Med,* 290, 1389-1393.

Fielding, A.K., Rowe, J.M., Richards, S.M., Buck, G., Moorman, A.V., Durrant, I.J., Marks, D.I., McMillan, A.K., Litzow, M.R., Lazarus, H.M., Foroni, L., Dewald, G., Franklin, I.M., Luger, S.M., Paietta, E., Wiernik, P.H., Tallman, M.S. & Goldstone, A.H. (2009) Prospective outcome data on 267 unselected adult patients with Philadelphia chromosome-positive acute lymphoblastic leukemia confirms superiority of allogeneic transplantation over chemotherapy in the pre-imatinib era: results from the International ALL Trial MRC UKALLXII/ECOG2993. *Blood,* 113, 4489-4496.

Fiere, D., Lepage, E., Sebban, C., Boucheix, C., Gisselbrecht, C., Vernant, J.P., Varet, B., Broustet, A., Cahn, J.Y., Rigal-Huguet, F. & et al. (1993) Adult acute lymphoblastic leukemia: a multicentric randomized trial testing bone marrow transplantation as postremission therapy. The French Group on Therapy for Adult Acute Lymphoblastic Leukemia. *J Clin Oncol,* 11, 1990-2001.

Forman, S.J. (1997) The role of allogeneic bone marrow transplantation in the treatment of high-risk acute lymphocytic leukemia in adults. *Leukemia,* 11 Suppl 4, S18-19.

Garcia-Manero, G. & Thomas, D.A. (2001) Salvage therapy for refractory or relapsed acute lymphocytic leukemia. *Hematol Oncol Clin North Am,* 15, 163-205.

Giebel, S., Stella-Holowiecka, B., Krawczyk-Kulis, M., Gokbuget, N., Hoelzer, D., Doubek, M., Mayer, J., Piatkowska-Jakubas, B., Skotnicki, A.B., Dombret, H., Ribera, J.M., Piccaluga, P.P., Czerw, T., Kyrcz-Krzemien, S. & Holowiecki, J. (2010) Status of minimal residual disease determines outcome of autologous hematopoietic SCT in adult ALL. *Bone Marrow Transplant,* 45, 1095-1101.

Giralt, S., Thall, P.F., Khouri, I., Wang, X., Braunschweig, I., Ippolitti, C., Claxton, D., Donato, M., Bruton, J., Cohen, A., Davis, M., Andersson, B.S., Anderlini, P., Gajewski, J., Kornblau, S., Andreeff, M., Przepiorka, D., Ueno, N.T., Molldrem, J. & Champlin, R. (2001) Melphalan and purine analog-containing preparative regimens: reduced-intensity conditioning for patients with hematologic malignancies undergoing allogeneic progenitor cell transplantation. *Blood,* 97, 631-637.

Goldman, F.D., Rumelhart, S.L., DeAlacron, P., Holida, M.D., Lee, N.F., Miller, J., Trigg, M. & Giller, R. (2000) Poor outcome in children with refractory/relapsed leukemia undergoing bone marrow transplantation with mismatched family member donors. *Bone Marrow Transplant,* 25, 943-948.

Goldstone, A.H., Richards, S.M., Lazarus, H.M., Tallman, M.S., Buck, G., Fielding, A.K., Burnett, A.K., Chopra, R., Wiernik, P.H., Foroni, L., Paietta, E., Litzow, M.R., Marks, D.I., Durrant, J., McMillan, A., Franklin, I.M., Luger, S., Ciobanu, N. & Rowe, J.M. (2008) In adults with standard-risk acute lymphoblastic leukemia, the greatest benefit is achieved from a matched sibling allogeneic transplantation in first complete remission, and an autologous transplantation is less effective than conventional consolidation/maintenance chemotherapy in all patients: final results of the International ALL Trial (MRC UKALL XII/ECOG E2993). *Blood,* 111, 1827-1833.

Gupta, V., Yi, Q.L., Brandwein, J., Minden, M.D., Schuh, A.C., Wells, R.A., Chun, K., Kamel-Reid, S., Tsang, R., Daly, A., Kiss, T., Lipton, J.H. & Messner, H.A. (2004) The role of allogeneic bone marrow transplantation in adult patients below the age of 55 years with acute lymphoblastic leukemia in first complete remission: a donor vs no donor comparison. *Bone Marrow Transplant,* 33, 397-404.

Herzig, R.H., Bortin, M.M., Barrett, A.J., Blume, K.G., Gluckman, E., Horowitz, M.M., Jacobsen, S.J., Marmont, A., Masaoka, T., Prentice, H.G. & et al. (1987) Bone-marrow transplantation in high-risk acute lymphoblastic leukaemia in first and second remission. *Lancet,* 1, 786-789.

Hoelzer D, G.N. (2002) Acute Lymphoblastic Leukemia in Adults. In: *Leukemia* (ed. by E.S. Henderson, Lister, T.A., Greaves, M.F.), pp. 621-654. Saunders.

Hoelzer, D. & Gokbuget, N. (2000) New approaches to acute lymphoblastic leukemia in adults: where do we go? *Semin Oncol,* 27, 540-559.

Horowitz, M.M., Gale, R.P., Sondel, P.M., Goldman, J.M., Kersey, J., Kolb, H.J., Rimm, A.A., Ringden, O., Rozman, C., Speck, B. & et al. (1990) Graft-versus-leukemia reactions after bone marrow transplantation. *Blood,* 75, 555-562.

Horowitz, M.M., Messerer, D., Hoelzer, D., Gale, R.P., Neiss, A., Atkinson, K., Barrett, A.J., Buchner, T., Freund, M., Heil, G. & et al. (1991) Chemotherapy compared with bone marrow transplantation for adults with acute lymphoblastic leukemia in first remission. *Ann Intern Med,* 115, 13-18.

Hunault, M., Harousseau, J.L., Delain, M., Truchan-Graczyk, M., Cahn, J.Y., Witz, F., Lamy, T., Pignon, B., Jouet, J.P., Garidi, R., Caillot, D., Berthou, C., Guyotat, D., Sadoun, A., Sotto, J.J., Lioure, B., Casassus, P., Solal-Celigny, P., Stalnikiewicz, L., Audhuy, B., Blanchet, O., Baranger, L., Bene, M.C. & Ifrah, N. (2004) Better outcome of adult acute lymphoblastic leukemia after early genoidentical allogeneic bone marrow transplantation (BMT) than after late high-dose therapy and autologous BMT: a GOELAMS trial. *Blood,* 104, 3028-3037.

Kantarjian, H.M., O'Brien, S., Smith, T.L., Cortes, J., Giles, F.J., Beran, M., Pierce, S., Huh, Y., Andreeff, M., Koller, C., Ha, C.S., Keating, M.J., Murphy, S. & Freireich, E.J. (2000) Results of treatment with hyper-CVAD, a dose-intensive regimen, in adult acute lymphocytic leukemia. *J Clin Oncol,* 18, 547-561.

Kebriaei, P. & Larson, R.A. (2003) Progress and challenges in the therapy of adult acute lymphoblastic leukemia. *Curr Opin Hematol,* 10, 284-289.

Kersey, J.H., Weisdorf, D., Nesbit, M.E., LeBien, T.W., Woods, W.G., McGlave, P.B., Kim, T., Vallera, D.A., Goldman, A.I., Bostrom, B. & et al. (1987) Comparison of autologous and allogeneic bone marrow transplantation for treatment of high-risk refractory acute lymphoblastic leukemia. *N Engl J Med,* 317, 461-467.

Khouri, I.F., Keating, M., Korbling, M., Przepiorka, D., Anderlini, P., O'Brien, S., Giralt, S., Ippoliti, C., von Wolff, B., Gajewski, J., Donato, M., Claxton, D., Ueno, N., Andersson, B., Gee, A. & Champlin, R. (1998) Transplant-lite: induction of graft-versus-malignancy using fludarabine-based nonablative chemotherapy and allogeneic blood progenitor-cell transplantation as treatment for lymphoid malignancies. *J Clin Oncol,* 16, 2817-2824.

Korbling, M., Przepiorka, D., Huh, Y.O., Engel, H., van Besien, K., Giralt, S., Andersson, B., Kleine, H.D., Seong, D., Deisseroth, A.B. & et al. (1995) Allogeneic blood stem cell transplantation for refractory leukemia and lymphoma: potential advantage of blood over marrow allografts. *Blood,* 85, 1659-1665.

Kottaridis, P.D., Milligan, D.W., Chopra, R., Chakraverty, R.K., Chakrabarti, S., Robinson, S., Peggs, K., Verfuerth, S., Pettengell, R., Marsh, J.C., Schey, S., Mahendra, P., Morgan, G.J., Hale, G., Waldmann, H., de Elvira, M.C., Williams, C.D., Devereux, S., Linch, D.C., Goldstone, A.H. & Mackinnon, S. (2000) In vivo CAMPATH-1H prevents graft-versus-host disease following nonmyeloablative stem cell transplantation. *Blood,* 96, 2419-2425.

Labar, B., Suciu, S., Zittoun, R., Muus, P., Marie, J.P., Fillet, G., Peetermans, M., Stryckmans, P., Willemze, R., Feremans, W., Jaksic, B., Bourhis, J.H., Burghouts, J.P. & de Witte, T. (2004) Allogeneic stem cell transplantation in acute lymphoblastic leukemia and non-Hodgkin's lymphoma for patients <or=50 years old in first complete remission: results of the EORTC ALL-3 trial. *Haematologica,* 89, 809-817.

Larson, R.A., Dodge, R.K., Linker, C.A., Stone, R.M., Powell, B.L., Lee, E.J., Schulman, P., Davey, F.R., Frankel, S.R., Bloomfield, C.D., George, S.L. & Schiffer, C.A. (1998) A randomized controlled trial of filgrastim during remission induction and consolidation chemotherapy for adults with acute lymphoblastic leukemia: CALGB study 9111. *Blood,* 92, 1556-1564.

Lee, S., Kim, D.W., Kim, Y.J., Park, Y.H., Min, C.K., Lee, J.W., Min, W.S. & Kim, C.C. (2002) Influence of karyotype on outcome of allogeneic bone marrow transplantation for adults with precursor B-lineage acute lymphoblastic leukaemia in first or second remission. *Br J Haematol,* 117, 109-118.

Locatelli, F., Zecca, M., Rondelli, R., Bonetti, F., Dini, G., Prete, A., Messina, C., Uderzo, C., Ripaldi, M., Porta, F., Giorgiani, G., Giraldi, E. & Pession, A. (2000) Graft versus host disease prophylaxis with low-dose cyclosporine-A reduces the risk of relapse in children with acute leukemia given HLA-identical sibling bone marrow transplantation: results of a randomized trial. *Blood,* 95, 1572-1579.

Marks, D.I., Aversa, F. & Lazarus, H.M. (2006) Alternative donor transplants for adult acute lymphoblastic leukaemia: a comparison of the three major options. *Bone Marrow Transplant,* 38, 467-475.

Martin, T.G. & Gajewski, J.L. (2001) Allogeneic stem cell transplantation for acute lymphocytic leukemia in adults. *Hematol Oncol Clin North Am,* 15, 97-120.

Mathe, G., Amiel, J.L., Schwarzenberg, L., Cattan, A., Schneider, M., Devries, M.J., Tubiana, M., Lalanne, C., Binet, J.L., Papiernik, M., Seman, G., Matsukura, M., Mery, A.M., Schwarzmann, V. & Flaisler, A. (1965) Successful Allogenic Bone Marrow Transplantation in Man: Chimerism, Induced Specific Tolerance and Possible Anti-Leukemic Effects. *Blood,* 25, 179-196.

Moorman, A.V., Harrison, C.J., Buck, G.A., Richards, S.M., Secker-Walker, L.M., Martineau, M., Vance, G.H., Cherry, A.M., Higgins, R.R., Fielding, A.K., Foroni, L., Paietta, E., Tallman, M.S., Litzow, M.R., Wiernik, P.H., Rowe, J.M., Goldstone, A.H. & Dewald, G.W. (2007) Karyotype is an independent prognostic factor in adult acute lymphoblastic leukemia (ALL): analysis of cytogenetic data from patients treated on the Medical Research Council (MRC) UKALLXII/Eastern Cooperative Oncology Group (ECOG) 2993 trial. *Blood,* 109, 3189-3197.

Mortuza, F.Y., Papaioannou, M., Moreira, I.M., Coyle, L.A., Gameiro, P., Gandini, D., Prentice, H.G., Goldstone, A., Hoffbrand, A.V. & Foroni, L. (2002) Minimal residual disease tests provide an independent predictor of clinical outcome in adult acute lymphoblastic leukemia. *J Clin Oncol,* 20, 1094-1104.

Mutis, T., Verdijk, R., Schrama, E., Esendam, B., Brand, A. & Goulmy, E. (1999) Feasibility of immunotherapy of relapsed leukemia with ex vivo-generated cytotoxic T lymphocytes specific for hematopoietic system-restricted minor histocompatibility antigens. *Blood,* 93, 2336-2341.

Nash, R.A., Pineiro, L.A., Storb, R., Deeg, H.J., Fitzsimmons, W.E., Furlong, T., Hansen, J.A., Gooley, T., Maher, R.M., Martin, P., McSweeney, P.A., Sullivan, K.M., Anasetti, C. & Fay, J.W. (1996) FK506 in combination with methotrexate for the prevention of graft-versus-host disease after marrow transplantation from matched unrelated donors. *Blood,* 88, 3634-3641.

Oh, H., Gale, R.P., Zhang, M.J., Passweg, J.R., Ino, T., Murakami, H., Ohno, R., Rowlings, P.A., Sobocinski, K.A., Tanimoto, M., Tomonaga, M., Weisdorf, D.J. & Horowitz, M.M. (1998) Chemotherapy vs HLA-identical sibling bone marrow transplants for adults with acute lymphoblastic leukemia in first remission. *Bone Marrow Transplant,* 22, 253-257.

Ohminami, H., Yasukawa, M. & Fujita, S. (2000) HLA class I-restricted lysis of leukemia cells by a CD8(+) cytotoxic T-lymphocyte clone specific for WT1 peptide. *Blood,* 95, 286-293.

Olavarria, E., Ottmann, O.G., Deininger, M., Clark, R.E., Bandini, G., Byrne, J., Lipton, J., Vitek, A., Michallet, M., Siegert, W., Ullmann, A., Wassmann, B., Niederwieser, D. & Fischer, T. (2003) Response to imatinib in patients who relapse after allogeneic stem cell transplantation for chronic myeloid leukemia. *Leukemia,* 17, 1707-1712.

Ottmann, O.G. & Pfeifer, H. (2009) Management of Philadelphia chromosome-positive acute lymphoblastic leukemia (Ph+ ALL). *Hematology Am Soc Hematol Educ Program,* 371-381.

Piccaluga, P.P., Martinelli, G., Isidori, A., Malagola, M., Rondoni, M., Paolini, S., Amabile, M., Iacobucci, I., Baccarani, M. & Visani, G. (2008) Long-term molecular complete remission with IFN-alpha in Ph+ adult acute lymphoid leukemia patients. *Leukemia,* 22, 1617-1618.

Piccaluga, P.P., Martinelli, G., Rondoni, M., Visani, G. & Baccarani, M. (2006) Advances and potential treatment for Philadelphia chromosome-positive adult acute lymphoid leukaemia. *Expert Opin Biol Ther,* 6, 1011-1022.

Piccaluga, P.P., Paolini, S. & Martinelli, G. (2007) Tyrosine kinase inhibitors for the treatment of Philadelphia chromosome-positive adult acute lymphoblastic leukemia. *Cancer,* 110, 1178-1186.

Popat, U., Carrum, G. & Heslop, H.E. (2003) Haemopoietic stem cell transplantation for acute lymphoblastic leukaemia. *Cancer Treat Rev,* 29, 3-10.

Porter, D.L. & Antin, J.H. (1999) The graft-versus-leukemia effects of allogeneic cell therapy. *Annu Rev Med,* 50, 369-386.

Pui, C.H. & Evans, W.E. (1998) Acute lymphoblastic leukemia. *N Engl J Med,* 339, 605-615.

Ribera, J.M., Ortega, J.J., Oriol, A., Bastida, P., Calvo, C., Perez-Hurtado, J.M., Gonzalez-Valentin, M.E., Martin-Reina, V., Molines, A., Ortega-Rivas, F., Moreno, M.J., Rivas, C., Egurbide, I., Heras, I., Poderos, C., Martinez-Revuelta, E., Guinea, J.M., del Potro, E. & Deben, G. (2007) Comparison of intensive chemotherapy, allogeneic, or autologous stem-cell transplantation as postremission treatment for children with very high risk acute lymphoblastic leukemia: PETHEMA ALL-93 Trial. *J Clin Oncol,* 25, 16-24.

Ribera, J.M., Ortega, J.J., Oriol, A., Fontanillas, M., Hernandez-Rivas, J.M., Brunet, S., Garcia-Conde, J., Maldonado, J., Zuazu, J., Gardella, S., Besalduch, J., Leon, P., Macia, J., Domingo-Albos, A., Feliu, E. & San Miguel, J.F. (1998) Late intensification chemotherapy has not improved the results of intensive chemotherapy in adult acute lymphoblastic leukemia. Results of a prospective multicenter randomized trial (PETHEMA ALL-89). Spanish Society of Hematology. *Haematologica,* 83, 222-230.

Rousseau, R.F., Bollard, C.M. & Heslop, H.E. (2001) Gene therapy for paediatric leukaemia. *Expert Opin Biol Ther,* 1, 663-674.

Rowe, J.M., Buck, G., Burnett, A.K., Chopra, R., Wiernik, P.H., Richards, S.M., Lazarus, H.M., Franklin, I.M., Litzow, M.R., Ciobanu, N., Prentice, H.G., Durrant, J., Tallman, M.S. & Goldstone, A.H. (2005) Induction therapy for adults with acute lymphoblastic leukemia: results of more than 1500 patients from the international ALL trial: MRC UKALL XII/ECOG E2993. *Blood*, 106, 3760-3767.

Rowe, J.M.B., G. Burnett, A.K. (2001) Induction therapy for adults with acute lymphoblastic leukemia (ALL): Results of nearly 1,400 patients from the International ALL Trial (MRC UKALL XII/ECOG E2993). *Blood*, 102, 224a.

Sebban, C., Lepage, E., Vernant, J.P., Gluckman, E., Attal, M., Reiffers, J., Sutton, L., Racadot, E., Michallet, M., Maraninchi, D. & et al. (1994) Allogeneic bone marrow transplantation in adult acute lymphoblastic leukemia in first complete remission: a comparative study. French Group of Therapy of Adult Acute Lymphoblastic Leukemia. *J Clin Oncol*, 12, 2580-2587.

Slavin, S., Nagler, A., Naparstek, E., Kapelushnik, Y., Aker, M., Cividalli, G., Varadi, G., Kirschbaum, M., Ackerstein, A., Samuel, S., Amar, A., Brautbar, C., Ben-Tal, O., Eldor, A. & Or, R. (1998) Nonmyeloablative stem cell transplantation and cell therapy as an alternative to conventional bone marrow transplantation with lethal cytoreduction for the treatment of malignant and nonmalignant hematologic diseases. *Blood*, 91, 756-763.

Snyder, D.S., Nademanee, A.P., O'Donnell, M.R., Parker, P.M., Stein, A.S., Margolin, K., Somlo, G., Molina, A., Spielberger, R., Kashyap, A., Fung, H., Slovak, M.L., Dagis, A., Negrin, R.S., Amylon, M.D., Blume, K.G. & Forman, S.J. (1999) Long-term follow-up of 23 patients with Philadelphia chromosome-positive acute lymphoblastic leukemia treated with allogeneic bone marrow transplant in first complete remission. *Leukemia*, 13, 2053-2058.

Storb, R., Yu, C., Deeg, H.J., Georges, G., Kiem, H.P., McSweeney, P.A., Nash, R.A., Sandmaier, B.M., Sullivan, K.M., Wagner, J.L. & Walters, M.C. (1998) Current and future preparative regimens for bone marrow transplantation in thalassemia. *Ann N Y Acad Sci*, 850, 276-287.

Thiebaut, A., Vernant, J.P., Degos, L., Huguet, F.R., Reiffers, J., Sebban, C., Lepage, E., Thomas, X. & Fiere, D. (2000) Adult acute lymphocytic leukemia study testing chemotherapy and autologous and allogeneic transplantation. A follow-up report of the French protocol LALA 87. *Hematol Oncol Clin North Am*, 14, 1353-1366, x.

Thomas, D.A., Kantarjian, H., Smith, T.L., Koller, C., Cortes, J., O'Brien, S., Giles, F.J., Gajewski, J., Pierce, S. & Keating, M.J. (1999) Primary refractory and relapsed adult acute lymphoblastic leukemia: characteristics, treatment results, and prognosis with salvage therapy. *Cancer*, 86, 1216-1230.

Thomas, E.D., Buckner, C.D., Banaji, M., Clift, R.A., Fefer, A., Flournoy, N., Goodell, B.W., Hickman, R.O., Lerner, K.G., Neiman, P.E., Sale, G.E., Sanders, J.E., Singer, J., Stevens, M., Storb, R. & Weiden, P.L. (1977) One hundred patients with acute leukemia treated by chemotherapy, total body irradiation, and allogeneic marrow transplantation. *Blood*, 49, 511-533.

Thomas, X., Boiron, J.M., Huguet, F., Dombret, H., Bradstock, K., Vey, N., Kovacsovics, T., Delannoy, A., Fegueux, N., Fenaux, P., Stamatoullas, A., Vernant, J.P., Tournilhac, O., Buzyn, A., Reman, O., Charrin, C., Boucheix, C., Gabert, J., Lheritier, V. & Fiere,

D. (2004) Outcome of treatment in adults with acute lymphoblastic leukemia: analysis of the LALA-94 trial. *J Clin Oncol,* 22, 4075-4086.

Uzunel, M., Mattsson, J., Jaksch, M., Remberger, M. & Ringden, O. (2001) The significance of graft-versus-host disease and pretransplantation minimal residual disease status to outcome after allogeneic stem cell transplantation in patients with acute lymphoblastic leukemia. *Blood,* 98, 1982-1984.

Vernant, J.P., Marit, G., Maraninchi, D., Guyotat, D., Kuentz, M., Cordonnier, C. & Reiffers, J. (1988) Allogeneic bone marrow transplantation in adults with acute lymphoblastic leukemia in first complete remission. *J Clin Oncol,* 6, 227-231.

Vey, N., Blaise, D., Stoppa, A.M., Bouabdallah, R., Lafage, M., Sainty, D., Cowen, D., Viens, P., Lepeu, G., Blanc, A.P. & et al. (1994) Bone marrow transplantation in 63 adult patients with acute lymphoblastic leukemia in first complete remission. *Bone Marrow Transplant,* 14, 383-388.

Visani, G., Martinelli, G., Piccaluga, P., Tosi, P., Amabile, M., Pastano, R., Cavo, M., Isidori, A. & Tura, S. (2000) Alpha-interferon improves survival and remission duration in P-190BCR-ABL positive adult acute lymphoblastic leukemia. *Leukemia,* 14, 22-27.

Wassmann, B., Pfeifer, H., Goekbuget, N., Beelen, D.W., Beck, J., Stelljes, M., Bornhauser, M., Reichle, A., Perz, J., Haas, R., Ganser, A., Schmid, M., Kanz, L., Lenz, G., Kaufmann, M., Binckebanck, A., Bruck, P., Reutzel, R., Gschaidmeier, H., Schwartz, S., Hoelzer, D. & Ottmann, O.G. (2006) Alternating versus concurrent schedules of Imatinib and chemotherapy as front-line therapy for Philadelphia-positive acute lymphoblastic leukemia (Ph+ALL). *Blood.*

Wassmann, B., Pfeifer, H., Scheuring, U., Klein, S.A., Gokbuget, N., Binckebanck, A., Martin, H., Gschaidmeier, H., Hoelzer, D. & Ottmann, O.G. (2002) Therapy with imatinib mesylate (Glivec) preceding allogeneic stem cell transplantation (SCT) in relapsed or refractory Philadelphia-positive acute lymphoblastic leukemia (Ph+ALL). *Leukemia,* 16, 2358-2365.

Wassmann, B., Pfeifer, H., Stadler, M., Bornhauser, M., Bug, G., Scheuring, U.J., Bruck, P., Stelljes, M., Schwerdtfeger, R., Basara, N., Perz, J., Bunjes, D., Ledderose, G., Mahlberg, R., Binckebanck, A., Gschaidmeier, H., Hoelzer, D. & Ottmann, O.G. (2005) Early molecular response to posttransplantation imatinib determines outcome in MRD+ Philadelphia-positive acute lymphoblastic leukemia (Ph+ ALL). *Blood,* 106, 458-463.

Zhang, M.J., Hoelzer, D., Horowitz, M.M., Gale, R.P., Messerer, D., Klein, J.P., Loffler, H., Sobocinski, K.A., Thiel, E. & Weisdorf, D.J. (1995) Long-term follow-up of adults with acute lymphoblastic leukemia in first remission treated with chemotherapy or bone marrow transplantation. The Acute Lymphoblastic Leukemia Working Committee. *Ann Intern Med,* 123, 428-431.

14

Treatment Options in Myelodysplastic Syndromes

Klara Gadó and Gyula Domján
Semmelweis University, Faculty of Medicine,
1st Department of Internal Medicine
Hungary

1. Introduction

Myelodysplastic syndromes are a heterogeneous group of clonal bone marrow disorders, which are considered to be cancers and have a tendency to transform into acute myeloid leukemia. MDS are characterised by the underproduction of normal blood cells. Disease-induced cytopenias result infections and bleeding complications.
The diverse pathobiology of the disease is manifested by a varied clinical course, with some patients having more indolent disease and longer life expectancy, and others presenting with aggressive variants that rapidly progress to AML.
The frequency of the disease is the highest in the elderly. Number of diagnoses has increased in the past decade as a consequence of the increased recognition, and the aging of the population.
Prognosis is influenced by the age and co-morbidities of the patient and also by the cytogenetic abnormalities.
Risk stratification is based on the number of myeloblast in the bone marrow, the number of cytopenias and the cytogenetic abnormalities. The most widely used prognostic systems are the International Prognostic Scoring System (IPSS) and the WHO classification-based Prognostic Scoring System (WPSS). The latter system incorporates transfusion burden as well (Greenberg, 1998).
From the standpoint of both disease biology and prognosis, it is important to distinguish primary MDS from therapy-related MDS; the latter is closely related to therapy-related AML and develops in the setting of prior exposure to chemotherapy (eg, alkylating agents, topoisomerase II inhibitors), radiotherapy, radiation accidents, benzene, or other toxins and is prognostically worse than primary MDS. Response to therapy is highly worse in this group of patients (Tefferi, 2010; Graubert, 2010).
Treatment of MDS in the past was restricted to supplementation of the missing cell type thus relieving the patient's symptoms. Advances over the last decade have given a multitude of treatment options. Therapeutic options now exist that not only reduce disease-related symptoms, improve quality of life, but alter the natural history of the disease as well.

2. Epidemiology

The incidence of newly diagnosed MDS exceeds 10.000 cases in the US annually. MDS are the most common in older patients. At the time of diagnosis, 86 % of patients are 60 years or

older. The incidence of MDS rises from 3,4 cases per 100.000 in the general population to 15 to 50 cases per 100.000 in people older than 75 years (Giralt et al., 2011).
The incidence and prevalence of MDS has increased significantly over the past 20 years as a result of increasing longevity of the population and increasing physician awareness (Goldberg et al., 2010).
Among patients diagnosed with MDS, 14% have been treated for other primary tumors prior to diagnosis (Ma et al., 2007; Sekeres, 2011a).

3. Survival

Patient outcomes vary depending on disease presentation. Among untreated patients with MDS, median survival varies from 5,7 years in low-risk patients to 0,4 years in patients with high risk disease. For patients with del(5q) as the only chromosomal aberration, medial survival is 73 months compared to 19,3 months for patients with more than one aberration. Transfusion need at diagnosis is the most independent parameter for survival, with transfusion-dependent patients having a median survival of 39 months versus 97 months for transfusion-independent patients.

4. Staging system, risk stratification

The IPSS was the first system to stratify patients according to their risk of death or evolution to AML. It is based on percentage of bone marrow blasts, cytogenetic abnormalities, and number of cytopenias. However, the IPSS does not consider the severity of anaemia and

Factors for determining IPSS score					
	Score value				
Prognostic variable	0	0.5	1.0	1.5	2.0
Blasts in marrow (%)	< 5	5-10	-	11-20	21-30
Karyotype	good	Intermediate	Poor	-	-
Cytopenias	0/1	2/3	-	-	-
Good = normal, -Y, del(5q), del(20q); poor = complex (3 abnormalities) or chromosome 7 anomalies; intermediate = other abnormalities.					

Risk of transformation and median survival			
Risk	**Total score**	**Progression to AML Without therapy (year)**	**Median survival Without therapy (years)**
Low	0	9.4	5.7
Int-1	0.5-1.0	3.3	3.5
Int-2	1.5-2.0	1.1	1.2
High	≥ 2.5	0.2	0.4

Int-1 = intermediate-1, Int-2 = intermediate-2

Table 1. The International Prognostic Score System (IPSS) for MDS

transfusion requirement, which are definitely associated with reduced survival in MDS. The newer WPSS has five prognostic groups with distinct survival times and probabilities of progression to AML. While the IPSS is limited to newly diagnosed patients with MDS, the WPSS can be applied at any time during the disease course. It has been shown that the WPSS improves survival prediction and identifies patients with very low-risk disease who may achieve long-term survival.

Variable	0	1	2	3	
WHO category	RA, RARS, 5q-	RCMD, RCMD-RS	RAEB-1	RAEB-2	
Karyotype	Good	Intermediate	Poor	–	
Transfusion requirement	No	Regular	–	–	

Abbreviations: RA = refractory anaemia; RAEB-1 = refractory anaemia with excess blasts type 1; RAEB-2 = refractory anaemia with excess blasts type 2; RARS = refractory anaemia with ringed sideroblasts; RCMD = refractory cytopenia with multilineage dysplasia; RCMD-RS = refractory cytopenia with multilineage dyslasia and ringed sideroblasts;

Table 2. The WHO Classification-Based Prognostic Score System (WPSS) for MDS

5. New prognostic factors

Conventional metaphase cytogenetic analysis is limited by the sensitivity of the assay, and therefore new technologies are being applied in MDS to detect genomic alterations. One of these approaches has been the use of single-nucleotide polymorphism arrays (SNP-A). Using SNP-A, several candidate genes have been described recently, and include c-CBL (encodes an E3 ubiquitin ligase that inhibits tyrosine kinase signalling) (Sanada et al., 2008), ASXL1, and TET-2 (which are significant because they may have a role in the control of epigenetic alterations in MDS.) (Boultwood et al., 2010; Jankowska et al., 2009; Garcia-Manero, 2010).

6. Treatment goals and options

The goals of treatment for patients with MDS are to prolong overall survival, reduce transfusion burden, and improve their quality of life (Scott, 2008).

Treatment options for individual patients with MDS depend on the disease classification, prognostic stage and the age and health status of the patient.

For patients with lower risk disease (IPSS: low- or intermediate-1-risk) options include lenalidomide, erythropoietin-stimulating agents and immunosuppressive therapy, as well as hypomethylating agents (azacitidine and decitabine). For patients with del(5q) syndrome lenalidomide is the treatment of choice. For patients with higher risk (IPSS: intermediate-2-, or high-risk), intensive therapy with allogeneic stem cell transplantation (alloSCT) remains the only potentially curative treatment option (Sekeres, 2009; Mufti & Chen, 2008).

For the majority of patients who are not candidates for alloSCT, treatment options include azacitidine or decitabine, clinical trials, or supportive care.

In addition, all patients receiving active therapy should receive continuous supportive care.

7. Considerations regarding to select the appropriate treatment options

Treatment consideration must take into account many factors, including the pathologic diagnosis, the prognosis based on the IPSS or WPSS, the unique disease features in that particular patient, feasibility of performing a clinical trial, the appropriateness of a bone marrow transplantation, and the philosophy of the patient and the family concerning his or her care (Malcovati & Nimer, 2008; Barzi & Sekeres, 2010; Greenberg, 2010).

In addition, if the patient has a secondary MDS, tolerability of therapy is probably worse because of previous exposure to DNA-damaging agents.

Older age per se has a negative impact on survival of MDS patients, in particular of those with low-risk disease. However, age affects the survival of high-risk patients indirectly as well, by limiting their eligibility to intensive treatments. In addition, aging is associated with an increasingly high risk of developing co-morbidity.

Co-morbidities have a significant impact on the outcome of patients with MDS. Risk stratification should also have to include the assessment of co-morbidities of the MDS patients.

One of the most important areas of research is the identification of biomarkers of response to therapy (Garcia-Manero, 2010). A critical example is the discovery by List *et al.* of the relationship between the presence of a chromosome 5 alterations and response to lenalidomide (List et al., 2006), which resulted in the development of targeted approaches to these patients, as well as an entirely new field of research with the discovery of RPS14 as a critical gene in 5q31 (Ebert et al, 2008). Predictors of responsiveness to immunosuppressive therapy are younger age, HLA-DR15 type, the interval between first transfusion and starting of immunosuppressive treatment. (Sloand et al., 2008; Barrett et al., 2006).

8. Patient selection

The selection of an adequate treatment option has to be preceded by thorough consideration, that have to take account of several points of view. The proper patient selection is the key element of successful treatment.

There is a significant role of the patient's intent beyond the severity of the illness in the embracement of the most appropriate treatment option in case of a given patient. Patient has to be committed to the treatment as he or she takes the risk of the adverse effects of the treatment, and also he or she undertakes the inconveniences of a long-term treatment process.

9. Treatment modalities in MDS

Taking into account the above mentioned aspects, there are several possibilities to choose: observation, supportive care, iron chelation, erythropoiesis-stimulating agents (EPOs), immunosuppressive treatment, intensive chemotherapy, and also disease-altering options: hypomethylating agents, lenalidomide, alloSCT, as well as entrance in a clinical trial (Barzi & Sekeres, 2010).

9.1 Observation

Many patients, especially those who are older and frail, benefit from a period of observation before any discussion about the need for therapy is made. Treatment should be reserved only if there are symptoms resulting from anaemia or other cytopenias or perhaps presymptomatic anaemia or severe thrombocytopenia.

9.2 Supportive therapy

Supportive care includes transfusion of red blood cells and platelets to minimize complication of cytopenias and to improve quality of life, as well as antibiotics to treat infections.

Patients with symptomatic anaemia should receive transfusion to relieve their symptoms. Platelet transfusion must be given in case of severe thrombocytopenia which cause bleeding.

9.3 Erythropoiesis-stimulating agents

Once an MDS patient become transfusion-dependent, erythropoietin-stimulating agents (ESAs) can be considered. These include epoietin (Eprex) and darbopoietin alfa (Aranesp). The clinical practice guideline of the American Society of Haematology suggests to use epoietin and darbepoietin in adults with lower risk MDS without chemotherapy. Erythropoiesis stimulating agents are used by more than 50% of patients (Sekeres, 2011a,).

Recently ESAs are increasingly used to treat anaemia of lower risk MDS, even before RBC transfusion requirement. Early introduction of ESA in lower risk MDS, may help to better avoid the consequences of anaemia and also to delay the need for RBC transfusions, hypothetically by slowing the disease course (Park et al., 2010).

EPOs are recommended particularly for those with low serum EPO level. Serum erythropoietin (EPO) level < 500 mU/ml may respond to EPO if relatively high doses are administered. The EPO dose required is 40.000-60.000 units 1-3 times a week subcutaneously. Erythroid response occur within 6 to 8 weeks of treatment (Hellström-Lindberg E, 2003; Jädersten et al., 2008).

In a phase 3 prospective randomized trial the efficacy and long-term safety of EPO with or without granulocyte colony-stimulating factor plus supportive care was evaluated versus supportive care alone for the treatment of anaemic patients with lower-risk MDS (Greenberg et al., 2009). In comparison with supportive care alone, patients receiving EPO with or without granulocyte colony-stimulating factor plus supportive care had improved erythroid responses, similar survival, and incidence of acute myeloid leukemia transformation. Responding patients had significantly lower serum EPO levels (45% vs 5% responses for levels < 200 mU/mL vs ≥ 200 mU/mL). (Rizzo et al., 2010).

9.4 Iron chelation therapy

Many MDS patients are dependent on red blood cell (RBC) transfusions for symptomatic management of refractory anaemia. Iron overload ensues when the iron acquired from transfused RBCs exceeds body storage capacity, thereby raising the risk for end organ damage (List, 2010).

The use of iron chelation in myelodysplastic syndrome (MDS) has generated much controversy recently (Steensma, 2009a). Transfusion dependency is associated with shortened overall survival and leukemia-free survival in MDS. The major question is whether this effect is mediated by transfusional iron overload itself or if need for red cell transfusion is simply a marker of disease severity (Malcovati et al., 2005).

Averting cardiac dysfunction in low-grade MDS patients who have sufficient longevity to experience deleterious cardiac effects of iron overload has been the major argument in favor of iron chelation. However, the role of iron chelation therapy in MDS remains controversial. Although there is significant evidence showing the adverse impact of transfusion dependency on survival in MDS, direct evidence linking tissue iron overload to poor survival or in particular to cardiac dysfunction is lacking.

Present evidence suggesting that the major benefit of iron chelation in MDS is not likely to come from reduction in cardiac and other end organ damage due to tissue iron overload, but from a potential favorable impact on 3 other outcomes namely: lowering infection risk, improving the outcome of allogeneic hematopoietic stem cell transplantation, and delaying leukemic transformation. These outcomes have particular relevance for patients with higher grades of MDS (Pullarkat, 2009).

Application of recent advances in the treatment of MDS can reduce or eliminate the need for transfusions, thus minimizing the risk of iron overload.

Iron chelation strategies include oral agents such as deferasirox (Exjade, Novartis), deferiprone (Ferriprox, Apotex Europe BV,) and parenteral administration of deferoxamine (Desferal, Novartis).

Considering the short survival of patients with MDS, the benefit of iron chelation is debatable. This procedure is advised for those who have received more than 25 units of packed red blood cells.

Unfortunately, iron chelation is expensive and, to date, has not shown a clear survival benefit in the MDS population.

9.5 Hypomethylating agents

MDS pathophysiology is complex, and still not completely understood. Structural alterations in DNA play a role in the pathogenesis of disease. Epigenetic changes in the form of modifications to the transcriptional capacity of the cell via processes, such as DNA methylation can also alter gene expression impacting disease biology. As such, hypermethylation of the promoters of certain tumor suppression genes is prevalent in MDS and secondary AML, and it is postulated that the DNA hypomethylation may result in the reactivation of silenced genes, restoring their cancer-suppressing functions, and inducing cellular differentiation providing a biologically rational therapeutic target for MDS.

Currently, both FDA- (U.S. Food and Drug Administration) approved hypomethylating agents (azacitidine and decitabine) have shown good clinical responses in patients with MDS. In the last two decades the use of epigenetic therapy has gained popularity, given its favorable side effect profile and its potential to improve survival.

9.5.1 Azacitidine (5-azacytidine; Vidaza)

Azacitidine (5-azacytidine; Vidaza; Pharmion Corporation) for injectable suspension received regular approval by the FDA for the treatment of all subtypes of MDS in 2004. It was the first agent that has been reported to prolong survival in MDS patients.

9.5.1.1 Mechanism of action

Azacitidine is a nucleosid analog that incorporates into RNA and requires the activity of ribonucleotide reductase to be incorporated to DNA. The incorporation of azacitidine into DNA leads to inhibition of DNA methyltransferase (DNMT) which has been known to hypermethylate cytosine residues at cytosine-guanine repeat sequences. By reducing the quantity of hypermethylated DNA produced, azacitidine can serve to limit the proliferation of malignant cells, while allowing an increased proportion of normal cells to differentiate. Although azacitidine falls into the category of hypomethylating agents, its mechanism of action is likely multifactorial (Yang et al, 2010).

9.5.1.2 Identification patients for azacitidine treatment

Azacitidine was approved by the US FDA on the basis of CALGB (Cancer and Leukemia Group B) 9221. This randomized trial compared azacitidine with best supportive care (BSC) targeting high-intermediate and high-risk disease and including low-intermediate risk patients with ongoing complications of cytopenias. The results demonstrated superiority of azacitidine over BSC in terms of quality of life, reduced transfusion needs, and delayed time to AML transformation or death for higher risk patients (Kurtin & Demakos. 2010,).

It was the AZA-001 trial (the second randomized trial, which was conducted to define whether an endpoint not of response but overall survival (OS) could be met with hypomethylation therapy) which revealed that azacitidine significantly improved OS compared to that of the conventionally treated patients (either BSC, low-dose cytosine-arabinoside, or intensive induction chemotherapy). There was no particular „winner" regarding the different subsets of patients treated. Response benefit was distributed fairly evenly across various subgroups of patients. This was true across younger and older patients, gender, performance status, FAB subtype, WPSS, IPSS score, cytogenetic risk category, and bone marrow blasts rate (Edlin et al., 2010).

The rate of complete response (CR) was rather low in the AZA-001 trials (17 % in the azacitidine arm). The low CR rate provides compelling data against oncology dogma that achievement of CR is required for survival benefit. Indeed, a retrospective subset analysis examined the survival impact of azacitidine excluding patients who achieved CR. One-year survival rates were superior for azacitidine treatment versus conventional treatment (68% vs 56 %, P= 0,015), thus achieving a CR is no longer a sufficient predictor for a therapy's ability to extend survival and alter course of MDS.

Critical to the success of azacitidine is the selection of appropriate patients. Higher risk MDS patients who are of reasonable performance status with adequate organ function are excellent candidates and also those lower risk patients who are transfusion dependent or have severe cytopenias.

9.5.1.3 Side effects

Adverse events are inevitably encountered. Haematological toxicities are common, particularly in the first one to two cycles. Infections have occured in 50 % of azacitidine-treated patients in AZA-001 trial. 58 % of patients experienced thrombocytopenia and/or neutropenia.

Common other adverse effects are gastrointestinal (vomiting, diarrhoea, anorexia), fevers, rigors, arthralgia, headache, and dizziness (San Miguel Amigo, 2011).

9.5.1.4 Dosage and administration

For azacitidine the most appropriate dosing schedule is 75 mg/m^2 subcutaneously for 7 days, every 4 weeks. It is also an important issue to keep on treatment for at least at six cycles. In case of the absence of unacceptable toxicity or the evidence of disease progression, therapy should be continued for the maintenance of the response.

9.5.1.5 Azacitidine for maintenance of complete remission after chemotherapy

A Phase II prospective study was performed to assess the feasibility and efficacy of maintenance azacitidine for older patients with high-risk MDS in CR after induction chemotherapy. Sixty patients were enrolled and treated by standard induction chemotherapy. Patients that reached CR started maintenance therapy with subcutaneous azacitidine, 5/28 days until relapse. Median overall survival was 20 months. A dose of 60 mg/m(2) was well tolerated. Grade III-IV thrombocytopenia and neutropenia occurred after 9.5 and 30% of the cycles, respectively, while haemoglobin levels increased during treatment. azacitidine treatment is safe, feasible and may be of benefit in a subset of patients (Grövdal et al., 2010).

9.5.2 Decitabine (dezocitidine, Dacogen, SuperGen, Inc., Dublin)

Decitabine is another hypomethylating agent approved by the US FDA in 2006 for the treatment of *de novo* and secondary MDS (for intermediate-1-, intermediate-2-, and high-risk MDS patients). It has been studied predominantly in higher-risk MDS patients.

The FDA approval was based on the randomized phase III trial (EORTC 06011) versus BSC using the European administration schedule. Decitabine-treated patients had improved quality of life, and reduced transfusion needs. Although there was a delayed time to AML transformation or death, this trial failed to shown any survival benefit for decitabine arm. The relatively short duration of therapy administered was the main limitation of the study design which may negatively affected outcomes (Garcia et al., 2010,).

Concerning studies with hypomethylating agents the best available evidence and consensus is that patients benefiting from therapy should continue treatment until progression or unacceptable toxicity.

The toxicity profile of the two AZA nucleotides is similar.

While the haematological response rates to both hypomethylating agents are similar, the survival advantage demonstrated by the FDA-approved dose schedule of azacitidine compared with conventional care, as well as the absence of such a survival advantage in response to the FDA-approved dose schedule of decitabine in two randomized studies, makes azacitidine the drug of choice for AZA nucleoside-naive, high-risk MDS patients. Perhaps the better-tolerated North American decitabine schedule would lead to similar survival advantage.

A comparative trial of azacitidine versus decitabine recently opened to accrual (http:// www.ClinicalTrials.gov; identifier: CT01011283). Unfortunately, the trial is designed with a primary endpoint of early response at 6 months and is not powered to examine the more relevant question of OS. (Blum, 2010)

9.5.2.1 Dosage and administration of decitabine

European schedule is 15 mg/m^2 intravenously over 3 hours, every 8 hours for 3 days, repeated every 6 weeks Kantarjian et al., 2006). In North America, the typical schedule is 20 mg/m^2 intravenously over 1 hour daily for 5 days, repeated every 4 weeks, based on promising high CR results of up to 39% in a single center (Kantarjian et al., 2007). A multicenter study of 99 patients with MDS using the same regimen, The Alternative Dosing for Outpatient Treatment (ADOPT) trial, showed a CR rate of 17%, with an overall response rate of 51% (Steensma et al., 2009b). The main point of administration is the number of cycles. The better OS results of azacitidine may be the consequence of the higher number of cycles. In the EORTC 06011 study the median number of cycles per patient was four, and 40% received two or fewer cycles. By comparison, the median number of cycles of AZA given in AZA-001 was nine.

9.6 Lenalidomide

Lenalidomide is an immunomodulatory drug that is FDA approved for lower risk patients with transfusion-dependent anaemia and an interstitial deletion of the long arm of chromosome 5 [del(5q)]. The approval was based on high rates of prolonged transfusion independence and complete cytogenetic response (Ximeri et al., 2010).

Treatment with a 10 mg daily dose for 21 days in every 4 weeks results transfusion independence in 67 % of patients with del(5q). Lenalidomide treatment does not increase risk of AML progression according to the preliminary result of MDS-004 study. It was a significant concern in Europe, preventing approval of the drug. Despite the frequent need of dose reduction due to myelosuppression, the starting dose of lenalidomide is suggested to be 10 mg (Sekeres et al., 2008).

Though lenalidomide is approved by FDA for del(5q) lower risk subset of MDS patients, it is reasonable to consider lenalidomide as frontline treatment in the few higher risk patients with MDS who have both isolated del(5q) and platelet count > 100.000/μL (Kurtin & Demakos, 2010).

In case of patients with lower risk and without del(5q) the ratio of patients reached transfusion independence was much lower and response duration was much shorter than that of patients with del(5q). Cytogenetic improvement was also inferior to that observed in the del(5q) patients. (Komrokji & List, 2010;, List et al., 2006).

9.6.1 Mechanism of action

The drug may have different mechanism of activity, depending on the disease type.

In del(5q), lenalidomide suppresses the malignant clone, but in non-del(5q) it appears to promote erythropoiesis.

There is now a better understanding of the mechanism of the karyotype-dependent drug action. In del(5q) patients, lenalidomide suppresses the clone by inhibiting the nuclear sequestration of the haplodeficient cell cycle regulatory protein cdc25c, thereby promoting selective G2 arrest and apoptosis (Wei et al., 2009). In non-del(5q) patients, lenalidomide enhances erythropoietin receptor signaling.

9.6.2 Combination treatment

The addition of lenalidomide to azacitidine may provide additional clinical benefit over e monotherapy.

A recent Phase I trial testing the lenalidomide and azacitidine combination yielded encouraging results. In this study, 18 higher-risk MDS patients were treated with the combination for seven cycles, after which lenalidomide was discontinued in eight patients who achieved a complete response, with azacitidine monotherapy continuing until disease progression. Three patients who relapsed on monotherapy with excess blasts at 12, 19, and 24 months, in whom lenalidomide was then resumed in combination with azacitidine. Each patient, one with normal cytogenetics at relapse; one with a 18 abnormality; and one with del(4q25), recaptured a complete response that was sustained for 5, 7, and 7+ months.

The addition of lenalidomide to azacitidine provides additional clinical benefit over azacitidine monotherapy. (Sekeres et al., 2011b)

9.7 Immunosuppressive treatment

An activated immune system has been observed in patients with myelodysplastic syndrome but its exact contribution to disease development and control is not fully clarified. On the one hand an activated and skewed T-cell repertoire has been reported, but on the other hand, decreased natural killer cell function has been found. Immune activation could reflect undesired autoimmune reactions against normal hematopoietic precursor cells as well as effective immune-surveillance against dysplastic clones (Chamuleau et al., 2009).

Autoreactive T-cell clones have a role in the apoptosis in the bone marrow of MDS, mainly in patients with hypocellular bone marrow. Immunosuppressive agents such as cyclosporine A (CsA)and anti-thymocyte globulin (ATG), or combination of the two agents can improve cytopenia and can reduce transfusion need in a subset of MDS patients. The anti-TNF agent etanercept can also be applied in combination with ATG. (Deeg et al., 2004).

Responses to immunosuppressive therapy in MDS remain among the most durable (and perplexing) of all available therapies in the disease. Typically, equine ATG is given at 40 mg/kg intravenously for four consecutive days, in conjunction with methylprednisone and CsA. Initial reports of its use demonstrated achievement of transfusion independence in 21 of 61 (34%) of patients. More importantly, the probability of continued transfusion independence after 5 years was 76%. However, the toxicity of the regimen has led to considerable apprehension in its use.

Most ATG studies in MDS have been single center experiences, and several reports in unselected patients have demonstrated lower response rates and significant toxicity with the regimen. Clearly, judicious patient selection for the therapy is critical for its success and productive implementation. Helpful guidance for the use of ATG in MDS was recently published. Outcomes for patients with MDS who were given equine ATG (with or without CsA) in sequential protocols at the National Heart, Lung, and Blood Institute between 1971 and 2003 were described. For 129 ATG-treated patients, 39 patients (30%) responded; 9% had CR. Serious infusion-related toxicities were infrequent, but meaningful because 9% of patients required temporary intensive care unit support. Responses included 18 of 74 (24%) treated with ATG alone, 20 of 42 (48%) treated with ATG_CsA, and 1 of 13 (8%) treated with CsA. Median response duration was 3 years (3 months to 10 years), and median survival in the cohort was 10,5 years. Notably, factors affecting response were younger age (< 60), HLA-DR15 positivity, and use of combination ATG+CsA. There was no association of response with pretreatment marrow cellularity, paroxysmal nocturnal hemoglobinuria clone, or absolute neutrophil count. Furthermore, a multivariate analysis of the IST-treated group compared with results from a control group of 816 patients with MDS from the International Myelodysplasia Risk Analysis Workshop database showed that survival was improved in younger patients, those treated with IST, and those with low/intermediate IPSS risk scores.

The role of IST in MDS treatment remains enigmatic, and the serious toxicities that can be encountered with its infusion and long-term immunosuppressive effects have properly led to hesitation to its use in community and academic practice. However, in a manner similar to the rigorous selection of patients for transplantation, appropriate selection of patients for IST affords some the opportunity for prolonged responses without the requirement for repeated maintenance chemotherapy (such as AZA or decitabine). IST should be considered instead of a hypomethylating agent for previously untreated, younger patients with MDS with low or Int-1 risk disease with HLADR15.

Predictors of responsiveness are younger age, HLA-DR15 type, the interval between first transfusion and starting of immunosuppressive treatment. (Sloand et al., 2008; Barrett et al., 2006).

Passweg *et al.* reported open-label randomized phase III trial on patients with MDS randomly assigned to 15 mg/kg of horse ATG for 5 days and oral CSA for 180 days (ATG+CSA) (45 patients) or best supportive care (43 patients), stratified by treatment center and International Prognostic Scoring System (IPSS) risk score. Primary end point was best hematologic response at 6 months. This trial demonstrated that ATG+CSA treatment seems to be associated with hematologic response in a subset of patients without apparent impact on transformation free survival and OS (Passweg et al., 2011).

Because of the toxicities associated with ATG/CsA, an alternative regimen with alemtuzumab in MDS was investigated (Sloand et al., 2010). It was a nonrandomized, off-label, pilot, phase I/II study of alemtuzumab monotherapy in patients with MDS who were judged likely to respond to immunosuppressive treatment based on the following criteria: HLA-DR15-negative patients whose age plus the number of months of RBC transfusion dependence (RCTD) was less than 58; and HLA-DR15-positive patients whose age plus RCTD was less than 72. Thirty-two patients were enrolled they received alemtuzumab 10 mg/d intravenously for 10 days. Primary end points were hematologic responses at 3, 6, and 12 months after alemtuzumab. Seventeen (77%) of 22 evaluable intermediate-1 patients and four (57%) of seven evaluable intermediate-2 patients responded to treatment with a median time to response of 3 months. Four of seven evaluable responders with cytogenetic abnormalities before treatment had normal cytogenetics by 1 year after treatment. Five (56%) of nine responding patients evaluable at 12 months had normal blood counts, and seven (78%) of nine patients were transfusion independent. Alemtuzumab seems to be safe and active in MDS and may be an attractive alternative to ATG in selected patients likely to respond to IST.

9.8 Intensive chemotherapy

About 25 % of patients with newly diagnosed MDS and 15-20 % of patients with established MDS have higher risk disease. These patients should be treated immediately, given the high likelihood of transformation to AML or death within 1.5 years. One of the treatment options is intensive chemotherapy (Bello et al., 2010; Gergis & Wissa, 2010).

Intensive chemotherapy means some kind of acute leukemia protocol. One of them is the FLAG-Ida regimen. The combination of fludarabine, high dose cytarabine and granulocyte colony stimulating factor (FLAG) with idarubicin. According to a single centre experience treated a total of 105 patients over a 4-year period with 59% achieving a complete remission. For patients responding to FLAG - Ida, the median event-free survival was 11 months at 5 years. Such patients proceeded either to further chemotherapy or a haematopoietic stem cell transplant (HSCT). The median EFS (13 months vs. 8 months) and projected 5-year survival (37% vs. 13%) of patients undergoing HSCT was significantly better than those who did not. The regimens were well tolerated, with the majority of patients experiencing grade 1 or less non-haematological toxicity (mainly nausea and vomiting). The median time to neutrophil and platelet recovery was 28 and 31 day, respectively. There was a 17% incidence of treatment-related deaths, of which 39% was caused by invasive aspergillus infection (Virchis et al., 2004).

To date, no prospective randomized study has evaluated hypomethylating agents against intensive chemotherapy. Proponents of hypomethylating agents argue that they have a lower toxicity profile, allowing patients to receive therapy on an outpatient basis, and that the lower complete response rates do not necessarily translate into lower survival rates. This may be particularly relevant for older MDS patients, who are at higher risk of morbidity from intensive chemotherapy. (Scott & Estey, 2008).

9.9 Autologous stem cell transplantation

Intensive chemotherapy with AML-like schedules followed by autologous stem cell transplantation may provide an alternative option for patients lacking a suitable donor. Intensive chemotherapy results in complete remission rates of 15-65%. The median remission duration without stem cell transplantation is usually short due to a high incidence of early relapses. In view of the high relapse rate after chemotherapy alone transplantation with autologous stem cells after remission induction and consolidation chemotherapy has been applied in various clinical studies.

Autologous SCT has been extremely investigated in MDS. It is limited to patients who have achieved a CR, can be harvested, and are candidates for the procedure. Autologous SCT after successful induction chemotherapy may increase the proportion of long-term survivors, thus improving CR duration in some patients with MDS, particularly in younger patients in remission. Results for older patients are unsatisfactory. The relapse rate is up to 75%, with a 2-year probability of disease-free survival of only 25% for patients 40-60 years of age (Meletis & Terpos, 2009).

Autologous peripheral stem cells (PB) result in faster hematopoietic recovery, but may be associated with a higher risk of relapse than bone marrow stem cells. In a study comparing 336 patients transplanted with either bone marrow (BM) (n=104) or PB (n=232). In the multivariate Cox model, the event-free survival was not different after PB or BM HSCT. The relapse risk after transplantation with stem cells from either source was similar. A significant interaction between age and the source of stem cells indicated a more favorable potential of autologous PB HSCT in young age groups. As autologous PB and BM HSCT result in equivalent outcomes, given the more rapid hematopoietic recovery PB is the preferred source of stem cells (de Witte et al., 2006).

According to the final results of a prospective randomized European Intergroup Trial (EORTC-06961) autologous stem cell transplantation does not provide longer survival than intensive chemotherapy (de Witte et al., 2010). Therefore, there is very limited enthusiasm for the future of autologous SCT in the management of MDS patients.

Patients with therapy-related MDS/AML had a significantly better disease-free survival than did those with the other categories of disease, even after adjustment for confounding factors in the Cox model, including interval between diagnosis and transplantation. The explanation of this unexpected outcome is not straightforward, but the contribution of patients with favorable cytogenetic characteristics might be relevant. We identified eight patients with t(8;21) or inversion 16 in an incomplete and ongoing analysis. The 3-year disease-free survival of these patients was 57% (de Witte et al., 2007).

9.10 Allogeneic stem cell transplantation

Despite the approval of three novel agents for MDS, allogeneic stem cell transplantation (alloSCT) is the only curative treatment modality for MDS patients. Allogeneic SCT replaces recipient dysplastic hemopoiesis with healthy donor haemopoiesis and immune system with an attendant graft-versus-leukemia (GvL) effect. Its applicability, however, is limited by the age of MDS patients, high rates of transplant-related mortality (TRM) and availability of a suitable HLA-matched donor. MDS is currently the third most common indication for alloSCT. As the disease is most frequent in the older population, the oftener use and acceptance of reduced-intensity conditioning (RIC) made the procedure more increased.

There are several unresolved issues in connection with transplantation. These are the impact of pre-transplant tumor debulking, the optimal timing and the transplantation of older patients and of those with co-morbidities, and also the issue of post-transplant maintaining therapy. The main difficulty is whether can we identify a subset of patients that could benefit from early transplantation and also a subset that are anticipated not going to benefit from alloSCT. Recent efforts to optimize the curative potential of transplant have focused on pretransplant therapy options, the use of predictive models to improve patient selection, and transplant modifications using reduced conditioning intensity.

Dependent upon disease status at the time of transplantation, 30% to 70% of patients can be expected to be cured of their disease and survive long term. However, posttransplant relapse and graft-versus-host disease (GvHD) remain problems (Bartenstein & Deeg, 2010). No prospective randomized controlled trials have directly compared non-transplant therapies with SCT for MDS.

As more than 50% of the patients will fail to benefit from this treatment approach either due to TRM or to relapse after transplantation being mindful of patient selection is a main issue in decision making.

9.10.1 Patient selection and decision making

However, as the majority of patients with MDS are in the seventh or eighth decade of life careful consideration must be made to determine whether the patient is a candidate for alloSCT. Only approximately 5-10% of higher risk MDS patients are alloSCT candidates.

Features to consider for alloSCT include the patient's age and IPSS score (moreover WPSS), performance status, co-morbid conditions, availability of a suitable donor (Warlick, 2010; Malcovati et al., 2005; Cutler, 2010a, 2010b;).

The proper assessment of co-morbidities becomes more important. A useful tool is hematopoietic cell transplantation co-morbidity index (HCTCI) developed by Sorror et a. (Sorror et al., 2005). It has been shown to have a predictive utility on survival also in MDS, even in case of patients getting only best supportive care (Zipperer et al., 2009). According to a study on 172 MDS patients assessing the impact of co-morbidities on survival and the prognostic utility of co-morbidity scores patients with an HCTCI of 0 had a median survival time of 68 months, those with an HCTCI of 1 or 2 lived for 34 months, and those with an HCTCI of ≥3 survived for 25 months. The HCTCI was able to further subdivide the IPSS intermediate-2 and high-risk groups. For patients in the IPSS intermediate-1 and low-risk groups, the HCTCI provided no additional prognostic information. The presence of pulmonary disease, gastrointestinal tract ulcers, cardiac disorders, and infection were independent prognostic factors for survival. HCTCI yielded prognostic information independent of the IPSS (Zipperer et al., 2009).

It is also important to determine whether the patient's marrow blast count is sufficiently low. Data suggest that patients with 5–20% marrow blasts have only 25-28% 5-year overall survival whereas lower risk patients (based on WPSS risk score) do well with alloSCT, with a 5-year overall survival of 80% (Alessandrino et al., 2008).

Risk groups	Patients	NRM (%)	Relapse (%)	OS (%)	RFS (%)
Group I (HCT-CI scores 0–2 and low-risk diseases)	Myeloablative (n=138)	11	14	78	75
	Nonmyeloablative (n=28)	4	33	70	63
Group II (HCT-CI scores 0–2 and intermediate and high-risk diseases)	Myeloablative (n=176)	24	34	51	43
	Nonmyeloablative (n=34)	3	42	57	56
Group III (HCT-CI scores ≥ 3 and low-risk diseases)	Myeloablative (n=52)	32	27	45	41
	Nonmyeloablative (n=19)	27	37	41	36
Group IV (HCT-CI scores ≥ 3 and intermediate and high-risk diseases)	Myeloablative (n=86)	46	34	24	20
	Nonmyeloablative (n=44)	29	49	29	23

Table 3. Two-year NRM, relapse, OS, and RFS incidences among 4 risk groups of nonmyeloablative and myeloablative patients with AML or MDS. Donors were either related (n=301) or unrelated (n=276) (Sorror et al., 2005).

9.10.2 Appropriate timing of alloSCT

The timing of transplantation has always been the most controversial topic of discussion for both patients and physicians. The lack of prospective data adds to the doubtfulness in this topic. Faced with the uncertainty of transplantation outcomes but the certainty of eventual MDS disease progression, decisions are often made based on patient preference. It is clear that there is inherent bias in these types of analyses, because the patients included are often selected and represent the best transplant candidates.

To address the shortcomings of these and other biased retrospective analyses, a Markov decision model was generated to best understand how treatment decisions would affect overall outcome in large cohorts of patients with newly diagnosed MDS (Cutler et al., 2004). The decision model was designed to determine if transplantation at the time of initial diagnosis, delayed a fixed number of years, or at the time of leukemic transformation was the optimal usage strategy for transplantation. Using data from several large, nonoverlapping databases, it was demonstrated that the optimal treatment strategy for patients with low- and intermediate-1-risk IPSS disease categories was to delay transplantation until the time of leukemic progression. Immediate transplantation was recommended for patients with high- and intermediate-2-risk IPSS scores (Cutler et al., 2004). However, a major limitation of this study is that it excluded patients over the age of 60 and focused solely on myeloablative conditioning.

According to the retrospective registry study of de Witte *et al.* low risk patients benefited by early transplantation (within one year after diagnosis) versus more than 12 months from diagnosis. Estimated four-year overall survival was 57 % versus 47% (de Witte et al., 2010).
Beside the IPSS there are also important clinical events such as a new transfusion requirement, recurrent infection, or recurrent bleeding episodes that could be considered triggers to move on to transplantation. Transfusion requirement is a very strong predictor of the advantage of early transplantation. WPSS incorporates transfusion dependence into the score, and enables a dynamic assessment of prognosis during the time-course of the disease. WPSS appears more useful than IPSS particularly with respect to patients with low-risk disease (Malcovati et al., 2007). Presumably, identification of newer prognostic markers, using cytogenetic, immunophenotypic, and molecular techniques will enable further improvement in terms of risk stratification and help to clarify the optimal timing for transplantation, particularly for patients with low-risk disease.

9.10.3 Conditioning regimen

Standard myeloablative conditioning (SMC) results in high treatment-related mortality. While a high proportion of patients can achieve long-term disease control when undergoing transplantation in early phases of the disease using SMC, transplantation-related deaths account for 20% to 30% of treatment failures with such type of conditioning regimens (Bearman et al., 1988; de Lima et al., 2004).
A combination of busulfan (Bu) and cyclophosphamide (Cy) has been used as a standard myeloablative regimen for alloSCT. Recent studies postulate that fludarabine (Flu) is a less toxic substitute for Cy. Lee *et al.* compared the two regimens (BuCy vs BuFlu) and showed that there was no significant intergroup difference in the time of engraftment, nausea/vomiting, acute/chronic graft-versus-host disease, hepatic veno-occlusive disease, or hemorrhagic cystitis. Moreover, the 2 groups showed no significant difference in the cumulative risk of relapse, event-free survival, or overall survival (Lee et al., 2010; Kindwall-Keller & Isola, 2009).
As the vast majority of MDS patients are over 60, using SMC is rarely an option. Conditioning regimens with less toxicity for alloSCT are being developed and should be available for a higher proportion of patients, particularly those who are elderly.
It has been well established that for many malignancies the curative potential of allogeneic transplantation is, in large part, due to the graft versus malignancy (GVM) effect. This has led to the development of less toxic, nonmyeloablative, and reduced intensity transplantation regimens that would provide donor cell engraftment and generation of a GVM effect. This approach has allowed treatment of older and debilitated patients who have been considered ineligible for transplantation using myeloablative regimens (de Lima et al., 2004; Gale & Champlin, 1984).
Reduced-intensity conditioning regimens (these regimens cannot be safely administered without stem cell support) usually involve a combination of a purine analog (primarily fludarabine) with an alkylating agent (usually melphalan or busulfan). These reduced-intensity regimens are generally considered to include less than 16 mg/kg busulfan or less than 10 Gy total body irradiation (Giralt et al., 1999, 2001). They have been usually associated with prompt engraftment of donor cells procured from both HLA-matched related and unrelated donors, while the truly nonablative regimens (NMA; can be given

routinely without stem cell support, with neutrophil recovery within 28 days) have been associated with a varying degree of mixed chimerism and a higher risk of primary and secondary graft failures (de Lima et al., 2004).

However, there are currently no completed prospective randomized controlled studies comparing outcomes of RIC to myeloablating conditioning. The best insight into differences in outcome comes from large registry-based retrospective studies. These studies demonstrate that the use of RIC is associated with a reduction in transplant-related mortality but an increased risk of disease relapse. (Horwitz, 2011).

Based on recent relevant data regarding RIC for transplantation (Laport et al., 2008; McClune et al., 2010), patient age and disease status generally dictate the type of conditioning to be utilized. For example, those relatively older patients (i.e. >50 or 60 years) with <10% marrow blasts would generally be recommended to receive RIC, whereas younger patients with a higher marrow blast burden would generally be recommended to receive standard conditioning. Regimens assessing novel approaches to RIC for MDS are being evaluated (Pagel et al, 2009).

At MD Anderson Cancer Center a retrospective analysis of transplantation outcomes was performed to determine whether in AML and MDS a reduced-intensity conditioning regimen would result in lower relapse rates than a truly nonablative regimen. Patients were included in this study if they had either AML or high-risk MDS and had undergone an alloSCT from an HLA-compatible donor with either a truly nonablative regimen of fludarabine, cytarabine (araC), and idarubicin (FAI) or with a RIC regimen with fludarabine in combination with melphalan 140 or 180 mg/m^2 (FM140 or FM180). FAI was intended to be the treatment of choice for older patients (> 55 years) with early disease, having sibling donors and who were in remission, and with a high risk of relapse. The 2 doses of melphalan were investigated in an attempt to minimize toxicities observed with FM180. The lower age limit for participation in the FM studies was 55 years, but younger patients with organ dysfunction that made them ineligible for high-dose treatment protocols were also eligible. Both related and unrelated donors were allowed. Patients with more advanced disease were preferably treated with FM in order to provide higher dose intensity in the preparative regimen. The 3-year cumulative incidence of non–relapse-related mortality was significantly higher after conditioning with FM than with FAI. Conversely, the 3-year cumulative incidence of relapse-related mortality was higher after FAI than after FM. The analysis suggests that disease recurrence is more frequent in patients receiving the truly nonablative regimen as compared with those receiving the reduced-intensity regimen of fludarabine and melphalan. for AML and MDS both cytoreduction of the preparative regimen as well as GVM contribute to disease control after allografting with reduced-intensity or nonablative conditioning regimens. The relative importance of these mechanisms, however, may differ with regimens other than those tested here. Truly nonablative regimens may be effective in minimal disease states or for diseases highly sensitive to GVM effects and may provide a platform for innovative cell therapy approaches that may obviate the need for direct cytoreduction of the malignancy (de Lima et al., 2004; Virchis et al., 2004).

There are studies intented to compare outcomes with different conditioning approaches. The relative efficacy of 3731 SMC transplant were compared with 1448 RIC/NMA procedures performed at 217 centers between 1997 and 2004 on AML/MDS patients. NMA conditioning resulted in inferior disease-free survival (DFS) and OS, but there was no difference in DFS and OS between RIC and SMC regimens. Late TRM negates early decreases in toxicity with RIC and NMA regimens. These data suggest that higher regimen intensity may contribute to optimal survival in patients with AML/MDS, suggesting roles for both regimen intensity and graft vs leukemia in these diseases (Luger et al., 2011).

The addition of alemtuzumab to a RIC regimen dramatically reduces the incidence of acute and chronic GVHD in patients with AML and MDS undergoing allogeneic transplantation, while TRM, relapse risk, OS and DFS are not affected (van Besien et al., 2009). In another study using alemtuzumab-based RIC alloSCT the 3-year non-relapse mortality (NRM) was 31%, DFS was 41% and overall survival (OS) was 46%. Comorbidity scoring and performance status have been suggested as strategies to guide dose adjustment and to identify fit 65- to 70-year-old patients who can tolerate myeloablative conditioning while those with comorbidities receive less intense regimens. HCT-CI was found to be an independent variable affecting 3-year NRM, DFS and OS, indicating that the HCT-CI provides an important means of stratifying patients with a high risk of inferior transplant outcomes (Lim et al., 2010a; Artz et al., 2006).

9.10.4 AlloSCT for the elderly

The median age of diagnosis for patients with MDS is 76 years of age with 86% of patients older than 60 years. The NCCN Practice Guidelines, as well as recent expert reviews, recommend alloSCT for patients with high-risk MDS without making any age restrictions (Greenberg, 2010). However, registry analysis show that few patients older than 65 years are actually undergoing this procedure.

There are no randomized controlled clinical trials evaluating the role of alloSCT for the treatment of elderly (> 60 years old) patients with intermediate- to high-risk MDS. The highest level of evidence is prospective cohort studies.

Two studies address the question of age and outcomes after transplantation.

Lim *et al.* recently published the results of their retrospective survey. Transplant outcomes of 1333 patients with MDS older than 50 years were reported to the European Group for Bone Marrow Transplant. The median age was 56 (range, 50 to 74) with 34% of patients being older than 60 years, 52% having advanced disease at the time of transplant and 62% receiving a RIC regimen. Overall survival for the whole group was 31% at 4 years, with 63% of patients dying of nonrelapse causes. The nonrelapse mortality rate was 36% at 4 years with no significant difference demonstrated in patients older or younger than 60 years. Patients receiving a myeloablative conditioning regimen had a higher risk of transplant-related mortality than those receiving an RIC regimen (44% *v* 32%). Relapse rates were higher in patients receiving RIC regimens than those receiving myeloablative conditioning (41% *v* 33%). In their multivariate analysis, age was not a significant factor for clinical outcomes of relapse or survival. Significant factors associated with nonrelapse mortality included type of conditioning, advanced disease at transplantation, and donor type. The 4-year overall survival estimate was not significantly different between those < 60 years old and those > 60 years old. However, relapse rate was increased in the > 60-year-old cohort (32% vs 41%). Finally, relapse rate and survival were significantly affected by poor-risk cytogenetics (Lim, et al., 2010b).

McClune *et al.* included 535 MDS patients and 545 AML patients in first complete remission registered with the Center for International Blood and Marrow Transplant Research from 1995 to 2005. All transplants were (RIC or nonmyeloablative (NMA) conditioning, in contrast to the prior study. Thirty-four percent of MDS patients and 36% of AML patients were > 60 years old. In both the MDS and AML cohorts, age did not significantly affect nonrelapse mortality, disease-free survival, or overall survival. Long-term disease control was seen in about one-third of MDS patients independent of age, supporting the curative potential of SCT for patients with MDS over the age of 60 (McClune et al., 2010).

Castagna *et al.* carried out a retrospective study of 63 patients >60 years with hematological malignancies and treated with RIC and alloSCT. Only the occurrence of aGVHD affected the TRM and OS. Acute GVHD is the main cause of TRM and more efforts should be made to reduce its incidence without sacrificing graft vs tumor effect (Castagna et al., 2010).

Umbilical cord blood (UCB) is feasible as an alternative donor source for RIC alloSCT among older patients with MDS who do not have suitable HLA-matched sibling donors (MSD). Majhail *et al.* compared outcomes of allo-SCT, using MSD or UCB among older patients (age over 55 years) with AML or MDS. All patients received a RIC regimen consisting of cyclophosphamid, fludarabine and 200 cGy TBI. Median age at alloSCT was 63 years for MSD and 61 years for UCB recipients. On multivariate analysis, donor source (MSD vs UCB) did not impact risks of OS, leukemia-free survival and relapse or treatment-related mortality (Majhail et al., 2011).

By means of the WPSS the impact of regular transfusion requirement (defined as requiring at least one transfusion every 8 weeks in a 4-month period) can be implicated into the risk stratification and those low-risk patients can be identified who can benefit of allo-SCT. Regular transfusion requirement is given the same regression weight as progressing to a higher cytogenetic risk group (Malcovati et al., 2007).

Based on the available evidence, transplantation for MDS in patients aged 60 or older is the only available curative therapy, but its benefit in terms of OS has not been demonstrated in randomized controlled trials. Careful patient selection may improve the results (Giralt et al., 2011; Ria et al,. 2009; Alatrash et al., 2011).

9.10.5 The issue of the stem cell source and donor selection

The patients with MDS who received peripheral blood stem cells (PBSC) had a faster engraftment and a lower relapse rate, resulting in improved DFS rates, than patients who received bone marrow stem cells (Guardiola et al., 2002). There is a shifting tendency towards to use PBSC rather than bone marrow stem cells.

Umbilical cord blood has increased access to hematopoietic cell transplantation (HCT) for patients without HLA-matched sibling donors (MSD).

A study was carried out through the Eurocord and European Group for Blood and Marrow Transplantation (EBMT) registries by Robin *et al.*, aiming to evaluate the outcomes and risk factors in adult patients who underwent single or double unrelated cord blood transplantation (UCBT) for myelodysplastic syndrome (MDS) or secondary acute myeloblastic leukemia (sAML). A total of 180 adults with MDS (n=39) or sAML (n=69) were analyzed. Median age was 43 (18-72) years. In all, 77 patients (71%) received a single UCBT. Myeloablative conditioning regimen (MAC) was given to 57 (53%) patients. A 2-year non-relapse mortality (NRM) was significantly higher after MAC (62 vs 34). A 2-year DFS and overall survival (OS) were 30 and 34%, respectively. In multivariate analysis, patients with high-risk disease (blasts >5% and International Prognostic scoring system (IPSS) intermediate-2 or high in MDS) had significant poorer DFS. In spite of high NRM, these data indicate that UCBT is an acceptable alternative option to treat adults with high-risk MDS or sAML, without a suitable human leukocyte antigen (HLA)-matched donor (Robin et al., 2011).

The rate of related versus unrelated donor is also changing. The percentage of older patients who are receiving unrelated donor transplants is increasing (Karanes et al., 2008).

A study compared allogeneic sibling-matched SCT data in MDS patients ≤60 years old to clinical outcomes of age-matched non-treated comparable stage MDS patients. Markov decision-making statistical analysis indicated that higher risk IPSS patients ≤60 years old should proceed to such human leucocyte antigen (HLA) identical sibling transplants at diagnosis, whereas for those lower risk MDS patients, delaying transplantation for several years and prior to disease progression would be beneficial (Cutler et al., 2004).

According to newer studies minor HLA disparity in unrelated compared to related donors could have a significant impact on transplant outcomes. To assess whether use of unrelated donors (URD) engenders more potent graft versus leukaemia effect in RIC alloSCT compared to matched related donors (MRD), a retrospective study has been performed Dana-Farber Cancer Institute on patients with different haematological malignancies including 66 MDS patients (Ho et al., 2010). Patients received uniformly fludarabine and intravenous busulfan conditioning, and GVHD prophylaxis with tacrolimus/mini-methroxate (mini-MTX) or tacrolimus/sirolimus ± mini-MTX. URD was associated with a lower risk of relapse (52% versus 65%).

9.10.6 Reducing relapse rate

The most important factors for TRM are age, co-morbidities, donor selection and intensity of the conditioning regimen. A careful donor selection and an appropriate choice of conditioning regimen will significantly reduce TRM.

On the other hand preferring RIC as preparative regimen, the incidence of relapse will be higher. So effords are made to reduce relapse rate. Either improving of the pretransplant remission status or by using post-transplant strategies, such as maintenance or consolidation therapies can play a role in reducing the risk of relapse (Kröger, 2008).

In addition to age, the cytogenetic risk score remained an independent prognostic factor for relapse. Lenalidomide has shown activity in terms of resolution of chromosomal abnormalities, especially in isolated 5q- but also in complex abnormalities involving 5q-. There are also some observations that MDS patients with poor cytogenetic abnormalities such as monosomy 7 may benefit from hypomethylating agents (Raj et al., 2007).

Besides improving pretransplant status of patients with MDS by inducing clinical and cytogenetic remission, further approaches to reduce the risk of relapse after alloSCT involve post-transplant modifications.

Donor lymphocyte infusion (DLI) can induce durable remission in relapsing patients with hematological malignancies after alloSCT. However, the experience of DLI in patients with MDS is limited. and the role of it is unclear. In Fred Hutchinson Cancer Research Center, Washington 16 patients treated with DLI for relapsed MDS after HCT between 1993 and 2004. CR with resolution of cytopenias and prior disease markers occurred in 3 of 14 patients who could be evaluated. Two patients survived without MDS for 68 and 65 months after DLI, respectively, but died with pneumonia. Grades II-IV acute GVHD and chronic

GVHD occurred after DLI in 6 (43%) and 5 (36%) patients, respectively. All three responders developed grades III-IV acute GVHD and extensive chronic GVHD after DLI. This result refers that DLI can result in CR in some patients with relapsed MDS after transplant, but long-term survival is infrequent (Campregher et al., 2007). Azacitidine combined with DLI for relapsed patients may also result remission (Czibere et al., 2010).
Maintenance therapy with azacitidine is another option. Low dose azacitidine (at 32 mg/m^2) given for 5 days is safe and can be administered after allogeneic transplant for at least 4 cycles to heavily pretreated AML/MDS patients. The trial also suggested that this treatment may prolong event-free and overall survival, and that more cycles may be associated with greater benefit (de Lima et al., 2010).

9.10.7 Treatment of relapse

Standard induction chemotherapy may be used for re-induction. The tolerability and potential of low dose azacitidine for treatment of relapsed MDS is supported by a Phase I trial (de Lima et al., 2010).
Immunotherapeutic strategies for example peptide vaccination targeting leukemia-associated antigens, such as the Wilm's Tumor protein (WT1) (Keilholz et al., 2009) and whole cell leukemia vaccination with CD80 and IL-2 genetically modified leukemic blasts are currently the subjects of Phase I clinical trial (Ingram et al., 2009).

9.10.8 The effect of iron oveload on the post-tranplantation survival

Pre-transplantation transfusion history and serum ferritin have significant prognostic value in patients with myelodysplastic syndrome undergoing myeloablative allogeneic stem cell transplantation, inducing a significant increase of non-relapse mortality. An elevated serum ferritin (>1000 μg/L) has been associated with reduced OS and increased risk of infection following HSCT. Given that ferritin is an acute phase reactant, the elevated ferritin serum level alone does not considered to be a satisfactory marker of iron overload. Other biomarkers of body iron load should be taken into account to estimate the effect of iron overload on the post-transplantation survival (Pullarkat, 2010). These results indicate that transfusion history should be considered in transplantation decision-making in patients with myelodysplastic syndrome. Outcome was significantly worse in subjects receiving more than 20 red cell units.
Elevated pre-transplantation liver iron content estimated by magnetic resonance imaging (MRI) is significantly associated with inferior post-transplantation survival. There is a strong correlation between pre-transplantation serum ferritin level and liver iron content which was mostly dependent on prior transfusion history (Armand).

9.11 New agents

The past several years have brought exciting new treatments strategies for MDS. However, despite the huge progress, no curative therapy does exist exclusive of alloSCT nowadays. Furthermore, *cure* is not necessarily essential in the presence of drug therapy that can effectively control disease symptoms and prevent disease-related mortality (Tefferi, 2008; Rajkumar, 2008). Since the knowledge of molecular genetics in AML and MDS has expanded recently, targeted therapeutics should offer new possibilities for advancement (Bryan et al., 2010).

The nucleoside analog clofarabine, and the alkylating agent cloretazine are now being tested. There are several clinical trials using new agents that act at a number of different levels. There are new agents with new therapeutic targets, for example tyrosine kinase inhibitors (FLT3 inhibitors), farnesyl-transferase inhibitors (Braun & Fenaux, 2008), which are thought to target mutant ras activity, agents with c-jun modificating effect, MAP-kinase inhibitors, histone deacetylase inhibitors, which cause epigenetic alterations.

Using combination of drugs with different and new mechanism of action may also be a novel approach. Azacitidine in combination with histone deacetylase inhibitors might offer better efficacy by modulating the methylation and acetylation states of silenced genes. Silverman, 2009, Leuk res

Single-agent tipifarnib (a farnesyltransferase inhibitor) in high-risk MDS patients achieves responses comparable to those of standard epigenetic therapies, including hypomethylating agents, and for this reason, farnesyltransferase inhibitors deserve further in MDS. (Braun & Fenaux, 2008; Grant, 2009).

10. Recommendations for treating lower risk (IPSS low, intermediate-1 or WPSS very low, low and Intermediate) MDS patients

Patients with del(5q) chromosomal abnormalities and symptomatic anaemia should receive lenalidomide.

Other patientswith symptomatic anaemia with EPO level ≤ 500 mU/ml, should be treated with recombinant human EPO or darbepoetin with or without granulocyte-colony stimulatin factor (G-CSF). Non-responders should be considered for treatment with azacitidine, decitabine or lenalidomide.

Those anaemic patients with EPO levels < 500 mU/ml should be evaluated to determine whether they have a good probability of responding to immunosuppressive therapy. For those with low probability treatment with azacitidine, decitabine or lenalidomide should be recommended.

Non-responders and those with severe cytopenias participation in a clinical trial, or allo-SCT should be an option.

11. Recommendations for treating higher risk (IPSS Intermediate-2, high-risk or WPSS high, very high) MDS patients

Treatment recommendations for higher risk patients depend on whether they are candidate for intensive therapy. Decision should be made by considering the patient's age, co-morbidities, physiosocial status and also the patient's preference and the availability of a suitable donor.

For eligible patients the first choice regarding the donor source has remained an HLA-matched sibling though results with HLA-matched unrelated donors have improved to a similar level.

Patients older than 55-60 years, particularly those with less than 10 % marrow myeloblasts, would be conditioning with RIC. Patients with high blast count pretransplant debulking therapy is generally needed. Younger patients, regardless of marrow blast burden, will generally receive high dose conditioning.

For patients eligible for intensive therapy lacking a suitable donor using an intensive induction chemotherapy should be recommended.

For higher risk patients who are not candidate for intensive therapy the use of azacitidine, decitabine or a participation in a clinical trial should be considered.

Good supportive care should be given to every MDS patients. This may be the only therapy for those with adverse clinical features or disease progression despite antitumor therapy.

12. Conclusion

Ten years ago, there was little to offer MDS patients other than transfusion support, but since then, the approach to treating patients with MDS has undergone a progress. New therapeutic options have added to the armamentarium of MDS treatments.

Historically, treatment goals have varied based on patient risk category, with modification of disease reserved for patients in the higher-risk population and response rates being a primary endpoint. In the light of recent data the treatment paradigm should be reevaluated to focus on prolonging time to leukemic transformation and extending survival while improving quality of life. Lenalidomide can modify disease activity, the methyltransferase inhibitors have the ability to prolong survival. These are shifting the focus away from response rates alone. Azacitidine has demonstrated a sustained impact on overall survival. In addition, data suggest that treatment with these therapeutics in patients with lower-risk disease may further extend survival by altering the biology of MDS. The new treatment paradigm should aim for prolonging leukemic-free transformation and extending survival while optimizing quality of life and maintaining hematologic and cytogenetic response.

The results of stem cell transplantation for MDS continue to improve together with the outlook of patients afflicted with myelodysplasia.

The incidence of MDS is increasing parallely with the increase of the average age of the population. That gives a particular importance to the development of MDS therapy has taken place in the last decade. In the close future some more possible progress are expected.

13. References

Advisory Committee of the International Bone Marrow Transplant Registry. 1989. Report from the International Bone Marrow Transplant Registry. *Bone Marrow Transplant*,48, 453-458,

Alatrash G, de Lima M, Hamerschlak N, Pelosini M, Wang X, Xiao L, Kerbauy F, Chiattone A, Rondon G, Qazilbash MH, Giralt SA, de Padua Silva L, Hosing C, Kebriaei P, Zhang W, Nieto Y, Saliba RM, Champlin RE & Andersson BS. 2011. Myeloablative Reduced-Toxicity i.v. Busulfan-Fludarabine and Allogeneic Hematopoietic Stem Cell Transplant for Patients with Acute Myeloid Leukemia or Myelodysplastic Syndrome in the Sixth through Eighth Decades of Life. *Biol Blood Marrow Transplant*, Febr. 18,

Alessandrino EP, Della Porta MG, Bacigalupo A, Van Lint MT, Falda M, Onida F, Bernardi M, Iori AP, Rambaldi A, Cerretti R, Marenco P, Pioltelli P, Malcovati L, Pascutto C, Oneto R, Fanin R & Bosi A; Gruppo Italiano Trapianto di Midollo Osseo (GITMO). 2008. WHO classification and WPSS predict posttransplantation outcome in patients with myelodysplastic syndrome: a study from the Gruppo Italiano Trapianto di Midollo Osseo (GITMO). *Blood*, 112,3,895-902,

Armand P, Kim HT, Rhodes J, Sainvil MM, Cutler C, Ho VT, Koreth J, Alyea EP, Hearsey D, Neufeld EJ, Fleming MD, Steen H, Anderson D, Kwong RY, Soiffer RJ & Antin JH.

2011. Iron Overload in Patients with Acute Leukemia or MDS Undergoing Myeloablative Stem Cell Transplantation. *Biol Blood Marrow Transplant*,6,852-60,

Artz AS, Pollyea D & Kocherginsky M. 2006. Performance status and comorbidity predict transplant related mortality after allogeneic hematopoietic cell transplantation. *Biol Blood and Marrow Transplant*,12,954–964,

Barrett J, Sloand E & Young N. 2006. Determining which patients with myelodysplastic syndrome will respond to immunosuppressive treatment. *Haematologica*, 91,5,583-4,

Bartenstein M & Deeg HJ. 2010. Hematopoietic stem cell transplantation for MDS. *Hematol Oncol Clin North Am*,24,2,407-22,

Barzi A & Sekeres MA. 2010. Myelodysplastic syndromes: a practical approach to diagnosis and treatment. *Cleve Clin J, Med*, 77,1,37-44,

Bearman SI, Appelbaum FR, Buckner CD, Petersen FB, Fisher LD, Clift RA & Thomas ED. 1988. Regimen-related toxicity in patients undergoing bone marrow transplantation. *J Clin Oncol*,6, 1562-1568,

Bello C, Yu D, Komrokji RS, Zhu W, Wetzstein GA, List AF & Lancet JE. 2010. Outcomes after induction chemotherapy in patients with acute myeloid leukemia arising from myelodysplastic syndrome. *Cancer*, Nov. 8,

van Besien K, Kunavakkam R, Rondon G, De Lima M, Artz A, Oran B & Giralt S. 2009. Fludarabine-melphalan conditioning for AML and MDS: alemtuzumab reduces acute and chronic GVHD without affecting long-term outcomes. *Biol Blood Marrow Transplant*,15,5,610-7,

Blum W. How much? How frequent? How long? 2010. A clinical guide to new therapies in myelodysplastic syndromes. *Hematology Am Soc Hematol Educ Program*,2010,314-21,

Boultwood J, Perry J & Pellagatti A. 2010. Frequent mutation of the polycomb-associated gene ASXL1 in the myelodysplastic syndromes and in acute myeloid leukemia. *Leukemia*,524,5,1062–1065,

Braun T & Fenaux P. 2008. Farnesyltransferase inhibitors and their potential role in therapy for myelodysplastic syndromes and acute myeloid leukaemia. *Br J Haematol*, 141,5,576-86,

Bryan J, Jabbour E, Prescott H, Garcia-Manero G, Issa JP & Kantarjian H. 2010. Current and future management options for myelodysplastic syndromes. *Drugs*, 70,11,1381-94,

Campregher PV, Gooley T, Scott BL, Moravec C, Sandmaier B, Martin PJ, Deeg HJ, Warren EH & Flowers ME. 2007. Results of donor lymphocyte infusions for relapsed myelodysplastic syndrome after hematopoietic cell transplantation. *Bone Marrow Transplant*,40,10,965-71,

Castagna L, Fürst S, Marchetti N, El Cheikh J, Faucher C, Mohty M, Bouabdallah R, Vey N, Stoppa AM, Esterni B & Blaise D. 2010. Retrospective analysis of common scoring systems and outcome in patients older than 60 years treated with reduced-intensity conditioning regimen and alloSCT. *Bone Marrow Transplant*, Oct 4,

Chamuleau ME, Westers TM, van Dreunen L, Groenland J, Zevenbergen A, Eeltink CM, Ossenkoppele GJ & van de Loosdrecht AA. 2009. Immune mediated autologous cytotoxicity against hematopoietic precursor cells in patients with myelodysplastic syndrome. *Haematologica*,94,4,496-506,

Cutler CS, Lee SJ & Greenberg P. 2004. A decision analysis of allogeneic bone marrow transplantation for the myelodysplastic syndromes: delayed transplantation for low-risk myelodysplasia is associated with improved outcome. *Blood*,104,579–585,

Cutler C. 2010. Patient selection for transplantation in the myelodysplastic syndromes. *Hematol Oncol Clin North Am*,24,2,469-76,

Cutler C. 2010. Allogeneic hematopoietic stem-cell transplantation for myelodysplastic syndrome. *Hematology Am Soc Hematol Educ Program*,2010,325-9,

Czibere A, Bruns I, Kröger N, Platzbecker U, Lind J, Zohren F, Fenk R, Germing U, Schröder T, Gräf T, Haas R & Kobbe G. 2010. 5-Azacytidine for the treatment of patients with acute myeloid leukemia or myelodysplastic syndrome who relapse after alloSCT: a retrospective analysis. *Bone Marrow Transplant*,45,5,872-6,

Deeg HJ, Jiang PY, Holmberg LA, Scott B, Petersdorf EW & Appelbaum FR. 2004. Hematologic responses of patients with MDS to antithymocyte globulin plus etanercept correlate with improved flow scores of marrow cells. *Leuk Res*, 28,1177-80,

Ebert BL, Pretz J & Bosco J. 2008. Identification of RPS14 as a 5q- syndrome gene by RNA interference screen. *Nature*,451,335–339,

Edlin R, Connock M, Tubeuf S, Round J, Fry-Smith A, Hyde C & Greenheld W. 2010. Azacitidine for the treatment of myelodysplastic syndrome, chronic myelomonocytic leukaemia and acute myeloid leukaemia. *Health Technol Assess*,14 Suppl 1,69-74,

Gale RP & Champlin RE. 1984. How does bone-marrow transplantation cure leukaemia? *Lancet*,2, 28-30,

Garcia JS, Jain N & Godley LA. 2010. An update on the safety and efficacy of decitabine in the treatment of myelodysplastic syndromes. *Onco Targets Ther*, 3,1-13,

Garcia-Manero G. 2010. Prognosis of myelodysplastic syndromes. *Hematology Am Soc Hematol Educ Program*,2010,330-7,

Gergis U & Wissa U. 2010. High-risk myelodysplastic syndromes: chemotherapy, transplantation, and beyond. *Curr Hematol Malig Rep*, 5,1,1-8,

Giralt SA, Horowitz M, Weisdorf D & Cutler C. 2011. Review of stem-cell transplantation for myelodysplastic syndromes in older patients in the context of the Decision Memo for Allogeneic Hematopoietic Stem Cell Transplantation for Myelodysplastic Syndrome emanating from the Centers for Medicare and Medicaid Services. *J Clin Oncol*,29,5,566-72,

Giralt S, Khouri I & Champlin R. 1999. Non-myeloablative "mini transplants." *Cancer Treat Res*,101, 97-108,

Giralt S, Thall P & Khouri I. 2001. Purine analogcontaining preparative regimens: reduced-intensity conditioning for patients with hematologic malignancies undergoing allogeneic progenitor cell transplantation. *Blood*,97, 631-637,

Goldberg SL, Chen E, Corral M, Guo A, Mody-Patel N, Pecora AL & Laouri M. 2010. Incidence and clinical complications of myelodysplastic syndromes among United States Medicare beneficiaries. *J Clin Oncol*,28,17,2847-52,

Greenberg PL. 1998. Risk factors and their relationship to prognosis in myelodysplastic syndromes. *Leuk Res*, 22 Suppl 1,S3-6,

Greenberg PL, Sun Z, Miller KB, Bennett JM, Tallman MS, Dewald G, Paietta E, van der Jagt R, Houston J, Thomas ML, Cella D & Rowe JM. 2009. Treatment of myelodysplastic syndrome patients with erythropoietin with or without granulocyte colony-stimulating factor: results of a prospective randomized phase 3 trial by the Eastern Cooperative Oncology Group (E1996). *Blood*,114,12,2393-400,

Greenberg PL. 2010. Current therapeutic approaches for patients with myelodysplastic syndromes. *Br J Haematol*,150,2,131-43,

Grant S. 2009. New agents for AML and MDS. *Best Pract Res Clin Haematol*,22,4,501-7,

Guardiola P, Runde V & Bacigalupo A. 2002. Subcommittee for Myelodysplastic Syndromes of the Chronic Leukaemia Working Group of the European Blood and Marrow Transplantation Group. Retrospective comparison of bone marrow and granulocyte colony-stimulating factor-mobilized peripheral blood progenitor cells for allogeneic stem cell transplantation using HLA identical sibling donors in myelodysplastic syndromes. *Blood*,99,4370–4378,

Graubert T. 2010. Therapy-related myelodysplastic syndrome: models and genetics. *Biol Blood Marrow Transplant*, 16,1, suppl,S45-S47,

Grövdal M, Karimi M, Khan R, Aggerholm A, Antunovic P, Astermark J, Bernell P, Engström LM, Kjeldsen L, Linder O, Nilsson L, Olsson A, Holm MS, Tangen JM, Wallvik J, Oberg G, Hokland P, Jacobsen SE, Porwit A & Hellström-Lindberg E. 2010. Maintenance treatment with azacytidine for patients with high-risk myelodysplastic syndromes (MDS) or acute myeloid leukaemia following MDS in complete remission after induction chemotherapy. *Br J Haematol*,150,3,293-302,

Hellström-Lindberg E. 2003. Approach to anemia associated with myelodysplastic syndromes. *Curr Hematol Rep*,2,2,122-9,

Ho VT, Kim HT, Aldridge J, Liney D, Kao G, Armand P, Koreth J, Cutler C, Ritz J, Antin JH, Soiffer RJ & Alyea EP. 2010. Use of Matched Unrelated Donors Compared with Matched Related Donors Is Associated with Lower Relapse and Superior Progression-Free Survival after Reduced-Intensity Conditioning Hematopoietic Stem Cell Transplantation. *Biol Blood Marrow Transplant*, 12,702,

Horwitz ME. 2011. Reduced intensity versus myeloablative allogeneic stem cell transplantation for the treatment of acute myeloid leukemia, myelodysplastic syndrome and acute lymphoid leukemia. Curr Opin *Oncol*,23,2,:197-202,

Ingram W, Chan L, Guven H, Darling D, Kordasti S, Hardwick N, Barber L, Mufti GJ & Farzaneh F. 2009. Human CD80/IL2 lentivirus-transduced acute myeloid leukaemia (AML) cells promote natural killer (NK) cell activation and cytolytic activity: implications for a phase I clinical study. *Br J Haematol*,145,6,749-60,

Jädersten M, Malcovati L, Dybedal I, Della Porta MG, Invernizzi R, Montgomery SM, Pascutto C, Porwit A, Cazzola M & Hellström-Lindberg E. 2008. Erythropoietin and granulocyte-colony stimulating factor treatment associated with improved survival in myelodysplastic syndrome. *J Clin Oncol*,26,21,3607-13,

Jankowska AM, Szpurka H & Tiu RV. 2009. Loss of heterozygosity 4q24 and TET2 mutations associated with myelodysplastic/myeloproliferative neoplasms. *Blood*,113,6403–6410,

Kantarjian H, Issa JP, Rosenfeld CS, Bennett JM, Albitar M & DiPersio J. 2006. Decitabine improves patient outcomes in myelodysplastic syndromes: results of a phase III randomized study. *Cancer*,106,1794 -1803,

Kantarjian H, Oki Y, Garcia-Manero G, Huang X, O'Brien S & Cortes J. 2007. Results of a randomized study of 3 schedules of low-dose decitabine in higher-risk myelodysplastic syndrome and chronic myelomonocytic leukemia. *Blood*,109,52–57,

Karanes C, Nelson GO, Chitphakdithai P, Agura E, Ballen KK, Bolan CD, Porter DL, Uberti JP, King RJ & Confer DL. 2008. Twenty years of unrelated donor hematopoietic cell transplantation for adult recipients facilitated by the National Marrow Donor Program. *Biol Blood Marrow Transplant*,14,9 Suppl,8-15,

Keilholz U, Letsch A, Busse A, Asemissen AM, Bauer S, Blau IW, Hofmann WK, Uharek L, Thiel E & Scheibenbogen C. 2009. A clinical and immunologic phase 2 trial of Wilms tumor gene product 1 (WT1) peptide vaccination in patients with AML and MDS. *Blood*,113,26,6541-8,

Kiladjian JJ, Visentin G, Viey E, Chevret S, Eclache V, Stirnemann J, Bourhis JH, Chouaib S, Fenaux P & Caignard A. 2008. Activation of cytotoxic T-cell receptor gammadelta T lymphocytes in response to specific stimulation in myelodysplastic syndromes. *Haematologica*,93,3,381-9,

Kindwall-Keller T & Isola LM. 2009. The evolution of hematopoietic SCT in myelodysplastic syndrome. *Bone Marrow Transplant*, 43,8,597-609,

Komrokji RS & List AF. 2010. Lenalidomide for treatment of myelodysplastic syndromes: current status and future directions. *Hematol Oncol Clin North Am*, 24,2,377-88,

Kröger N. 2008. Epigenetic modulation and other options to improve outcome of stem cell transplantation in MDS. *Hematology Am Soc Hematol Educ Program*,60-7,

Kurtin SE &, Demakos EP. 2010. An update on the treatment of myelodysplastic syndromes. *Clin J Oncol Nurs*,14,3,E29-44,

Laport GG, Sandmaier BM, Storer BE, Scott BL, Stuart MJ, Lange T, Maris MB, Agura ED, Chauncey TR, Wong RM, Forman SJ, Petersen FB, Wade JC, Epner E, Bruno B, Bethge WA, Curtin PT, Maloney DG, Blume KG & Storb RF. 2008. Reduced-intensity conditioning followed by allogeneic hematopoietic cell transplantation for adult patients with myelodysplastic syndrome and myeloproliferative disorders. Biol Blood Marrow Transplant, 14,2,246-55.

Lee JH, Choi J, Kwon KA, Lee S, Oh SY, Kwon HC, Kim HJ, Han JY & Kim SH. 2010. Fludarabine-based myeloablative regimen as pretransplant conditioning therapy in adult acute leukemia/myelodysplastic syndrome: comparison with oral or intravenous busulfan with cyclophosphamide. *Korean J Hematol*, 45,2,102-8,

Lim Z, Brand R & Martino R. 2010. Allogeneic hematopoietic stem-cell transplantation for patients 50 years or older with myelodysplastic syndromes or secondary acute myeloid leukemia. *J Clin Oncol*,28,405–411,

de Lima M, Giralt S, Thall PF, de Padua Silva L, Jones RB, Komanduri K, Braun TM, Nguyen HQ, Champlin R & Garcia-Manero G. 2010. Maintenance therapy with low-dose azacitidine after allogeneic hematopoietic stem cell transplantation for recurrent acute myelogenous leukemia or myelodysplastic syndrome: a dose and schedule finding study. *Cancer*,116,23,5420-31,

de Lima M, Anagnostopoulos A, Munsell M, Shahjahan M, Ueno N, Ippoliti C, Andersson BS, Gajewski J, Couriel D, Cortes J, Donato M, Neumann J, Champlin R & Giralt S. 2004. Nonablative versus reduced-intensity conditioning regimens in the treatment of acute myeloid leukemia and high-risk myelodysplastic syndrome: dose is relevant for long-term disease control after allogeneic hematopoietic stem cell transplantation. *Blood*,104,3,865-72,

List A, Dewald G & Bennett J. 2006. Lenalidomide in the myelodysplastic syndrome with chromosome 5q deletion. *N Engl J Med*,355,1456–1465,

List AF. 2010. Iron overload in myelodysplastic syndromes: diagnosis and management. *Cancer Control*, 17 Suppl,2-8,

Litzow MR, Tarima S, Pérez WS, Bolwell BJ, Cairo MS, Camitta BM, Cutler CS, de Lima M, Dipersio JF, Gale RP, Keating A, Lazarus HM, Luger S, Marks DI, Maziarz RT, McCarthy PL, Pasquini MC, Phillips GL, Rizzo JD, Sierra J, Tallman MS & Weisdorf DJ 2010. Allogeneic transplantation for therapy-related myelodysplastic syndrome and acute myeloid leukemia. *Blood*,115,9,1850-7,

Luger SM, Ringdén O, Zhang MJ, Pérez WS, Bishop MR, Bornhauser M, Bredeson CN, Cairo MS, Copelan EA, Gale RP, Giralt SA, Gulbas Z, Gupta V, Hale GA, Lazarus HM, Lewis VA, Lill MC, McCarthy PL, Weisdorf DJ & Pulsipher MA. 2011. Similar outcomes using myeloablative vs reduced-intensity allogeneic transplant preparative regimens for AML or MDS. *Bone Marrow Transplant*, Mar 28.

Ma, X., Does, M., Raza, A., & Mayne, S.T. 2007. Myelodysplastic syndromes: Incidence and survival in the United States. *Cancer*, 109, 1536–1542,

Majhail NS, Brunstein CG, Shanley R, Sandhu K, McClune B, Oran B, Warlick ED, Wagner JE & Weisdorf DJ. 2011. Reduced-intensity hematopoietic cell transplantation in older patients with AML/MDS: umbilical cord blood is a feasible option for patients without HLA-matched sibling donors. *Bone Marrow Transplant*, 14,3,282-9,

Malcovati L, Della Porta MG & Pascutto C. 2005. Prognostic factors and life expectancy in myelodysplastic syndromes classified according to WHO criteria: a basis for clinical decision making. *J Clin Oncol*, 23,30,7594-7603,

Malcovati L, Germing U, Kuendgen A, Della Porta MG, Pascutto C, Invernizzi R, Giagounidis A, Hildebrandt B, Bernasconi P, Knipp S, Strupp C, Lazzarino M, Aul C & Cazzola M. 2007. Time-dependent prognostic scoring system for predicting survival and leukemic evolution in myelodysplastic syndromes. *J Clin Oncol*,25,23,3503-10,

Malcovati L & Nimer SD. 2008. Myelodysplastic syndromes: diagnosis and staging. *Cancer Control*, 15 Suppl,4-13,

McClune BL, Weisdorf DJ & Pedersen TL. 2010. Effect of age on outcome of reduced-intensity hematopoietic cell transplantation for older patients with acute myeloid leukemia in first complete remission or with myelodysplastic syndrome. *J Clin Oncol*,28,1878–1887,

McCormack SE & Warlick ED. 2010. Epigenetic approaches in the treatment of myelodysplastic syndromes: clinical utility of azacitidine. *Onco Targets Ther*, 3,157-65,

Meletis J & Terpos E. 2009. Transplantation strategies for the management of patients with myelodysplastic syndromes. *J BUON*,14,4,551-64,

Mufti GJ & Chen TL. 2008. Changing the treatment paradigm in myelodysplastic syndromes. *Cancer Control*, 15 Suppl,14-28,

National Comprehensive Cancer Network (2010) The NCCN 2.2010 Myelodysplastic Syndromes Clinical Practice Guidelines in Oncology (National Comprehensive Cancer Network, Fort Washington, PA).

Park S , Kelaidi C, Sapena R, Vassilieff D, Beyne-Rauzy O, Coiteux V, Vey N, Ravoet C, Cheze S, Rose C, Legros L, Stamatoullas A, Escoffre-Barbe M, Guerci A, Chaury MP, Fenaux P & Dreyfus F. 2010. Early introduction of ESA in low risk MDS patients may delay the need for RBC transfusion: a retrospective analysis on 112 patients. *Leuk Res*, 34,11,1430-6,

Passweg JR, Giagounidis AA, Simcock M, Aul C, Dobbelstein C, Stadler M, Ossenkoppele G, Hofmann WK, Schilling K, Tichelli A & Ganser A. 2011. Immunosuppressive therapy for patients with myelodysplastic syndrome: a prospective randomized multicenter phase III trial comparing antithymocyte globulin plus cyclosporine with best supportive care--SAKK 33/99. *J Clin Oncol*,29,3,303-9,

Pullarkat V. 2010. Iron overload in patients undergoing hematopoietic stem cell transplantation. *Adv Hematol,* 2010. pii: 345756,

Pullarkat V. 2009. Objectives of iron chelation therapy in myelodysplastic syndromes: more than meets the eye? *Blood,*114,26,5251-5,

Raj K, John A & Ho A. 2007. CDKN2B methylation status and isolated chromosome 7 abnormalities predict responses to treatment with 5-azacytidine. *Leukemia,*;21,1937–1944,

Rajkumar SV. 2008. Treatment of myeloma: cure vs control. *Mayo Clin Proc,*83,1142-1145,

Ria R, Moschetta M, Reale A, Mangialardi G, Castrovilli A, Vacca A & Dammacco F. 2009. Managing myelodysplastic symptoms in elderly patients. *Clin Interv Aging,* 4,413-23,

Rizzo JD, Brouwers M, Hurley P, Seidenfeld J, Arcasoy MO, Spivak JL, Bennett CL, Bohlius J, Evanchuk D, Goode MJ, Jakubowski AA, Regan DH & Somerfield MR. 2010. American Society of Hematology and the American Society of Clinical Oncology Practice Guideline Update Committee, American Society of Hematology/American Society of Clinical Oncology clinical practice guideline update on the use of epoetin and darbepoetin in adult patients with cancer. *Blood,*116,20,4045-59,

Robin M, Sanz GF, Ionescu I, Rio B, Sirvent A, Renaud M, Carreras E, Milpied N, Mohty M, Beguin Y, Bordigoni P, de Witte T, Picardi A, Purtill D, Gluckman E, Kroger N & Rocha V. 2011. Unrelated cord blood transplantation in adults with myelodysplasia or secondary acute myeloblastic leukemia: a survey on behalf of Eurocord and CLWP of EBMT. *Leukemia,*25,1,75-81,

Sanada M, Suzuki T & Shih LY. 2009. Gain-of-function of mutated C-CBL tumour suppressor in myeloid neoplasms. *Nature,*460,904–908,

San Miguel Amigo L, Franco Osorio R, Mercadal Vilchez S & Martínez-Francés A. 2011. Azacitidine adverse effects in patients with myelodysplastic syndromes. *Adv Ther,* 28 Suppl 4,6-11,

Scott BL & Estey E 2008. Management of myelodysplastic syndromes: 2008 update. *Oncology (Williston Park),* 22,12,1344-52,

Sekeres MA, Maciejewski JP, Giagounidis AA, Wride K, Knight R & Raza A. 2008. Relationship of treatment-related cytopenias and response to lenalidomide in patients with lower-risk myelodysplastic syndromes. *J Clin Oncol,*26, 5943–5949,

Sekeres MA. 2009. Treatment of MDS: something old, something new, something borrowed.... *Hematology Am Soc Hematol Educ Program,*656-63,

Sekeres MA. 2011a. Epidemiology, natural history, and practice patterns of patients with myelodysplastic syndromes in 2010. *J Natl Compr Canc Netw,*9,1,57-63,

Sekeres MA, List AF, Paulic K, Afable M, Englehaupt R & Maciejewski JP. 2011b. Demonstration of additional benefit in adding lenalidomide to azacitidine in patients with higher-risk myelodysplastic syndromes. *Am J Hematol,*86,1,102-3.

Silverman LR. 2009. Hypomethylating agents in myelodysplastic syndromes changing the inevitable: the value of azacitidine as maintenance therapy, effects on transfusion and combination with other agents. *Leuk Res*,33 Suppl 2,S18-21,

Sloand EM, Olnes MJ, Shenoy A, Weinstein B, Boss C, Loeliger K, Wu CO, More K, Barrett AJ, Scheinberg P & Young NS. 2010. Alemtuzumab treatment of intermediate-1 myelodysplasia patients is associated with sustained improvement in blood counts and cytogenetic remissions. *J Clin Oncol*,28,35,5166-73,

Sloand EM, Wu CO, Greenberg P, Young N & Barrett J. 2008. Factors affecting response and survival in patients with myelodysplasia treated with immunosuppressive therapy. *J Clin Oncol*,26,15,2505-11,

Sloand EM, Olnes MJ, Shenoy A, Weinstein B, Boss C, Loeliger K, Wu CO, More K, Barrett AJ, Scheinberg P & Young NS. 2010. Alemtuzumab treatment of intermediate-1 myelodysplasia patients is associated with sustained improvement in blood counts and cytogenetic remissions. *J Clin Oncol*,28,35,5166-73,

Sorror ML, Maris MB & Storb R. 2005. Hematopoietic cell transplantation (HCT)-specific comorbidity index: a new tool for risk assessment before allogeneic HCT. *Blood*,106,2912–2919,

Steensma DP. 2009. Myelodysplasia paranoia: iron as the new radon. *Leuk Res*,33,9,1158-63,

Steensma DP, Baer MR, Slack JL, Buckstein R, Godley LA & Garcia-Manero G. 2009. Multicenter study of decitabine administered daily for 5 days every 4 weeks to adults with myelodysplastic syndromes: the alternative dosing for outpatient treatment (ADOPT) trial. *J Clin Oncol*,27,3842–3848,

Tefferi A. 2008. Is "cure" essential in the treatment of cancer [letter]? *Mayo Clin Proc*,83,1413-1414,

Tefferi A. 2010. Myelodysplastic syndromes--many new drugs, little therapeutic progress. *Mayo Clin Proc*, 85,11,1042-5,

Virchis A, Koh M, Rankin P, Mehta A, Potter M, Hoffbrand AV & Prentice HG. 2004. Fludarabine, cytosine arabinoside, granulocyte-colony stimulating factor with or without idarubicin in the treatment of high risk acute leukaemia or myelodysplastic syndromes. *Br J Haematol*, 124,1,6-32,

Vigil CE, Martin-Santos T & Garcia-Manero G. 2010. Safety and efficacy of azacitidine in myelodysplastic syndromes. *Drug Des Devel Ther*, 4,221-9,

Warlick ED. 2010. Optimizing stem cell transplantation in myelodysplastic syndromes: unresolved questions. *Curr Opin Oncol*,22,2,150-4,

Wei S, Chen X, Rocha K, Epling-Burnette PK, Djeu JY, Liu Q, Byrd J, Sokol L, Lawrence N, Pireddu R, Dewald G, Williams A, Maciejewski J & List A. 2009. A critical role for phosphatase haplodeficiency in the selective suppression of deletion 5q MDS by lenalidomide. *Proc Natl Acad Sci U S A*,106(31),12974-9,

Wijermans P, Suciu S, Baila L, Platzbecker U, Giagounidis A & Selleslag D. 2009. Low dose decitabine versus best supportive care in elderly patients with intermediate or high risk MDS not eligible for intensive chemotherapy: final results of the randomized phase III study (06011) of the EORTC Leukemia and German MDS Study Groups. [abstract 226]. *Blood*, 114,22,

de Witte T, Brand R, van Biezen A, Delforge M, Biersack H, Or R, Meloni G, Bandini B, Sierra J, Kroger N, Gratwohl A & Niederwieser D. 2006. MDS subcommittee of the EBMT Chronic Leukemia Working Party. The role of stem cell source in autologous hematopoietic stem cell transplantation for patients with myelodysplastic syndromes. *Haematologica,*91,6,750-6,

de Witte T, Suciu S, Brand R, Muus P & Kröger N. 2007. Autologous stem cell transplantation in myelodysplastic syndromes. *Semin Hematol,*44,4,274-7,

de Witte T, Hagemeijer A, Suciu S, Belhabri A, Delforge M, Kobbe G, Selleslag D, Schouten HC, Ferrant A, Biersack H, Amadori S, Muus P, Jansen JH, Hellström-Lindberg E, Kovacsovics T, Wijermans P, Ossenkoppele G, Gratwohl A, Marie JP & Willemze R. 2010. Value of allogeneic versus autologous stem cell transplantation and chemotherapy in patients with myelodysplastic syndromes and secondary acute myeloid leukemia. Final results of a prospective randomized European Intergroup Trial. *Haematologica,*95,10,1754-61,

Ximeri M, Galanopoulos A, Klaus M, Parcharidou A, Giannikou K, Psyllaki M, Symeonidis A, Pappa V, Kartasis Z, Liapi D, Hatzimichael E, Kokoris S, Korkolopoulou P, Sambani C, Pontikoglou C & Papadaki HA. 2010. Hellenic MDS Study Group, Effect of lenalidomide therapy on hematopoiesis of patients with myelodysplastic syndrome associated with chromosome 5q deletion. *Haematologica,*95,3,406-14,

Yang X, Lay F, Han H & Jones PA. 2010. Targeting DNA methylation for epigenetic therapy. *Trends Pharmacol Sci,* 31,11,536-46,

Zipperer E , Pelz D, Nachtkamp K, Kuendgen A, Gattermann N, Haas R & Germing U. 2009. The hematopoietic stem cell transplantation comorbidity index is of prognostic relevance for patients with myelodysplastic syndrome. *Haematologica,*94,5,729-32,

Zy Ingram W, Brand R, Ho A, Kenyon M, Devereux S, Marsh J, Mufti GJ & Pagliuca. 2010. A Impact of pretransplant comorbidities on alemtuzumab-based reduced-intensity conditioning allogeneic hematopoietic SCT for patients with high-risk myelodysplastic syndrome and AML. *Bone Marrow Transplant,*45,4,633-9

15

Mantle Cell Lymphoma: Decision Making for Transplant

Yener Koc and Taner Demirer
Medical Park Hospital, Antalya and Ankara University Medical School, Ankara Turkey

1. Introduction

1.1 Definition and clinical characteristics

Mantle Cell Lymphoma (MCL) is a relatively rare type of mature B-cell lymphoma that comprises 5% of Non-Hodgkin's Lymphomas(NHL)[1-3]. MCL was added to the Revised European–American Lymphoma classification in 1994. Having both indolent and incurable features associated with aggressive clinical course, MCL is most frequently seen in 6th decade of life, with male dominance 3 to 4:1[2]. Malignant origin of MCL cells appear to derive from an antigen-naive pregerminal center cell[4, 5].

1. **B-cell neoplasms**
   1. Precursor B-cell
   2. Mature B-cell
      - Chronic lymphocytic leukemia/small lymphocytic lymphoma
      - Lymphoplasmacytic lymphoma
      - Splenic marginal zone lymphoma
      - Extranodal marginal zone B-cell lymphoma of MALT
      - Nodal marginal zone B-cell lymphoma
      - Follicular lymphoma
      - *Mantle cell lymphoma*
      - Diffuse large B-cell lymphoma
      - Mediastinal (thymic) large B-cell lymphoma
      - Intravascular large B-cell lymphoma
      - Primary effusion lymphoma
      - Burkitt's lymphoma/leukemia
   3. B-cell proliferations of uncertain malignant potential
2. **T-Cell and NK-Cell neoplasms**

Table 1. World Health Organization Classification of Lymphomas[1].

At the time of diagnosis, patients tend to have more extranodal disease and low serum albumin[6]. Although MCL has differential diagnosis with Chronic Lymphocytic Leukemia (CLL) and low-grade NHL, mimicking malignancies with indolent behavior, it follows an aggressive course with 10% to 15% long-term survivors[7] despite administration of standard chemotherapy courses commonly used in NHLs[8], corresponding to a median survival of 3 to 5 years.

2. Diagnosis

Diagnosis of MCL can be made by lymph node or bone marrow biopsy, or analysis of malignant cells obtained from peripheral blood, if the disease is in the leukemic phase[9, 10]. Differential diagnosis with CLL is important since both MCL and CLL cell have co-expression of CD5 and CD19/20[11]. Malignant cells are negative for CD10, CD23 and BCL6. Although absence of CD23 antigen expression on malignant cell population strongly favor a diagnosis of MCL[12, 13], presence of cyclinD1 expression by immunohistochemical staining[14] or determination of t(11;14) translocation by molecular analysis[15-18] is required to confirm the diagnosis[19]. Cyclin D1 is overexpressed in MCL as a result of the landmark t(11;14)(q13;q32) translocation[20]. Cyclin D1 complex with cyclin dependent kinases 4 and 6 (Cdk4 and Cdk6) and cyclin E-Cdk2, leading to phosphorylation of retinoblastoma protein (Rb), irreversibly inducing progression of the cell from G1 to S phase, which is not the only biologic dysregulation on the way to malignant transformation[21-24].

Disease	**CD5**	**CD10**	**CD23**	**CD43**	**Cyclin D1**	**Ig class**
FL	-	+	+/-	-	-	IgM, IgG
MCL	+	-	-	+	+	IgM/IgD
CLL/SLL	+	-	+	+	-	IgM/IgD
LPL	-	-	-	+/-	-	IgM (c)
MALT	-	-	-	+/-	-	IgM (c, s)
SMZL	-	-	-	-	-	IgM/IgD
HCL	-	-	-	-	-/+	IgG

FL, follicular lymphoma; MCL, mantle cell lymphoma; CLL/SLL, chronic lymphocytic leukemia/small lymphocytic lymphoma; LPL, Lymphoplasmacytic lymphoma; MALT, marginal zone lymphoma of MALT type; SMZL, splenic marginal zone lymphoma; HCL, hairy cell leukemia; Ig class, most commonly expressed heavy chain classes; c, cytoplasmic Ig; s, surface Ig.

Table 2. Differential Diagnosis of "Small" B-Cell Lymphomas[25].

3. Prognostic parameters

Prognosis can be estimated by using MIPI (mantle cell lymphoma international prognostic index, Figure 1)[26] which seems to be more efficient than international prognostic index (IPI)[27] or follicular lymphoma international prognostic index (FLIPI)[28, 29], which includes leukemic phase[10, 30, 31], besides other clinical parameters used in the IPI.

Points	Age, y	ECOG	LDHULN	WBC, 10^9/L
0	<50	0-1	<0.67	< 6.700
1	50-59	—	0.67-0.99	6.700-9.999
2	60-69	2-4	1.000 -1.49	1.000-14.999
3	≥70	—	≥1.5000	≥15000

Table 3. Simplified MIPI[26].

For each prognostic factor, 0 to 3 points were given to each patient and points were summed up to a maximum of 11. Patients with 0 to 3 points in summary were classified as low risk, patients with 4 to 5 points as intermediate risk, and patients with 6 to 11 points as high risk.

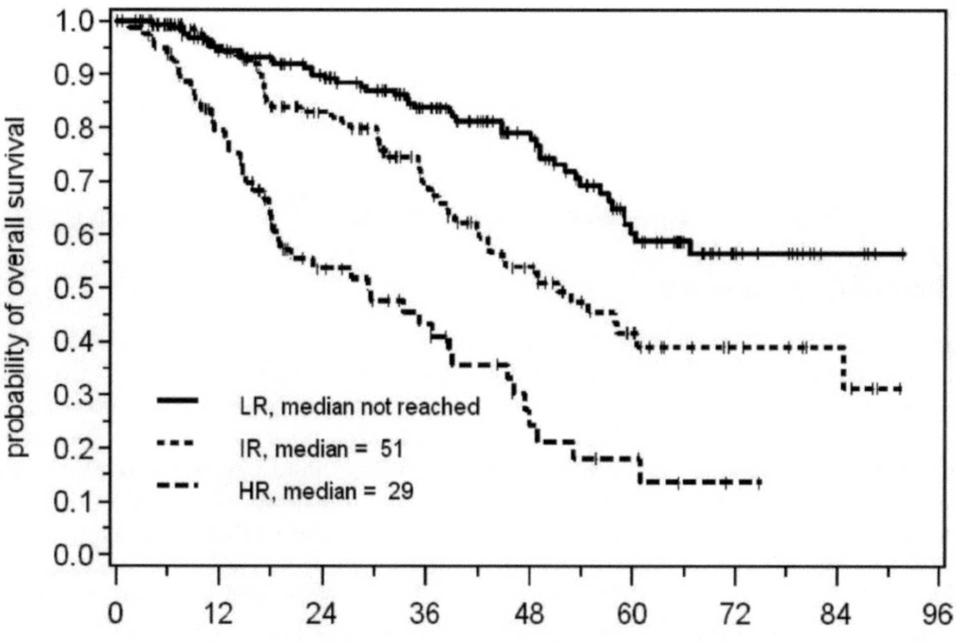

Fig. 1. Overall survival according to the mantle cell lymphoma international prognostic index (MIPI).

Although Ki-67 has previously been shown to predict prognosis (Figure 2)[32-34], analysis of Ki-67 did not substantially change the regression coefficient of the MIPI score and served as an important biologic marker with strong additional prognostic relevance[26].
Recently, proliferation gene expression signature has been reported to be the best molecular predictor of survival in patients with MCL[21], leading to a prognostic model defined as an optimized survival predictor composed of five genes: RAN, MYC, TNFRSF10B, POLE2, and SLC29A2[35]. Furthermore, this model was validated for application in formalin-fixed paraffin-embedded tissue samples and appeared superior to the immunohistochemical marker Ki-67[35].

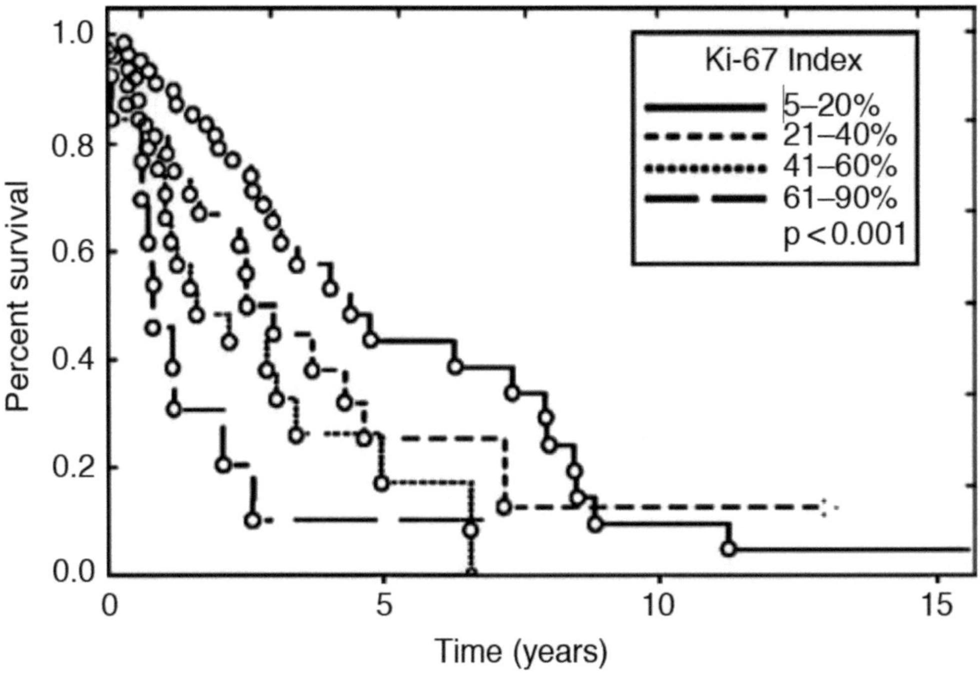

Fig. 2. Overall survival of patients with MCL according to Ki-67 proliferation index[33].

4. Role of chemotherapy

When it comes to assessment of long term overall survival, MCL has the worst prognosis among all lymphoma types[36], a watch-and-wait strategy should be avoided. Retrospective analyses of administration of standard chemotherapy have not shown any improvement in overall survival in patients with MCL[6, 37]. Most regiments induce around 80% response rates with complete remissions up to 30% of cases[6, 8]. While more aggressive regimens were used[38], reliable cure rates with conventional treatment has not been reported until now[36]. Compared to historical controls, median survival with conventional chemotherapy CVP and CHOP have not been improved[37]. In a randomized trial, these two regimens were associated with similar response rates (84% and 89%) and median overall survivals (32 and 37 months)[39]. Only one third of patients with untreated[40] or relapsed[41] MCL will respond to rituximab as single agent. Rituximab may also be used for purging tumor cells either with standard chemotherapy[42] or prior to high dose therapy regimens with stem cell support[43]. But in a randomized trial, addition of rituximab to CHOP regimen failed to improve overall survival (Figure 3)[44]. Maintenance interferon-alfa therapy following induction regimen has not been proven to improve survival either[45, 46].

Fludarabine can also be used in patients with lymphoma with or without rituximab combination[47]. Fludarabine alone in previously treated MCL can induce temporary responses in a third of patients lasting between 4 to 8 months[48]. In newly diagnosed patients, fludarabine appears to be more active, inducing responses in 60% , half of which is complete response[49]. Combining fludarabine with idarubicin or cyclophosphamide may improve induction of complete responses[50, 51]. Fludarabine and cyclophosphamide combination(FC) can induce a higher response rate of 63% in patients recurrent MCL[51]. Previously untreated patients have response rate up to 100% with a complete response rate of 70% and 28 months of progression free survival[51].

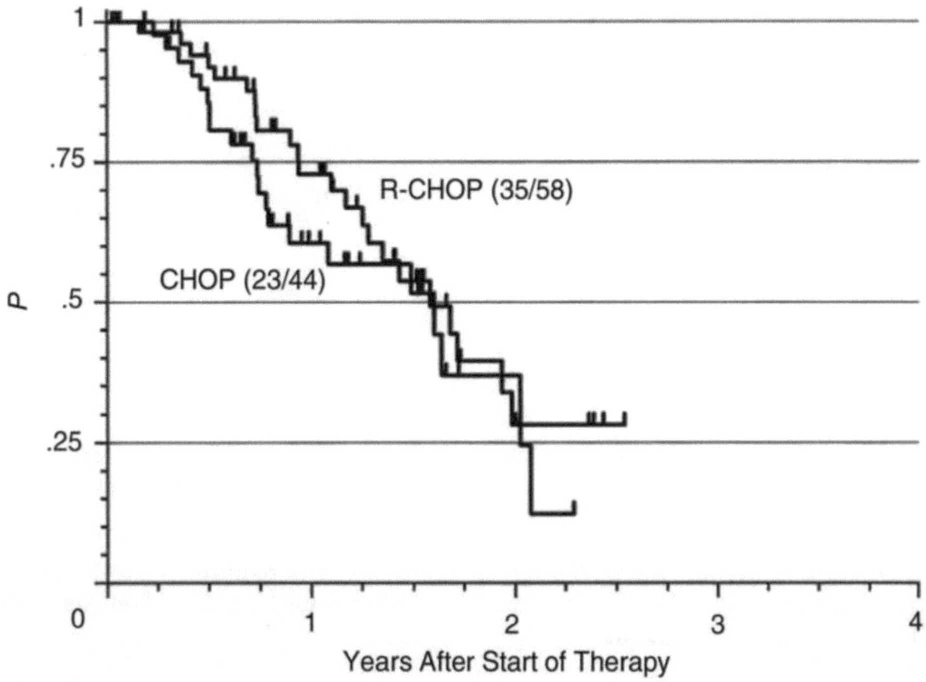

Patients at Risk						
R-CHOP	58	45	28	15	5	1
CHOP	44	35	17	10	3	

Fig. 3. Progression-free survival after cyclophosphamide, doxorubicin, vincristine, and prednisone (CHOP) and rituximab and CHOP (R-CHOP)[44].

In a randomized study testing addition of rituximab to fludarabine, cyclophosphamide and mitoxantrone (FCM), patients in the rituximab arm were found to have significant improvement in disease free and overall survival[52-54]. Fludarabine combined with cyclophosphamide and rituximab is a highly effective regimen in patients relapsing from previously received CHOP regimen[55]. Addition of rituximab to the chemotherapy regimens does not appear to increase toxicity[56].

Although tolerating patients treated with the hyper-CVAD regimen had excellent response and survival rates of greater than 90% at 3 years and 4.6 years of time to treatment failure (TTF at 8yr , 16% if age>65, 46% if age<65) following 10 year observation period[7, 38, 57, 58], patients treated with conventional therapies have also reported to have similar 3 year survival rates[59, 60]. The hyper-CVAD regimen without stem cell transplantation was not associated with a plateau in the survival curves (Figure 4)[57]. In this study, beta-2-microglobulin levels and IPI/MIPI scores were found to predict survival[57]. Hyper-CVAD regimen data may suggest high efficacy of the high-dose Ara-C (HIDAC) regimen, which needs to be tested in further clinical trials.

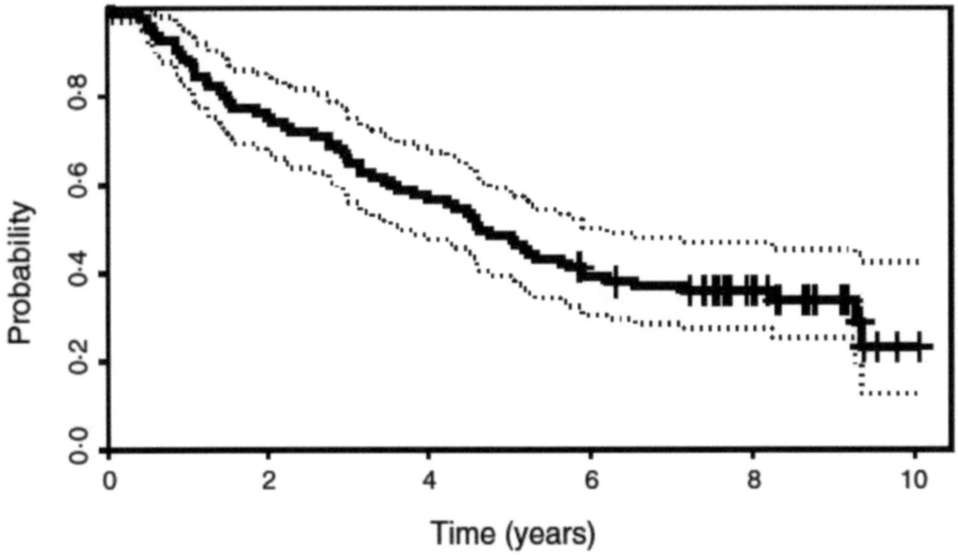

Fig. 4. Overall survival and time to failure in 97 patients treated with R hyper-CVAD alternating with R M/A[57].

Due to advanced age at the time of presentation of MCL, there is a concern about intensive therapy regimens being associated with higher toxicity rates requiring patient selection and patients tolerating intensive therapies are being selected for high-dose therapies with stem cell support.

5. New agents

Recently, bortezomib[61-63], lenalidomide[64-66], bendamustine[67-69], pixantrone[70], azaepothilone ixabepilone[71] and mTOR inhibitor temsirolimus[72-74] has been shown to demonstrate activity alone and in combination with rituximab and mitoxantrone, and may induce complete responses in relapsed or refractory MCL[61, 64, 69, 75, 76]. Along with rituximab, these biotherapy agents can be used frontline or may also be introduced into post transplant maintenance to prevent or treat relapses[61, 68, 77-81]. There may be an advantage of combining rituximab with bendamustine compared to combining with the conventional CHOP regimen in terms of disease free survival[82]. Recently, it was reported that frontline use of cladribine and rituximab can induce an overall response rate of 87%, with 61% of patients achieving complete remission, suggesting use of this combination for the initial treatment of MCL[83]. Radioimmunotherapy with ^{90}Y-labeled anti-CD20 monoclonal antibody (ibritumomab tiuxetan) or ^{131}I-rituximab is now being tested in phase I-II trials, including stem cell transplant setting[84-87].

6. High dose chemotherapy with autologous stem cell support (ASCT)

Since conventional and dose intense chemotherapy regimens have failed to induce cure or a plateau phase in time to treatment failure curves, high dose therapy with autologous stem cell support have been studied in the relapse setting[88] as well as consolidating complete responses following frontline chemotherapy regimens[88-94].

In a randomized trial testing ASCT versus IFN-alfa consolidation following response to CHOP or CHOP like regimens with or without rituximab, consolidation with BEAM/ASCT improved progression free survival (PFS) but has not been shown to improve overall survival (OS)(Figure 5)[93]. When patients responding to R-CHOP regimen received HIDAC therapy followed with autologous tranplantation with the BEAM regimen, also a plateau was not observed in time to treatment failure (TTF) and OS curves[94].

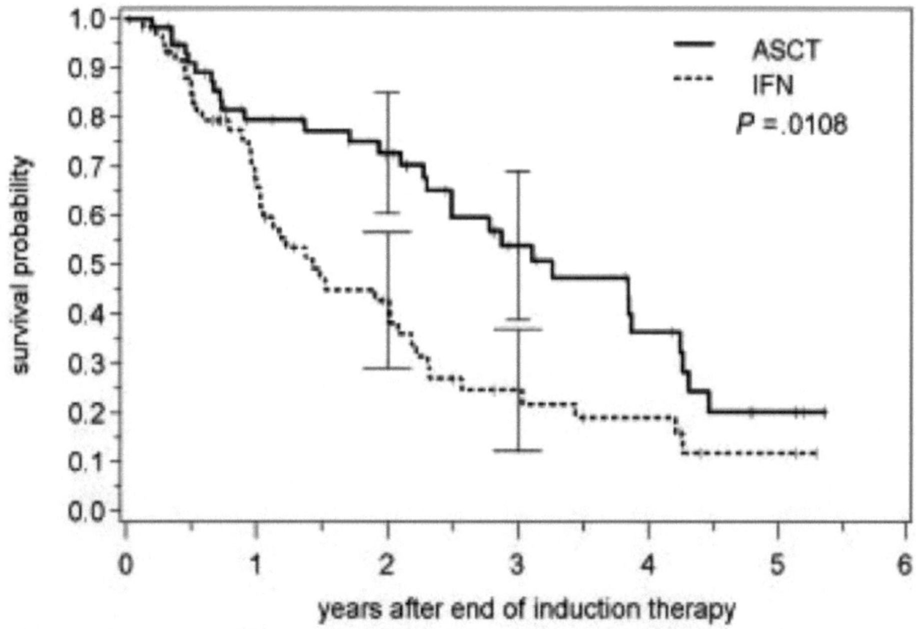

numbers of patients at risk

ASCT	62	38	31	17	10	3
IFN	60	33	19	9	6	2

Fig. 5. Progression-free survival after high-dose radiochemotherapy followed by ASCT and IFN-alfa maintenance in MCL[93].

Autologous transplantation with conventional conditioning regimens without in vivo purging induced a median survival of 47 months[88]. Analysis of this patient group did not reveal a plateau in the survival curve[88].

Addition of rituximab to ASCT regimen leads to better event free survival (EFS) curves in patients with first remission, not affecting progression in patients with relapsed/refractory disease[91].

The R-HDS regimen consisting of high-dose sequential chemotherapy (including intravenous administration of high-dose cyclophosphamide, high-dose cytarabine, high-dose melphalan, and high-dose mitoxantrone plus melphalan) and in vivo purging with rituximab resulted in OS and EFS rates at 54 months were 89% and 79%, respectively[96]. These results compare favorably with the 42% OS rate and the 18% EFS rate observed in 35 age-matched historic controls treated with standard-dose chemotherapy at the participating centers (Figure 6)[96].

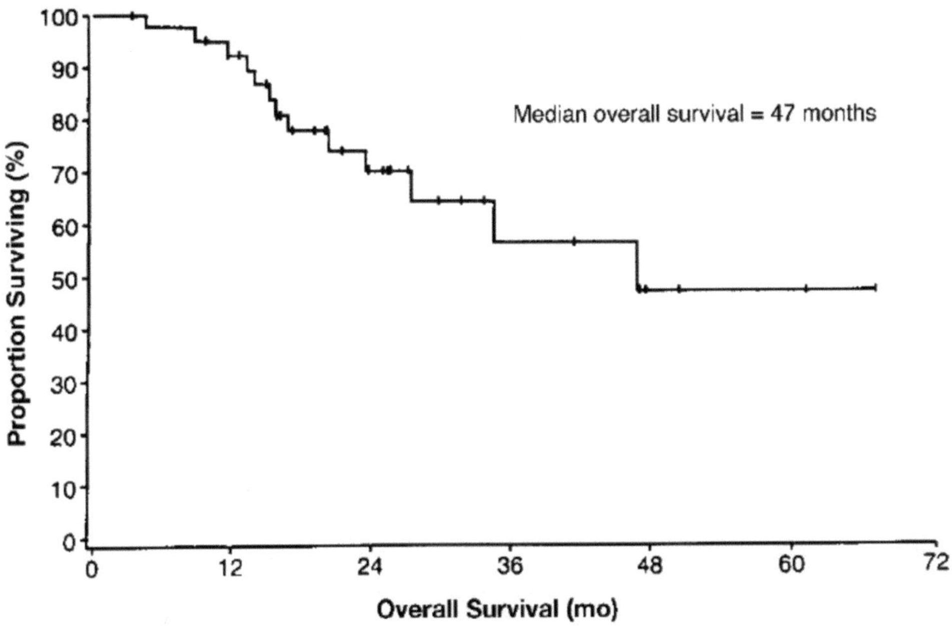

Fig. 6. Overall survival for mantle cell lymphoma patients after autologous transplantation without in vivo purging[88].

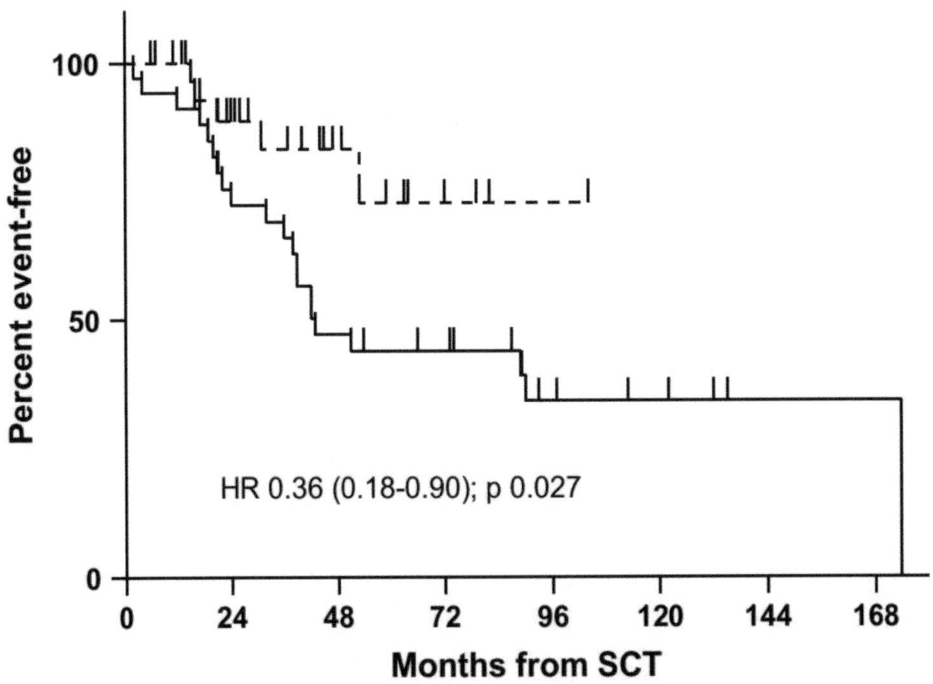

Fig. 7. EFS following ASCT after myeloablative therapy with (broken line) or without peritransplant rituximab (solid line)[95].

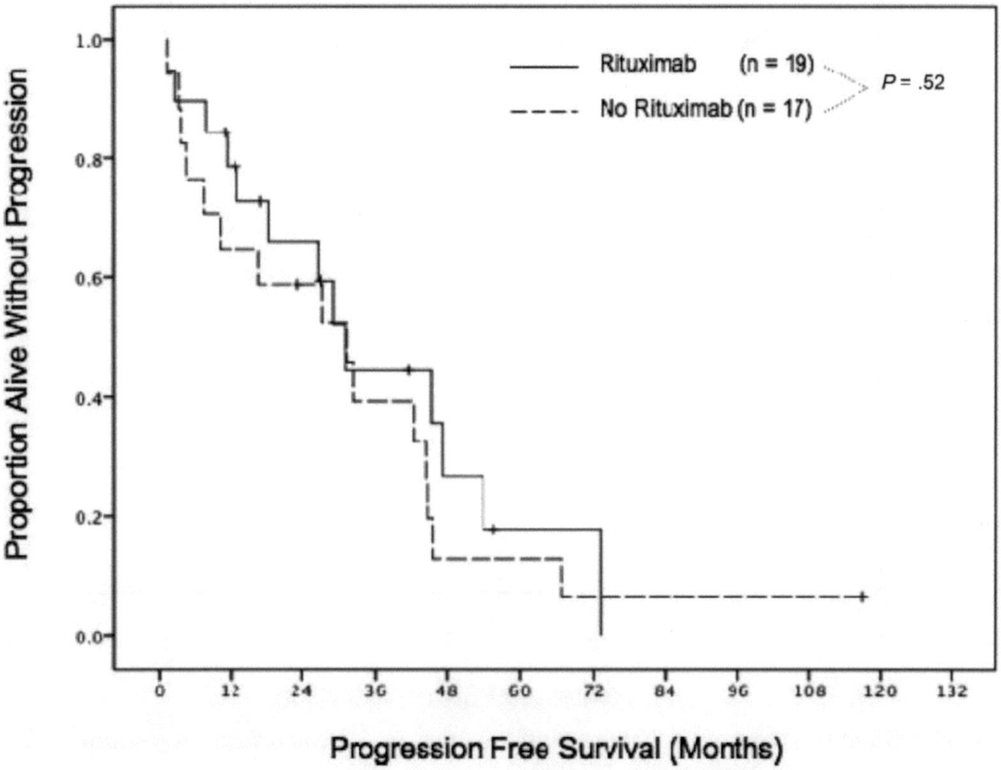

Fig. 8. PFS for patients receiving ASCT for relapsed/refractory disease[91].

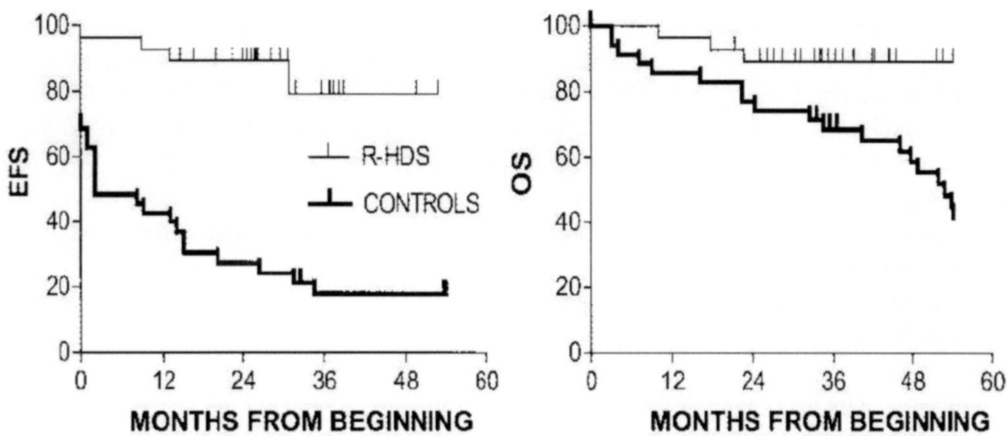

Fig. 9. Overall survival and event-free survival of patients treated with R-HDS versus conventional chemotherapy[96].

Heavily treated patients with multiple recurrences may undergo an effective salvage with high-dose radioimmunotherapy (RIT) coupled with autologous stem cell support[97].

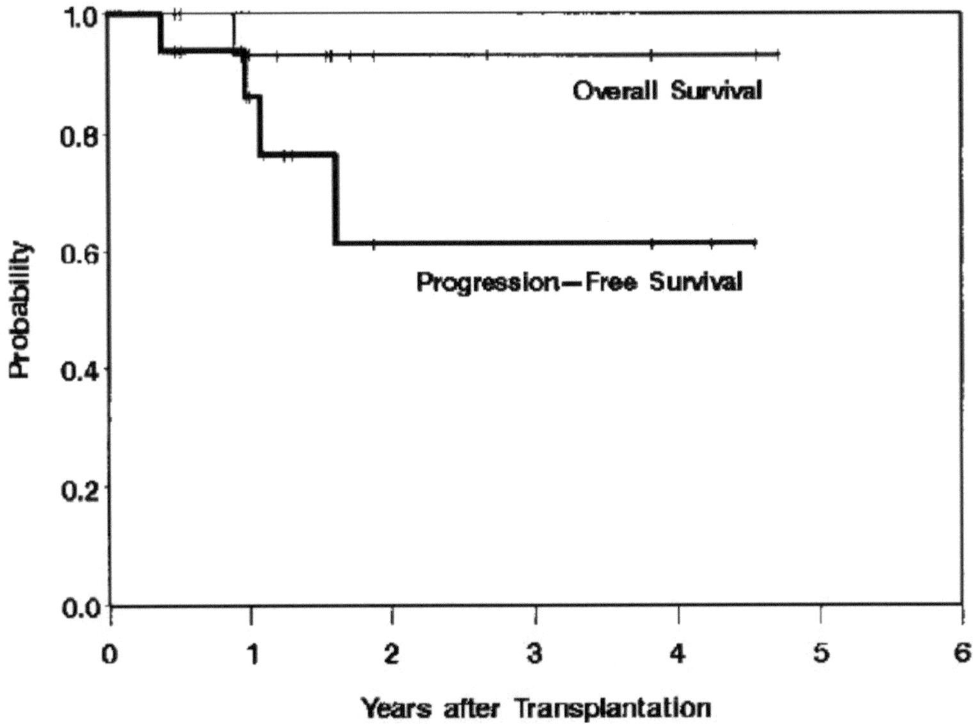

Fig. 10. OS and PFS obtained in patients with relapsed MCL treated with ^{131}I-tositumomab, etoposide, cyclophosphamide, and ASCT[97].

7. Allogeneic transplantation for achieving cure

Myeloablative or non-myeloablative allo-HSCT is generally performed in patients with lymphoma relapsing following auto-HSCT, since patients need tumor-free grafts that can induce a graft-versus-lymphoma (GVL) effect.

Physicians should consider the need for an allogeneic transplant when there is a need for a GVL effect (high risk of relapse) which can be predicted by the presence of;

- A high MIPI score
- Aggressive clinical behavior characterized by not achieving a satisfactory response to chemotherapy regimens
- remaining PET positive following ASCT[98]
- Multiple relapses

and;

- If the patient is young (<55 years) or,
- Autologous stem cells cannot be mobilized

in patients considered to be eligible following pretransplant screening tests.

Evidence for presence of graft versus lymphoma (GVL) effect has been reported in patients with MCL who underwent allo-HSCT[99]. CVL effect may be observed as conversion to pcr negativity for t(11;14) or achieving CR in the presence of GVHD[99], or observing lower relapse rates in allo-HSCT recipients compared to patients undergoing autologous transplantation[100-102]. A healthy comparison of autologous and allogeneic transplants cannot be made at present, due to lack of randomized trials and the different prognostic groups undergoing each transplant type. By using myeloablative regimens for relapsed patients, allogeneic transplants can induce three year event free survivals (EFS)around 50%[99, 101-104]. However, if performed in first CR or PR, EFS induced by allo transplants at three years can be as high as 70%[103].

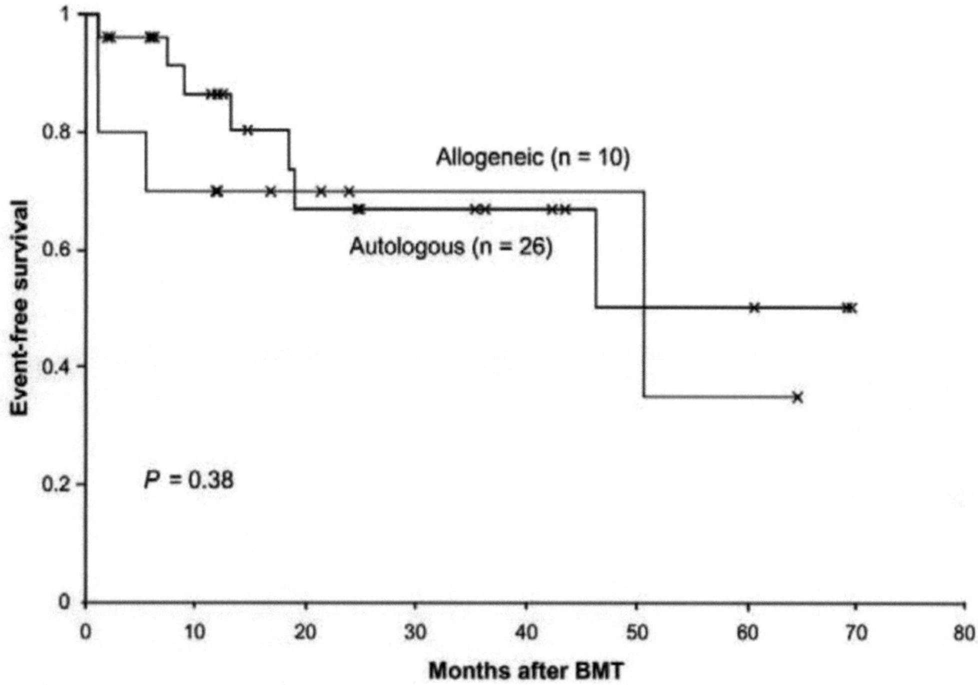

Fig. 11. Event-free survival according to type of BMT[103].

8. Reduced intensity allogeneic stem cell transplantation

In the allogeneic transplant setting, benefit of cure over transplant related mortality can be positively affected by using reduced intensity (RIC) or non-myeloablative (NMA) conditioning regimens[105-116]. Using donor lymphocyte infusions for GVL effect in patients undergoing allogeneic transplantation with RIC or NMA regimens, low transplant mortality rates (<10%) and higher OS (73 to 85%) and EFS rates (73 to 82%) can be achieved[105, 117, 118].

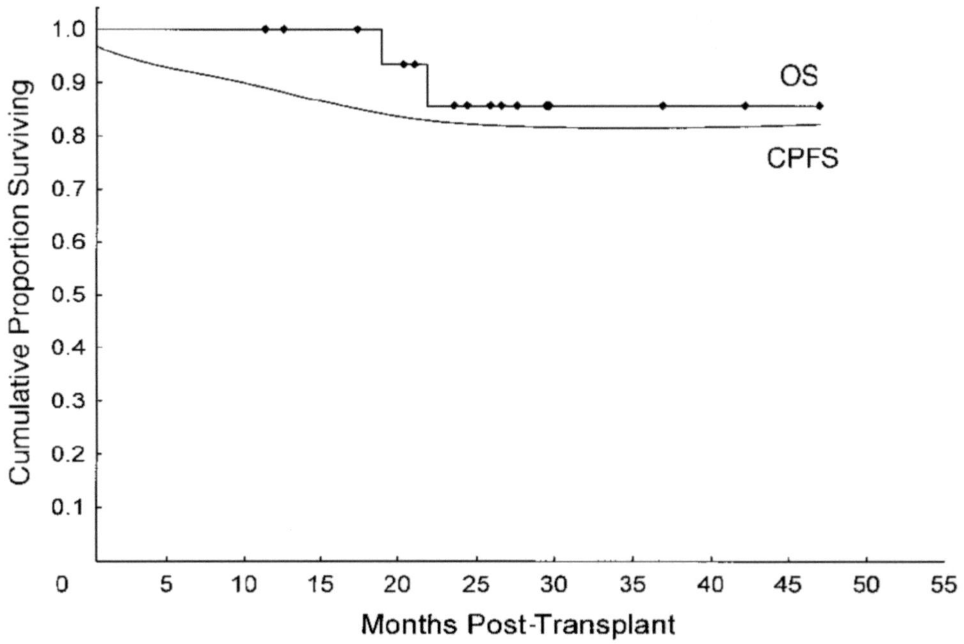

Fig. 12. Kaplan-Meier overall survival (OS) and current progression-free survival (CPFS) accounting for salvage post-DLI in with relapsed MCL who received NMA allogeneic transplantation[105].

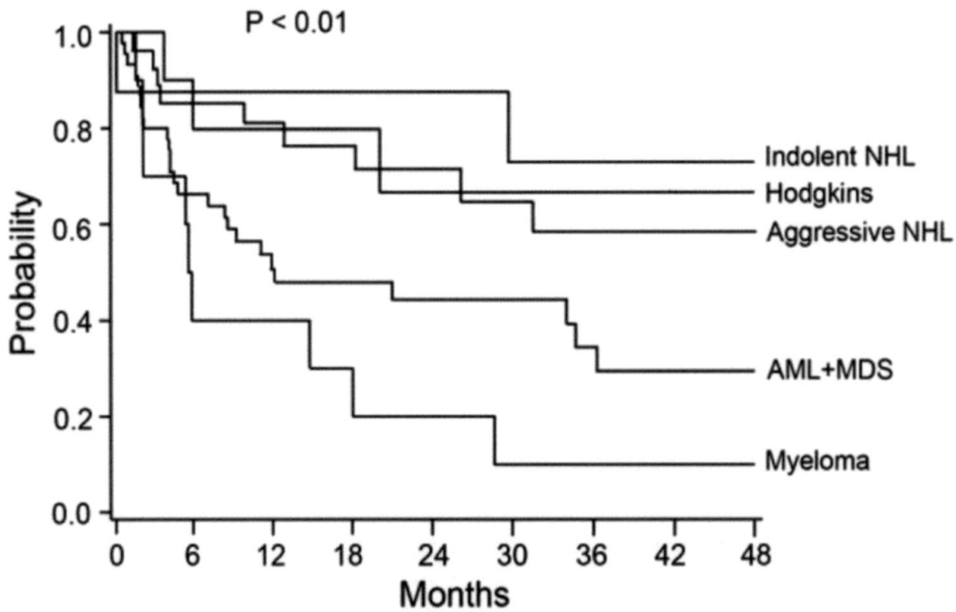

Fig. 13. Favorable overall survival following reduced intensity allogeneneic transplants by disease group[117].

9. Decision making for transplant options

While standard chemotherapy regimens do not offer long term progression free survival, intensive therapy followed by ASCT should be considered for each newly diagnosed patient[92], especially if the MIPI score is translating into poor prognosis.
Multiply relapsed patients will not do well after autologous transplants[91, 103]. Best time to perform an autologous transplant is following CR[91, 92, 119, 120] obtained after high-dose therapy with hyperCVAD or HDS regimen, coupled with rituximab[91], including in vivo purging prior to stem cell collection[81, 92, 96]. With this approach, patients eligible for these sequential intensive regimens may enjoy long term disease free survival, although cure cannot be achieved[121].

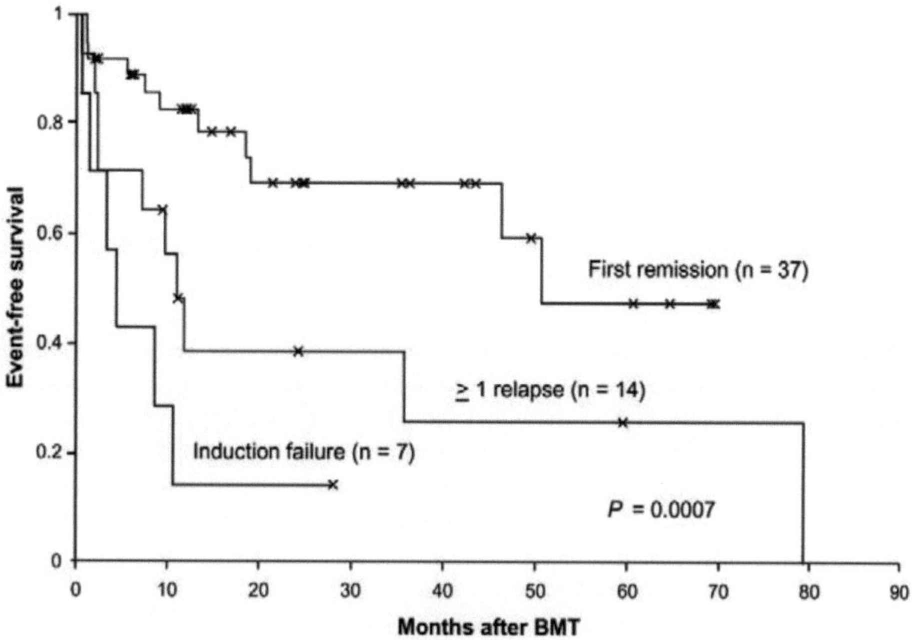

Fig. 14. Event-free survival according to remission status at BMT[103].

Newly diagnosed and relatively younger patients less than 60 years of age who are not eligible for ASCT or relapsing after ASCT, or patients with recurrence following chemotherapy, can still be cured by allogeneic SCT following an effective salvage regimen inducing CR or near CR[121].
As the outcome of transplantation is most promising in the newly diagnosed patients without chemorefractory disease, graft contamination and the lack of a survival plateau following autologous transplantation, allogeneic transplantation deserve investigation as the upfront therapy in the management of patients with MCL[92].
Patients over age 65, and individuals who are not eligible for ASCT with poor prognostic MIPI scores remain as a major challenge for the hematologist. In this group of patients, promising new drugs such as bendamustine, cladribine and bortezomib in combination with synergizing agents such as rituximab, may be offered to patients in the context of clinical trials. One should remember that ASCT with intermediate dose melphalan at doses

between 100 to 140 mg/m^2 can still be utilized in patients over age 65, as well as non-myeloablative allogeneic stem cell transplants, immediately after achieving a clinical complete response in experienced transplant centers. These approaches have to be tested in clinical trials in the near future.

10. Rational approaches in managing patients with MCL

Due to heterogeneity of disease, paucity of randomized trials, availability of a wide range of chemotherapy regimens including new agents, and altering prognostic factors among patients, an individualized treatment approach should be adapted for each patient diagnosed with MCL. As in therapy of patients with myeloma, long term survival may be achieved following sequential and risk-adapted use of available therapies such as intensive or nucleoside analogue based chemotherapy, followed by autologous and/or reduced intensity allogeneic transplantation, including radioimmunotherapy.

Rational strategy at present can be outlined as;

1. After assessing the MIPI score, low risk patients can be followed without transplantation strategy, because of late age of onset, which is over 60 in most patients. This group of patients will do well with standard chemotherapy regimens utilized in NHLs, such as R-CHOP or new generation combination regimens including a purine analogue and rituximab, such as R-FCM or FCR.
2. Due to its poor prognosis and failure to achieve improved survival curves with conventional chemotherapy regimens, consideration should be given to upfront autologous transplantation in patients with high MIPI scores. To induce remission FCR, R-hyperCVAD and R-CHOP-14 regimens seem to be effective approaches. Inducing high quality complete responses -molecular remission for t(11;14) translocation- prior to autologous transplantation may further improve disease free and overall survival curves. Success of this strategy has been shown in comparison to historical controls, and also has to be proven in randomized clinical trials.
3. Young patients with intermediate MIPI scores may choose to continue with the same approach with the ones with high scores as described above. Elderly patients may receive R-CHOP, R-FCM or FCR as the initial therapy.
4. Patients having recurrence following conventional chemotherapy or an autologous transplant and who are determined to be fit for an allogeneic transplant following a careful evaluation and screening tests should be given a chance to have this treatment option to achieve cure.
5. Regardless of the MIPI score, patients who are not eligible for high dose therapy and autologous stem cell support and relapsing after conventional chemotherapy regimens and not a candidate for an allogeneic transplant may be treated with bendamustine and cladribine containing regimens in the context of clinical trials.

Transplant strategy can be summarized as;

Patients with predicted poor survival by MIPI score should undergo intensive induction regimens if eligible, to achieve CR, followed by ASCT preferably in first remission. If cure is targeted, especially in young patients, allogeneic SCT may be performed following an excellent cytoreduction with ASCT or intensive chemotherapy regimen including high dose Ara-C and rituximab, such as R-HDS or hyper-CVAD regimens.

11. Emerging new drugs to improve results in the near future

New agents which may have potential to improve outcome of relapsed or refractory patients with MCL include chemicals targeting cyclin D1 and the cell cycle regulatory proteins (cdk), inhibitors of mammalian target of rapamycin, the proteasome, and proapoptotic family members[122].

Both histone deacetylase (HDAC) and mTOR inhibitors may downregulate cyclin D1 reducing cell cycle drive, which is upregulated in patients with MCL. Cyclin D1 is the key factor for upregulating cellular proliferation rate[22]. HDAC inhibitor sueroylanilide hydroxamic acid (vorinostat) has ability to reduce intracellular cyclin D1 levels in MCL cells, leading to inhibition of progression to S-phase in the cell cycle, and may induce response in MCL patients, currently being tested in phase I clinical trials[123]. Cyclin D1 protein itself can be used as a target for sensitized T-cells, generating a potential treatment option by cellular therapy[124].

Other than overexpression of cyclin D1, activation of mammalian target of rapamycin (mTOR) also plays a major role in growth and proliferation in MCL cells[125]. Inhibitiors of mTOR (rapamycin, temsirolimus, everolimus) may increase apoptosis and decrease proliferation rate by causing G1 arrest in the cell cycle[126]. Single agent temsirolimus has been shown to have clinical activity in phase II trials, and a randomized phase III trial[127-129]. Alternative mTOR inhibitor everolimus is currently in phase II trials[130, 131].

Deficiencies of Noxa and Bim proteins and aberrant expression of Bcl-2 may reduce apoptosis and lead to increased resistance to chemotherapeutic agents[132]. Proteasome inhibitors such as bortezomib and flavopiridol may function as accumulating cdk inhibitors (p21/p27), causing cell cycle arrest in malignant cells[132-134]. Accumulation of proapoptotic Noxa protein has been consistently shown in MCL cells treated with bortezomib[135].

There is a rationale combining bortezomib with HDAC inhibitor vorinostat. Vorinostat may be able to turn off cyclin D1, whereas bortezomib turns on cdk inhibitors p27 and p21, targeting two complementary genetic lesions involved in the cell cycle aberration in MCL cells which may lead to induction of apoptosis[136]. Bortezomib may also be combined with cyproheptadine, recently identified inhibitor of cyclin D1, synergistic in inducing apoptosis in vitro[137].

Development of new drugs for MCL will further improve the recent achievements in the management options published in the last decade, in a disease currently not curable by medical therapy except allogeneic transplantation.

12. References

[1] Jaffe ESHNL, Stein H., et al:. Tumours of haematopoietic and lymphoid tissues *In: Jaffe ES, et al ed. World Health Organization Classification of Tumours, Lyon, France: IARC Press* 2001: 1-352.

[2] Lenz G, Dreyling M, Hiddemann W. Mantle Cell Lymphoma. *In: Canellos G.P. e al ed. The Lymphomas. 2nd Edition Philadelphia, USA: Elsevier Inc.* 2006: 397-405.

[3] Velders GA, Kluin-Nelemans JC, De Boer CJ, Hermans J, Noordijk EM, Schuuring E *et al.* Mantle-cell lymphoma: a population-based clinical study. *J Clin Oncol* 1996; 14(4): 1269-74.

[4] Walsh SH, Rosenquist R. Immunoglobulin gene analysis of mature B-cell malignancies: reconsideration of cellular origin and potential antigen involvement in pathogenesis. *Med Oncol* 2005; 22(4): 327-41.
[5] Welzel N, Le T, Marculescu R, Mitterbauer G, Chott A, Pott C *et al.* Templated nucleotide addition and immunoglobulin JH-gene utilization in t(11;14) junctions: implications for the mechanism of translocation and the origin of mantle cell lymphoma. *Cancer Res* 2001; 61(4): 1629-36.
[6] Vandenberghe E, De Wolf-Peeters C, Vaughan Hudson G, Vaughan Hudson B, Pittaluga S, Anderson L *et al.* The clinical outcome of 65 cases of mantle cell lymphoma initially treated with non-intensive therapy by the British National Lymphoma Investigation Group. *Br J Haematol* 1997; 99(4): 842-7.
[7] Dreyling M. Hyper-CVAD in mantle-cell lymphoma: really "hyper" or just hype? *Leuk Lymphoma* 2008; 49(6): 1017-8.
[8] Zucca E, Roggero E, Pinotti G, Pedrinis E, Cappella C, Venco A *et al.* Patterns of survival in mantle cell lymphoma. *Ann Oncol* 1995; 6(3): 257-62.
[9] Singleton TP, Anderson MM, Ross CW, Schnitzer B. Leukemic phase of mantle cell lymphoma, blastoid variant. *Am J Clin Pathol* 1999; 111(4): 495-500.
[10] Schlette E, Lai R, Onciu M, Doherty D, Bueso-Ramos C, Medeiros LJ. Leukemic mantle cell lymphoma: clinical and pathologic spectrum of twenty-three cases. *Mod Pathol* 2001; 14(11): 1133-40.
[11] Bentz M, Plesch A, Bullinger L, Stilgenbauer S, Ott G, Muller-Hermelink HK *et al.* t(11;14)-positive mantle cell lymphomas exhibit complex karyotypes and share similarities with B-cell chronic lymphocytic leukemia. *Genes Chromosomes Cancer* 2000; 27(3): 285-94.
[12] Zukerberg LR, Medeiros LJ, Ferry JA, Harris NL. Diffuse low-grade B-cell lymphomas. Four clinically distinct subtypes defined by a combination of morphologic and immunophenotypic features. *Am J Clin Pathol* 1993; 100(4): 373-85.
[13] Pittaluga S, Wlodarska I, Stul MS, Thomas J, Verhoef G, Cassiman JJ *et al.* Mantle cell lymphoma: a clinicopathological study of 55 cases. *Histopathology* 1995; 26(1): 17-24.
[14] de Boer CJ, Schuuring E, Dreef E, Peters G, Bartek J, Kluin PM *et al.* Cyclin D1 protein analysis in the diagnosis of mantle cell lymphoma. *Blood* 1995; 86(7): 2715-23.
[15] Fan H, Gulley ML, Gascoyne RD, Horsman DE, Adomat SA, Cho CG. Molecular methods for detecting t(11;14) translocations in mantle-cell lymphomas. *Diagn Mol Pathol* 1998; 7(4): 209-14.
[16] Samaha H, Dumontet C, Ketterer N, Moullet I, Thieblemont C, Bouafia F *et al.* Mantle cell lymphoma: a retrospective study of 121 cases. *Leukemia* 1998; 12(8): 1281-7.
[17] Espinet B, Sole F, Woessner S, Bosch F, Florensa L, Campo E *et al.* Translocation (11;14)(q13;q32) and preferential involvement of chromosomes 1, 2, 9, 13, and 17 in mantle cell lymphoma. *Cancer Genet Cytogenet* 1999; 111(1): 92-8.
[18] Li JY, Gaillard F, Moreau A, Harousseau JL, Laboisse C, Milpied N *et al.* Detection of translocation t(11;14)(q13;q32) in mantle cell lymphoma by fluorescence in situ hybridization. *Am J Pathol* 1999; 154(5): 1449-52.
[19] Lai R, Medeiros LJ. Pathologic diagnosis of mantle cell lymphoma. *Clin Lymphoma* 2000; 1(3): 197-206; discussion 207-8.

[20] de Boer CJ, van Krieken JH, Kluin-Nelemans HC, Kluin PM, Schuuring E. Cyclin D1 messenger RNA overexpression as a marker for mantle cell lymphoma. *Oncogene* 1995; 10(9): 1833-40.
[21] Rosenwald A, Wright G, Wiestner A, Chan WC, Connors JM, Campo E *et al.* The proliferation gene expression signature is a quantitative integrator of oncogenic events that predicts survival in mantle cell lymphoma. *Cancer Cell* 2003; 3(2): 185-97.
[22] Wiestner A, Tehrani M, Chiorazzi M, Wright G, Gibellini F, Nakayama K *et al.* Point mutations and genomic deletions in CCND1 create stable truncated cyclin D1 mRNAs that are associated with increased proliferation rate and shorter survival. *Blood* 2007; 109(11): 4599-606.
[23] Garcia-Conde J, Cabanillas F. Mantle cell lymphoma: a lymphoproliferative disorder associated with aberrant function of the cell cycle. *Leukemia* 1996; 10 Suppl 2: s78-83.
[24] Hartwell LH, Kastan MB. Cell cycle control and cancer. *Science* 1994; 266(5192): 1821-8.
[25] Feldman AL, Pittaluga S, Jaffe ES. Classification and Histopathology of the Lymphomas. *In: Canellos G.P. et al. The Lymphomas. 2nd Edition Philadelphia, USA: Elsevier Inc.* 2006: 2-38.
[26] Hoster E, Dreyling M, Klapper W, Gisselbrecht C, van Hoof A, Kluin-Nelemans HC *et al.* A new prognostic index (MIPI) for patients with advanced-stage mantle cell lymphoma. *Blood* 2008; 111(2): 558-65.
[27] Shipp MA. Prognostic factors in aggressive non-Hodgkin's lymphoma: who has "high-risk" disease? *Blood* 1994; 83(5): 1165-73.
[28] Moller MB, Pedersen NT, Christensen BE. Mantle cell lymphoma: prognostic capacity of the Follicular Lymphoma International Prognostic Index. *Br J Haematol* 2006; 133(1): 43-9.
[29] Solal-Celigny P, Roy P, Colombat P, White J, Armitage JO, Arranz-Saez R *et al.* Follicular lymphoma international prognostic index. *Blood* 2004; 104(5): 1258-65.
[30] Tiemann M, Schrader C, Klapper W, Dreyling MH, Campo E, Norton A *et al.* Histopathology, cell proliferation indices and clinical outcome in 304 patients with mantle cell lymphoma (MCL): a clinicopathological study from the European MCL Network. *Br J Haematol* 2005; 131(1): 29-38.
[31] Geisler CH, Kolstad A, Laurell A, Raty R, Jerkeman M, Eriksson M *et al.* The Mantle Cell Lymphoma International Prognostic Index (MIPI) is superior to the International Prognostic Index (IPI) in predicting survival following intensive first-line immunochemotherapy and autologous stem cell transplantation (ASCT). *Blood* 2010; 115(8): 1530-3.
[32] Determann O, Hoster E, Ott G, Wolfram Bernd H, Loddenkemper C, Leo Hansmann M *et al.* Ki-67 predicts outcome in advanced-stage mantle cell lymphoma patients treated with anti-CD20 immunochemotherapy: results from randomized trials of the European MCL Network and the German Low Grade Lymphoma Study Group. *Blood* 2008; 111(4): 2385-7.

[33] Katzenberger T, Petzoldt C, Holler S, Mader U, Kalla J, Adam P *et al.* The Ki67 proliferation index is a quantitative indicator of clinical risk in mantle cell lymphoma. *Blood* 2006; 107(8): 3407.

[34] Klapper W, Hoster E, Determann O, Oschlies I, van der Laak J, Berger F *et al.* Ki-67 as a prognostic marker in mantle cell lymphoma-consensus guidelines of the pathology panel of the European MCL Network. *J Hematop* 2009.

[35] Hartmann E, Fernandez V, Moreno V, Valls J, Hernandez L, Bosch F *et al.* Five-gene model to predict survival in mantle-cell lymphoma using frozen or formalin-fixed, paraffin-embedded tissue. *J Clin Oncol* 2008; 26(30): 4966-72.

[36] Wilson WH, Armitage JO. Non Hodgkin's Lymphoma. *In: Abeloff, M.D. et al ed. Abeloff's Clinical Oncology* 2008: 2371-2404.

[37] Zucca E, Fontana S, Roggero E, Pedrinis E, Pampallona S, Cavalli F. Treatment and prognosis of centrocytic (mantle cell) lymphoma: a retrospective analysis of twenty-six patients treated in one institution. *Leuk Lymphoma* 1994; 13(1-2): 105-10.

[38] Khouri IF, Romaguera J, Kantarjian H, Palmer JL, Pugh WC, Korbling M *et al.* Hyper-CVAD and high-dose methotrexate/cytarabine followed by stem-cell transplantation: an active regimen for aggressive mantle-cell lymphoma. *J Clin Oncol* 1998; 16(12): 3803-9.

[39] Meusers P, Engelhard M, Bartels H, Binder T, Fulle HH, Gorg K *et al.* Multicentre randomized therapeutic trial for advanced centrocytic lymphoma: anthracycline does not improve the prognosis. *Hematol Oncol* 1989; 7(5): 365-80.

[40] Foran JM, Rohatiner AZ, Cunningham D, Popescu RA, Solal-Celigny P, Ghielmini M *et al.* European phase II study of rituximab (chimeric anti-CD20 monoclonal antibody) for patients with newly diagnosed mantle-cell lymphoma and previously treated mantle-cell lymphoma, immunocytoma, and small B-cell lymphocytic lymphoma. *J Clin Oncol* 2000; 18(2): 317-24.

[41] Coiffier B, Haioun C, Ketterer N, Engert A, Tilly H, Ma D *et al.* Rituximab (anti-CD20 monoclonal antibody) for the treatment of patients with relapsing or refractory aggressive lymphoma: a multicenter phase II study. *Blood* 1998; 92(6): 1927-32.

[42] Howard OM, Gribben JG, Neuberg DS, Grossbard M, Poor C, Janicek MJ *et al.* Rituximab and CHOP induction therapy for newly diagnosed mantle-cell lymphoma: molecular complete responses are not predictive of progression-free survival. *J Clin Oncol* 2002; 20(5): 1288-94.

[43] Gianni AM, Cortelazzo S, Magni M, Martelli M. Rituximab: enhancing stem cell transplantation in mantle cell lymphoma. *Bone Marrow Transplant* 2002; 29 Suppl 1: S10-3.

[44] Lenz G, Dreyling M, Hoster E, Wormann B, Duhrsen U, Metzner B *et al.* Immunochemotherapy with rituximab and cyclophosphamide, doxorubicin, vincristine, and prednisone significantly improves response and time to treatment failure, but not long-term outcome in patients with previously untreated mantle cell lymphoma: results of a prospective randomized trial of the German Low Grade Lymphoma Study Group (GLSG). *J Clin Oncol* 2005; 23(9): 1984-92.

[45] Hiddemann W, Unterhalt M, Herrmann R, Woltjen HH, Kreuser ED, Trumper L *et al.* Mantle-cell lymphomas have more widespread disease and a slower response to

chemotherapy compared with follicle-center lymphomas: results of a prospective comparative analysis of the German Low-Grade Lymphoma Study Group. *J Clin Oncol* 1998; 16(5): 1922-30.

[46] Teodorovic I, Pittaluga S, Kluin-Nelemans JC, Meerwaldt JH, Hagenbeek A, van Glabbeke M *et al.* Efficacy of four different regimens in 64 mantle-cell lymphoma cases: clinicopathologic comparison with 498 other non-Hodgkin's lymphoma subtypes. European Organization for the Research and Treatment of Cancer Lymphoma Cooperative Group. *J Clin Oncol* 1995; 13(11): 2819-26.

[47] Lenz G, Hiddemann W, Dreyling M. The role of fludarabine in the treatment of follicular and mantle cell lymphoma. *Cancer* 2004; 101(5): 883-93.

[48] Decaudin D, Bosq J, Tertian G, Nedellec G, Bennaceur A, Venuat AM *et al.* Phase II trial of fludarabine monophosphate in patients with mantle-cell lymphomas. *J Clin Oncol* 1998; 16(2): 579-83.

[49] Zinzani PL, Magagnoli M, Moretti L, Battista R, Ronconi F, De Renzo A *et al.* Fludarabine-based chemotherapy in untreated mantle cell lymphomas: an encouraging experience in 29 patients. *Haematologica* 1999; 84(11): 1002-6.

[50] Zinzani PL, Magagnoli M, Moretti L, De Renzo A, Battista R, Zaccaria A *et al.* Randomized trial of fludarabine versus fludarabine and idarubicin as frontline treatment in patients with indolent or mantle-cell lymphoma. *J Clin Oncol* 2000; 18(4): 773-9.

[51] Cohen BJ, Moskowitz C, Straus D, Noy A, Hedrick E, Zelenetz A. Cyclophosphamide/fludarabine (CF) is active in the treatment of mantle cell lymphoma. *Leuk Lymphoma* 2001; 42(5): 1015-22.

[52] Forstpointner R, Dreyling M, Repp R, Hermann S, Hanel A, Metzner B *et al.* The addition of rituximab to a combination of fludarabine, cyclophosphamide, mitoxantrone (FCM) significantly increases the response rate and prolongs survival as compared with FCM alone in patients with relapsed and refractory follicular and mantle cell lymphomas: results of a prospective randomized study of the German Low-Grade Lymphoma Study Group. *Blood* 2004; 104(10): 3064-71.

[53] Forstpointner R, Unterhalt M, Dreyling M, Bock HP, Repp R, Wandt H *et al.* Maintenance therapy with rituximab leads to a significant prolongation of response duration after salvage therapy with a combination of rituximab, fludarabine, cyclophosphamide, and mitoxantrone (R-FCM) in patients with recurring and refractory follicular and mantle cell lymphomas: Results of a prospective randomized study of the German Low Grade Lymphoma Study Group (GLSG). *Blood* 2006; 108(13): 4003-8.

[54] Hiddemann W, Dreyling M, Unterhalt M. Rituximab plus chemotherapy in follicular and mantle cell lymphomas. *Semin Oncol* 2003; 30(1 Suppl 2): 16-20.

[55] Thomas DW, Owen RG, Johnson SA, Hillmen P, Seymour JF, Wolf MM *et al.* Superior quality and duration of responses among patients with mantle-cell lymphoma treated with fludarabine and cyclophosphamide with or without rituximab compared with prior responses to CHOP. *Leuk Lymphoma* 2005; 46(4): 549-52.

[56] Eve HE, Linch D, Qian W, Ross M, Seymour JF, Smith P *et al.* Toxicity of fludarabine and cyclophosphamide with or without rituximab as initial therapy for patients

with previously untreated mantle cell lymphoma: results of a randomised phase II study. *Leuk Lymphoma* 2009; 50(2): 211-5.

[57] Romaguera JE, Fayad LE, Feng L, Hartig K, Weaver P, Rodriguez MA *et al.* Ten-year follow-up after intense chemoimmunotherapy with Rituximab-HyperCVAD alternating with Rituximab-high dose methotrexate/cytarabine (R-MA) and without stem cell transplantation in patients with untreated aggressive mantle cell lymphoma. *Br J Haematol* 2010; 150(2): 200-8.

[58] Romaguera JE, Khouri IF, Kantarjian HM, Hagemeister FB, Rodriguez MA, McLaughlin P *et al.* Untreated aggressive mantle cell lymphoma: results with intensive chemotherapy without stem cell transplant in elderly patients. *Leuk Lymphoma* 2000; 39(1-2): 77-85.

[59] Dreyling MH, Hoster E, Hermine O, Kluin-Nelemans JC, Walewski J, Trneny M *et al.* European MCL Network: An Update on Current First Line Trials. *Blood* 2007; 110: 388 Abstract.

[60] Martin P, Chadburn A, Christos P, Furman R, Ruan J, Joyce MA *et al.* Intensive treatment strategies may not provide superior outcomes in mantle cell lymphoma: overall survival exceeding 7 years with standard therapies. *Ann Oncol* 2008; 19(7): 1327-30.

[61] Friedberg JW, Vose JM, Kelly JL, Young F, Bernstein SH, Peterson D *et al.* The combination of bendamustine, bortezomib, and rituximab for patients with relapsed/refractory indolent and mantle cell non-Hodgkin lymphoma. *Blood* 2011; 117(10): 2807-12.

[62] Ruan J, Martin P, Furman RR, Lee SM, Cheung K, Vose JM *et al.* Bortezomib plus CHOP-rituximab for previously untreated diffuse large B-cell lymphoma and mantle cell lymphoma. *J Clin Oncol* 2011; 29(6): 690-7.

[63] Lamm W, Kaufmann H, Raderer M, Hoffmann M, Chott A, Zielinski C *et al.* Bortezomib combined with rituximab and dexamethasone is an active regimen for patients with relapsed and chemotherapy-refractory mantle cell lymphoma. *Haematologica* 2011; 96(7): 1008-14.

[64] Habermann TM, Lossos IS, Justice G, Vose JM, Wiernik PH, McBride K *et al.* Lenalidomide oral monotherapy produces a high response rate in patients with relapsed or refractory mantle cell lymphoma. *Br J Haematol* 2009; 145(3): 344-9.

[65] Richardson SJ, Eve HE, Copplestone JA, Dyer MJ, Rule SA. Activity of thalidomide and lenalidomide in mantle cell lymphoma. *Acta Haematol* 2010; 123(1): 21-9.

[66] Qian Z, Zhang L, Cai Z, Sun L, Wang H, Yi Q *et al.* Lenalidomide synergizes with dexamethasone to induce growth arrest and apoptosis of mantle cell lymphoma cells in vitro and in vivo. *Leuk Res* 2011; 35(3): 380-6.

[67] Weide R, Hess G, Koppler H, Heymanns J, Thomalla J, Aldaoud A *et al.* High anti-lymphoma activity of bendamustine/mitoxantrone/rituximab in rituximab pretreated relapsed or refractory indolent lymphomas and mantle cell lymphomas. A multicenter phase II study of the German Low Grade Lymphoma Study Group (GLSG). *Leuk Lymphoma* 2007; 48(7): 1299-306.

[68] Rummel MJ, Al-Batran SE, Kim SZ, Welslau M, Hecker R, Kofahl-Krause D *et al.* Bendamustine plus rituximab is effective and has a favorable toxicity profile in the

treatment of mantle cell and low-grade non-Hodgkin's lymphoma. *J Clin Oncol* 2005; 23(15): 3383-9.

[69] Rummel MJ, Gregory SA. Bendamustine's emerging role in the management of lymphoid malignancies. *Semin Hematol* 2011; 48 Suppl 1: S24-36.

[70] Abbott BL. Diagnosis and management of lymphoma. *Clin Lymphoma Myeloma* 2006; 7(1): 30-2.

[71] O'Connor OA, Portlock C, Moskowitz C, Straus D, Hamlin P, Stubblefield M *et al.* A multicentre phase II clinical experience with the novel aza-epothilone Ixabepilone (BMS247550) in patients with relapsed or refractory indolent non-Hodgkin lymphoma and mantle cell lymphoma. *Br J Haematol* 2008; 143(2): 201-9.

[72] Ansell SM, Tang H, Kurtin PJ, Koenig PA, Inwards DJ, Shah K *et al.* Temsirolimus and rituximab in patients with relapsed or refractory mantle cell lymphoma: a phase 2 study. *Lancet Oncol* 2011; 12(4): 361-8.

[73] Galimberti S, Petrini M. Temsirolimus in the treatment of relapsed and/or refractory mantle cell lymphoma. *Cancer Manag Res* 2010; 2: 181-9.

[74] Hoy SM, McKeage K. Temsirolimus: In relapsed and/or refractory mantle cell lymphoma. *Drugs* 2010; 70(14): 1819-29.

[75] Garnock-Jones KP. Bendamustine: a review of its use in the management of indolent non-Hodgkin's lymphoma and mantle cell lymphoma. *Drugs* 2010; 70(13): 1703-18.

[76] Zhang L, Qian Z, Cai Z, Sun L, Wang H, Bartlett JB *et al.* Synergistic antitumor effects of lenalidomide and rituximab on mantle cell lymphoma in vitro and in vivo. *Am J Hematol* 2009; 84(9): 553-9.

[77] Ohmachi K, Ando K, Ogura M, Uchida T, Itoh K, Kubota N *et al.* Multicenter phase II study of bendamustine for relapsed or refractory indolent B-cell non-Hodgkin lymphoma and mantle cell lymphoma. *Cancer Sci* 2010; 101(9): 2059-64.

[78] Ogura M, Ando K, Taniwaki M, Watanabe T, Uchida T, Ohmachi K *et al.* A Feasibility and Pharmacokinetic Study of Bendamustine Hydrochloride in Combination With Rituximab in Relapsed or Refractory Aggressive B-Cell Non-Hodgkin Lymphoma. *Cancer Sci* 2011.

[79] Ogura M. Current treatment strategy and new agents in mantle cell lymphoma. *Int J Hematol* 2010; 92(1): 25-32.

[80] Cortelazzo S, Ponzoni M, Ferreri AJ, Dreyling M. Mantle cell lymphoma. *Crit Rev Oncol Hematol* 2011.

[81] Oyan B, Koc Y, Kansu E. Successful salvage with high-dose sequential chemotherapy coupled with in vivo purging and autologous stem cell transplantation in 2 patients with primary refractory mantle cell lymphoma presenting in the leukemic phase. *Int J Hematol* 2005; 81(2): 155-8.

[82] Keating GM. Rituximab: a review of its use in chronic lymphocytic leukaemia, low-grade or follicular lymphoma and diffuse large B-cell lymphoma. *Drugs* 2010; 70(11): 1445-76.

[83] Spurgeon SE, Pindyck T, Okada C, Chen Y, Chen Z, Mater E *et al.* Cladribine plus rituximab is an effective therapy for newly diagnosed mantle cell lymphoma. *Leuk Lymphoma* 2011.

[84] Weigert O, Illidge T, Hiddemann W, Dreyling M. Recommendations for the use of yttrium-90 ibritumomab tiuxetan in malignant lymphoma. *Cancer* 2006; 107(4): 686-95.

[85] Bethge WA, Lange T, Meisner C, von Harsdorf S, Bornhaeuser M, Federmann B *et al.* Radioimmunotherapy with yttrium-90-ibritumomab tiuxetan as part of a reduced-intensity conditioning regimen for allogeneic hematopoietic cell transplantation in patients with advanced non-Hodgkin lymphoma: results of a phase 2 study. *Blood* 2010; 116(10): 1795-802.

[86] Leahy MF, Turner JH. Radioimmunotherapy of relapsed indolent non-Hodgkin lymphoma with 131I-rituximab in routine clinical practice: 10-year single-institution experience of 142 consecutive patients. *Blood* 2011; 117(1): 45-52.

[87] Krishnan A, Nademanee A, Fung HC, Raubitschek AA, Molina A, Yamauchi D *et al.* Phase II trial of a transplantation regimen of yttrium-90 ibritumomab tiuxetan and high-dose chemotherapy in patients with non-Hodgkin's lymphoma. *J Clin Oncol* 2008; 26(1): 90-5.

[88] Vose JM, Bierman PJ, Weisenburger DD, Lynch JC, Bociek Y, Chan WC *et al.* Autologous hematopoietic stem cell transplantation for mantle cell lymphoma. *Biol Blood Marrow Transplant* 2000; 6(6): 640-5.

[89] Jacobsen E, Freedman A. An update on the role of high-dose therapy with autologous or allogeneic stem cell transplantation in mantle cell lymphoma. *Curr Opin Oncol* 2004; 16(2): 106-13.

[90] Murali S, Winton E, Waller EK, Heffner LT, Lonial S, Flowers C *et al.* Long-term progression-free survival after early autologous transplantation for mantle-cell lymphoma. *Bone Marrow Transplant* 2008; 42(8): 529-34.

[91] Tam CS, Bassett R, Ledesma C, Korbling M, Alousi A, Hosing C *et al.* Mature results of the M. D. Anderson Cancer Center risk-adapted transplantation strategy in mantle cell lymphoma. *Blood* 2009; 113(18): 4144-52.

[92] Kasamon YL. Blood or marrow transplantation for mantle cell lymphoma. *Curr Opin Oncol* 2007; 19(2): 128-35.

[93] Dreyling M, Lenz G, Hoster E, Van Hoof A, Gisselbrecht C, Schmits R *et al.* Early consolidation by myeloablative radiochemotherapy followed by autologous stem cell transplantation in first remission significantly prolongs progression-free survival in mantle-cell lymphoma: results of a prospective randomized trial of the European MCL Network. *Blood* 2005; 105(7): 2677-84.

[94] van 't Veer MB, de Jong D, MacKenzie M, Kluin-Nelemans HC, van Oers MH, Zijlstra J *et al.* High-dose Ara-C and beam with autograft rescue in R-CHOP responsive mantle cell lymphoma patients. *Br J Haematol* 2009; 144(4): 524-30.

[95] Dreger P, Laport GG. Controversies in lymphoma: the role of hematopoietic cell transplantation for mantle cell lymphoma and peripheral T cell lymphoma. *Biol Blood Marrow Transplant* 2008; 14(1 Suppl 1): 100-7.

[96] Gianni AM, Magni M, Martelli M, Di Nicola M, Carlo-Stella C, Pilotti S *et al.* Long-term remission in mantle cell lymphoma following high-dose sequential chemotherapy and in vivo rituximab-purged stem cell autografting (R-HDS regimen). *Blood* 2003; 102(2): 749-55.

[97] Gopal AK, Rajendran JG, Petersdorf SH, Maloney DG, Eary JF, Wood BL *et al.* High-dose chemo-radioimmunotherapy with autologous stem cell support for relapsed mantle cell lymphoma. *Blood* 2002; 99(9): 3158-62.
[98] Svoboda J, Andreadis C, Elstrom R, Chong EA, Downs LH, Berkowitz A *et al.* Prognostic value of FDG-PET scan imaging in lymphoma patients undergoing autologous stem cell transplantation. *Bone Marrow Transplant* 2006; 38(3): 211-6.
[99] Khouri IF, Lee MS, Romaguera J, Mirza N, Kantarjian H, Korbling M *et al.* Allogeneic hematopoietic transplantation for mantle-cell lymphoma: molecular remissions and evidence of graft-versus-malignancy. *Ann Oncol* 1999; 10(11): 1293-9.
[100] Nademanee A. Transplantation for non-Hodgkin lymphoma. *Expert Rev Hematol* 2009; 2(4): 425-42.
[101] Popplewell L, Nademanee A, Carter N, al. e. Autologous vs allogeneic cell transplantation for mantle cell lymphoma (MCL): outcomes over a 10-year period at City of Hope. *Blood* 2004; 104: abstract 894.
[102] Ganti AK, Bierman PJ, Lynch JC, Bociek RG, Vose JM, Armitage JO. Hematopoietic stem cell transplantation in mantle cell lymphoma. *Ann Oncol* 2005; 16(4): 618-24.
[103] Kasamon YL, Jones RJ, Diehl LF, Nayer H, Borowitz MJ, Garrett-Mayer E *et al.* Outcomes of autologous and allogeneic blood or marrow transplantation for mantle cell lymphoma. *Biol Blood Marrow Transplant* 2005; 11(1): 39-46.
[104] Laudi N, Arora M, Burns L, McGlave P, Miller J, Bohac G *et al.* Efficacy of high-dose therapy and hematopoietic stem cell transplantation for mantle cell lymphoma. *Am J Hematol* 2006; 81(7): 519-24.
[105] Khouri IF, Lee MS, Saliba RM, Jun G, Fayad L, Younes A *et al.* Nonablative allogeneic stem-cell transplantation for advanced/recurrent mantle-cell lymphoma. *J Clin Oncol* 2003; 21(23): 4407-12.
[106] Corradini P, Tarella C, Olivieri A, Gianni AM, Voena C, Zallio F *et al.* Reduced-intensity conditioning followed by allografting of hematopoietic cells can produce clinical and molecular remissions in patients with poor-risk hematologic malignancies. *Blood* 2002; 99(1): 75-82.
[107] Maris MB, Sandmaier BM, Storer BE, Chauncey T, Stuart MJ, Maziarz RT *et al.* Allogeneic hematopoietic cell transplantation after fludarabine and 2 Gy total body irradiation for relapsed and refractory mantle cell lymphoma. *Blood* 2004; 104(12): 3535-42.
[108] Morris E, Thomson K, Craddock C, Mahendra P, Milligan D, Cook G *et al.* Outcomes after alemtuzumab-containing reduced-intensity allogeneic transplantation regimen for relapsed and refractory non-Hodgkin lymphoma. *Blood* 2004; 104(13): 3865-71.
[109] Robinson SP, Goldstone AH, Mackinnon S, Carella A, Russell N, de Elvira CR *et al.* Chemoresistant or aggressive lymphoma predicts for a poor outcome following reduced-intensity allogeneic progenitor cell transplantation: an analysis from the Lymphoma Working Party of the European Group for Blood and Bone Marrow Transplantation. *Blood* 2002; 100(13): 4310-6.

[110] Le Gouill S, Mohty M, Guillaume T, Gastinne T, Moreau P. Allogeneic stem cell transplantation in mantle cell lymphoma: where are we now and which way should we go? *Semin Hematol* 2011; 48(3): 227-39.

[111] Sorror ML. Comorbidities and hematopoietic cell transplantation outcomes. *Hematology Am Soc Hematol Educ Program* 2010; 2010: 237-47.

[112] Baron F, Storb R, Storer BE, Maris MB, Niederwieser D, Shizuru JA *et al.* Factors associated with outcomes in allogeneic hematopoietic cell transplantation with nonmyeloablative conditioning after failed myeloablative hematopoietic cell transplantation. *J Clin Oncol* 2006; 24(25): 4150-7.

[113] Sorror ML, Maris MB, Storer B, Sandmaier BM, Diaconescu R, Flowers C *et al.* Comparing morbidity and mortality of HLA-matched unrelated donor hematopoietic cell transplantation after nonmyeloablative and myeloablative conditioning: influence of pretransplantation comorbidities. *Blood* 2004; 104(4): 961-8.

[114] Hill BT, Bolwell BJ, Rybicki L, Dean R, Kalaycio M, Pohlman B *et al.* Nonmyeloablative second transplants are associated with lower nonrelapse mortality and superior survival than myeloablative second transplants. *Biol Blood Marrow Transplant* 2010; 16(12): 1738-46.

[115] Tomblyn M, Brunstein C, Burns LJ, Miller JS, MacMillan M, DeFor TE *et al.* Similar and promising outcomes in lymphoma patients treated with myeloablative or nonmyeloablative conditioning and allogeneic hematopoietic cell transplantation. *Biol Blood Marrow Transplant* 2008; 14(5): 538-45.

[116] Dreger P, Brand R, Milligan D, Corradini P, Finke J, Lambertenghi Deliliers G *et al.* Reduced-intensity conditioning lowers treatment-related mortality of allogeneic stem cell transplantation for chronic lymphocytic leukemia: a population-matched analysis. *Leukemia* 2005; 19(6): 1029-33.

[117] Warlick ED, Tomblyn M, Cao Q, Defor T, Blazar BR, Macmillan M *et al.* Reduced-intensity conditioning followed by related allografts in hematologic malignancies: long-term outcomes most successful in indolent and aggressive non-Hodgkin lymphomas. *Biol Blood Marrow Transplant* 2011; 17(7): 1025-32.

[118] Shea T, Johnson J, Westervelt P, Farag S, McCarty J, Bashey A *et al.* Reduced-Intensity Allogeneic Transplantation Provides High Event-Free and Overall Survival in Patients with Advanced Indolent B Cell Malignancies: CALGB 109901. *Biol Blood Marrow Transplant* 2011.

[119] Till BG, Gooley TA, Crawford N, Gopal AK, Maloney DG, Petersdorf SH *et al.* Effect of remission status and induction chemotherapy regimen on outcome of autologous stem cell transplantation for mantle cell lymphoma. *Leuk Lymphoma* 2008; 49(6): 1062-73.

[120] Evens AM, Winter JN, Hou N, Nelson BP, Rademaker A, Patton D *et al.* A phase II clinical trial of intensive chemotherapy followed by consolidative stem cell transplant: long-term follow-up in newly diagnosed mantle cell lymphoma. *Br J Haematol* 2008; 140(4): 385-93.

[121] Tam CS, Khouri IF. Autologous and allogeneic stem cell transplantation: rising therapeutic promise for mantle cell lymphoma. *Leuk Lymphoma* 2009; 50(8): 1239-48.

[122] Diefenbach CS, O'Connor OA. Mantle cell lymphoma in relapse: the role of emerging new drugs. *Curr Opin Oncol* 2010; 22(5): 419-23.
[123] Watanabe T, Kato H, Kobayashi Y, Yamasaki S, Morita-Hoshi Y, Yokoyama H *et al.* Potential efficacy of the oral histone deacetylase inhibitor vorinostat in a phase I trial in follicular and mantle cell lymphoma. *Cancer Sci* 2010; 101(1): 196-200.
[124] Wang M, Sun L, Qian J, Han X, Zhang L, Lin P *et al.* Cyclin D1 as a universally expressed mantle cell lymphoma-associated tumor antigen for immunotherapy. *Leukemia* 2009; 23(7): 1320-8.
[125] Rudelius M, Pittaluga S, Nishizuka S, Pham TH, Fend F, Jaffe ES *et al.* Constitutive activation of Akt contributes to the pathogenesis and survival of mantle cell lymphoma. *Blood* 2006; 108(5): 1668-76.
[126] Peponi E, Drakos E, Reyes G, Leventaki V, Rassidakis GZ, Medeiros LJ. Activation of mammalian target of rapamycin signaling promotes cell cycle progression and protects cells from apoptosis in mantle cell lymphoma. *Am J Pathol* 2006; 169(6): 2171-80.
[127] Ansell SM, Inwards DJ, Rowland KM, Jr., Flynn PJ, Morton RF, Moore DF, Jr. *et al.* Low-dose, single-agent temsirolimus for relapsed mantle cell lymphoma: a phase 2 trial in the North Central Cancer Treatment Group. *Cancer* 2008; 113(3): 508-14.
[128] Witzig TE, Geyer SM, Ghobrial I, Inwards DJ, Fonseca R, Kurtin P *et al.* Phase II trial of single-agent temsirolimus (CCI-779) for relapsed mantle cell lymphoma. *J Clin Oncol* 2005; 23(23): 5347-56.
[129] Hess G, Herbrecht R, Romaguera J, Verhoef G, Crump M, Gisselbrecht C *et al.* Phase III study to evaluate temsirolimus compared with investigator's choice therapy for the treatment of relapsed or refractory mantle cell lymphoma. *J Clin Oncol* 2009; 27(23): 3822-9.
[130] Witzig TE, Reeder CB, LaPlant BR, Gupta M, Johnston PB, Micallef IN *et al.* A phase II trial of the oral mTOR inhibitor everolimus in relapsed aggressive lymphoma. *Leukemia* 2011; 25(2): 341-7.
[131] Yee KW, Zeng Z, Konopleva M, Verstovsek S, Ravandi F, Ferrajoli A *et al.* Phase I/II study of the mammalian target of rapamycin inhibitor everolimus (RAD001) in patients with relapsed or refractory hematologic malignancies. *Clin Cancer Res* 2006; 12(17): 5165-73.
[132] Marshansky V, Wang X, Bertrand R, Luo H, Duguid W, Chinnadurai G *et al.* Proteasomes modulate balance among proapoptotic and antiapoptotic Bcl-2 family members and compromise functioning of the electron transport chain in leukemic cells. *J Immunol* 2001; 166(5): 3130-42.
[133] Perez-Galan P, Roue G, Villamor N, Campo E, Colomer D. The BH3-mimetic GX15-070 synergizes with bortezomib in mantle cell lymphoma by enhancing Noxa-mediated activation of Bak. *Blood* 2007; 109(10): 4441-9.
[134] Goy A, Kahl B. Mantle cell lymphoma: The promise of new treatment options. *Crit Rev Oncol Hematol* 2010.
[135] Rummel MJ, de Vos S, Hoelzer D, Koeffler HP, Hofmann WK. Altered apoptosis pathways in mantle cell lymphoma. *Leuk Lymphoma* 2004; 45(1): 49-54.

[136] Paoluzzi L, Gonen M, Gardner JR, Mastrella J, Yang D, Holmlund J *et al.* Targeting Bcl-2 family members with the BH3 mimetic AT-101 markedly enhances the therapeutic effects of chemotherapeutic agents in in vitro and in vivo models of B-cell lymphoma. *Blood* 2008; 111(11): 5350-8.
[137] Paoluzzi L, Scotto L, Marchi E, Seshan VE, O'Connor OA. The anti-histaminic cyproheptadine synergizes the antineoplastic activity of bortezomib in mantle cell lymphoma through its effects as a histone deacetylase inhibitor. *Br J Haematol* 2009; 146(6): 656-9.

16

Autologous Peripheral Blood Purified Stem Cells Transplantation for Treatment of Systemic Lupus Erythematosus

Ledong Sun and Bing Wang
Department of Dermatology
Zhujiang Hospital
Southern Medical University, Guangzhou,
China

1. Introduction

"Lupus", from the Latin in the 19th century, had appeared in the West before and after the medical literature. In 1828, the French dermatologist Bette (Biett) firstly reported such a patient: facial skin has erythema the same as the the wolf bites. To the mid-19th century, there was a doctor named Carson Musharraf formally using the "lupus" (lupus erythematosus, LE) as the medical terminology, which only refered to skin lesions of discoid lupus erythematosus then. With the development of medical science and clinical practice, more and more doctors found that lupus is not only skin damage, but also associated with kidney, brain, heart, lungs, nerves, muscles and joints, blood and other systemic diseases. To the 1890s, American doctor Osler (Osler) proposed the disease as "systemic lupus erythematosus" (systemic lupus erythematosus, SLE) [1]. Systemic lupus erythematosus (SLE), a common autoimmune disease without explicit etiology, has a high incidence of secondary infection, treatment difficulties and high mortality, arousing the focus of domestic and foreign scholars. According to the prevalence rate from 70 to 100/10 million people [2], China has reached to 1.12 million patients which is constituted mainly by young women. In addition to involving the outer skin, the disease violates various body organs, including vital organs such as the nervous system, heart, kidney and so on, the central nervous system such as meningitis, encephalitis, acute cerebrovascular disease, spinal cord inflammation and subarachnoid hemorrhage, even to death. It is the most common when occurring cardiac involvement, such as pericarditis, pericardial effusion heart attack, and life-threatening acute cardiac tamponade, which affect the quality of life. Involving the kidneys, it may be kidney failure. Respiratory system involvement includes bronchial pneumonia, pleurisy, atelectasis and respiratory failure. Secondary infection is a common complication of SLE, causing death at worst.

SLE is considered a typical multi-organ involvement autoimmune disease, as a result of the disorder of immune system[3,4]. Before the 1960s, the 5-year survival rate was below 50%. With the progress of the treatment and the development of therapeutic drugs, especially corticosteroids and cytotoxic drugs for SLE, the survival rate has improved significantly, has reached 93% to the 1990s, while 10 and 20 year survival rate has reached 92%, 68%, respectively[5,6].

In treatment, we begin with corticosteroid and cytotoxic at large doses the most, in order to suppress the disease, in 1-2 weeks, if necessary, plus corticosteroid pulse therapy or immunosuppressive therapy. Glucocorticoid inhibits T lymphocytes mainly through the production of growth factors and Fc receptor expression [7,8]. However, the dosage of corticosteroids varies with the insensitivity of the patients. In addition, due to large amount of medication for a long time, it prones to side effects and complications, such as: infection (viruses, bacteria, fungi, etc.), induced diabetes, peptic ulcer, perforation or gastrointestinal bleeding, fractures and avascular necrosis, mental disorders, Cushing's syndrome, acne, hirsutism, etc. More than 60% of patients died of hormonal side effects, which has bad impact on life quality [7]. Therefore, it is still a medical problem without effective cure. The latest research reports, it is an effective and new treatment for SLE by hematopoietic stem cell transplantation [4], the evaluation of long-term efficacy of which remains to be seen. Since the pathological immune T, B lymphocytes from the common lymphoid stem / progenitor cells, it may be an effective way to cure the disease by the application of certain tools (such as chemotherapy or radiotherapy) to destroy pathological immune and re-establish normal immune system by hematopoietic stem cell transplant[3,7]. In recent years, it gradually become a hot topic in international research for autologous peripheral blood stem cell transplantation for treatment of SLE and other autoimmune diseases,whose exact mechanism is not fully clear, only defined to the level of the original autoimmune cell clones completely destroyed, while normal one re-established [7 10]. Autoimmunity and self-tolerance is a dynamic equilibrium between the process, in which, we use immunosuppressive agents to weaken the strength of their immune, while hematopoietic stem cell transplantation to support the power of self-tolerance and eventually make the body's immune system normal [11]. Theoretically, it is possible to cure SLE and other autoimmune disease, and may achieve the purpose of healing.

2. Materials and methods

2.1 Clinical data

A total of 16 patients consistent with the diagnostic criteria for SLE by American College of Rheumatology in 1982 were recruited, for whom corticosteroids and immunosuppressive therapy ineffective. There are 4 males and 12 females, aging from 11-37 years old, and the type of renal pathology, II, III, IV, V were3, 3, 6, 4 cases, respectively. SLEDAI scores are 14-35 points.

2.2 Reagents

$CD34^+$ cell isolation kit, Germany Miltenyi Biotec Company; Lymphocyte separation medium, Beijing Ding Guo; Antibody (mouse IgG1-PE, mouse anti-human CD34-PE), American Gene Company; 3 - aminopropyl triethoxy silane solution, a company in Kyoto, Japan

2.3 Main instruments

Immunomagnetic separation system, company Miltenyi Biotec, Germany; Blood cell separator,U.S. BAXTER Company; BD's flow cytometry, U.S.; Low-speed desktop centrifuge,Beijing Medical Centrifuge FactoryHettich, Germany's low-temperature high-speed centrifuge

2.4 Hematopoietic stem cell mobilization, collection, purification and cryopreservation

Mobilization plan: cyclophosphamide + granulocyte colony stimulating factor. We use cyclophosphamide intravenous $4g/m^2$, divided into 2 to 3 days, and urine alkalization and hydration to protect the heart, liver and kidney. When WBC <1.0×10^9 / L, give Recombinant Human Granulocyte Colony-stimulating Factor Injection 5ug/kg, until WBC> 5.58×10^9 / L and $CD34^+$> 2% we start collecting with CS3000Plus blood cell separator (BAXTER products). After dilution and volume adjustment, plus CD34 monoclonal antibody in cells in 19 ~ 25 ℃ for 30 minutes, then wash 2 times by PBS buffer to pre-sort on MiniMACS separator, at last cool and liquid nitrogen froze immediately.

2.5 Pre-treatment

Cyclophosphamide 50mg/kg daily, intravenous infusion, once every 4 days (-5 to -2 days); ATG 2.5mg/kg.d, intravenous infusion, once every 3 days, while urine alkalization and hydration to protect the heart, liver and kidney.

2.6 Hematopoietic stem cell reinfusion

Transplant day 0, recover the frozen stem cells in 37 ℃ ~ 40 ℃water bath, then rapid transfusion through subclavian vein cannulation.

2.7 Follow-up indicators

Dynamic monitoring hematopoietic and immune indicators of clinical manifestations of SLE patients after transplantation.

2.8 Statistical analysis

We use software package SPSS10.0 processing repeated measurements analysis of variance of autoantibodies, immune function monitoring indicators in patients with SLE who underwent autologous peripheral blood stem cell transplantation.

3. Results

3.1 Autologous peripheral blood stem cell mobilization

The total number of collected $CD34^+$ cells ranged from $(2.06\text{-}9.9) \times 10^8$, higher than the required 2×10^6/kg, is the success of hematopoietic stem cell collection for the 16 patients.

3.2 Mobilization-related complications

The total of 16 patients have leukopenia and gastrointestinal adverse reactions, and all with varying degrees of fever, but blood cultures were negative. 1 case has severe edema, but alleviate after the supplement of albumin and furosemide. 1 case has atopic dermatitis (specific reasons unknown), the anti-histamine treatment effective. All the patients were in stable condition after mobilization without the original mobilization-related illness or death.

3.3 Changes in clinical transplantation

Within 1 month after transplantation, the facial rash, joint pain and other symptoms disappeared completely in all patients. 13 patients with proteinuria disappeared completely in three months, 3 cases of the four V-type, still urinary protein in a year, but the 24-hour urine protein continued to decline. One patient, 1 year after transplantation, had the recurrence of oral ulcers, arthralgia, leukopenia and elevated titers of autoantibodies. The other patient, 2 and a half years after transplantation, had facial erythema, leukopenia and elevated titers of autoantibodies.

3.4 Hematopoietic reconstitution after transplantation

The number of peripheral blood leukocytes is > 1.0×10^9 / L after 7 to 15 days, platelets is > 20×10^9 / L 0 to 21 days.

3.5 Transplant-related complications

All patients suffered from uneven serum sickness-like reactions, 1 case severe renal failure and heart failure, 3 cases hemorrhagic cystitis, 1 case psychogenic mental disorders, 1 case perineal candidiasis, 1 case acute pulmonary edema, 1 case septic shock, but all recovered after active treatment. 2 patients had fever after transplantation because of tuberculosis infection (see Chapter VII). One case with fever in October after transplantation, had chest CT showing: 1, multiple new lung nodules; 2, left lung consolidation; 3, pericardial thickening, small left pleural effusion. According to CT, it was much as the invasive nature of purulent bacterial or fungal infection, in combination with other manifestations, at last diagnosed as fungal infections, alleviated after itraconazole injection . 12 cases of cytomegalovirus infection became negative after ganciclovir treatment (see Chapter VI). 3 cases of herpes zoster, 1 case of generalized herpes zoster, were recovered after treatment by ganciclovir. 1 case in the purification of autologous peripheral blood stem cell transplantation in August, appeared Evans syndrome (also known as idiopathic thrombocytopenic purpura with autoimmune hemolytic anemia). Check ENA polypeptide: ANA 70.50u/ml, ds-DNA 18.65u/ml, RNP 49.90u/ml, Sm 15.44u/ml, Coomb's test showed: more than a single specific anti-human globulin (++), specific anti-C3 (+), monospecific anti-IgG (+). Bone marrow pathology: bone marrow dysfunction, three lines of hematopoietic cells can be seen, the proportion of red tablets was normal, megakaryocyte number and morphology were without exception. Be double filtration plasmapheresis and methylprednisolone injection pulse therapy, Coomb's test showed: more specific anti-human globulin (+), single-specific anti-C3 (+), I (-). Plus immunosuppressive agents cyclosporine and vincristine, while reducing the hormone dosage, it is stable in clinical and laboratory condition.

3.6 Post-transplant changes in autoantibody

After transplantation ANA, anti Ds-DNA antibodies, anti-Sm antibodies, anti-RNP antibody titers decreased than those before transplantation, but among the different phases they showed no significant ($P > 0.05$), while ANA, anti-RNP antibody titers were significantly lower than before transplantation ($P < 0.05$), specifically in chapter IV.

3.7 Post-transplant immune reconstitution

It is different for $CD3^+$, $CD45RA^+$ $CD4^+$, $CD45RA^+$ $CD8^+$, $CD45RO^+$ $CD4^+$ expression when 3 months,6 months and 1 year after transplantation ($P < 0.05$). (1) 3,6 month after treatment, CD3 + expression levels were significantly lower than that before transplantation (P <0.05); (2) 6,12months after transplantation, CD45RA + CD4 + expression levels were significantly higher than that 3 months after transplantation (P <0.05), but no significant difference than before treatment (P> 0.05); (3) 3,6 months after transplantation when CD45RA + CD8 + expression levels were significantly lower than before transplantation (P <0.05). (4) 3,6, 12 months after transplantation, the CD45RO + CD4 + expression levels were significantly lower than that before transplantation (P <0.05). Detailed in Table 1. and Chapter.

time	CD_3^+	$CD_{45}RA^+CD_4^+$	$CD_{45}RA^+CD_8^+$	$CD_{45}RO^+CD_4^+$
Before transplantion	74.49±9.79	5.80±3.96	28.95±11.28	16.66±5.60
3 months later	58.04±20.61*	3.41±4.46	20.29±11.11*	8.43±3.32*
6 months later	62.75±17.09*	6.33±4.87$^{\Delta}$	18.89±8.18*	10.26±4.84*
12 months later	67.28±14.16	7.84±3.63$^{\Delta}$	23.14±10.15	11.55±5.25*
F	3.727	4.457	4.539	8.325

Mauchly test of sphericity, P> 0.05, the spherical assumption is not rejected, without correction with correction factor ε freedom.

* Compared with the pre-transplant, P <0.05; Δ compared with 3 months after transplant, P <0.05

Table 1. Alternation of immune function in patients with SLE before and after AHSCT

4. Discussion

Systemic lupus erythematosus is an autoimmune disease, without clear cause and difficult to cure. In recent years, with the increasing levels of health care, new drugs and treatments are emerging for treatment. Particularly glucocorticoids and cytotoxic drugs for SLE, the survival rate has improved significantly. To the 1990s, 5,10,20-year survival rate has reached 93%, 1 92%, 68%,respectively [5,6].

Though corticosteroid treatment with cytotoxic drugs is still the preferred solution, However, due to the large amount of medication for a long time, prone to side effects and complications and make 60% of patients eventually die from side effects o [7]. Therefore, it is still medical problem at home and abroad.

In 1985, Ikehara confirmed Allo-BMT to treat mice autoimmune disease. In 1991 Knaan-Shanzer confirmed their bone marrow transplant for adjuvant arthritis in rats. At the same time,it was discovered that receiving Allo-BMT or autologous bone marrow transplantation can achieve remission for autoimmune disease such as the blood system or other malignancies [12]. Karussis et al [13] firstly used cyclophosphamide and granulocyte colony-stimulating factor to stimulate the bone marrow to release hematopoietic stem cells into the blood, and isolated, purified and separate mouse hematopoietic stem cells in vitro. After sorting the total number of $CD34^+$ cells greater than 2.0×10^6/kg, use large doses of immunosuppressive agents in the second step for abnormal immune cells elimination, then reinfuse the $CD34^+$ cells into mice. After the reconstruction of immune system, and long term follow up, it was found that it has achieved remarkable results in animal models.

In recent years, it is effective that autologous peripheral blood stem cell transplantation for autoimmune diseases and becoming a hot topic in international research. Firstly, we use doses of immunosuppressive agents or radiation in patients to finish a strong immune suppression, and then rebuild the patient's immune system by autologous hematopoietic stem cell transplantation. The use of peripheral blood transplantation, hematopoietic stem cells has no anesthesia, trauma and has an earlier immune reconstitution compared to bone marrow transplantation. The patient's immune regulation in the reconstruction process can reach a new equilibrium and (or) immune tolerance, may complete remission or even completely cured[7 ~ 11].

Hematopoietic stem cell transplantation (HSCT) for the treatment of SLE has opened up a new way, however, there is a certain risk, which confined its use. Combined with existing research and the experience, the performance of at least one of the following may be considered for HCST treatment; ① epileptic seizures or psychiatric symptoms; ② lung involvement at one of the following: pulmonary hemorrhage, infiltration and no infection is present, within the last 6 months is greater than 15% lower forced vital capacity, pulmonary hypertension; ③ refractory hemolytic anemia, reticulocyte count is greater than 3%, hemoglobin less than 100g / L; or the decreaseing of other fatal blood cells; ④ SLE disease activity index greater than 16 points; ⑤ serious anti-cardiolipin antibody syndrome. Nephritis and has one of the following persons: proteinuria greater than 1g / d and albumin less than 30 g / L, serum creatinine greater than 15mg / L, and hematuria, cellular casts, renal biopsy with acute proliferative damage. Patients also need the following therapy: oral prednisone 0.5mg/kg at least 2 months, last 6 months or intravenous methyl poured nylon 1g, 3 times; application CTX 500 mg/m^2, 3 times/month [3,7].

Stem cells can self-renew,replicate to mature blood cells, has CD34 antigen expressed, including hematopoietic progenitor / unique cell populations. The cceptable number of $CD34^+$ cell transplantation cellsis in the range (2 ~ 5) × 106/kg, but the number of peripheral blood stem cells often less than 0.1%. Therefore, we need a way to increase the concentration of peripheral blood stem cells, which is called stem cell mobilization. The key method is chemotherapy + granulocyte colony-stimulating factor or single granulocyte colony-stimulating factor. The former may alleviate the disease, while the later may deteriorate it. Joint mobilization is better than single granulocyte-colony stimulating factor, cyclophosphamide $4g/m^2$ than $2g/m^2$ [3,14,15]. The group of 16 patients were treated with cyclophosphamide $4g/m^2$ combined granulocyte colony-stimulating factor 5ug/kg to make a success rate of 100%. But in the process of collecting peripheral blood stem cells, there are two important problems to be solved, that is, when to start and when to end. We need to keep abreast of peripheral blood mononuclear cells and stem cells numbers timely for clinical decisions.$CD34^+$ cell count is the most widely used in the determination of graft stem / progenitor cells of the indicators. The absolute count of which is closely related to content, to predict the effect of acquisition. The peripheral blood WBC, MNC count and the number of harvested $CD34^+$ cells were not significantly correlated, indicating which does not accurately reflect the level of peripheral blood stem cells, so it became the key that the absolute count of $CD34^+$ collection . In general, the best collection time should be when the peripheral blood $CD34^+$ cells is from 20 to 85 / uL, according to its cell surface antigen and labeled antibody to the tiny beads binded specifically. Some scholars [16] reported that $CD34^+$ cell graft up to a median of 49% by positive selection, compared with pre-separation concentration as 33 times. The main purpose is to remove the graft in the immune cells to reduce the chance of a relapse transplant. We used the German company Miltenyi Biotec MACS immunomagnetic system in $CD34^+$ sorting [17].

Mobilization of patients with SLE-related complications included fever, hypotension, abdominal pain, joint pain, seizures, myocardial infarction, pulmonary edema [17]. The total of 16 patients mobilized had both leukopenia and gastrointestinal adverse reactions, with varying degrees of fever, while blood cultures negative, after $CD34^+$ cells collected the temperature returned to normal but its mechanism remains unclear, needing further study.

Pretreatment is one of the important part of autologous hematopoietic stem cell transplantation. Its main purpose is to destroy the abnormal cells in vivo in patients, minimize recurrence, and provided the necessary space to implant. Regimen should also have the function to suppress hematopoietic and immune, which often require multiple medications and / or radiotherapy combined.

Currently, there are no uniform for conditioning regimen for SLE. In general, the efficacy of pre-clinical and pre-treatment intensity have some relevance, for example, by cyclophosphamide / busulfan, cyclophosphamide or other drugs combined with systemic lymph node irradiation, the disease recurrence rate is low, but transplant-related mortality also increased. Because of the transplanted organ damage associated with SLE patients, most scholars do not advocate the use of intense myeloablative program, but the lymphatic clearance program. The current use of cyclophosphamide (200mg/kg) + anti-thymocyte globulin (ATG, 90mg/kg) is more common, while some individual case reports cyclophosphamide (120 ~ 150mg/kg) + systemic lymph node irradiation (400 ~ 600cGy, lung shielding to 400cGy) or cyclophosphamide (120mg/kg) + busulfan (16 mg / kg) and the BEAM program (BCNU 300mg / m^2, VP16 400 ~ 800mg / m^2, cytarabine 800 ~ 1600mg / m^2, melphalan 140mg / m^2) [10,15,18]. The whole body and lymph node irradiation can induce more tumor-related complications, should be taken seriously enough [15].

Conditioning regimen containing cyclophosphamide, 5% patients emerged life-threatening cardiac toxicity and severe cardiac toxicity, mainly showing low-voltage ECG, progressive heart failure or pericarditis with the drug doses. One case in our group occurred serious heart failure in use of drugs to support and pericardiocentesis when necessary.

Over 90% patients had the incidence of oral mucositis after hematopoietic stem cell transplantation, approximately the date of reinfusion, continuing until blood implanted. Some patients have severe pain and require total parenteral parenteral nutrition support and make fungal, bacterial, viral and other infections common.

Almost all patients have varying degrees of gastrointestinal reactions, mainly nausea, vomiting, or diarrhea. We can use 5 - hydroxy tryptophan-blockers to control gastrointestinal adverse reactions before chemotherapy.

Renal dysfunction is a common complication with amounts of pathogenic factors, such as direct drug toxicity, hypovolemia and use of nephrotoxic drugs. By disabling nephrotoxic drugs, paying attention to water and electrolyte balance, most patients can return to normal.

Another prominent side effect caused by cyclophosphamide is hemorrhagic cystitis, mainly due to as cyclophosphamide metabolites acrolein from urine, causing extensive mucosal ulceration, necrosis and bleeding. The effective prevention measures include alkaline urine and application Gomez sodium, which can be combined with acrolein to reduce the cell toxicity in a total dose of 1.6 times of that of cyclophosphamide and divided in 4 times.

Animal experiments confirmed that the graft of T lymphocytes related to disease recurrence after transplantation [3,7,11]. At present, most foreign scholars are using cell sorting system to remove T-cell line in vitro, only can remove about 3 logarithmic T cells, which can be re-entered to arouse the body's own response to the disease clone. ATG is a kind of polyclonal immunoglobulin antibodies using the human thymus cells immunized animals to collect the serum extract, selective against human T lymphocytes. Anti-thymocyte globulin in the body can bind to T cells, swallowed by circulating monocytes, and fixed macrophages components, through the activation of complement to remove T cells [19]. We applied plenty of ATG in vivo to remove the residual T lymphocytes and memory cells to enhance the success rate and reduce the relapse rate.

The exact mechanism of autologous peripheral blood stem cell transplantation for autoimmune diseases is not fully clear, what can be certain is that the original autoimmune cell clones were completely destroyed while re-established a normal immune system [7-10]. By high dose systemic immunosuppressants or radiation, we eliminate its own mature immune cells, while stem cells, due to a large quantity of aldehyde dehydrogenase, can resist to cyclophosphamide. Pretreatment process can avoid the impact of memory T cell so that the new stem cells can develop into non-anti-self lymphocytes, inducing tolerance to the autoantigen. In addition, the mobilization and pre-processing stage eliminate autoimmune lymphocytes, reduce antibodies and immune complexes to allow healing of damaged tissue and normal immune.

Autoimmunity and self-tolerance is a dynamic equilibrium between the process. By using immunosuppressive agents and hematopoietic stem cell transplantation can eventually balance the body's immune system back to normal [11]. Theoretically, it is possible to cure autoimmune disease.

Since 1997, Marmont using autologous bone marrow stem cell transplantation for treatment of SLE has been successful, there have been a lot of research on bone marrow and peripheral blood stem cell transplantation for treatment of SLE at home and abroad, and have achieved encouraging clinical results. In 2003, ASH meeting reported the U.S. total number of cases is 681, of which 127 cases were systemic lupus erythematosus, mainly in remission. Van lear have found that 45 SLE patients undergoing transplantation, 27 of which have significantly improved, 14 recurred after the initial alleviation, 7 deaths, for 5 have relation with the transplantation and the overall transplant-related mortality rate is 11% [20]. Traynor et al from North America [10] found 15 patients with SLE completing the study of autologous transplantation were alive, but 2 died after mobilization. In more than 1 year's follow-up, 12 patients, 8 cases completely withdrawaled, 2 have small doses of hormones to maintain, 2 had disease activity requiring the addition of immunosuppressive agents. Our studies show that it has an ideal short-term effect about autologous peripheral blood purified CD34 + cell transplantation for SLE, for all patients had hematopoietic reconstitution well, most of the clinical manifestations and related immunological parameters disappeared and were markedly improved in survival quality. But there are still many issues worth further reflection and research [3], such as: SLE multiple genetic abnormalities exist in the stem cell gene level, the transplanted stem cells will differentiate into the same self-reactive immune cells; whether disease recurrence are related to stem cells derived from the input; if the new input stem cells in the relatively normal environment can develop into self-tolerance T, B lymphocytes, and so on.

5. References

[1] Sun Le-dong . The origin of the name for systemic lupus erythematosus .Paper of family doctor, 2005.04.11.

[2] Sun Le-dong; Zeng Kang; Zhou Zai-gao. Progress in treatment of systemic lupus erythematosus. *Southern China Journal of Dermato-Venereology*, 2000,7（4）:41-45.

[3] Sun Le-dong; Zeng Kang; Zhou Zai-gao, et al. Progress in treatment of systemic lupus erythematosus by autologous stem cells transplantation. *Journal of Dalian Medical University*, 2006,28:421-422,427.

[4] Shlomchik MJ, Craft JE, Mamula MJ. From T to B and back again：positive feedback in systemic autoimmune disease. *Nat Rev Immunol*, 2001, 1：147-153

[5] Cervera R, Khamashta MA, Font J, et al. Morbidity and mortality in systemic lupus erythematosua during a 10-year period：a comparison of early and late manifestations in a cohort of 1000 patients. *Medicine*, 2003, 82：299-308.

[6] Lukac J, Rovensky J, Lukacova O, et al. Systemic lupus erythematodes. *Vnitr Lek*, 2006,52):702-711.

[7] Diao You-tao, Zeng Kang, SUN Jing ,SUN Le-dong,et al.Autologous peripheral blood stem cell transplantation for the treatment of systemic lupus erythematosus. *Chinese Journal of Dermatology*, 2006,39：317-319.

[8] Zeng Kang , Peng Xue-biao, Sun Jing , et al. Transplantation of purified CD34~+ stem cells from autologous peripheral blood for treatment of systemic lupus erythematosus. *Journal of First Military Medical University*, 2002, 22：1052.

[9] Sibilia J. Treatment of systemic lupus erythematosus in 2006. *Joint Bone Spine*. 2006, 73:591-598.

[10] Tyndall A, Daikeler T. Autologous hematopoietic stem cell transplantation for autoimmune diseases. *Acta Haematol*. 2005,114:239-247.

[11] Lehnhardt FG, Scheid C, Holtik U, et al.Autologous blood stem cell transplantation in refractory systemic lupus erythematodes with recurrent longitudinal myelitis and cerebral infarction. *Lupus*. 2006,15:240-243.

[12] Marmont AM, van-Lint MT, Gualandi F.Autologous marrow stem cell transplantation for severe systemic lupus erythematosus of long duration.Lupus, 1997, 6：545-548.

[13] Karussis D, Slavin S.Hematopoietic stem cell transplantation in multiple sclerosis: experimental evidence to rethink the procedures.*J Neurol Sci*, 2004, 223：59-64.

[14] Rojas Ramos E, Martinez Jimenez NE, Reyes Salina A. Hematopoietic steam cell transplantation in the management of systemic lupus erythematosus resistant to treatment. *Rev Alerg Mex*, 2004,51：61-65.

[15] Jayne D.. Stem cell transplantation in systemic lupus erythematosus. *Best Pract Res Clin Haematol*. 2004,17:291-304.

[16] Ikehara S.New strategies for BMT, organ transplantation and regeneration therapy.*Hematology*,2003, 8：77-81.

[17] Chen Bing; Ouyang J ian; Li Cuiping; et al. Study on the peripheral blood CD 34~+ cell mobilization, harvesting and purification on patients with auto-immune diseases. *Jiangsu Medical Journal*, 2003, 29：247-249.

[18] Lisukov IA, Sizikova SA, Kulagin AD, et al. High-dose immunosuppression with autologous stem cell transplantation in severe refractory systemic lupus erythematosus. Lupus, 2004, 13 : 89-94.

[19] Durez P, Toungouz M, Schandene L, et al.Remission and immune reconstitution after T-cell-depleted stem -cell transplantation for reeumatoid arthritis. *Lancet,* 1998, 352 : 881

[20] Van Laar JM, Tyndall A. Intense immunosuppression and stem-cell transplantation for patients with severe rheumatic autoimmune disease: a review[J]. *Cancer Control.* 2003, 10 : 57-65.

17

Allogeneic Hematopoietic Cell Transplantation for Paroxysmal Nocturnal Hemoglobinuria

Markiewicz Miroslaw, Koclega Anna, Sobczyk-Kruszelnicka Malgorzata, Dzierzak-Mietla Monika, Zielinska Patrycja, Frankiewicz Andrzej, Bialas Krzysztof and Kyrcz-Krzemien Slawomira
Medical University of Silesia, Department of Hematology and BMT, Katowice, Poland

1. Introduction

Paroxysmal nocturnal hemoglobinuria /PNH/ is an acquired, clonal disorder of hematopoietic stem cell, characterized by intravascular hemolysis and manifested by episodes of hemoglobinuria, life-threatening venous thromboses, bone marrow failure and smooth muscle dystonias. The triad of hemolytic anemia, pancytopenia and thrombosis makes PNH a truly unique clinical syndrome. The disease can arise de novo or in the setting of acquired aplastic anemia (Robert A. Brodsky, 2008). The name of the disease refers to the occurrence of hemoglobinuria responsible for red or dark brown urine. Hemoglobinuria in patients with PNH very often appears after awakening and is due to intravascular lysis of red blood cells that are abnormally sensitive to the complement attack (Bessler and Hiken, 2008). PNH is unique as it is an acquired hemolytic disorder resulting from abnormality of the red cell. Almost all other acquired hemolytic conditions are due to an extrinsic attack on the red cells, eg. antibody- mediated or mechanical hemolysis (Hillmen, 2008). So it is not surprising that this rare disease (the estimated incidence of 2 to 5 new cases per million inhabitants (Brodsky, 2008) has been fascinating scientists for almost 150 years. The first clear and detailed clinical description of "intermittent haematuria" in an anemic patient was given by Wiliam Gull in 1866. Dr. Paul Strubling recognized the urinary pigment as haemoglobin and called this sign "paroxysmal haemoglobinuria" in 1882. He has developed the hypothesis that the erythrocytes of PNH are abnormally sensitive to hemolysis when the plasma is acidified during sleep because of carbon dioxide and lactic acid accumulation as a result of slower circulation (Parker, 2008). His work was largely ignored until Marchiafava reported a case in Italy 29 years later. In 1931 Michelli published further observations and used the term " splenomegalic hemolytic anemia with hemoglobinuria and hemosiderinuria Marchiafava-Michelli type"- hence the name Marchiafava-Michelli disease. Early in the 20th century Van der Bergh showed that erythrocytes from PNH patients underwent lysis in acidified serum and he suspected that complement played a role in this abnormality. In 1944, Dacie noted a possible association with aplasia and suggested that PNH is an acquired clonal dysfunction affecting hematopoietic stem cells occurring due to their somatic mutation (Johnson RJ et al, 2002). The mechanism of clonal dominance in PNH remains unknown.

2. Pathogenesis and classification

PNH is not only interesting for hematologists, but it is also very important problem for neurologists, nephrologists, gastroenterologists, pediatricians (about 10% of patients are younger than 21 years) and gynecologists (25% of the cases are diagnosed during pregnancy (Parker et al. 2005).

PNH is caused by an acquired somatic mutation of the PIG-A gene located on the short arm of the X chromosome in multipotent hematopoietic stem cell. Most PIG-A mutations are small insertions or deletions that result in a frame-shift in the coding region and consequently a shortened non-functional GPI product (Brodsky, 2008). Due to its localization on the X chromosome and due to X inactivation in the female somatic cells, only one mutation is required in either males or females to abolish the expression GPI-linked proteins. To date over 180 mutations have been identified in the PIG-A gene in blood cells from patients with PNH (Hillmen, 2008). Small to moderate PNH clones are found in up to 70% of patients with acquired aplastic anemia, demonstrating a patho-physiological link between these disorders (Schrezenmeier et al. 1995; Dunn DE et al. 1999; Mukhima GL et al. 2001; Wang H et al, 2002). What is more, many of aplastic anemia patients exhibit expansion of the PIG-A mutant clone and progress to clinical PNH (Brodsky, 2008). GPI- anchored proteins deficient cells have also been reported in patients with myelodysplastic syndrome (MDS) (Dunn DE et al. 1999; Mukhina Gl et al. 2001; Wang H et al. 2002). Patients with MDS possessing small PNH populations tend to be classified as having refractory anemia and often have a hypocellular marrow with HLA-DR15 positivity, normal cytogenetics, moderate to severe thrombocytopenia and high likelihood of response to immunosuppressive therapy (Wang H et al. 2002; Nagarajan S et al. 1995; Sugimori C et al. 2006).

PIG-A gene encodes complement-protecting surface proteins which are normally tethered to the cell membrane of erythrocytes, neutrophils and platelets by the GPI anchor (Rosse WF et al. 2004). Two proteins play particularly important role:

- CD55 (DAF- Decay Accelerating Factor) which controls the first steps of the complement cascade by regulating the activity of C3 and C5 convertases (Rosse WF et al. 2004; Johnson RJ et al. 2002; Parker C et al. 2005).
- CD59 (MIRL-Membrane Inhibitor of Reactive Lysis) which inhibits the formation of the Terminal Complement Complex by preventing the incorporation of C9 in the C5b-8 complex, thereby preventing the formation of Membrane Attack Complex (MAC or C5b-9) (Rosse WF et al. 2004; Johnson RJ et al. 2002; Parker C et al. 2005).

Defect in PIG-A gene leads to partial or complete deficiency of these protective factors (CD55 and CD59) on the cell surface responsible for susceptibility to the lytic effects of activated complement and intravascular lysis of erythrocytes.

Flow cytometry analysis reveals that all hematopoietic lineages: myeloid, erythroid and lymphoid are involved (Brodsky, 2008).

Glycosylphosphatidyloinositol (GPI) protein deficiency on PNH red cells can be complete (PNH type III cels) or partial (PNH type II cells). Cells with normal levels of GPI proteins on their surface are referred to as type I cells (Fujioka & Yamada, 1994). The analysis of GPI-anchored proteins on the surface of the hematopoietic cells, particularly red cells in PNH reveals that approximately 40% of patients have a combination of types I, II, III PNH cells (Hill et al. 2007).

Current classification system has recently been proposed, which divides patients with PNH clones into the following subgroups (Parker C et al. 2005):

- Classic PNH- patients have clinical evidence of intravascular hemolysis: signs of marrow stimulation- reticulocytosis, high levels of LDH and direct or indirect bilirubin, low concentration of serum haptoglobin, without evidence of another defined bone marrow abnormality.
- PNH in the setting of another bone marrow disorder- patients in this subcategory have clinical and laboratory evidence of hemolysis, but also have or have had a history of a defined underlying marrow abnormality (eg. PNH/ aplastic anemia, PNH/ refractory anemia-MDS).
- Subclinical PNH- patients have no clinical or laboratory evidence of hemolysis. PNH-sc is observed in association with bone marrow failure syndromes, particularly aplastic anemia and refractory anemia-MDS.

The course of the disease is marked by exacerbations of hemolysis of varying severity on top of the chronic hemolysis which exists in these patients. Repeated paroxysmal episodes often necessitate a transfusion. The paroxysms may be triggered by infections, stress, trauma, surgery, pregnancy and drugs.

Intravascular hemolysis is responsible for most symptoms observed in PNH patients. Free hemoglobin released from PNH red cells during intravascular hemolysis can completely deplete haptoglobin and then overflow into the urine resulting in hemoglobinuria (Tabbara, 1992). Once the capacity of this hemoglobin scavenging protein is exceeded, (Parker et al. 2005) consumption of endogenous nitric oxide (NO) ensues. This reduced bioavailability of NO could also promote platelet and endothelial activation, leading to intravascular thrombosis (Rosse WF et al. 2004; Parker C et al. 1996; Rother RP et al. 2005). The occurrence of thrombotic events can sometimes lead to the diagnosis. Venous thrombosis may be recurrent and often occur in uncommon sites (Gralnick HR et al. 1995) such as pre-hepatic veins (Budd-Chiari syndrome), splenic, portal, renal, cerebral or mesenteric veins or inferior vena cava. An increased incidence of arterial thrombosis has also been reported.

Nitric oxide depletion also causes disturbances in the regulation of smooth muscle tone (especially during PNH exacerbations), resulting in abdominal pain, dysphagia and erectile dysfunction- it has been well established that local NO deficiency is one of the major factors responsible for erectile dysfunction (Corbin et al. 2002). Moreover, about 50% of patients with hemolytic PNH have elevated pulmonary artery systolic pressure (Hill et al 2006).

About 35% of PNH patients are presenting symptoms of anemia. It should be distinguished how much of the anemia is a consequence of hemolysis and how much is due to impaired erythropoiesis. Review of the complete blood count is informative, because evidence of thrombocytopenia, leucopenia or both suggests stem cell dysfunction. Marrow failure is likely to be a contributing factor in a patient with PNH who has anemia with an inappropriately low reticulocyte count.

3. Diagnostics

As PNH is a very rare disease, appropriate choice of patients for PNH screening is a very important issue. It should include patients with hemoglobinuria, Coombs negative intravascular hemolysis (based on abnormally high serum LDH, especially with iron deficiency), venous thrombosis involving unusual sites, aplastic anemia and refractory anemia. In 1939 Ham and Dingle's historic paper described the first diagnostic test for PNH utilizing the finding that PNH red cells were exquisitely sensitive to hemolysis in the setting of acidified serum, now known as the Ham's test. It has been used to diagnose PNH for decades. Other traditional diagnostic tests were sucrose and thrombin lysis tests, cheap and simple to perform, but with decreased sensitivity due to the short half-life of circulating PNH red blood cells (Bessler and Hiken, 2008).

Nowadays flow cytometry is the gold standard method as it is the most sensitive and yields the most quantitative and qualitative information (Parker C et al. 2005). This technique uses specific antibodies directed against antigens normally tethered by GPI to the surface of hematopoietic cells (erythrocytes, granulocytes and monocytes). Analysis of erythrocytes should either precede transfusions or be performed during a period of transfusion abstinence. The results of the analysis for erythrocytes should include the percentage of type I, type II, and type III PNH cells (Parker C et al. 2005). Analysis of granulocytes provides the best estimate of the size of the PNH clone and it is unaffected by red cells transfusion (Bessler and Hiken 2008). Another method used for PNH diagnosis is lack of FLAER (fluorescently labeled inactive toxin aerolysin) binding to GPI anchor on peripheral blood granulocytes. The method cannot be used for the analysis of red blood cells or platelets, thus it might be difficult to perform in severe aplastic anemia due to very low number of circulating granulocytes. PIGA gene mutation analysis, although very specific, is not used for diagnosing PNH. Supportive laboratory tests confirming presence of intravascular hemolysis are helpful, although not diagnostic: increased lactate dehydrogenase (LDH), low haptoglobin, increased unconjugated bilirubin, hemoglobinuria, hemosiderinuria, reticulocytosis and erythroid hyperplasia in the bone marrow.

Median survival in PNH is 10 to 15 years after diagnosis, with a wide range (Brodsky, 2009). Men and women are affected equally and there is no familial tendency. Thrombophilia is a major cause of morbidity and mortality in PNH (Hillmen et al. 1995; Socie G et al. 1996). About 40% of patients with PNH experience a thrombotic event during the course of disease (Hillmen et al. 1995; Socie G et al. 1996; Ware RE et al. 1992). Risk of developing thrombosis is proportional to the size of the PNH clone (Hillmen P et al. 2007; Hall C et al. 2003; Moyo V et al. 2004; Nishimura J 2004). Population of PNH granulocytes >50% predicts an increased risk of thrombosis (Hall C et al. 2003). The cause of thrombotic tendency in PNH is multi-factorial. Free plasma hemoglobin may contribute to platelets activation and thrombosis. The effect of free hemoglobin on platelet function is probably through the scavenging of nitric oxide (NO). Reduction of NO generation leeds to increased platelet aggregation (Broekman et al. 1991, Schafer et al. 2004).

Another complication in PNH is bone marrow insufficiency (Hillmen et al. 1995). It has long been known that there is a close correlation between PNH and aplastic anemia.

Renal failure has been recognized as a complication of PNH for many years. Between 8% and 18 % of deaths in PNH have been reported to have renal failure as at least a contributing factor (Nishimura JI et al. 2004; Clark DA et al. 1981). The cause of renal failure is probably damage of proximal renal tubules by iron released during intravascular hemolysis (Hillmen 2008).

PNH is only rarely associated with acute myeloid leukemia. It occurs in approximately 5% of patients with aplastic anemia. Marrow aplasia predisposes to PNH and to the leukemic transformation into acute leukemia. In most leukemia patients with PNH who have been studied the leukemic clone has been derived from the PNH clone (Hillmen P et al. 1995). A significant proportion of patients survive for prolonged periods (approximately 25% surviving over 25 years with blood transfusion support only) and about 10% to 15% experiences a spontaneous recovery from PNH with no sequelae attributable to the disease . A possible hypothesis to explain occasional spontaneous recovery is that intensity of aplastic process that is positively selecting PNH clones reduces with time. Recovery would depend on the presence of normal stem cells capable of regenerating the bone marrow (Hillmen P et al. 1995)

Before initiating treatment of PNH, it is useful to stratify patients into classical or hypoplastic PNH. This can usually be accomplished by obtaining a complete blood and reticulocyte count, LDH determination, peripheral blood flow cytometry and bone marrow examination. Regardless of how patients with small PNH clones are classified (eg. AA/PNH, MDS/PNH or subclinical PNH), they do not have to and should not be treated, unless symptoms appear. Patients with PNH clones less than 10% rarely require clinical intervention, nevertheless they should be monitored closely because expansion of the clone may occur. Monitoring of granulocyte and erythrocyte clone size is sometimes performed every 6 to 12 months. Patients with PNH have an impaired quality of life in comparison with the general population. They suffer from chronic fatique, which seriously compromises their professional and personal lives (Moyo VM et al. 2004). Patients with chronic hemolysis complain of lethargy, malaise, myalgia (Clark DA et al. 1981). The dysphagia and male impotence of PNH appear related to hemolysis (Rosse WF 2000).

4. Treatment

PNH patients with either chronic hemolysis or with acute hemolytic exacerbations may be treated with corticosteroids (Rosse WF 1982; Zhao M et al. 2002; Bourants K 1994; Issaragrisil S et al. 1987). Some patients respond rapidly and dramatically to prednisone given in the dose range of 0,25-1,0 mg/kg per day, but chronic use of corticosteroids may cause adverse effects such as immunosuppression, osteopenia, myopathy, iatrogenic Cushing syndrome, hypertension or in diabetes (Rosse WF 1982). Androgen therapy either alone or combination with steroids has been used successfully to treat anemia in PNH (Rosse WF 1982; Hartman RC 1966), especially caused by the underlying bone marrow failure. Although starting dose of 400 mg twice a day is recommended, a lower dose (200-400mg/24h) may be adequate to control chronic hemolysis. Potential complications of androgen treatment include liver toxicy (cholestatic jaundice, peliosis hepatic) prostatic hypertrophy and virilizing effects.

Transfusion therapy, despite improvement of blood hemoglobin, reduces hemolysis by suppressing erythropoiesis. The iron overload that follows repeated transfusion is prevented by the urinary loss of iron due to the hemoglobinuria and hemosiderinuria connected with intravascular hemolysis (Parker C et al. 2005). To prevent transfusion reaction arising from the interaction between donor leukocytes and recipient's antibodies hemofiltration is recommended. Folates supplementation (5 mg/24h) is recommended to compensate for their utilization associated with increased erythropoiesis consequencing from hemolysis (Parker et al. 2005). Chronic hemosiderinuria and hemoglobinuria may lead to severe iron deficiency despite repeated transfusions, thus PNH patients often require iron supplementation. PNH patients with cytopenias often respond to immunosuppressive therapy with anti-lymphocyte globulin and/or ciclosporin, because of associated aplastic anemia.

As the hemolysis of PNH is a consequence of complement- mediated cytolysis, inhibition of complement is a logical approach to therapy (Parker CJ 2003). Eculizumab is a humanized monoclonal antibody that targets complement protein C5, thereby preventing assembly of the terminal complement complex (also called the Membrane Attack Complex or MAC) during complement activation. Eculizumab is an IgG kappa immunoglobulin with engineered Fc portion that is a hybrid of IgG2 and IgG4 designed to have no downstream activity- thus eculizumab is a purely "blocking" antibody that prevents the cleavage and thus activation of C5 (Hillmen, 2008). Eculizumab is highly effective in reducing intravascular hemolysis in PNH, but it does not stop extravascular hemolysis, and it does not treat the bone marrow failure. Thus treatment with eculizumab is most effective in patients with classic PNH- it decreases or eliminates the need of blood transfusion, reduces the risk of thrombosis and improves quality of life (Hillmen et al. 2007; Hillmen et al. 2006; Brodsky RA 2008). Eculizumab has been approvated by FDA for PNH treatment in 2007. Thrombosis is an absolute indication for initiating treatment with eculizumab (Brodsky 2009). Some authors recommend eculizumab for patients with disabling fatigue, transfusion dependency, frequent painful paroxysms, renal insufficiency. Watchful waiting is appropriate for asymptomatic patients or those with mild symptoms (Brodsky and Jones, 2005; Young 2002). Eculizumab therapy shouldn't be initiated in patients with unresolved serious Neisseria menigitidis infection and who are not currently vaccinated against it. Eculizumab is administered intravenously at a dose of 600mg weekly for first 4 weeks, then 900mg biweekly starting on week 5 (Brodsky 2009). Eculizumab is generally well tolerated, but the most frequent adverse events observed in clinical studies were: headache, nasopharyngitis, back pain, nausea and fatique. Eculizumab increases risk of meningococcal infections, so ideally patients should be vaccinated 2 weeks before commencing eculizumab and re-vaccinated every 2 to 3 years. Patients also need to be instructed to contact doctor immediately if they develop any symptoms suggesting meningococcal infection, such as very high temperatures, a hemorrhagic rash or neck stiffness (Hillmen 2008). Most patients with PNH who are treated with eculizumab will cease to require transfusions. This leads to an increase in the proportion of PNH red cells as they are not destroyed by complement nor diluted by transfusions (Socie G et al. 2007). In some patients the proportion of PNH red cells may exceed 90% so if eculizumab is stopped for any reason, it may be followed by a sudden fall in hemoglobin and brisk hemoglobinuria. Patients who receive eculizumab must be educated that they cannot simply miss a scheduled dose of therapy (Hillmen, 2008).

Prophylaxis against thromboembolic events in patients with PNH is an issue of debate among members of the International PNH Interests Group (Moyo VM et al. 2004; Nishimura et al. 2004; Clark DA et al 1981; Sloand EM, 2000). The relatively high risk of a thromboembolic complications justifies prophylaxis with warfarin in PNH when over 50% granulocytes are GPI-AP deficient and there are no contraindications to anticoagulation (Nishimura et al. 2004). Patients with PNH who experience a thromboembolic event should receive high-intensity warfarin therapy (International Normalized Ratio INR 3.0-4.0) or subcutaneous low molecular weight heparin (Parker et al. 2005). Prophylactic anticoagulation with low molecular weight heparin should be initiated in PNH patients with elevated D-dimer levels, during pregnancy and puerperal period and in the perioperative period (Brodsky, 2009) . Patients with acute thrombotic events require anticoagulation with heparin. Thrombolytic therapy (McMullin et al. 1994; Sholar et al. 1985) or radiologic intervention (Griffith et al. 1996) should be strongly considered in patients with acute onset of Budd-Chiari syndrome.

5. Hematopoietic cell transplantation

Nowadays, the availability of option to prevent complement-dependent hemolysis by use of eculizumab has decreased use of allogeneic hematopoietic cell transplantation (allo-HCT) in PNH. Allo-HCT is the only curative and durable therapy for PNH which is still recommended for PNH patients with life-threatening cytopenias, hemolysis or thrombosis, that is not controlled with eculizumab therapy. The decision to perform allo-HCT should weigh disease prognosis by incorporating known adverse prognostic factors such as previous history of thrombosis and/or evolution to pancytopenia, against the risk of transplant-related complications. Multiple transplant-related risk factors for hemolysis can be managed without exacerbation of hemolysis during the procedure (Markiewicz et al. 2006). Selection of the appropriate candidate and, equally important, the right time to perform allo-HCT are important questions that need to be answered (Matozs-Fernandez et al. 2009). The fact that the majority of patients undergoing allo-HCT usually already had an extensive transfusion history is an important issue. PNH patients with peripheral cytopenias meeting criteria for severe aplastic anemia are considered good candidates for allo-HCT. Allo-HCT should not be offered as initial therapy for most patients with classical PNH. Exceptions are patients with PNH in countries where eculizumab therapy is not available, primarily due to considerable cost of treatment. Although allogeneic hematopoietic cell transplantation (allo-HCT) offers complete cure of PNH with complete eradication of the PNH clone, outcomes presented in the past were controversial. Thus it is important to re-evaluate results of allo-HCT in PNH. There are few reports on the use of allo-HCT for PNH and nearly all of them include small numbers of patients. Small series and reporting biases make assessment of transplant outcomes difficult. There are few risk factors associated with worse survival: occurrence of thrombosis, progression to pancytopenia, transformation to myelodysplastic syndrome or acute leukemia and thrombocytopenia at diagnosis. Transplants from siblings and especially from unrelated donors performed in the past, before implementation of complete allelic HLA-typing, were based on suboptimal donor-recipient matching. Moreover, improvements of HLA typing, growing experience of transplant teams and better either myeloablative or reduced intensity conditioning and concomitant therapy with introduction of novel antimicrobial agents enabled achieving better outcomes of allo-HCT in PNH. Similar to other allo-HCT transplantation patients, younger age and the availability of a fully HLA-matched sibling donor are favorable prognostic factors.

Proper choice of conditioning regimen is a very important component of treatment with allo-HCT. Eradication of the PNH clone has been achieved with both myeloablative and reduced-intensity conditioning regimens (Matozs-Fernandez et al. 2009; Suenaga et al. 2001). Myeloablative regimens used in PNH consisted of busulfan and cyclophosphamide or fludarabine, high-dose cyclophosphamide alone or together with total body irradiation (TBI) (Raiola et al. 2000; Woodard et al. 2001). This type of conditioning has been associated with high treatment-related mortality, high-dose cyclophosphamide used alone did not exert a significant effect on PNH clone in the long term (Cho SG et al. 2003). In reduced-intensity conditioning fludarabine has been used with or without TBI or cyclophosphamide, and less often with cladribine, thiotepa, busulfan or melphalan in various combinations (Grosskreutz C et al. 2003). In addition to conditioning anti-thymocyte globulin (ATG, Thymoglobuline) and alemtuzumab (Campath) are used in unrelated allo-HCT setting.

High-dose myeloablative conditioning regimen is not required. Non-myeloablative regimens, capable to completely eradicate the PNH clone and thus cure PNH, may be preferable in younger patients seeking to maintain fertility or in patients with moderate organ dysfunction who may not tolerate myeloablative doses of chemotherapy. Non-myeloablative syngeneic bone marrow transplantation did not cure the disease, suggesting that there is an important "graft versus PNH" effect of transplantation. The conditioning regimen of cyclophosphamide/ATG is recommended for patients with PNH/aplastic anemia. For patients with classic PNH a more myeloablative regimen is indicated, eg.

radiation- or busulfan- based (Parker C et al. 2005). The rationale for use of treosulfan-based conditioning has been its favorably low non-hematological toxicity profile and myeloablative efficiency demonstrated earlier in a dose-finding study (Casper J et al. 2010). It has been reported to be safe and efficient conditioning treatment for PNH (Markiewicz et al. 2005).

The primary objective of the reduced intensity conditioning (RIC) is to achieve stable engraftment of donor cells rather than bone marrow ablation. It provides the potentially curative option for patients with advanced PNH (Hegenbart U et al. 2003). The non-myeloablative conditioning regimens consisting of fludarabine with either TBI (Grosskreutz C et al. 2003; Hegenbart U et al. 2003), or cyclophosphamide (Takahashi Y et al. 2004) were reported to have adequate potential to eliminate the PNH clone and to allow engraftment. Fludarabine seems to be a highly immunosuppressive nucleoside analogue that has been used in several reduced-intensity conditioning regimens, yielding high engraftment rates with minimal conditioning-related morbidity (Sirinivasan R et al. 2006). Numerous evidence suggests that donor alloimmune effects play a critical role in the eradication of PNH cells following reduced-intensity conditioning with fludarabine. Importantly, in vitro and in vivo data from patients undergoing fludarabine-based reduced-intensity conditioning have shown that allo-reactive donor T-cells recognizing minor histocompatibility antigens expressed on both normal and GPI-anchored protein negative recipient cells can eradicate PNH (Takahashi Y et al. 2004). At first PNH cells persist following conditioning chemotherapy, then gradual decline of GPI-negative cells takes place followed by conversion to full donor T-cell chimerism and disappearance of residual PNH cells after CsA withdrawal. Minor histocompatibility antigen-specific donor T-cells can be expanded from patients post-transplant with in vitro cytotoxicity against GPI-negative cells of recipient origin. Addition of fludarabine to cyclophosphamide and ATG provides additional host immunosuppression that helps to achieve engraftment in patients at high risk for rejection. It also enables to reduce the traditional dose of cyclophosphamide (200mg/kg) without compromising engraftment. Excellent engraftment and long-term survival has been reported in patients with PNH undergoing this reduced-intensity transplant approach (Sirinivasan R, 2006), especially suitable for heavily transfused and/or alloimmunized PNH patients.

Overall survival for unselected PNH patients who undergo transplantation using a HLA-matched sibling donor is in the range of 50 to 60% (Parker C et al. 2005). Acute GVHD and extensive chronic GVHD are reported as a major complications that may increase treatment-related mortality after alloHCT in PNH. Chronic graft-versus host disease (GVHD) is a major complication of patients after allo-HCT. Acute and chronic GVHD after RIC are also seen and may not differ substantially from those occurring after fully ablative transplantation (Grosskreutz C et al. 2003). The combination of methotrexate and cyclosporine A appears to be superior to either agent alone in preventing acute GVHD (Strob R et al. 1986).

6. Transplant centers' experience

There are few reports about survival after allogenic transplantation in PNH.

The International Bone Marrow Transplantation Registry reported 2-year survival in recipients of HLA-identical sibling transplantation in years 1978-1995 (median age of recipients was 28 years) (Saso R, 1999). Saso et al. described 57 patients who underwent transplantation with myeloablative conditioning from related (n=50) und unrelated (n=7) donors. 19 (33,3%) died within the first 4 months from graft failure (35%), infection (15%), GvHD (15%), bleeding (5%) and interstitial pneumonitis or acute respiratory distress syndrome (30%). The majority of the deaths in this study occurred within one year of transplantation. Long-term survival was 56% after related and 14% after unrelated transplantation.

The EBMT (European group for Blood and Marrow Transplantation) has reported 70% survival after allo-HCT for PNH, but only 54% met the criteria of classical PNH (Latour RP et al. 2008). Graft failure occurred in 6% of patients. Acute and chronic GVHD occurred in 15% and 20% of patients, respectively. Uni-variant analysis showed better outcome in younger patients (less than 30 years of age) and in patients transplanted less than 2 years from the date of diagnosis (Piccin A et al. 2009).

The GITMO (Gruppo Italiano Trapianto Midollo Osseo) has reported a retrospective study of 26 patients (median age 32 years) with PNH who received allo-HCT between 1988 and 2006 (Santaronr et al. 2010). Median time from diagnosis to transplant was 33 months (range: 3-208 months). HLA-matched sibling donor were used for 22 patients, HLA-matched unrelated donor for 1 patient and 3 patients received cells from HLA-mismatched donors (2 related and 1 unrelated). Myeloablative conditioning regimen (busulfan and cyclophosphamide) has been used in 15 patients and non-myeloablative regimen (treosulfan and cyclophosphamide) in 11 patients. The 10-year probability of survival was 57% for all patients with a median follow-up of 131 months. There has been one primary graft failure after myeloablative conditioning regimen and one secondary graft failure in a patient who received RIC allo-HCT. Both died from complications of a second transplant. Acute GVHD greater than stage 2 occurred in 2 of the 26 patients. The cumulative incidence of chronic GVHD in 20 evaluable patients was 50%, in 4 of them chronic GVHD was extensive. Transplant related mortality at one year was 25% in patients after myeloablative conditioning and 63% in patients after RIC. Three patients died of multiorgan failure after RIC allo-HCT from non-identical donor. 15 patients are alive with complete hematologic recovery and no evidence of PNH.

21 patients with PNH or AA/PNH have received allo-HCT after treosulfan-based conditioning in Department of Hematology and BMT of Medical University of Silesia in Katowice, Poland, in years 2004-2011 (see Table 1).

19 allo-HCTs were from 10/10 (15 pts) or 9/10 (4 pts) HLA-matched unrelated donors and 2 from HLA-matched siblings. Median age of recipients was 28 (range 20-51) years and donors 33 (19-53) years, median time from diagnosis to allo-HCT was 16 (2-97) months. Median size of PNH clone was 90% (4%-98%) granulocytes. Indication for allo-HCT was aplastic/hypoplastic bone marrow (11 pts), MDS (1 pt), severe course of PNH with hemolytic crises (5 pts) and transfusion-dependency (in remaining 4 pts). Additional risk factors were Budd-Chiari syndrome and hepatosplenomegaly (1 pt), history of renal insufficiency requiring hemodialyses (2 pts) and chronic hepatitis B (1 pt). The preparative regimen consisted of treosulfan 3x14 g/m2 plus fludarabine 5x30 mg/m2 (19 pts) or treosulfan 2x10 g/m2 plus cyclophosphamide 4x40 mg/kg (2 pts). Standard GVHD prophylaxis consisted of cyclosporine-A, methotrexate and pre-transplant ATG or thymoglobulin in MUD-HCT. Source of cells was bone marrow (9 pts) or peripheral blood (12 pts) with median 2.9 or 11.0 x10(8)NC/kg, 2.4 or 6.2 x10(6)CD34+cells/kg, 26.2 or 266.9 x10(6)CD3+cells/kg, respectively. Results of alloHCT are presented in Table 2.

Patient characteristics	Mean (range)
Age (years)	
Patients	28 (20–51)
Donors	33 (19-53)
Time from diagnosis to allo-HCT (months)	16 (2-97)
Number of transplanted cells /kg	
Bone marrow	
x10(8) NC	2.9 (0.5-3.9)
x10(6) CD34+cells	2.4 (1.2-6.3)
x10(6) CD3+cells	26.2 (1.6-266.2)
Peripheral blood	
x10(8) NC	11.0 (4.3-16.5)
x10(6) CD34+cells	6.2 (4.1-14.8)
x10(6) CD3+cells	266.9 (21.9-712.0)
	Number of pts
Sex	
Patients	
Men	15
Women	6
Donors	
Men	12
Women	9
Primary indication for allo-HCT	
Aplastic/hypoplastic bone marrow	11
Myelodysplasia	1
Severe course of PNH with hemolytic crises	5
Transfusion-dependency	4
Source of transplant	
Sibling	2
Unrelated donor	19
Source of cells	
Bone marrow	9
Peripheral blood	12
Regimen	
treosulfan 3x14 g/m^2, fludarabine 5x30 mg/m^2	19
treosulfan 2x10 g/m^2, cyclophosphamide 4x40 mg/kg	2

Table 1. Characteristics of donor/recipient pairs and allo-HCT procedures for PNH in Department of Hematology and BMT of Silesian Medical University in Katowice, Poland, experience in years 2004 to 2011.

Results	Median (range)
Absolute agranulocytosis <0.1 G/l (days)	9 (6-13)
Number of transfusions (units)	
RBC	7 (0-12)
Platelets	6 (3-15)
Regeneration of hematopoiesis (days)	
Granulocytes 1.0 G/l	17 (13-39)
Platelets 50 G/l	17 (9-39)
Hb 10 g/dl	30 (16-50)
LDH decrease at 30 days post-transplant	80% (5%-91%)
	Number of pts
Survival at 26 months (2-82) post-transplant	
Alive	20 (95.2%)
Died	1 (previously hemodialysed pt)
GVHD	
Acute grade I	8
grade II	5
grade III	1
Chronic, limited	7
Other complications	
FUO	5
CMV reactivation	3
Mucositis	3
VOD	1
Neurotoxicity	1
Venal thrombosis	1
Hemorrhagic cystitis	1
Achieving 100% donor chimerism	21 (100%)
PNH clone disappearance	21 (100%)

Table 2. Results of allo-HCT for PNH in Department of Hematology and BMT of Silesian Medical University in Katowice, Poland, experience in years 2004 to 2011.

Myeloablation was complete in all pts with median 9 (6-13) days of absolute agranulocytosis <0.1 G/l. Median number of transfused RBC and platelets units was 7(0-12) and 6(3-15). All pts engrafted, median counts of granulocytes 1.0 G/l, platelets 50 G/l and Hb 10 g/dl were achieved on days 17(13-39), 16(9-39) and 30(16-50). Acute GVHD grade I,II and III was present in 8, 5 and 1 pt, limited chronic GVHD in 7 pts. LDH decreased by 80%(5%-91%) in first 30 days indicating disappearance of hemolysis. 100% donor chimerism was achieved in all pts. 1 previously hemodialysed pt died on day +102 in a consequence of nephrotoxicity complicating adenoviral/CMV hemorrhagic cystitis. Complications in survivors were FUO (5 pts), CMV reactivation (3), VOD (1), neurotoxicity (1), venal thrombosis (1), hemorrhagic cystitis (1) and mucositis (3). 20 pts (95.2%) are alive 16 months (2-61) post-transplant and are doing well without treatment. Complete disappearance of PNH clone was confirmed by flow cytometry in all surviving patients. We concluded that allo-HCT with treosulfan-based conditioning is effective and well tolerated curative therapy in PNH.

Another novel theoretically interesting approach is related to selection of normal circulating CD34+/CD59+ cells with flow cytometry from PNH patients who have a mixed population of normal and abnormal cells for the purpose of performing autologous hematopoietic cell transplantation (AHCT). As yet, no case of AHCT for PNH has been reported (Xiao et al. 2010).

7. Conclusion

In conclusion, allo-HCT can be offered as an effective treatment option for PNH patients. Transplant-related risk factors for hemolysis can be managed without exacerbation of hemolysis during the procedure. Treosulfan, fludarabine and antithymocyte globulin treatment can be safely and effectively used for conditioning in PNH.

Since the first report of allo-HCT performed in PNH from sibling in 1973, several questions related to PNH and transplantation remain unresolved. What is the optimal time for alloHCT in PNH? What type of the conditioning therapy– conventional or reduced- enables to achieve durable cure without exacerbation of possible complications? And for what extend benefit of cure offered by alloHCT will defend recommendations for this method in the era of lifelong treatment of the disease with complement inhibitors? We are still looking for answers to these questions.

8. References

Bessler M, Hiken J. The Pathophysiology of Disease in Patients with Paroxysmal Nocturnal Hemoglobinuria. Hematology 2008, 104-110.

Bourants K. High dose recombinant human erythropoietin and low dose corticosteroids for treatment of anemia in paroxysmal nocturnal hemoglobinuria. Acta Haematol. 1994; 91: 62-65.

Brodsky RA, Jones RJ. Aplastic anemia. Lancet 2005; 365: 1647-1656.

Brodsky RA. Paroxysmal Nocturnal Hemoglobinuria: the physiology of complement-related hemolytic anemia. Ann Intern Med. 2008; 148: 587-595.

Brodsky RA, Young NS, Antonioli E, et al. Multicenter phase 3 study of the complement inhibitor eculizumab for the treatement of patients with paroxysmal nocturnal hemoglobinuria. Blood.2008; 111: 1840-1847.

Brodsky RA. How I treat paroxysmal nocturnal hemoglobinuria. Blood 2009, 113, 26; 6522-6527.

Broekman, M.J., Eiora, A.M & Marcus, A.J. Inhibition of human platelet reactivity by endothelium- derived relaxing factor from human umbilical vein endothelial cells in suspension: blockade of aggregation and secretion by aspirin-insensitive mechanism. Blood 1991, 78, 1033-1040.

Casper J, wolff D, Knauf W, Blau IW, Rutuu T, Volin L, Wandt H, Schafer-Eckart K, Hołowiecki J, Giebel S, Aschan J, Zander AR, Kroger N, Hilgendorf I, Baumgart J,

Mylius HA, Pichlmeier U, Freund M. Allogenic hematopoietic stem-cell transplantation in patients with hematologic malignancies after dose-escaleted treosulfan-fludarabine conditioning. J.Clin. Oncol.2010 Jul 10; 28(20): 3344-3351. Epub 2010 May 24. Erratum in: J Clin Oncol. 2010 Aug 10 10;28(23): 3797.

Cho SG, Lim J, Kim Y, Eom HS, Jin CY, Han CW, Kim CC. Conditioning with high-dose cyclophosphamide may not be sufficient to provide a long term remission of paroxysmal nocturnal hemoglobinuria following syngeneic peripheral blood stem cell transplantation. Biol Blood Marrow Transplant. 2003 Jul; 9(7): 453-459.

Corbin, J.D., Francis, S.H & Webb, D.J. Phosphodiesterase type 5 as a pharmacologic target in erectile dysfunction. Urology 2002, 60, 4-11.

Dunn DE, Tanawattanacharoen P, Boccuni P, Nagakura S, Green SW, Kirby MR, Kumar MS, Rosenfeld S, Young NS. " Paroxysmal nocturnal hemoglobinuria cells in patients with bone marrow failure syndromes. Ann Intern Med. 1999; 131: 401-408.

Epner E, Witherspoon R, Blume K, Storb R. Hematopoietic cell transplantation from related and unrelated donors after minimal conditioning as a curative treatment modality for severe paroxysmal nocturnal hemoglobinuria. Biol Blood Marrow Transplant. 2003 Nov: 9(11): 689-97

Gralnick H.R, Vail, M., McKeown, L.P., Merryman, P., Wilson, O., Chu, I & Kimball, J . Activated platelets in paroxysmal nocturnal hemoglobinuria BJH 1995; 91: 697-702.

Griffith JF, Mahmoud AE, Cooper S, Elias E, West RJ, Olliff SP. Radiological intervention in Budd-Chiari syndrome: techniques and outcome in 18 patients. Clin Radiol. 1996; 51: 775-784.

Grosskreutz C, Ross V, Scigliano E, Fruchtman S, Isola L. Low dose total body irradiation, fludarabine and antithymocyte globulin conditioning for nonmyeloablative allogeneic transplantation. Biol Blood Marrow Transplant. 2003; 9(7): 453-459

Hall ST, Rosse WF. " The use of monoclonal antibodies and flow cytometry in the diagnosis of paroxysmal nocturnal hemoglobinuria. Blood 1996; 12: 5332-5340.

Hall C, Richard S, Hillmrn P. Primary prophylaxis with warfarin prevents thrombosis in paroxysmal nocturnal hemoglobinuria" Blood 2003: 102: 3587-3591.

Hartmann RC, Jenkins DE Jr, McKee LC, Heyssel RM. Paroxysmal nocturnal hemoglobinuria: clinical and laboratory studies relating to iron metabolism and therapy with androgen and iron. Medicine (Baltimore). 1966; 45: 331-363.

Hegenbart U, Niederwieser D, Forman S, Holler E, Leiblein S, Johnston L, Ponisch Clark DA, Butler SA, Braren V, Hartmann RC, Jenkins DE. The kidneys in paroxysmal nocturnal hemoglobinuria. Blood 1981: 57: 83-89.

Hill A., Wang, X., Sapsford, R.J., McGawley, G.M., Oxborough, D.L., Richards, S.J., Kroon, H., Rother, R.P., Gladwin, M.T., & Hillmen, P. High prevalence of pulmonary hypertension and hemolysis associated nitric oxide depletion in patients with paroxysmal nocturnal hemoglobinuria (Abstract). Hematologica/ The Hematology Journal 2006, 91, 192.

Hill A, Richards SJ, Hillmen P. Recent developments in the understanding and management of paroxysmal nocturnal haemoglobinuria. British Journal of Haematology 2007, 137; 181-192.

Hillmen P, S.M. Lewis, M. Bessler, L. Luzzatto, and J.V. Dacie. Natural History of Paroxysmal Nocturnal Hemoglobinuria. The New England Journal of Medicine 1995, Vol. 333 No. 19. 1237 -1241.

Hillmen, P., Young, N.S, Schubert, J., Brodsky R.A., Socie, G., Muus, P., Roth, A., Szer, J., Elebute, M.O, Nakamura, R., Browne, P., Risitano, A.M., Hill, A., Schrezenmeier, H., Fu, C.L., Maciejewski, J., Rollins, S.A., Mojcik, C.F., Rother, R.P & Luzatto, L. The complement inhibitor eculizumab in paroxysmal nocturnal hemoglobinuria. New England Journal of Medicine, 2006; 355, 1233-1243.

Hillmen P, Muus P, Duhrsen U, Risitano AM, Schubert J, Luzzattio L, Schrezenmeier H, Szer J, Brodsky RA, Hill A, Socie G, Bessler M, Rollins SA, Bell L, Rother RP, Young NS. Effect of the complement inhibitor eculizumab on thromboembolism in patients with paroxysmal nocturnal hemoglobinuria Blood 2007, vol 110, No 12, 4123-4128.

Hillmen P. The role of Complement Inhibition in PNH. Am Soc Hematol Educ Program. 2008: 116-123.

Issaragrisil S, Piankijagum A, Tang-naitrisorana Y. Corticosteroids therapy in paroxysmal nocturnal hemoglobinuria. Am J Hematol.1987; 25: 77-83.

Johnson RJ, P. Hillmen. Paroxysmal nocturnal hemoglobinuria: nature's gene therapy? Clin Pathol 2002; 55: 145-152.

Latour RP, Mary JY, Salanoubat C, Terriou L, Etienne G, Mohty M, Roth S, de Guibert S, Maury S, Cahn JY, Socie G, French Society of Hematology; French Association of Young Hematologists. Paroxysmal nocturnal hemoglobinuria: natural history of disease subcategories. Blood 2008; 112(8): 3099-3106. Epub 2008 Jun 5

Lee JL, Lee JH, Lee Jh, Choi SJ, Kim S, Seol M, Lee YS, Chi Hs, Park Cj, Kim WK, Lee JS, Lee KH. Allogenic hematopoietic cell transplantation for paroxysmal nocturnal hemoglobinuria. Eur. J Haematol. 2003. 71(2) 114-118.

Markiewicz M, Wojciechowska M, Wyleżoł I, Giebel S, Wożniczka K, Wojnar J, Mendek-Czajkowska E, Hołowiecki J. First two successful unrelated donor bone marrow transplantations for paroxysmal nocturnal hemoglobinuria in Poland Annals of Transplantation 2005, 10(3), 26-30.

Markiewicz M, Wojciechowska M, Wyleżoł I, Woźniczka K, Giebel S, Wojnar J, Krawczyk-Kuliś M, Mendek-Czajkowska E, Hołowiecki J: Unrelated donor bone marrow transplantation with treosulfan-based myeloablative conditioning for paroxysmal nocturnal hemoglobinuria- successful treatment despite multiple transplant-related risk factors for hemolysis including major Kidd group incompatibility. Bone Marrow Transplant 2006, 37(2), 231–232.

Matozs-Fernandez NA, Abou Mourad YR, Caceres W, Kharfan-Dabaja MA. Current status of allogenic hematopoietic stem cell transplantation for paroxysmal nocturnal hemoglobinuria. Biol Blood Marrow Transplant. 2009. 15(6): 656-661.

McMullin MF, Hillmen P, Jackson J, Ganly P, Luzatto L, Tissue plasminogen activator for hepatic vein thrombosis in paroxysmal nocturnal hemoglobinuria. J. Intern Med. 1994: 235: 85-89.

Moyo VM, Mukhina GL, Garret ES, Brodsky RA, "Natural History of paroxysmal nocturnal hemoglobinuria using modern diagnostic assays" Br. J. Haematol. 2004; 126: 133-138

Mukhina GL, Buckley JT, Barber JP, Jones RJ, Brodsky RA. "Multilineage glycosylphosphatidylinositol anchor deficient hematopoiesis in untreated aplastic anemia. Br J Haematol. 2001; 115: 476-482

Nagarajan S, Brodsky RA, Young NS, Medof ME. Genetic defects underlying paroxysmal nocturnal hemoglobinuria that arises out of aplastic anemia Blood. 1995; 86: 4656-4661.

Nishimura JI, Kanakura Y, Ware RE, Schichishima T, Nakakuma H, Ninomiya H., Decastro CM, Hall S, Kanamaru A, Sullivan Km, Mizoquchi H, Omine M, Kinishita T, Rosse WF. Clinical course and flow cytometric analysis of paroxysmal nocturnal hemoglobinuria In the United States and Japan. Medicine 2004; 83: 193-207.

Parker CJ. An overview of the development of specific inhibitors of complement: opportunites for therapy of paroxysmal nocturnal hemoglobinuria. In: Omnie M, Kinoshita T, eds. Paroxysmal nocturnal hemoglobinuria and Related Disorders: Molecular Aspects of Pathogenesis. Tokyo: Springer; 2003: 11-23.

Parker C, Mitsuhiro O, Stephen Richards, Jun-ichi Nishimura, Monica Bessler, Russler Ware, Peter Hillmen, Lucio Luzatto, Neal Young, Taroh Kinoshita, Wendell Rose, and Gerard Socie, for the International PNH Interest Group. Diagnosis and management of paroxysmal nocturnal hemoglobinuria Blood 2005, vol 106, No 12; 3699-3708.

Parker Charles J. Paroxysmal Nocturnal Hemoglobinuria: an Historical Overview. Hematology 2008, 93-103.

Piccin A McCann S, Socie G, Oneto R, Bacigalupo A, Locasciulli A, Marsh J, Schrezenmeier H, Tichelli A, Hand E, Lawler M, Passweg J, Aplastic Anemia Party of the European Group for Blood and Marrow transplantation. Survival of patients with documented autologus recovery after SCT for severe aplastic anemia, a study by the WPSAA of the EBMT. Bone Marrow Transplant. 2010. 45(6): 1008-1013.

Raiola AM, Van Lint MT, Lamparelli T, Gualandi F, Benventuro F, Figari O, Mordini N, Berissa G, Bregante S, Frassoni F, Bacigalupo A. Bone marrow transplantation for paroxysmal nocturnal hemoglobinuria. Haematologica. 2000. 85(1) 59-62.

Rosse WF. Treatment of paroxysmal nocturnal hemoglobinuria Blood.1982; 60-23.

Rosse WF. Paroxysmal nocturnal hemoglobinuria. In Hoffman R, Benz EJ, Shattil SJ, et al. eds. Hematology: Basic principials and Practice. New York: Churhill Livingstone; 2000:331-342

Rosse WF, Peter Hillmen, Alan D. Schreiber. Immune mediated hemolytic anemia. Hematology 2004:48-6219.

Russell P, Rother, PhD; Leonard bell, MD; Peter Hillmen, MB; PhD, Mark T, Gladwin MD. The clinical sequeale of intravascular hemolysis and extracellular plasma hemoglobin. A novel mechanism of human disease. JAMA 2005; 293: 1653-1662.

Santaronr S, Bacigalupo A, Risitano AM, Tagliaferri E, Di Bartolomeo E, Iori AP, Rambaldi A, Angelucci E, Spagnoli A, Papineschi F, Tammiazo S, Di Nicola M, Di Bartolomeo P. Hematopoietic stem cell transplantation for paroxysmal nocturnal hemoglobinuria: long-term results of a retrospective study on behalf of the Gruppo Italiano Trapianto Midollo Osseo (GITMO). Haematologica. 2010; 95(6): 983-988.

Saso R, Marsh J, cevreska L, Szer J, Gale RP, Rowlings PA, Passweg JR, Nugent ML, Luzzatto L, Horwitz MM, Gordon-Smith EC. Bone Marrow transplants for paroxysmal nocturnal haemoglobinuria. Br J Haematol. 1999; 104(2): 396-396.

Schafer, A., Wiesman, F., Neubauer, S., Eigenthaler, M., Bauersachs, J. & Channon, K.M. (2004) Rapid regulation of platelet activation in vivo by nitric oxide. Circulation, 109, 1819-1822.

Schrezenmeier H, Hertenstein B, Wagner B, Raghavachar A, Heimpel H, A pathogenetic link between aplastic anemia and paroxysmal nocturnal hemoglobinuria is suggested by a high frequency of aplastic anemia patients with deficiency of phosphatidylinositol glycan anchored proteins" Exp. Hematol. 1995:23: 81-87

Sholar PW, Bell WR. Thrombolytic therapy for inferior vena cava thrombosis in paroxysmal nocturnal hemoglobinuria. Ann Intern Med. 1985; 103: 539-541

Sirinivasan R, Takahashi Y, McCoy JP, Espinoza- Delgado I, Dorrance C, Igarashi T, Lundqvist A, Barrett AJ, Young NS, Geller N, Childs RW. Overcoming graft rejection in heavily transfused and allo-immunised patients with bone- marrow failure syndromes using fludarabine-based haematopoietic cell transplantation. Br. J. Haematol. 2006: 133(3): 305-314.

Sloand EM, Young NS. Thrombotic complications in PNH. In: Young NS, Moss J, eds. Paroxysmal Nocturnal Hemoglobinuria and Glycosylphosphatidylinositol- Linked Proteins. San- Diego, CA Academic Press; 2000: 101-112.

Socie G, Marry JY, de Gramont A, Rio, B., Leporrier, M., Rose, C., Heudier, P., Rochant, H., Cahn, J.Y. &Gluckman, E. " Proxysmal nocturnal hemoglobinuria: long-term follow-up and prognostic factors" French Society of Haematology. Lancet. 1996; 348: 573-577.

Socie G, Hillmen P, Muus P,et al. Sustained improvements in transfusion requirements, fatique and thrombosis with eculizumab treatement in paroxysmal nocturnal hemoglobinuria. Blood 2007: 110 Abstract 3672.

Strob R, Deeg HJ, Farwell V, Doney K, Appelbaum F, Baetty P, Bensinger W, Buskner CD, Clift R, Hansen J, et al. Marrow Transplantation for severe aplastic anemia: metothrexate alone compared with combination of metothrexate and cyclosporine for prevention of acute graft-versus host disease. Blood 1986 Jul;68(1): 119-25

Suenaga K, Kanda Y, Niiya H, Nakai K, Satio A, Ohnishi M, Takeuchi T, Tanosaki R, Makimoto A, Miyawaki S, Ohnishi T, Kanai S, Tobinai K, Takaue Y, Mineishi S, Successful application of nonmyeloablative transplantation for paroxysmal nocturnal hemoglobinuria. Exp. Hematol 2001 May 29(5): 639-642

Sugimori C Chuhjo T, Feng X, et al. Minor population of CD55- blood cells predicts response to immunosuppressive therapy and prognosis in patients with aplastic anemia. Blood 2006; 107: 1308-1314.

Tabbara, I.A. (1992) Hemolytic anemias. Diagnosis and management. Medical Clinics of North America, 76, 649-668.

Takahashi Y, McCoy JP Jr, Carvallo C, Rivera C, igarashi T, Srinivasan R, Young NS, Childs RW. In vitro and in vivo evidence of PNH cell sensitivity to immune attack after nonmyeloablative allogeneic hematopoietic cell transplantation. Blood.2004, 103(4): 1383-90.

Wang H, Chuhjo T, Yasue S, Omine M, Nakao S. Clinical significance of a minor population of paroxysmal nocturnal hemoglobinuria- type cells in bone marrow failure syndrome. Blood 2002; 100: 3897-3902.

Ware RE, Hall SE, Rosse WF. " Paroxysmal nocturnal hemoglobinuria with onset in childhood and adolescence. N Engl J Med. 1992; 325: 991-996.

Woodard P, Wang W, Pitts N, Benaim E, Horwitz E, Cunningham J, Bowman L. Succesful unrelated donor bone marrow transplantation for paroxysmal nocturnal hemoglobinuria. Bone Marrow Transplant. 2001 Mar. 27(6) 589-92.

Xiao J, Han B, Wu YJ, Zhong YP, Sun WL. Ex vivo expansion and long-term hematopoietic reconstitution ability of sorted CD34+, CD59+ cells from patients with paroxysmal nocturnal hemoglobinuria. Int J Hematol. 2010. 92(1): 58-67.

Young NS. Acquired aplastic anemia. Ann Intern Med. 2002; 136: 534-546.

Zhao M, Shao Z. Li K, et al. Clinical analysis of 78 cases of paroxysmal nocturnal hemoglobinuria diagnosed in the past ten years Chin Med J (Engl) 2002; 115: 398-401.

18

Intensified Chemotherapy with Stem Cell Support for Solid Tumors in Adults: 30 Years of Investigations Can Provide Some Clear Answers?

Paolo Pedrazzoli[1], Giovanni Rosti[2], Simona Secondino[1], Marco Bregni[3] and Taner Demirer[4]

[1]*Dipartimento Onco-Ematologico, Fondazione IRCCS Policlinico S. Matteo, Pavia,*
[2]*Oncologia Medica, Ospedale Ca' Foncello, Treviso,*
[3]*Dipartimento Ematologico, Istituto Scientifico San Raffaele, Milano,*
[4]*Department of Hematology, Ankara University Medical School, Cebeci Hospital, Ankara,*
[1,2,3]*Italy*
[4]*Turkey*

1. Introduction

High-dose chemotherapy (HDC) with autologous hematopoietic stem cell transplantation (AHSCT) was first introduced as an experimental treatment for solid tumours in the late seventies (1). The basic line for such treatment was extremely simple: can grams work where milligrams fail? Initial data were from pediatric tumors, but soon after an incredible amount of data came from germ cell tumor (GCT), breast carcinoma (BC), small cell lung cancer (SCLC), ovarian cancer and nearly all tumor types have been treated, many of them reported as anecdotal.

After more than thirty years of investigations we are now in the position of giving a lookout of what happened and why some questions received an answer and why others did not. The "rush to transplantation" in the early times led to thousands of patients (pts) treated with such modality, BC being the most preferred indication with nearly 2,000 pts per year in the late nineties only in Europe (2). Unfortunately, the vast majority of those pts have been treated outside clinical trials, so the level of evidence for years remained at the lowest level. Nevertheless some important studies have been conducted and at least for some tumor types some robust data have been provided.

This article will try to clarify what happened and where we are now in this still fascinating and probably not exhausted field. We will first consider BC due to the huge amount of data available, and then other tumors for which HDC has been widely utilized.

2. Breast cancer

Several nonrandomized studies conducted in the eighties and early nineties demonstrated considerable improvements for pts with BC receiving HDC (3-5). This led to the premature acceptance of HDC as a treatment option for high risk adjuvant and metastatic BC, and the number of transplants performed worldwide consistently increased during the nineties (2). By the mid-1990s, BC had become the most common indication for AHSCT in North America and Europe. There followed a phase of disillusion after premature reports of some randomized studies not showing significant overall survival (OS) benefit of HDC and after a case of scientific misconduct (6). The scene was set for the demise of HDC in BC, and in more recent years the number of procedures has diminished and in fact abandoned by the vast majority of Centers (7). In more recent years, new information provided from mature phase III trials and a metaanalysis (8-10) suggest that HDC may still have a role in subgroups of pts with BC.

2.1 Adjuvant setting

The body of evidence consists of 15 randomized studies including more than 6,200 pts (10,11) Three of those have to be considered pilot or feasibility studies with less than 100 pts each (12-14). The Scandinavian Study by Bergh et al (15) is the only one with a poorer survival of the HDC arm. However, because of the unusual design of the study (chemotherapy (CT) in the control arm tailored to individual tolerance), pts in the control arm received a significantly higher total dose of CT and had a higher incidence of treatment-related leukemia and myelodysplastic syndromes.

Three important trials conducted in the era of peripheral blood stem cells, and comparing HDC with appropriate control have been recently published. The transplant-related mortality (TRM) in these studies ranged from 0 to 1%. The IBCSG reported a PFS advantage of intensified doxorubicin-cyclophosphamide regimen over conventional antracyclin-based CT pts with high risk of relapse harboring ER+ tumors (16). The largest study of HDC in BC, recently updated with a follow up of 7 years (8), has shown a clear advantage in favour of HDC in RFS in women with > 9 positive nodes and in OS in pts with > 3 positive nodes and Her2 negative tumors. Finally, Nitz et al (9) demonstrated a significant RFS and OS improvement in pts with high-risk BC (>9 positive nodes) with a tandem high-dose regimen with no treatment-related mortality.

Some have argued that HDC have been so far compared with "old fashion" conventional treatments. Among new antineoplastic agents introduced in the last two decades, taxanes, the only ones to show additional benefit to conventional antracycline-based combinations, produce limited survival advantage, if any, in the higher risk population for nodal status, i.e. > 3 LN (17,18). Moreover, taxanes are more effective in the HER2 positive pts (17), which is the population not likely to benefit from HDC alkylating agents. The only targeted therapy currently utilized in the adjuvant setting of BC (trastuzumab) is effective (and approved) in HER2+ disease. It is true that taxanes and targeted therapies were not included in the control arms. But they were not included in the HDC arms as well

In summary, the following considerations in the adjuvant setting can be drawn:

1. Most of the studies of HDC show an advantage in RFS, regarded as the primary end-point in the adjuvant setting by oncologists, while an OS benefit was observed in two modern trials comparing HDC with appropriate control and utilizing blood stem cells.
2. The Dutch study (8) and the recent metaanalysis by Berry at al. (10) provide evidence that only pts with Her2 negative tumors do benefit from HDC. This observation confirms the conclusion of several retrospective analyses that Her2 positive tumors should probably not receive HD alkylating CT. Moreover, if one assumes that 25-30% of

pts in randomized studies have HER2 positive disease, these pts will not benefit from HDC and will in fact do worse if effective antracycline-based CT is withheld, then any benefit from HDC for the HER2 negative tumors will be invisible in the final outcome.

3. While difficult to perform due to the limited interest of the pharmaceutical companies, future studies on HDC should concentrate on populations likely to benefit most from this form of therapy and on regimens with low mortality rate.

2.2 Metastatic BC

Individual studies have suggested that age/menopausal status, hormone-receptor status or HER-2/*neu* expression, tumor load, and chemosensitivity may be predictive of the benefits of HDC (19-22).

When looking at prospective randomized studies An equivalent or better EFS has been shown in all trials but, only one published as full report could demonstrate an OS advantage (11). A recent metaanalysis of phase III studies failed to demonstrate a survival advantage of HDC in MBC (23). Also for MBC it would be important to identify pts, if any, that are likely to benefit most from HDC. In particular, HDC appears to be effective in stage IV pts who were rendered free of macroscopic disease by previous therapy or in patients with oligometastatic disease (19,22). It clearly comes out from two large retrospective analysis (2,24) that a significant proportion of patients undergoing HDC after achieving response from conventional treatments are long term (>5 years) disease free survivors. Such results suggest that HDC with AHSCT can cure a subset of patients with MBC.

Finally, as observed in the adjuvant setting, also stage IV pts with HER2 positive tumors derive no benefit from high-dose alkylating agents (19). It requires to be demonstrated whether the use of non-alkylating drugs, including taxanes, antracycline and mitoxantrone (24, 32), in HDC regimens might prove effective in subset of pts with HER2 positive or with other biological tumor characteristics (29).

Also in the setting of metastatic disease, randomized studies in selected patient populations are necessary to define the exact role of HDC.

2.3 Small cell lung carcinoma (SCLC)

Probably due to its high chemosensitivity and its poor prognosis, SCLC has been the first adult tumor to be tested with HDC in a relative large number of pts. The first attempts have been very pioneristic, with pts usually grafted after first or subsequent lines with poor performance status and in a chemorefractory phase of their diseases.

After several small studies conducted in the eighties and early nineties showing that HDC might have a role in selected patients (25), The EBMT launched the first and so far only randomized study where pts with limited and extensive disease ≤ 2 metastatic sites were given either conventional CT or a multiple transplant program, i.e. three high-dose ICE (ifosfamide, carboplatin and etoposide) shots (26). This international study, enrolling 145 pts (half of the planned number), showed no significantly improved 3-year PFS or OS benefit of increasing the dose-intensity, the peak-dose or the total dose of ICE. It is important to note that 9% of pts in the HDC arm died of therapy-related toxicity, a data that other than being unacceptable today, can *per se* explain the lack of favourable results of this study.

At present the number of pts receiving HDC worldwide is anecdotal. The role of dose intensification in the treatment of SCLC would deserve further evaluation, taking into consideration also the very poor progress obtained in such tumor in the last decades and the so far pessimistic scenario of target therapy in this disease (27).

2.4 Germ cell tumors (GCT)

Because of the extremely high chemosensitivity of GCTs, the concept of HDC in this disease has been rapidly developed worldwide and intensively investigated. Clinical trials have been performed in a variety of settings, ranging from resistant or absolute refractory disease to chemosensitive relapse. The role of dose-intensification with stem cell support has also been explored as a part of first-line strategy for patients at higher risk of recurrence. This issue has been recently comprehensively reviewed by Simonelli et al (28) who suggest that the role of HDC in GCTs remains controversial mainly due to the heterogeneity of patient population and treatment approaches, the lack of well-defined prognostic variables and the limited number of randomized trials conducted.

HDC with stem cell rescue cannot be proposed for poor risk patients, neither as front-line therapy nor as consolidation, in patients achieving response by conventional CT. The two randomized studies conducted in this patient population (29,30) failed to demonstrate an OS benefit. Albeit supported by limited data, first line "consolidation" HDCT may be considered in selected patients with chemosensitive primary mediastinal disease (31,32). Intensified treatments has been more widely investigated as a salvage therapy for patients with an incomplete response to initial CT and for those with relapsed GCTs. Despite the robust data from the Indiana group (33) and from other retrospective/phase II studies (28), the role of HDCT as second line treatment for relapsed GCTs, remains today uncertain. Also in view of the very recent data produced by Lorch et al (34), the most pressing issues in GCT treatment are defining standards of HDC and optimizing outcomes of salvage treatment. The recently proposed TIGER study, comparing four cycles of conventional dose TIP versus paclitaxel/ifosfamide followed by multiple HD-CE as first line salvage treatment in refractory/relapsed GCT patients, goes in this direction (35). HDC should be considered a treatment option for patients that are (primary) refractory to platinum-based CT or for those with a first or further relapse (33,36,37). Multiple intensified cycles with carboplatin/etoposide (33,38) is recommended as the standard HD treatment also due to concern that using a three-drug regimen would require dose reductions of the two most active drugs in this setting.

2.5 Ovarian carcinoma

Early reports of HDC for ovarian carcinoma dealt with pts with resistant or refractory disease. Several small phase II studies have shown activity of HDC, but responses were generally short lasting (39). These early studies were extremely heterogeneous in terms of pts' selection, use of chemotherapeutic agents and schedule of administration.

Stiff et al reported a single centre experience in 100 pts with relapsed or persistent disease treated with various HDC regimens (40). The median OS were 9,6 and 23.1 months for pts with platinum-resistant and platinum-sensitive disease, respectively, leading to the conclusion that platinum sensitivity and tumor bulk are the two most important predictors of survival following HDC. Two large retrospective analyses including data from the European and American registries database (251 and 421 patients, respectively) suggested a survival benefit in pts receiving HDC as consolidation following conventional first line CT (41) or being in remission at transplant (42).

In late 1998 the EBMT launched an international randomised trial of multi-cycle HDC in optimally debulked pts. Control arm consisted of 6 courses of standard dose carboplatin-paclitaxel. Later on this study merged with a similar German trial. The final analysis of this combined study included 149 patients, half of the planned population providing evidence of no benefit of HDC on survival (43). Two other randomised trials, conducted by the Gynaecological Oncology Group and by a Finnish group closed early due to poor recruitment. At present no HDC studies are ongoing in ovarian cancer.

2.6 Sarcoma

In *soft tissue sarcoma* (STS) several phase I and II studies have been conducted (39, 44) but, because of the heterogeneity of study population and histology, no evidence-based conclusions could be extrapolate. Recently, the only randomized study to date has been published by the French sarcoma group. This trial failed to show an OS advantage for advanced STS patients treated with dose-intensified chemotherapy with stem cell support (45). In the highly chemosensitive *Ewing family of tumors* (EFT), a rare disease in adults, data are far more convincing both in locally advanced and metastatic disease (46-48). First line EFT pts with good responding disease, i.e. complete-partial response, as well as patients with sensitive relapse are good candidates to be considered for HDC, whenever possible within controlled studies.

Finally, evidence of HDC role in *rhabdomyosarcomas* and *osteosarcomas* is still missing (49,50).

3. Conclusion

What we can take from the bulk of clinical studies performed over the last 30 years, which unfortunately include a limited number of randomized studies, is that greatly increasing the total dose of CT may prove effective in subgroup of pts with defined clinical and biological characteristics, as suggested for BC. Many oncologists believe that is not quite enough and suggest that this approach should cease while we entered the era of targeted therapies (51). However, such a conclusion could be just as premature and thoughtless as the uncritical use of HDC that was so common 15 years ago.

HDC with AHSCT has become a safe and reasonably well-tolerated treatment modality (52) that can even be administered in the outpatient setting. Moreover, the prognosis of solid tumors discussed in this article has changed very little in the past decades, as novel targeted therapies had a clear impact only in the subset of pts with BC overexpressing HER2 (53,54).

We believe that, instead of simply giving up on a potential treatment modality, it is more logical and practical to refine and improve this existing therapy in addition to developing new approaches in the clinical trial setting (55). Improvement of treatment of solid tumor may well come, in the future, by integrating intensified CT, being *per-se* capable of remarkable and rapid tumor regression, with novel treatment strategies (i.e. immunotherapy or target-specific therapy) for their potential to eliminate residual disease (56,57)

In conclusion, data available to date do not support, outside controlled trials, the use of HDC with AHSCT for solid tumors in adults, with the possible exception of highly selected, well informed pts with GCTs, BC and EFT.

4. References

[1] Skipper HE. Dose intensity versus total dose of chemotherapy: An experimental basis. In: De Vita VT, Hellman S, Rosenberg SA eds. Important Advances in Oncology. Philadelphia Lippincott, 1990, pp 43-64.

[2] Pedrazzoli P, Ferrante P, Kulekci A, Schiavo R, De Giorgi U, Carminati O, et al: Autologous hematopoietic stem cell transplantation for breast cancer in Europe: critical evaluation of data from the European Group for Blood and Marrow Transplantation (EBMT) Registry 1990-1999. Bone Marrow Transplant 32: 489-494, 2003.
[3] Eder J, Antman K, Peters W, Henner WD, Elias A, Shea T, et al: High-dose combination alkylating agent chemotherapy with autologous bone marrow support for metastatic breast cancer. J Clin Oncol 4:1592-1597, 1986
[4] Peters WP, Ross M, Vredenburgh JJ, Meisenberg B, Marks LB, Winer E, et al: High-dose chemotherapy and autologous bone marrow support as consolidation after standard-dose adjuvant therapy for high-risk primary breast cancer. J Clin Oncol 11:1132-1143, 1993
[5] Gianni AM, Siena S, Bregni M, Di Nicola M, Orefice S, Cusumano F, et al: Efficacy, toxicity, and applicability of high-dose sequential chemotherapy as adjuvant treatment in operable breast cancer with 10 or more involved axillary nodes: five-year results. J Clin Oncol 15:2312-21, 1997
[6] Weiss RB, Rifkin RM, Stewart FM, Theriault RL, Williams LA, Hermann AA, et al: High-dose chemotherapy for high-risk primary breast cancer: an on-site review of the Bezwoda study. Lancet 355:999-1003, 2000
[7] Gratwohl A, Schmid O, Baldomero H, Horisberger B, Urbano-Ispizua A: Haematopoietic stem cell transplantation (HSCT) in Europe 2002. Changes in indication and impact of team density. A report of the EBMT activity survey. Bone Marrow Transpl 34:855-875, 2004
[8] Rodenhuis S, Bontenbal M, van Hoesel QG, Smit WM, Nooij MA, Voest EE, et al: Efficacy of high-dose alkylating chemotherapy in HER2/neu-negative breast cancer. Ann Oncol 17:588-596, 2006
[9] Nitz UA, Mohrmann S, Fisher J, Lindemann W, Berdel WE, Jackisch C, et al: Comparison of rapidly cycled tandem high-dose chemotherapy plus peripheral-blood stem-cell support versus dose-dense conventional chemotherapy for adjuvant treatment of high-risk breast cancer: results of a multicentre phase III trial. Lancet 366:1935-1944, 2005
[10] Berry DA, Ueno NT, Johnson MM et al. High-dose chemotherapy with autologous stem cell support as adjuvant therapy in breast cancer: overview of 15 randomized trials J Clin Oncol, in press
[11] Pedrazzoli P, Ledermann JA, Lotz JP, Leyvraz S, Aglietta M, Rosti G et al: High dose chemotherapy with autologous hematopoietic stem cell support for solid tumors other than breast cancer in adults. Ann Oncol 17:1479-1488, 2006
[12] Hanrahan EO, Broglio K, Frye D, Buzdar AU, Theriault RL, Valero V, et al: Randomized trial of high-dose chemotherapy and autologous hematopoietic stem cell support for high-risk primary breast carcinoma: follow-up at 12 years. Cancer 106:2327-36, 2006.
[13] Schrama JG, Faneyte IF, Schronagel JH, Baars JW, Pertese JL, van de Vijver MJ, et al: Randomized trial of high-dose chemotherapy and hematopoietic progenitor-cell support in operable breast cancer with extensive lymph node involvement: final analysis with 7 years of follow-up. Ann Oncol 13:689-698. 2002.

[14] Tokuda Y, Tajima T, Narabayashi M, Takeyama T, Watanabe T, Fukutomi T, et al: Randomized phase 3 study of high-dose chemotherapy (HDC) with autologous stem cell support as consolidation in high-risk postoperative breast cancer: Japan Clinical Oncology (JCOG9208). Proc ASCO 20:38a, 2001 (abstr)

[15] Bergh J, Wiklund T, Erikstein B, Lidbrink E, Lindman H, Malmström P, et al: Tailored fluorouracil, epirubicin and cyclophosphamide compared with marrow-supported high-dose chemotherapy as adjuvant treatment for high-risk breast cancer: a randomized trial. Scandinavian Breast Group 9401 study. Lancet 356:1384-1391,2000

[16] Basser RL, O'Neill A, Martinelli G, Green MD, Peccatori F, Cinieri S, et al: Multicycle dose-intensive chemotherapy for women with high-risk primary breast cancer: results of International Breast Cancer Study Group Trial 15-95. J Clin Oncol 24:370-8, 2006

[17] Martin M, Pienkowski T, Mackey J Pawlicki M, Guastalla JP, Weaver C, et al: Adjuvant docetaxel for node-positive breast cancer. N Engl J Med 352:2302-2313, 2005

[18] Henderson IC, Berry DA, Demetri GD et al. Improved outcomes from adding sequential Paclitaxel but not from escalating Doxorubicin dose in an adjuvant chemotherapy regimen for patients with node-positive primary breast cancer. J Clin Oncol. 2003 Mar 15;21(6):976-83.

[19] Nieto Y, Nawaz S, Jones RB, Shpall EJ, Cagnoni PJ, Mcweeney PA, et al: Prognostic model for replapse after high-dose chemotherapy with autologous stem-cell transplantation for stage IV ologometastatic breast cancer. J Clin Oncol 20: 707-718, 2002

[20] Kurian S, Qazilbash M, Fay J, Wolff S, Herzig R, Hobbs G, et al: Complete Response after High-Dose Chemotherapy and Autologous Hemopoietic Stem Cell Transplatation in Metastatic Breast Cancer Results in Survival Benefit. Breast J 12:531-235, 2006

[21] Diallo-Danebrock R, Ting E, Gluz O, Herr A, Mohrmann S, Geddert H, et al. Protein expression profiling in high-risk breast cancer patients treated with high-dose or conventional dose-dense chemotherapy. Clin Cancer Res 13: 488-497, 2007

[22] Montemurro F, Rondón G, Ueno NT et al. Factors affecting progression-free survival in hormone-dependent metastatic breast cancer patients receiving high-dose chemotherapy and hematopoietic progenitor cell transplantation: role of maintenance endocrine therapy. Bone Marrow Transplant 29: 2002; 861-866.

[23] DA Berry, NT Ueno, Johnson MM et al. High-dose chemotherapy with autologous hematopoietic stem cell transplantation versus standard-dose chemotherapy: Overview of individual patient data from six randomized metastatic breast cancer trials. J Clin Oncol, in press

[24] Martino M, Ballestrero A, Zambelli A, et al, on behalf of GITMO. Long term survival in patients with metastatic breast cancer receiving intensified chemotheapy and stem cell rescue: data from the italian registry. Submitted

[25] Elias A. Hematopoietic stem cell transplantation for small cell lung cancer. Chest 116: (s6):531S-538S,1999

[26] Leyvraz S, Pampallona S, Martinelli G, et al: A threefold dose intensity treatment with ifosfamide, carboplatin, and etoposide for patients with small cell lung cancer: a randomized trial. J Natl Cancer Inst. 100: 533-41, 2008.

[27] Rosti G, Carminati O, Monti M, et al. Chemotherapy advances in small cell lung cancer. Ann Oncol 17(s5):v99-102, 2006

[28] M Simonelli M, G Rosti G, GL Banna GL, P Pedrazzoli P. Intensified chemotherapy with stem cell rescue in germ-cell tumors. Ann Oncol, in press

[29] Motzer RJ, Nichols CJ, Margolin KA et al. Phase III randomized trial of conventional-dose chemotherapy and autologous hematopoietic stem-cell rescue as first-line treatment for patients with poor-prognosis metastatic germ cell tumors. J Clin Oncol 25: 247-256, 2007

[30] Daugaard G, Skoneczna IA, Aass N, et al. A randomized phase III study comparing standard dose BEP with sequential high-dose cisplatin, etoposide, ifosfamide (VIP) plus stem cell support in males with poor prognosis germ cell cancer (GCC): An intergroup study of EORTC, GTCSG, and Grupo Germinal (EORTC 30974). Proc Am Soc Clin Oncol 2010 28: 15s, (suppl; abstr 4512).

[31] Bokemeyer C, Schleucher N, Metzner B et al. First-line sequential high-dose VIP chemotherapy with autologous transplantation for patients with primary mediastinal nonseminomatous germ cell tumours: a prospective trial. Br J Cancer 89: 29–35, 2003.

[32] Rosti G, De Giorgi U, Wandt H et al. First-line high-dose chemotherapy for patients with poor prognosis extragonadal germ cell tumors: the experience of the European Bone Marrow Transplantation (EBMT) Solid Tumors Working Party. Bone Marrow Transplant 34: 1033–1037, 2004

[33] Einhorn LH, Williams SD, Chamness BA et al. High-dose chemotherapy and stem cell rescue for metastatic germ-cell tumors. N Engl J Med 357; 340-348, 2007.

[34] Lorch A, Bascoul-Mollevi C, Kramar A, et al. Conventional-dose versus high-dose chemotherapy as first salvage treatment in male patients with metastatic germ cell tumors: evidence from a large international database. J Clin Oncol. 29: 2178-84, 2011

[35] Pedrazzoli P, De Giorgi U, Bregni M. High-dose chemotherapy (HDCT) and autologous stem cell transplantation in relapsed germ cell tumors: do we need a randomized study? BMT 45 (suppl 2)S30, 2010.

[36] Vaena DA, Abonour R, Einhorn LH: Long-term survival after high-dose salvage chemotherapy for germ cell malignancies with adverse prognostic variables. J Clin Oncol 21: 4100–4104, 2003.

[37] Lorch A, Neubauer A, Hackenthal M, et al. High-dose chemotherapy (HDCT) as second-salvage treatment in patients with multiple relapsed or refractory germ-cell tumors. Ann Oncol 21: 820-825, 2010.

[38] Kondagunta GV, Bacik J, Sheinfeld J et al. Paclitaxel plus ifosfamide followed by high-dose carboplatin plus etoposide in previously treated germ cell tumors. J Clin Oncol 25: 85-90, 2007.

[39] Pedrazzoli P, Ledermann JA, Lotz JP, et al: High dose chemotherapy with autologous hematopoietic stem cell support for solid tumors other than breast cancer in adults. Ann Oncol 17:1479-1488, 2006

[40] Stiff PJ, Bayer R, Kerger C, Potkul RK, Malhotra D, Peace DJ, et al: High-dose chemotherapy with autologous transplantation for persistent/relapsed ovarian cancer: a multivariate analysis of survival for 100 consecutively treated patients. J Clin Oncol 15: 1309-1317, 1997

[41] Lederman JA, Herd R, Maraninchi D, Viens P, Buclon M, Philip T, et al: High-dose chemotherapy for ovarian carcinoma: long term results from the Solid Tumour Registry of the European Group for Blood and Marrow Transplantation (EBMT). Ann Oncol 12: 639-699, 2001

[42] Stiff P, Veum-Stone J, Lazarus HM, Ayash L, Edwards JR, Keating A, et al: High-dose chemotherapy and autologous stem-cell transplantation for ovarian cancer: an autologous blood and marrow transplant registry. Ann Intern Med 133: 504-515, 2000

[43] Möbus V, Wandt H, Frickhofen N, et al Phase III trial of high-dose sequential chemotherapy with peripheral blood stem cell support compared with standard dose chemotherapy for first-line treatment of advanced ovarian cancer: intergroup trial of the AGO-Ovar/AIO and EBMT. J Clin Oncol. 25:4187-93, 2007.

[44] Peinemann F, Smith LA, Kromp M, et al. Autologous hematopoietic stem cell transplantation following high-dose chemotherapy for non-rhabdomyosarcoma soft tissue sarcomas. Cochrane Database Syst Rev. 2011 Feb 16;(2): CD008216.

[45] Bui-Nguyen B, Ray-Coquard I, Chevreau C, et al.High-dose chemotherapy consolidation for chemosensitive advanced soft tissue sarcoma patients: an open-label, randomized controlled trial. Ann Oncol. 2011 Jun 7. [Epub ahead of print]

[46] Laurence V, Pierga JY, Barthier S, Long-term follow up of high-dose chemotherapy with autologous stem cell rescue in adults with Ewing tumor. Am J Clin Oncol 28:301-9, 2005

[47] Ladenstein R, Pötschger U, Le Deley MC, Primary disseminated multifocal Ewing sarcoma: results of the Euro-EWING 99 trial. J Clin Oncol 28:3284-91, 2010.

[48] Ferrari S, Sundby Hall K, et al. Nonmetastatic Ewing family tumors: high-dose chemotherapy with stem cell rescue in poor responder patients. Results of the Italian Sarcoma Group/Scandinavian Sarcoma Group III protocol. Ann Oncol 22:1221-7, 2011.

[49] Admiraal R, van der Paardt M, Kobes J, et al. High-dose chemotherapy for children and young adults with stage IV rhabdomyosarcoma. Cochrane Database Syst Rev. 2010 Dec 8;(12):CD006669.

[50] Fagioli F, Biasin E, Mereuta OM et al. Poor prognosis osteosarcoma: new therapeutic approach. Bone Marrow Transplant 41 Suppl 2:S131-4, 2008.

[51] Hortobagyi GN: What is the role of high-dose chemotherapy in the era of targeted therapies? J Clin Oncol 22: 2263-2266, 2004

[52] Piccart-Gebhart MJ, Procter M, Leyland-Jones B, Goldhirsch A, Untch M, Smith I, et al: Trastuzumab after adjuvant chemotherapy in HER2-positive breast cancer. N Engl J Med 353:1659-1672, 2005.

[53] Geyer CE, Forster J, Lindquist D, Chan S, Romieu CG, Pienkowski T, et al: Lapatinib plus Capecitabine for HER2-Positive Advanced Breast Cancer. N Engl J Med 355:2733-2743, 2006.

[54] Marino P, Roché H, Moatti JP; PEGASE Group. High-dose chemotherapy for patients with high-risk breast cancer: a clinical and economic assessment using a quality-adjusted survival analysis. Am J Clin Oncol 31:117-124, 2008

[55] Cheng YC, Ueno NT. Is high-dose chemotherapy with autologous hematopoietic stem cell transplantation in breast cancer patients a done deal? Womens Health 6:481-485, 2010

[56] Siena S: Towards the integration of hematopoietic stem cells into therapy of breast cancer? Haematologica 84:865-867,1999
[57] Cheng YC, Rondón G, Sanchez LF et al. Interleukin-2 and granulocyte-macrophage-colony-stimulating factor immunomodulation with high-dose chemotherapy and autologous hematopoietic stem cell transplantation for patients with metastatic breast cancer. Int J Hematol 90: 627-634, 2009.

19

Hematopoietic Stem Cell Transplantation for Malignant Solid Tumors in Children

Toshihisa Tsuruta
Department of Clinical Examination
National Hospital Organization Kumamoto Medical Center
Japan

1. Introduction

The prognosis of malignant solid tumors in children has been improved by multidisciplinary therapy involving surgery, radiotherapy and chemotherapy. High-dose chemotherapy (HDC) followed by autologous (Auto) or allogeneic (Allo) hematopoietic stem cell transplantation (HSCT) has contributed significantly to improvements in the prognosis of high-risk chemosensitive tumors. It has been utilized in many kinds of pediatric malignancies, not only hematopoietic tumors but also solid tumors such as neuroblastoma, brain tumors, Ewing's sarcoma, osteosarcoma, Wilms' tumor, rhabdomyosarcoma, hepatoblastoma, and retinoblastoma. Fig, 1 shows the numbers of cases in which HSCT was used for pediatric solid tumors in Japan. When HSCT is utilized in patients with malignant solid tumors, there are several treatment objectives. First, it can be intended as a curative treatment acting on the minimal residual malignant cells after

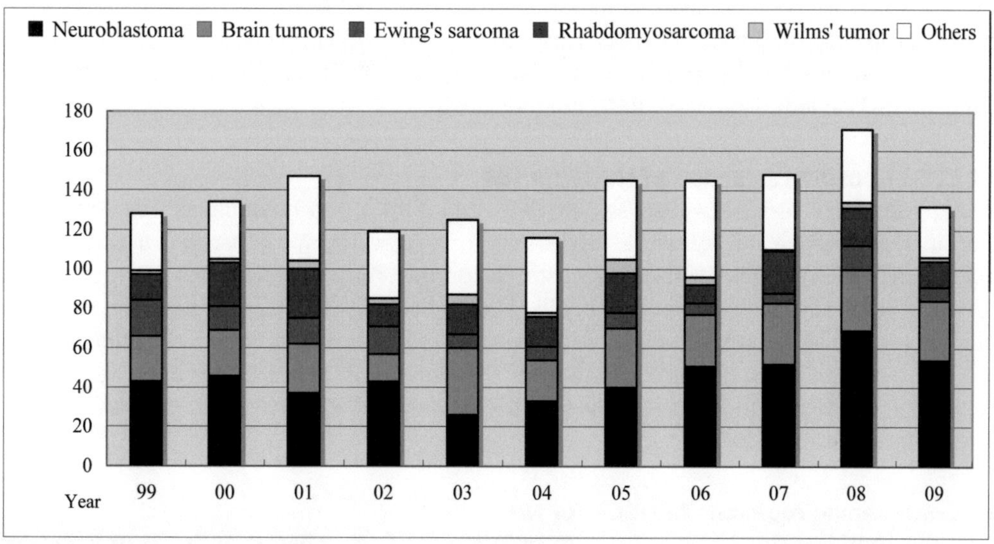

Fig. 1. The number of hematopoietic stem cell transplants for pediatric solid tumors in Japan.

conventional therapies. Second, it can be used in refractory or relapsed disease when conventional doses of anti-tumor drugs may have not reached or killed tumor cells. Third, it can be used for the immunological effect of Allo HSCT. Allo HSCT grafts sometimes have immunological anti-tumor effects called graft versus tumor (GVT) effects. Non-myeloablative conditioning regimens with Allo HSCT for solid tumors are one such example. Furthermore, tandem HSCT to increase dose intensity and HSCT with targeted radionuclide therapy have been studied clinically. In this chapter, I review the literature on HSCT for malignant solid tumors, especially non-hematopoietic tumors in children according to tumor type and consider future therapy.
Fig. 1. The Japan Society for Hematopoietic Cell Transplantation Office of Nationwide Survey, (March 2011), *The Japan Society for Hematopoietic Cell Transplantation Annual Report of Nationwide Survey 2010*, p. 60

2. Neuroblastoma

Neuroblastoma (NB) is the most common extracranial solid tumor of childhood. The prognostic indicators of NB are age at diagnosis, tumor stage as classified by the International NB Staging System (INSS) (Brodeur et al., 1993), amplification of the *MYCN* oncogene (Brodeur et al., 1984), and histological findings (Shimada et al., 1999). HSCT has been performed for high-risk NB, characterized as stage IV by INSS, stage III with other negative prognostic factors e.g. unfavorable histology (Shimada system), or any stage with amplification of the *MYCN* oncogene (Matthay et al., 1999; Weinstein et al., 2003).

2.1 Stem cell sources for HSCT in NB

Bone marrow transplantation (BMT) and peripheral blood stem cell transplantation (PBSCT) following mobilization by granulocyte colony-stimulating factor (G-CSF) are available for Auto HSCT in patients with NB (Fish & Grupp, 2008). Since NB cells infiltrate the bone marrow (BM) more frequently than peripheral blood (Cohn et al., 1997; Trager et al., 2003), and PB stem cells provide faster hematopoietic recovery, PB stem cells are used more often than BM cells in Auto HSCT (Fish & Grupp, 2008). Allo umbilical cord blood transplantation (CBT) from an unrelated donor, as well as Allo BMT or PBSCT, have also been performed in NB patients (Vanlemmens et al., 1992; Wagner, 1993).

2.2 CD34+ positive selection of HSCT for NB

Since minimal residual disease (MRD) in PBSC, as evaluated by anti-GD2 immunofluorescence or tyrosine hydroxylase reverse transcriptase-polymerase chain reaction (RT-PCR), is one of the prognostic factors of NB, the positive selection of CD34+ Auto PB stem cells has been studied in NB (Marabelle et al., 2011). MRD has been shown to exist in the CD34 negative fraction with most NB cells removed by the selection. CD34+ positive selection does not affect long-term hematopoiesis although immune recovery is delayed. Treatment with 13-*cis*-retinoic acid therapy after HSCT is thought to be effective for MRD (Marabelle et al., 2011).

2.3 Conditioning regimens of HSCT for NB

As a preparative regimen prior to Auto or Allo HSCT for patients with NB, fractionated total body irradiation (TBI) or local irradiation is often utilized (Kushner et al., 1991). TBI-containing regimens to treat NB have acceptable toxicity and are as effective as non-TBI preparative regimens (Kamani et al., 1996; McCowage et al., 1995). Patients who receive TBI have long-term adverse effects such as growth failure (Hovi et al., 1999) and severe disturbances of dental development (Holtta et al., 2002) more often than patients who receive non-TBI-containing regimens. Many kinds of regimens, especially melphalan-based regimens with or without TBI, have been studied as conditioning regimens for Auto HSCT in high-risk NB patients (Yalcin et al., 2011). Recently, a study of 36 patients with high-risk NB (median age at transplantation: 3 years, range: 0.7–14 years) who received high-dose busulfan and melphalan as the conditioning regimen for Auto HSCT was reported. No toxic deaths were observed. With a median follow-up of 55 months, progression-free survival (PFS) was 57 ± 9% for the whole group and patients who achieved early complete remission (CR) following transplantation (3 months post-transplantation) had a better outcome than those who did not (PFS: 91 ± 6% vs. 9 ± 8%). They also showed younger patients with no 1p deletions had better outcomes (Molina et al., 2011).

2.4 Randomized studies of HSCT for NB

The Children's Cancer Group (CCG) in the US first reported a randomized trial involving Auto HSCT. High-risk NB patients between 1 to 18 years of age who received five cycles of initial chemotherapy (doxorubicin 30 mg/m^2, etoposide 100 mg/m^2 daily for 2 days, and cyclophosphamide 1000 mg/m^2 daily for 2 days) who showed no disease progression were randomized to myeloablative therapy (carboplatin 1000 mg/m^2, etoposide 640 mg/m^2, melphalan 140 mg/m^2), TBI (333 cGy daily for 3 days), or to three cycles of continuation chemotherapy (cisplatin 160 mg/m^2, etoposide 500 mg/m^2 and doxorubicin 40 mg/m^2). The mean event-free survival (Fangusaro et al.) rate three years after randomization was significantly better among the 189 patients who were assigned to undergo transplantation than among the 190 patients assigned to receive conventional chemotherapy (34 ± 4% vs. 22 ± 4%, $p = 0.034$) (Matthay et al., 1999) .

There are two reported randomized trials from Europe. The European NB Study Group (ENSG) demonstrated the value of melphalan myeloablative therapy. In their study, 167 children with stage III or IV (44 stage III and stage IV from 6 to 12 month old at diagnosis, 123 stage IV > 1 years old) were treated by initial induction chemotherapy consisting of vincristine, cisplatin, epipodophyllotoxin, and cyclophosphamide and surgical excision of the primary tumor. Ninety patients were achieved complete or good partial responses, and then 65 (72%) children were actually randomized either to high-dose melphalan (180 mg/m^2) with Auto bone marrow support or to no further treatment. The 5-year EFS with median follow-up from randomization of 14.3 years was 38% (95% confidence interval (CI): 24–54%) in the melphalan-treated group and 27% (95% CI: 12–42%) in the group who did not receive melphalan. Although this difference was not statistically significant, the outcome of 48 randomized stage IV patients >1 year at diagnosis was significantly better in the melphalan-treated group (5-year EFS 33% vs. 17%, $p = 0.01$) (Pritchard et al., 2005).

In Germany, a randomized study involving myeloablative HDC with Auto HSCT versus oral maintenance chemotherapy was conducted. Two hundred and ninety-five patients with high-risk NB (i.e. stage IV or *MYCN* amplified tumors) were randomly assigned to myeloablative HDC (melphalan, etoposide, and carboplatin) with Auto HSCT (n = 149) or oral maintenance chemotherapy with cyclophosphamide (n = 146). The 3-year EFS in the HDC group (47%, 95% CI: 38–55%) was better than that (30%, 95% CI: 45–62%) in the oral maintenance therapy group ($p = 0.0221$), but there was no difference in overall survival (OS) (62%; 95% CI: 54–70% vs. 53%; 95% CI: 45–62%; $p = 0.0875$). In the HDC group, five patients died from acute complications related to HDC, but no patients on oral maintenance therapy died from acute treatment-related toxic effects (Berthold et al., 2005).

2.5 *MYCN* Amplification in NB and HSCT

In patients with stage II or III neuroblastoma with *MYCN* amplification, the prognosis of patients who received high-dose busulfan (600 mg/m^2) and melphalan (140 mg/m^2) (n = 12) is better than that of patients who received conventional chemotherapy (carboplatin, etoposide, vincristine, cyclophosphamide, and doxorubicin), radiotherapy, or both (n = 20). The 6-year OS was significantly different (25 ± 10% vs. 83 ± 11%; $p = 0.004$) (Laprie et al., 2004).

2.6 Tandem HSCT for NB

Scheduled multiple cycles of (tandem) Auto HSCT for high-risk NB were reported first by Philip et al. They used two different regimens for advanced NB and harvested bone marrow twice before and after first BMT. The harvested bone marrow cells were purged *in vitro* by an immunomagnetic technique. The first regimen was a combination of tenoposid, carmustine, and cisplatinium (or carboplatin). The second regimen consisted of vincristine, melphalan, and TBI. This double HSCT achieved a prolonged relapse free survival (Philip et al., 1993). There are several reports of double Auto HSCT with one course of TBI (George et al., 2006; Sung et al., 2007) or without TBI (Grupp et al., 2000). Central nervous system relapse and secondary malignancies are rare in patients who receive tandem HSCT (George et al., 2006), although there is an increase in treatment-related deaths after the second HSCT (Grupp et al., 2000; Sung et al., 2007) and a higher rate of late adverse effects including growth hormone deficiency, dental problems, osteochondromas, and hearing deficiencies (Hobbie et al., 2008). Children's Healthcare of Atlanta performed a study comparing single versus tandem HSCT for high-risk NB. There were 28 patients who received a single HSCT and 56 who received tandem HSCT. Tandem HSCT had a significantly improved 4-year EFS (59.3 ± 6.7% vs. 26.8 ± 9.2%, $p = 0.01$) and OS (70.6 ± 9.2% vs. 44.7 ± 11.2%, $p = 0.06$) when compared with single HSCT (Qayed et al., 2011).

Triple-tandem HDC with PBSCT for high-risk NB was reported by two different groups. The conditioning regimens of the first first consisted of two courses of carboplatin (667 mg/m^2 daily for 2 days) and etoposide (1000 mg/m^2 daily for 2 days) following one course of thiotepa (300 mg/m^2 daily for 3 days) and cyclophosphamide (60 mg/kg daily for 3 days). Seventeen patients were able to complete all three cycle of HSCT. The 3-year EFS and survival were 57 ± 11% and 79 ± 10%, respectively (Kletzel et al., 2002). The study by the Children's Oncology Group in the US involved patients with stage IV NB receiving three cycles of interpatient dose escalating carboplatin (800 to 1000 mg/m^2), cyclophosphamide (4 g/m^2) and etoposide (600 mg/m^2). The overall 3-year EFS and OS rates were 20 ± 10% and 26 ± 11%, respectively (Bensimhon et al., 2010).

2.7 Allo HSCT for NB

Allo HSCT with TBI for NB has been attempted more than a quarter century ago (August et al., 1984). Although the graft contains no contamination of tumor cells, initial studies of conventional Allo HSCT failed to show a clear clinical benefit with a high ratio of treatment-related mortality (TRM) and graft versus host diseases (GVHD) (Evans et al., 1994; Ladenstein et al., 1994; Matthay et al., 1994). However, the development of supportive care and the establishment of Allo HSCT for leukemia had renewed interest in Allo immunity for various malignant cells that is described as a graft versus tumor effect (GVT) (Pedrazzoli et al., 2002). Cytotoxic T-lymphocytes can target antigens found on malignant cells. These antigens include linage-restricted antigens found on tumors of similar origin and on related normal cells (e.g., GD2 ganglioside expressed in NB), antigens found on tumors of different origins but not on normal tissue, and tumor-specific antigens produced by mutant genes within the tumor (Navid et al., 2009).

Clinical GVT was first described in a patient with NB who received HSCT from a HLA haplo-identical donor. Although the patient received further chemotherapy after the Allo HSCT and the response was not correlated to GVHD, the patient entered and sustained complete remission for 4 years (Inoue et al., 2003). A more recent report shows clinical GVT for NB is correlated with GVHD in a patient who underwent a reduced intensity Allo HSCT (Marabelle et al., 2007). In mice, efficient GVT for NB could be induced by donor lymphocyte infusion or immunomodulation with dendritic cells after Allo HSCT in the absence of GVHD (Ash et al., 2009; Ash et al., 2010). Allo CBT for NB was also performed since the early 1990s in three patients with advanced NB with highly amplified *MYCN* (Vanlemmens et al., 1992; Wagner, 1993). After receiving Auto HSCT followed by Allo CBT, they have maintained disease-free survival (DFS) for 37 to 60 months without severe acute complications (Goi et al., 2011). Recently, a reduced-intensity conditioning regimen (cyclophosphamide 50 mg/kg, fludarabine 40 mg/m^2 daily for two days, TBI 200 cGy, and rabbit anti-thymocyte globulin 2.5 mg/kg daily for two days) followed by unrelated CBT was studied for patients with relapsed NB. Although all patients progressed after transplant (median: 55 days, range: 26 to 180 days), natural killer (NK) cell counts were normal within 2 months with no evidence of GVHD, whereas the T-cell recovery was slower. These results show the possibility of treating patients with MRD after transplant with less intensive immunosuppression and adding NK-cell based post-transplant immunotherapy (Jubert et al., 2011).

2.8 Radioactive Iodine therapy with HSCT for NB

The norepinephrine analog metaiodobenzylguanidine (MIBG) is selectively accumulated in sympathetic nervous tissue, and several clinical studies of high dose iodine-131-labeled MIBG (^{131}I-MIBG) in combination with myeloablative chemotherapy and Auto HSCT for the treatment of NB were performed. Although this regimen has acceptable toxicity (Yanik et al., 2002), six of 22 assessable patients had complete or partial remission and the 3-year EFS and survival rates were 31% ± 10% and 58% ± 10% (Yanik et al., 2002). A case of ^{131}I-MIBG therapy with reduced intensity Allo HSCT in recurrent extensive neuroblastoma was also reported. This patient achieved remission which was maintained for 3 months after transplantation (Takahashi et al., 2008). A study of high-dose ^{131}I-MIBG with T cell-depleted haplo-identical HSCT and post-transplant immunotherapy in children with relapsed or refractory NB shows durable remission. All five patients in this study had no acute GVHD and four patients who received additional treatment with donor lymphocyte infusion (DLI) developed controllable chronic GVHD after DLI. Analysis of immunologic recovery showed a fast reappearance of NK- and T-cells. Two patients are alive with no evidence of disease 40 and 42 months after transplantation (Toporski et al., 2009). A phase I trial in which patients with refractory or relapsed NB received double ^{131}I-MIBG infusions (target red marrow radiation index ranged from 22 to 50 mCi/kg) at 2-week intervals with Auto HSCT 2 weeks after the second dose was reported. This regimen had no dose-limiting toxicities and dose intensification of ^{131}I-MIBG. The PFS at 6 and 12 months was 57 ± 11% and 33 ± 11%, respectively (Matthay et al., 2009).

3. Brain tumors

Brain tumors are the most common pediatric malignant tumors with various types and various grades of malignancy. Recent advances in multidisciplinary treatment have improved the OS of pediatric brain tumors, especially chemotherapy-sensitive high-grade malignant tumors such as medulloblastoma (MB) or primitive neuroectodermal tumors (PNET). Many studies of HSCT were undertaken for high-grade pediatric brain tumors that are both chemotherapy-sensitive and resistant. Initial HSCT studies for high-grade glioma and ependymoma, which are relatively chemotherapy-resistant, failed to show an efficient clinical effect (Grill et al., 1996; Kalifa et al., 1992), although HSCT for patients with high-risk MB played an important role in improving survival (Gajjar & Pizer, 2010). Recently, molecular targeted therapies for adult high-grade gliomas and various grades of pediatric glioma or ependymoma are being studied (Tsuruta et al., 2011a). The combination of molecular therapy and HSCT may be of special interest in such chemotherapy-resistant cases.

3.1 Conditioning regimens & stem cell sources of HSCT for brain tumors

Until the early 1990's, high-dose 1,3-bis(2-chloroethyl)-1-nitrosourea (BCNU) based regimens followed by Auto BMT were utilized for high-risk brain tumors, especially for high-grade glioma in adults. In 1979, the first use of intensive BCNU and Auto BMT for brain tumors was reported with doses of 900 mg/m^2 and 1500 mg/m^2 (Phillips et al., 1979). Subsequent reports of single treatment BCNU showed the limiting dose of BCNU ranges from 600 to 1,200 mg/m^2 (Hochberg et al., 1981; Phillips et al., 1983). Although several studies of high-dose BCNU treatment for high-grade glioma showed responses (Hochberg et al., 1981) or prolonged survival when compared with historical control groups (Johnson et al., 1987), the effects were thought to be limited by late toxicity (Phillips et al., 1986). In France, an analysis of high-dose BCNU (800 mg/m^2) single treatment followed by Auto HSCT (BMT = 84, PBSCT = 30) and local irradiation (60 Gy) for patients (median age, 44 years) with supratentorial high-grade glioma (glioblastoma (GB) = 78, anaplastic astrocytoma (AA) = 24, anaplastic oligodendroglioma (OD) = 12) was performed. With 89 months of median follow-up, the OS was 12 months for GB, 37 months for OD, and 81 months for AA. There were some long-term survivors, although OS was comparable to that described for other treatments (Durando et al., 2003). A BCNU combination regimen (BCNU 600 mg/m^2, thiotepa 900 mg/m^2, and etoposide 1,500 or 750 mg/m^2) followed by BMT and local irradiation was studied in 42 relatively young patients (median age, 12.2 years) with high-risk brain tumors at 10 academic oncology centers in the US, UK, and Australia. Of 25 (GB = 20) evaluable newly diagnosed patients, 20% achieved CR and 4% partial remission (PR), while 28% remained in continuing complete remission (CCR) and 44% remained with stable disease prior to radiation therapy Of eight (GB = 3) evaluable patients with recurrent disease, one (GB = 1) achieved CR and two (GB = 1) PR, while one (GB = 1) remained in CCR and four with stable disease for 1 to 110.2 months (Papadakis et al., 2000).

A Phase II study of high-dose busulfan (150 mg/m^2 daily for 4 days) and thiotepa (350 mg/m^2 daily for 3 days) followed by BMT in children with recurrent brain tumors was designed in France. In 20 children (median age, 6 years), five partial responses (3/6 MB, 1/5 ependymoma, 1/2 PNET), three objective responses, ten cases of stable disease, and one case of progressive disease were observed, although one patient died due to toxicity. There were high levels of toxicity in terms of aphasia and cutaneous, hepatic, and neurological complications (Kalifa et al., 1992). Subsequent studies of regimens of thiotepa + etoposide, cyclophosphamide + melphalan, busulfan + melphalan, carboplatin + etoposide, and thiotepa + melphalan followed by BMT (or PBSCT) in patients with high-risk brain tumors reported some long-term survivors, including some patients with recurrent tumors (Finlay et al., 1996; Graham et al., 1997; Hara et al., 1998; Mahoney et al., 1996). In patients with recurrent tumors, long-term survivors usually have local recurrence, and patients with disseminated tumors have very poor prognosis (Graham et al., 1997)

Recently, a study of low-dose continuous intravenous etoposide (day −22 to day −2) along with high-dose carboplatin (667 mg/m^2 daily or area under the curve (AUC) = 9 mg/ml/min for 3 days) and thiotepa (300 mg/m^2 daily for 3 days) followed by HSCT conducted in patients with recurrent malignant brain tumors reported that some patients had long-term survival (Grodman et al., 2009).

3.2 HSCT & type of brain tumors

HSCT was often considered in patients with high-risk (e.g. metastatic or disseminated) or recurrent MB. The Children's Cancer Group in the US reported 23 patients with recurrent MB who were treated with Auto HSCT and a conditioning regimen consisting of carboplatin, thiotepa, and etoposide. This resulted in long-term survivors; the EFS and OS were 34 ± 10% and 46 ± 11%, respectively, at 36 months post-HSCT (Dunkel et al., 1998). In Spain, 19 patients with high-risk (n = 13) and recurrent (n = 6) MB and supratentorial PNET were treated with high-dose busulfan and melphalan followed by Auto HSCT. The 2-year EFS was 38 ± 14% in all patients and 57 ± 15% for the high-risk group (Perez-Martinez et al., 2005). In France, high-dose busulfan and thiotepa with HSCT followed by posterior fossa irradiation for local MB recurrence or progression after conventional chemotherapy was studied in 39 children (median age at diagnosis, 31 months) and was showed to yield a high OS rate (68.8% at 5 years; 95% CI: 53.0–81.2%) (Ridola et al., 2007). A risk-adapted craniospinal radiotherapy (23.4–39.6 Gy) followed by high-dose HSCT (St Jude Medulloblastoma-96) performed at St. Jude Children's Research Hospital also showed an improved outcome (Gajjar et al., 2006). Patients aged < 3 years at diagnosis had a better outcome (Shih et al., 2008). Pre-relapse definitive radiation therapy is thought to be one of the prognostic factors for recurrent MB. HDC is not effective in patients with recurrent MB who received cranial radiation therapy prior to recurrence (Gururangan et al., 2008; Massimino et al., 2009).

In the 1990s, two studies of HDC (thiotepa, busulfan, etoposide, and carboplatin) followed by HSCT for patients with recurrent ependymoma failed to show clinical efficacy (Grill et al., 1996; Mason et al., 1998). On the other hand, for patients with relapsed or progressive central nervous system germ cell tumors, thiotepa-based HDC regimens followed by HSCT were studied. Patients with germinomas have better outcomes than patients with nongerminomatous germ cell tumors (Modak et al., 2004).

Central nervous system atypical teratoid/rhabdoid tumors (CNS AT/RT) are rare tumors of childhood with a very poor prognosis. Median survival of CNS AT/RT is less than 1 year with conventional therapy. One study involving high-dose carboplatin, thiotepa, and etoposide followed by HSCT was performed in 13 patients with CNS AT/RT after tumor resection and conventional chemotherapy consisting of cisplatin, vincristine, cyclophosphamide, and etoposide with or without high-dose methotrexate. Three of seven patients who received the high-dose methotrexate combined HSCT regimen had long-term remission (42+, 54+, and 67+ months), although they received no radiation therapy (Gardner et al., 2008). Trilateral retinoblastoma is also a rare brain tumor in the pineal region with very poor prognosis. HDC for trilateral retinoblastoma has some effect (De Ioris et al., 2010; Dunkel et al., 2010b), even if the tumor has relapsed (Tsuruta et al., 2011b).

3.3 HSCT for young children with brain tumors

HSCT is often considered for young children with newly diagnosed MB instead of radiation therapy because cranial irradiation in young children is thought to incur late complications (Kim et al., 2010). In the US, the Memorial Sloan-Kettering Cancer Center study of HDC followed by Auto BMT for young children with newly diagnosed malignant brain tumors shortened the period of maintenance chemotherapy and reduced the risk of radiation therapy and had some long survivors (Mason et al., 1998). The neuropsychological findings of patients who received BMT revealed within low average range of intelligence (Sands et al., 1998). They also studied Auto BMT for young children with recurrent malignant tumors and showed that HDC with BMT followed by additional external-beam irradiation is effective for some young children with recurrent malignant tumors (Guruangan et al., 1998). Patients with supratentorial PNET have poor outcomes when compared to MB. HDC followed by HSCT in young patients with newly diagnosed supratentorial PNET provides improved EFS and OS and radiation therapy for these patients can sometimes be deferred or eliminated (Fangusaro et al., 2008).

3.4 Tandem & allo HSCT for brain tumors

Tandem HSCT is performed for brain tumors as well as neuroblastoma. Two courses of high-dose melphalan (100 mg/m^2) with PBSCT in 16 patients with cerebral PNET safely resulted in a high response rate, i.e. 11 PR in 14 patients with measurable disease (Vassal et al., 2001). A study of children with newly diagnosed MB and supratentorial PNET who were treated with four cycles of high-dose cyclophosphamide (4,000 mg/m^2), cisplatin (75 mg/m^2), and vincristine (1.5 mg/m^2 x 2) with 49 of 53 patients completing all four cycles and HSCT showed a 2-year PFS rate of 73.7 ± 10.5% for high-risk patients (residual tumor ≥ 1.5 cm^2 after conventional therapy) and 93.6 ± 4.7% for average-risk patients (Strother et al., 2001). In Korea, 18 patients with high-risk solid tumors including brain tumors received two courses of HDC (cyclophosphamide, melphalan, thiotepa, carboplatin, etoposide, etc.) with HSCT. In this study, all patients with MB were alive and disease-free, although patients with other types of brain tumors died (Sung et al., 2003). Our case report series showed that double HDC consisting of melphalan and thiotepa may cure patients with primary disseminated MB; however, a patient with high expression of erythroblastic leukemia viral oncogene homolog 2 (*ERBB2*) relapsed (Aihara et al., 2010). Recently, there is a report that single HSCT regimens have greater toxicity than ≥ 3 tandem HSCT which use milder HDC per HSCT cycle (Panosyan et al., 2011).

Allo HSCT for brain tumors is still experimental, although there are case reports from 1992 (Lundberg et al., 1992; Tsuruta et al., 2011b). An antitumor GVT effect associated with chronic GVHD was observed in a patient with metastatic MB after Allo HSCT, (Secondino et al., 2008).

4. Ewing's sarcoma

Studies of Auto BMT for patients with Ewing's sarcoma (ES) were reported in both adults and children since the early 1980s. In initial studies of Auto BMT for ESwing, melphalan-based conditioning regimens were used for disseminated or relapsed patients, with some achieving remission (Cornbleet et al., 1981; Graham-Pole et al., 1984). Subsequent studies of Auto HSCT for ES involved regimens combined with TBI or busulfan.

4.1 TBI combined regimens for ES

Until the 1990s, TBI-combined regimens were often utilized. Regimens consisting of TBI (12 Gy), melphalan, and etoposide with or without carboplatin for multi-focal or relapsed ES showed better outcomes (relapse-free survival 45 ± 12% at 6 years) when compared with historical controls (Burdach et al., 1993). For patients with high-risk ES in first CR, HSCT with or without TBI was associated with a DFS of 62.7 ± 11% overall and 40 ± 21.9% for those with metastatic disease (Madero et al., 1998). One study of HSCT with versus without TBI for patients with primary metastatic (stage IV) ES (n = 171) showed improved prognosis (overall EFS 40% vs. 19%), although patients with pulmonary or skeletal metastases had especially poor outcome (EFS 27% vs. 0%) (Paulussen et al., 1998). HSCT with TBI for patients with ES did not improve the prognosis of patients with metastases to bone or bone marrow; the majority experienced relapse and died with progressive disease (Kushner & Meyers, 2001; Meyers et al., 2001). Recurrent ES also has a poor prognosis. However, in secondary remission HSCT can improve the prognosis of recurrent ES (Barker et al., 2005).

4.2 Busulfan & melphalan regimens for ES

Since the late 1990s busulfan has been often utilized instead of TBI. Studies of melphalan-based conditioning regimens including busulfan were reported from several groups. In 21 children with high-risk ES resistant to conventional therapy and recurrent disease, PBSCT with melphalan-based regimens did not show any benefit (Drabko et al., 2005). In a study of 33 recurrent or progressive ES patients treated with regimens of high-dose busulfan or melphalan with or without TBI, some patients achieved long-term survival although the treatment was associated severe toxicity (McTiernan et al., 2006). In 2006, in Europe the benefits of high-dose busulfan (600 mg/m^2) and melphalan (140 mg/m^2) for ES with metastasis to lung or bone without bone marrow involvement were shown (Oberlin et al., 2006). The subsequent Euro-EWING-99 trial reported an estimated 3-year EFS of 45% in 46 children younger than 14 years with primary disseminated multifocal ES (Ladenstein et al., 2010). Recently, an Italian and Scandinavian group showed the effectiveness of HSCT with busulfan and melphalan for patients with non-metastatic, poor-response ES (5-year EFS of HSCT and non-HSCT group was 72% and 33%) (Ferrari et al., 2011).

4.3 Tandem HSCT for ES

The European Intergroup Cooperative Ewing Sarcoma Study (EICESS) compared the regimen of HyperME (TBI 12 Gy, melphalan 120 to 180 mg/m^2, etoposide 40 to 60 mg/kg, and carboplatin maximum dose 1,800mg) to TandemME (two cycles of melphalan 120 to

140mg/m^2 and etoposide 60mg/kg). The 5-year EFS of Hyper ME and TandemME were 22 ± 8% and 29 ± 9%, respectively, while the lethal complication rate was 23% in HyperME and 4% in TandemME (Burdach et al., 2003). A study using two cycles of alkylating agent-based HDC for primary metastatic bulky disease or recurrent ES showed acceptable toxicities and OS of 45% and EFS of 47% at 3 years, although only 65% of patients were able to proceed to the second cycle (Rosenthal et al., 2008).

4.4 Allo HSCT for ES

From EICESS, 26 cases of Auto HSCT and 10 cases of Allo HSCT for multifocal or recurrent advanced ES were reported. Although patients who received Allo HSCT failed to show allogeneic immunological effects e.g. GVT, eight of 26 cases with Auto HSCT received systemic interleukin-2 (IL-2) therapy after HSCT. The IL-2 group had a better EFS rate (60 ± 18%) compared to the groups that received Allo HSCT (20 ± 13%) alone or Auto HSCT alone (22 ± 13%) (Burdach et al., 2000). A case report of a 6 year-old girl with multifocal ES who received Allo HSCT from her HLA-matched mother showed a shrinkage of her pulmonary tumors and grade 1 GVHD during a taper of the immunosuppression (Lucas et al., 2008). Recently, reduced- versus high-intensity conditioning for Allo HSCT was compared. There was no improvement in survival with reduced intensity conditioning due to an increased rate of death caused by disease and relapse (Thiel et al., 2011).

5. Osteosarcoma

The prognosis of metastatic or relapsed osteosarcoma is very poor. High-dose methotrexate has been used since the 1970s, but HSCT for osteosarcoma remains experimental. There are fewer reports of HSCT for osteosarcoma than for ES since patients with metastatic or recurrent disease have < 20% chance of long-term survival despite aggressive treatment (Chou et al., 2008). However, many kinds of new agents including antifolate compounds and muramyl tripeptide phosphatidylethanolamine are under investigation (Jaffe, 2009).

The initial study of HSCT in two cases of advanced osteosarcoma failed to reveal a clinical effect. In 2001, the Cooperative German-Austrian-Swiss Osteosarcoma Study Group reported the results of HSCT in 15 patients with relapsed osteosarcoma. Three of the 15 patients died of toxic complications and the treatment outcome was not better than with conventional therapy (Sauerbrey et al., 2001). A study using tandem HSCT for relapsed osteosarcoma by the Italian Sarcoma Group Study showed two points: only patients who are chemosensitive to induction treatment can obtain complete remission after HSCT, and the length of remission is short (Fagioli et al., 2002). From Japan, two case reports of Allo HSCT for osteosarcoma were presented. In the first case, non-myeloablative chemotherapy followed by PBSCT for a third CR showed a GVT effect in reduction of the metastatic lung lesion with GVHD (Kounami et al., 2005). In the second report, a patient with progressive osteosarcoma and multiple lung and bone metastases received TBI combined with myeloablative HCT (thiotepa 600 mg/m^2 and etoposide 1,000mg/m^2) followed by Allo BMT. The metastatic lesions successfully disappeared (Goi et al., 2006).

6. Wilms' tumor

Although multidisciplinary therapies including chemotherapy improved the cure rate of Wilms' tumor (WT) to approximately 80–85%, the EFS in recurrent disease was less than

15% before HDC with HSCT was developed (Grundy et al., 1989). Furthermore, a number of adverse prognostic factors related to poor outcomes during relapse have been identified and are used as inclusion criteria for studies involving HSCT (Dallorso et al., 2008) e.g. loss of heterozygosity in chromosomes 1p and 16q (Grundy et al., 2005).
The European Bone Marrow Transplantation for Solid Tumor Registry reported a study of 25 children with WT (12 stage IV, 5 stage III, 3 stage II, and 1 stage I) who were treated with Auto BMT and melphalan-based or other regimens. Of 21 recurrent cases, 13 achieved second or subsequent CR after Auto BMT and eight patients were event-free at 14–90 months (Garaventa et al., 1994). High-dose melphalan, etoposide, and carboplatin (MEC) followed by Auto HSCT in high-risk recurrent WT patients with chemotherapy-responsive disease was evaluated prospectively by the French Society of Pediatric Oncology. The DFS and OS at 3 years were 50 ± 17% and 60 ± 18%, respectively (Pein et al., 1998). A report of 23 patients with high-risk WT, including 20 recurrent cases, treated with HDC including MEC followed by Auto HSCT in the German Cooperative Wilms Tumor Studies showed an estimated survival and EFS are 60.9% and 48.2% (Kremens et al., 2002). At the Children's Memorial Hospital in Chicago, one or two cycles of HSCT were tried in patients with relapsed WT. The EFS and OS of this study at 4 years were 60% and 73%, respectively (Campbell et al., 2004). A study in Italy also showed the benefit of Auto HSCT for high-risk relapsed WT with a 3-year DFS approaching 50% (Spreafico et al., 2008). A meta-analysis of 6 studies with or without HDC followed by Auto HSCT revealed that patients who only relapsed in the lung have a higher survival rate than those who relapsed in other locations. The 4-year survival rate among stage I or II patients was about 30% higher in the no HSCT group than the HSCT group, but the 4-year survival rate was similar in HSCT vs. no-HSCT in stage III and IV patients. These findings suggest that salvage chemotherapy is typically the better choice for relapsed WT patients, and that HSCT could be considered for stage III or IV patients with relapse only in the lungs (Presson et al., 2010). There is a case report describing a refractory case of WT with lung metastasis being cured by CBT (Massimino et al., 2010). There are rare case reports of WT successfully treated by HSCT, e.g. a case of third remission (Brown et al., 2010), a bilateral Wilms' tumor treated without any irradiation (Saarinen-Pihkala et al., 1998), and an anephric pediatric case with multiple recurrent WT treated in conjunction with hemodialysis (Dagher et al., 1998).

7. Rhabdomyosarcoma

Since the end of 1990s, the prognosis of patients with localized rhabdomyosarcoma (RS) has improved with an EFS rate of approximately 70%. Although HDC with or without TBI followed by Auto or Allo HSCT has been performed for patients with metastatic or recurrent diseases by several groups, a retrospective multi-center analysis with 36 patients by the German and Austrian Pediatric Bone Marrow Transplantation Group in 1997 showed that the effect of HSCT was uncertain. In this study, the EFS rate at 2 years after HDC in 26 patients who were treated with MEC with or without TBI regimen was 36 ± 7%, and none of the 5 patients who received Allo BMT alive (Koscielniak et al., 1997). The European Intergroup MMT4-89 (including 4 cycles of standard chemotherapy) and MMT4-91 (melphalan-based HDC followed by HSCT after three courses of standard chemotherapy) studies on childhood metastatic RS suggested that there was prolonged survival with the MMT4-91 regimen (EFS and OS at 3 years were 29.7% and 40%, respectively) when compared with the MMT4-89 regimen (EFS and OS at 3 year were 19.2% and 27.7%, respectively) (Carli et al., 1999). However, their final report concluded there is no evidence that consolidation HDC improves outcomes (Carli et al., 2004). The analysis of 177 patients with stage IV RS from 22 studies who received HSCT also concluded that HSCT for patients with relapsed or refractory high-risk RS does not provide a significant advantage (Weigel et al., 2001). A recent meta-analysis of three non-randomized controlled studies on survival showed 3-year OS ranged from 22% to 55% in the HSCT group versus 18% to 55% in the control group, and also showed no differences between treatments (Peinemann et al., 2011). On the other hand, an analysis of HSCT for high-risk RS according to histological type suggested that for unfavorable histologies (alveolar or undifferentiated subtypes) HSCT after relapse may have some benefit because long-term survivors were sometimes seen (Stiff et al., 2010).

8. Hepatoblastoma

HCT followed by Auto or Allo HSCT for patients with high-risk hepatoblastoma (HB) was tried by several groups. The initial report described three patients with stage II or III HB who achieved CR after receiving Auto HSCT, and one patient with a third relapse showed PR (Hara et al., 1998). HSCT for MRD may improve survival (Yoshinari et al., 1998). Subsequent reports of tandem HSCT for three high-risk patients (metastatic or recurrent disease) showed that they failed to remain in remission (Katzenstein et al., 2002). Recent reports by the Japanese Study Group for Pediatric Liver Tumors also showed that the prognosis of HDC with HSCT for metastatic HB was poor (5-year OS of 43.9%) (Hishiki et al., 2011). On the other hand, a report of a patient with recurrent metastatic HB who received non-myeloablative HSCT from a HLA-matched unrelated donor showed a probable GVT effect with GVHD (Inaba et al., 2006).

9. Retinoblastoma

The prognosis of retinoblastoma improved to a 5-year survival rate of approximately 90% through early diagnosis and proper treatment, including enucleation, chemotherapy, or radiation therapy. However, the outcome of progressive disease, i.e. optic nerve invasion, extra-ocular extension, or metastatic disease, especially into central nervous system (CNS), remains poor (Ali et al., 2011; Chantada et al., 2007; Cozza et al., 2009). Since the 1980s, HDC with HSCT with or without TBI has been performed with immunomagnetic purging (Saleh et al., 1988), or monitoring of MRD in the BM or PB by monoclonal antibody of 3A7 (Saarinen et al., 1991) or protein gene product 9.5 gene expression (Yamane et al., 1999) because retinoblastoma frequently infiltrates into these sites (Karcioglu et al., 1997). Several trials of Auto HSCT for metastatic retinoblastoma with or without CNS involvement were reported. Patients without CNS involvement were successfully treated by HDC followed by HSCT with or without local irradiation (Dunkel et al., 2000; Kremens et al., 2003; Matsubara et al., 2005). A patient with recurrent disseminated retinoblastoma in the lymph nodes, bone, and BM was successfully treated by HDC with CD34-selected Auto HSCT (Hertzberg et al., 2001). High-dose carboplatin (250 or 350 mg/m^2 daily for 5 days), etoposide (350 mg/m^2 daily for 5 days), and cyclophosphamide (1.6 g/m^2 daily for 4 days), followed by HSCT, can improve survival in retinoblastoma involving bone or BM but not the CNS (Namouni et al., 1997). A patient with cerebrospinal fluid metastasis who received HSCT with high-dose/short-infusion cyclosporine to inhibit P-glycoprotein, which is expressed by most metastatic retinoblastomas, achieved long-term remission (Dimaras et al., 2009). A recent study of HDC with Auto HSCT for stage 4b (CNS involvement) retinoblastoma showed some benefit for some patients with CNS metastasis (Dunkel et al., 2010a). Furthermore, a study of HDCT in South America for metastatic retinoblastoma with CNS involvement also showed that 2 of 4 patients with CNS involvement were alive and disease free at 39 and 27 months (Dunkel et al., 2010a). Tandem HSCT was attempted to treat advanced bilateral retinoblastoma. Nine patients with bilateral retinoblastoma received two cycles of HDC with regimens of MEC + BM (busulfan and melphalan) or CTE (carboplatin, thiotepa, and etoposide) + CM (cyclophosphamide and melphalan) followed by Auto HSCT in order to avoid external-beam radiation therapy. All patients had at least one functional eye and in two patients both eyes were preserved (Lee et al., 2008).

10. References

Aihara Y.; Tsuruta, T.; Kawamata, T., et al. (2010). Double High-Dose Chemotherapy Followed by Autologous Peripheral Blood Stem Cell Transplantation for Primary Disseminated Medulloblastoma: A Report of 3 Cases, *J Pediatr Hematol Oncol*, Vol.32, 2, pp. e70-74

Ali M.J.; Reddy, V.A.; Honavar, S.G., et al. (2011). Orbital Retinoblastoma: Where Do We Go from Here?, *J Cancer Res Ther*, Vol.7, 1, pp. 11-14

Ash S.; Gigi, V.; Askenasy, N., et al. (2009). Graft Versus Neuroblastoma Reaction Is Efficiently Elicited by Allogeneic Bone Marrow Transplantation through Cytolytic Activity in the Absence of GVHD, *Cancer Immunol Immunother*, Vol.58, 12, pp. 2073-2084

Ash S.; Stein, J.; Askenasy, N., et al. (2010). Immunomodulation with Dendritic Cells and Donor Lymphocyte Infusion Converge to Induce Graft Vs Neuroblastoma Reactions without GVHD after Allogeneic Bone Marrow Transplantation, *Br J Cancer*, Vol.103, 10, pp. 1597-1605

August C.S.; Serota, F.T.; Koch, P.A., et al. (1984). Treatment of Advanced Neuroblastoma with Supralethal Chemotherapy, Radiation, and Allogeneic or Autologous Marrow Reconstitution, *J Clin Oncol*, Vol.2, 6, pp. 609-616

Barker L.M.; Pendergrass, T.W.; Sanders, J.E., et al. (2005). Survival after Recurrence of Ewing's Sarcoma Family of Tumors, *J Clin Oncol*, Vol.23, 19, pp. 4354-4362

Bensimhon P.; Villablanca, J.G.; Sender, L.S., et al. (2010). Peripheral Blood Stem Cell Support for Multiple Cycles of Dose Intensive Induction Therapy Is Feasible with Little Risk of Tumor Contamination in Advanced Stage Neuroblastoma: A Report from the Childrens Oncology Group, *Pediatr Blood Cancer*, Vol.54, 4, pp. 596-602

Berthold F.; Boos, J.; Burdach, S., et al. (2005). Myeloablative Megatherapy with Autologous Stem-Cell Rescue Versus Oral Maintenance Chemotherapy as Consolidation Treatment in Patients with High-Risk Neuroblastoma: A Randomised Controlled Trial, *Lancet Oncol*, Vol.6, 9, pp. 649-658

Brodeur G.M.; Pritchard, J.; Berthold, F., et al. (1993). Revisions of the International Criteria for Neuroblastoma Diagnosis, Staging, and Response to Treatment, *J Clin Oncol*, Vol.11, 8, pp. 1466-1477

Brodeur G.M.; Seeger, R.C.; Schwab, M., et al. (1984). Amplification of N-Myc in Untreated Human Neuroblastomas Correlates with Advanced Disease Stage, *Science*, Vol.224, 4653, pp. 1121-1124

Brown E.; Hebra, A.; Jenrette, J., et al. (2010). Successful Treatment of Late, Recurrent Wilms Tumor with High-Dose Chemotherapy and Autologous Stem Cell Rescue in Third Complete Response, *J Pediatr Hematol Oncol*, Vol.32, 6, pp. e241-243

Burdach S.; Jurgens, H.; Peters, C., et al. (1993). Myeloablative Radiochemotherapy and Hematopoietic Stem-Cell Rescue in Poor-Prognosis Ewing's Sarcoma, *J Clin Oncol*, Vol.11, 8, pp. 1482-1488

Burdach S.; Meyer-Bahlburg, A.; Laws, H.J., et al. (2003). High-Dose Therapy for Patients with Primary Multifocal and Early Relapsed Ewing's Tumors: Results of Two Consecutive Regimens Assessing the Role of Total-Body Irradiation, *J Clin Oncol*, Vol.21, 16, pp. 3072-3078

Burdach S.; van Kaick, B.; Laws, H.J., et al. (2000). Allogeneic and Autologous Stem-Cell Transplantation in Advanced Ewing Tumors. An Update after Long-Term Follow-up from Two Centers of the European Intergroup Study Eicess. Stem-Cell Transplant Programs at Dusseldorf University Medical Center, Germany and St. Anna Kinderspital, Vienna, Austria, *Ann Oncol*, Vol.11, 11, pp. 1451-1462

Campbell A.D.; Cohn, S.L.; Reynolds, M., et al. (2004). Treatment of Relapsed Wilms' Tumor with High-Dose Therapy and Autologous Hematopoietic Stem-Cell Rescue: The Experience at Children's Memorial Hospital, *J Clin Oncol*, Vol.22, 14, pp. 2885-2890

Carli M.; Colombatti, R.; Oberlin, O., et al. (2004). European Intergroup Studies (MMT4-89 and MMT4-91) on Childhood Metastatic Rhabdomyosarcoma: Final Results and Analysis of Prognostic Factors, *J Clin Oncol*, Vol.22, 23, pp. 4787-4794

Carli M.; Colombatti, R.; Oberlin, O., et al. (1999). High-Dose Melphalan with Autologous Stem-Cell Rescue in Metastatic Rhabdomyosarcoma, *J Clin Oncol*, Vol.17, 9, pp. 2796-2803

Chantada G.L.; Dunkel, I.J.; Antoneli, C.B., et al. (2007). Risk Factors for Extraocular Relapse Following Enucleation after Failure of Chemoreduction in Retinoblastoma, *Pediatr Blood Cancer*, Vol.49, 3, pp. 256-260

Chou A.J.Geller, D.S. & Gorlick, R. (2008). Therapy for Osteosarcoma: Where Do We Go from Here?, *Paediatr Drugs*, Vol.10, 5, pp. 315-327

Cohn S.L.; Moss, T.J.; Hoover, M., et al. (1997). Treatment of Poor-Risk Neuroblastoma Patients with High-Dose Chemotherapy and Autologous Peripheral Stem Cell Rescue, *Bone Marrow Transplant*, Vol.20, 7, pp. 543-551

Cornbleet M.A.; Corringham, R.E.; Prentice, H.G., et al. (1981). Treatment of Ewing's Sarcoma with High-Dose Melphalan and Autologous Bone Marrow Transplantation, *Cancer Treat Rep*, Vol.65, 3-4, pp. 241-244

Cozza R.; De Ioris, M.A.; Ilari, I., et al. (2009). Metastatic Retinoblastoma: Single Institution Experience over Two Decades, *Br J Ophthalmol*, Vol.93, 9, pp. 1163-1166

Dagher R.; Kreissman, S.; Robertson, K.A., et al. (1998). High Dose Chemotherapy with Autologous Peripheral Blood Progenitor Cell Transplantation in an Anephric Child with Multiply Recurrent Wilms Tumor, *J Pediatr Hematol Oncol*, Vol.20, 4, pp. 357-360

Dallorso S.; Dini, G.; Faraci, M., et al. (2008). SCT for Wilms' Tumour, *Bone Marrow Transplant*, Vol.41 Suppl 2, pp. S128-130

De Ioris M.A.; Fidani, P.; Munier, F.L., et al. (2010). Successful Treatment of Trilateral Retinoblastoma with Conventional and High-Dose Chemotherapy Plus Radiotherapy: A Case Report, *J Pediatr Hematol Oncol*, Vol.32, 8, pp. e343-345

Dimaras H.; Heon, E.; Budning, A., et al. (2009). Retinoblastoma CSF Metastasis Cured by Multimodality Chemotherapy without Radiation, *Ophthalmic Genet*, Vol.30, 3, pp. 121-126

Drabko K.; Zawitkowska-Klaczynska, J.; Wojcik, B., et al. (2005). Megachemotherapy Followed by Autologous Stem Cell Transplantation in Children with Ewing's Sarcoma, *Pediatr Transplant*, Vol.9, 5, pp. 618-621

Dunkel I.J.; Aledo, A.; Kernan, N.A., et al. (2000). Successful Treatment of Metastatic Retinoblastoma, *Cancer*, Vol.89, 10, pp. 2117-2121

Dunkel I.J.; Boyett, J.M.; Yates, A., et al. (1998). High-Dose Carboplatin, Thiotepa, and Etoposide with Autologous Stem-Cell Rescue for Patients with Recurrent Medulloblastoma. Children's Cancer Group, *J Clin Oncol*, Vol.16, 1, pp. 222-228

Dunkel I.J.; Chan, H.S.; Jubran, R., et al. (2010a). High-Dose Chemotherapy with Autologous Hematopoietic Stem Cell Rescue for Stage 4B Retinoblastoma, *Pediatr Blood Cancer*, Vol.55, 1, pp. 149-152

Dunkel I.J.; Jubran, R.F.; Gururangan, S., et al. (2010b). Trilateral Retinoblastoma: Potentially Curable with Intensive Chemotherapy, *Pediatr Blood Cancer*, Vol.54, 3, pp. 384-387

Durando X.; Lemaire, J.J.; Tortochaux, J., et al. (2003). High-Dose BCNU Followed by Autologous Hematopoietic Stem Cell Transplantation in Supratentorial High-Grade Malignant Gliomas: A Retrospective Analysis of 114 Patients, *Bone Marrow Transplant*, Vol.31, 7, pp. 559-564

Evans A.E.; August, C.S.; Kamani, N., et al. (1994). Bone Marrow Transplantation for High Risk Neuroblastoma at the Children's Hospital of Philadelphia: An Update, *Med Pediatr Oncol*, Vol.23, 4, pp. 323-327

Fagioli F.; Aglietta, M.; Tienghi, A., et al. (2002). High-Dose Chemotherapy in the Treatment of Relapsed Osteosarcoma: An Italian Sarcoma Group Study, *J Clin Oncol*, Vol.20, 8, pp. 2150-2156

Fangusaro J.; Finlay, J.; Sposto, R., et al. (2008). Intensive Chemotherapy Followed by Consolidative Myeloablative Chemotherapy with Autologous Hematopoietic Cell Rescue (AuHCR) in Young Children with Newly Diagnosed Supratentorial Primitive Neuroectodermal Tumors (sPNETs): Report of the Head Start I and Ii Experience, *Pediatr Blood Cancer*, Vol.50, 2, pp. 312-318

Ferrari S.; Sundby Hall, K.; Luksch, R., et al. (2011). Nonmetastatic Ewing Family Tumors: High-Dose Chemotherapy with Stem Cell Rescue in Poor Responder Patients. Results of the Italian Sarcoma Group/Scandinavian Sarcoma Group III Protocol, *Ann Oncol*, Vol.22, 5, pp. 1221-1227

Finlay J.L.; Goldman, S.; Wong, M.C., et al. (1996). Pilot Study of High-Dose Thiotepa and Etoposide with Autologous Bone Marrow Rescue in Children and Young Adults with Recurrent CNS Tumors. The Children's Cancer Group, *J Clin Oncol*, Vol.14, 9, pp. 2495-2503

Fish J.D. & Grupp, S.A. (2008). Stem Cell Transplantation for Neuroblastoma, *Bone Marrow Transplant*, Vol.41, 2, pp. 159-165

Gajjar A.; Chintagumpala, M.; Ashley, D., et al. (2006). Risk-Adapted Craniospinal Radiotherapy Followed by High-Dose Chemotherapy and Stem-Cell Rescue in Children with Newly Diagnosed Medulloblastoma (St Jude Medulloblastoma-96): Long-Term Results from a Prospective, Multicentre Trial, *Lancet Oncol*, Vol.7, 10, pp. 813-820

Gajjar A. & Pizer, B. (2010). Role of High-Dose Chemotherapy for Recurrent Medulloblastoma and Other CNS Primitive Neuroectodermal Tumors, *Pediatr Blood Cancer*, Vol.54, 4, pp. 649-651

Garaventa A.; Hartmann, O.; Bernard, J.L., et al. (1994). Autologous Bone Marrow Transplantation for Pediatric Wilms' Tumor: The Experience of the European Bone Marrow Transplantation Solid Tumor Registry, *Med Pediatr Oncol*, Vol.22, 1, pp. 11-14

Gardner S.L.; Asgharzadeh, S.; Green, A., et al. (2008). Intensive Induction Chemotherapy Followed by High Dose Chemotherapy with Autologous Hematopoietic Progenitor Cell Rescue in Young Children Newly Diagnosed with Central Nervous System Atypical Teratoid Rhabdoid Tumors, *Pediatr Blood Cancer*, Vol.51, 2, pp. 235-240

George R.E.; Li, S.; Medeiros-Nancarrow, C., et al. (2006). High-Risk Neuroblastoma Treated with Tandem Autologous Peripheral-Blood Stem Cell-Supported Transplantation: Long-Term Survival Update, *J Clin Oncol*, Vol.24, 18, pp. 2891-2896

Goi K.; Inukai, T.; Honna, H., et al. (2011). Successful Tandem (Autologous-Cord Blood) SCT in Advanced Neuroblastomas with Highly Amplified Mycn, *Bone Marrow Transplant*, Vol.46, 6, pp. 835-839

Goi K.; Sugita, K.; Tezuka, T., et al. (2006). A Successful Case of Allogeneic Bone Marrow Transplantation for Osteosarcoma with Multiple Metastases of Lung and Bone, *Bone Marrow Transplant*, Vol.37, 1, pp. 115-116

Graham M.L.; Herndon, J.E., 2nd; Casey, J.R., et al. (1997). High-Dose Chemotherapy with Autologous Stem-Cell Rescue in Patients with Recurrent and High-Risk Pediatric Brain Tumors, *J Clin Oncol*, Vol.15, 5, pp. 1814-1823

Graham-Pole J.; Lazarus, H.M.; Herzig, R.H., et al. (1984). High-Dose Melphalan Therapy for the Treatment of Children with Refractory Neuroblastoma and Ewing's Sarcoma, *Am J Pediatr Hematol Oncol*, Vol.6, 1, pp. 17-26

Grill J.; Kalifa, C.; Doz, F., et al. (1996). A High-Dose Busulfan-Thiotepa Combination Followed by Autologous Bone Marrow Transplantation in Childhood Recurrent Ependymoma. A Phase-II Study, *Pediatr Neurosurg*, Vol.25, 1, pp. 7-12

Grodman H.Wolfe, L. & Kretschmar, C. (2009). Outcome of Patients with Recurrent Medulloblastoma or Central Nervous System Germinoma Treated with Low Dose Continuous Intravenous Etoposide Along with Dose-Intensive Chemotherapy Followed by Autologous Hematopoietic Stem Cell Rescue, *Pediatr Blood Cancer*, Vol.53, 1, pp. 33-36

Grundy P.; Breslow, N.; Green, D.M., et al. (1989). Prognostic Factors for Children with Recurrent Wilms' Tumor: Results from the Second and Third National Wilms' Tumor Study, *J Clin Oncol*, Vol.7, 5, pp. 638-647

Grundy P.E.; Breslow, N.E.; Li, S., et al. (2005). Loss of Heterozygosity for Chromosomes 1p and 16q Is an Adverse Prognostic Factor in Favorable-Histology Wilms Tumor: A Report from the National Wilms Tumor Study Group, *J Clin Oncol*, Vol.23, 29, pp. 7312-7321

Grupp S.A.; Stern, J.W.; Bunin, N., et al. (2000). Rapid-Sequence Tandem Transplant for Children with High-Risk Neuroblastoma, *Med Pediatr Oncol*, Vol.35, 6, pp. 696-700

Guruangan S.; Dunkel, I.J.; Goldman, S., et al. (1998). Myeloablative Chemotherapy with Autologous Bone Marrow Rescue in Young Children with Recurrent Malignant Brain Tumors, *J Clin Oncol*, Vol.16, 7, pp. 2486-2493

Gururangan S.; Krauser, J.; Watral, M.A., et al. (2008). Efficacy of High-Dose Chemotherapy or Standard Salvage Therapy in Patients with Recurrent Medulloblastoma, *Neuro Oncol*, Vol.10, 5, pp. 745-751

Hara J.; Osugi, Y.; Ohta, H., et al. (1998). Double-Conditioning Regimens Consisting of Thiotepa, Melphalan and Busulfan with Stem Cell Rescue for the Treatment of Pediatric Solid Tumors, *Bone Marrow Transplant*, Vol.22, 1, pp. 7-12

Hertzberg H.; Kremens, B.; Velten, I., et al. (2001). Recurrent Disseminated Retinoblastoma in a 7-Year-Old Girl Treated Successfully by High-Dose Chemotherapy and CD34-Selected Autologous Peripheral Blood Stem Cell Transplantation, *Bone Marrow Transplant*, Vol.27, 6, pp. 653-655

Hishiki T.; Matsunaga, T.; Sasaki, F., et al. (2011). Outcome of Hepatoblastomas Treated Using the Japanese Study Group for Pediatric Liver Tumor (JPLT) Protocol-2: Report from the JPLT, *Pediatr Surg Int*, Vol.27, 1, pp. 1-8

Hobbie W.L.; Moshang, T.; Carlson, C.A., et al. (2008). Late Effects in Survivors of Tandem Peripheral Blood Stem Cell Transplant for High-Risk Neuroblastoma, *Pediatr Blood Cancer*, Vol.51, 5, pp. 679-683

Hochberg F.H.; Parker, L.M.; Takvorian, T., et al. (1981). High-Dose BCNU with Autologous Bone Marrow Rescue for Recurrent Glioblastoma Multiforme, *J Neurosurg*, Vol.54, 4, pp. 455-460

Holtta P.; Alaluusua, S.; Saarinen-Pihkala, U.M., et al. (2002). Long-Term Adverse Effects on Dentition in Children with Poor-Risk Neuroblastoma Treated with High-Dose Chemotherapy and Autologous Stem Cell Transplantation with or without Total Body Irradiation, *Bone Marrow Transplant*, Vol.29, 2, pp. 121-127

Hovi L.; Saarinen-Pihkala, U.M.; Vettenranta, K., et al. (1999). Growth in Children with Poor-Risk Neuroblastoma after Regimens with or without Total Body Irradiation in Preparation for Autologous Bone Marrow Transplantation, *Bone Marrow Transplant*, Vol.24, 10, pp. 1131-1136

Inaba H.; Handgretinger, R.; Furman, W., et al. (2006). Allogeneic Graft-Versus-Hepatoblastoma Effect, *Pediatr Blood Cancer*, Vol.46, 4, pp. 501-505

Inoue M.; Nakano, T.; Yoneda, A., et al. (2003). Graft-Versus-Tumor Effect in a Patient with Advanced Neuroblastoma Who Received HLA Haplo-Identical Bone Marrow Transplantation, *Bone Marrow Transplant*, Vol.32, 1, pp. 103-106

Jaffe N. (2009). Osteosarcoma: Review of the Past, Impact on the Future. The American Experience, *Cancer Treat Res*, Vol.152, pp. 239-262

Johnson D.B.; Thompson, J.M.; Corwin, J.A., et al. (1987). Prolongation of Survival for High-Grade Malignant Gliomas with Adjuvant High-Dose Bcnu and Autologous Bone Marrow Transplantation, *J Clin Oncol*, Vol.5, 5, pp. 783-789

Jubert C.; Wall, D.A.; Grimley, M., et al. (2011). Engraftment of Unrelated Cord Blood after Reduced-Intensity Conditioning Regimen in Children with Refractory Neuroblastoma: A Feasibility Trial, *Bone Marrow Transplant*, Vol.46, 2, pp. 232-237

Kalifa C.; Hartmann, O.; Demeocq, F., et al. (1992). High-Dose Busulfan and Thiotepa with Autologous Bone Marrow Transplantation in Childhood Malignant Brain Tumors: A Phase II Study, *Bone Marrow Transplant*, Vol.9, 4, pp. 227-233

Kamani N.; August, C.S.; Bunin, N., et al. (1996). A Study of Thiotepa, Etoposide and Fractionated Total Body Irradiation as a Preparative Regimen Prior to Bone Marrow Transplantation for Poor Prognosis Patients with Neuroblastoma, *Bone Marrow Transplant*, Vol.17, 6, pp. 911-916

Karcioglu Z.A.; al-Mesfer, S.A.; Abboud, E., et al. (1997). Workup for Metastatic Retinoblastoma. A Review of 261 Patients, *Ophthalmology*, Vol.104, 2, pp. 307-312

Katzenstein H.M.; Rigsby, C.; Shaw, P.H., et al. (2002). Novel Therapeutic Approaches in the Treatment of Children with Hepatoblastoma, *J Pediatr Hematol Oncol*, Vol.24, 9, pp. 751-755

Kim S.Y.; Sung, K.W.; Hah, J.O., et al. (2010). Reduced-Dose Craniospinal Radiotherapy Followed by High-Dose Chemotherapy and Autologous Stem Cell Rescue for Children with Newly Diagnosed High-Risk Medulloblastoma or Supratentorial Primitive Neuroectodermal Tumor, *Korean J Hematol*, Vol.45, 2, pp. 120-126

Kletzel M.; Katzenstein, H.M.; Haut, P.R., et al. (2002). Treatment of High-Risk Neuroblastoma with Triple-Tandem High-Dose Therapy and Stem-Cell Rescue: Results of the Chicago Pilot II Study, *J Clin Oncol*, Vol.20, 9, pp. 2284-2292

Koscielniak E.; Klingebiel, T.H.; Peters, C., et al. (1997). Do Patients with Metastatic and Recurrent Rhabdomyosarcoma Benefit from High-Dose Therapy with Hematopoietic Rescue? Report of the German/Austrian Pediatric Bone Marrow Transplantation Group, *Bone Marrow Transplant*, Vol.19, 3, pp. 227-231

Kounami S.; Nakayama, K.; Yoshiyama, M., et al. (2005). Non-Myeloablative Allogenic Peripheral Blood Stem Cell Transplantation in a Patient with Refractory Osteosarcoma, *Pediatr Transplant*, Vol.9, 3, pp. 342-345

Kremens B.; Gruhn, B.; Klingebiel, T., et al. (2002). High-Dose Chemotherapy with Autologous Stem Cell Rescue in Children with Nephroblastoma, *Bone Marrow Transplant*, Vol.30, 12, pp. 893-898

Kremens B.; Wieland, R.; Reinhard, H., et al. (2003). High-Dose Chemotherapy with Autologous Stem Cell Rescue in Children with Retinoblastoma, *Bone Marrow Transplant*, Vol.31, 4, pp. 281-284

Kushner B.H. & Meyers, P.A. (2001). How Effective Is Dose-Intensive/Myeloablative Therapy against Ewing's Sarcoma/Primitive Neuroectodermal Tumor Metastatic to Bone or Bone Marrow? The Memorial Sloan-Kettering Experience and a Literature Review, *J Clin Oncol*, Vol.19, 3, pp. 870-880

Kushner B.H.; O'Reilly, R.J.; Mandell, L.R., et al. (1991). Myeloablative Combination Chemotherapy without Total Body Irradiation for Neuroblastoma, *J Clin Oncol*, Vol.9, 2, pp. 274-279

Ladenstein R.; Lasset, C.; Hartmann, O., et al. (1994). Comparison of Auto Versus Allografting as Consolidation of Primary Treatments in Advanced Neuroblastoma over One Year of Age at Diagnosis: Report from the European Group for Bone Marrow Transplantation, *Bone Marrow Transplant*, Vol.14, 1, pp. 37-46

Ladenstein R.; Potschger, U.; Le Deley, M.C., et al. (2010). Primary Disseminated Multifocal Ewing Sarcoma: Results of the Euro-Ewing 99 Trial, *J Clin Oncol*, Vol.28, 20, pp. 3284-3291

Laprie A.; Michon, J.; Hartmann, O., et al. (2004). High-Dose Chemotherapy Followed by Locoregional Irradiation Improves the Outcome of Patients with International Neuroblastoma Staging System Stage II and III Neuroblastoma with Mycn Amplification, *Cancer*, Vol.101, 5, pp. 1081-1089

Lee S.H.; Yoo, K.H.; Sung, K.W., et al. (2008). Tandem High-Dose Chemotherapy and Autologous Stem Cell Rescue in Children with Bilateral Advanced Retinoblastoma, *Bone Marrow Transplant*, Vol.42, 6, pp. 385-391

Lucas K.G.Schwartz, C. & Kaplan, J. (2008). Allogeneic Stem Cell Transplantation in a Patient with Relapsed Ewing Sarcoma, *Pediatr Blood Cancer*, Vol.51, 1, pp. 142-144

Lundberg J.H.; Weissman, D.E.; Beatty, P.A., et al. (1992). Treatment of Recurrent Metastatic Medulloblastoma with Intensive Chemotherapy and Allogeneic Bone Marrow Transplantation, *J Neurooncol*, Vol.13, 2, pp. 151-155

Madero L.; Munoz, A.; Sanchez de Toledo, J., et al. (1998). Megatherapy in Children with High-Risk Ewing's Sarcoma in First Complete Remission, *Bone Marrow Transplant*, Vol.21, 8, pp. 795-799

Mahoney D.H., Jr.; Strother, D.; Camitta, B., et al. (1996). High-Dose Melphalan and Cyclophosphamide with Autologous Bone Marrow Rescue for Recurrent/Progressive Malignant Brain Tumors in Children: A Pilot Pediatric Oncology Group Study, *J Clin Oncol*, Vol.14, 2, pp. 382-388

Marabelle A.; Merlin, E.; Halle, P., et al. (2011). Cd34+ Immunoselection of Autologous Grafts for the Treatment of High-Risk Neuroblastoma, *Pediatr Blood Cancer*, Vol.56, 1, pp. 134-142

Marabelle A.; Paillard, C.; Tchirkov, A., et al. (2007). Graft-Versus-Tumour Effect in Refractory Metastatic Neuroblastoma, *Bone Marrow Transplant*, Vol.39, 12, pp. 809-810

Mason W.P.; Grovas, A.; Halpern, S., et al. (1998). Intensive Chemotherapy and Bone Marrow Rescue for Young Children with Newly Diagnosed Malignant Brain Tumors, *J Clin Oncol*, Vol.16, 1, pp. 210-221

Massimino M.Cohen, K.J. & Finlay, J.L. (2010). Is There a Role for Myeloablative Chemotherapy with Autologous Hematopoietic Cell Rescue in the Management of Childhood High-Grade Astrocytomas?, *Pediatr Blood Cancer*, Vol.54, 4, pp. 641-643

Massimino M.; Gandola, L.; Spreafico, F., et al. (2009). No Salvage Using High-Dose Chemotherapy Plus/Minus Reirradiation for Relapsing Previously Irradiated Medulloblastoma, *Int J Radiat Oncol Biol Phys*, Vol.73, 5, pp. 1358-1363

Matsubara H.; Makimoto, A.; Higa, T., et al. (2005). A Multidisciplinary Treatment Strategy That Includes High-Dose Chemotherapy for Metastatic Retinoblastoma without Cns Involvement, *Bone Marrow Transplant*, Vol.35, 8, pp. 763-766

Matthay K.K.; Quach, A.; Huberty, J., et al. (2009). Iodine-131--Metaiodobenzylguanidine Double Infusion with Autologous Stem-Cell Rescue for Neuroblastoma: A New Approaches to Neuroblastoma Therapy Phase I Study, *J Clin Oncol*, Vol.27, 7, pp. 1020-1025

Matthay K.K.; Seeger, R.C.; Reynolds, C.P., et al. (1994). Allogeneic Versus Autologous Purged Bone Marrow Transplantation for Neuroblastoma: A Report from the Childrens Cancer Group, *J Clin Oncol*, Vol.12, 11, pp. 2382-2389

Matthay K.K.; Villablanca, J.G.; Seeger, R.C., et al. (1999). Treatment of High-Risk Neuroblastoma with Intensive Chemotherapy, Radiotherapy, Autologous Bone Marrow Transplantation, and 13-Cis-Retinoic Acid. Children's Cancer Group, *N Engl J Med*, Vol.341, 16, pp. 1165-1173

McCowage G.B.; Vowels, M.R.; Shaw, P.J., et al. (1995). Autologous Bone Marrow Transplantation for Advanced Neuroblastoma Using Teniposide, Doxorubicin, Melphalan, Cisplatin, and Total-Body Irradiation, *J Clin Oncol*, Vol.13, 11, pp. 2789-2795

McTiernan A.; Driver, D.; Michelagnoli, M.P., et al. (2006). High Dose Chemotherapy with Bone Marrow or Peripheral Stem Cell Rescue Is an Effective Treatment Option for Patients with Relapsed or Progressive Ewing's Sarcoma Family of Tumours, *Ann Oncol*, Vol.17, 8, pp. 1301-1305

Meyers P.A.; Krailo, M.D.; Ladanyi, M., et al. (2001). High-Dose Melphalan, Etoposide, Total-Body Irradiation, and Autologous Stem-Cell Reconstitution as Consolidation Therapy for High-Risk Ewing's Sarcoma Does Not Improve Prognosis, *J Clin Oncol*, Vol.19, 11, pp. 2812-2820

Modak S.; Gardner, S.; Dunkel, I.J., et al. (2004). Thiotepa-Based High-Dose Chemotherapy with Autologous Stem-Cell Rescue in Patients with Recurrent or Progressive Cns Germ Cell Tumors, *J Clin Oncol*, Vol.22, 10, pp. 1934-1943

Molina B.; Alonso, L.; Gonzalez-Vicent, M., et al. (2011). High-Dose Busulfan and Melphalan as Conditioning Regimen for Autologous Peripheral Blood Progenitor Cell Transplantation in High-Risk Neuroblastoma Patients, *Pediatr Hematol Oncol*, Vol.28, 2, pp. 115-123

Namouni F.; Doz, F.; Tanguy, M.L., et al. (1997). High-Dose Chemotherapy with Carboplatin, Etoposide and Cyclophosphamide Followed by a Haematopoietic Stem Cell Rescue in Patients with High-Risk Retinoblastoma: A Sfop and Sfgm Study, *Eur J Cancer*, Vol.33, 14, pp. 2368-2375

Navid F.Armstrong, M. & Barfield, R.C. (2009). Immune Therapies for Neuroblastoma, *Cancer Biol Ther*, Vol.8, 10, pp. 874-882

Oberlin O.; Rey, A.; Desfachelles, A.S., et al. (2006). Impact of High-Dose Busulfan Plus Melphalan as Consolidation in Metastatic Ewing Tumors: A Study by the Societe Francaise Des Cancers De L'enfant, *J Clin Oncol*, Vol.24, 24, pp. 3997-4002

Panosyan E.H.; Ikeda, A.K.; Chang, V.Y., et al. (2011). High-Dose Chemotherapy with Autologous Hematopoietic Stem-Cell Rescue for Pediatric Brain Tumor Patients: A Single Institution Experience from Ucla, *J Transplant*, Vol.2011, pp. e740673

Papadakis V.; Dunkel, I.J.; Cramer, L.D., et al. (2000). High-Dose Carmustine, Thiotepa and Etoposide Followed by Autologous Bone Marrow Rescue for the Treatment of High

Risk Central Nervous System Tumors, *Bone Marrow Transplant*, Vol.26, 2, pp. 153-160

Paulussen M.; Ahrens, S.; Burdach, S., et al. (1998). Primary Metastatic (Stage IV) Ewing Tumor: Survival Analysis of 171 Patients from the Eicess Studies. European Intergroup Cooperative Ewing Sarcoma Studies, *Ann Oncol*, Vol.9, 3, pp. 275-281

Pedrazzoli P.; Da Prada, G.A.; Giorgiani, G., et al. (2002). Allogeneic Blood Stem Cell Transplantation after a Reduced-Intensity, Preparative Regimen: A Pilot Study in Patients with Refractory Malignancies, *Cancer*, Vol.94, 9, pp. 2409-2415

Pein F.; Michon, J.; Valteau-Couanet, D., et al. (1998). High-Dose Melphalan, Etoposide, and Carboplatin Followed by Autologous Stem-Cell Rescue in Pediatric High-Risk Recurrent Wilms' Tumor: A French Society of Pediatric Oncology Study, *J Clin Oncol*, Vol.16, 10, pp. 3295-3301

Peinemann F.; Kroger, N.; Bartel, C., et al. (2011). High-Dose Chemotherapy Followed by Autologous Stem Cell Transplantation for Metastatic Rhabdomyosarcoma--a Systematic Review, *PLoS One*, Vol.6, 2, pp. e17127

Perez-Martinez A.; Lassaletta, A.; Gonzalez-Vicent, M., et al. (2005). High-Dose Chemotherapy with Autologous Stem Cell Rescue for Children with High Risk and Recurrent Medulloblastoma and Supratentorial Primitive Neuroectodermal Tumors, *J Neurooncol*, Vol.71, 1, pp. 33-38

Philip T.; Ladenstein, R.; Zucker, J.M., et al. (1993). Double Megatherapy and Autologous Bone Marrow Transplantation for Advanced Neuroblastoma: The LMCE2 Study, *Br J Cancer*, Vol.67, 1, pp. 119-127

Phillips G.L.; Fay, J.W.; Herzig, G.P., et al. (1983). Intensive 1,3-Bis(2-Chloroethyl)-1-Nitrosourea (BCNU), NSC #4366650 and Cryopreserved Autologous Marrow Transplantation for Refractory Cancer. A Phase I-II Study, *Cancer*, Vol.52, 10, pp. 1792-1802

Phillips G.L.; Fay, J.W.; Herzig, G.P., et al. (1979). Intensive 1,3-Bis(2-Chloroethyl)-1-Nitrosourea (BCNU) Autologous Bone Marrow Transplantation Therapy of Refractory Cancer: A Preliminary Report, *Exp Hematol*, Vol.7 Suppl 5, pp. 372-383

Phillips G.L.; Wolff, S.N.; Fay, J.W., et al. (1986). Intensive 1,3-Bis (2-Chloroethyl)-1-Nitrosourea (BCNU) Monochemotherapy and Autologous Marrow Transplantation for Malignant Glioma, *J Clin Oncol*, Vol.4, 5, pp. 639-645

Presson A.Moore, T.B. & Kempert, P. (2010). Efficacy of High-Dose Chemotherapy and Autologous Stem Cell Transplant for Recurrent Wilms' Tumor: A Meta-Analysis, *J Pediatr Hematol Oncol*, Vol.32, 6, pp. 454-461

Pritchard J.; Cotterill, S.J.; Germond, S.M., et al. (2005). High Dose Melphalan in the Treatment of Advanced Neuroblastoma: Results of a Randomised Trial (ENSG-1) by the European Neuroblastoma Study Group, *Pediatr Blood Cancer*, Vol.44, 4, pp. 348-357

Qayed M.; Chiang, K.Y.; Ricketts, R., et al. (2011). Tandem Stem Cell Rescue as Consolidation Therapy for High-Risk Neuroblastoma, *Pediatr Blood Cancer*, doi: 10.1002/pbc.23155. [Epub]

Ridola V.; Grill, J.; Doz, F., et al. (2007). High-Dose Chemotherapy with Autologous Stem Cell Rescue Followed by Posterior Fossa Irradiation for Local Medulloblastoma

Recurrence or Progression after Conventional Chemotherapy, *Cancer*, Vol.110, 1, pp. 156-163

Rosenthal J.; Bolotin, E.; Shakhnovits, M., et al. (2008). High-Dose Therapy with Hematopoietic Stem Cell Rescue in Patients with Poor Prognosis Ewing Family Tumors, *Bone Marrow Transplant*, Vol.42, 5, pp. 311-318

Saarinen U.M.Sariola, H. & Hovi, L. (1991). Recurrent Disseminated Retinoblastoma Treated by High-Dose Chemotherapy, Total Body Irradiation, and Autologous Bone Marrow Rescue, *Am J Pediatr Hematol Oncol*, Vol.13, 3, pp. 315-319

Saarinen-Pihkala U.M.Wikstrom, S. & Vettenranta, K. (1998). Maximal Preservation of Renal Function in Patients with Bilateral Wilms' Tumor: Therapeutic Strategy of Late Kidney-Sparing Surgery and Replacement of Radiotherapy by High-Dose Melphalan and Stem Cell Rescue, *Bone Marrow Transplant*, Vol.22, 1, pp. 53-59

Saleh R.A.; Gross, S.; Cassano, W., et al. (1988). Metastatic Retinoblastoma Successfully Treated with Immunomagnetic Purged Autologous Bone Marrow Transplantation, *Cancer*, Vol.62, 11, pp. 2301-2303

Sands S.A.van Gorp, W.G. & Finlay, J.L. (1998). Pilot Neuropsychological Findings from a Treatment Regimen Consisting of Intensive Chemotherapy and Bone Marrow Rescue for Young Children with Newly Diagnosed Malignant Brain Tumors, *Childs Nerv Syst*, Vol.14, 10, pp. 587-589

Sauerbrey A.; Bielack, S.; Kempf-Bielack, B., et al. (2001). High-Dose Chemotherapy (HDC) and Autologous Hematopoietic Stem Cell Transplantation (ASCT) as Salvage Therapy for Relapsed Osteosarcoma, *Bone Marrow Transplant*, Vol.27, 9, pp. 933-937

Secondino S.; Pedrazzoli, P.; Schiavetto, I., et al. (2008). Antitumor Effect of Allogeneic Hematopoietic Sct in Metastatic Medulloblastoma, *Bone Marrow Transplant*, Vol.42, 2, pp. 131-133

Shih C.S.; Hale, G.A.; Gronewold, L., et al. (2008). High-Dose Chemotherapy with Autologous Stem Cell Rescue for Children with Recurrent Malignant Brain Tumors, *Cancer*, Vol.112, 6, pp. 1345-1353

Shimada H.; Ambros, I.M.; Dehner, L.P., et al. (1999). The International Neuroblastoma Pathology Classification (the Shimada System), *Cancer*, Vol.86, 2, pp. 364-372

Spreafico F.; Bisogno, G.; Collini, P., et al. (2008). Treatment of High-Risk Relapsed Wilms Tumor with Dose-Intensive Chemotherapy, Marrow-Ablative Chemotherapy, and Autologous Hematopoietic Stem Cell Support: Experience by the Italian Association of Pediatric Hematology and Oncology, *Pediatr Blood Cancer*, Vol.51, 1, pp. 23-28

Stiff P.J.; Agovi, M.A.; Antman, K.H., et al. (2010). High-Dose Chemotherapy with Blood or Bone Marrow Transplants for Rhabdomyosarcoma, *Biol Blood Marrow Transplant*, Vol.16, 4, pp. 525-532

Strother D.; Ashley, D.; Kellie, S.J., et al. (2001). Feasibility of Four Consecutive High-Dose Chemotherapy Cycles with Stem-Cell Rescue for Patients with Newly Diagnosed Medulloblastoma or Supratentorial Primitive Neuroectodermal Tumor after Craniospinal Radiotherapy: Results of a Collaborative Study, *J Clin Oncol*, Vol.19, 10, pp. 2696-2704

Sung K.W.; Lee, S.H.; Yoo, K.H., et al. (2007). Tandem High-Dose Chemotherapy and Autologous Stem Cell Rescue in Patients over 1 Year of Age with Stage 4 Neuroblastoma, *Bone Marrow Transplant*, Vol.40, 1, pp. 37-45

Sung K.W.; Yoo, K.H.; Chung, E.H., et al. (2003). Successive Double High-Dose Chemotherapy with Peripheral Blood Stem Cell Rescue Collected During a Single Leukapheresis Round in Patients with High-Risk Pediatric Solid Tumors: A Pilot Study in a Single Center, *Bone Marrow Transplant*, Vol.31, 6, pp. 447-452

Takahashi H.; Manabe, A.; Aoyama, C., et al. (2008). Iodine-131-Metaiodobenzylguanidine Therapy with Reduced-Intensity Allogeneic Stem Cell Transplantation in Recurrent Neuroblastoma, *Pediatr Blood Cancer*, Vol.50, 3, pp. 676-678

Thiel U.; Wawer, A.; Wolf, P., et al. (2011). No Improvement of Survival with Reduced- Versus High-Intensity Conditioning for Allogeneic Stem Cell Transplants in Ewing Tumor Patients, *Ann Oncol*, Vol.22, 7, pp. 1614-1621

Toporski J.; Garkavij, M.; Tennvall, J., et al. (2009). High-Dose Iodine-131-Metaiodobenzylguanidine with Haploidentical Stem Cell Transplantation and Posttransplant Immunotherapy in Children with Relapsed/Refractory Neuroblastoma, *Biol Blood Marrow Transplant*, Vol.15, 9, pp. 1077-1085

Trager C.; Kogner, P.; Lindskog, M., et al. (2003). Quantitative Analysis of Tyrosine Hydroxylase mRNA for Sensitive Detection of Neuroblastoma Cells in Blood and Bone Marrow, *Clin Chem*, Vol.49, 1, pp. 104-112

Tsuruta T.; Aihara, Y.; Kanno, H., et al. (2011a). Shared Molecular Targets in Pediatric Gliomas and Ependymomas, *Pediatr Blood Cancer*, doi: 10.1002/pbc.23009. [Epub]

Tsuruta T.; Aihara, Y.; Kanno, H., et al. (2011b). High-Dose Chemotherapy Followed by Autologous and Allogeneic Peripheral Blood Stem Cell Transplantation for Recurrent Disseminated Trilateral Retinoblastoma, *Childs Nerv Syst*, Vol.27, 6, pp. 1019-1024

Vanlemmens P.; Plouvier, E.; Amsallem, D., et al. (1992). Transplantation of Umbilical Cord Blood in Neuroblastoma, *Nouv Rev Fr Hematol*, Vol.34, 3, pp. 243-246

Vassal G.; Tranchand, B.; Valteau-Couanet, D., et al. (2001). Pharmacodynamics of Tandem High-Dose Melphalan with Peripheral Blood Stem Cell Transplantation in Children with Neuroblastoma and Medulloblastoma, *Bone Marrow Transplant*, Vol.27, 5, pp. 471-477

Wagner J.E. (1993). Umbilical Cord Blood Stem Cell Transplantation, *Am J Pediatr Hematol Oncol*, Vol.15, 2, pp. 169-174

Weigel B.J.; Breitfeld, P.P.; Hawkins, D., et al. (2001). Role of High-Dose Chemotherapy with Hematopoietic Stem Cell Rescue in the Treatment of Metastatic or Recurrent Rhabdomyosarcoma, *J Pediatr Hematol Oncol*, Vol.23, 5, pp. 272-276

Weinstein J.L.Katzenstein, H.M. & Cohn, S.L. (2003). Advances in the Diagnosis and Treatment of Neuroblastoma, *Oncologist*, Vol.8, 3, pp. 278-292

Yalcin B.; Kremer, L.C.; Caron, H.N., et al. (2011). High-Dose Chemotherapy and Autologous Haematopoietic Stem Cell Rescue for Children with High-Risk Neuroblastoma, *Cochrane Database Syst Rev*, Vol.2011, 5, pp. CD006301

Yamane S.; Shirai, C.; Arimoto, A., et al. (1999). Disseminated Retinoblastoma Successfully Treated with Myeloablative Chemotherapy--Implication for Molecular Detection of Minimal Residual Disease, *Bone Marrow Transplant*, Vol.23, 9, pp. 971-974

Yanik G.A.; Levine, J.E.; Matthay, K.K., et al. (2002). Pilot Study of Iodine-131-Metaiodobenzylguanidine in Combination with Myeloablative Chemotherapy and Autologous Stem-Cell Support for the Treatment of Neuroblastoma, *J Clin Oncol*, Vol.20, 8, pp. 2142-2149

Yoshinari M.; Imaizumi, M.; Hayashi, Y., et al. (1998). Peripheral Blood Stem Cell Transplantation for Hepatoblastoma with Microscopical Residue: A Therapeutic Approach for Incompletely Resected Tumor, *Tohoku J Exp Med*, Vol.184, 3, pp. 247-254

20

Stem Cells in Ophthalmology

Sara T. Wester[1] and Jeffrey Goldberg[1,2]
[1]*Department of Ophthalmology, Bascom Palmer Eye Institute*
[2]*Interdisciplinary Stem Cell Institute*
University of Miami Miller School of Medicine, Miami, Florida
USA

1. Introduction

Stem cell research in the field of ophthalmology over the past two decades has advanced knowledge and treatment dramatically. Stem cells are functionally defined as (a) cells that can self-renew and provide ongoing populations of identical daughter cells with the same

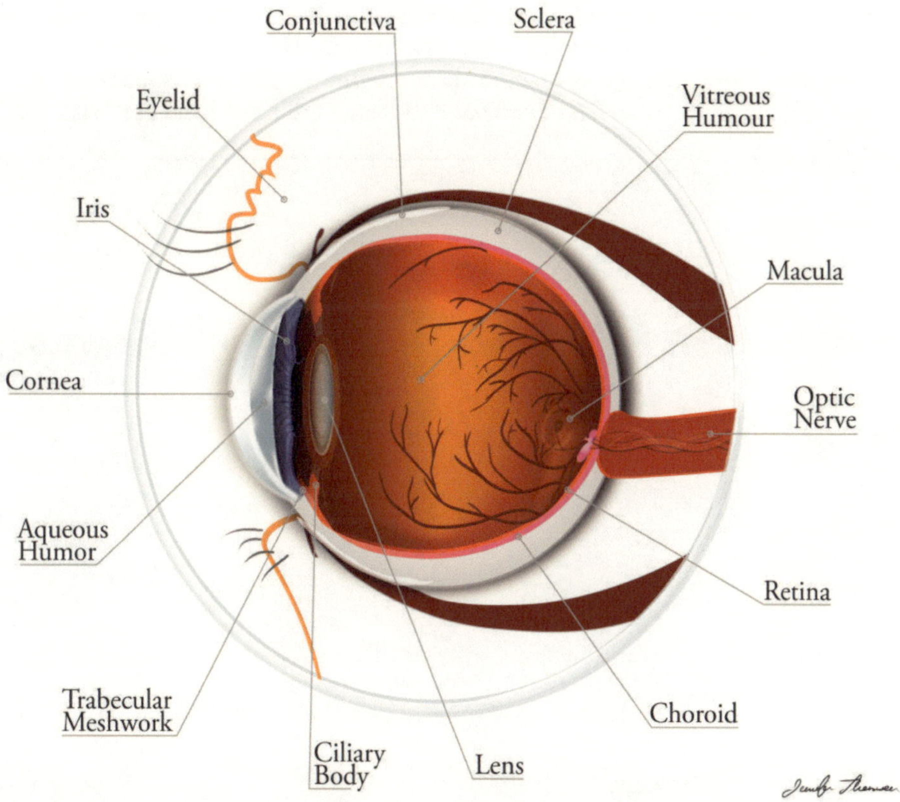

Fig. 1. Anatomy of the Eye and Extraocular Tissues (Illustration by Jennifer Thomson Ph D., Bascom Palmer Eye Institute, University of Miami Miller School of Medicine)

unrestricted proliferation potential and (b) cells that are multipotent and able to give rise to all cell lineages in a particular tissue. Stem cells repopulate damaged or lost tissue either through differentiation into tissue appropriate cells and /or by releasing paracrine signaling molecules to recruit inflammatory cells and other tissue progenitor cells. As such, these cells have a key role in therapeutic tissue regeneration (Lau, 2009). From the establishment of the location of corneal epithelial stems cells to promising developments in diagnosis and treatment of ophthalmic diseases, an enhanced understanding of stem cells in ophthalmology offers promising diagnostic and therapeutic advances.

The purpose of this chapter is to outline our current knowledge and recent advances in stem cell research in ophthalmology. We will focus on the potential clinical applications of stem cell research in six main areas: Cornea, Conjunctiva, Orbit, Eyelid, Trabecular Meshwork and Retina.

2. Cornea

2.1 Background

The human cornea is highly specialized. The refractive index of a normal cornea is 1.376 and is essential to maintaining clarity of visual images on the retina. The cornea receives nutrition via glucose from the aqueous humour and oxygen through the tear film, although the peripheral cornea gets oxygen from the corneoscleral limbus. The cornea is composed of five layers: the epithelium, Bowman's layer, the stroma, Descemet's membrane and the endothelium (see Figure 2). The epithelium is composed of stratified squamous epithelial cells. These cells make up about 5% of corneal thickness. Perilimbal basal epithelial cells give

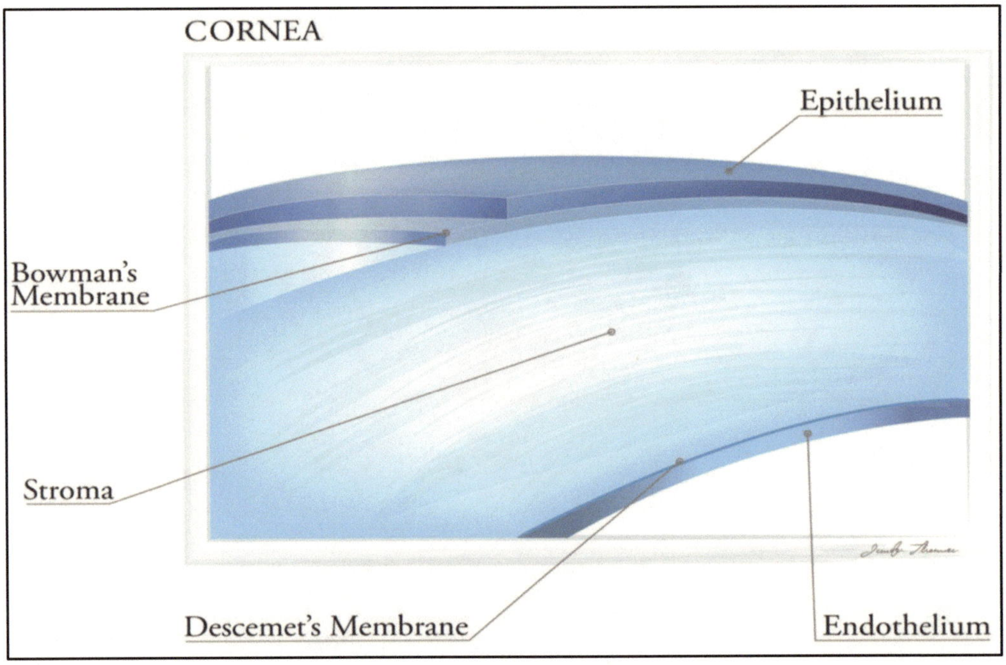

Fig. 2. Anatomy of the Cornea (Illustration by Jennifer Thomson Ph D., Bascom Palmer Eye Institute, University of Miami Miller School of Medicine)

rise to superficial cells which then desquamate into the tear film over a period of one to two weeks. Constant renewal of the epithelial cells is essential to the cornea's visually essential role of light transmittance and refraction. With each blink, superficial epithelial cells may be shed into the tear pool, and these cells must be constantly repopulated from epithelial stem cells.

The stroma is also essential to the maintenance of clear vision. It is composed of keratocytes in a lattice arrangement, which reduces light scattering by destructive interference. The lattice is smaller than the wavelength of visible light, an essential characteristic to maintaining corneal clarity. Corneal transparency also depends on corneal hydration which depends on the endothelial pump. The endothelium is especially susceptible to damage as endothelial cells do not proliferate and cell loss leads to enlargement and spread of neighboring cells to cover the defect.

Thus, each layer of the cornea is essential to maintaining clarity of visual images. The discovery of the location of corneal epithelial stems cells at the limbus (the junctional zone between the cornea and conjunctiva) was one of the first major stem cell advances in the field of ophthalmology, and was proposed in the early 1970s (Davanger and Evenson, 1971).

2.2 Stem cell localization

An extensive review of corneal stem cells will be documented in the chapter Limbal Stem Cell Transplantation and Corneal Neovascularization by Katikireddy and Ula. The purpose of this section of the chapter is to provide a brief overview of corneal stem cells in the context of other stem cell research in ophthalmology, such as in extensive ocular surface disease.

The corneal limbus was first proposed in 1971 as the site of corneal stem cells based on the observation that pigment migrated from the limbus to the central cornea in healing corneal defects (Davanger & Evenson, 1971). Subsequent studies demonstrated that slow-cycling limbal epithelial basal cells retained tritiated thymidine (Cotsarelis et al, 1989), supporting the notion that this was the location of stem cells. In addition, keratinocyte growth factor (KGF) is known to stimulate limbal epithelial precursor cells to grow (Wilson et al, 1994), and limbal epithelial cells exhibit higher levels of KGF compared to similar corneal epithelial cells (Li & Tseng, 1997). Interestingly, corneal epithelial cells grown on limbal stroma also revert to a more proliferative cell type (Espana et al, 1987). In addition, stem cell markers such as p63 (Pelligrini et al, 2001) and transporter ABCG2 (Watanabe et al, 2004) have been identified in the limbal cell population. Even studies on the distribution of stem cells in the human corneal stroma and endothelium which found precursor cells distributed throughout the cornea show that the peripheral cornea has the large majority of the stem cell population (Yamagami et al, 2007).

How does this translate into the clinical setting? In culture, limbal basal cells have the highest proliferative capacity, diseases affecting the limbus result in delayed healing and conjunctival overgrowth, and limbal transplants can regenerate corneal-like epithelium. These findings all support the premise that the corneal surface is renewed by limbal stem cells in both normal physiological conditions and after injury.

What are the key characteristics of corneal stem cells? Corneal stem cells are characterized by small size, low rate of replication, and expression of a side population (SP) phenotype (Watanabe et al, 2004), a common feature of stem cells in other organ systems. These cells represent only a minor fraction of the cell population in the cornea. In the mouse model, these limbal stem cells only represent from <1% (Budak et al, 2005) to 3.8% (Krulova et al, 2008) of the limbal cell population. In rat and human cornea, flow cytometry and analysis of certain stem cell markers (such as ABCG2 expression) has shown that these cells range from 0.20% (Watanabe et al, 2004) to 10% (Umemoto et al, 2005) of the total limbal epithelial cell population. The small fraction they represent is important in understanding techniques for extraction and cultivation of stem cells.

What about the stem cell niche? Even in tissues with a high turnover rate, such as the cornea, there are a proportion of stem cells among this population that are quiescent and reactivate only in specific situations such as injury or trauma (Wang et al, 2010). This stem cell "niche" hypothesis maintains that upon stem cell division one cell remains a stem cell and returns to replenish the stem cell supply while the other becomes differentiated to repair and resupply lost cells (Schofield, 1983). In addition, the niche in which stem cells reside provides ongoing nutritional support and protection. For example, in the cornea the stem cells at the limbus are close to blood vessels which can provide essential nutrients and allow for ongoing proliferation. The properties and environment which lead to support and proliferation of stem cells as well as maintenance of supply are key factors to the successful treatment of diseases with limbal stem cell deficiency.

2.3 Pathology

The protective microenvironment which maintains the stem cell population and allows for self-renewal is extremely important. Any disruption in this environment can damage the stem cells and cause pathologic changes with significant consequences, such as occurs in limbal stem cell deficiency. There are congenital and acquired forms of limbal stem cell deficiency. For example, aniridia is a congenital ocular disease associated with limbal stem cell deficiency. More commonly, limbal stem cell deficiency develops due to local tissue injury. Stevens Johnson syndrome (see Figure 3), ocular cicatricial pemphigoid, chemical or thermal burns, severe contact lens keratopathy, surgical damage to the limbus and other cicatrizing disease can all lead to a deficiency in these important cells.

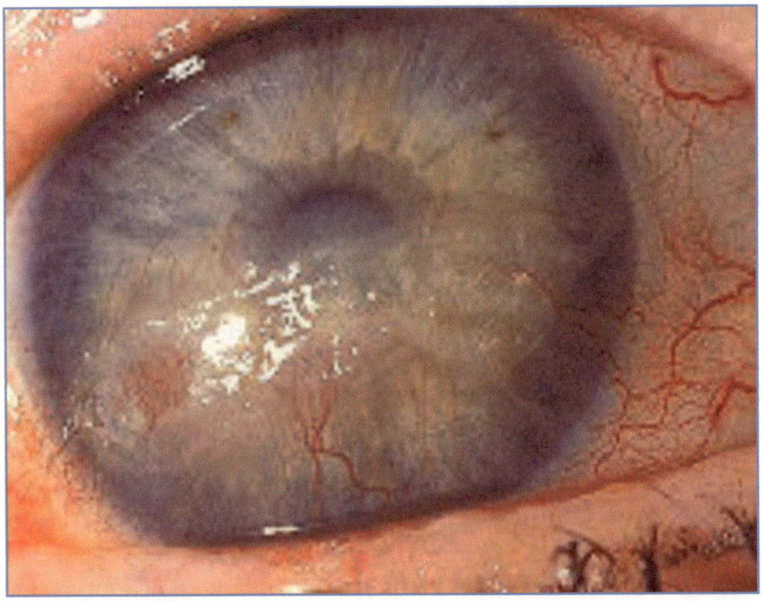

Fig. 3. Corneal scarring from Stevens Johnson Syndrome (Courtesy of Miami VA Hospital Photography Department)

Typically, when limbal stem cell deficiency occurs, the neighboring conjunctival cells (normally hindered by limbal cells) migrate over the cornea and cause conjunctivalization (corneal neovascularization, appearance of goblet cells and an irregular epithelium) (Daniels et al, 2001). The result of this overgrowth is poor vision, corneal opacity, surface irregularity and patient discomfort. In the cornea, the identification of limbal stem cells has enabled advances in not only the diagnosis of corneal disease but also in treatment options for patients with severe limbal stem cell deficiency.

2.4 Treatment

In certain severe ocular diseases with corneal compromise, treatment of limbal stem cell deficiency was initially demonstrated to be effective via transplantation of limbal allografts, autografts or keratoepithelioplasty (Frucht-Pery et al, 1998) (Kenyon & Tseng, 1989). Limbal allografts, however, carry a high risk of rejection because of the presence of specific antigens which are more likely to stimulate an immune response (Pels & van der Gaag, 1984). As a result, they require long term immunosuppression. In addition, cadaver grafting, which in cases of severe bilateral disease is sometimes the only option, has a high failure rate compared to living related donor in the treatment of total limbal stem cell deficiency (Miri, Al-Deiri et al, 2010). This form of grafting has been associated with poor clinical outcomes.

Prolonged inflammation due to the ongoing disease process can also destroy corneal stem cells and substantially increases the risk of any surgical intervention in these cases. As such, limbal transplantation is contraindicated in patients with severe dry eye given the low chance of survival of the graft (Dua et al, 2010). Given these factors, it is important that an extensive ocular evaluation is performed including eyelid position and function, ocular surface lubrication and presence or absence of dry eye syndrome, visual potential of the eye, condition of the fellow eye and systemic health of the patient (especially when allografts are planned and immunosuppression is required). Treatment of ocular or systemic co-morbidities is essential to the long-term success of stem cell transplantation in the eye, just as in other organ systems.

The following general algorithm for surgical intervention in cases of stem cell deficiency was proposed (Dua et al, 2010):

1. Partial stem cell deficiency with no involvement of the visual axis, aggressive lubrication
2. Partial stem cell deficiency with visual involvement, sequential sector conjunctival epitheliectomy (SSCE)
3. Partial stem cell deficiency with pannus, SSCE with a sector limbal transplant
4. Total unilateral stem cell deficiency, autolimbal transplantation
5. Total bilateral stem cell deficiency use of allografts from a living related or cadaver donor

For partial stem cell deficiency, amniotic membrane transplantation with superficial keratectomy is also a viable treatment option (Anderson et al, 2001). The details of the surgical procedures are beyond the scope of this chapter, but the extensive algorithm summarizes the significant advances that have taken place in terms of our understanding of the anatomy, pathology and treatment options for stem cell deficiency in the cornea.

2.5 Further stem cell applications and new advances

Limitations of the aforementioned treatment modalities have led to treatment of unilateral or bilateral total stem cell deficiency with ex vivo expansion of limbal derived cells with or without a substrate (Dua et al, 2010). Ex vivo transplantation uses cells from a small biopsy sample which are cultured to generate cohesive sheets of corneal epithelium. This epithelium can then be transplanted (Pelligrini et al, 1997). Several techniques are used for producing ex vivo cultured limbal epithelial cells for transplantation. The explants culture system and the suspension culture system have been shown to have similar success rates (Shortt et al, 2007). Ex vivo cultured oral mucosal epithelium has also been used to treat limbal stem cell deficiency (Inatomi et al, 2006). While ex vivo expansion has the advantage of limited donor site morbidity because it only requires a small biopsy, it necessitates specialized techniques, takes time to generate the limbal stem cells and is very costly.

What about the niche environment? The success of limbal stem cell transplantation, whether ex-vivo or in-vivo, is also not surprisingly dependent on the limbal stem cell niche environment. Processing of cultured limbal stem cells may become more efficient and treatments more efficacious as these factors are elucidated. In fibroblast cultures from the cornea, limbus and sclera of donor ocular tissue, only corneal and limbal fibroblasts supported limbal epithelial precursor culture establishment; limbal fibroblasts also had significantly better growth (Ainscough et al, 2011). Scleral fibroblasts had increased myofibroblast formation, which inhibited limbal epithelial stem cell growth, possibly preventing proliferation of limbal epithelial stem cells towards the sclera (Ainscough et al, 2011). These surrounding fibroblasts thus regulate the stem cell enrivonment. In addition, in the rabbit model, corneal stromal inflammation was associated with poor graft outcome (Tsai, & Tseng 1995). The environment in which the cultured epithelial graft cells are placed is therefore important for the long term therapeutic effect.

How does this translate into clinical applications? Improved success of cultured limbal epithelial grafts may be obtained by "debriding" the wound of myofibroblasts. Stromal tissue substitutes that provide a good scaffold and hinder growth of myofibroblasts may lead to improved clinical outcomes (Ainscough et al, 2011). Transplantation with surrounding conjunctival tissue may lead to improved clinical results even in cadaver grafts (Dua et al, 2010). It is possible this frill of conjunctiva possesses factors which enhance clinical outcomes of transplantation. Co-transfer of limbal stem cells with other stem cells, such as mesenchymal cells, may also lead to improved outcomes by inhibiting local inflammation and improving healing through immunosuppressive properties of these cells (Zajicova et al, 2010). Furthermore, human amniotic membrane may serve as a niche for limbal stem cells, possibly through the suppression of apoptosis and expression of certain growth factors (Tseng et al, 1998). As such, amniotic membrane may contribute to the long term survival of cultured limbal stem cell transplantation (Tsai RJ & Tsai RY, 2010). Understanding the milieu into which stems cells are to be transplanted for maximal success is essential to the advancement of these treatment options.

Therapy focused on reconstruction of the ocular surface in disease such as severe ocular burns, ocular cicatricial pemphigoid, Stevens Johnson Syndrome and others has thus been dramatically improved with limbal stem cell and amniotic membrane transplantation, with improved corneal clarity and clinical outcomes. As new growth factors are elucidated and stem cell culture techniques are developed, the treatment of corneal disease with stem cells will continue to advance. Further analysis of the location and level of markers such as CK3, CK19, p63, ABCG2, and vimentin (Ghoubay-Benallaoua et al, 2011) will enhance understanding of stem cell deficiency as well as stem cell treatment in the cornea and other organ systems. In addition, emerging advances in nanotechnology for stem cell application in the eye have shown that electrospun nanofibers can serve as a scaffold to grow and transfer limbal stem cells in the treatment of limbal stem cell deficiency (Zajicova et al, 2010). As biological discoveries identify key factors to stem cell survival and technological advances provide new biomaterials, the potential for improved treatment is dramatic.

3. Conjunctiva

3.1 Background

The conjunctiva is essential to ocular health as it contains essential support cells, provides for passage of fluids and nutrients and is a barrier to pathogen entrance and fluid loss. It functions to provide mucin for the tear film, immune defense (via production of antimicrobial peptides such as defensins as well as the conjunctiva associated lymphoid tissue - CALT), and a mechanical barrier. It is a thin transparent mucous membrane that is made up of a non-keratinized epithelium and a vascularized stroma. The epithelium varies from stratified cuboidal cells over the tarsus to columnar in the fornices to squamous on the globe. The basal cell layer contains goblet cells, which secrete mucin. The membrane covers the posterior surface of the eyelids and the anterior surface of the globe and is made up of the tarsal conjunctiva, forniceal conjunctiva, bulbar conjunctiva and the semilunar fold. The tarsal conjunctiva adheres firmly to the underlying tarsus. In contrast, the bulbar conjunctiva is loosely adherent to Tenon's capsule and has multiple folds, allowing for unrestricted ocular movement.

Therefore, the conjunctiva is essential to ocular health, providing a mechanical and immunological barrier to injury and infection, as well as contributing to the production and stability of the tear film through goblet cell secretion of mucin (Qi et al, 2010). Diseases of the conjunctiva can therefore have significant ocular morbidity.

Interestingly, there has been considerable debate regarding the role of the conjunctiva in corneal regeneration, and some have proposed that corneal and conjunctival epithelia are equipotent (Majo et al, 2010). Despite these findings and its essential role in ocular health, conjunctival stem cell biology has not been as well investigated as that of corneal stem cells (Ang & Tan, 2004).

3.2 Stem cell localization

The location of conjunctival stem cells has been proposed to include the fornix (Wei, 1995), palpebral and bulbar conjunctiva (Pelligrini et al, 1999), bulbar conjunctiva (Qi et al, 2010; Chen et al, 2003), and mucocutaneous junction (Wirstschafter et al, 1999). Human bulbar and forniceal conjunctival cells have been found to have equivalent proliferative capacity, suggesting that conjunctival epithelial stem cells may be evenly distributed between the bulbar and forniceal conjunctiva (Pelligrini et al, 1999). Various keratins and molecules associated with stem cells were found in undifferentiated basal cells in the human bulbar conjunctival epithelium, supporting the localization of progenitor cells in this region (Qi et al, 2010). Research has focused on not only the location of these progenitor cells but also their differentiation potential. Interestingly, both conjunctival non-goblet and goblet cells are thought to be derived from a common progenitor (Pelligrini et al, 1999). This suggests that differentiation into goblet cells may occur late and these cells may be generated from transient amplifying cells (Schrader et al, 2010).

In pathological processes, a decrease in the conjunctival stem cell population has been demonstrated through staining for stem cell markers such as p63 in humans, rats and mice, which had reduced expression in damaged conjunctiva and conjunctivalized cornea (Moore et al, 2002). These discoveries and others have enhanced our understanding of conjunctival epithelial stem cells, which enables more appropriate surgical and medical treatment of conjunctival disease through advances such as tissue engineering.

3.3 Pathology

Cicatrizing ocular processes that affect the cornea can also affect the conjunctiva, such as ocular cicatricial pemphigoid, Stevens Johnson syndrome, graft versus host disease, autoimmune disease, post surgical changes and myriad other causes. Conjunctival disease and inflammation can lead to conjunctival thickening, forniceal foreshortening, symblepharon, severe dry eye due to poor blink dynamic, and subsequent keratinization of the ocular surface epithelia. In addition, damage to the goblet cells, ineffective blink and keratinization of the ocular surface epithelium can lead to disruption of the tear film and dry eye syndrome. Forniceal cicatrization can lead to cicatricial entropion and secondary trichiasis which damages the already susceptible corneal surface epithelium and makes the eye more prone to recurrent erosions, with the risk of secondary bacterial infection.

3.4 Treatment

Prolonged conjunctival inflammation can also deplete the limbal epithelial stem cell population, causing decreased vision through the mechanisms highlighted above. Given the interposition of conjunctival disease on corneal dysfunction, treatment of severe scarring, or cicatrizing disease, is fraught with associated complications, as discussed previously. Traditional treatment regimens have been staged, multi-specialty approaches. Prior to any corneal reconstruction, the ocular surface must be optimized so as to reduce the risk of subsequent failure (Gomes et al, 2003). Surgical reconstruction of the fornix can improve conjunctival function and lead to better long term success, with improved vision and comfort for the patient. The frequent ongoing inflammatory disease, however, can hinder long-term benefits.

Treatment modalities are all aimed at excision of scar tissue with placement of a tissue substitute to recreate the fornices and ocular epithelial surface. This can be achieved through mucous membrane grafting, conjunctival autografts, and amniotic membrane. Amniotic membrane transplantation has been used for severe cicatrizing conjunctival diseases, such as ocular cicatricial pemphigoid (Barabino et al, 2003). Inflammation at the time of grafting can have deleterious effects on outcomes (Tsai & Tseng, 1995). Thus, transplantation is often performed in a staged surgical process whereby amniotic membrane transplantation is performed first to be followed by limbal transplantation in order to achieve the optimal result (Barabino et al, 2003). In cases of chemical burns with partial and total limbal stem cell deficiency, amniotic membrane transplantation has been shown to be an effective component of ocular surface reconstruction (Gomes et al, 2003). As previously highlighted, amniotic membrane may hinder apoptosis of essential cells and promote growth of stem cells essential to graft success (Tseng et al, 1998). When performed in conjunction with limbal stem cell transplantation, it is effective at improving the ocular surface in cases of limbal stem cell deficiency (Barabino et al, 2003).

Other bioengineered synthetic material also has the potential to help in cases of severe ocular surface disease. Acellular polymer (such as porous collagen-glycosaminoglycan copolymer matrix) grafting can minimize fornix contracture after injury by inhibiting early scar formation (Schrader et al, 2009). Engineered tissue substitutes are limited, however, because they do not have a donor epithelium and are often not elastic enough for forniceal placement. As with any surgical intervention in these cases, outcomes may be dependent on the level of inflammation. Features of the ideal transplant material include: low immunogenicity, high tolerance, easy surgical manipulation and elasticity, with an associated layer of donor epithelium available in cases of conjunctival stem cell deficiency (Schrader, 2009).

3.5 Further stem cell applications and new advances

While current transplantation techniques such as conjunctival autografts, oral mucous membrane grafts, nasal turbinate mucosa grafts and amniotic membrane are often used with benefit in patients with conjunctival stem cell deficiency and scarring, they are limited by complications secondary to the underlying stem cell deficiency and ongoing inflammatory disease. Long term success in patients with diffuse loss of conjunctival epithelial stem cells may require that the transplanted tissue contains epithelial stem cells (Pelligrini et al, 1999).

How are conjunctival epithelial stem cells transplanted? Conjunctival epithelial cells have been cultured in serum-free media. These cells have been shown to have a similar in vivo proliferative capability as those cultured in serum with 3T3 feeder cells, mitigating the need for use of animal serum (Ang, Tan, Beuerman et al, 2004). A conjunctival epithelial equivalent has been developed using this serum-free culture system (Ang, 2004). Subsequently, autologous cultivated conjunctival cells grown in this manner on amniotic membrane have been successfully transplanted in primary pterygium (Ang et al, 2005). As in the cornea, conjunctival epithelial cell differentiation and growth on amniotic membrane in vitro has been demonstrated (Meller et al, 1999). Other synthetic materials, such as ultrathin poly(epsilon-caprolactone) membrane substrate, have been studied for possible transplantation and found to be biocompatible and able to support conjunctival epithelial cell proliferation (Ang et al, 2006).

What about the surrounding environment? Understanding the matrix and appropriate substrates for growth of these cells is an integral component of tissue transplantation success. The goal of treatment should be not only tissue replacement but also stimulation of new growth and proliferation of healthy tissue. Both the extracellular matrix and the co-culture environment are essential to the proliferation and differentiation of conjunctival epithelial cells (Tsai & Tseng, 1988; Tsai et al, 1994). Certain growth factors can also modulate the outcomes of stem cells. For example, when placed in the appropriate environment and with associated factors such as nerve growth factor (Li W et al, 2010), progenitor cells can differentiate into goblet cells (Ang, Tan, Phan et al, 2004). This is not surprising given the previous suggestion of the common origin of these cells (Pelligrini et al, 1999). Due to the essential role of goblet cells in the stability of the tear film, this is an integral part of restoring the ocular surface. Similar to limbal stem cells, support is also required in the stroma for conjunctival cell growth. Serum free co-culture of conjunctival epithelial cells with mitotically active subconjunctival fibroblasts stimulates the maintenance of conjunctival cells with stem cell characteristics (Ainscough et al, 2011). Thus, the conjunctival stromal scaffold and surrounding growth factors may also need manipulation in order to best support epithelial grafts.

All of these advances will continue to be translated into clinical application, allowing for improved treatment of severe conjunctival disease and it's potentially blinding sequelae. Cultivated conjunctival cells on amniotic membrane or biocompatible membranes can help improve the long term outcomes of surgical interventions in these difficult cases by stimulating new growth and creation of a new niche environment for stem cell proliferation.

4. Orbit

4.1 Background

The orbit provides the support framework for the eye. Each orbit is a bony cavity that contains the eye, optic nerve, extraocular muscles, nerves, fat, lacrimal gland, and blood vessels. Orbital anatomy is intricate and any disruption to the orbital tissue via inflammation, infection, trauma, scarring or neoplasm can cause significant visual consequences. In addition, multiple apertures and fossa connect the orbit to the surrounding structures of the cranium and sinus cavities. The proximity of the orbit to the cranial vault and nasal cavity can lead to neurological or nasal sequelae from orbital disease. In cases of more aggressive malignant tumors, tumor spread into the brain can lead to devastating neurological consequences. In addition, infectious processes, such as sinusitis, can spread from the sinuses into the orbital cavity and cause blindness. These infections can also spread into the brain and become life threatening.

Processes within the orbit are further complicated by the confined bony walls of the orbital cavity. For example, in cases of benign orbital tumors in the intraconal space such as cavernous hemangiomas, as the tumors grow they can sometimes compress the optic nerve and cause visual loss. In addition, all of the rectus muscles in the orbit originate from the annulus of zinn, a fibrous ring formed by their origin which encircles the optic foramen and the central portion of the superior orbital fissure. Diseases such as thyroid eye disease (TED) can lead to blindness by enlargement of these extraocular muscles causing compression of the optic nerve. For purposes of brevity, this chapter will focus on orbital adipose tissue and fibroblasts and one of the most common diseases affecting the orbit, thyroid eye disease (TED).

4.2 Stem cell localization

As stem cell research elsewhere has discovered, the human adipose tissue is a source of multipotent stem cells. Processed lipoaspirate cells can differentiate into osteogenic, adipogenic, myogenic, neurogenic and chondrogenic lineages and have many similar features to mesenchymal stem cells (Zuk et al, 2002). While adipose stem cells have been studied extensively throughout the body, research on the orbital adipose stem cell population is limited.

Interestingly, orbital adipocytes (which are continuous with the nasal fat pad) are derived from neural crest cells, versus adipose tissue throughout the body which is generally mesodermal in origin (Johnston et al, 1979). The central preaponeurotic fat has been presumed to be more similar to systemic adipose tissue and derived from mesoderm. The eyelid, therefore, provides a unique place to compare orbital stem cells (nasal fat pad) with fat more similar to systemic adipose tissue (central fat pads) (Korn et al, 2009). Both nasal and central adipose depots have been found to have the potential to differentiate into adipocytes, smooth muscle and neuronal/glial lineages (Korn et al, 2009). Interestingly, the nasal adipose stem cells expressed a higher level of antigens such as CD34 correlated with neurogenic differentiation potential. Conversely, central cells expressed more neuronal/glial associated antigens under the neurogenic environment used, suggesting that the pathway for differentiation is modulated at several levels (Korn et al, 2009).

Further study of orbital adipose tissue is needed to understand the factors which lead to differentiation of this tissue and the neurogenic potential. Orbital adipose stem cell ability to differentiate into smooth muscle, neuronal lineages and adipocytes may also provide novel therapeutic treatments in disease affecting nerves, soft tissue, and muscles in the orbit. In addition, understanding the factors necessary for adipogenesis and expansion of the orbital adipose compartment in diseases such as TED has significant clinical applications. The ease of access to orbital adipose tissue makes it an interesting new target for stem cell research.

4.3 Pathology

Thyroid eye disease is an autoimmune inflammatory disease that has significant orbital consequences. Clinical signs of thyroid eye disease can vary from mild to severe and include proptosis (bulging of the eye), strabismus (double vision), bulging eyelids, exposure keratopathy, eyelid retraction and optic neuropathy. Enlargement of the extraocular muscles and expansion of the orbital fat from stimulation of orbital fibroblasts and adipogenesis in the fixed volume of the bony orbit can lead to these symptoms. Increased volume of the retroocular tissue and fibrotic changes can compress the optic nerve, cause extraocular muscle or eyelid dysfunction, or push the eye forward leading to corneal compromise.

The pathology of thyroid eye disease involves an autoantibody cross-reaction from the thyroid-stimulating hormone (TSH) receptor with orbital fibroblasts. This leads to orbital fibroblast stimulation of lymphocytes, release of various cytokines and stimulation of orbital fibroblast production of glycosaminoglycans that increase volume by osmotic edema. In addition, preadipocyte fibroblasts in this environment can differentiate into adipocytes, leading to expansion of the orbital fat cavity. De novo adipogenesis within orbital tissues is an important factor in the pathology of TED (Kumar et al, 2003). Proliferation of fibroblasts and adipocytes and secondary hyaluronan accumulation (among other aminoglycosides) causes tissue volume expansion, orbital congestion, exophthalmos, and compressive optic neuropathy (Kahaly et al, 1998). Differentiation of fibroblasts to adipocytes and proliferation of orbital fibroblasts are thus critical steps in the development of TED.

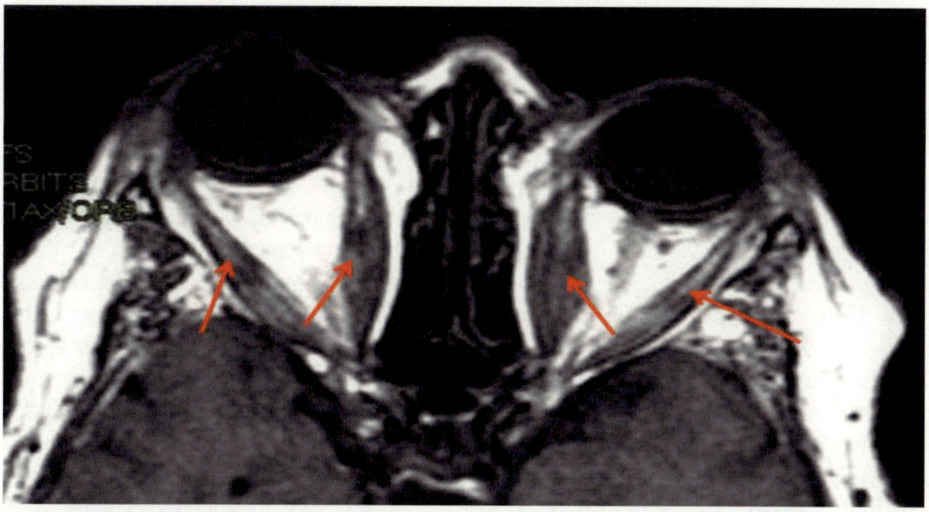

Fig. 4. Orbital Congestion from extraocular muscle enlargement in TED Note: arrows pointing to enlarged extraocular muscles

Fibroblast proliferation and infiltration into muscle fibers leads to extraocular muscle fibrosis and disabling diplopia, which some classify as the first subtype of TRO, characterized by prominent extraocular muscle enlargement (see Figure 4) and diplopia (Hiromatsu et al, 2000). Fibroblast differentiation into adipocytes leads to accumulation of fatty tissue in the orbit, which has been classified by some as type two TED, associated with expansion of the orbital fat compartment and proptosis (Hiromatsu et al, 2000). Thus, a key aspect of the clinical progression of TED is the ability of these fibroblasts to differentiate into adipocytes and propagate at a pathological level.

4.4 Treatment

Traditional treatment of TED focuses on the ocular sequelae. Proptosis or eyelid retraction can lead to corneal exposure. Mild cases may be treated with medical management, such as aggressive lubrication, elevation of the head while sleeping, patching at night, and the use of eye shields. Patients with persistent symptoms despite these treatments may necessitate tarsorrhaphy (surgical closure of part of the eyelid) to help protect the cornea. In severe cases, orbital decompression (removing some of the bones of the orbit cavity to expand orbital volume) is sometimes indicated. Lid retraction and resultant exposure keratopathy are treated with aggressive lubrication, patching, botulinum toxin to the Müllers muscle, and sometimes a tarsorrhaphy in the near term; levator recession and lower lid retraction repair via a number of different techniques is used for permanent treatment of lid retraction. In cases of optic neuropathy from compression at the apex, steroids and surgical treatment with orbital decompression may be used. For disabling diplopia, radiation therapy and steroids may be of benefit. Surgical intervention is indicated once disease stability has been documented (with prisms and patching often recommended in the interim if the above treatment modalities are not sufficient). The order of any possible surgical intervention is important, with decompression prior to strabismus surgery followed by eyelid surgery, in order to achieve the best long-term outcome.

Despite various treatment options available, the ophthalmopathy may be progressive during times of thyroid dysfunction making surgical intervention more risky. In addition, some patients are not good candidates for surgery. Steroids are used as treatment in many cases, but the side effect profile of steroids is significant. As such, treatment advances have more recently been focused on alternative forms of immunomodulation (Salvi et al, 2006).

4.5 Further stem cell applications and new advances

Localization of orbital stem cells may help provide important therapeutic targets for TED. Preadipocyte fibroblasts in the adipose tissues are the precursor cells that differentiate into adipocytes in a process controlled by multiple regulatory genes and transcription factors (Sorisky, 1999). Preadipocyte fibroblast differentiation is associated with an increase in TSH receptor mRNA expression in patients with TED (Feliciello et al, 1993). Studies have also shown that cultured orbital preadipocytes can be induced in vitro to differentiate into adipocytes with functional TSH receptors (Valyasevi, 1992). Significant progress has been made into understanding the interactions between autologous T-lymphocytes and proliferation of fibroblasts (Feldon et al, 2005; Valyasevi et al, 1992). Understanding the factors that lead to differentiation of these fibroblasts into adipocytes as well as their proliferation in the orbit may help provide interventions to prevent this cell differentiation and stimulation. In addition, if orbital adipose stem cells are able to differentiate into neuronal, adipose and smooth muscle lineages, they may offer treatment options for myriad ocular conditions.

Orbital stem cells have been studied much less extensively than those in other areas of ophthalmology. The potential clinical applications for research in this area are broad. Orbital adipose tissue and the reservoirs of stem cells within are routinely excised during blepharoplasty surgery. As such, further research into these progenitor cells may provide insight on pathological processes affecting the orbit, such as TED. In addition, treatment modalities may be advanced through harvesting and differentiation of these stem cells.

5. Eyelid

5.1 Background

The eyelid is made up of an anterior and posterior lamella, both of which are essential to maintaining appropriate lid height, eyelid closure, blink dynamic, functioning of the nasolacrimal drainage system and protection of the globe. The skin of the eyelid is the thinnest in the body, with no underlying subcutaneous fat layer. Underlying the skin are the muscles of protraction, the orbicularis oculi, whose function can become altered by diseases such as seventh nerve palsies or benign essential blepharospasm. The orbital septum is the next layer which is a thin, multilayered sheet of fibrous tissue that separates preseptal from postseptal or orbital tissue planes. The fat lies posterior to the septum and anterior to the muscles of retraction, the levator palpebrae and the müllers muscle. Inferior to the müllers

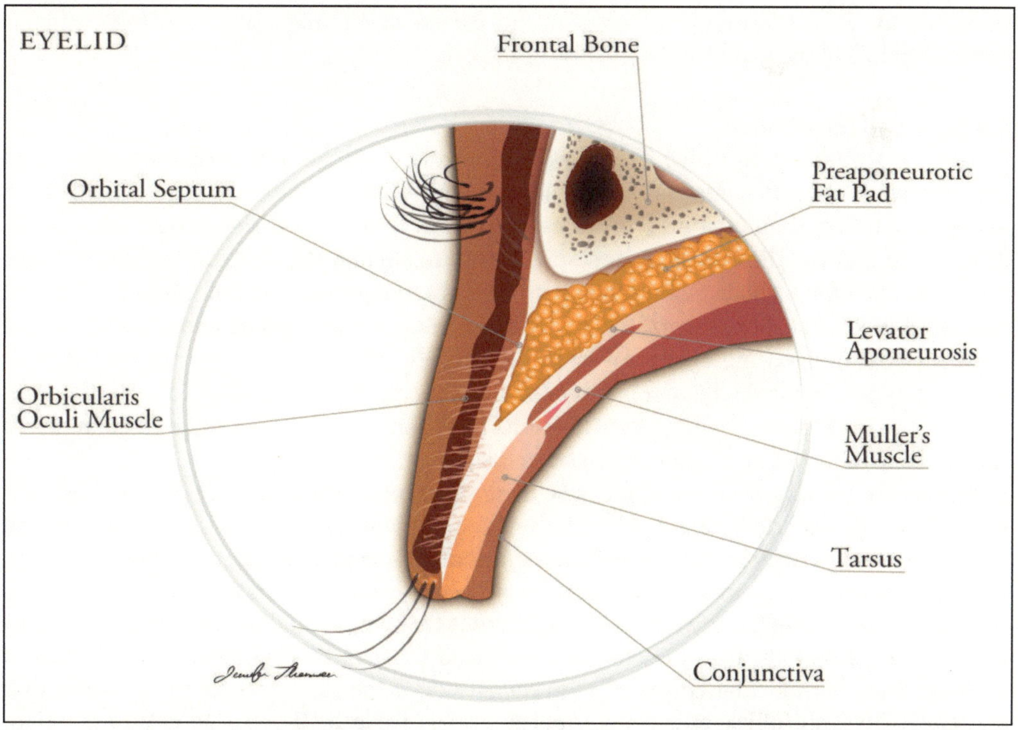

Fig. 5. Anatomy of the eyelid (Illustration by Jennifer Thomson Ph D., Bascom Palmer Eye Institute, University of Miami Miller School of Medicine)

muscle is the tarsus, a firm, dense plate of connective tissue that serves as the "backbone" of the eyelid. The most posterior layer is the conjunctiva which extends from the mucocutaneous junction to the fornix (see Figure 5).

The purpose of this section is to discuss potential stem cell applications in cutaneous wound repair in the eyelid as well as neurogenic causes of poor orbicularis tone, namely facial nerve palsy. While stem cell applications may also extend to the muscles of retraction, the tarsus and posterior lamellar reconstruction, the chapter will focus on these two principal areas.

5.2 Pathology

Pathologic processes affecting the cutaneous layer of the eyelid are extensive. The primary relevance of stem cell research to the eyelid skin relate to tissue reengineering in the cases of loss of tissue that most often arise from cutaneous malignancy, orbital malignancy, or trauma. While only 15-20% of eyelid neoplasms are malignant, malignancy is becoming an increasingly common cause for severe loss of eyelid tissue. Basal cell carcinoma, squamous cell carcinoma, melanoma, sebaceous cell carcinoma as well as other much less frequent malignancies can invade locally, necessitating extensive resection. In cases of recurrent malignancies in patients who have undergone previous radiation, tissue shortage and poor vascular supply compound this issue.

In addition, traumatic and post-surgical cases of anterior lamellar deficiency can have significant potential ocular repercussions. Restoration of eyelid function is essential to the eye, as the eyelid functions to lubricate and protect the globe and eyelid opening is necessary for vision. Improper repair of eyelid defects or chronic wound healing problems can lead to loss of the vision and even loss of the eye.

5.3 Stem cell localization

Cutaneous wound repair in the eyelid, as in other places of the body, has been researched extensively. Wound healing and subsequent scar formation occurs through a cascade of events, including inflammatory, proliferative (granulation tissue deposition) and tissue remodeling phases. The details of these processes are beyond the scope of this chapter (see Lau et al, 2009 for a good review), but stem cells are integral to each of the phases. In the process of wound healing, several factors lead to a relatively hypoxic environment, in which bone marrow-derived mesenchymal stem cells thrive (Ren, 2006). Hypertrophic scars or keloids can develop from dysregulation of some of the processes involved in wound repair. Conversely, inadequate healing after radiation, infection, or in cases of poor vascular status often seen in diabetics or smokers can lead to insufficient wound repair.

Where are the stem cells? The major components of the epidermis with the potential for self-renewal are the interfollicular epidermis, sebaceous gland and hair follicle. These are all repopulated by the differentiation of their own stem cells in physiologic conditions (Ko et al, 2011). In cases of skin injury, the stem cells become more pluripotent and are able to repopulate one another. For example, keratinocytes that reepithelialize a wound are derived from both the interfollicular epithelial stem cells and the hair follicle bulge stem cells (Lau et al, 2009). The interfollicular epithelial stem cells reside in the basal layer of the interfollicular epidermis. The hair follicle bulge epidermal stem cells are located in the outer root sheath of the hair follicle and play an essential role in epidermal healing in the early phases of wound repair (Nowak et al, 2008). In addition, the bulge epidermal stem cells retain a larger reservoir with age as opposed to the interfollicular stem cells which decline with age (Nowak et al, 2008). This tissue is easily isolated, allowing for easier procurement of these stem cells.

What about the surrounding environment? Angiogenesis plays a key role in wound healing, and is likewise dependent on surrounding stem cells. Precursor cells such as the "tissue-resident endothelial precursors" and "pluripotent nestin expressing cells" (Lau et al, 2009) are important components to this process. In addition, as previously highlighted, adipose stem cells may also play a role in wound healing through promotion of angiogenesis. As with mesenchymal stem cells, adipose-derived stromal cells proliferate and support angiogenesis under the hypoxic conditions of wound healing (Lee et al, 2009). Chemokines also play an integral role as they induce wound repair via the accumulation of mesenchymal derived stem cells both in vitro and in vivo in the mouse model (Sasaki et al, 2008). It is thought that this process is modulated through paracrine signaling and the release of growth factors such as fibroblast growth factor and endothelial growth factors. Stem cells therefore contribute to wound healing both through repopulation of the cells that are essential for wound repair (i.e. keratinocytes) as well as modulating the wound repair process.

5.4 Treatment

Treatment of eyelid pathology is dependent on the underlying etiology and subsequent defect. Eyelid defects are repaired by sub-specialty trained Oculoplastic surgeons to reconstruct an anterior and posterior lamella best able to simulate normal anatomy. The myriad techniques for repair are beyond the scope of this chapter, but each is aimed at achieving the optimal functional and cosmetic result. In cases of severe trauma or tumor resection, the absence of an adequate amount of tissue for anterior lamellar reconstruction can be a complicating factor. Tissue expanders and other advances are helpful techniques in these types of patients, but tissue engineering with viable stem cell transplantation has potential significant applications.

5.5 Further stem cell applications and new advances

In patients with large eyelid defects from cutaneous malignancy requiring reconstruction, stem cells may help augment wound repair. In many patients with recurrent malignancy that have previously undergone radiation therapy, the risk of tissue necrosis is increased and stem cell promotion of angiogenesis may have an integral role. Patients with radiation damage may also have decreased functionality of their own stem cells. Certain factors integral to attracting stem cells to the wound, thus essential to the wound healing process, can also become deficient with age and diabetes (Ko et Al, 2010).

Because of the deficiency in stem cell functionality in cases of radiation, diabetes and age, stem cell application to these wounds may accelerate healing. Stem cell application to wounds using artificial dermis in patients with radiation injuries was shown to improve fat angiogenesis and dermal reconstitution, and minimize inflammatory epidermal damage (Akita et al, 2010). In addition, mesenchymal stem cells applied to wounds were found to accelerate wound closure and increase proangiogenic factors in both normal and diabetic mice (Wu et al, 2007). Translating these findings into clinical application, bone marrow-derived stem cells were applied topically to non-healing wounds and led to clinical closure of the wound and histological evidence of reduced scarring (Badiavas & Falanga, 2003).

Accelerated closure of wounds was also demonstrated by topical application of autologous bone marrow-derived mesenchymal stem cells in culture via a fibrin spray system (Falanga et al, 2007). Biological skin substitutes with stem cells have also been found to be suitable for wound bed repair for future autografting in cases of full-thickness skin defects caused by burn or trauma in a Phase I/II study (Schurr et al, 2009). Such advances and the application of other technologies and biomaterials for the application of cells and growth factors to the wound all have potential clinical application in repairing severe eyelid defects.

In addition, as hair growth stimulation has been supported elsewhere by adipose stem cells, it is possible that these stem cells could also play a role in treatment of hypotrichosis and/or eyelash loss after eyelid reconstruction. Bimatoprost (Latisse™) has been shown to promote hair growth in several studies (Wester et al, 2010) and has been found to be beneficial in cases of hypotrichosis, but it requires ongoing application of the solution. In cases of severe hypotrichosis, it is possible that stem cells applied topically may have a benefit as well. Won et al studied hair growth from topical application and injection of adipose stem cells and suggest that these cells may promote hair growth by increasing the proliferation of dermal papilla cells through cell cycle modulation and activation of the anagen phase (Won, 2010). This may be important in hair loss surrounding surgical wounds and in treatment of hypotrichosis.

Stem cell research in the eyelid is in its nascency, but findings in other organ systems can be translated into wound repair. Future research may focus on application of cutaneous stem cell treatments to periocular regions. Research in this area may provide key therapeutic options for patients with significant soft tissue defects which can lead to inadequate ocular lubrication and blindness if not treated appropriately.

6. Facial nerve

6.1 Background

A number of peripheral nerves greatly impact ocular function. For example, the facial nerve, cranial nerve seven, arises in the pons and enters the petrous temporal bone into the internal auditory meatus (in close proximity to the inner ear), then runs a tortuous course through the facial canal. It travels the longest distance through a bony canal of any nerve in the body. After passing through the facial canal it emerges from the stylomastoid foramen and passes through the parotid gland, where it divides into five major branches (see Figure 6). The five major branches that exit the parotid gland include the temporal, zygomatic, buccal, marginal mandibular and cervical branch, all of which are essential to facial movement. The facial nerve also functions as the efferent limb of the corneal reflex and the blink reflex.

6.2 Stem cell localization

Stem cells in the area of denervated axons have been found to have an essential role in the attachment and growth of the regenerating axons. These cells contribute to the up-regulation of multiple genes and downstream release of cytokines and growth factors which provides a stimulatory environment for nerve repair (Walsh 2009). Schwann cells also initially up-regulate multiple factors used for the attachment and growth of regenerating axons.

Over time, however, in the absence of axonal contact these cells slowly lose their capacity to proliferate and lose the ability to support axonal regeneration (Chen, 2010). Chronic denervation leads to a decrease in this capacity and a decline in the Schwann cell population, hindering further axonal regeneration (De Medinaceli & Rawlings, 1987). The clinical implication is that axonal regeneration must occur during a "critical window" to maximize success of reinnervation.

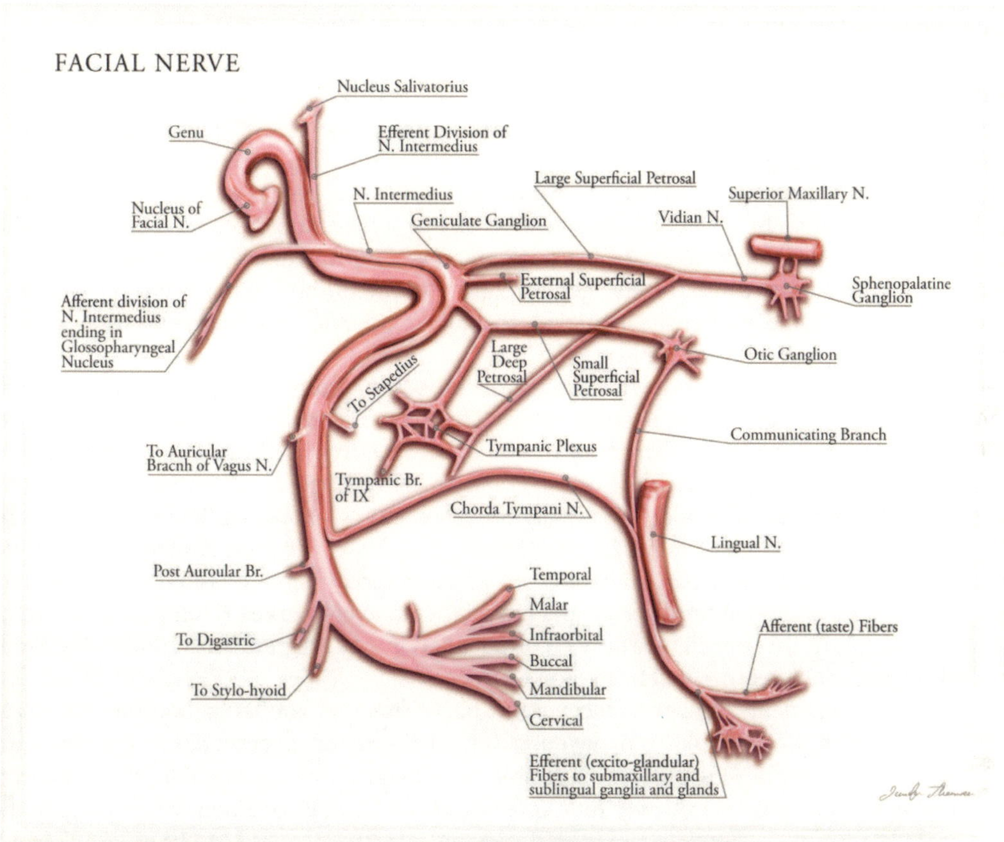

Fig. 6. Facial Nerve Anatomy (Illustration by Jennifer Thomson Ph D., Bascom Palmer Eye Institute, University of Miami Miller School of Medicine)

6.3 Pathology

Seventh nerve injury can affect the periocular tissue and eye and lead to chronic irritation, corneal infection, and even blindness. Seventh nerve palsy is a cause of paralytic ectropion that causes neurogenic orbicularis dysfunction and chronic ocular symptoms. Upper eyelid lagophthalmos from orbicularis dysfunction and poor blinking and eyelid closure causes chronic ocular irritation and potential infection. Chronic stimulation of reflex secretors causes secondary tearing, but poor blink dynamic and lid closure can lead to severe exposure keratopathy and infection. The etiology ranges from viral or other infectious causes to congenital or neoplastic causes, and trauma, toxic exposures, or iatrogenic causes. In addition, because of the long course and narrow caliber of the canal, inflammatory processes involving the central nervous system and facial nerve, and traumatic or other injuries to the temporal bone and surrounding structures can affect the function of the nerve.

The most common cause of unilateral facial paralysis is Bell's palsy, idiopathic facial paralysis, which accounts for the majority of cases of acute unilateral facial paralysis. The etiology has been attributed to several possible causes including viral (such as HSV), inflammatory, autoimmune, and ischemic etiologies. In the large majority of cases, Bell's palsy resolves spontaneously within the first few months (Ropper & Samuels, 2009), but others may persist longer and have more significant consequences.

6.4 Treatment

Even the non-ophthalmic literature highlights that the most important primary consideration in the treatment of facial paralysis is protection of the eye and vision (Meltzer & Alam, 2010). Temporary treatment includes artificial tears, ointment, eyelid taping, and moisture chambers. In cases without resolution, or other etiologies of seventh nerve dysfunction which are less likely to have spontaneous improvement, long-term treatment includes tarsorrhaphy, canthoplasty, gold weight placement, horizontal eyelid tightening, fascia lata or silicone suspension sling of the lower lid and full thickness hard palate mucosal grafting to the lower eyelid.

In cases of damage to a nerve via transection or other trauma, surgical repair is often indicated. Gaps or scars are particularly complicated to repair as the regenerating axons face a gap between the injury and the target (which becomes more difficult to overcome the longer the distance). Interpositional nerve grafts are often used as a treatment for longer gaps or scar segments to provide a pathway for regenerating axons, while direct repair can be used for shorter areas of denervation (Millesi et al, 1972). Current strategies use cable grafting in most cases of facial paralysis when primary repair is not possible, using greater auricular, sural or medial and lateral antebrachial cutaneous nerves (Humphrey & Kreit, 2008). Importantly, these grafts are only recommended if they are performed within the time period during which reinnervation is possible.

Despite these treatments, however, recovery of nerve function is often suboptimal in cases of seventh nerve palsy. Prolonged degeneration in the distal denervated nerve leads to changes that render the Schwann cells less able to promote axon regeneration, thus limiting the success of these procedures over time (Meltzer et al, 2010). In addition, while peripheral nerves do regenerate better than the central nervous system, recovery can be incomplete or misdirected, even when surgical treatment is attempted. In the cases of facial nerve palsy, synkinesis (miswiring of nerves after injury) can lead to post paretic sequelae such as Marin Amat syndrome and crocodile tears. Chronic facial spasm can also develop after resolution of the facial nerve palsy. Surgical procedures in cases of chronic paralysis are often focused on static suspension procedures, regional myofascial procedures and free tissue transfer with the goal of achieving maximal symmetry. Ipsilateral and contralateral nerve transfers, static slings (such as used in the brow), dynamic muscle transfers, and gracilis microneurovascular free tissue transfer are all also used in the treatment of facial palsy, but have associated morbidities (Meltzer et al, 2010).

6.5 Further stem cell applications and new advances

Adipose and skin derived precursor cells have been shown to have the potential to differentiate and proliferate into Schwann and glial-like cells when treated with the appropriate medium (Zavan et al, 2010). While autologous nerve grafts remain the gold standard, harvesting of autologous stem cells and differentiation into neurogenic lineages has tremendous clinical potential in peripheral nerve injuries. The accessibility of skin and adipose derived stem cells may "bridge" the gap and allow translation of this into clinical applications in such areas as facial nerve palsies. Hair follicle bulge cells used to repair a peripheral nerve gap contributed to recovery of function by differentiating into glial cells (Amoh, 2005). Skin-derived stem cells injected into a nerve distal to a crush injury were found to associate with axons, express stem cell markers, and eventually express a myelinating phenotype (Walsh, 2009). Nerve repair in patients may be improved in the future by the use of autologous, cultured stem cells to help promote nerve regeneration. Skin and adipose derived stem cells have the added benefit that they can be easily harvested and have been shown to have neurogenic potential, as previously outlined.

7. Retina

The retina is developmentally an outgrowth from the forebrain, and like the rest of the central nervous system, exhibits neurodegenerative diseases that prove devastating to affected patients. Retinal degenerations often end with the death of retinal neurons, such as rod and cone photoreceptors in age-related macular degeneration and retinitis pigmentosa (discussed in section 7.1 below) and retinal ganglion cells in optic neuropathies such as glaucoma (discussed in section 7.2 below). Although it may be possible to salvage these cells before they die, for example, using neurotrophic factors or stem cells or gene therapy, for the many patients who have lost these cells, the challenge remains to figure out a way to replace them and thereby restore vision.

7.1 Outer retina: Photoreceptors and retinal pigment epithelium

7.1.1 Background

Photoreceptors are the retinal cells that initiate vision by converting photons of light into chemical signals. In humans, photoreceptors are divided into rods and cones, which are responsible for dim, achromatic and brighter color vision, respectively. The center of the retina is more cone-dominant and the periphery more rod-dominant. Throughout the retina, photoreceptors are in close, critical contact with an adjacent layer of supportive cells, the retinal pigment epithelium (RPE). The RPE is responsible for phagocytosis of photoreceptor outer segments, critical to the constant replenishment of the membranous discs that house the photoreceptors' photon-conversion machinery. The RPE is also responsible for the replenishment of critical photopigments in the photoreceptor outer segments.

7.1.2 Pathology

Primary degeneration of either the photoreceptors or RPE leads to eventual degeneration of the other, and in either case degenerative loss of vision. Such pathologies are observed in a number of diseases, including common diseases like macular degeneration, and less common diseases such as retinitis pigmentosa.

Age-related macular degeneration (ARMD) is one of the most common causes of blindness worldwide (Klein et al., 2011). ARMD usually affects older adults and occurs in "dry" and "wet" forms. In dry ARMD, deposits called drusen accumulate between the retina and the choroid, and slow degeneration of RPE and later photoreceptors leads to localized "geographic" atrophy and central vision loss. In wet ARMD, new, leaky blood vessels grow up from the choroid behind the retina, leading to acute vision loss (Wong et al., 2008).

Retinitis pigmentosa is a collection of genetic diseases that generally start with night blindness and then continues on to tunnel vision, reflecting a pathology that affects rod photoreceptors in the peripheral retina earlier in the disease (Sahel et al.). Hundreds of different mutations in dozens of different primary genes lead to similar pathologies, including RPE hypertrophy, migration and proliferation and photoreceptor death (Daiger et al., 2007). Interestingly, these mutations may be found in either the photoreceptor or RPE cells, reinforcing the mutual interdependence of these two cell types.

7.1.3 Current treatment

Wet ARMD is currently treated with success using anti-VEGF antibodies (Mitchell, 2011); dry ARMD is slowed slightly by nutritional supplementation, at least in the later stages. Both of these sets of degenerative processes have in recent years become the target of a number of newer therapeutic approaches designed to slow, halt, or even reverse the disease process. Neurotrophic factors such as ciliary neurotrophic factor (CNTF) may provide neuroprotection (Sieving et al., 2006; Tao et al., 2002; Zhang et al., 2011). Gene therapy delivered with viral vectors may replace the defective gene (Ashtari et al., 2011; Cideciyan, 2010; Simonelli et al., 2009). However, both forms of the disease eventually lead to loss of photoreceptors and RPE cells, and these are not replaced through endogenous reparative mechanisms once lost. Thus there is considerable motivation to bring stem cell therapies to treat these diseases.

7.1.4 Stem cell research and opportunities

Significant steps forward have been made in using stem cells to replace photoreceptors and retinal pigment epithelium to address these diseases. Stem cells may prove therapeutic in two regards. As more commonly conceived, they may differentiate into the cell type lost in the disease, take up that cell's connections, and thereby replace that cell's function. However, it turns out stem cells also have remarkable properties of neuroprotection.

How do stem cells protect other cells from degenerating? One key way likely involves their expression and secretion of neurotrophic factors (Crigler et al., 2006) and immune system modulators that deter invading immune cells from creating neuronal death (Pluchino et al., 2005). Their neuroprotective activity can be further enhanced by transducing them with vectors that further increase their expression of specific neurotrophic survival factors (see section 7.2.4). An attraction to harnessing the neuroprotective effects of stem cell transplantation is that it does not require any specific differentiation and integration: the transplanted cells need only persist locally, for example in the vitreous or subretinal space. Success with such approaches has been demonstrated in rat and mouse models of retinitis pigmentosa (Arnhold et al., 2007; Inoue et al., 2007).

A major advantage in thinking about cell replacement therapy approaches for these diseases is that photoreceptors and RPE cells only have to interact with local, neighboring cells to carry out their full roles in retinal physiology (Lamba et al., 2009b). A number of stem cell therapies are approaching human clinical investigation now, including trials in the United Kingdom, Israel and the United States. All three of these are directed at the use of stem cell-derived RPE cells implanted subretinally to resume the function of lost RPE and thereby protect photoreceptors from further degeneration (Lu et al., 2009; Lund et al., 2006). Similar approaches for photoreceptor replacement have demonstrated some success (Lamba et al., 2009a) and may be aided by treatments that manipulate the host retinal environment to encourage grafting and integration, for example through steroid treatment (Singhal et al.), matrix metalloproteinase manipulation (Zhang et al., 2007), or disruption of the outer limiting membrane of the retina (West et al., 2008).

A number of cell sources may prove valuable for replacing photoreceptors, although direct comparison of different cell sources has not yet been performed. One study recently examined retinal progenitors isolated from different developmental stages, and found that donor cells from retinas at an age close to photoreceptors' peak birthdate *in vivo* demonstrated the greatest integration after transplantation (Hernit-Grant & Macklis; MacLaren et al.). , and postmitotic rod precursors reintegrated synaptically in the photoreceptor layer better than multipotent progenitor cells. Transplantation of immature rods improved visually evoked potentials in mouse models of retinitis pigmentosa, either by direct synaptic input or by improving the function of the residual photoreceptors (MacLaren et al.).

Can we harvest retinal stem cells from patients? Recently, it was demonstrated that in adult humans, a small population of stem cells reside at the edge of the retina, called the ciliary marginal zone (Coles et al., 2004; Tropepe et al., 2000). It is unclear where these cells come from during development, or whether they have any normal role to play in the retina. Importantly, however, if you remove them from the eye into cell cultures, they can proliferate and be induced to differentiate into all the cell types in the retina, including rods and cones (Coles et al., 2004; Tropepe et al., 2000). Such a technique could benefit patients by offering a source of retinal progenitor cells that are autologous (from the patient themselves), or allogeneic (from a matched donor), for example. However, many questions remain, including how to optimally support these progenitor cells' proliferative capacity, and how to guide their differentiation into specific retinal neurons such as photoreceptors or retinal ganglion cells.

7.2 Inner retina and optic nerve

7.2.1 Background

Retinal ganglion cell axons originate at the retinal ganglion cell bodies in the inner layer of the retina. Retinal ganglion cells receive synaptic signals from inner retinal bipolar and amacrine cells, and project this visual information down their axon in the nerve fiber layer. The optic nerve begins at the optic nerve head or optic disk, and from there through the orbit (Wolff, 1948) and optic canal (Maniscalco & Habal, 1978) to the chiasm, optic tract and into the brain.

7.2.2 Pathology

The inner retina and optic nerve are commonly involved in disease, resulting in retinopathy and optic neuropathy, and thus visual loss. Vein occlusions, in which the inner retinal circulation is compromised, are the most common cause of inner retinal injury, damaging retinal ganglion cells as well as amacrine cells but sparing the outer retina including photoreceptors (Laouri et al., 2011). Less common, retinal artery occlusions have the same effects (Mangat, 1995). More specific damage to retinal ganglion cells occurs when the ischemic insult is observed at the optic nerve in arteritic and non-arteritic anterior ischemic optic neuropathies. This is generally accompanied by a pale disk and loss of optic nerve axons. Ischemia in either location may result from hypotension or from occlusive or embolic disease (Connolly et al., 1994; Johnson et al., 1987).

Glaucoma is the most common optic neuropathy and the most common cause of irreversible blindness worldwide (Quigley & Broman, 2006). Glaucomatous optic neuropathy is distinguished by a distinct morphology of progressive excavation of the nerve head without significant pallor of the remaining neuroretinal rim. Within the retina, there are decreased numbers of retinal ganglion cell bodies in glaucoma (Giles & Soble, 1971; Minckler, 1989; Quigley, 1995), and this likely reflects death by apoptosis (Dkhissi et al., 1999; Kerrigan et al., 1997; Okisaka et al., 1997; Quigley et al., 1995). The number of retinal ganglion cells lost correlates with the visual field deficit (Quigley et al., 1989). In addition to the retinal ganglion cell body loss, there is loss of the ganglion cell axons, manifested by segmental loss of the nerve fiber layer (Airaksinen et al., 1984; Drance, 1985; Hoyt et al., 1973; Iwata et al., 1985; Quigley et al., 1980), increased cup-to-disk ratio, thinning of the optic nerve (Stroman et al., 1995) and chiasm (Iwata et al., 1997), changes in post-synaptic cell counts within the lateral geniculate nucleus (Chaturvedi et al., 1993; Vickers et al., 1997; Weber et al., 2000; Yücel et al., 2000) (the main target of retinal ganglion cell axons in higher animals), and even the cerebral cortex (Crawford et al., 2001; Crawford et al., 2000).

Other optic neuropathies include optic neuritis, ischemic optic neuropathy and compressive optic neuropathy, commonly associated with a tumor or aneurysm. Examples of these tumors include cavernous hemangioma, hemangiopericytoma, fibrous histiocytoma, lymphoma, and schwannoma. In addition, enlargement of the orbital extraocular muscles themselves, as in Grave's ophthalmopathy, may compress the optic nerve and lead to death of retinal ganglion cells (see Section 5). Traumatic optic neuropathy and optic nerve transection are less commonly observed in humans but have been the best studied causes of retinal ganglion cell death in animal models. In humans, direct transection or even concussive injury to the optic nerve injures axons and leads to retinal ganglion cell death back in the retina. Optic neuritis is the most common inflammatory optic neuropathy in young adults (1997; Rizzo & Lessell, 1988). Demyelination itself does not immediately cause loss of retinal ganglion cells; many rounds of inflammation and demylination eventually lead to axon loss (Evangelou et al., 2000; Perry & Anthony, 1999; Trapp et al., 1998), which again clinically appears as optic atrophy and loss of the nerve fiber layer (MacFadyen et al., 1988).

What mechanisms induce retinal ganglion cell death in glaucoma and other optic neuropathies? Glaucoma, for example, has been extensively modeled in animals (reviewed in (Morrison et al., 2005; Whitmore et al., 2005). Studies of tissue from human patients with glaucoma and non-human primates and other mammals with experimental glaucoma confirm changes at the optic nerve head, such as with respect to bowing out of the lamina cribrosa, intra-axonal accumulation of organelles (consistent with blocked axonal transport), and Wallerian degeneration distal to the lamina cribrosa (Fontana et al., 1998; Quigley & Addicks, 1980; Quigley et al., 1981). Whether due to mechanical trauma of axons (Allcutt et al., 1984a; Allcutt et al., 1984b; Barron et al., 1986; Berkelaar et al., 1994), ischemia (Cioffi & Sullivan, 1999), generation of nitric oxide (Neufeld, 1999; Neufeld et al., 1997), or other causes, axonal injury causes changes in retinal ganglion cells, eventually resulting in death. Increased intraocular pressure perturbs rapid anterograde and retrograde axonal transport at the lamina cribrosa (Anderson & Hendrickson, 1974; Minckler et al., 1977; Quigley et al., 1979; Radius, 1983). This may cause retinal ganglion cells to be deprived of neurotrophic factors or other survival signals produced by brain targets. Studies in experimental animals have shown that injury to the optic nerve, for example from increased intraocular pressure in experimental glaucoma, blocks the retrograde transport of the neurotrophic factor brain-derived neurotrophic factor (BDNF) along with its associated receptor, TrkB (Johnson et al., 2000; Pease et al., 2000; Quigley et al., 2000). Interestingly, one of the genes associated with a hereditary form of glaucoma, optineurin (Rezaie et al., 2002), may in its mutated form contribute to retinal ganglion cell sensitivity to optic nerve insults by failing to adequately participate in retrograde transport of neurotrophin signaling complexes, or by redox dysregulation (Anborgh et al., 2005; Chalasani et al., 2007; De Marco et al., 2006; Park et al., 2006).

7.2.3 Current treatment

Current treatment for optic neuropathies and optic nerve and retinal ischemias is directed almost exclusively at reducing risk factors for disease progression. For example, primary open-angle glaucoma therapy is focused on reducing the only known modifiable risk factor, intraocular pressure (discussed in section 8, below). Retinal and optic nerve ischemia treatment is directed at reducing identifiable risk factors such as uncontrolled diabetes or hypertension. All of these treatments, however, are directed at reducing disease progression or preventing recurrence. There are no current treatments able to restore function already lost once retinal ganglion cells die or their axons are severed in the optic nerve. Thus, there is considerable clinical need for novel therapeutics directed at enhancing function through restoration of lost cells and connectivity, as well as improved approaches to neuroprotection, similar to that discussed for the photoreceptor/RPE complex above.

7.2.4 Stem cell research and opportunities

There is considerable promise regarding the use of stem cells for glaucoma and other optic neuropathies, although most of this work has focused on the use of stem cells as neuroprotection delivery agents. Mesenchymal stem cells transplanted into the vitreous body in a rodent model of glaucoma showed considerable capacity to protect retinal ganglion cell degeneration (Johnson et al., 2010; Yu et al., 2006). This did not require any differentiation or integration of the transplanted stem cells. Similar results including RGC neuroprotection have been demonstrated in models of retinal ischemia (Li et al., 2009).

As a cell replacement therapy for glaucoma, stem cells face a much tougher set of hurdles. Transplanted stem cells thus far have shown limited success at integrating into and restoring function elsewhere in the central nervous system in preclinical models (Calford et al., 2005; Sohur et al., 2006). Optimal cellular integration may be limited by immune reactivity, response to injury, or factors found in development but missing in the adult microenvironment. For replacing retinal ganglion cells, stem cells would have to properly differentiate into retinal ganglion cells, integrate with their presynaptic partners in the retina to receive visual information, and project their axons down the optic nerve and properly wire up with their targets in the lateral geniculate nucleus of the thalamus and other regions of the brain (Goldberg, 2003b).

Considerable progress on understanding how retinal ganglion cells differentiate from their precursors has been made. For example, two well-studied transcription factors, Math5 and Brn3, are required for initial and terminal differentiation, respectively. In the absence of Math5 expression, retinal ganglion cells fail to differentiate from retinal progenitors during development (Brown et al., 2001; Wang et al., 2001). Math5 normally activates the expression of Brn3 genes (Mu et al., 2005), but in the absence of Brn3 expression, retinal ganglion cells die shortly after their generation in the retina (Badea et al., 2009; Wang et al., 2002). Despite these and other steps forward in our understanding of retinal ganglion cell differentiation, we know little about how to force stem cells to differentiate into retinal ganglion cells with high efficiency.

Considerable recent progress has been made in our understanding of how to encourage retinal ganglion cells to regenerate their axons through the optic nerve towards the brain (Moore & Goldberg, 2011). A combination of inhibitory molecules in the optic nerve (Berry et al., 2008), insufficient neurotrophic factors available to promote axon growth (Aguayo et al., 1996), and a decreased intrinsic regenerative capacity of retinal ganglion cells after an early period of developmental growth (Liu et al., 2011) all contribute to regenerative failure, but overcoming these alone or in combination enhances regenerative ability (Benowitz & Yin, 2007). A major question remains as to whether retinal ganglion cells derived from stem cells will also need to be treated to overcome these barriers to optic nerve axon growth and reconnection to the brain.

8. Trabecular meshwork

8.1 Background

The trabecular meshwork is a tissue in the angle between the cornea and iris in the anterior segment of the eye that is responsible for drainage of aqueous fluid. The ciliary body secretes aqueous fluid into the anterior chamber of the eye, which bathes and feeds the lens and cornea with critical nutrients (See Fig. 7). The balance between aqueous secretion and outflow is a major determinant of intraocular pressure (IOP), which is a major risk factor for the development of glaucoma (Goldberg, 2003a).

Fig. 7. Anterior chamber of the eye, with arrow demonstrating flow of aqueous humour (Illustration by Jennifer Thomson Ph D., Bascom Palmer Eye Institute, University of Miami Miller School of Medicine)

The trabecular meshwork is comprised of endothelial cells and extracellular matrix that separate the anterior chamber from an outflow track called Schlemm's canal. These trabecular meshwork cells may be divided into endothelial cells, juxtacanalicular cells, and Schlemm's canal cells, although the relationships between these anatomical distinctions is not well understood. The interaction between trabecular meshwork cells and the extracellular matrix, including the ability of the cells to phagocytose debris and thereby prevent clogging (Buller et al., 1990), is thought to create the mechanical relationships that determine outflow (WuDunn, 2009).

8.2 Pathology

In glaucoma, RGCs and their axons in the optic nerve demonstrate a typical pattern of damage and degeneration associated with permanent loss of vision, as discussed above. IOP is a major risk factor for glaucoma, although most people with IOP above population averages never develop glaucoma, and many people with "normal" IOP but particularly susceptible RGCs develop glaucoma. Randomized clinical trials have demonstrated that glaucoma patients with either "high" or "normal" starting IOPs can slow the progression of their disease by lowering IOP (reviewed elsewhere). Thus there is strong motivation to understand the dynamics of aqueous fluid production and outflow in the determining IOP and devising treatment strategies to lower IOP.

In glaucoma, changes in the cellular and molecular milieu of the trabecular meshwork have been observed. There is an age-dependent (Grierson & Howes, 1987) and glaucoma-dependent (Alvarado et al., 1984; Alvarado et al., 1981) loss of trabecular meshwork cells. Cultured trabecular meshwork cells (Gasiorowski & Russell, 2009) have provided a rich source of experimental evidence that their biology likely plays a role in determining outflow and thus IOP-associated glaucoma risk .

8.3 Current treatment

Current treatment for glaucoma is focused on lowering IOP, the only modifiable risk factor for primary open-angle glaucoma. Topical and oral medications designed to decrease aqueous production or increase outflow are typically first-line treatments, followed by laser trabeculoplasty and a number of incisional surgical approaches . Interestingly, laser trabeculoplasty may work by stimulating local cell replacement in the trabecular meshwork. Explanted human donor eyes treated with laser trabeculoplasty demonstrated increased cell division in the trabecular meshwork and in a line of cells located just anterior to the trabecular meshwork called insert cells (Acott et al., 1989; Alexander & Grierson, 1989; Alexander et al., 1989; Dueker et al., 1990). Cells from this anterior zone were observed to migrate into the trabecular meshwork, raising the hypothesis that these were stem cells or progenitor cells for this tissue.

8.4 Stem cell research and opportunities

Human trabecular meshwork cells can now be isolated and cultured, and a subset of these cells in culture appear to behave like stem cells (or, trabecular meshwork progenitor cells) . Whether this proliferative population is derived from insert cells or a subpopulation of trabecular meshwork cells remains to be determined. trabecular meshwork progenitor cells can be expanded in three-dimensional sphere-like cultures (Gonzalez et al., 2006). Cells from these cultures express markers and gene expression profiles associated with monolayer cultures of TM stem cells (Gonzalez et al., 2006).

This raises the question, would transplanting trabecular meshwork cells or trabecular meshwork progenitor cells back to the trabecular meshwork enhance outflow, thereby lowering IOP and protecting RGCs from damage in glaucoma? There is little data yet exploring this question, but the observation that laser trabeculoplasty enhances both cellularity and trabecular meshwork outflow suggests that additional cells, if added in the right fashion, may remodel the trabecular meshwork to similarly reduce IOP. A number of groups are working on this hypothesis, and the relatively easy accessibility of the anterior segment makes this an attractive approach to consider cell therapies in the eye (Kelley et al., 2009).

9. Conclusion

In summary, there is considerable interest in and scientific basis for pursuing stem cell therapies in ophthalmology. From epithelial to mesenchymal and even endothelial cell-based approaches, the eye and periocular tissues present a range of diseases with immense clinical need for novel therapeutics. We look forward to the advances in both pre-clinical and translational research into cell therapies for ophthalmology, and predict a rapid and encouraging step into a new generation of stem cell-based therapeutics throughout the field.

10. Acknowledgements

The authors gratefully acknowledge support from the Department of Defense, the Hope for Vision foundation, the National Eye Institute (P30 EY014081, Miami), and an unrestricted grant from Research to Prevent Blindness. Illustration by Jennifer Thomson Ph D., Bascom Palmer Eye Institute, University of Miami Miller School of Medicine.

11. References

Acott TS, Samples JR, Bradley JM, et al. Trabecular repopulation by anterior trabecular meshwork cells after laser trabeculoplasty. *Am J Ophthalmol.* 107, (1989), 1-6.

Aguayo AJ, Clarke DB, Jelsma TN, et al. Effects of neurotrophins on the survival and regrowth of injured retinal neurons. *Ciba Foundation Symp.* 196, (1996), 135-44.

Ainscough S, Linn M, Barnard Z et al. Effects of fibroblast origin and phenotype on the proliferative potential of limbal epithelial progenitor cells. *Experimental Eye Research,* Vol. 92 (2011), pp. 10-19.

Airaksinen PJ, Drance SM, Douglas GR, et al. Diffuse and localized nerve fiber loss in glaucoma. *Am J Ophthalmol.* 98, (1984), 566-71.

Akhtiari P, Djalilian A. Update on Limbal Stem Cell Transplantation. *Middle East Afr J Ophthalmol.* Vol.17, No.1, (2010 Jan), pp. 9-14.

Akita S, Akino K, Hirano A et al. Mesenchymal Stem Cell Therapy for Cutaneous Radiation Syndrome. *Health Phys,* Vol.98, No. 6, (2010), pp.858–862.

Akpek EK, Foster SC. Limbal stem-cell transplantation. *Int Ophthalmol Clin,* Vol.39, (1999), pp.79-82.

Alexander RA, Grierson I. Morphological effects of argon laser trabeculoplasty upon the glaucomatous human meshwork. *Eye (Lond).* 3 (Pt 6), (1989), 719-26.

Alexander RA, Grierson I, Church WH. The effect of argon laser trabeculoplasty upon the normal human trabecular meshwork. *Graefes Arch Clin Exp Ophthalmol.* 227, (1989), 72-7.

Allcutt D, Berry M, Sievers J. A qualitative comparison of the reactions of retinal ganglion cell axons to optic nerve crush in neonatal and adult mice. *Brain Res.* 318, (1984a), 231-40.

Allcutt D, Berry M, Sievers J. A quantitative comparison of the reactions of retinal ganglion cells to optic nerve crush in neonatal and adult mice. *Brain Res.* 318, (1984b), 219-30.

Alvarado J, Murphy C, Juster R. Trabecular meshwork cellularity in primary open-angle glaucoma and nonglaucomatous normals. *Ophthalmology.* 91, (1984), 564-79.

Alvarado J, Murphy C, Polansky J, Juster R. Age-related changes in trabecular meshwork cellularity. *Invest Ophthalmol Vis Sci.* 21, (1981), 714-27.

Amoh Y, Li L, Campillo R, Kawahara K, et al. Implanted hair follicle stem cells form Schwann cells that support repair of severed peripheral nerves. *Proc Natl Acad Sci USA*, Vol. 102, (2005), pp.17734–17738.

Anborgh PH, Godin C, Pampillo M, et al. Inhibition of metabotropic glutamate receptor signaling by the huntingtin-binding protein optineurin. *J Biol Chem.* 280, (2005), 34840-8.

Anderson D, Ellies P, Pires R. Amniotic membrane transplantation for partial limbal stem cell deficiency. *Br J Ophthalmol,* Vol 85, (2001), pp.567–575.

Anderson DR, Hendrickson A. Effect of intraocular pressure on rapid axoplasmic transport in monkey optic nerve. *Invest Ophthalmol.* 13, (1974), 771.

Ang L, Cheng Z, Beuerman R et al. The development of a serum-free derived bioengineered conjunctival epithelial equivalent using an ultrathin poly(epsilon-caprolactone) membrane substrate. *Invest Ophthalmol Vis Sci.* Vol 47, No.1 (2006 Jan), pp.105-112.

Ang LP, Tan DT. Ocular surface stem cells and disease: current concepts and clinical applications. *Ann Acad Med Singapore.* Vol 33, No. 5, (2004 Sep), pp.576-580.

Ang LP, Tan DT, Beuerman RW, Lavker RM. Development of a conjunctival epithelial equivalent with improved proliferative properties using a multistep serum-free culture system. *Invest Ophthalmol Vis Sci.* Vol. 45, No.6, (2004), pp. 1789-1795.

Ang L, Tan D, Cajucom-Uy H et al. Autologous cultivated conjunctival transplantation for pterygium surgery. *Am J Ophthalmol.* Vol. 139, (2005), pp.611-619.

Ang L, Tan D, Phan T et al. The in vitro and in vivo proliferative capacity of serum-free cultivated human conjunctival epithelial cells. *Curr Eye Res.* Vol. 28, (2004), pp.307-317.

Arnhold S, Absenger Y, Klein H, et al. Transplantation of bone marrow-derived mesenchymal stem cells rescue photoreceptor cells in the dystrophic retina of the rhodopsin knockout mouse. *Graefes Arch Clin Exp Ophthalmol.* 245, (2007), 414-22.

Ashtari M, Cyckowski LL, Monroe JF, et al. The human visual cortex responds to gene therapy-mediated recovery of retinal function. *J Clin Invest.* 121, (2011), 2160-8.

Badea TC, Cahill H, Ecker J, et al. Distinct roles of transcription factors brn3a and brn3b in controlling the development, morphology, and function of retinal ganglion cells. *Neuron.* 61, (2009), 852-64.

Badiavas EV, Falanga V. Treatment of chronic wounds with bone marrow-derived cells. *Arch Dermatol.* Vol. 139, No. 4, (2003), pp. 510-516.

Barabino S, Rolando M, Bentivoglio G. Role of amniotic membrane transplantation for conjunctival reconstruction in ocular-cicatricial pemphigoid. *Ophthalmology* Vol.110, No. 3, (2003 Mar), pp.474-480.

Barron KD, Dentinger MP, Krohel G, et al. Qualitative and quantitative ultrastructural observations on retinal ganglion cell layer of rat after intraorbital optic nerve crush. *J Neurocytol*. 15, (1986), 345-62.

Benowitz LI, Yin Y. Combinatorial treatments for promoting axon regeneration in the cns: Strategies for overcoming inhibitory signals and activating neurons' intrinsic growth state. *Dev Neurobiol*. 67, (2007), 1148-65.

Berkelaar M, Clarke DB, Wang YC, et al. Axotomy results in delayed death and apoptosis of retinal ganglion cells in adult rats. *J Neurosci*. 14, (1994), 4368-74.

Berry M, Ahmed Z, Lorber B, et al. Regeneration of axons in the visual system. *Restor Neurol Neurosci*. 26, (2008), 147-74.

Brown NL, Patel S, Brzezinski J, Glaser T. Math5 is required for retinal ganglion cell and optic nerve formation. *Development*. 128, (2001), 2497-508.

Budak MT, Alpdogan OS, Zhou M et al. Ocular surface epithelia contain ABCG2-depending side population cells exhibiting features associated with stem cells. *J Cell Sci*. Vol. 118, (2005), pp.1715-1724.

Buller C, Johnson DH, Tschumper RC. Human trabecular meshwork phagocytosis. Observations in an organ culture system. *Invest Ophthalmol Vis Sci*. 31, (1990), 2156-63.

Calford MB, Chino YM, Das A, et al. Neuroscience: Rewiring the adult brain. *Nature*. 438, (2005), E3; discussion E3-4.

Chalasani ML, Radha V, Gupta V, et al. A glaucoma-associated mutant of optineurin selectively induces death of retinal ganglion cells which is inhibited by antioxidants. *Invest Ophthalmol Vis Sci*. 48, (2007), 1607-14.

Chaturvedi N, Hedley-Whyte ET, Dreyer EB. Lateral geniculate nucleus in glaucoma. *Am J Ophthalmol*. 116, (1993), 182-8.

Cideciyan AV. Leber congenital amaurosis due to rpe65 mutations and its treatment with gene therapy. *Prog Retin Eye Res*. 29, (2010), 398-427.

Cioffi GA, Sullivan P. The effect of chronic ischemia on the primate optic nerve. *Eur J Ophthalmol*. 9 Suppl 1, (1999), S34-6.

Coles BL, Angenieux B, Inoue T, et al. Facile isolation and the characterization of human retinal stem cells. *Proc Natl Acad Sci U S A*. 101, (2004), 15772-7.

Cotsarelis SZ, Cheng G, Dong G, et al. Existence of slow-cycling limbal epithelial basal cells that can be preferentially stimulated to proliferate: implications on epithelial stem cells. *Cell*. Vol 57, (1989), pp.201–209.

Connolly SE, Gordon KB, Horton JC. Salvage of vision after hypotension-induced ischemic optic neuropathy. *Am J Ophthalmol*. 117, (1994), 235-42.

Crawford ML, Harwerth RS, Smith EL, 3rd, et al. Experimental glaucoma in primates: Changes in cytochrome oxidase blobs in v1 cortex. *Invest Ophthalmol Vis Sci*. 42, (2001), 358-64.

Crawford ML, Harwerth RS, Smith EL, 3rd, et al. Glaucoma in primates: Cytochrome oxidase reactivity in parvo- and magnocellular pathways. *Invest Ophthalmol Vis Sci.* 41, (2000), 1791-802.

Crigler L, Robey RC, Asawachaicharn A, et al. Human mesenchymal stem cell subpopulations express a variety of neuro-regulatory molecules and promote neuronal cell survival and neuritogenesis. *Exp Neurol.* 198, (2006), 54-64.

Daiger SP, Bowne SJ, Sullivan LS. Perspective on genes and mutations causing retinitis pigmentosa. *Arch Ophthalmol.* 125, (2007), 151-8.

Daniels J, Dart J, Tuft S et al. Corneal stem cells in review. *Wound Rep Reg,* Vol. 9, (2001), pp.483–494.

Davanger M, Evensen A. Role of the pericorneal papillary structure in renewal of corneal epithelium. *Nature,* 1971; 229: 560–1.

Daya SM, Watson A, Sharpe JR et al. Outcomes and DNA analysis of ex vivo expanded stem cell allograft for ocular surface reconstruction. *Ophthalmology,* Vol. 112, (2005), 470–477.

De Marco N, Buono M, Troise F, Diez-Roux G. Optineurin increases cell survival and translocates to the nucleus in a rab8-dependent manner upon an apoptotic stimulus. *J Biol Chem.* 281, (2006), 16147-56.

De Medinaceli L, Rawlings RR: Is it possible to predict the outcome of peripheral nerve injuries? A probability model based on prospects for regenerating neurites. *Biosystems,* Vol.20, (1987), pp.243–258.

Dkhissi O, Chanut E, Wasowicz M, et al. Retinal tunel-positive cells and high glutamate levels in vitreous humor of mutant quail with a glaucoma-like disorder. *Invest Ophthalmol Vis Sci.* 40, (1999), 990-5.

Drance SM. The early structural and functional disturbances of chronic open-angle glaucoma. Robert n. Shaffer lecture. *Ophthalmology.* 92, (1985), 853-7.

Dua H, Miri A, Said D. Contemporary limbal stem cell transplantation - a review. *Clinical and Experimental Ophthalmology,* Vol. 38, (2010), pp. 104–117.

Dueker DK, Norberg M, Johnson DH, et al. Stimulation of cell division by argon and nd:Yag laser trabeculoplasty in cynomolgus monkeys. *Invest Ophthalmol Vis Sci.* 31, (1990), 115-24.

Espana E, Ti S, Grueterich M, et al. Corneal stromal changes following reconstruction by ex vivo expanded limbal epithelial cells in rabbits with total limbal stem cell deficiency. *Br. J. Ophthalmol,* Vol. 87, (2003), pp.1509-14.

Evangelou N, Esiri MM, Smith S, et al. Quantitative pathological evidence for axonal loss in normal appearing white matter in multiple sclerosis. *Ann Neurol.* 47, (2000), 391-5.

Falanga V, Iwamoto S, Chartier M, et al. Autologous bone marrow-derived cultured mesenchymal stem cells delivered in a fibrin spray accelerate healing in murine and human cutaneous wounds. *Tissue Eng,* Vol.13, (2007), pp.1299-1312.

Feldon S, Park DJ, O'Loughlin C, et al. Autologous T-Lymphocytes stimulate proliferation of orbital fibroblasts derived from patients with graves ophthalmopathy. *Invest Ophthal and Vis Sci,* Vol. 46, No. 11, (2005), pp. 3913-3921.

Feliciello A, Porcellini A, Ciullo L et al. Expression of thyrotropin receptor mRNA in healthy and Graves disease retro-orbital tissue. *Lancet,* Vol. 342, (1993), pp.337-338.

Fontana L, Bhandari A, Fitzke FW, Hitchings RA. In vivo morphometry of the lamina cribrosa and its relation to visual field loss in glaucoma. *Curr Eye Res.* 17, (1998), 363-9.

Frucht-Pery J, Siganos CS, Solomon A. Limbal cell autograft transplantation for severe ocular surface disorders. *Graefes Arch Clin. Exp Ophthalmol,* Vol. 236, (1998), 582–587.

Gasiorowski JZ, Russell P. Biological properties of trabecular meshwork cells. *Exp Eye Res.* 88, (2009), 671-5.

Ghoubay-Benallaoua D, Basli E, Goldschmidt P et al. Human epithelial cell cultures from superficial limbal explants. Mol Vis, Vol. 17, (2011), pp. 341–354.

Giles CL, Soble AR. Intracranial hypertension and tetracycline therapy. *Am J Ophthalmol.* 72, (1971), 981-2.

Goldberg I. Relationship between intraocular pressure and preservation of visual field in glaucoma. *Surv Ophthalmol.* 48 Suppl 1, (2003a), S3-7.

Goldberg JL. How does an axon grow? *Genes Dev.* 17, (2003b), 941-58.

Gomes J, Serapião dos Santos M, Cunha M et al. Amniotic membrane transplantation for partial and total limbal stem cell deficiency secondary to chemical burn. *Ophthalmology,* Vol. 110, No. 3, (2003), pp.466-73.

Gonzalez P, Epstein DL, Luna C, Liton PB. Characterization of free-floating spheres from human trabecular meshwork (htm) cell culture in vitro. *Exp Eye Res.* 82, (2006), 959-67.

Grierson I, Howes RC. Age-related depletion of the cell population in the human trabecular meshwork. *Eye (Lond).* 1 (Pt 2), (1987), 204-10.

Hernit-Grant CS, Macklis JD. Embryonic neurons transplanted to regions of targeted photolytic cell death in adult mouse somatosensory cortex re-form specific callosal projections. *Exp Neurol.* 139, (1996), 131-42.

Hiromatsu Y, Yang D, Bednarczuk T, et al. Cytokine profiles in eye muscle tissue and orbital fat tissue from patients with thyroid-associated ophthalmopathy. *J Clin Endocrinol Metab,* Vol. 85, (2000), pp.1194-1199.

Hoyt WF, Frisen L, Newman NM. Fundoscopy of nerve fiber layer defects in glaucoma. *Invest Ophthalmol.* 12, (1973), 814-29.

Humphrey CD, Kreit JD. Nerve repair and cable grafting for facial paralysis.*FPS,* Vol. 24, (2008),pp.170–176.

Inatomi T, Nakamura T, Koizumi N. Midterm results on ocular surface reconstruction using cultivated autologous oral mucosal epithelial transplantation. *Am J Ophthalmol,* Vol. 141, (2006), pp. 267–275.

Inoue Y, Iriyama A, Ueno S, et al. Subretinal transplantation of bone marrow mesenchymal stem cells delays retinal degeneration in the rcs rat model of retinal degeneration. *Exp Eye Res.* 85, (2007), 234-41.

Iwata F, Patronas NJ, Caruso RC, et al. Association of visual field, cup-disc ratio, and magnetic resonance imaging of optic chiasm. *Arch Ophthalmol.* 115, (1997), 729-32.

Iwata K, Kurosawa A, Sawaguchi S. Wedge-shaped retinal nerve fiber layer defects in experimental glaucoma preliminary report. *Graefes Arch Clin Exp Ophthalmol.* 223, (1985), 184-9.

Johnson EC, Deppmeier LMH, Wentzien SKF, et al. Chronology of optic nerve head and retinal responses to elevated intraocular pressure. *Invest Ophthalmol Vis Sci.* 41, (2000), 431-442.

Johnson MW, Kincaid MC, Trobe JD. Bilateral retrobulbar optic nerve infarctions after blood loss and hypotension. A clinicopathologic case study. *Ophthalmology.* 94, (1987), 1577-84.

Johnson TV, Bull ND, Hunt DP, et al. Neuroprotective effects of intravitreal mesenchymal stem cell transplantation in experimental glaucoma. *Invest Ophthalmol Vis Sci.* 51, (2010), 2051-9.

Johnston MC, Noden DM, Hazelton RD et al. Origins of avian ocular and periocular tissues. Exp Eye Res 1979; 29: 27-43.

Kahaly J, Forster G, Hansen C. Glycosaminoglycans in thyroid eye disease. *Thyroid,* Vol.8, (1998), pp.429-432.

Kelley MJ, Rose AY, Keller KE, et al. Stem cells in the trabecular meshwork: Present and future promises. *Exp Eye Res.* 88, (2009), 747-51.

Kenyon KR, Tseng SC. Limbal autograft transplantation for ocular surface disorders. *Ophthalmology,* Vol.96, (1989), pp.709–722.

Kerrigan LA, Zack DJ, Quigley HA, et al. Tunel-positive ganglion cells in human primary open-angle glaucoma. *Arch Ophthalmol.* 115, (1997), 1031-5.

Klein R, Chou CF, Klein BE, et al. Prevalence of age-related macular degeneration in the us population. *Arch Ophthalmol.* 129, (2011), 75-80.

Korn B, Kikkawa D, Hicok K. Identification and characterization of adult stem cells from human orbital adipose tissue. *Ophthal Plast Reconstr Surg,* Vol. 25, No. 1, (2009), pp. 27- 32.

Ko S, Nauta A, Wong V. The role of stem cells in cutaneous wound healing: what do we really know? *Plast Reconstr Surg,* Vol. 127, (2011), pp.10S-20S.

Krulova M, Pokorna K, Lencova A et al. A Rapid Separation of Two Distinct Populations of Mouse Corneal Epithelial Cells with Limbal Stem Cell Characteristics by Centrifugation on Percoll Gradient. *Invest Ophthalmol Vis Sci,* Vol. 49, (2008), pp. 3903–3908.

Kumar S, Coenen M, Scherer P et al. Evidence for enhanced adipogenesis in the orbits of patients with graves ophthalmopathy. *Journ Clinic Endocrin Metab,* Vol.89, (2003), pp.930-935.

Lamba DA, Gust J, Reh TA. Transplantation of human embryonic stem cell-derived photoreceptors restores some visual function in crx-deficient mice. *Cell Stem Cell.* 4, (2009a), 73-9.

Lamba DA, Karl MO, Reh TA. Strategies for retinal repair: Cell replacement and regeneration. *Prog Brain Res.* 175, (2009b), 23-31.

Laouri M, Chen E, Looman M, Gallagher M. The burden of disease of retinal vein occlusion: Review of the literature. *Eye (Lond),* (2011).

Lau K, Paus R, Tiede S et al. Exploring the role of stem cells in cutaneous wound healing. *Experimental Dermatology,* Vol. 18, (2009), pp.921–933.

Lee EY, Xia Y, Kim WS et al. Hypoxia-enhanced wound healing function of adipose-derived stem cells: Increase in stem cell proliferation and up-regulation of VEGF and bFGF. *Wound Repair Regen*, Vol.17, (2009), pp.540-547.

Li D, Tseng S. Differential regulation of keratinocyte growth factor and hepatocyte growth factor/scatter factor by different cytokines in human corneal and limbal fibroblasts. *J. Cell. Physiol*, Vol. 172, (1997), pp.361-72.

Li N, Li XR, Yuan JQ. Effects of bone-marrow mesenchymal stem cells transplanted into vitreous cavity of rat injured by ischemia/reperfusion. *Graefes Arch Clin Exp Ophthalmol*. 247, (2009), 503-14.

Li W, Sun X, Wang Z et al. The effect of nerve growth factor on differentiation of corneal limbal epithelial cells to conjunctival goblet cells in vitro. *Mol Vis*, Vol. 16, (2010 Dec 15), pp.2739-2744.

Liu K, Tedeschi A, Park KK, He Z. Neuronal intrinsic mechanisms of axon regeneration. *Annu Rev Neurosci*. 34, (2011), 131-52.

Lu B, Malcuit C, Wang S, et al. Long-term safety and function of rpe from human embryonic stem cells in preclinical models of macular degeneration. *Stem Cells*. 27, (2009), 2126-35.

Lund RD, Wang S, Klimanskaya I, et al. Human embryonic stem cell-derived cells rescue visual function in dystrophic rcs rats. *Cloning Stem Cells*. 8, (2006), 189-99.

MacFadyen DJ, Drance SM, Douglas GR, et al. The retinal nerve fiber layer, neuroretinal rim area, and visual evoked potentials in ms. *Neurology*. 38, (1988), 1353-8.

MacLaren RE, Pearson RA, MacNeil A, et al. Retinal repair by transplantation of photoreceptor precursors. *Nature*. 444, (2006), 203-7.

Majo F, Rochat A, Nicolas M et al. Location of corneal epithelial stem cells. *Nature*, Vol 463, (2010), pp. E10.

Mangat HS. Retinal artery occlusion. *Surv Ophthalmol*. 40, (1995), 145-56.

Maniscalco JE, Habal MB. Microanatomy of the optic canal. *J Neurosurg*. 48, (1978), 402-6.

Meller D, Pires RT, Tseng SC. Ex vivo preservation and expansion of human limbal epithelial stem cells on amniotic membrane cultures. *Br J Ophthalmol*, Vol. 86; (2002), pp. 463–471.

Meller D, Tseng S. Conjunctival epithelial cell differentiation on amniotic membrane. *Invest Ophthalmol Vis Sci*, Vol. 40, (1999), pp. 878-86.

Meltzer N, Alam D. Facial paralysis rehabilitation: state of the art. *Curr Opin Otolaryngol Head Neck Surg*, Vol. 18, (2010), pp.232-7.

Millesi H, Meissl G, Berger A. The interfascicular nerve-grafting of the median and ulnar nerves. *J Bone Joint Surg Am*, Vol. 54, No. 4, (1972 Jun), pp.727-750.

Minckler DS. Histology of optic nerve damage in ocular hypertension and early glaucoma. *Surv Ophthalmol*. 33 Suppl, (1989), 401-11.

Minckler DS, Bunt AH, Johanson GW. Orthograde and retrograde axoplasmic transport during acute ocular hypertension in the monkey. *Invest Ophthalmol Vis Sci*. 16, (1977), 426-41.

Miri A, Mathew M, Dua HS. Quality of life after limbal transplants. *Ophthalmology*, Vol.117, No. 3, (2010), pp.638, 638.e1-3.

Miri A, Al-Deiri B, Dua HS. Long-term outcomes of Autolimbal and Allolimbal transplants. *Ophthalmology,* Vol. 117, No. 6, (2010), pp.1207-1213.

Mitchell P. A systematic review of the efficacy and safety outcomes of anti-vegf agents used for treating neovascular age-related macular degeneration: Comparison of ranibizumab and bevacizumab. *Curr Med Res Opin.* 27, (2011), 1465-75.

Moore DL, Goldberg JL. Four steps to optic nerve regeneration. *J Neuroophthalmol.* 30, (2011), 347-60.

Moore J, Mcmullen C, Mahon G et al. The Corneal Epithelial Stem Cell. *DNA And Cell Biology,* Vol. 21, (2002), pp. 443–51.

Morrison JC, Johnson EC, Cepurna W, Jia L. Understanding mechanisms of pressure-induced optic nerve damage. *Prog Retin Eye Res.* 24, (2005), 217-40.

Mu X, Fu X, Sun H, et al. A gene network downstream of transcription factor math5 regulates retinal progenitor cell competence and ganglion cell fate. *Dev Biol.* 280, (2005), 467-81.

Neufeld AH. Nitric oxide: A potential mediator of retinal ganglion cell damage in glaucoma. *Surv Ophthalmol.* 43 Suppl 1, (1999), S129-35.

Neufeld AH, Hernandez MR, Gonzalez M. Nitric oxide synthase in the human glaucomatous optic nerve head. *Arch Ophthalmol.* 115, (1997), 497-503.

Nowak JA, Polak L, Pasolli HA et al. Hair follicle stem cells are specified and function in early skin morphogenesis. *Cell Stem Cell,* Vol. 3, (2008), pp.33-43.

Okisaka S, Murakami A, Mizukawa A, Ito J. Apoptosis in retinal ganglion cell decrease in human glaucomatous eyes. *Jpn J Ophthalmol.* 41, (1997), 84-8.

Optic Neuritis Study Group. The 5-year risk of ms after optic neuritis. Experience of the optic neuritis treatment trial. *Neurology.* 49, (1997), 1404-13.

Park BC, Shen X, Samaraweera M, Yue BY. Studies of optineurin, a glaucoma gene: Golgi fragmentation and cell death from overexpression of wild-type and mutant optineurin in two ocular cell types. *Am J Pathol.* 169, (2006), 1976-89.

Pease ME, McKinnon SJ, Quigley HA, et al. Obstructed axonal transport of bdnf and its receptor trkb in experimental glaucoma. *Invest Ophthalmol Vis Sci.* 41, (2000), 764-774.

Pellegrini G, Dellambra E, Golisano O, et al. p63 identifies keratinocyte stem cells. *Proc Natl Acad Sci U S A.*, Vol.98, (2001), pp. 3156–3161.

Pellegrini G. Golisano O, Paterna P, et al. Location and clonal analysis of stem cells and their differentiated progeny in the human ocular surface. *J Cell Biol,* Vol. 145, (1999), pp.769-82.

Pellegrini G, Traverso C, Franzi A et al. Long-term restoration of damaged corneal surfaces with autologous cultivated corneal epithelium. *Lancet,* Vol. 349, (1997), pp. 990-993.

Pels E, van der Gaag R. HLA-A,B,C, and HLA-DR antigens and dendritic cells in fresh and organ preserved corneas. *Cornea,* Vol. 3, (1984), pp.231–239.

Perry VH, Anthony DC. Axon damage and repair in multiple sclerosis. *Philos Trans R Soc Lond B Biol Sci.* 354, (1999), 1641-7.

Pluchino S, Zanotti L, Rossi B, et al. Neurosphere-derived multipotent precursors promote neuroprotection by an immunomodulatory mechanism. *Nature*. 436, (2005), 266-71.

Qi hong, Zheng Xiaofen, Yuan Xioyong et al. Potential localization of putative stem/progenitor cells in human bulbar conjunctival epithelium. *J Cell Physiol,* Vol. 225, (2010), pp.180-185.

Quigley HA. Ganglion cell death in glaucoma: Pathology recapitulates ontogeny. *Aust N Z J Ophthalmol*. 23, (1995), 85-91.

Quigley HA, Addicks EM. Chronic experimental glaucoma in primates. Ii. Effect of extended intraocular pressure elevation on optic nerve head and axonal transport. *Invest Ophthalmol Vis Sci*. 19, (1980), 137-52.

Quigley HA, Addicks EM, Green WR, Maumenee AE. Optic nerve damage in human glaucoma. Ii. The site of injury and susceptibility to damage. *Arch Ophthalmol*. 99, (1981), 635-49.

Quigley HA, Broman AT. The number of people with glaucoma worldwide in 2010 and 2020. *Br J Ophthalmol*. 90, (2006), 262-7.

Quigley HA, Dunkelberger GR, Green WR. Retinal ganglion cell atrophy correlated with automated perimetry in human eyes with glaucoma. *Am J Ophthalmol*. 107, (1989), 453-64.

Quigley HA, Guy J, Anderson DR. Blockade of rapid axonal transport. Effect of intraocular pressure elevation in primate optic nerve. *Arch Ophthalmol*. 97, (1979), 525-31.

Quigley HA, McKinnon SJ, Zack DJ, et al. Retrograde axonal transport of bdnf in retinal ganglion cells is blocked by acute iop elevation in rats. *Invest Ophthalmol Vis Sci*. 41, (2000), 3460-6.

Quigley HA, Miller NR, George T. Clinical evaluation of nerve fiber layer atrophy as an indicator of glaucomatous optic nerve damage. *Arch Ophthalmol*. 98, (1980), 1564-71.

Quigley HA, Nickells RW, Kerrigan LA, et al. Retinal ganglion cell death in experimental glaucoma and after axotomy occurs by apoptosis. *Invest Ophthalmol Vis Sci*. 36, (1995), 774-86.

Radius RL. Pressure-induced fast axonal transport abnormalities and the anatomy at the lamina cribrosa in primate eyes. *Invest Ophthalmol Vis Sci*. 24, (1983), 343.

Rezaie T, Child A, Hitchings R, et al. Adult-onset primary open-angle glaucoma caused by mutations in optineurin. *Science*. 295, (2002), 1077-9.

Ren H, Cao Y, Zhao Q et al. Proliferation and differentiation of bone marrow stromal cells under hypoxic conditions. *Biochem Biophys Res Commun,*Vol. 347, (2006), pp.12-21.

Rizzo JFd, Lessell S. Risk of developing multiple sclerosis after uncomplicated optic neuritis: A long-term prospective study. *Neurology*. 38, (1988), 185-90.

Ropper AH, Samuels MA. (2009). Bell's palsy section of Diseases of the cranial nerves, *In Adams and Victor's Principles of Neurology*, (9th ed.), pp. 1330-1331. New York: McGraw-Hill.

Sahel J, Bonnel S, Mrejen S, Paques M. Retinitis pigmentosa and other dystrophies. *Dev Ophthalmol.* 47, 160-7.

Salvi M, Vannucchi G, Campi I. Efficacy of rituximab treatment for thyroid-associated ophthalmopathy as a result of intraorbital B-cell depletion in one patient unresponsive to steroid immunosuppression. *Eur Journ Endocrinol,* Vol. 154, (2006 Apr), pp. 511-7.

Sasaki M, Abe R. Fujita Y et al. Mesenchymal stem cells are recruited into wounded skin and contribute to wound repair by transdifferentiation into multiple skin cell type. *J Immunol,* Vol.180, (2008), pp. 2581-2587.

Schofield R. The stem cell system. *Biomed Pharmacother,* Vol. 37, (1983), pp. 375–380.

Schrader S, Notara M, Beaconsfield M et al. Tissue engineering for conjunctival reconstruction: established methods and future outlooks. *Current Eye Res,* Vol. 34, No. 11, (2009), pp. 913-24.

Schrader S, Notara M, Tuft S et al. Simulation of an in vitro niche environment that preserves conjunctival progenitor cells. *Regen Med,* Vol. 5, No. 6, (2010 Nov), pp. 877-889.

Schurr MJ, Foster KN, Centanni JM et al. Phase I/II clinical evaluation of Stratagraft: A consistent, pathogen-free human skin substitute. *J Trauma,* Vol. 66, (2009), pp. 866-873,4.

Shortt AJ, Secker GA, Lomas RJ et al. The effect of amniotic membrane preparation method on its ability to serve as a substrate for the ex-vivo expansion of limbal epithelial cells. *Biomaterials,* Vol. 30, (2009), pp. 1056–1065.

Sieving PA, Caruso RC, Tao W, et al. Ciliary neurotrophic factor (cntf) for human retinal degeneration: Phase i trial of cntf delivered by encapsulated cell intraocular implants. *Proc Natl Acad Sci U S A.* 103, (2006), 3896-901.

Simonelli F, Maguire AM, Testa F, et al. Gene therapy for leber's congenital amaurosis is safe and effective through 1.5 years after vector administration. *Mol Ther.* 18, (2009), 643-50.

Singhal S, Lawrence JM, Salt TE, et al. Triamcinolone attenuates macrophage/microglia accumulation associated with nmda-induced rgc death and facilitates survival of muller stem cell grafts. *Exp Eye Res.* 90, 308-15.

Sohur US, Emsley JG, Mitchell BD, Macklis JD. Adult neurogenesis and cellular brain repair with neural progenitors, precursors and stem cells. *Philos Trans R Soc Lond B Biol Sci.* 361, (2006), 1477-97.

Sorisky A. From preadipocyte to adipocyte: differentiation-directed signals of insulin from the cell surface to the nucleus. *Crit Rev Clin Lab Sci,* Vol.36, (1999), pp.1–34

Stroman GA, Stewart WC, Golnik KC, et al. Magnetic resonance imaging in patients with low-tension glaucoma. *Arch Ophthalmol.* 113, (1995), 168-72.

Tao W, Wen R, Goddard MB, et al. Encapsulated cell-based delivery of cntf reduces photoreceptor degeneration in animal models of retinitis pigmentosa. *Invest Ophthalmol Vis Sci.* 43, (2002), 3292-8.

Trapp BD, Peterson J, Ransohoff RM, et al. Axonal transection in the lesions of multiple sclerosis. *N Engl J Med.* 338, (1998), 278-85.

Tropepe V, Coles BL, Chiasson BJ, et al. Retinal stem cells in the adult mammalian eye. *Science*. 287, (2000), 2032-6.

Tsai RJ, Ho YS, Chen J. The effects of fibroblasts on the growth and differentiation of human bulbar conjunctival epithelial cells in an in vitro conjunctival equivalent. *Invest Ophthalmol Vis Sci,* Vol. 35, (1994), pp. 2865-2875.

Tsai RJ, Tsai RY. Ex vivo expansion of corneal stem cells on amniotic membrane and their outcome. *Eye and Contact Lens,* Vol. 36, No. 5, (2010), pp. 305-309.

Tsai R, Tseng S. Substrate modulation of cultured rabbit conjunctival epithelial cell differentiation and morphology. *Invest Ophthalmol Vis Sci,* Vol. 10, (1988 Oct), pp.1565-1576.

Tsai R, Tseng S. Effect of stromal inflammation on the outcome of limbal transplantation for corneal surface reconstruction. *Cornea,* Vol. 14, (1995), pp. 439–449.

Tseng SC, Prabhasawat P, Barton K, et al. Amniotic membrane transplantation with or without limbal allografts for corneal surface reconstruction in patients with limbal stem cell corneal epithelium deficiency. *Arch Ophthalmol,* Vol. 116, (1998), pp.431-441.

Umemoto T, Yamato M, Nishida K, et al. Rat limbal epithelial side population cells exhibit a distinct expression of stem cell markers that are lacking in side population cells from the central cornea. *FEBS Lett,* Vol. 579; (2005), pp. 6559–6574.

Valyasevi R, Harteneck D, Dutton C and Bahn R. Stimulation of adipogenesis, peroxisome proliferator-activated receptor-y (ppary) and thryotropin receptor by ppary agonist in human orbital preadipocyte fibroblasts. *J Clin Endocrinol Metab,* Vol. 87, (1992), pp. 2353-2358.

Vickers JC, Hof PR, Schumer RA, et al. Magnocellular and parvocellular visual pathways are both affected in a macaque monkey model of glaucoma. *Aust N Z J Ophthalmol*. 25, (1997), 239-43.

Walsh S, Midha R. Use of Stem Cells to Augment Nerve Injury Repair. *Neurosurgery,* Vol. 65, (2009), pp. A80–A86.

Wang C, Jacob B, Nah G et al. Runx family genes, niche, and stem cell quiescence. *Blood Cells, Molecules, and Diseases,* Vol. 44, (2010), pp. 275–286.

Wang SW, Kim BS, Ding K, et al. Requirement for math5 in the development of retinal ganglion cells. *Genes Dev*. 15, (2001), 24-9.

Wang SW, Mu X, Bowers WJ, et al. Brn3b/brn3c double knockout mice reveal an unsuspected role for brn3c in retinal ganglion cell axon outgrowth. *Development*. 129, (2002), 467-77.

Watanabe K, Nishida K, Yamato M et al. Human limbal epithelium contains side population cells expressing the ATP-binding cassette transporter ABCG2. *FEBS Letters,* Vol. 565, (2004), pp. 6–10.

Weber AJ, Chen H, Hubbard WC, Kaufman PL. Experimental glaucoma and cell size, density, and number in the primate lateral geniculate nucleus. *Invest Ophthalmol Vis Sci*. 41, (2000), 1370-9.

Wei ZG, Cotsarelis G, Sun TT et al. Label-retaining cells are preferentially located in forniceal epithelium: Implications on conjunctival epithelial homeostasis. *Invest Ophthalmol Vis Sci,* Vol. 36, (1995), pp.236-246.

West EL, Pearson RA, Tschernutter M, et al. Pharmacological disruption of the outer limiting membrane leads to increased retinal integration of transplanted photoreceptor precursors. *Exp Eye Res.* 86, (2008), 601-11.

Wester ST, Lee WW, Shi W. Eyelash growth from application of bimatoprost in gel suspension to the base of the eyelashes. *Ophthalmology.*Vol.117, No. 5, (2010 May), pp. 1024-1031.

Whitmore AV, Libby RT, John SW. Glaucoma: Thinking in new ways-a role for autonomous axonal self-destruction and other compartmentalised processes? *Prog Retin Eye Res.* 24, (2005), 639-62.

Wilson S, He Y, Zieske J, et al. Effect of epidermal growth factor, hepatocyte growth factor, and keratinocyte growth factor, on proliferation, motility and differentiation of human corneal epithelial cells. *Exp. Eye Res.* Vol. 59, (1994), pp. 665-678.

Wirstschafter JD, Ketcham JM, Weinstock RJ et al. Mucocutaneous junction as the major source of replacement palpebral conjunctival epithelial cells. *Invest Ophthalmol Vis Sci,* Vol. 40, (1999), pp.3138-3146.

Wolff E. The anatomy of the eye and orbit. Philadelphia: Blakiston, 1948.

Won CH, Yoo HG, Kwon OS. Hair growth promoting effects of adipose tissue-derived stem cells. *J Dermatol Sci.* Vol. 57, No. 2, (2010), pp. 134-137.

Wong TY, Chakravarthy U, Klein R, et al. The natural history and prognosis of neovascular age-related macular degeneration: A systematic review of the literature and meta-analysis. *Ophthalmology.* 115, (2008), 116-26.

Wu Y, Chen L, Scott PG, Tredgett EE. *Stem Cells,* Vol. 25, (2007), pp. 2648-2659.

WuDunn D. Mechanobiology of trabecular meshwork cells. *Exp Eye Res.* 88, (2009), 718-23.

Yamagami S, Yokoo S, Mimura T et al. Distribution of Precursors in Human Corneal Stromal Cells and Endothelial Cells. *Ophthalmology,* Vol.114, No. 3 (2007 Mar), pp. 433–439

Yu S, Tanabe T, Dezawa M, et al. Effects of bone marrow stromal cell injection in an experimental glaucoma model. *Biochem Biophys Res Commun.* 344, (2006), 1071-9.

Yücel YH, Zhang Q, Gupta N, et al. Loss of neurons in magnocellular and parvocellular layers of the lateral geniculate nucleus in glaucoma. *Arch Ophthalmol.* 118, (2000), 378-84.

Zajicova A, Pokorna K, Lencova A et al. Treatment of Ocular Surface Injuries by Limbal and Mesenchymal Stem Cells Growing on Nanofiber Scaffolds. *Cell Transplantation,* Volume 19, Number 10, (2010) , pp. 1281-1290.

Zavan B, Michelotto L, Lancerotto L et al. Neural potential of a stem cell population in the adipose and cutaneous tissues. *Neurological Research,* Vol. 32, No. 1, (2010), pp. 47-54.

Zhang K, Hopkins JJ, Heier JS, et al. Ciliary neurotrophic factor delivered by encapsulated cell intraocular implants for treatment of geographic atrophy in age-related macular degeneration. *Proc Natl Acad Sci U S A.* 108, (2011), 6241-5.

Zhang Y, Klassen HJ, Tucker BA, et al. Cns progenitor cells promote a permissive environment for neurite outgrowth via a matrix metalloproteinase-2-dependent mechanism. *J Neurosci.* 27, (2007), 4499-506.

Zuk P, Zhu M, Ashjian P et al. Human Adipose Tissue Is a Source of Multipotent Stem Cells. *Molecular Biology of the Cell,* Vol. 13, (2002), pp. 4279-4295.

21

Limbal Stem Cell Transplantation and Corneal Neovascularization

Kishore Reddy Katikireddy[1] and Jurkunas V. Ula[1,2]
[1]*Schepens Eye Research Institute*
[2]*Massachusetts Eye and Ear Infirmary, Harvard Medical School*
USA

1. Introduction

The human ocular surface spans from the conjunctiva to the cornea and plays a critical role in visual perception. Cornea, the anterior portion of the eye, is transparent and provides the eye with two-thirds of its focusing power and protection of ocular integrity. The cornea consists of five main layers, namely, corneal epithelium, Bowman's layer, corneal stroma, Descemet's membrane and corneal endothelium.

The outermost layer of the cornea, which is exposed to the external environment, is the corneal epithelium. Corneal epithelial integrity and transparency are maintained by somatic stem cells (SC) that reside in the limbus. The limbus, an anatomical structure 1-2 mm wide, circumscribes the peripheral cornea and separates it from the conjunctiva (Cotsarelis *et al.*, 1989, Davanger and Evensen, 1971) (**Figure 1**). Any damage to the ocular surface by burns, or various infections, can threaten vision. The most insidious of such damaging conditions is limbal stem cell deficiency (LSCD). Clinical signs of LSCD include corneal vascularization, chronic stromal inflammation, ingrowth of conjunctival epithelium onto the corneal surface and persistent epithelial defects (Lavker *et al.*, 2004). Primary limbal stem cell deficiency is associated with aniridia and ectodermal dysplasia. Acquired limbal stem cell deficiency has been associated with inflammatory conditions (Stevens–Johnson syndrome (SJS), ocular cicatricial pemphigoid), ocular trauma (chemical and thermal burns), contact lens wear, corneal infection, neoplasia, peripheral ulcerative corneal disease and neurotrophic keratopathy (Dua *et al.*, 2000, Jeng *et al.*, 2011). Corneal stem cells and/or their niche are known to play important anti-angiogenic and anti-inflamatory roles in maintaining a normal corneal microenvironment, the destruction of which in LSCD, tips the balance toward pro-angiogenic conditions (Lim *et al.*, 2009).

For a long time, the primary treatment for LSCD has been transplantation of healthy - keratolimbal tissue from autologous, allogenic, or cadaveric sources. In the late 1990s, cultured, autologous, limbal epithelial cell implants were used successfully to improve vision in two patients with chemical injury-induced LSCD (Pellegrini *et al.*, 1997). Since then, transplantation of cultivated epithelial (stem) cells has become a treatment of choice for numerous LSCD patients worldwide. While the outcomes are promising, the variability of methodologies used to expand the cells, points to an underlying need for better standardization of *ex vivo* cultivation-based therapies and their outcome measures (Sangwan *et al.*, 2005, Ti *et al.*, 2004, Grueterich *et al.*, 2002b, Kolli *et al.*, 2010).

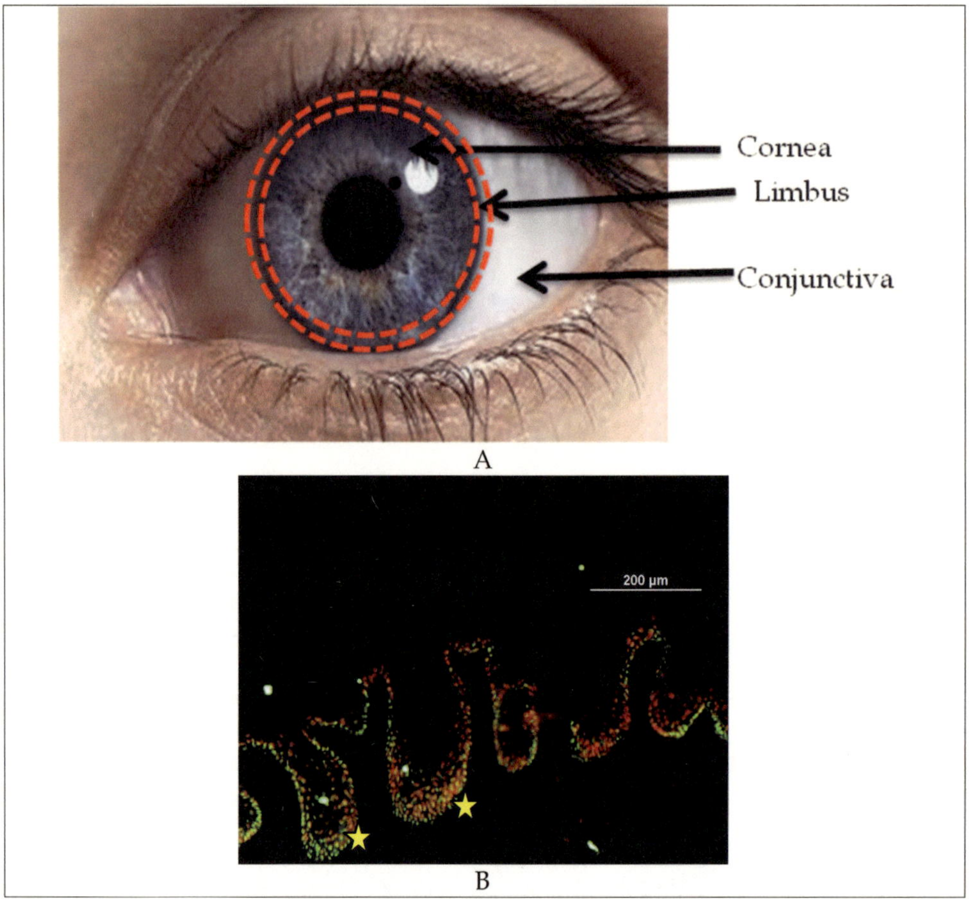

Fig. 1. (A) Photograph of the front of the human eye. The cornea, which is clear, can be seen at the front of the eye highlighted by the light reflections. The cornea is continuous with the limbus, the narrow band of tissue found encircling the cornea. The limbus is continuous with the conjunctiva and its underlying sclera, together forming the anterior portion of the eye. (B) Cross section of human limbal palisades of Vogt.

This chapter provides a general update on LSCD pathogenesis, the key role of stem cells and their niche in maintaining corneal homeostasis and avascularity and a general update on LSCD pathogenesis. An overview of different methodologies used for *ex vivo* stem cell expansion will shed some light on the future of cell-based therapy for LSCD patients.

2. What defines limbal stem cell niche

The ocular surface is anatomically composed of the entire epithelial surface of the cornea, limbus and conjunctiva, and is physically continuous with the eyelids and adenexa. The limbal epithelium, which separates the corneal epithelium from the conjunctival epithelium, is made up of a nonkeratinizing, stratified squamous epithelium, which is much thicker, up to 10 cell layers, than the central corneal epithelium. Moreover, limbal epithelium is thought to contain the source of the stem cells (SC) that serve as the source of corneal epithelial cell renewal and provide a barrier between conjunctival and corneal epithelia. These cells are known as corneal epithelial SCs or limbal stem cells (LSCs). In addition to intrinsic factors (i.e., characteristics inherent to the stem cells), extrinsic influences from the microenvironment surrounding the stem cells may also play a role in corneal cell renewal (Barros *et al.*, 1995, Morrison *et al.*, 1997, Tseng, 1996), as discussed below.

2.1 Limbal microenvironment and stem cell properties

Within the limbus, LSCs are thought to reside in a stem cell niche, which maintains them in their undifferentiated state. This stem cell niche is anatomically denoted as the palisades of Vogt. The palisades of Vogt are radial infoldings (Figure 1 B) located at the limbo-corneal junction, extending outwards of 1-2 mm from that junction (Goldberg and Bron, 1982); a anatomical feature unique to the human eye that can be used as a clinical marker indicating the presence of corneal epithelial stem cells. Davenger and Evensen (1971) and Bron (1973) reported the significant characterizations of the palisades of Vogt in 1971 (Davanger and Evensen, 1971, Bron, 1973). The authors observed heavily pigmented epithelium migrating in lines from the limbus to the central cornea in response to corneal epithelial defects.

2.1.1 Characterization and microenvironment of the palisades of Vogt

The limbal basement membrane differs from that of the cornea that undulates, with pegs of stroma extending upward and interconnecting with anchoring fibrils that link to the basement membrane (Gipson, 1989). This structure could provide resident stem cells with an adherent niche, protecting them from injury and movement within their microenvironment. Differences in the composition of the limbal and central corneal basement membranes have been observed, and these are thought to play a role in the maintenance of their respective populations of epithelial cells. Moreover the LSC niche lies beneath a number of cell layers where oxygen tension is likely to be lower. Limbal cells are in close proximity to a blood supply and have access to survival factors such as keratinocyte growth factor (KGF), IL-6 and components of basement membrane unavailable to cells of the central cornea. Interestingly, hypoxic *ex vivo* conditions have been found to produce larger, less differentiated limbal epithelial cell colonies, suggesting that low oxygen levels may induce selective proliferation of undifferentiated cells (Miyashita *et al.*, 2007).

Photomicrographic, angiographic and histological studies have demonstrated the fibrovascular nature of the palisades and the presence of "ridges of thickened epithelium" in the interpalisade zone (Goldberg and Bron, 1982, Townsend, 1991). Confocal microscopy along with scanning electron microscopy (SEM) provides the opportunity to optically section the corneal limbus and create 3D reconstructions of the tissue (Romano *et al.*, 2003, Shortt *et al.*, 2007). This approach allows identification of previously unrecognized candidates for the LSC niche, limbal crypts (LCs) and focal stromal projections (FSPs), and has significantly advanced understanding of the structure of this adult stem cell niche.

2.1.2 Asymmetric cell division

Adult SCs exist in an optimal microenvironment or 'niche' that promotes their maintenance in an undifferentiated state (Fuchs *et al.*, 2004). When SCs undergo asymmetric division, only one of the daughter cells can re-enter the niche to replenish the SC population (**Figure 2**). The other cell loses the protection of the niche and is destined to differentiate and become a transient amplifying cell (TAC). The role of the TAC is to divide at an exponential rate in

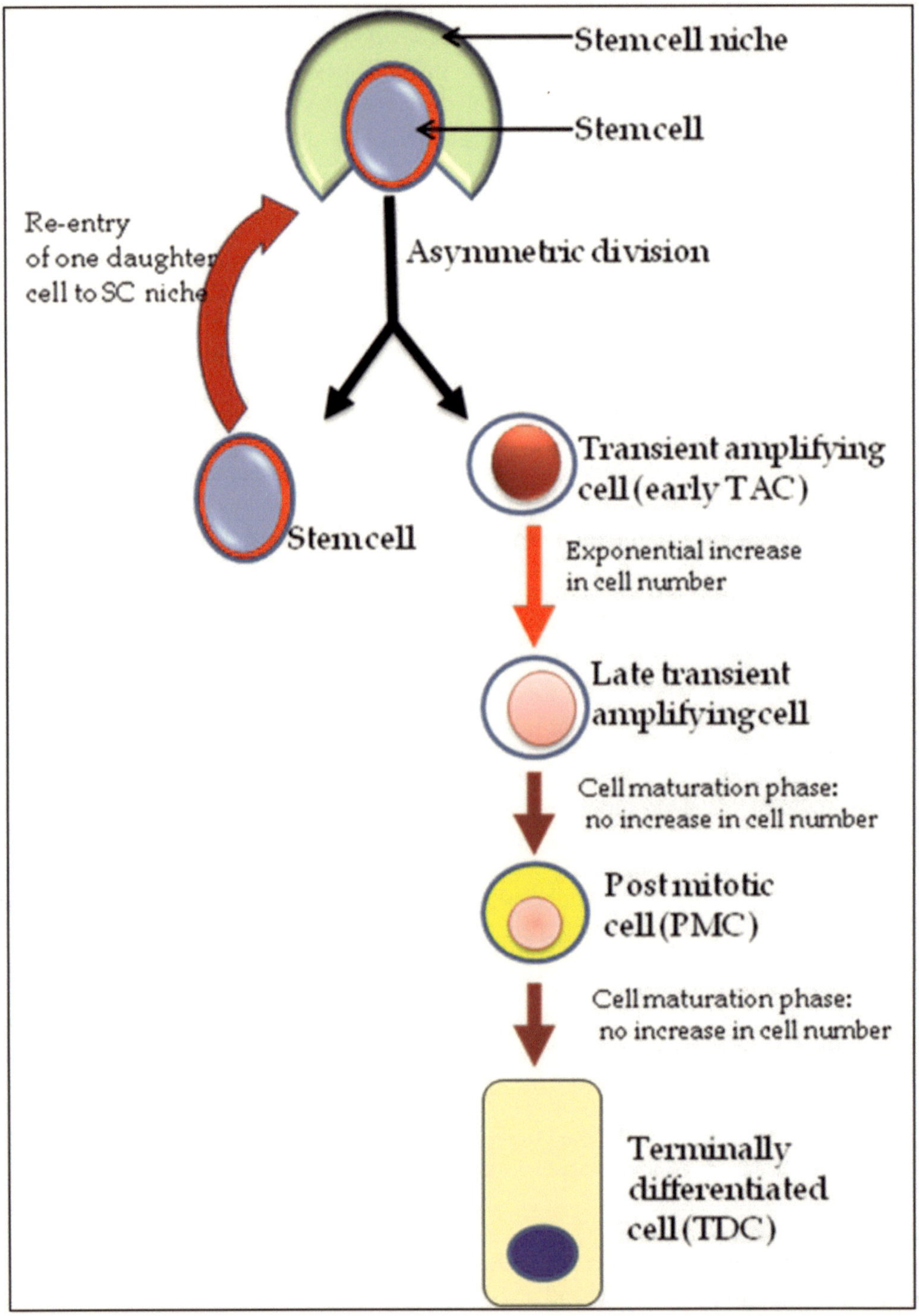

Fig. 2. Schematic diagram of the SC hierarchy. Modified from (Drug Discovery Today. Volume 15,Number 7/8.April 2010.

order to provide increased cell numbers. The ability of the TAC to multiply is limited and will eventually differentiate into a post-mitotic cell that can no longer multiply (Cotsarelis *et al.*, 1989). The PMCs are committed to cellular differentiation and mature to form terminally differentiated cells that represent the final phenotypic expression of the tissue type. These mechanisms are essential for the understanding of normal stem cell function and homeostasis. Moreover the LSC population is positioned to be influenced by a wide variety of cells. These influences include cell-to-cell contact, cell–extracellular matrix contact, and paracrine signalling factors and their receptors, and they from adjacent conjunctival epithelial cells and fibroblasts, corneal epithelium and fibroblasts, limbal blood vessels, limbal melanocytes, corneal nerves and Langerhans cells. The limbal fibroblasts in the underlying stroma are heterogeneous and express secreted protein acidic and rich in cysteine (SPARC), which may contribute to LSC adhesion (Shimmura *et al.*, 2006).

2.1.3 Cell signalling

Cytokine of the cornea and limbus implied specific differences in their microenvironments, suggesting fibroblast-mediated paracrine regulation of LSCs. For example, corneal fibroblasts release hepatocyte growth factor (HGF), whereas limbal fibroblasts secrete keratinocyte growth factor (KGF) (Sotozono *et al.*, 1994). Most importantly, epithelial cells at the basal layer of the limbus express high levels of the KGF-R, in contrast to central corneal epithelium, which expresses fewer of this receptor. KGF is a potent stimulator of proliferation in epithelial cells, and its presence in the limbus results in the proliferation of LSCs and TACs (Sotozono *et al.*, 1994). HGF is known to stimulate the migration of epithelial cells, and it has been proposed that HGF aids the migration of new TACs produced at the limbus through the action of KGF (Wilson *et al.*, 1994). The basement membrane separates the LSC from stroma and, thus modulates growth factors and cytokines involved in LSC regulation and function (Klenkler and Sheardown, 2004).

Sonic hedgehog, Wnt/β-catenin, TGF-β and Notch signalling pathways have all been implicated in niche control of stem cells; however, little is known of their potential roles in the LSC niche. Lack of Dkk2, leads to increased Wnt/β-catenin signalling in limbal stroma. This demonstrates the importance of limbal niche control over LSC differentiation during development. PAX6 expression is also lost in the corneal epithelial cells of mutant mice, suggesting the control downstream of Dkk2 (Mukhopadhyay *et al.*, 2006). PAX6 deficiency leads to aniridia, resulting in turn to impaired corneal epithelial function and eventual LSC failure, which may be due to altered niche development. A more recent study by Schlotzer-Schrehardt *et al.* found patchy immunolocalization of the laminin γ3 chain, BM40/SPARC and tenancin C, which were also found to colocalize with ABCG2/p63/K19-positive cell clusters. These factors may be involved in retaining cell stemness (Schlotzer-Schrehardt *et al.*, 2007).

3. Insult to the niche and corneal limbal stem cell deficiency

Corneal blindness is a condition currently affecting about 10 million people worldwide (Klenkler *et al.*, 2005). Corneal damage and disease can be a result of LSC deficiency, which can be heritable or acquired. Limbal stem cell deficiency (LSCD) results from the loss or dysfunction of LSC, most often because of injury or inflammation (Dua *et al.*, 2000). When the corneal limbal stem cells are depleted below a certain threshold, the cornea becomes covered by an abnormal conjunctiva-like epithelium, a process termed "conjunctivalization." As a result of LSC deficiency, neovascularization, chronic inflammation, recurrent erosions, ulceration and stromal scarring can occur, causing painful vision loss (Holland and Schwartz, 1996, Kenyon and Tseng, 1989).

LSCD is classified as either primary or secondary. Primary LSCD is characterized by the absence of identifiable external factors and an insufficient microenvironment to support the LSCs. Primary LCSD is seen in hereditary aniridia, congenital erythrokeratoderma, keratitis with multiple endocrine deficiencies and inadequate nutrition or cytokine supply, neurotrophic keratopathy, peripheral inflammation, and sclerocornea (Puangsricharern and Tseng, 1995). Dysfunction/poor regulation of stromal microenvironment of limbal epithelial SCs results in gradual loss of the SC population or TAC generation and amplification.

Secondary LSCD occurs following the destruction of LSCs by external factors such as trauma, chemical-acid or alkali or thermal injuries (Tseng, 1985), or ultraviolet and ionizing radiation (Fujishima *et al.*, 1996); systemic conditions such as Stevens-Johnson Syndrome (SJS), or ocular cicatricial pemphigoid (OCP) (Pfister, 1994).

3.1 Clinical conditions

The corneal epithelium cannot be maintained or renewed in eyes with LSCD, and leads to chronic epithelial defects. As the condition progresses, punctate epithelial keratopathy may develop and lead to severe epithelial defect. Due to decreased healing ability, corneal epithelial defects may become persistent, which can lead to stromal scarring, ulceration, and perforation (Puangsricharern and Tseng, 1995). Conjunctivalization of the cornea may occur, where the corneal epithelium is replaced with a conjunctival epithelial phenotype (Huang and Tseng, 1991). Loose intercellular connections result in stippled or late staining with fluorescein, and a lack of normal hemidesmosomal attachments at the base of corneal epithelial cells (Dua and Forrester, 1990). The cause of the LSCD often dictates whether the disease is unilateral or bilateral (i.e., affecting one eye or both).

3.1.1 Pathology of limbal stem cell deficiency and histologic, ultrastructural studies

Currently, diagnosis of LSCD is based on clinical evaluation, although clinical signs and symptoms of LSCD are nonspecific and common to several ocular surface diseases. The disease manifests as epithelial defects, chronic inflammation, keratitis, vascularization, and fibrosis, ultimately resulting in corneal blindness. In the absence of specific markers for limbal stem cells, identification of conjunctival goblet cells on the corneal surface is used to demonstrate conjunctival epithelial ingrowth or growth of epithelia on the corneal surface and to confirm the diagnosis of LSCD (Tseng, 1989, Dua and Forrester, 1990, Pfister, 1994).

Normal limbal architecture, with rows of palisades and a perilimbal vascular arcade, is usually best defined at the superior and inferior limbus. Definition of the palisades is less distinct nasally and temporally. Delicate changes such as staining of conjunctiva-derived cells on and across the limbus, which may be associated with a persistent epithelial defect, and other changes such as scarring, vascularization, or limbal hyperaemia, indicating chronic inflammation, can be seen in the early stages. SC deficiency following mild injury, a superficial or a disease process is slowly progressive, and loss of a segment of limbal epithelium may occur without significant damage to the substratum. A sheet of conjunctival/metaplastic epithelium without any notable vascularization consequently covers the cornea (Dua, 1998). Other pathological conditions resulting from LSCD are abnormal conjunctival/metaplastic epithelium; tags of loose epithelium filaments with mucus, and recurrent erosions, with unstable tear film that readily takes up fluorescein dye (Huang and Tseng, 1991).

In moderate to severe cases of SC deficiency, superficial and/or deep vascularization of the cornea occurs. Initially it is restricted to the segment with SC deficiency, but in later stages, the entire circumference may become involved. However, in moderate to severe cases of SC deficiency, epithelial cover of the denuded cornea is associated with encroachment of fibrovascular tissue of varying thicknesses (Kenyon and Tseng, 1989). In human eyes, ocular surface insult from chemical burns may go beyond damage to the limbal tissue, leading to long-term ocular surface damage. This damage may include distortion and dysfunction of proximal or distal bulbar conjunctiva, tarsal conjunctiva, and anterior orbital tissues, resulting in abnormalities in surface mucous secretion and wetting, cicatrizing of conjunctiva, symblepharon, and entropion (McCulley, 1990, McCulley *et al.*, 1983, Grant and Kern, 1955).

In summary, alkali burns are the commonest cause of severe LSCD, producing long-standing and persistent changes, like epithelial hyperplasia, fibrosis, and inflammation of the ocular surface. Presence of goblet cells on the cornea confirms the diagnosis of LSCD, but their absence does not rule it out, and correlates presence or absence with prolonged interval and squamous metaplasia of the entire surface in such cases (Fatima *et al.*, 2008).

3.1.2 Accompanying neovascularization

The cornea is avascular, but a wide variety of insults can cause capillary invasion from the limbal vascular plexus. This process of new blood vessel formation is termed as corneal neovascularization (NV) (Chang *et al.*, 2001, Takahashi *et al.*, 1999). The three major categories of corneal NV are superficial vascularization, fibrovascular pannus, and deep stromal vascularization. Superficial vascularization rarely causes a decrease in vision; however, the latter two types of corneal NV can lead to significant loss of vision if they involve the visual axis. The most frequent causes of corneal angiogenesis are injury and defective limbal stem cells.

4. Putative mechanisms of LSCD and corneal neovascularization

There is a delicate balance between angiogenic and anti-angiogenic factors in the cornea, which determines its vascularity or lack thereof. The normal cornea is avascular, and corneal epithelial stem cells may play a significant role in maintaining the angiogenic balance in favour of avascularity. Nevertheless, there are numerous pro-angiogenic factors that play key roles in pathologic corneal NV. A comprehensive review of all the factors was described by Lim *et.al*, and included key factors and their possible mechanisms of action (Lim *et al.*, 2009).

4.1 Angiogenic factors and anti-angiogenic factors

Vascular endothelial growth factor (VEGF) is a protein that acts as a mitogen for vascular endothelial cells, stimulating these cells to divide and multiply. The most important member of this group is VEGF-A, which has been shown to stimulate endothelial cell proliferation and migration. It also increases microvascular permeability. In rats with surgically induced LSCD, high levels of VEGF correlated with inflammation and corneal NV. The corneal NV was suppressed with the addition of anti-VEGF antibody (Amano *et al.*, 1998). Other members of this family include VEGF-B, VEGF-C, VEGF-D, and placenta growth factor.

Inflammatory cytokines, chemokines, and cell adhesion molecules (CAMs) play a significant role in inflammatory cell infiltration and angiogenesis. The inflammatory cytokine interleukin-1 (IL-1) interacts directly with endothelial cells to stimulate migration and proliferation. IL-6 can stimulate corneal and inflammatory cells, in a paracrine manner, to secrete VEGF (Biswas *et al.*, 2006). Tumor necrosis factor-α (TNF-α) enhances leukocyte infiltration (Yoshida *et al.*, 1997). Chemokines are a family of secreted proteins that act as chemoattractants for inflammatory cells through activation of chemokine receptors. They may also enhance endothelial cell chemotaxis and proliferation (Berger *et al.*, 1999). The chemokines IL-8 (Strieter *et al.*, 1992, Koch *et al.*, 1992) and monocyte chemoattractant-1 (MCP-1) (Goede *et al.*, 1999) have been shown to induce corneal NV in rabbit corneas. Cell adhesion molecules (CAMs) are proteins located on the cell surface involved in binding with other cells or with the extracellular matrix. They facilitate inflammatory cell infiltration into the corneal stroma by mediating the rolling, adhesion, or trans-endothelial migration of leukocytes (Vaporciyan *et al.*, 1993). Fibroblast growth factors (FGF) are a family of heparin-binding peptides that are expressed in tissues during angiogenesis, cellular differentiation, mitogenesis, and wound repair. Basic fibroblast growth factor (bFGF) is an angiogenic factor that exhibits differential binding of vascular basement membranes in newly formed corneal vessels, based on degree of maturation of the vessels (Soubrane *et al.*, 1990).

Several anti-angiogenic factors are produced by corneal epithelial cells, and a variety of vascular endothelial cell inhibitors including angiostatin, endostatin, restin, neostatin, and thrombospondins. Several of these endogenous anti-angiogenic factors are derived from the proteolysis of larger extracellular matrix proteins that have no intrinsic anti-angiogenic activity. Angiostatin, for example, is derived from the proteolysis of plasminogen. Endostatin is another example, produced by the cleavage of collagen XVIII. It inhibits proliferation (Hanai *et al.*, 2002) and migration (Dhanabal *et al.*, 1999) of endothelial cells (EC), and induces EC apoptosis. It may block the activation and catalytic activity of matrix metalloproteinases (MMPs) (Kim *et al.*, 2000) and interfere with the binding of VEGF to its receptors. Arrestin, canstatin, and tumstatin, derived from collagen IV, also inhibit the proliferation and migration of ECs and induce EC apoptosis. Restin, which is related to endostatin, is produced by the proteolytic cleavage of collagen XV and is believed to inhibit EC migration (Ramchandran *et al.*, 1999). Neostatin, derived from collagen XVIII, and thrombospondin are other inhibitors of EC proliferation.

Avascular tissue-derived factors, such as pigment epithelium-derived factor (PEDF), and angiogenic factor antagonists, such as angiopoetin-2, have been found in the cornea (Karakousis *et al.*, 2001). PEDF and angiopoetin-2 inhibit angiogenesis by inhibiting endothelial cell migration and inducing EC apoptosis.

4.1.1 Role of matrix metalloproteinases (MMPs)

Matrix metalloproteinases (MMPs) are a family of zinc-dependent proteases. They are commonly categorized as collagenases, gelatinases, matrilysins, or membrane-type MMPs, and they play a role in the proteolytic processes involved in angiogenesis (Chang *et al.*, 2001). Their action is inhibited by tissue inhibitors of matrix metalloproteinases (TIMPs). Some are involved in the cleavage process that generates the anti-angiogenic factors angiostatin and endostatin. Plasminogen activator inhibitors inhibit tissue plasminogen activator (tPA) and urokinase plasminogen activator (uPA), the activators of plasminogen, thereby preventing fibrinolysis and the physiologic breakdown of blood clots.

In response to corneal injury, lipid mediators are released from cell membranes. Platelet activating factor (PAF) is a lipid mediator that facilitates endothelial cell migration through the cornea via increased expression of VEGF (Tao *et al.*, 1996), MMP-9 (Tao *et al.*, 1995), and uPA (Tao *et al.*, 1996). Another important lipid mediator is 12(R)-HETrE, an arachidonic acid metabolite that binds to limbal endothelial cells and causes endothelial mitogenesis and neovascularization (Stoltz and Schwartzman, 1997).

4.1.2 Animal disease models

Development of experimental animal models to study ocular NV have been useful in understanding its mechanisms and pathogenesis, as well as to developing potential therapeutic modalities. For example, this basic research led to the discovery of new therapeutic targets for pathologic ocular NV (Barouch and Miller, 2006). In the field of neovascular research, the testing of angiogenic and anti-angiogenic substances relies heavily on the sensitivity and specificity of *in vivo* and *ex vivo* bioassays.

Several corneal angiogenesis models in the rabbit have been described, including direct intrastromal injections of substances like heparin-copper complex (Alessandri *et al.*, 1983), induced chemical or thermal injury (Deutsch and Hughes, 1979, Korey *et al.*, 1977), and intrastromal tumor implantation (Gimbrone *et al.*, 1974). The cornea suture model created in mice and rabbits simulate a clinical scenario and are not expensive. However, models inducing an inflammatory component have disadvantages that are surgery dependent and yield highly variable responses. Recently developed micropocket assay models are not expensive, less time consuming; however, they are surgery dependent and yield highly variable responses (Rogers *et al.*, 2007, Loughman *et al.*, 1996).

A conjunctivalized corneal epithelium lacks the smoothness and cohesion of normal corneal epithelium, making it optically inferior and prone to erosions. It is also heavily vascularized. To study inflammatory response, an important component of NV, Moromizato *et al.* produced burn and chemical cornea model lesions by corneal cauterization or local injection into the cornea of small amounts of acid (hydrochloric acid, 0.1 N) or alkali (sodium hydroxide, 0.1 N) (Moromizato *et al.*, 2000). Complex pathophysiology is a disadvantage of this model.

Animal models will continue to be needed for evaluation of safety and efficacy of combination therapies (Kim *et al.*, 2006). Each model has strengths and weaknesses; thus, in order to gain the most useful information about the disease process and responses to therapy, the best approach may be combination of different models. Therefore, search for better, more reliable and reproducible animal models should continue.

5. Treatment modalities

A number of therapeutic strategies have been adopted to treat LCSD. In patients with partial limbal stem cell deficiency, mechanical debridement of the conjunctivalized epithelium can be performed. This prevents conjunctival epithelium from crossing the limbus and allows the corneal epithelium to heal the defect. A groundbreaking report by Kenyon and Tseng, describes case studies where autologous limbal stem cell transplantation proved to be a powerful and effective surgical procedure to restore severely damaged ocular surfaces (Kenyon and Tseng, 1989). However, in later studies, where donor tissue was harvested in a "blinded" fashion, two surgical risks were identified. First, if the donor limbal grafts do not contain adequate numbers of stem cells, the surgery will either fail immediately or lack longevity. Second, if too many stem cells are removed, the donor eye will develop problems associated with LSCD (Copeland and Char, 1990, Holland and Schwartz, 1996, Holland, 1996, Frucht-Pery *et al.*, 1998).

5.1 Amniotic membrane transplantation

Human amniotic membrane (HAM) has several properties that render it extremely useful as a biomaterial for surgical purposes and it promotes epithelialization, inhibits fibrosis, has anti-inflammatory, anti-angiogenic, antimicrobial and antiviral properties, and has a high hydraulic conductivity (van Herendael *et al.*, 1978, Gomes *et al.*, 2005, Fernandes *et al.*, 2005) Moreover, HAM shows low or no immunogenicity (Akle *et al.*, 1985, Akle *et al.*, 1981). Over the past decade, HAM's abilities to reduce scarring and inflammation, and enhance wound healing, as well as its epithelialization properties, has led to an increase in its use as a biomaterial in ophthalmic surgery, particularly for ocular surface reconstruction. HAM has also been used to promote healing in corneal diseases such as chemical and thermal burns, neurotrophic ulcers, persistent epithelial defects, shield ulcers, microbial keratitis, band keratopathy and bullous keratopathy (Dua *et al.*, 2004, Tosi *et al.*, 2005a, Tosi *et al.*, 2005b). Furthermore, the use of HAM as a biomaterial has not been confined to ophthalmology—it is also widely used in general surgery and wound treatment, e.g., for burned skin, bedsores, ulcers (Faulk *et al.*, 1980), and in head and neck surgery (Zohar *et al.*, 1987). More recently, HAM has been used as a culture substrate for *ex vivo* cultured LSC transplantation.

5.1.1 Corneal limbal autografts and allografts

The current treatment for LSCD is limbal transplantation using auto- or allograft limbal tissues, each of which have their associated risks and benefits. In 1989, Kenyon and Tseng, treated LSCD by transplanting healthy limbal tissue on diseased limbus, procedure termed as conjunctival limbal autograft (CLAU) and living-related conjunctival limbal autograft (lr-CLAL) (Kenyon and Tseng, 1989). The main disadvantage of taking limbal tissue from the patient's own contralateral healthy eye or from that of a living related donor is the risk of inducing LSCD in the donor eye, if a large amount of tissue is required (Jenkins *et al.*, 1993). In bilateral cases with total LSCD, or in a one-eyed patient who develops total LSCD in the seeing eye, an allograft limbal transplant utilizing donor tissue from a cadaver or a living relative is the only option. A high risk of immune rejection is associated with such transplantation because of the vascularity of the limbus, the high immunogenic stimulus of the limbal transplant relative to the abundance of Langerhan's cells and HLA-DR antigens. Hence, an effective immunosuppressant is essential indefinitely or until the graft is viable. Moreover, the dosage of immunosuppression can be increased or decreased if it proves ineffective or causes adverse effects, respectively (Tseng *et al.*, 1998, Coster *et al.*, 1995, Santos *et al.*, 2005). For this reason, transplantation of auto-or allograft limbal tissues is not often a viable treatment option.

5.1.2 *Ex vivo* expansion of corneal limbal stem cells

A novel method of transplanting limbal stem cells is *ex vivo* expansion of LSCs. This technique is based on Rheinwald and Green's pioneering work in skin (Green *et al.*, 1977). Skin epithelial cells have been grown successfully on a feeder layer of 3T3 fibroblasts that have been rendered mitotically inactive by irradiation or treatment with mitomycin C (Rheinwald, 1980). In the late 1990s, cultured autologous limbal epithelial cells were successfully used to improve vision in two patients with chemical injury-induced LSCD. From 1–2 mm^2 limbal tissue, epithelial cells were isolated and expanded in the laboratory on tissue culture plastic, in the presence of growth-arrested 3T3 mouse fibroblast feeders, before transfer to the eye on a petrolatum gauze or a soft contact lens (Pellegrini *et al.*, 1997). This technique has theoretical advantages over conventional treatments. Its proposed advantage over CLAU and lr-CLAL is that the required size of the limbal biopsy is substantially smaller. This minimizes the risk of precipitating stem cell failure in the donor eye and provides the option of taking a further biopsy if required.

Since 1997's landmark report, a variety of culture techniques have been developed to produce contiguous epithelial cell sheets for transplantation. These techniques can be broadly defined as either explant culture in which cells migrate out from limbal tissue attached to a surface (Grueterich *et al.*, 2002b, Koizumi *et al.*, 2001a, Koizumi *et al.*, 2001b, Sangwan *et al.*, 2006) or suspension culture in which cells are released from enzymatically digested extracellular matrix before culture (Daya *et al.*, 2005, Nakamura *et al.*, 2006, Pellegrini *et al.*, 1997). The aforementioned methods have been used in studies to culture limbal epithelial cells successfully, on either a growth-arrested 3T3 fibroblast feeder layer or an amniotic membrane (AM), with varying results (Kim *et al.*, 2004, Zito-Abbad *et al.*, 2006).

5.1.3 Explant culture method

The explant culture method involves plating a whole piece of limbal tissue or limbal explant, sized approximately 1 mm by 1 mm. One variation of the explant culture system uses AM, which acts as both a substrate and a carrier for the cultured cells. To attain growth from limbal explants the limbal biopsy is placed on the basement membrane surface of the amniotic membrane and allowed to adhere to it. Once attached, the biopsy and amniotic membrane are submerged in culture medium (**Figure 3**). This contains nutrients and mitogens that stimulate LSCs to proliferate and migrate out of the biopsy and cover the surface of the AM, which occurs over a period of 14 to 28 days.

Recently, exciting discussions have led to the question of whether or not AM epithelial cells are necessary to cultivate limbal epithelial cells. Several groups have cultured limbal epithelial cells on intact AM (AM with amniotic epithelium) and have reported that these limbal epithelial cells contained slow-cycling and label-retaining characteristic cells, which did not express corneal epithelial differentiation proteins cytokeratin 3 (CK3) or cytokeratin 12 (CK12) and connexin 43. The amniotic epithelial cells were killed by cryopreservation and then removed by enzymatic digestion, chemical treatment, or physical scraping of the membrane prior to use (Koizumi *et al.*, 2001a, Koizumi *et al.*, 2001b, Nakamura *et al.*, 2004b, Sangwan *et al.*, 2005). Grueterich *et al.* demonstrated that culturing LSC on amnion with intact amniotic epithelium may result in a more stem cell-like phenotype than with de-epithelialized amnion (Grueterich *et al.*, 2002b). Their study showed that, while cells on both the substrates were well attached to the AM stroma, morphologically superior, better stratified and/or differentiated limbal cells could be cultured on denuded AM compared to those cells cultured on intact AM. As a result, these studies hypothesized that denuded AM

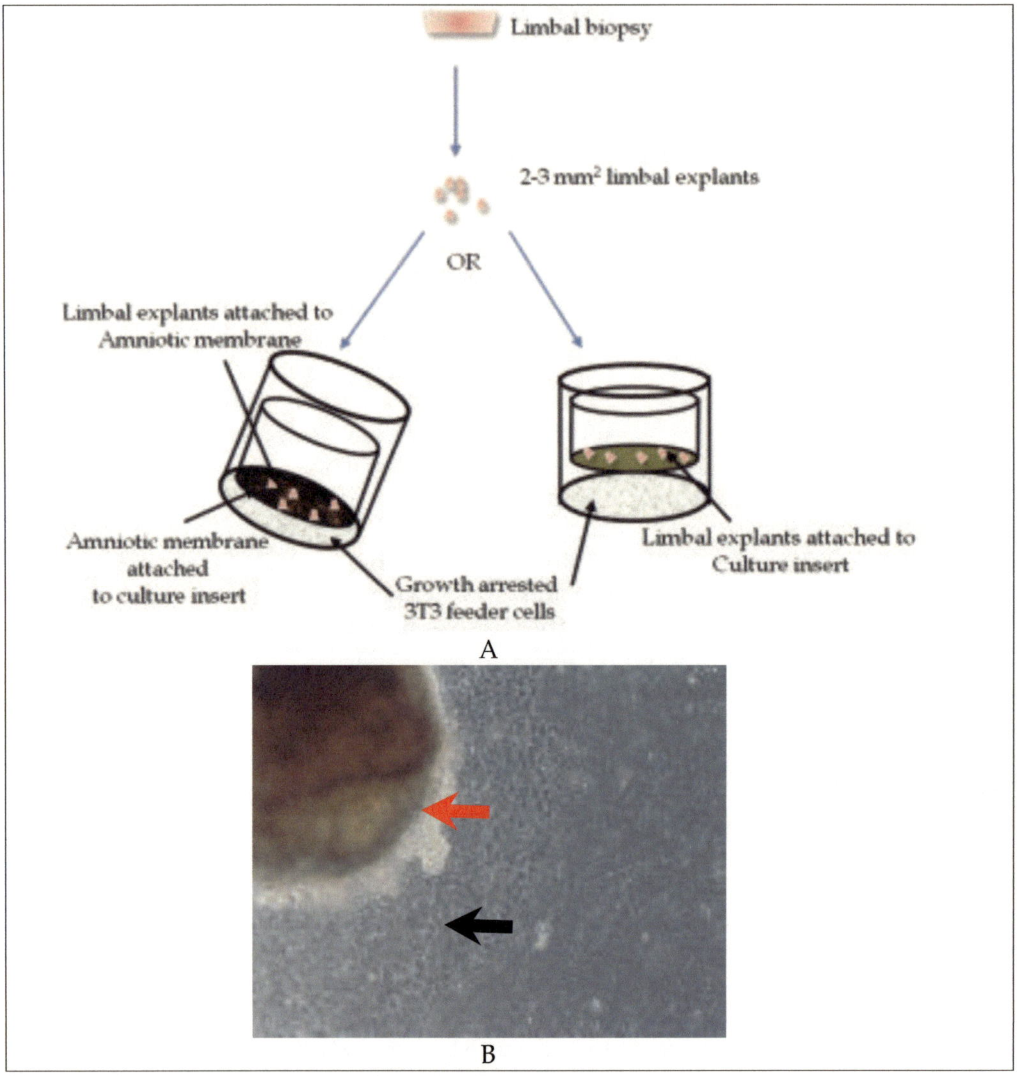

Fig. 3. The explant culture system. This method employs amniotic membrane as the carrier for cultured cells (**A**, Left); or cell culture insert (**A**-Right) as a substrate. Phase contrast micrograph reveals LSC (black arrow) migrating out of the explants (red arrow) and covering the surface of the cell culture insert over the course of 14 to 28 days (**B**).

is probably a more practical carrier for human limbal epithelial cell cultures in a cell-suspension culture system. Kolli *et al.* analyzed the outgrowths from human cadaveric limbal explants cultured on AM by dividing the explant outgrowths into three zones—inner, middle, and outer, depending on proximity to the explant. This yielded a successive decline in colony-forming efficiency (CFE), and ΔNp63α and ABCG2 expression, and an increase in expression of CK3 in zones further away from the explants. These Results support the importance of putative niche environment in maintaining the undifferentiated state of the limbal stem cells during explant outgrowth (Kolli *et al.*, 2008).

5.1.4 Suspension culture system

The cell suspension culture system employs the enzymes dispase, which digests basement membrane collagen and separates epithelial cells from the stroma, and trypsin, which separates clumps of limbal epithelial cells into a suspension of single cells (**Figure 4**). The single cells are then seeded either onto AM or onto a plastic tissue culture dish containing a feeder layer of growth-arrested 3T3 fibroblasts. Culture medium is then added, and the cells are incubated for 14 to 21 days. When confluent, the epithelial sheet is transferred to the ocular surface using either a contact lens or fibrin gel (Pellegrini *et al.*, 1997, Pellegrini *et al.*,

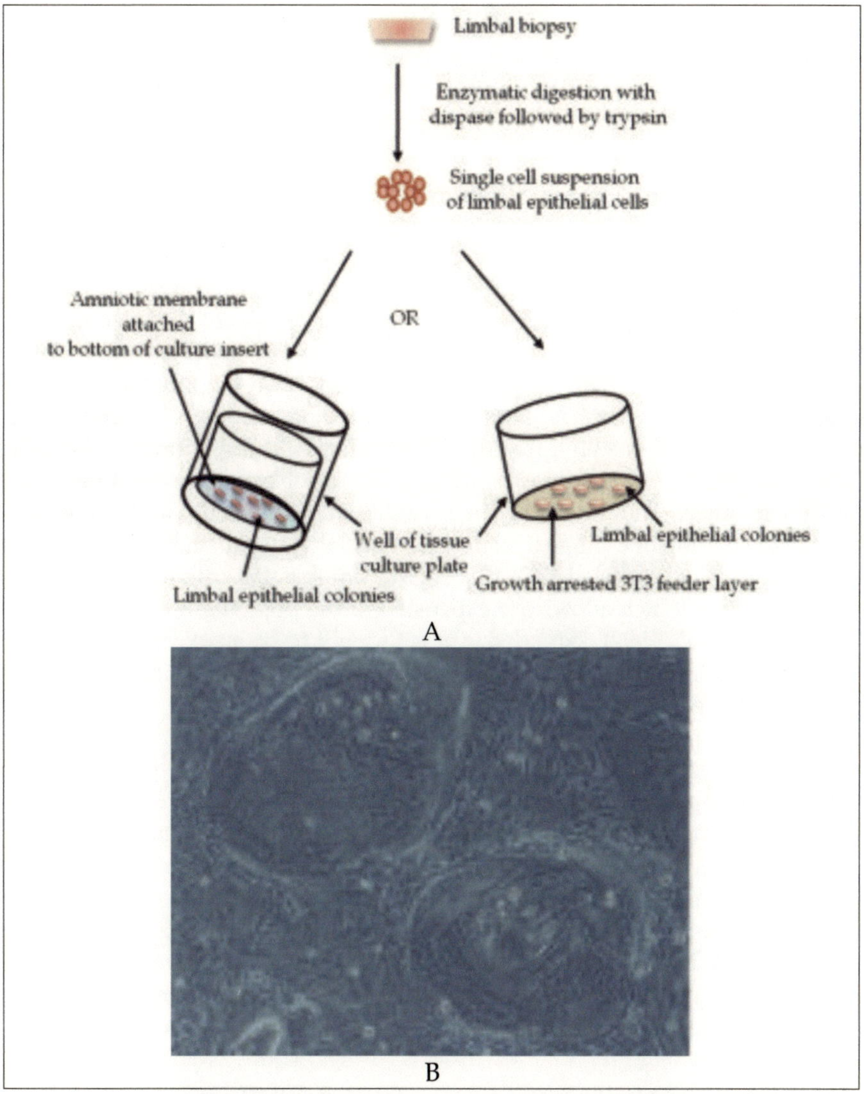

Fig. 4. The suspension culture system. This method employs the enzymes dispase and trypsin. This suspension is seeded onto either amniotic membrane (**A**, Left) or a plastic tissue culture dish that contains a feeder layer of growth-arrested 3T3 fibroblasts (**A**, Right). Phase contrast micrograph of limbal epithelial colonies on 3T3 cells (**B**).

1999). Recently, Di Girolamo *et al.* cultured limbal epithelial cells in autologous serum, on lotrafilcon A and balafilcon A contact lenses. The cells cultured on the lotrafilcon A contact lenses proliferated, migrated, showed a corneal phenotype and microvillae on the apical surface; whereas cells cultured on balafilcon A contact lenses showed no growth (Di Girolamo *et al.*, 2007).
Various substrates have been used in suspension culture systems for proliferation of limbal epithelial cells, for example, reconstituted collagen substrate membrane from HAM (Merrett *et al.*, 2008), a chitosan cross-linked collagen membrane (Auxenfans *et al.*, 2009), which has good adherence, retains epithelial morphology and supports the nontoxic nature of the membrane. Reconstituted human amniotic collagen membrane has been used to act as collagen scaffolding for the culture of limbal epithelial cells (Doillon *et al.*, 2003, Griffith *et al.*, 1999, Li *et al.*, 2005). Rama *et al.* have provided data demonstratng that the use of fibrin as a carrier supports stem cell maintenance (Rama *et al.*, 2001).

5.1.5 Variations in explant and suspension cultures

The variations in explant and cell suspension culture systems are airlifting and/or (with or without the addition) of 3T3 feeder cells. In airlifting cultures, the level of culture medium in the dish is lowered to the surface level of the epithelium, promoting stratification and differentiation of the epithelium (Ban *et al.*, 2003, Cooper *et al.*, 2004, Koizumi *et al.*, 2001a, Nakamura *et al.*, 2003b). This method can, therefore, be used for investigating the stratification and differentiation potential of cultured limbal epithelial cells with high expression of CK3 and CK12. The latter culture method, adding 3T3 feeder cells as an additional feeder layer of growth-arrested 3T3 fibroblasts in the bottom of the cell culture well is a duplex system (Miyashita *et al.*, 2008). 3T3 fibroblasts are primitive cells isolated from embryonic mice. They have a high proliferative capacity and have been used extensively in the culture of epithelial stem cells especially in skin and cornea. Both amniotic membrane and growth-arrested 3T3 fibroblasts inhibit the differentiation of corneal epithelial cells *ex vivo*, which allows the expansion of the population of LSCs (Grueterich *et al.*, 2003, Pellegrini *et al.*, 1999). Denuded AM is thought to be an excellent substrate for the corneal limbal cell culture used for ocular surface reconstruction, even though AM epithelial cells contain various growth factors with 3T3 feeder cells. This theory, however, remains to be fully investigated.
The variation in suspension culture techniques are use of novel temperature-responsive culture surfaces. Temperature-responsive polymers chemically immobilized to thin films on cell-culture surfaces facilitate cell adhesion and growth of cells in normal culture conditions, at 37°C. Temperatures below 30°C can reversibly alter cell hydration properties with respect to temperature, hydration and swelling—prompting complete detachment of adherent cells without the use of proteolytic enzymes or treatment with EDTA (Nishida *et al.*, 2004b, Nishida *et al.*, 2004a). The advantage of this system is that no enzymes are needed to free the epithelial sheet from the culture membrane; however, application of temperature-responsive polymers in a clinical setting remains questionable for safety and efficacy.

6. *Ex vivo* expansion of alternative sources of epithelium

Although adult stem cells hold considerable promise for the treatment of a number of diseases in regenerative medicine, the second major obstacle has been to obtain sufficient number of autologous or allogeneic stem cells.

6.1 Oral mucosal cells

In an attempt to overcome the problems inherent in the transplantation of allogeneic tissue, there has been recent interest in the possibility of using alternative autologous epithelial cells. There are several potential sources of non-keratinizing, stratified squamous epithelium in the adult human, including oral mucosal, conjunctival, nasal, esophageal, vaginal and rectal epithelia. The main concept is to use the smooth surface of the oral mucosa, with its stem cell properties, to reconstruct the ocular surface. Oral mucosae are thought to be at a lower stage of differentiation than epidermal keratinocytes (Collin *et al.*, 1992, Schermer *et al.*, 1986) because they divide rapidly and can be maintained in culture for prolonged periods without keratinization (Hata *et al.*, 1995). Various antimicrobial peptides (AMPs) are known to be present on the epithelial cells of ocular and oral surfaces (Haynes *et al.*, 1999). Moreover, keratin 3 is expressed by both corneal epithelium (Sangwan *et al.*, 2003b) and oral mucosa (Collin *et al.*, 1992, Juhl *et al.*, 1989), suggesting that gene expression in oral and corneal epithelium may be similar. Wet-surfaced epithelia, produce a group of highly glycosylated, protective membrane glycoproteins termed mucins (Gipson, 2007). Although the function of these mucins in the oral cavity remains to be elucidated, it is possible that they contribute to the epithelial protective mucin layer and act as receptors initiating one or more intracellular signal transduction pathways (Argueso *et al.*, 2003). At the ocular surface, at least three membrane-associated mucins (MUC1, -4, and -16) and two secreted mucins (MUC5 and -7) are expressed (Hori *et al.*, 2007).

More extensive studies have been performed to check the feasibility of using cultivated oral mucosal epithelium (COME) for this purpose, as it is readily available and could be harvested without invasive surgery. A small biopsy of the patient's oral mucosa is harvested, connective tissue is dissected and, after enzymatic digestion, an epithelial cell suspension is prepared. The oral epithelial cells are cultivated on the substrate with a feeder layer. The sheet of cultivated cells is then transplanted onto the diseased ocular surface (**Figure 5**). These studies suggest that oral mucosal epithelium is a feasible alternative for

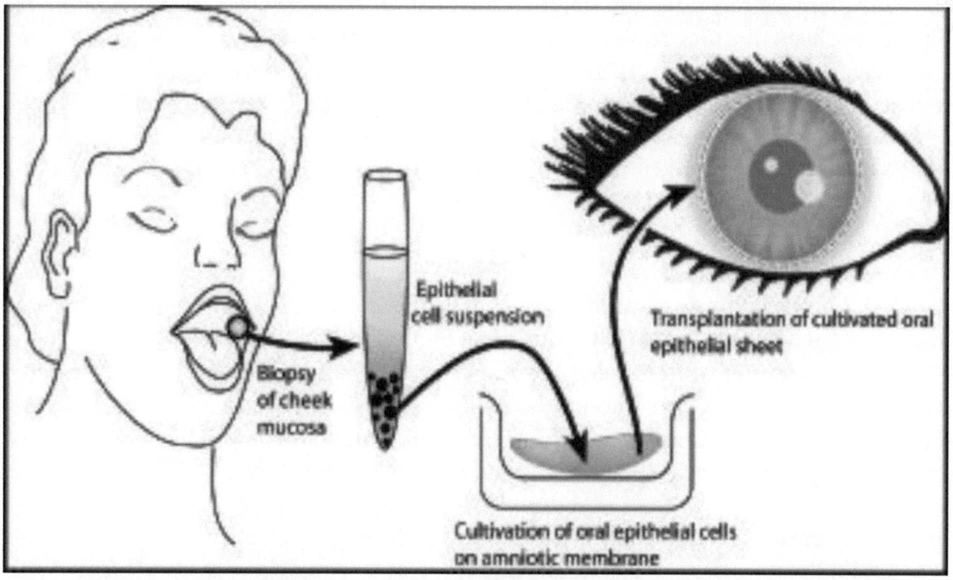

Fig. 5. Systematic representation of cultivated oral mucosal epithelium transplantation. *Adopted from (Lim et al., 2009).*

allogenous limbal transplants (Nakamura *et al.*, 2003b, Nakamura *et al.*, 2004a, Inatomi *et al.*, 2006, Nishida *et al.*, 2004b, Hayashida *et al.*, 2005). Oral mucosal epithelium cultured on human amniotic membrane with the help of feeder cells has been characterized extensively, and has been used to reconstruct the ocular surface in rabbits (Nakamura *et al.*, 2003a) as well as humans with chemical injury and SJS (Nakamura *et al.*, 2004a). In the lengthiest study reported so far, cultured oral mucosal epithelial cells were transplanted into patients with LSCD and followed up with for up to 34 months (Inatomi *et al.*, 2006) where the followups continued until tissue became viable.

The main advantages of using oral mucosa are the absence of tissue rejection and, therefore, avoidance of immunosuppression; easy access to the oral cavity; inconspicuous location of the resultant small scar, and repeatability. This treatment method has not, however, provided satisfactory long-term results. The oral mucosal epithelial cell grafts seldom transdifferentiate to a corneal epithelium phenotype, e.g., expression of keratin 12 (Krt12) (Inatomi *et al.*, 2006). This new approach, however, provides an exciting possibility for treatment of this difficult group of patients with blindness from bilateral total LSCD and warrants further study.

6.1.2 Dental pulp stem cells

In an attempt to prevent rejection, isolated human immature dental pulp stem cells (hIDPSC) from deciduous teeth have been used as an autologous alternative epithelium source for LSCD treatment or corneal reconstruction. hIDPSCs are known to express both mesenchymal stem cell markers (SH2, SH3 and SH4) and human embryonic stem cell markers (OCT 4, NANOG, SSEA-3 and SSEA-4) (Gomes *et al.*, 2010). Furthermore, hIDPSCs have a normal karyotype and show the capacity for multilineage differentiation into neurons, smooth and skeletal muscle, cartilage, bone, and other cell types *ex vivo* and *in vivo* (Kerkis *et al.*, 2006, Lavagnolli *et al.*, 2009). Limited studies have been performed to test dental pulp stem cells as potential treatment for LSCD. Monteiro *et al.* demonstrated, using immunohistochemistry and reverse transcription polymerase chain reaction, that hIDPSCs express markers in common with LSCs, such as ABCG2, integrin β1, vimentin, p63, connexin 43 and cytokeratins 3/12. These have been shown to be capable of reconstructing the eye surface after induction of unilateral total LSCD in rabbits (Monteiro *et al.*, 2009). Further studies are required to understand the long-term outcome of LSCD treated using hIDPSCs.

6.1.3 Hair follicle stem cells

Hair has the potential to regenerate from stem cells located at the lower part of the follicle called the bulge (Oshima *et al.*, 2001). Additionally, hair follicle stem cells (HFSCs) can give rise to epidermis and sebaceous gland tissues. Recently, Blazejewska *et al.* demonstrated that murine HFSCs can transform into epithelia-like cells when cultured in limbal fibroblast-derived medium (Blazejewska *et al.*, 2009). Although, these findings need further substantiation using *in vivo* functional studies in animal models, they provide the first step towards the design of protocols that could use human autologous hair follicle stem cells to replace corneal epithelium in therapeutic applications. Due to their multipotency, easy accessibility, and high proliferation rate *ex vivo*, hair follicles are an attractive source of autologous adult stem cells and a promising therapeutic tool for ocular surface reconstruction.

6.1.4 Embryonic stem cells

Embryonic stem cells (ESCs) are pluripotent (i.e., they can give rise to any cell type in the body) and are derived from the inner cell mass of a blastocyst. Differentiation of ESCs to a corneal epithelial lineage could be achieved by replication of the LSC niche environment. Transplantation of epithelial cells derived from mouse ESCs have been used to reconstruct chemically injured mouse corneas (Homma *et al.*, 2004). Moreover, human ESCs express cornea-like epithelial markers such as CK3 and CK12 when cultured in conditions similar to those of the limbal stem cell niche. *Ex vivo* replication of the LSC niche environment has been achieved by culturing hESCs on an extracellular matrix of collagen IV and fed with medium conditioned with limbal fibroblasts (Ahmad *et al.*, 2007). The differentiated human ESCs demonstrate phenotypic differences to cultured human limbal epithelial cells, such as differences in size and cilia length, which are characteristically much longer than those of limbal epithelial cells. These data suggest that some differences exist between LSCs and hESC-derived corneal progenitors (Ahmad *et al.*, 2007). The translation of this approach to human therapeutic use requires further work to overcome problems associated with functionality, complete hESC differentiation and immune rejection as well as ethical concerns.

7. Future considerations for cultivated epithelial transplants

Ex vivo expansion of limbal epithelium has been performed by several unrelated groups in a number of countries and used to treat a variety of ocular surface disorders that are thought to be the result of limbal stem cell failure. Despite a substantial number of experimental models of this technique and an ever-growing body of laboratory data on limbal epithelial stem cell biology, scientific basis for this procedure is poorly understood. Future considerations in the field of limbal epithelial cell culture and transplantation methods are discussed below.

7.1 Culture methodologies

The culture of limbal epithelium is essential for furthering our understanding of limbal stem cell biology. *Ex vivo* expansion of LSCs for the treatment of LSCD is one of the more recent stem cell treatments available in the field of regenerative medicine. Although the technique of *ex vivo* expansion of LSCs is still in its infancy, it is starting to be used by several groups for clinical purposes, and different approaches have been attempted. There are many varying techniques for culturing limbal epithelial cells, such as different media and sera; culture with 3T3 fibroblasts or amniotic membrane, or both; removal of amniotic membrane epithelial cells from the membrane, or not; use of explant or suspension methods; and airlifting of the culture, or not (Shortt *et al.*, 2007, Grueterich *et al.*, 2003). Whichever culture method is used, the end result, often after a period of weeks, is a cultivated epithelial sheet composed of LSCs. The ease of explanting culture technique could be adapted without initial dispase treatment for transplantation studies. However, further studies on the interaction between epithelial progenitor and autologous stromal cells *ex vivo,* and complete characterization of limbal cells in this culture system, including cell junction proteins, are required.

7.1.2 Autologous sources and xenobiotic-free conditions

With growing concerns regarding the potential transmission of opportunistic agents such as prions and animal viruses, it would be preferable to culture cells for human transplantation under xenobiotic-free conditions. In limbal corneal epithelial cultures, the use of autologous human serum in the medium and/or human amniotic membrane would be appropriate as it would reduce the need for animal cells or products in the culture (Kolli *et al.*, 2010, Nakamura *et al.*, 2004a, Nakamura *et al.*, 2006). To avoid contamination from xeno-feeder layers, different human feeder layers have also been developed and analyzed (Chen *et al.*, 2007). Moreover, cells have been cultured on feeder layers during the proliferative phase and further separated using robotic technology (Schneider *et al.*, 2008). Limbal epithelial cells grown with MRC–5 human embryonic fibroblasts can sustain the stem cell phenotype (Notara *et al.*, 2007). The successful use of autologous serum–derived oral epithelial equivalent to treat severe ocular surface disease was reported in 2006 (Ang *et al.*, 2006, Nakamura *et al.*, 2006). This represents an important advance in the pursuit of completely autologous xenobiotic-free bioengineered ocular equivalents for clinical transplantation.

In conclusion, a specific culture system that mimics natural tissue architecture and its niche closely by enabling formation of an autofeeder layer, thereby eliminating the need for xeno-feeder layers for *ex vivo* expansion of epithelial progenitor cells is important.

7.1.3 Combination techniques with anti-angiogenic treatments

Angiogenic and anti-angiogenic cytokines are involved in cell migration and differentiation in wound healing and tumor progression (even in the absence of angiogenesis). The ever-increasing use of anti-angiogenic therapy for corneal neovascularization brings great promise for the treatment of a wide variety of corneal conditions, including LSCD. TGF-β, IL-1, and VEGF are the key pro-angiogenic factors involved in the process of angiogenesis; and are targets for anti-angiogenic therapies. The ultimate goal for research on corneal angiogenesis is to discover effective treatment for pathological corneal neovascularization; however, to date, most anti-angiogenic therapies remain experimental.

Angiostatin, a 38 kDa proteolytic fragment of plasminogen, is a potent anti-angiogenic factor. Recombinant angiostatin has been used successfully to suppress tumor growth and metastasis in animal model systems (Shin *et al.*, 2000, Ambati *et al.*, 2002). Gabison and associates have demonstrated the involvement of angiostatin in corneal avascularity after wounding. They confirmed that angiostatin-like molecules are expressed in the corneal epithelium and in cultured corneal epithelial cells (Gabison *et al.*, 2004). Corneal neovascularization was observed after excimer laser keratectomy when anti-angiostatin antibodies were injected into the cornea; this was significantly higher than when plasmin B chain antibodies were injected. These studies suggest that angiostatin may contribute to the maintenance of corneal avascularity after excimer laser keratectomy.

The corneal epithelium may be the source of anti-angiogenic molecules. It has long been demonstrated that corneal epithelium inhibits angiogenesis (Kaminski and Kaminska, 1978). Ferreras and associates demonstrated the generation of endostatin-like fragments by cleaving collagen XVIII with MMP-2, -3, -9, -12, -13, or -14 (Ferreras *et al.*, 2000). Upregulation of matrix-derived anti-angiogenic factors such as endostatin (Kato *et al.*, 2003) and restin (Saika *et al.*, 2004), along with increased anti-inflammatory factor IL-1ra may play an important role in LSC and AM transplantation-mediated anti-angiogenic effect. Recently, Cursiefen *et al.* demonstrated a critical mechanism that contributed to corneal avascularity by VEGF receptor 3, which is normally present on lymphatic and proliferating blood vascular endothelium, is strongly constitutively expressed by corneal epithelium and is mechanistically responsible for suppressing inflammatory corneal angiogenesis (Cursiefen *et al.*, 2006). Knowledge gained from using epithelia–matrix interaction to regulate corneal angiogenesis will enable us to optimize the anti-angiogenic effect of the cultivated cells like oral mucosal epithelial cells or mesehchymal stem cells for future ocular surface reconstruction.

8. Clinical trials and outcomes: present and future

LSCD by any cause may result in poor corneal epithelialization, persistent epithelial defects, corneal vascularization, corneal scarring, and so-called conjunctivalization of the cornea. These problems may in turn lead to decreased vision, ocular discomfort and pain, as well as an unstable ocular surface. Limbal stem cells may be partially or totally depleted, resulting in varying degrees of stem cell deficiency and resulting abnormalities on the corneal surface. The first use of *ex vivo* expanded LSCs for the treatment of LSCD in human subjects was described by Pellegrini and co-workers, who used a culture system of LSCs grown on mouse J2-3T3 fibroblasts with fetal calf serum (FCS) supplemented media (Pellegrini *et al.*, 1997). Since then, several studies have been reported on the transplantation of *ex vivo* cultured LSCs to treat LSCD (Schwab *et al.*, 2000, Tsai *et al.*, 2000, Koizumi *et al.*, 2001b, Grueterich *et al.*, 2002a). Schwab *et al.* reported the results of treating 19 eyes of 18 patients (Schwab, 1999). In 2003, Sangwan *et al.* reported the largest study to date, which enrolled 125 patients (Sangwan *et al.*, 2003b). In this study *ex vivo* cultured limbal and conjunctival epithelial cells were grown on AM. In 2006, Sangwan *et al.* published a more detailed report on clinical outcomes for 88 eyes of 86 patients (Sangwan *et al.*, 2006). In 2001, Rama *et al.* reported treatment outcomes for 18 eyes of 18 patients (Rama *et al.*, 2001). The same group recently reported a 10 year follow-up with permanent restoration of transparent corneal epithelium in 76.6% of eyes, with few 63 instances of inflammation out of 113 eyes treated (Rama *et al.*, 2010). The outcome of cultured limbal stem cells transplants differed significantly depending on whether the transplanted cultures contained more than 3% p63-bright holoclone-forming stem cells or 3% or less – the success rates were 78% with the larger number of stem cells and 11% increase with the smaller number of p63 cells (Rama *et al.*, 2001, Rama *et al.*, 2010), suggesting the importance of a high percentage of p63 in cultured limbal stem cells. The remaining studies each treated between 2 and 14 patients (Kolli *et al.*, 2010, Grueterich *et al.*, 2002b, Nakamura *et al.*, 2004a, Sangwan *et al.*, 2003a).

In total, current studies reported data to enable analysis of outcomes in 506 eyes (Baylis *et al.*, 2011) treated using *ex vivo*, cultured LSC transplants and 27 eyes treated with *ex vivo* cultured oral mucosal epithelium. To date, the limited periods (mean follow-up was 16.74 months, median 14 months, range 6 to 29.5 months), the use of subjective outcome measures, and the absence of a reliable method for detecting transplanted (stem) cells shows that a long term success rate has yet to be established.

Ex vivo expansion and transplantation of limbal epithelium have been performed by several unrelated groups in a number of countries, with uncertainty in the limbal stem cell cultivation, with different variables, including source of donor limbal tissue, culture method (e.g., suspension versus explant culture), culture composition (e.g., 3T3 fibroblasts, amniotic membrane and fibrin), and variation in culture time. Overall the success rate for cultured limbal epithelial transplantation is 76%. This is based on clinical restoration of the corneal epithelium. It is interesting that, despite the different methodologies employed, the success rates and number of significant clinical outcomes are remarkably high. It would, therefore, appear that, as long as viable limbal epithelial stem cells are transferred, the method used to achieve this is relatively unimportant.

The future directions in LSC transplantation involve modifications of existing technologies to allow improved safety and efficacy of techniques at hand. The studies on the molecular mechanism of limbal stem cells and their niche will enhance our knowledge of how cell therapy can be further modified to closely mimic *in vivo* conditions. Identification of key niche factors controlling limbal stem cell behaviour would allow these conditions to replicate *ex vivo* and, thus, make the process of culturing limbal stem cells safer and more efficient. Newly developed transplantation techniques using tissue-engineered epidermal adult stem cells, immature dental pulp stem cells (Gomes *et al.*, 2010), and hair follicle bulge-derived stem cells (Blazejewska *et al.*, 2009) were reportedly successful for the reconstruction of corneal epithelium in an animal model of severe LSCD. The recombinant human cross-linked collagen scaffold and a Food and Drug Administration-approved contact lens are also promising new techniques for successfully achieving ocular surface reconstruction (Di Girolamo *et al.*, 2007). There are main challenges in improving the established techniques, in elimination of animal-derived products and minimization of allogeneic human tissue use. Since cell-based therapies are being scrutinized by ever-increasing regulatory requirements, many investigators that have been successfully performing ex vivo cultivation in the past are reformulating their methodologies to conform to these guidelines (Daniels *et al.*, 2006, Kolli *et al.*, 2010). The future of this methodology lies in the standardization of the stem cell preparation and cultivation techniques with clearly defined endpoints and outcome measures of treatment success.

9. Acknowledgment

We would like to thank Ms. Sandhya M. Sharma (Schepens Eye Research Institute, Harvard University) for providing human limbus cross section in Figure 1 (B).

10. References

Ahmad, S., Stewart, R., Yung, S., et al. (2007) Differentiation of human embryonic stem cells into corneal epithelial-like cells by in vitro replication of the corneal epithelial stem cell niche, *Stem Cells*, 25, 1145-55.

Akle, C., McColl, I., Dean, M., et al. (1985) Transplantation of amniotic epithelial membranes in patients with mucopolysaccharidoses, *Exp Clin Immunogenet*, 2, 43-8.

Akle, C. A., Adinolfi, M., Welsh, K. I., et al. (1981) Immunogenicity of human amniotic epithelial cells after transplantation into volunteers, *Lancet*, 2, 1003-5.

Alessandri, G., Raju, K. and Gullino, P. M. (1983) Mobilization of capillary endothelium in vitro induced by effectors of angiogenesis in vivo, *Cancer Res*, 43, 1790-7.

Amano, S., Rohan, R., Kuroki, M., et al. (1998) Requirement for vascular endothelial growth factor in wound- and inflammation-related corneal neovascularization, *Invest Ophthalmol Vis Sci*, 39, 18-22.

Ambati, B. K., Joussen, A. M., Ambati, J., et al. (2002) Angiostatin inhibits and regresses corneal neovascularization, *Arch Ophthalmol*, 120, 1063-8.

Ang, L. P., Nakamura, T., Inatomi, T., et al. (2006) Autologous serum-derived cultivated oral epithelial transplants for severe ocular surface disease, *Arch Ophthalmol*, 124, 1543-51.

Argueso, P., Spurr-Michaud, S., Russo, C. L., et al. (2003) MUC16 mucin is expressed by the human ocular surface epithelia and carries the H185 carbohydrate epitope, *Invest Ophthalmol Vis Sci*, 44, 2487-95.

Auxenfans, C., Builles, N., Andre, V., et al. (2009) [Porous matrix and primary-cell culture: a shared concept for skin and cornea tissue engineering], *Pathol Biol (Paris)*, 57, 290-8.

Ban, Y., Cooper, L. J., Fullwood, N. J., et al. (2003) Comparison of ultrastructure, tight junction-related protein expression and barrier function of human corneal epithelial cells cultivated on amniotic membrane with and without air-lifting, *Exp Eye Res*, 76, 735-43.

Barouch, F. C. and Miller, J. W. (2006) Potential future targets for treating ocular neovascularization, *Ophthalmol Clin North Am*, 19, 401-9.

Barros, E. J., Santos, O. F., Matsumoto, K., et al. (1995) Differential tubulogenic and branching morphogenetic activities of growth factors: implications for epithelial tissue development, *Proc Natl Acad Sci U S A*, 92, 4412-6.

Baylis, O., Figueiredo, F., Henein, C., et al. (2011) 13 years of cultured limbal epithelial cell therapy: a review of the outcomes, *J Cell Biochem*, 112, 993-1002.

Berger, O., Gan, X., Gujuluva, C., et al. (1999) CXC and CC chemokine receptors on coronary and brain endothelia, *Mol Med*, 5, 795-805.

Biswas, P. S., Banerjee, K., Kinchington, P. R. and Rouse, B. T. (2006) Involvement of IL-6 in the paracrine production of VEGF in ocular HSV-1 infection, *Exp Eye Res*, 82, 46-54.

Blazejewska, E. A., Schlotzer-Schrehardt, U., Zenkel, M., et al. (2009) Corneal limbal microenvironment can induce transdifferentiation of hair follicle stem cells into corneal epithelial-like cells, *Stem Cells*, 27, 642-52.

Bron, A. J. (1973) Vortex patterns of the corneal epithelium, *Trans Ophthalmol Soc U K*, 93, 455-72.

Chang, J. H., Gabison, E. E., Kato, T. and Azar, D. T. (2001) Corneal neovascularization, *Curr Opin Ophthalmol*, 12, 242-9.

Chen, Y. T., Li, W., Hayashida, Y., et al. (2007) Human amniotic epithelial cells as novel feeder layers for promoting ex vivo expansion of limbal epithelial progenitor cells, *Stem Cells*, 25, 1995-2005.

Collin, C., Ouhayoun, J. P., Grund, C. and Franke, W. W. (1992) Suprabasal marker proteins distinguishing keratinizing squamous epithelia: cytokeratin 2 polypeptides of oral masticatory epithelium and epidermis are different, *Differentiation*, 51, 137-48.

Cooper, L. J., Fullwood, N. J., Koizumi, N., et al. (2004) An investigation of removed cultivated epithelial transplants in patients after allocultivated corneal epithelial transplantation, *Cornea*, 23, 235-42.

Copeland, R. A., Jr. and Char, D. H. (1990) Limbal autograft reconstruction after conjunctival squamous cell carcinoma, *Am J Ophthalmol*, 110, 412-5.

Coster, D. J., Aggarwal, R. K. and Williams, K. A. (1995) Surgical management of ocular surface disorders using conjunctival and stem cell allografts, *Br J Ophthalmol*, 79, 977-82.

Cotsarelis, G., Cheng, S. Z., Dong, G., et al. (1989) Existence of slow-cycling limbal epithelial basal cells that can be preferentially stimulated to proliferate: implications on epithelial stem cells, *Cell*, 57, 201-9.

Cursiefen, C., Chen, L., Saint-Geniez, M., et al. (2006) Nonvascular VEGF receptor 3 expression by corneal epithelium maintains avascularity and vision, *Proc Natl Acad Sci U S A*, 103, 11405-10.

Daniels, J. T., Secker, G. A., Shortt, A. J., et al. (2006) Stem cell therapy delivery: treading the regulatory tightrope, *Regen Med*, 1, 715-9.

Davanger, M. and Evensen, A. (1971) Role of the pericorneal papillary structure in renewal of corneal epithelium, *Nature*, 229, 560-1.
Daya, S. M., Watson, A., Sharpe, J. R., et al. (2005) Outcomes and DNA analysis of ex vivo expanded stem cell allograft for ocular surface reconstruction, *Ophthalmology*, 112, 470-7.
Deutsch, T. A. and Hughes, W. F. (1979) Suppressive effects of indomethacin on thermally induced neovascularization of rabbit corneas, *Am J Ophthalmol*, 87, 536-40.
Dhanabal, M., Ramchandran, R., Waterman, M. J., et al. (1999) Endostatin induces endothelial cell apoptosis, *J Biol Chem*, 274, 11721-6.
Di Girolamo, N., Chui, J., Wakefield, D. and Coroneo, M. T. (2007) Cultured human ocular surface epithelium on therapeutic contact lenses, *Br J Ophthalmol*, 91, 459-64.
Doillon, C. J., Watsky, M. A., Hakim, M., et al. (2003) A collagen-based scaffold for a tissue engineered human cornea: physical and physiological properties, *Int J Artif Organs*, 26, 764-73.
Dua, H. S. (1998) The conjunctiva in corneal epithelial wound healing, *Br J Ophthalmol*, 82, 1407-11.
Dua, H. S. and Forrester, J. V. (1990) The corneoscleral limbus in human corneal epithelial wound healing, *Am J Ophthalmol*, 110, 646-56.
Dua, H. S., Gomes, J. A., King, A. J. and Maharajan, V. S. (2004) The amniotic membrane in ophthalmology, *Surv Ophthalmol*, 49, 51-77.
Dua, H. S., Saini, J. S., Azuara-Blanco, A. and Gupta, P. (2000) Limbal stem cell deficiency: concept, aetiology, clinical presentation, diagnosis and management, *Indian J Ophthalmol*, 48, 83-92.
Fatima, A., Iftekhar, G., Sangwan, V. S. and Vemuganti, G. K. (2008) Ocular surface changes in limbal stem cell deficiency caused by chemical injury: a histologic study of excised pannus from recipients of cultured corneal epithelium, *Eye (Lond)*, 22, 1161-7.
Faulk, W. P., Matthews, R., Stevens, P. J., et al. (1980) Human amnion as an adjunct in wound healing, *Lancet*, 1, 1156-8.
Fernandes, M., Sridhar, M. S., Sangwan, V. S. and Rao, G. N. (2005) Amniotic membrane transplantation for ocular surface reconstruction, *Cornea*, 24, 643-53.
Ferreras, M., Felbor, U., Lenhard, T., et al. (2000) Generation and degradation of human endostatin proteins by various proteinases, *FEBS Lett*, 486, 247-51.
Frucht-Pery, J., Siganos, C. S., Solomon, A., et al. (1998) Limbal cell autograft transplantation for severe ocular surface disorders, *Graefes Arch Clin Exp Ophthalmol*, 236, 582-7.
Fuchs, E., Tumbar, T. and Guasch, G. (2004) Socializing with the neighbors: stem cells and their niche, *Cell*, 116, 769-78.
Fujishima, H., Shimazaki, J. and Tsubota, K. (1996) Temporary corneal stem cell dysfunction after radiation therapy, *Br J Ophthalmol*, 80, 911-4.
Gabison, E., Chang, J. H., Hernandez-Quintela, E., et al. (2004) Anti-angiogenic role of angiostatin during corneal wound healing, *Exp Eye Res*, 78, 579-89.
Gimbrone, M. A., Jr., Cotran, R. S., Leapman, S. B. and Folkman, J. (1974) Tumor growth and neovascularization: an experimental model using the rabbit cornea, *J Natl Cancer Inst*, 52, 413-27.
Gipson, I. K. (1989) The epithelial basement membrane zone of the limbus, *Eye (Lond)*, 3 (Pt 2), 132-40.

Gipson, I. K. (2007) The ocular surface: the challenge to enable and protect vision: the Friedenwald lecture, *Invest Ophthalmol Vis Sci*, 48, 4390; 4391-8.

Goede, V., Brogelli, L., Ziche, M. and Augustin, H. G. (1999) Induction of inflammatory angiogenesis by monocyte chemoattractant protein-1, *Int J Cancer*, 82, 765-70.

Goldberg, M. F. and Bron, A. J. (1982) Limbal palisades of Vogt, *Trans Am Ophthalmol Soc*, 80, 155-71.

Gomes, J. A., Geraldes Monteiro, B., Melo, G. B., et al. (2010) Corneal reconstruction with tissue-engineered cell sheets composed of human immature dental pulp stem cells, *Invest Ophthalmol Vis Sci*, 51, 1408-14.

Gomes, J. A., Romano, A., Santos, M. S. and Dua, H. S. (2005) Amniotic membrane use in ophthalmology, *Curr Opin Ophthalmol*, 16, 233-40.

Grant, W. M. and Kern, H. L. (1955) Action of alkalies on the corneal stroma, *AMA Arch Ophthalmol*, 54, 931-9.

Green, H., Rheinwald, J. G. and Sun, T. T. (1977) Properties of an epithelial cell type in culture: the epidermal keratinocyte and its dependence on products of the fibroblast, *Prog Clin Biol Res*, 17, 493-500.

Griffith, M., Osborne, R., Munger, R., et al. (1999) Functional human corneal equivalents constructed from cell lines, *Science*, 286, 2169-72.

Grueterich, M., Espana, E. and Tseng, S. C. (2002a) Connexin 43 expression and proliferation of human limbal epithelium on intact and denuded amniotic membrane, *Invest Ophthalmol Vis Sci*, 43, 63-71.

Grueterich, M., Espana, E. M., Touhami, A., et al. (2002b) Phenotypic study of a case with successful transplantation of ex vivo expanded human limbal epithelium for unilateral total limbal stem cell deficiency, *Ophthalmology*, 109, 1547-52.

Grueterich, M., Espana, E. M. and Tseng, S. C. (2003) Ex vivo expansion of limbal epithelial stem cells: amniotic membrane serving as a stem cell niche, *Surv Ophthalmol*, 48, 631-46.

Hanai, J., Gloy, J., Karumanchi, S. A., et al. (2002) Endostatin is a potential inhibitor of Wnt signaling, *J Cell Biol*, 158, 529-39.

Hata, K., Kagami, H., Ueda, M., et al. (1995) The characteristics of cultured mucosal cell sheet as a material for grafting; comparison with cultured epidermal cell sheet, *Ann Plast Surg*, 34, 530-8.

Hayashida, Y., Nishida, K., Yamato, M., et al. (2005) Ocular surface reconstruction using autologous rabbit oral mucosal epithelial sheets fabricated ex vivo on a temperature-responsive culture surface, *Invest Ophthalmol Vis Sci*, 46, 1632-9.

Haynes, R. J., Tighe, P. J. and Dua, H. S. (1999) Antimicrobial defensin peptides of the human ocular surface, *Br J Ophthalmol*, 83, 737-41.

Holland, E. J. (1996) Epithelial transplantation for the management of severe ocular surface disease, *Trans Am Ophthalmol Soc*, 94, 677-743.

Holland, E. J. and Schwartz, G. S. (1996) The evolution of epithelial transplantation for severe ocular surface disease and a proposed classification system, *Cornea*, 15, 549-56.

Homma, R., Yoshikawa, H., Takeno, M., et al. (2004) Induction of epithelial progenitors in vitro from mouse embryonic stem cells and application for reconstruction of damaged cornea in mice, *Invest Ophthalmol Vis Sci*, 45, 4320-6.

Hori, Y., Sugiyama, H., Soma, T. and Nishida, K. (2007) Expression of membrane-associated mucins in cultivated human oral mucosal epithelial cells, *Cornea*, 26, S65-9.

Huang, A. J. and Tseng, S. C. (1991) Corneal epithelial wound healing in the absence of limbal epithelium, *Invest Ophthalmol Vis Sci,* 32, 96-105.

Inatomi, T., Nakamura, T., Koizumi, N., et al. (2006) Midterm results on ocular surface reconstruction using cultivated autologous oral mucosal epithelial transplantation, *Am J Ophthalmol,* 141, 267-275.

Jeng, B. H., Halfpenny, C. P., Meisler, D. M. and Stock, E. L. (2011) Management of focal limbal stem cell deficiency associated with soft contact lens wear, *Cornea,* 30, 18-23.

Jenkins, C., Tuft, S., Liu, C. and Buckley, R. (1993) Limbal transplantation in the management of chronic contact-lens-associated epitheliopathy, *Eye (Lond),* 7 (Pt 5), 629-33.

Juhl, M., Reibel, J. and Stoltze, K. (1989) Immunohistochemical distribution of keratin proteins in clinically healthy human gingival epithelia, *Scand J Dent Res,* 97, 159-70.

Kaminski, M. and Kaminska, G. (1978) Inhibition of lymphocyte-induced angiogenesis by enzymatically isolated rabbit cornea cells, *Arch Immunol Ther Exp (Warsz),* 26, 1079-82.

Karakousis, P. C., John, S. K., Behling, K. C., et al. (2001) Localization of pigment epithelium derived factor (PEDF) in developing and adult human ocular tissues, *Mol Vis,* 7, 154-63.

Kato, T., Chang, J. H. and Azar, D. T. (2003) Expression of type XVIII collagen during healing of corneal incisions and keratectomy wounds, *Invest Ophthalmol Vis Sci,* 44, 78-85.

Kenyon, K. R. and Tseng, S. C. (1989) Limbal autograft transplantation for ocular surface disorders, *Ophthalmology,* 96, 709-22; discussion 722-3.

Kerkis, I., Kerkis, A., Dozortsev, D., et al. (2006) Isolation and characterization of a population of immature dental pulp stem cells expressing OCT-4 and other embryonic stem cell markers, *Cells Tissues Organs,* 184, 105-16.

Kim, H. S., Jun Song, X., de Paiva, C. S., et al. (2004) Phenotypic characterization of human corneal epithelial cells expanded ex vivo from limbal explant and single cell cultures, *Exp Eye Res,* 79, 41-9.

Kim, I. K., Husain, D., Michaud, N., et al. (2006) Effect of intravitreal injection of ranibizumab in combination with verteporfin PDT on normal primate retina and choroid, *Invest Ophthalmol Vis Sci,* 47, 357-63.

Kim, Y. M., Jang, J. W., Lee, O. H., et al. (2000) Endostatin inhibits endothelial and tumor cellular invasion by blocking the activation and catalytic activity of matrix metalloproteinase, *Cancer Res,* 60, 5410-3.

Klenkler, B. and Sheardown, H. (2004) Growth factors in the anterior segment: role in tissue maintenance, wound healing and ocular pathology, *Exp Eye Res,* 79, 677-88.

Klenkler, B. J., Griffith, M., Becerril, C., et al. (2005) EGF-grafted PDMS surfaces in artificial cornea applications, *Biomaterials,* 26, 7286-96.

Koch, A. E., Polverini, P. J., Kunkel, S. L., et al. (1992) Interleukin-8 as a macrophage-derived mediator of angiogenesis, *Science,* 258, 1798-801.

Koizumi, N., Inatomi, T., Suzuki, T., et al. (2001a) Cultivated corneal epithelial stem cell transplantation in ocular surface disorders, *Ophthalmology,* 108, 1569-74.

Koizumi, N., Inatomi, T., Suzuki, T., et al. (2001b) Cultivated corneal epithelial transplantation for ocular surface reconstruction in acute phase of Stevens-Johnson syndrome, *Arch Ophthalmol,* 119, 298-300.

Kolli, S., Ahmad, S., Lako, M. and Figueiredo, F. (2010) Successful clinical implementation of corneal epithelial stem cell therapy for treatment of unilateral limbal stem cell deficiency, *Stem Cells,* 28, 597-610.

Kolli, S., Lako, M., Figueiredo, F., et al. (2008) Loss of corneal epithelial stem cell properties in outgrowths from human limbal explants cultured on intact amniotic membrane, *Regen Med,* 3, 329-42.

Korey, M., Peyman, G. A. and Berkowitz, R. (1977) The effect of hypertonic ointments on corneal alkali burns, *Ann Ophthalmol,* 9, 1383-7.

Lavagnolli, T. M., Fonseca, S. A., Serafim, R. C., et al. (2009) Presumptive germ cells derived from mouse pluripotent somatic cell hybrids, *Differentiation,* 78, 124-30.

Lavker, R. M., Tseng, S. C. and Sun, T. T. (2004) Corneal epithelial stem cells at the limbus: looking at some old problems from a new angle, *Exp Eye Res,* 78, 433-46.

Li, D. Q., Chen, Z., Song, X. J., et al. (2005) Partial enrichment of a population of human limbal epithelial cells with putative stem cell properties based on collagen type IV adhesiveness, *Exp Eye Res,* 80, 581-90.

Lim, P., Fuchsluger, T. A. and Jurkunas, U. V. (2009) Limbal stem cell deficiency and corneal neovascularization, *Semin Ophthalmol,* 24, 139-48.

Loughman, M. S., Chatzistefanou, K., Gonzalez, E. M., et al. (1996) Experimental corneal neovascularisation using sucralfate and basic fibroblast growth factor, *Aust N Z J Ophthalmol,* 24, 289-95.

McCulley, J. P. (1990) Ocular hydrofluoric acid burns: animal model, mechanism of injury and therapy, *Trans Am Ophthalmol Soc,* 88, 649-84.

McCulley, J. P., Whiting, D. W., Petitt, M. G. and Lauber, S. E. (1983) Hydrofluoric acid burns of the eye, *J Occup Med,* 25, 447-50.

Merrett, K., Fagerholm, P., McLaughlin, C. R., et al. (2008) Tissue-engineered recombinant human collagen-based corneal substitutes for implantation: performance of type I versus type III collagen, *Invest Ophthalmol Vis Sci,* 49, 3887-94.

Miyashita, H., Higa, K., Kato, N., et al. (2007) Hypoxia enhances the expansion of human limbal epithelial progenitor cells in vitro, *Invest Ophthalmol Vis Sci,* 48, 3586-93.

Miyashita, H., Shimmura, S., Higa, K., et al. (2008) A novel NIH/3T3 duplex feeder system to engineer corneal epithelial sheets with enhanced cytokeratin 15-positive progenitor populations, *Tissue Eng Part A,* 14, 1275-82.

Monteiro, B. G., Serafim, R. C., Melo, G. B., et al. (2009) Human immature dental pulp stem cells share key characteristic features with limbal stem cells, *Cell Prolif,* 42, 587-94.

Moromizato, Y., Stechschulte, S., Miyamoto, K., et al. (2000) CD18 and ICAM-1-dependent corneal neovascularization and inflammation after limbal injury, *Am J Pathol,* 157, 1277-81.

Morrison, S. J., Shah, N. M. and Anderson, D. J. (1997) Regulatory mechanisms in stem cell biology, *Cell,* 88, 287-98.

Mukhopadhyay, M., Gorivodsky, M., Shtrom, S., et al. (2006) Dkk2 plays an essential role in the corneal fate of the ocular surface epithelium, *Development,* 133, 2149-54.

Nakamura, T., Endo, K., Cooper, L. J., et al. (2003a) The successful culture and autologous transplantation of rabbit oral mucosal epithelial cells on amniotic membrane, *Invest Ophthalmol Vis Sci,* 44, 106-16.

Nakamura, T., Inatomi, T., Sotozono, C., et al. (2004a) Transplantation of cultivated autologous oral mucosal epithelial cells in patients with severe ocular surface disorders, *Br J Ophthalmol,* 88, 1280-4.

Nakamura, T., Inatomi, T., Sotozono, C., et al. (2006) Transplantation of autologous serum-derived cultivated corneal epithelial equivalents for the treatment of severe ocular surface disease, *Ophthalmology,* 113, 1765-72.

Nakamura, T., Inatomi, T., Sotozono, C., et al. (2004b) Successful primary culture and autologous transplantation of corneal limbal epithelial cells from minimal biopsy for unilateral severe ocular surface disease, *Acta Ophthalmol Scand,* 82, 468-71.

Nakamura, T., Koizumi, N., Tsuzuki, M., et al. (2003b) Successful regrafting of cultivated corneal epithelium using amniotic membrane as a carrier in severe ocular surface disease, *Cornea,* 22, 70-1.

Nishida, K., Yamato, M., Hayashida, Y., et al. (2004a) Functional bioengineered corneal epithelial sheet grafts from corneal stem cells expanded ex vivo on a temperature-responsive cell culture surface, *Transplantation,* 77, 379-85.

Nishida, K., Yamato, M., Hayashida, Y., et al. (2004b) Corneal reconstruction with tissue-engineered cell sheets composed of autologous oral mucosal epithelium, *N Engl J Med,* 351, 1187-96.

Notara, M., Haddow, D. B., MacNeil, S. and Daniels, J. T. (2007) A xenobiotic-free culture system for human limbal epithelial stem cells, *Regen Med,* 2, 919-27.

Oshima, H., Rochat, A., Kedzia, C., et al. (2001) Morphogenesis and renewal of hair follicles from adult multipotent stem cells, *Cell,* 104, 233-45.

Pellegrini, G., Golisano, O., Paterna, P., et al. (1999) Location and clonal analysis of stem cells and their differentiated progeny in the human ocular surface, *J Cell Biol,* 145, 769-82.

Pellegrini, G., Traverso, C. E., Franzi, A. T., et al. (1997) Long-term restoration of damaged corneal surfaces with autologous cultivated corneal epithelium, *Lancet,* 349, 990-3.

Pfister, R. R. (1994) Corneal stem cell disease: concepts, categorization, and treatment by auto- and homotransplantation of limbal stem cells, *Clao J,* 20, 64-72.

Puangsricharern, V. and Tseng, S. C. (1995) Cytologic evidence of corneal diseases with limbal stem cell deficiency, *Ophthalmology,* 102, 1476-85.

Rama, P., Bonini, S., Lambiase, A., et al. (2001) Autologous fibrin-cultured limbal stem cells permanently restore the corneal surface of patients with total limbal stem cell deficiency, *Transplantation,* 72, 1478-85.

Rama, P., Matuska, S., Paganoni, G., et al. (2010) Limbal stem-cell therapy and long-term corneal regeneration, *N Engl J Med,* 363, 147-55.

Ramchandran, R., Dhanabal, M., Volk, R., et al. (1999) Antiangiogenic activity of restin, NC10 domain of human collagen XV: comparison to endostatin, *Biochem Biophys Res Commun,* 255, 735-9.

Rheinwald, J. G. (1980) Serial cultivation of normal human epidermal keratinocytes, *Methods Cell Biol,* 21A, 229-54.

Rogers, M. S., Birsner, A. E. and D'Amato, R. J. (2007) The mouse cornea micropocket angiogenesis assay, *Nat Protoc,* 2, 2545-50.

Romano, A. C., Espana, E. M., Yoo, S. H., et al. (2003) Different cell sizes in human limbal and central corneal basal epithelia measured by confocal microscopy and flow cytometry, *Invest Ophthalmol Vis Sci,* 44, 5125-9.

Saika, S., Okada, Y., Miyamoto, T., et al. (2004) Protein expression pattern of collagen type XV in mouse cornea, *Graefes Arch Clin Exp Ophthalmol,* 242, 432-6.

Sangwan, V. S., Matalia, H. P., Vemuganti, G. K., et al. (2006) Clinical outcome of autologous cultivated limbal epithelium transplantation, *Indian J Ophthalmol,* 54, 29-34.

Sangwan, V. S., Murthy, S. I., Vemuganti, G. K., et al. (2005) Cultivated corneal epithelial transplantation for severe ocular surface disease in vernal keratoconjunctivitis, *Cornea,* 24, 426-30.

Sangwan, V. S., Vemuganti, G. K., Iftekhar, G., et al. (2003a) Use of autologous cultured limbal and conjunctival epithelium in a patient with severe bilateral ocular surface disease induced by acid injury: a case report of unique application, *Cornea,* 22, 478-81.

Sangwan, V. S., Vemuganti, G. K., Singh, S. and Balasubramanian, D. (2003b) Successful reconstruction of damaged ocular outer surface in humans using limbal and conjuctival stem cell culture methods, *Biosci Rep,* 23, 169-74.

Santos, M. S., Gomes, J. A., Hofling-Lima, A. L., et al. (2005) Survival analysis of conjunctival limbal grafts and amniotic membrane transplantation in eyes with total limbal stem cell deficiency, *Am J Ophthalmol,* 140, 223-30.

Schermer, A., Galvin, S. and Sun, T. T. (1986) Differentiation-related expression of a major 64K corneal keratin in vivo and in culture suggests limbal location of corneal epithelial stem cells, *J Cell Biol,* 103, 49-62.

Schlotzer-Schrehardt, U., Dietrich, T., Saito, K., et al. (2007) Characterization of extracellular matrix components in the limbal epithelial stem cell compartment, *Exp Eye Res,* 85, 845-60.

Schneider, A., Spitkovsky, D., Riess, P., et al. (2008) "The good into the pot, the bad into the crop!"--a new technology to free stem cells from feeder cells, *PLoS One,* 3, e3788.

Schwab, I. R. (1999) Cultured corneal epithelia for ocular surface disease, *Trans Am Ophthalmol Soc,* 97, 891-986.

Schwab, I. R., Reyes, M. and Isseroff, R. R. (2000) Successful transplantation of bioengineered tissue replacements in patients with ocular surface disease(1), *Am J Ophthalmol,* 130, 543-4.

Shimmura, S., Miyashita, H., Higa, K., et al. (2006) Proteomic analysis of soluble factors secreted by limbal fibroblasts, *Mol Vis,* 12, 478-84.

Shin, S. H., Kim, J. C., Chang, S. I., et al. (2000) Recombinant kringle 1-3 of plasminogen inhibits rabbit corneal angiogenesis induced by angiogenin, *Cornea,* 19, 212-7.

Shortt, A. J., Secker, G. A., Munro, P. M., et al. (2007) Characterization of the limbal epithelial stem cell niche: novel imaging techniques permit in vivo observation and targeted biopsy of limbal epithelial stem cells, *Stem Cells,* 25, 1402-9.

Sotozono, C., Kinoshita, S., Kita, M. and Imanishi, J. (1994) Paracrine role of keratinocyte growth factor in rabbit corneal epithelial cell growth, *Exp Eye Res,* 59, 385-91.

Soubrane, G., Jerdan, J., Karpouzas, I., et al. (1990) Binding of basic fibroblast growth factor to normal and neovascularized rabbit cornea, *Invest Ophthalmol Vis Sci,* 31, 323-33.

Stoltz, R. A. and Schwartzman, M. L. (1997) High affinity binding sites for 12(R)-Hydroxyeicosatrienoic acid [12(R)-HETrE] in microvessel endothelial cells, *J Ocul Pharmacol Ther,* 13, 191-9.

Strieter, R. M., Kunkel, S. L., Elner, V. M., et al. (1992) Interleukin-8. A corneal factor that induces neovascularization, *Am J Pathol,* 141, 1279-84.

Takahashi, T., Kalka, C., Masuda, H., et al. (1999) Ischemia- and cytokine-induced mobilization of bone marrow-derived endothelial progenitor cells for neovascularization, *Nat Med,* 5, 434-8.

Tao, Y., Bazan, H. E. and Bazan, N. G. (1995) Platelet-activating factor induces the expression of metalloproteinases-1 and -9, but not -2 or -3, in the corneal epithelium, *Invest Ophthalmol Vis Sci,* 36, 345-54.

Tao, Y., Bazan, H. E. and Bazan, N. G. (1996) Platelet-activating factor enhances urokinase-type plasminogen activator gene expression in corneal epithelium, *Invest Ophthalmol Vis Sci,* 37, 2037-46.

Ti, S. E., Grueterich, M., Espana, E. M., et al. (2004) Correlation of long term phenotypic and clinical outcomes following limbal epithelial transplantation cultivated on amniotic membrane in rabbits, *Br J Ophthalmol,* 88, 422-7.

Tosi, G. M., Massaro-Giordano, M., Caporossi, A. and Toti, P. (2005a) Amniotic membrane transplantation in ocular surface disorders, *J Cell Physiol,* 202, 849-51.

Tosi, G. M., Traversi, C., Schuerfeld, K., et al. (2005b) Amniotic membrane graft: histopathological findings in five cases, *J Cell Physiol,* 202, 852-7.

Townsend, W. M. (1991) The limbal palisades of Vogt, *Trans Am Ophthalmol Soc,* 89, 721-56.

Tsai, R. J., Li, L. and Chen, J. (2000) Reconstruction of damaged corneas by transplantation of autologous limbal epithelial cells(1), *Am J Ophthalmol,* 130, 543.

Tseng, S. C. (1985) Staging of conjunctival squamous metaplasia by impression cytology, *Ophthalmology,* 92, 728-33.

Tseng, S. C. (1989) Concept and application of limbal stem cells, *Eye (Lond),* 3 (Pt 2), 141-57.

Tseng, S. C. (1996) Regulation and clinical implications of corneal epithelial stem cells, *Mol Biol Rep,* 23, 47-58.

Tseng, S. C., Prabhasawat, P., Barton, K., et al. (1998) Amniotic membrane transplantation with or without limbal allografts for corneal surface reconstruction in patients with limbal stem cell deficiency, *Arch Ophthalmol,* 116, 431-41.

van Herendael, B. J., Oberti, C. and Brosens, I. (1978) Microanatomy of the human amniotic membranes. A light microscopic, transmission, and scanning electron microscopic study, *Am J Obstet Gynecol,* 131, 872-80.

Vaporciyan, A. A., DeLisser, H. M., Yan, H. C., et al. (1993) Involvement of platelet-endothelial cell adhesion molecule-1 in neutrophil recruitment in vivo, *Science,* 262, 1580-2.

Wilson, S. E., He, Y. G., Weng, J., et al. (1994) Effect of epidermal growth factor, hepatocyte growth factor, and keratinocyte growth factor, on proliferation, motility and differentiation of human corneal epithelial cells, *Exp Eye Res,* 59, 665-78.

Yoshida, S., Ono, M., Shono, T., et al. (1997) Involvement of interleukin-8, vascular endothelial growth factor, and basic fibroblast growth factor in tumor necrosis factor alpha-dependent angiogenesis, *Mol Cell Biol,* 17, 4015-23.

Zito-Abbad, E., Borderie, V. M., Baudrimont, M., et al. (2006) Corneal epithelial cultures generated from organ-cultured limbal tissue: factors influencing epithelial cell growth, *Curr Eye Res,* 31, 391-9.

Zohar, Y., Talmi, Y. P., Finkelstein, Y., et al. (1987) Use of human amniotic membrane in otolaryngologic practice, *Laryngoscope,* 97, 978-80.

22

Bone Marrow Stromal Cells for Repair of the Injured Spinal Cord

D. S. Nandoe Tewarie Rishi[1],
Oudega Martin[2] and J. Ritfeld Gaby[2]
[1]*Slotervaart Hospital, Department of Neurosurgery, Amsterdam,*
[2]*University of Pittsburgh, Department of Physical Medicine & Rehabilitation*
[1]*Netherlands*
[2]*USA*

1. Introduction

In 1927 Harvey Cushing described the outcome for soldiers with spinal cord injury (SCI) sustained during World War I: "Fully 80 percent died in the first few weeks in consequence of infection from bedsores and catheterization. Only those cases survived in which the spinal cord lesion was a partial one" (Cushing 1927). Nowadays, this has been reversed. In well-organized systems of care for trauma and SCI and due to improved critical-care medicine most patients survive the initial hospitalization. At present, there is no treatment available that effectively re-establishes disrupted axonal circuitries that are necessary to restore injury-induced functional deficits. Due to the lack of a cure and the improved health care, the number of wheelchair bound people increases steadily each year. Currently, in the United States there are an estimated 400,000 people with SCI, with an annual incidence of 11,000 (The National Spinal Cord Injury Statistical Center, Birmingham, AL). In Western European countries similar leading causes of SCI are obtained as in the United States, with vehicular crashes and falls as leading causes of SCI and predominantly young males are affected (Ditunno & Formal, 1994; Ackery et al., 2004). In contrast to the developed countries, in the less developed countries a shift of etiology can be observed towards falls (Hoque et al., 1999) and violence (Da Paz 1992, Dincer 1992, Hart 1994).
Following the first medical care in a hospital, continuing medical care is necessary to maintain the SCI patient's health and quality of life. This does not lead to functional repair. Repair-promoting pharmaceutical and/or surgical interventions will be necessary to significantly change the functional outcome after SCI. Transplantation of repair-supporting cells is considered a candidate repair approach. A bone marrow stromal cell (BMSC) transplant has shown great promise for spinal cord repair. This chapter will give an overview of clinical consequences, assessments, and treatments of SCI and will then focus on stem cells and in particular BMSC as a therapy for SCI.

2. Clinical consequences of SCI

A direct force to the vertebral column can cause damage to bony and soft tissue structures. Torn ligaments or fractures can cause instability of the vertebral column with potential risk of additional damage. Fracture dislocation and hematomas can directly compress the spinal cord and cause immediate neural cell death, axon damage and demyelination, resulting in instant loss of motor and sensory function. After the first destructive events, a sequence of molecular and cellular pathophysiological events, including an aggressive inflammatory response within the damaged tissue leads to additional tissue loss at the injury epicenter and at distant sites (secondary injury; Blight 1992; Hagg & Oudega, 2006). The functional consequences of SCI are highly variable and depend on the degree of tissue damage, which in turn depends on the impact severity. In patients with SCI with a relatively small amount of tissue damage, some endogenous recovery of function can be observed, which is most likely resulting from plasticity of the spinal nervous tissue (Dietz & Harkema, 2004; Dobkin et al, 2007). In people with SCI with extensive tissue damage the neurological deficits are generally major and permanent. There are very few reports of people with a large injury that regain motor function to a degree that independence can be achieved.

Over 95% of SCI patients survive their initial hospitalization. The relatively young age when SCI occurs, improved medical care, and lack of effective therapies are responsible for the continually increasing number of paralyzed people with SCI. This puts a high financial burden on the patient, his/her family, and society (Fiefler et al., 1999; Ackery et al, 2004). The psychological consequences of SCI should not be underestimated and appropriate guidance of patient and family should have an important place in the management of SCI (Boekamp et al., 1996; Widerström-Noga et al., 1999/2007). Patients need time to accept their deficits. One can expect an initial period of denial and/or inability to fully comprehend the consequences of the paralysis caused by the injury. After the patient realizes his/her fate to the fullest extent, a period of acceptance will have to run its course (Boekamp et al., 1996). After that, the patient needs to learn to live with his/her disabilities, and this may be accompanied by bouts of depression. The mental state of the patient can have its effect on his/her medical treatments (Widerström-Noga et al., 1999).

SCI is the second most expensive condition to treat in the United States after respiratory distress syndrome in infants and is ranked third in medical conditions requiring the longest stay in hospitals (Winslow et al., 2002). The costs of lifetime care for a SCI patient varies between 1 and 3 million dollars. The Center for Disease Control in the United States estimated that about 10 billion dollars are spent yearly on SCI treatment excluding the management of pressure ulcers, a common adverse effect of SCI, which adds another billion dollars per year (McKinley et al., 1999).

2.1 Clinical assessment of SCI

The American Spinal cord Injury Association (ASIA) impairment scale is widely used and provides clinicians with a standard way of grading the functional severity of a spinal cord injury. The scale has one grade for complete injuries (ASIA A), three other grades for incomplete injuries (ASIA B-D), and another for no impairment from the injury (ASIA E). To categorize the grades for incomplete injuries, clinicians determine the degree of muscle strength, being active movement against full resistance, of the key muscles below the neurological level of the injury. The assignment is based on the extent to which more than half of the key muscles have a muscle strength grade of 3 or higher. Classifications of SCI can also be achieved by implementing tests that measure functional ability (Ditunno et al., 2005). A widely used scale for such measurements is the Functional Independence Measure (FIM), which uses a 7-point scale to measure 18 items concerning mobility, locomotion, self-care, bowel and/or bladder function, communication, and social cognition (McKinley et al., 2004; Ditunno et al., 2007). Other functional assessment scales are the Quadriplegic Index of Function (QIF), Modified Barthel Index (MBI), Walking Index for SCI (WISCI), Capabilities of Upper Extremity Instrument (CUE), Spinal Cord Independence Measure (SCIM), and the Canadian Occupational Performance Measure (COPM). These functional tests become more important in the rehabilitation phase of the patient when it is important to analyze the limitations in daily life.

2.2 Treatment of SCI

An acute and a chronic phase can be distinguished after SCI. Since SCI is often a consequence of severe accidents, initial treatment is generally focused on stabilization of the patient. There is insufficient evidence that would support standards of care during the acute phase of SCI. It is advised to maintain patients in an intensive care unit for close monitoring of respiratory and hemodynamic complications. For adequate spinal perfusion, which can be at risk due to injury-induced edema, a mean arterial pressure of 85-90 mmHg should be maintained (Botel et al., 1997). Depending on the type of injury, surgical interventions should be considered to decompress the spinal cord and or stabilize the spinal column (Brodkey et al., 1980; Fehlings & Perrin, 2006). Decompression surgeries may accelerate functional improvements and result in shorter hospitalization and rehabilitation periods (Papadopoulos et al., 2002; McKinley et al., 2004). However, it does not result in an improved functional outcome (Chen et al., 1998). A lack of consensus of care during the acute phase of SCI is in part due to the large variability among injuries and makes its early management complicated. If bone fragments continue to compress the spinal cord, early surgery may be vital to prevent exacerbation of spinal cord tissue destruction. However, in cases without a clear sign of such urgency there is no consensus on whether and what type of early surgical/clinical interventions must be implemented (Fehlings & Perrin, 2006). The type of surgical intervention should be considered on a case-to-case basis, which makes it complicated to study the efficacy of intervention in the acute phase after SCI in randomized and controlled clinical trials.

Besides surgical interventions, pharmacological treatments to limit the secondary injury after SCI are often considered. The best-known treatment is a high dose of the glucocorticosteroid, methylprednisolone sodium succinate (MPSS) within 8 hours after the injury (Bracken et al., 1990/ 1997/ 2002). Experimentally it was demonstrated that a high dose of MPSS reduces the inflammatory response and limit tissue loss after damage to the spinal cord. The effects of MPSS in patients with SCI were investigated in 3 consecutive National Acute Spinal Cord Injury Studies (NASCIS; Bracken et al., 1990/ 1997/ 2002). The results demonstrated that MPSS treatment in the acute phase of SCI resulted in neurological improvements up to 6 months after injury. After a thorough review of the results from the NASCIS studies and a more comprehensive assessment of the benefits and risks involved in high dose MPSS treatment, the therapeutic benefits are now disputed (Nesutherai, 1998/ 2001; Lee et al., 2007). Especially in patients with complete SCI high dose steroid treatment can lead to adverse effects such as myopathy and wound infection that may negatively influence functional outcome and in some cases may be life-threatening (Qian et al., 2005; Lee et al., 2007). Currently, many SCI clinics worldwide have discontinued the 'standard' acute administration of MPSS after SCI.

Treatment paradigms in the chronic stage after SCI are multidisciplinary and intensive. Different complications may occur that each demands specific interventions. For instance, SCI can lead to pain (Widerstrom et al., 1999/ 2004), decreased fertility (Patki et al., 2007), and autonomic dysreflexia with loss of bladder and bowel control (Weaver et al., 2006). It has to be taken into consideration that many SCI patients get accustomed to the specific injury-related pain they experience and as a result reveal their distress to their physician often at a late stage (Mariano, 1992; Sawatzky et al., 2008). For some SCI-related conditions, such as decreased fertility, it is the patient's personal desire that should guide the physician's actions. Other common problems that arise after SCI are septicemia, respiratory insufficiency, and pneumonia due to muscle atrophy. These complications may cause clinical deterioration and could eventually result in death. They often occur without typical symptoms. It is imperative that SCI patients receive annual screenings and long-term follow-ups to prevent these secondary complications. It is advised to treat patients on a regular basis with pneumococcal and influenza vaccine to prevent opportunistic infections. Monitoring the skin and urinary tract and implementing aggressive treatments against pressure ulcers and urinary tract infections is needed to reduce the risk of septicemia. Appropriate nutrition and exercise should also be incorporated in the (new) lifestyle. Rehabilitation programs should be implemented to reduce the risk of cardiovascular disease (Kennedy et al., 2003; Strauss et al., 2006).

3. A stem cell; the origin of life?

Stem cells are defined by their capacity to divide in one cell that remains the stem cell without signs of ageing (self-renewal), and another cell that begins the process of differentiation, thus providing an unlimited source for implantation. Moreover, a stem cell is able to differentiate into different lineages (pluripotency). It may also be possible for a stem cell to cease proliferation, entering a quiescent phase, in which it is not an actual stem cell, but still has the potential since it can re-enter the cycle of self-renewal (Potten et al., 1990).

3.1 The origin of life

Aristotle (384-322 BC) developed the concept of spontaneous generation, which was first mentioned during ancient Roman times, and described that earth, air, fire, and water (the four elements) mixed with an essence known as "quintessence" or "ether" would give 'life' after it was brought in contact with 'pneuma' (or `soul`). In the middle ages it became generally believed that the concept of living was based on "spontaneous generation", hypothesizing that the embryo was derived from the mother's menstrual blood, based on the concept that living animals arose from slime or decaying matter. It was not until 1855 that zoologist and comparative anatomist, Leydig pronounced that life could only arise from preexisting life *(omne vivum ex vivo)*. Virchow extended this and postulated that all cells in an organism are derived from preexisting cells *(omnis cellula e cellula)*; all cells of the human body arise from the fertilized egg. The theoretical proof for his hypothesis was provided by the French chemist and biologist, Louis Pasteur (1822-1895) in his essay "Mémoire sur les corpuscules organisés qui existent dans l'atmosphère" of 1861 in which he described a series of experiments showing that germs do not form spontaneously in sterilized flasks. Nineteenth century pathologists first hypothesized the presence of stem cells in the adult as "embryonal rests" to try to explain the cellular origin of cancer cells. Cancer results from an imbalance between the rate at which cells are produced and the rate of their apoptosis. Nowadays, it is generally agreed that stem cells reside in the adult body as well, possibly in a quiescent state. Certain cues can cause them to re-enter the cycle of cell proliferation and differentiation (Sell, 2004).

3.2 Determination

The ultimate stem cell is the fertilized egg, a totipotent cell, which can differentiate in each cell type. After a few divisions, when the dividing cells are called blastomeres (32 to 64 cells), the cells are determined to become specific for one of the three germ layers; the ectodermal layer, which will give rise to skin and neural tissue, the mesodermal layer, which will give rise to connective tissue, muscle, bone and blood cells, and the endodermal layer, which will give rise to gastrointestinal tract and internal glandular organ cells. In classic embryology, this 'determination' of stem cells is thought to be an irreversible process. As the cell mass develops, the daughter cells begin to acquire properties different from one another, so that different regions destined to become different components of the embryo are formed. Recently, it has been suggested that the determined stem cell is in fact phenotypically plastic and is able to give rise to cells from different germ layers, a process known as transdifferentiation (Morrison et al., 1997; Johansson et al., 2003). The process of transdifferentiation can also result in the formation of abnormal phenotypes having no counterpart in the normal body (dysplasia). These dysplastic cells may be the first step in carcinogenesis (Slack, 2007). The possibility of transdifferentiation has raised the interest in using of easy to harvest (adult) stem cells and try to transdifferentiate them into a more desirable cell type.

3.3 Embryonic versus adult

Embryonic stem (ES) cells are obtained from the undifferentiated inner cell mass of the blastocyst, which in humans forms 4 to 5 days after fertilization. At this stage the inner cell mass consists of around 50-150 cells. ES cells are pluripotent cells and can develop into more than 200 cell types of the adult body when given the appropriate stimuli for differentiation (Morrison et al., 1997). They are known to proliferate rapidly in culture. Because of this plasticity some people prefer to use embryonic stem cells for research purposes. On the other hand, it recently has been shown that the ES cells need to be genetically modified and extensively manipulated in vitro before they can be transplanted safely. Direct transplant of ES cells are known to give rise to teratomas and uncontrollable cell proliferation (Slack, 2007). To avoid immune rejection a life-long use of immunosuppressive drugs after transplantation is necessary or the ES cells have to be tissue-matched from a bank of stem cells created from 'spare' human embryos; a procedure prone to failure and morally objectionable to many, including scientists.

By contrast, adult stem cells can be transplanted directly without genetic modification or pre-treatments. There is no problem with immune rejection because the cells can readily be isolated from the patient's bone marrow, skin, fat, umbilical cord blood and other sources *(autografting)*. Although the adult stem cells show less plasticity in vitro, they do show high degrees of genomic stability during culture. They have the potential to differentiate in response to cues from the surrounding tissues and do not show uncontrollable proliferation resulting in tumors. Finally, since it is not necessary to take cells from embryos there are less ethical concerns for the use of this kind of stem cell.

3.4 Ethical and social perspectives

The involvement of human cells gives rise to ethical and social concerns, mostly concerning the status of the human embryo. Although there is a general consensus that the early human embryo is worthy of some measure of respect as an organized assembly of human cells, opinions vary as to the measure of respect deserved. The Roman Catholic Church teaches that "from its conception, the child has a right to life" and "it must be treated from conception as a person" (Roman Catholic Church, 1994). On the other hand, there are those that consider the early human embryo as merely an assembly of human cells and not a human individual at all at this stage. The embryo should be considered an individual once it becomes viable. Over the last few decades this viability of the embryo has shifted as a result of advanced medical techniques to the 20th week of gestation (Goldenring, 1985; Peterfy, 1995). The use of human embryo's for research or therapeutic purposes is still the subject of controversy.

The conversion of stem cells into gametes (sperm or egg cells) in itself presents, if any, few ethical hurdles (Testa & Harris, 2005). On the positive side, stem cells as a potential source of oocytes for clinical applications, such as in vitro fertilization to induce pregnancy in childless couples, have proven successful. Many of the ethical concerns arise from the potential for misuse rather than the use of such gametes (Westpal, 2003). The ability to derive oocytes from stem cells of male origin, inducing X and Y chromosomes and fertilized oocytes would raise the possibility of in vitro fertilization of such oocytes and the production of a child with two male biological parents (Hubner et al., 2003; Geijsen et al., 2004; Azim, 2004). Most extreme, the possibility of producing oocytes from the male stem cells and fertilizing these oocytes with his own sperm would open the avenue for single-parent families and in theory would be the most incestuous relationship. Obviously, there still has to be a female surrogate mother to bring any offspring to term. The offspring can be expected, if it survived, to exhibit many developmental defects resulting from acquisition of pairs of (recessive) genes for anomalies (Hubner et al., 2003, Whittaker, 2007).

Therapeutic cloning and genetic manipulation of stem cells have ethical and social concerns. Cloning of cells, genetically matched for the host can contribute to the field of organ transplantation, where a shortage of donor organs and rejection of not perfectly matched grafted organs by the host are major concerns. However, the destruction of human embryos for this application opposes obvious problems in itself and is in general considered unethical. Genetic manipulation may perhaps be one of the most significant applications of converting embryonic stem cells into gametes, opening the possibility for germ line gene therapy (GLGT) (Newson & Smajdor, 2005). GLGT offers the possibility for eliminating the transfer of a gene mutation from a parent to their children and their children's children, and as a result eliminating the aberrations in the gene pattern. Genetic manipulation could, even more controversially, open the possibilities for inserting 'desirable' genes into the germ line, resulting in the so-called "designer babies".

Stem cell-based therapies are still at a very early stage and the associated risks are still unclear. On the other hand, when a patient is suffering from a disabling and sometimes life-threatening disease, a case might be made for lowering the ethical barriers to enable treatment. As with any medical interventions, the questions to be asked are whether this approach is the most likely to achieve success and whether the potential benefits outweight the potential risks.

4. Stem cells for central nervous system repair

The promise of neural stem/progenitor cells (NSC) for repair of traumatic injuries of central nervous tissue lies in their ability to differentiate into the three neural cell types; neurons, astrocytes, and oligodendrocytes (Fig. 1). Besides NSC, other types of stem cells may also be beneficial for replacement-based repair approaches if they can be successfully manipulated into the neural lineage (i.e., transdifferentiation).

Although, following an injury to the central nervous system, endogenous NSC are typically recruited, they generally do not lead to significant repair. A potential approach to benefit more from NSC for neural repair is to isolate the cells from donor tissue and prepare them ex vivo for transplantation. The advantages of ex vivo isolation of cells are the large numbers of cells that can be cultured, and the opportunity of committing the cells towards a certain cell type. Experimental evidence has demonstrated that neural stem cells after transplantation into damaged nervous tissue are more likely to differentiate into astrocytes than into neurons or oligodendrocytes. In some cases this may be the objective, but most likely there is a larger need for neurons or oligodendrocytes. Therefore, it is more effective to transplant cells already committed to become a certain neural cell type such as oligodendrocyte-precursor cells (OPCs) and glial-restricted precursor cells (GRPCs).

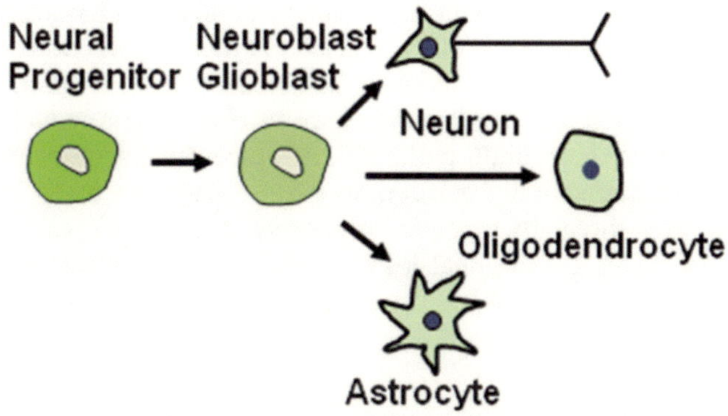

Fig. 1. Neural lineage differentiation: Neural stem cells differentiate into neuroblasts, which give rise to neurons, and in glioblasts, which give rise to astrocytes and oligodendrocytes.

Replacement of lost neurons with NSC-derived neurons may provide the cellular foundation for restoring axonal connections between affected regions in the nervous system (Fig. 2). The prerequisite for such repair is that the transplanted neurons grow their axon towards the original (or new) target cells and form functional synaptic connections. Replacing lost oligodendrocytes by ones generated from stem/progenitor cells may benefit remyelination of axons that have survived the initial and secondary damage but have lost their insulating myelin sheaths due to oligodendrocyte death (Fig. 2). In addition, oligodendrocyte replacement may help myelination of newly generated axons from the transplanted neurons as well as regenerated axons from original neurons. Research has demonstrated that oligodendrocyte replacement using OPCs is a powerful tool to elicit (re)myelination and as a result improve functional recovery. Currently, OPCs are being tested in a clinical trial for their potential to repair the spinal cord. Replacement of lost astrocytes can also be of great value. Astrocytes play a crucial role in a number of events such as neuronal functioning, blood-spinal cord-barrier maintenance, and more. Astrocyte replacement is often undervalued mainly because astrocytes are known to be an integral component of the glial scar that forms at an injury site. This glial scar presents a molecular and cellular barrier for axon growth and thus for functional restoration.

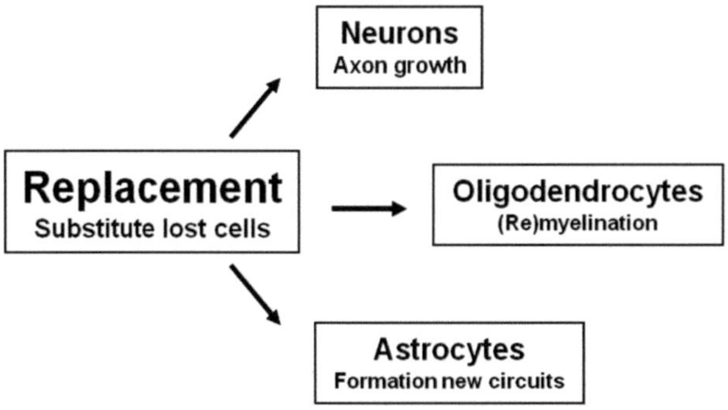

Fig. 2. Replacement strategies for nervous tissue repair. Neural progenitor cells could be employed to replace lost cells. This may result in formation of new neurons, oligodendrocytes or astrocytes.

The cellular consequences after SCI in time can generally be characterized in 5 stages, starting with 1) immediate neural cell death and axonal damage accompanied by motor and sensory function loss due to the trauma; 2) progressive loss of tissue resulting in the development of cystic cavities; 3) infiltration of macrophages to remove cellular debris; 4) formation of a glial scar surrounding the injury epicenter limiting and/or preventing endogenous axonal sprouting; 5) dieback of damaged axon stumps away from the injury.
In general, there are three therapeutic platforms for spinal cord repair (Fig. 3); neuroprotection (i.e., limiting additional loss of nervous tissue; *rescue tissue*), regeneration (i.e., promoting axonal growth and/or myelination, nervous tissue formation/modeling; *restore tissue*), and plasticity (i.e., exploiting axonal circuits that were left intact; reuse *tissue*).

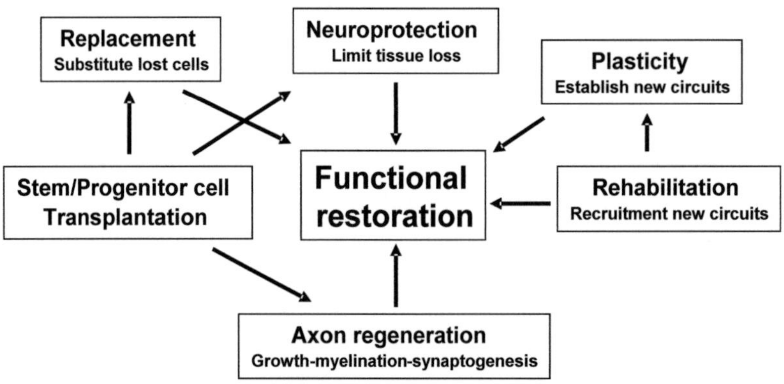

Fig. 3. Repair platforms for the injured spinal cord. Spinal cord repair may be achieved through a variety of opportunities that involve tissue rescue (neuroprotection), tissue restoration (axonal regeneration/myelination), and tissue reuse (plasticity). Stem cells may influence all of these platforms as well as replace lost cells, which has its own repair effects. All can lead to functional recovery.

The consensus in the field of spinal cord injury/repair is that different interventions addressing the different repair platforms need to be combined in an approach to obtain significant functional restoration. The components of such combination strategies will need to be applied simultaneously and/or successively depending on each individual purpose thereby creating optimal conditions for histological and functional repair. Intensive rehabilitation to develop and stabilize new axonal circuits will be necessary either as one of the components in the repair strategy or as an 'anchoring'-mechanism of the outcomes obtained by neuroprotection- and regeneration-based interventions.

5. Bone marrow stromal cells

Mesenchymal stem cells from bone marrow (BMSC) have therapeutic potential for the injured spinal cord (Nandoe Tewarie et al., 2006). BMSC were shown to differentiate into bone, fat, tendon and cartilage cells (Pittenger et al., 1999). Although still debated, it has been reported that BMSC can transdifferentiate in vitro into liver cells (Petersen et al., 1999), skeletal cells (Wakitani et al., 1995), cardiac muscle cells (Orlic et al., 2001), and neural cells (Petersen et al., 1999, Mezey et al., 2000). Besides this ability, BMSC are also known to produce different types of growth factors that could potentially influence nervous tissue repair positively. Together, these abilities make BMSC interesting for repair strategies for the injured spinal cord.

Several other aspects make BMSC interesting candidates for cell-based approaches for central nervous system repair. Firstly, BMSC are relatively easy to obtain from a fairly routine bone marrow extraction followed by a quick centrifuge and culture procedure to remove the hematopoietic cells. Secondly, BMSC are easy to culture as they do not need complicated growth media or special culture circumstances. Basic cell culture equipment is sufficient to successfully culture millions of BMSC. Thirdly, BMSC are easy to transduce with viral vectors which, if necessary, may be helpful to boost the overall reparative abilities of the cells. The use of viral vectors to genetically modify cells prior to transplantation has not yet become mainstream as there are some biological and ethical issues that need to be resolved. Finally, BMSC do not have the ethical concerns that embryonic or fetal stem cells have, and therefore circumvent public rejection as a possible treatment for neural and non-neural trauma and disorders.

At this time, there is no irrefutable evidence that BMSC transplanted into the damaged nervous tissue differentiate into neural cells that successfully replace lost cells. Also, there is no convincing evidence that neural cells derived from grafted BMSC contributed to functional improvements after transplantation. As long as the potential of BMSC for differentiation into neural cells is in debate, the ability to produce and secrete different types of growth-promoting molecules, which include several neurotrophins and cytokines, is the more interesting and more likely characteristic of BMSC that makes these cells important candidates for spinal cord repair approaches. By releasing these molecules, BMSC can positively influence all consequences of spinal cord injury and support anatomical and functional repair.

6. Rodent models of SCI

Promising therapies for spinal cord injury are typically tested in rodent models, and mostly in rats. Similar as in humans, a SCI in the rat results in progressive loss of the grey and white matter creating large fluid filled cysts. Proliferation and activation of astrocytes result in formation of scar tissue, which acts as a barrier for axonal regeneration. Importantly, as in humans, there is no spontaneous regeneration in the injured spinal cord in rodents. The histological similarity between human and rat spinal cord injury has made the rat an extensively studied model for experimental therapeutic strategies, including BMSC transplantation.

The most widely used model of spinal cord injury involves a spinal cord contusion inflicted by an impactor device. A contusion is clinically the most frequently occurring type of spinal cord injury; approximately 75% of all human injuries are contusions. The consequences of a contusive injury in rats are similar as the known consequences in the contused human spinal cord. Figure 4 shows the rat model system for spinal cord contusive injury.

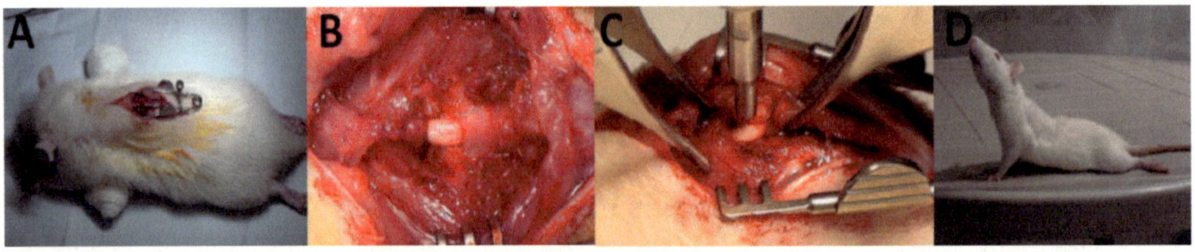

Fig. 4. Rat spinal cord contusion model. A. A laminectomy is performed exposing the underlying spinal cord. **B.** Enlarged view of the exposed spinal cord segment. **C.** A computerized impactor is used to contuse the spinal cord. The piston is attached to a sensor to record velocity, force and displacement to ensure consistency. **D.** A moderate contusion results in loss of function at and below the level of injury and loss of bladder function.

An alternative model for a contusion-like spinal cord injury is the clip compression model. The main difference between the impactor-inflicted contusion and the clip-inflicted compression is time. With an impactor the spinal cord is compressed for a brief moment of time while with a clip the spinal cord is compressed for a longer, regulatable, time. The clip model is clinically more relevant as most spinal cord injuries are inflicted by a lasting compression rather than a brief one.

There are a number of other, non-contusive, spinal cord injury models employed in laboratories around the world to test treatment paradigms. These are valuable in their own right to investigate the underlying mechanisms and/or validity of certain approaches. Partial transections of specific regions in the spinal cord are used especially to study the effects of treatments that aim to promote axonal regeneration; specific descending or ascending pathways can be damaged with relatively small local knife cuts and the regeneration response quantified at later time points. The involvement of specific axonal pathways in locomotor function can also be investigated using partial transections. The main disadvantages of partial transections are the low clinical relevance and the possible misinterpretation of results due to compensatory sprouting, i.e., other previously non-involved axonal pathways become involved in particular functions.

Another model that has been used is the complete transection of the spinal cord. Although this is not often seen in the clinic, complete transections are particularly advantageous to study cell types for their ability to promote regeneration of damaged axons without contaminating sprouting of undamaged pathways and to serve as bridging material between spinal cord stumps. This model is also suitable to study the efficacy of synthetic or natural biomaterials for their efficacy to serve as carrier of cells or drugs. A disadvantage besides the low clinical relevance is that rats with a completely transected spinal cord are more laborious to maintain.

7. BMSC preparation and injection

It is difficult to provide standard guidelines for cell preparation because every cell type requires special conditions and circumstances for optimal isolation and culturing. Cell injection procedures may vary but are essentially similar. The standard procedures to harvest, culture and genetically modify BMSC with lentiviral vectors encoding for green fluorescent protein (GFP) to enable easy identification in vivo, as well as to inject BMSC as used in our laboratory are depicted in Figure 5. The length of the culture (preparation) time for BMSC depends on how many cells are needed to fill the damaged area. Thus, the number of BMSC necessary depends on the overall loss of tissue which, in turn, depends on the severity of the initial insult and on the time between insult and transplantation. Imaging techniques may provide the necessary information to guide the decisions on damaged tissue volumes and number of cells.
There are a number of studies that have explored injection paradigms other than straight acute injections into the injury site. BMSC have been infused systemically or into the 4th ventricle (Ohta et al, 2004), or transplanted acutely into the cervical (Lu et al, 2004) or thoracic spinal cord (Hofstetter et al, 2002; Ankeny et al, 2004) or into the chronically injured cord (Zurita and Vaquero, 2004).

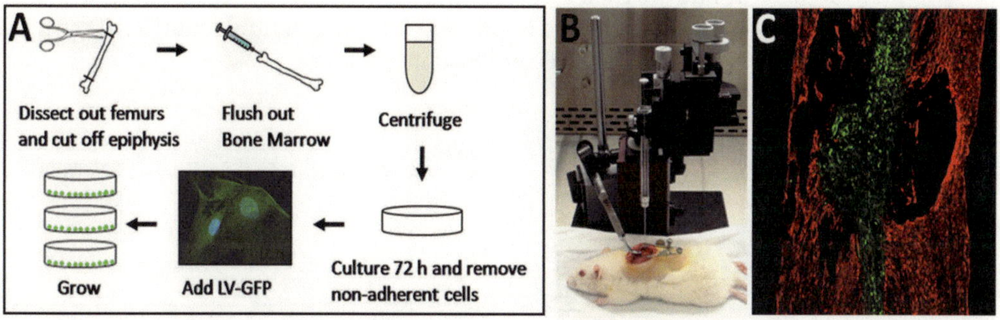

Fig. 5. Transplantation of BMSC. **A.** BMSC are isolated from femurs of rats by cutting off the epiphyses and flushing out the bone marrow. Cells are plated onto plastic culture dishes. Non-adherent hematopoietic stem cells are removed and the plastic-adherent BMSC are infected with LV-GFP. **B.** Cells are injected into the spinal cord contusion epicenter using a Hamilton syringe with a pulled glass needle attached, held within a micromanipulator. **C.** Appearance of transplanted BMSC (green) in the contused rat spinal cord seven days post transplantation (20 µm thick section at 2.5 x magnification). The red color represents immunohistochemically stained glial fibrillary acidic protein (GFAP), a commonly used marker for astrocytes.

8. BMSC transplantation improves function after spinal cord contusion

BMSC have been transplanted into rodent models of spinal cord injury by various research groups and with generally promising results. Considerable variation exists between injury models, treatments paradigms, and analytical methods used by the various investigators, and consequently, different results have been found by different groups. The injection

paradigms vary considerably in time, number, location, survival times, etc. Given these differences between the approaches, it is difficult to compare the respective results and thus to properly value the regeneration-promoting abilities of BMSCs.
However, it was observed by most groups and it is clear that some histological and/or functional beneficial effects can be expected after transplantation of BMSC into the injured spinal cord. These effects include improvements in locomotion, sensorimotor function, sensory function, promotion of axonal regeneration, and preservation of neural tissue (reviewed by Nandoe Tewarie et al., 2006).
Improvements in hind limb locomotor recovery have been reported after BMSC transplantation in the acutely, subacutely, and chronically injured spinal cord. Hind limb function is typically evaluated using the open field BBB-test, which scores for joint movements, paw placement, weight support, and coordination between fore limbs and hind limbs. Although a valid way to test hind limb function, the BBB test has limitations; the scoring is subjective and difficult for observation of coordination. This may affect the assessment of hind limb motor performance. Other sensorimotor tests such as foot print, grid and beam walking, and analysis of gait using the CatWalk® provide a more complete measurement of hind limb function (Figure 6). Recently, Ritfeld et al. (2011) performed an extensive battery of functional tests after BMSC transplantation into a spinal cord contusion injury rat model, including BBB open field testing, performance on horizontal ladder, footprint analysis, mechanical allodynia and thermal hyperalgesia. Improvements were found in the BBB-subscore, horizontal ladder performance, base of support, angle of paw rotation, and mechanical and thermal hypersensitivity. No improvements were found in BBB open field locomotion, illustrating the importance of using multiple functional outcome assays to detect treatment-mediated effects on functional recovery.

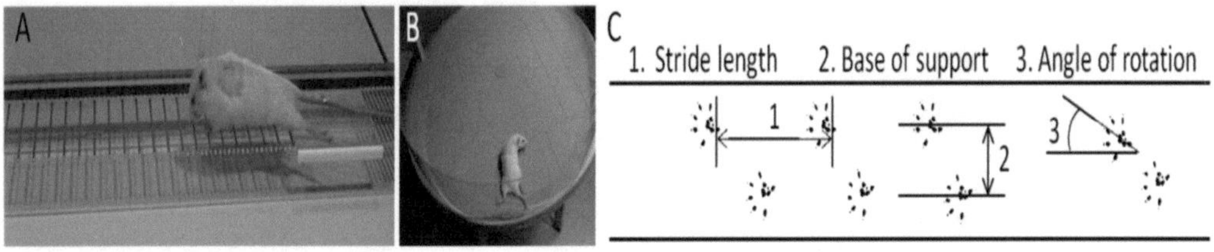

Fig. 6. **A.** Performance on a horizontal ladder is used as an indicator of sensorimotor function. The number of slips made on the rungs is counted by playing the videotaped runs back in slow motion. **B.** Testing of BBB open field locomotion includes scoring for joint movements, paw placement, weight support, and fore/hind limb coordination. **C.** In footprint analysis stride length, base of support and angle of paw rotation are determined.

The finding that BMSC transplantation into the contused spinal cord leads to improvements in functional restoration holds promise for future BMSC-based repair strategies for the injured spinal cord. However, knowledge of the underlying mechanisms is essential to increase the overall outcomes. Which repair-mechanisms are positively influenced by the BMSC transplant? Which molecules and receptors are involved? How do BMSC transplant-mediated events elicit functional recovery? Initial reports on BMSC-initiated functional improvements suggested that beneficial effects could be due to replacement of lost cells by BMSC. The first study that provided evidence that BMSC can differentiate into neural-like cells *in vivo* was from Mezey and colleagues (2000). Using a mouse model they transplanted male BMSC into the peritoneal cavity of female recipients. The grafted BMSC preparation did not contain neuron- or glia-like cells at the time of transplantation, although it should be noted that about 18% of the cells expressed the neural precursor cell marker nestin, when cultured for several weeks. Using in situ hybridization techniques, Y chromosome-containing neurons were located in the brain of the host, suggesting that the grafted BMSC had crossed the blood-brain barrier and formed neurons within the CNS.

Interestingly, Cogle and colleagues (2004) also demonstrated Y chromosome-containing neurons that were nicely integrated in the hippocampus of three female humans that had received transplants of male BMSC up to six years earlier. It should be mentioned that a fusion between a grafted BMSC and a host cell could provide false-positive results. In several studies it has been reported that BMSC can spontaneously fuse with other cells in vitro. Whereas this is a real possibility, Cogle and coworkers (2004) used fluorescence in situ hybridization techniques to reveal the presence of only one X chromosome, concluding that the neurons found in their study had not fused with BMSC. The Y chromosome containing, transgender cells accounted for approximately 1 % of all neurons and 1 -2 % of all astrocytes and microglial cells in the hippocampus. These studies have provided exciting data suggesting that BMSC can differentiate into neural cells in the mature CNS. However, to date this possible transdifferentiation of BMSC is intensely debated and conclusive evidence of this possibility is lacking. The data showing functional and histological improvements after BMSC transplantation into spinal cord contusion injury models point to repair mechanisms other than cell replacement of lost cells by BMSC. Within spinal cord injury models, few investigators have shown expression of neural markers by BMSC transplanted in the spinal cord. Functionality of BMSC-derived neurons (e.g. synapse formation, firing of action potentials, release of neurotransmitters, myelination) and astrocytes has not been demonstrated. Thus whether transplanted BMSC can transdifferentiate into neural cells in the damaged spinal cord is highly questionable. Whether BMSC-derived neural cells are involved in the observed functional improvements after BMSC transplantation in the injured spinal cord is even more improbable. This raises the question in what manner transplanted BMSC contribute to the functional improvements.

9. BMSC and neuroprotection

One possible repair mechanism that could be influenced by a BMSC transplant is neuroprotection. Neuroprotection aims to protect neural cells from loss due to pathophysiological events secondary to the initial injury resulting in more spinal cord nervous tissue which may lead to functional gains. Neuroprotective approaches are considered the first line of defense against the devastating life-debilitating consequences of spinal cord injury. The sooner neuroprotective strategies are implemented after injury the more effective they will be. If functional recovery is not achieved with neuroprotective strategies there may still be important benefits because a larger volume of nervous tissue is present to serve as a substrate for regenerative or plasticity-promoting interventions. In general, stem/progenitor cell transplants may elicit neuroprotection through different direct and indirect mechanism. After transplantation the cells may secrete molecules that prevent/limit the loss of neural cells due to apoptosis (direct mechanism), decrease the invasion of macrophages and thus macrophage-mediated cell death (indirect mechanism), or prevent the breakdown of blood vessels leading to blood-spinal cord-barrier permeability (indirect mechanism).

Our current understanding of BMSC-mediated repair points towards a mechanism that involves secretion of trophic factors by BMSC resulting in decreased loss of neural tissue (i.e., neuroprotection). Ritfeld and colleagues (2011) demonstrated a strong correlation between improved function and volumes of spared tissue in rats with BMSC transplanted in their contused spinal cord at three days post-injury. The mechanisms underlying BMSC-mediated tissue sparing in the injured spinal cord are still under investigation. A possible mechanism is that BMSC transplants positively influence blood vessel presence in the injury site either by blood vessel stabilization/rescuing or formation which results in tissue sparing. This in turn may involve sparing of descending and ascending axons mediating those functions that were improved (Ritfeld et al., 2011).

BMSC have been shown to secrete various growth factors, including brain-derived neurotrophic factors (BDNF), glial-derived neurotrophic factor (GDNF), vascular endothelial growth factor (VEGF), fibroblast growth factor 2 (FGF-2), nerve growth factor (NGF) and neurotrophin-3. NGF and BDNF increase survival and decrease apoptotic death of neurons and oligodendrocytes. BDNF also increases oligodendrocyte proliferation. GDNF has been implicated in the rescue of motor neurons, possibly by activating MAP kinase and Bcl-2, an anti-apoptotic regulator. FGF-2 is known to positively affect tissue sparing and promote neuronal survival and angiogenesis following spinal cord injury. VEGF is a potent angiogenic factor and as such could positively affect tissue preservation. Thus, based on the secretion profile, BMSC may contribute to neuroprotection in a direct manner (rescuing neural cells) and/or in an indirect manner (promoting angiogenesis).

The secretion of these molecules is thought to be the main factor behind the observed functional improvements after BMSC transplantation in spinal cord injury models (Figure 7). However, which factors or combinations of factors are specifically necessary and the precise mechanisms by which these factors elicit their effect remain to be elucidated in future investigations. Such mechanistic information is imperative as it can provide us with valuable insights on how to increase the overall reparative effects of BMSC transplants. For

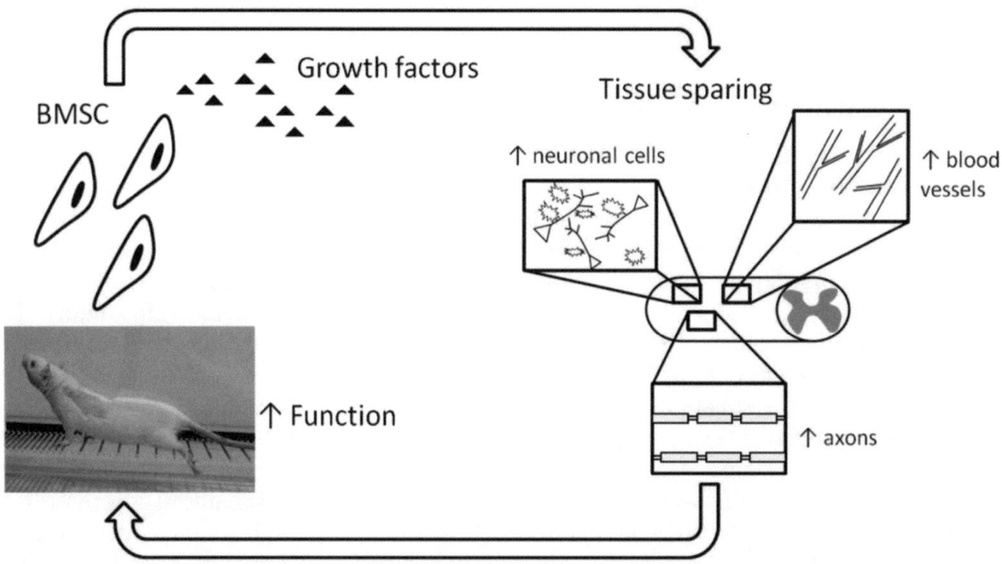

Fig. 7. BMSC secrete various growth factors, including BDNF, VEGF, NGF and NT-3. These factors are thought to limit the loss of tissue in the injured spinal cord, contributing to the increased functional outcomes after BMSC transplantation.

example, increasing the production and secretion of certain growth factors by viral transduction of the cells before transplantation is currently used to gain mechanistic insights and to optimize the therapeutic effect of BMSC.

10. BMSC and axonal regeneration

BMSC transplantation may also lead to improved functional outcomes by promoting axonal regeneration. Spinal cord injury leaves axons cut and/or demyelinated which prevents or limits the conduction of information to their target cells, including signals essential for the generation and control of motor behaviors. Eliciting axonal growth after injury may lead to the reestablishment of previously existing axonal circuits that were involved in motor function or in the formation of new axonal circuits that could be recruited for motor functions. Remyelination is an integral part of axonal regeneration and crucial for proper functioning of axons. A major obstacle in axonal regeneration after spinal cord injury is the distance between injury, where the proximal end of the cut axon is located, and the target area, where synaptic connections need to be established. Many axonal regeneration therapies have shown to promote axonal regeneration but fail to elicit lengthy regeneration. The relative short regenerated sprouts may in the best scenario form synaptic connections with neurons close to the injury site, but not with original target neurons or with neurons nearby or in the target area. As a result, the resulting behavioral effects may be limited or absent, despite the axonal regeneration response.

In general, stem/progenitor cell transplants could promote axonal regeneration (i.e., growth and/or (re-)myelination) by producing and secreting molecules that initiate the growth-process in damaged neurons and promote oligodendrocytes to form myelin sheaths around new or existing undamaged but demyelinated axons. Secreted molecules that promote the differentiation of local oligodendrocyte progenitor cells into mature oligodendrocytes could also contribute effectively to the axonal regeneration process. Alternatively, stem/progenitor cells may elicit an axonal growth response by diminishing the growth-inhibitory nature of the injury environment especially that of the glial scar. This latter effect could set the stage for successful axon growth initiated by the stem/progenitor cells and/or endogenous axon sprouting.

There are few studies that have focused on the axonal regeneration promoting abilities of BMSC. Lu and co-workers (2004) demonstrated that transplantation of native BMSCs into the contused spinal cord promoted modest sensory and motor axon regeneration. In another study from Lu and colleagues (2005) it was shown that BMSC modified to produce and secrete higher levels of BDNF also promoted axonal regeneration. BMSCs produce and secrete several growth factors including GDNF, NGF, and BDNF that could positively affect axonal growth and/or myelination. To date, most studies that investigated the effects of BMSC transplants in the injured spinal cord have limited focus on axonal regeneration. In those studies that did focus on axonal regeneration lengthy regeneration was not demonstrated and the involvement of the responding axons in behavioral recovery not proven. Clearly, further investigations are necessary to establish the axonal regeneration-promoting abilities of BMSC transplants in the injured spinal cord. Especially effective in such studies could be techniques that silence the production of particular molecules in BMSC, for instance approaches based on small interfering RNA molecules, to elucidate which of the molecules if any are in fact needed for BMSC to elicit an axonal growth response.

11. BMSC and plasticity

BMSC transplants may lead to functional recovery after spinal cord injury by promoting plasticity within circuits that have been spared. Most spinal cord injuries in humans are anatomically incomplete; connections between brain and below-level spinal cord segments exist but are ineffective. Locomotor patterns can be generated by a complex network of axonal circuits which is present in the lumbar spinal cord segments and known as the central pattern generator (CPG). The existence of ineffective axonal connections and the CPG offers the possibility to elicit plasticity within these circuitries after spinal cord after injury possibly resulting in motor activities. Plasticity may occur at the molecular level resulting in changes in neuronal properties (e.g., excitability), or at the structural level resulting in new synaptic connections. Eliciting plasticity has gained much attention in the last decade as an interventional goal with the potential to elicit functional recovery after spinal cord injury.

Intraspinal stem/progenitor cell transplants may evoke plasticity via secreting molecules that could elicit changes in the functioning of neurons or modulations in synaptic organization. Although plasticity within the damaged spinal cord has been in the spotlight for some time, studies that involve stem/progenitor cells in general or BMSC in particular and address plasticity in the injured spinal cord are sparse. There are, however, many BMSC studies that have demonstrated some degree of functional restoration without clear evidence for underlying mechanisms. Possibly, plasticity was among the BMSC-mediated effects in the damaged spinal cord resulting in improved functional outcomes.

12. BMSC survival in damaged nervous tissue

Although functional and histological improvements are observed after BMSC transplantation in the injured rodent spinal cord, BMSC survival is typically poor, with the majority of cells dying within the first week after transplantation (Nandoe Tewarie et al., 2009). Possible cytotoxic factors include an extensive inflammatory response, free radical accumulation due to initial cell loss, lack of oxygen and nutrients due to the loss of blood vessels, and lack of a survival-promoting substrate. The question arises how BMSC transplant exert their repair effects if their survival is poor. To date, the answer to this question is not known but the most likely answer is that their relative short presence in the injury site is sufficient to bring about repair-supporting processes. This would point at neuroprotection as the main mechanism underlying BMSC-mediated repair because axonal regeneration and plasticity may require their presence in later stages after injury. Evidence that supports this idea is that BMSC transplantation at seven days post-injury or later does not lead to tissue sparing whereas transplantation immediately or three days after injury does (Nandoe Tewarie et al., 2009). An interesting associated question is whether BMSC are in fact necessary during later stages after injury. Future investigations will need to be conducted to obtain the right answers to these questions which will have major impact on BMSC-based spinal cord repair approaches.

The growth factors secreted by BMSCs may affect BMSC survival and/or proliferation *in vivo* through autocrine actions. Poor survival then may reflect inadequate amounts of these factors to positively affect their own survival within an extremely harsh injury milieu with many cells and factors that negatively influence survival. Timing of BMSC transplantation, combination with anti-inflammatory and/or immune-modulatory drugs, and transplantation within a surviving -promoting scaffold have been studied for their effects on BMSC survival.

13. Timing of transplantation

In an experiment by Nandoe Tewarie and colleagues (2006), BMSC were transplanted into a moderately contused adult rat spinal cord at 15 min, and at 3, 7, and 21 day post-injury and BMSC survival was closely assessed both during the transplantation procedure and up to four weeks after transplantation. In addition, the effect of the timing of BMSC transplantation on tissue sparing was determined. BMSC were collected from culture dishes, kept on ice, and passed through a glass pulled needle for injection into the contusion site. This procedure resulted in a majority (67 %) of the BMSC intended to be transplanted being present in the contusion at 15 min after transplantation. Thereafter, BMSC numbers rapidly decreased. The rate at which cell death occurs is different when transplanting acutely or delayed. In an acute transplantation paradigm (15 min post-contusion) and sub-acute transplantation paradigm (3 days post-injury) BMSC survival is better than in a delayed transplantation paradigm (7 days or 21 days post-injury). The percentages of BMSC in the contusion at seven days after transplantation are 32% and 52% for acute and sub-acute transplantation, respectively, and 9% for delayed transplantation. Four weeks after transplantation, almost no BMSC can be found in either paradigm (see figure 8). Interestingly, the presence of BMSC for this short period of time is sufficient to elicit tissue sparing. Acute and subacute transplantation, but not delayed transplantation results in neuroprotection, and tissue volumes in these paradigms are strongly correlated with the number of BMSC present (Nandoe Tewarie et al., 2009). These results indicate that timing of BMSC transplantation is important for optimal survival and neuroprotective effect, with acute and subacute transplantation being superior to delayed transplantation. However,

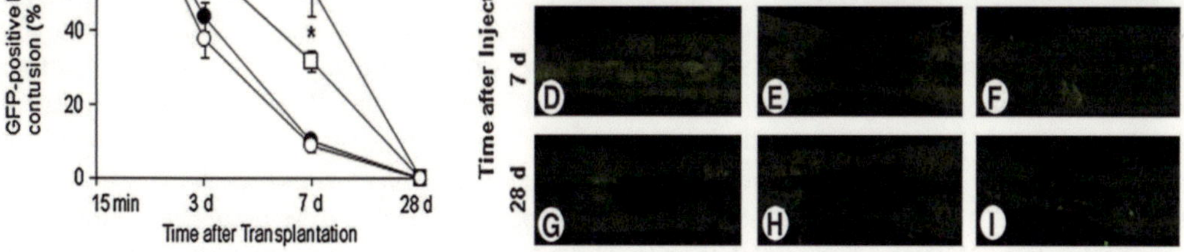

Fig. 8. **A.** BMSC numbers within a moderate contusion in the adult rat thoracic spinal cord decrease during 28 days post-injection. The rate at which cell death occurs is higher when BMSC are transplanted 7 or 21 days post-contusion, compared to BMSC transplantation 15 min or 3 days after contusion. **B.** The decreasing transplant is shown at 15 min (A–C), 7 days (D–F), and 28 days (G–I) after an injection at 15 min (acute), 7 days, and 21 days, respectively, post-injury. All microphotographs are from horizontal cryostat sections. (A) Scale bar, 600 mm in A-I.

because of the clinical relevance of delayed treatment, it seems imperative to find strategies to improve BMSC survival in delayed paradigms.
Previously, using a rat contusion injury model, Hofstetter and colleagues (2002) showed that more BMSCs survived when transplanted one week after injury compared to immediately after injury. The surviving cells were located within trabeculae that span the injury site. These data are in disagreement with those from the Nandoe Tewarie study (2009) although long-term results were in agreement with only 1% of the cells (about 3000 total) surviving at 4 weeks after grafting.. The difference in early survival between the two studies may be that Hofstetter and co-workers (2002) injected the BMSC not only into the contusion but also rostral and caudal thereof into the spinal cord nervous tissue. Possibly, the surviving cells were located nearby but not in the contusion epicenter. Most studies have reported a poor survival of BMSC. Nandoe Tewarie and colleagues (2009) demonstrated that the contusion milieu is less detrimental during the first week after injury than the second and fourth week after injury. What factors are important for BMSC survival *in vivo*? BMSCs are cultured in medium containing 10-20% serum. Factors other than present in serum are not essential for their survival and proliferation within the culture dish. In fact, addition of growth factors such as BDNF, FGF-2, or NT-3 instigates differentiation of the BMSCs into neural-like cells rather than affect survival. To date, the factors that may promote BMSC survival in vivo are unknown and further investigations are necessary to reveal them.

14. Anti-inflammatory drugs

It has been proposed that one of the mechanisms underlying death of cells transplanted into the spinal cord is injury-induced inflammation. The cellular and molecular components of the inflammatory response could initiate cell death, which would also explain improved survival with delayed grafting paradigms. Application of immune-modulatory molecules could possibly support better transplanted cell survival. A role of macrophages in death of transplanted cells has been suggested for a number of cell types and anti-inflammatory drugs often improve the outcome. A direct relationship between number of macrophages and transplanted cell survival is less clear in case of BMSC. For instance, if macrophage number plays a role in survival of BMSC then delayed transplantation paradigms at time points when macrophage numbers are lower would lead to better survival. This was not demonstrated by Nandoe Tewarie and colleagues (2009). On the contrary, delayed grafting into the adult rat contused spinal cord resulted in decreased survival rates of transplanted BMSC.
Ritfeld and colleagues (2010) studied the effect of pharmacologically decreasing the number of macrophages in the contused rat spinal cord on transplanted BMSC survival. Systemic treatment with the anti-inflammatory agents methylprednisolone, minocycline, or cyclosporine, all showed an effective decrease in macrophage presence in the contused spinal cord at three days post-contusion, the time point of BMSC transplantation. However, the decreased number of macrophages did not significantly improve BMSC survival. A trend towards improved survival was seen, however, for cyclosporine, and accordingly, other studies have reported beneficial effects of cyclosporine on BMSC survival. Conflictingly, BMSC have been reported to have the ability to evade immune responses due to low expression of MHC I and no expression of MHC II molecules. As such, BMSC are being explored for their potential use as immune suppressing agents in combination with other cell transplants. At present, no consensus exists on the necessity, type or dose of immune-suppressing agents for BMSC transplantation.

15. Survival promoting scaffolds

Another means of improving BMSC survival in the injured spinal cord is the use of biomaterials as scaffolds for BMSC transplants. Hydrogels, extracellular matrix-based materials (e.g.fibronectin), agarose scaffolds and fibrin scaffold are among the materials being used as scaffolds for BMSC transplantation. Typically, BMSC survival can be improved by use of such materials, either by providing the cells with a substrate for survival/growth and/or by protecting the cells from detrimental immune responses.

16. Clinical application of BMSC

BMSC have several features that make them appealing candidates for transplantation after spinal cord injury in the human. BMSC can be easily isolated under local anesthetics, rapidly and extensively expanded in cell culture and there is no evidence of tumorigenicity in vivo, even after immortalization to ensure an unlimited source of self-renewal ex vivo. In addition, these cells have demonstrated capacity for tissue repair, and secrete growth factors that enhance histological and functional repair. Clearly, BSMC are a promising candidate for transplantation into the injured spinal cord. However, since considerable variation consists among donors regarding gender, genetics, and general health, specific parameters need to be found that allow rapid and reliable selection of BMSC with therapeutic potential. In addition, it is also clear that our current understanding of BSMC function is not yet sufficient to provide us with the so-called "silver-bullet", the one therapy that will promote regeneration and restore function in the injured spinal cord. In fact, it is generally accepted in the field of spinal cord injury research that such a therapy does not exist. Spinal cord injury results in a particularly complex cascade of histopathological events and a treatment aimed at functional repair requires that these events are dealt with in a timely, most likely sequential, fashion. Despite our incomplete knowledge, several clinical trials are being conducted in Korea, Mexico, Columbia, Brazil, and India. In a Korean study, 35 complete spinal cord injured patients received a BMSC transplant, combined with granulocyte macrophage-colony stimulating factor (GM-CSF) administration within 2 weeks, between 2 weeks and 8 weeks or after 8 weeks of spinal cord injury. Patients were assessed using the American Spinal Injury Association Impairment Scale (ASIA), electrophysiology, and magnetic resonance imaging (MRI). No adverse histological were observed on MRI at four months and no serious clinical adverse effects were reported. In 30.4 % of the acute and subacute treated group ASIA score increased to ASIA B or ASIA C after a median follow-up period of 10.4 months after injury. The chronic treatments group showed no improvements. Although these results are promising, neuropathic pain and tumor formation remain to be evaluated and a long-term and large scale multicenter clinical study is required for more accurate assessment of efficacy. Other studies have reported no adverse effects of BMSC transplants, but efficacy remains to be revealed.

17. Conclusion

Stem cells have gained attraction over the last years in the field of neuroscience. In vitro it has been shown, although still disputed, that BMSC can transdifferentiate into cells of neural lineage. This has made this adult stem cell type interesting for neural transplantation paradigms. We have shown that after transplantation of BMSC in the injured spinal cord most cells die. Nevertheless, especially in early transplantation, cells have a neuroprotective effect on the host tissue. This effect may well be the result of secretion of growth factors. Further studies are needed to investigate the true potential of BMSC.

18. References

Ackery A, Tator C, Krassioukov A (2004). A global perspective on spinal cord injury epidemiology. *J Neurotr* 21 (10): 1355-70.

Akiyama Y, Radtke C & Kocsis JD (2002). Remyelination of the rat spinal cord by transplantation of identified bone marrow stromal cells. *J Neurosci* 22(15): 6623-6630.

Ankeny DP, McTigue DM, Jakeman LB (2004). Bone marrow transplants provide tissue protection and directional guidance for axons after contusive spinal cord injury in rats. *Exp Neurol* 190:17-31.

Azim S (2004). How to make eggs and sperm. *Nature* 427: 106-7.

Bjornson CR, Rietze RL, Reynolds BA et al. (1999). Turning brain into blood: a hematopoietic fate adopted by adult neural stem cells in vivo. *Science* 283: 534-537.

Blight AR (1992). Macrophages and inflammatory damage in spinal cord injury. *J Neurotr* 9: S83-91.

Boekamp JR, Overholser JC, Schubert DS (1996). Depression following a spinal cord injury. *Int J Psychiatry Med* 26: 329-349.

Botel U, Glaser E & Niedeggen A (1997). The surgical treatment of acute spinal paralyzed patients. *Spinal Cord* 35: 420-428.

Bracken MB, Shepard MJ, Collins WF, et al. (1990). A randomized, controlled trial of methylprednisolone or naloxone in the treatment of acute spinal-cord injury. Results of the Second National Acute Spinal Cord Injury Study. *N Eng J Med* 322: 1405-1411.

Bracken MB, Shepard MJ, Holford TR, et al. (1997). Administration of methylprednisolone for 24 or 48 hours or tirilazad mesylate for 48 hours in the treatment of acute spinal cord injury. Results of the Third National Acute Spinal Cord Injury Randomized Controlled Trial. National Acute Spinal Cord Injury Study. *JAMA* 277: 1597-1604.

Bracken MB (2002). Steroids for acute spinal cord injury. *Cochrane Database Syst Rev* 3: CD001046.

Brodkey JS, Miller CF & Harmody RM (1980). The syndrome of acute central cervical spinal cord injury revisited. *Surg Neurol* 14: 251-257.

Chen TY, Dickman CA, Eleraky M, et al. (1998). The role of decompression for acute incomplete cervical spinal cord injury in cervical spondylosis. *Spine* 23: 2398-2403.

Chopp M, Zhang XH, Li Y, et al. (2000). Spinal cord injury in rat: Treatment with bone marrow stromal cell transplantation. *Neuroreport* 11(13): 3001-5.

Cogle CR, Yachnis AT, Laywell ED, et al. (2004). Bone marrow transdifferentiation in brain after transplantation: A retrospective study. *Lancet* 363: 1432-37.

Corti S, Locatelli F, Donadoni C, et al. (2002). Neuroectodermal and microglial differentiation of bone marrow cells in the mouse spinal cord and sensory ganglia. *J Neurosci Res* 70(6): 721-733.

Cushing H (1927). Organization and activities of the neurological service American expeditionary forces. In: Ireland MW, ed. The medical department of the United

States Army in the world war. Washington, DC: Government Printing Office, 1927: 749-58.

Cyranoski D (2006). Patients warned about unproven spinal surgery. *Nature* 2006; 440: 850-851.

Da Paz AC, Beraldo PS, Almeida MC, et al. (1992). Traumatic injury to the spinal cord. Prevalence in Brazilian hospitals. *Paraplegia* 30 (9): 636-40.

Dietz V & Harkema SJ (2004). Locomotor activity in spinal cord-injured persons. *J Appl Physiol* 96: 1954-60.

Dincer F, Oflazer A, Beyazova M, et al. (1992). Traumatic spinal cord injuries in Turkey. *Paraplegia* 30 (9): 641-6.

Ditunno JF & Formal CS (1994). Chronic spinal cord injury. *N Eng J Med* 330: 550-56.

Ditunno JF, Burns AS, Marino RJ (2005). Neurological and functional capacity outcome measures: essential to spinal cord injury clinical trials. *J Rehabil Res Dev* 42: S35-41.

Ditunno JF, Barbeau H, Dobkin BH, et al. (2007). Spinal Cord Injury Locomotor Trial Group. Validity of walking scale for spinal cord injury and other domains of function in a multicenter clinical trial. *Neurorehabil Neural Repair* 21 (6): 539-50.

Dobkin BH, Barbeau H, DeForge D, et al. (2007). Spinal Cord Injury Locomotor Trial Group. The evolution of walking-related outcomes over the first 12 weeks of rehabilitation for incomplete traumatic spinal cord injury: the multicenter randomized Spinal Cord Injury Locomotor Trial. *Neurorehabil Neural Repair* 21 (1): 25-35.

Fehlings MG & Perrin RG. The timing of surgical intervention in the treatment of spinal cord injury: a systematic review of recent clinical evidence. *Spine* 15: S28-35.

Feron F, Perry C, Cochrane J, et al. (2005). Autologous olfactory ensheating cell transplantation in human spinal cord injury. *Brain* 128: 2951-60.

Fiefler IG, Laud PW, Maiman DJ, et al. (1999). Economics of managed care in spinal cord injury. *Arch Phys Med Rehabil* 80: 1441-49.

Geijsen N, Horoschak M, Kim K, et al. (2004). Derivation of embryonic germ cells and male gametes from embryonic stem cells. *Nature* 427: 148-54.

Geisler FH, Coleman WP, Benzel E, et al. (2002). Spinal cord injury. *Lancet* 360: 1883.

Goldenring JM (1985). The brain-life theory: towards a consistent biological definition of humanness. *J Med Ethics* 11: 198-204.

Hagg T, Oudega M (2006). Degenerative and spontaneous regenerative processes after spinal cord injury. *J Neurotr* 23: 264-80.

Hart C, Williams E (1994). Epidemiology of spinal cord injuries: a reflection of changes in South African society. *Paraplegia* 32 (11): 709-14.

Hofstetter CP, Schwarz EJ, Hess D, et al. (2002). Marrow stromal cells form guiding strands in the injured spinal cord and promote recovery. *Proc Natl Acad Sci* 99(4): 2199-2204.

Hoque MF, Grangeon C, Reed K (1999). Spinal cord lesions in Bangladesh: an epidemiological study 1994 - 1995. *Spinal Cord* 37 (12): 858-61.

Horwitz EM, Prockop DJ, Fitzpatrick LA, et al. (1999). Transplantability and therapeutic effects of bone marrow-derived mesenchymal cells in children with osteogenesis imperfecta. *Nature Med* 5(3): 309-13.

Hsu SP, Shih YH, Huang MC, et al. (2004). Repair of multiple cervical root avulsion with sural nerve graft. *Injury* 35: 896-907.

Hubner K, Fuhrmann G, Christenson LK, et al. (2003). Derivation of oocytes from mouse embryonic stem cells. *Science* 300: 1251-6.

Inoue M, Honmou O, Oka S, et al. (2003). Comparative analysis of remyelinating potential of focal and intravenous administration of autologous bone marrow cells into the rat demyelinated spinal cord. *Glia* 44(2): 111-18.

Johansson CB (2003). Mechanism of stem cells in the central nervous system. *J Cell Physiol* 196: 409-418.

Kennedy P, Evans MJ, Berry C, et al. (2003a). Comparative analysis of goal achievement during rehabilitation for older and younger adults with spinal cord injury. *Spinal Cord* 41 (1): 44-52.

Kennedy P, Berry C, Coggrave M, et al. (2003b). The effect of a specialist seating assessment clinic on the skin management of individuals with spinal cord injury. *J Tissue Viability* 13 (3): 122-5.

Knoller N, Auerbach G, Fulga V, et al. Clinical experience using incubated autologous macrophages as a treatment for complete spinal cord injury: phase 1 study results. *J Neurosurg Spine* 3: 173-81.

Krassioukov AV, Furlan JC, Fehlings MG. (2003). Medical co-morbidities, secondary complications and mortality in elderly with acute spinal cord injury. *J Neurotr* 20: 391-99.

Lee J, Kuroda S, Shichinohe H, et al. (2003). Migration and differentiation of nuclear fluorescence labeled bone marrow stromal cells after transplantation into cerebral infarct and spinal cord injury in mice. *Neuropathol* 23: 169-80.

Lee HC, Cho DY, Lee WY, et al. (2007). Pitfalls in treatment of acute cervical spinal cord injury using high dose of methylprednisolone: a retrospect audit of 111 patients. *Surg Neurol* 68: S37-42.

Lima C, Pratas-Vital J, Escada P, et al. Olfactory mucosa autografts in human spinal cord injury: a pilot clinical study. J Spin Cord Med 2006; 29: 191-206.

Lu P, Yang H, Jones LL, et al. (2004). Combinational therapy with neurotrophins and cAMP promotes axonal regeneration beyond sites of spinal cord injury. *J Neurosci* 24(28): 6402-6409.

Lu P, Jones LL, Tuszynski MH (2005). BDNF-expressing marrow stromal cells support extensive axonal growth at sites of spinal cord injury. *Exp Neurol* 191(2): 344-360.

Mariano AJ (1992). Chronic pain and spinal cord injury. *Clin J Pain* 8: 87-92.

McKinley WO, Jackson AB, Cardenas DD, et al. (1999). Long-term medical complications after traumatic spinal cord injury: a regional model systems analysis. *Arch Phys Med Rehabil* 80 (11): 1402-10.

McKinley WO, Tewksbury MA, Sitter P, et al. (2004). Assistive technology and computer adaptations for indiciduals with spinal cord injury. *Neurorehabil* 19: 141-146.

Mezey E, Chandross KJ, Harta G, et al. (2000). Turning blood into brain: Cells bearing neuronal antigens generated in vivo from bone marrow. *Science* 290: 1779-80.

Morrison SJ, Shah NM, Anderson DJ (1997). Regulatory mechanism in stem cell biology. *Cell* 88: 287-98.

Nandoe Tewarie RDS, Hurtado A, Levi AD, et al. (2006). Bone marrow stromal cells for repair of the spinal cord: towards clinical application. *Cell Transplantation* 15(7): 563-577.

Nandoe Tewarie RDS, Hurtado A, Ritfeld GJ, et al. (2009). Bone marrow stromal cells elicit tissue sparing after acute but not delayed transplantation into the contused adult rat thoracic spinal cord. *J Neurotr* 26 (12): 2313-22.

Nesutherai S (1998). Steroids and spinal cord injury: revisiting the NASCIS 2 and NASCIS 3 trials. *J Trauma* 45: 1088-93.

Nesutherai S (2001). The role of methylprednisolone in acute spinal cord injuries. *J Trauma* 51: 421-23.

Newson AJ & Smajdor A (2005). Artificial gametes: new paths for parenthood. *J Med Ethics* 31: 184-86.

Ohta M, Suzuki Y, Noda T, et al. (2004). Bone marrow stromal cells infused into the cerebrospinal fluid promote functional recovery of the injured rat spinal cord with reduced cavity formation. *Exp Neurol* 187 (2): 266-78.

Orlic D, Kajstura J, Chimenti S, et al. (2001). Transplanted adult bone marrow cells repair myocardial infarcts in mice. *Ann Acad Sci* 938: 221-29.

Papadopoulos SM, Selden NR, Quint DJ, et al. (2002). Immediate spinal cord decompression for cervical spinal cord injury : feasibility and outcome. *J Trauma* 52: 323-32.

Patki P, Hamid R, Shah J, et al. (2007). Fertility following spinal cord injury: a systematic review. *Spinal Cord* 45: 187.

Peterfy A (1995). Fetal viability as a threshold to personhood. A legal analysis. *J Leg Med* 16: 607-36.

Petersen BE, Bowen WC, Patrene KD, et al. (1999). Bone marrow as a potential source of hepatic oval cells. *Science* 284 (5417): 1168-70.

Pittenger MF, Mackay AM, Beck SC, et al. (1999). Multilineage potential of adult human mesenchymal stem cells. *Science* 284 (5411): 143-47.

Potten CS, Loeffler M (1990). Stem cells: Attributes, cycles, spirals, pitfalls, and uncertainties. Lessons for and from the crypt. *Development* 110: 1001-20.

Qian T, Guo X, Levi AD, et al. (2005). High-dose methylprednisolone may cause myopathy in acute spinal cord injury patients. *Spinal Cord* 43: 199-203.

Ritfeld GJ, Nandoe Tewarie RD, Rahiem ST, et al. (2010). Reducing macrophages to improve bone marrow stromal cell survival in the contused spinal cord. *Neuroreport* 21 (3): 21-6.

Roman Catholic Church (1994). Cathechism of the Catholic Church (paragraphs 2274; 2322). Vatican City by Libreria Editrice Vaticana. 2nd English translation.

Satake K, Lou J, Lenke LG (2004). Migration of mesenchymal stem cells through cerebrospinal fluid into injured spinal cord tissue. *Spine* 29 (18): 1971-79.

Sawatzky B, Bishop CM & Miller WC (2008). SCIRE Research team: Classification and measurement of pain in the spinal cord-injured population. *Spinal Cord* 46: 2-10.

Schwartz M & Yoles E (2006). Immune-based therapy for spinal cord repair: autologous macrophages and beyond. *J Neurotr* 23: 360-70.

Sell S (2004). Stem Cells Handbook, Totowa, NJ: Humana Press Inc. ISBN 1-58829-113-8.

Sigurjonsson OE, Perreault MC, Egeland T, et al. (2005). Adult human hematopoietic stem cells produce neurons efficiently in the regenerating chicken embryo spinal cord. *Proc Natl Acad Sci* 102 (14): 5227-32.

Slack JM (2007). Metaplasia and transdifferentiation: from pure biology to the clinic. *Nat Rev Mol Cell Biol* 8 (5): 369-78.

Strauss DJ, DeVivo MJ, Paculdo DR, et al. (2006). Trends in life expectancy after spinal cord injury. *Arch Phys Med Rehabil* 87: 1079-85.

Testa G, Harris J (2005). Ethics and synthetics gametes. *Bioethics* 9: 146-66.

Wakitani S, Saito T, Caplan AI (1995). Myogenic cells derived from rat bone marrow mesenchymal stem cells exposed to 5-azacytidine. *Muscle Nerve* 18 (12): 1417-26.

Weaver LC, Marsh DR, Gris D, et al. (2006). Autonomic dysreflexia after spinal cord injury: central mechanisms and strategies for prevention. *Prog Brain Res* 152: 245-63.

Westpal SP (2003). The next IVF revolution? *New Scientist* 178: 3.

Whittaker P (2007). Stem cells to gametes: how far should we go? *Human Fertility* 10 (1): 1-5.

Widerstrom-Noga EG, Felipe-Cuervo E, Broton JG, et al. (1999). Perceived difficulty in dealing with consequences of spinal cord injury. *Arch Phys Med Rehabi*l 80: 580-86.

Widerstrom-Noga EG, Felix ER, Cruz-Almeida Y, et al. (2007). Psychosocial subgroups in persons with spinal cord injuries and chronic pain. *Arch Phys Med Rehabil* 88: 1628-35.

Winslow C, Bode RK, Felton D, et al. (2002). Impact of respiratory complications on length of stay and hospital costs in acute cervical spine injury. *Chest* 121 (5): 1548-54.

Wu S, Suzuki Y, Ejiri Y, et al. (2003). Bone marrow stromal cells enhance differentiation of cocultured neurosphere cells and promote regeneration of injured spinal cord. *J Neurosci Res* 72 (3): 343-51.

Yoon SH, Shim YS, Park YH (2007). Complete spinal cord injury treatment using autologous bone marrow cell transplantation and bone marrow stimulation with granulocyte macrophage-colony stimulating factor: Phase I/II clinical trial. *Stem cells* 25 (8): 2066-73.

Zurita M & Vaquero J (2004). Functional recovery in chronic paraplegia after bone marrow Stromal cells transplantation. *Neuroreport* 15 (7): 1105-08.

23

What Do We Know About the Detailed Mechanism on How Stem Cells Generate Their Mode of Action

Peter Riess[1] and Marek Molcanyi[2]
[1]*Clinic of Trauma and Orthopedic Surgery, HELIOS Klinikum Siegburg / Institute for Research in Operative Medicine (IFOM), University of Witten/Herdecke, Cologne-Merheim Medical Center (CMMC), Cologne*
[2]*Clinic of Neurosurgery, University of Cologne*
Germany

1. Introduction

Transplantation of stem cells may provide cures for the damaged Central nervous system (CNS). They hold an enormous potential in cell replacement therapy following traumatic brain injury (TBI) and received a great scientific and public interest in recent years. TBI remains the leading cause of long-term neurological disabilities, including cognitive, sensory, motor and emotional impairments among children and young adults. It has been suggested that stem cells hold great potential for the repair of the damaged nervous system. The therapeutic potential of stem cells has been examined in experimental brain injury using a variety of approaches. Although these results emphasized their potential therapeutic role in traumatic brain injury, crucial mechanism on how stem cells take effect, e.g. timing of stem cell implanation, stem cell survival and integration, effect of brain microenvironment and local trophic support, will be explained in this Overview of Different In-vivo/In-vitro Experimental Settings.

TBI is the leading cause of death and disability worldwide (Bruns 2003). Thus, TBI is a highly relevant medical and socio-economic problem of modern society. During the last two decades, improvements in acute pre- and inhospital care, time management, diagnostic procedures, and rehabilitation strategies have substantially improved the level of care and outcome following TBI (Maegele 2007). But still, to date, no therapeutic approach has been proven effective in reversing the pathologic cellular sequelae underlying the progression of cell loss and in improving neurobehavioral outcome. As the brain has limited capacity for self-repair, restorative approaches with focus on replacement and repair of dysfunctional and dead cells by transplantation of cells (eg stem cells) into the traumatically injured brain have been studied. In the last 15 years various types of cells have been tested for their potential to restore the function in animal models after TBI. These cells include fetal cells, adult stem cells isolated from bone marrow as well as pluripotent stem cells (see review Schouten 2004).

In the traumatically injured brain, several experimental studies have been performed using engrafted hNT cells. These are post-mitotic human neurons (NT2N cells, commercially known as hNT cells, or LBS-neurons; Layton Bioscience Inc., Palo Alto, CA), derived from a human embryonal teratocarcinoma line (NT2 cell line) and differentiated into an exclusively neuronal phenotype by retinoic acid treatment *in vitro* (Trojanowski 1993). These cells appear to possess many of the key features of normal developing and mature human neurons, and survive up to 1 year in the brain after transplantation in immunodeficient neonatal or adult mice (Trojanowski et al., 1997). Muir (1999) first transplanted hNT cells into the injured cortex at 24 h after lateral Fluid Percussion (FP) brain injury in adult rats. The hNT transplant remained viable for up to 2 weeks, although no significant effect on acute posttraumatic neurologic motor function was observed. A follow-up study reported long-term (4-week) survival of hNT cells transplanted into the injured cortex 24 h following lateral FP brain injury in non-immunosuppressed adult rats. Integration of the graft in the peri-injured cortex was noted, again without significant improvement in motor or cognitive function (Philips 1999). More recently, hNT cells genetically engineered *ex vivo* to express NGF, transplanted into the medial septum at 24 h following Controlled Cortical Impact (CCI) brain injury in mice, attenuated cognitive dysfunction for up to 4 weeks post-transplant (Watson 2003; Longhi 2005).

Bone marrow stromal cells (BMSCs) have been evaluated in the CCI model of TBI in adult rats, administered by intra-arterial or intravenous injection at 24 h post-injury were subsequently found in multiple organs, including the brain. In these studies, improvement of neurological outcome and cellular expression of both the neuronal marker NeuN and the astrocytic marker GFAP in the engrafted cells were observed at 1 and 2 weeks post-administration (Lu 2001; Mahmood 2001). Intraparenchymal transplantation of whole bone marrow into the pericontusional tissue at 24 h after CCI brain injury in rats also resulted in improved functional outcome and differentiation of transplanted cells into populations of cells expressing neuronal and glial markers up to 4 weeks post-transplantation (Mahmood 2001).The mechanism by which BMSCs limit damage or promote repair has been suggested to be either via cell replacement by proliferation and differentiation of transplanted BMSCs into the phenotype of the damaged and/or lost cells, via trophic support, or via manipulation of the environment to stimulate endogenous regeneration (Hofstetter 2002). However, Breitbach showed, that the developmental fate of BM-derived cells is not restricted by the surrounding tissue and that the MSC fraction may underlie extended bone formation. These findings seriously question the biologic basis and clinical safety of using BM and in particular MSCs to treat nonhematopoietic disorders (Breitbach 2007).

Immortalized cells, eg HiB5 cells, derived from the embryonic rat hippocampus and conditionally immortalized, were first transplanted into the neonatal brain, where they acquired neuronal and glial morphologies appropriate to the site of transplantation. Stable transduction of HiB5 cells to secrete (Nerve Growth Factor) NGF in vivo have been observed up to 9 months following transplantation into the medial septum, resulting in a prevention or reversal of cholinergic neuronal atrophy and related behavioral impairments normally occurring with age (Martinez-Serrano 1996, 1998). In non-immunosuppressed rats subjected to lateral FP brain injury, HiB5-NGF cells were transplanted in the injured cortex, 24 h after injury (Philips 2001). Thereby brain-injured animals receiving either the HiB5-

NGF cells or untransduced HiB5 cells showed significant improvements in neuromotor function and spatial learning, but hippocampal cell death was significantly reduced only in the HiB5-NGF cell transplant group, indicating that the transplantation of HiB5 cells genetically engineered to secrete trophic factors may have behavioral and neuroprotective effects after brain injury (Philips 2001).

The C17.2 cell is a clonal multipotent progenitor cell derived from the external germinal layer of the neonatal murine cerebellum, immortalized by retroviral transduction of the avian gene myc (Ryder 1990). The cells were also marked with a second retrovirus for expression of bacterial X-galactosidase. A recent study form our working group has evaluated the behavioral effects of engrafted C17.2 cells following experimental TBI. In adult mice subjected to CCI brain injury, transplantation of C17.2 cells into the cortex (either ipsilateral or contralateral to the injury) at 3 days after injury significantly improved motor function but not cognitive function over a 12-week post-transplantation period (Riess 2002). Following engraftment, the C17.2 cells expressed either neuronal or astrocytic markers when transplanted ipsilateral to the lesion, while after contralateral transplantation only neuronal differentiation was observed.

Pluripotent murine embryonic stem cells (ESC) have been shown to survive in the implanted healthy and hypoxic injured brain. Furthermore they differentiate into neural cell types following transplantation into rat brains in an experimental stroke model. Additionally it was shown, that they migrate along the corpus callosum to the ventricular walls and they populate the border zone of the damaged brain tissue on the hemisphere opposite to the implantation site, indicating that ES cells have high migrational dynamics (Hoehn 2002, Erdoe 2003). Based on these results, our working group has implanted these undifferentiated, eGFP-expressing ES-D3 cells into the ipsi- or contra-lateral cortex of rat brains following the induction of a moderate lateral fluid-percussion (FP) brain injury (Riess 2007, Molcanyi 2007).

As one result we were able to show, that the differentiation and integration of transplanted ES-D3 cells was barely observed at any time point. But sporadic tumor formation was observed. However, ES-D3 cell grafted animals demonstrated a significant improvement in functional outcome on a range of behavioral tasks. Motor function improvements were observed as early as one week following implantation. Other observers are reporting that transplantation of pre-differentiated ES cells improved behavioural outcome on sensorimotor and locomotor tests but failed to improve cognitive function during memory tasks assessed by Morris water maze (Hoane 2004). An improvement of motor function was also demonstrated when motoneuron enriched neural cells derived from mouse embryonic stem cells were transplanted after cryogenic brain injury (Chiba 2003).

However, the success of TX depends not only on cell type and their differentiation state, but also on several other important parameters, such as inflammatory response after injury, microenvironment of the host tissue, time window for TX, immune response to cell TX, immunosuppression, site of TX.

2. Location for TX

Miyazono (Miyazono 1995) demonstrated that the proliferation, differentiation and survival of implanted cells were modulated by the anatomical target site into which the grafts were placed. The implantation of undifferentiated Ntera-derived cells into rat cortex resulted into formation of lethal tumors within 70 days posttransplantation. In contrast, the same cells implanted into caudoputamen ceased to proliferate and progressively differentiated into postmitotic neuron-like cells. Following transplantation of C17.2 NSCs following CCI brain injury, our working group was able to show, that cell differentiation depends on the location of the implantation site. Thirteen weeks after the transplantation under the contusion cavity, NSCs were found in the ipsilateral hippocampus and in the cortical parenchyma adjacent to the injury cavity. NSCs transplanted into the contralateral hemisphere were also detected in the contralateral hippocampus and the contralateral cortex. The NSCs in the injured cortex were found to express neuronal and astrocytic markers, implicating differentiation into neurons and glia, whereas the NSCs transplanted into the uninjured contralateral cortex showed an almost exclusively neuronal phenotype. The reasons underlying these apparent variations in differentiation remain unclear, although the demand for both neurons and glial cell replacement may be far greater in the areas of the injured hemisphere that have sustained significant tissue loss.

3. Timepoint for TX

Beyond cell specific effects on functional improvement following TX Soares and coworkers suggested that an optimal time window for TX may exist (Soares 1991). However, studies evaluating optimal time for transplantation in models of experimental TBI are still pending. In most studies cells were implanted between 24 and 72 hours following TBI. However this relatively early time point for transplantation may be appropriate for avoiding the peak of any inflammatory reaction and may allow the integration, migration and differentiation of cells prior to formation of the astroglial scar that may counteract these processes. Glial scar formed by CNS injury is considerd as the main inhibitory barrier of nerve regeneration. Many efforts have been made to prevent scarring and to pomote axonal regeneration after injury (see review: Sofroniew 2010). In their experiment, Soares and coworkers demonstrated that there is a temporal window in which fetal cortical transplants can attenuate glial scarring as well as be successfully incorporated into host brains following FP injury. They showed that cells transplanted two days, one week and two weeks after injury survived and were incorporated into the host tissue whereby the cells transplanted at later time points (4 weeks) failed to incorporate. This phenomenon was explained by the fact that there was a little evidence of a glial scar formation at day two and one week time points, whereas the scarring raised significantly thereafter (Soares 1991). More evidence for a optimal time window for TX is shown by comparing the following experiments in experimental Spinal Cord Injury. Han (Han 2002) reported that immediately TX of neuronal restricted progenitors (NRPs) after injury gives rise to neurons and survived for at least one month. However, after transplantation 10 days after spinal cord injury, NRPs did not give rise to neurons, suggesting that the environment of the injured spinal cord influences the implanted cells (Cao 2002). The influence of ES cells due to host environment was also proven in the following in vitro experiment. Hereby, we have shown that the extract derived from rat brains in the acute phase following TBI impairs survival of undifferentiated murine ES cells and induces rapid differentiation of surviving cells. This observation suggests that during the early acute phase of traumatic injury the cerebral environment contains detrimental as well as protective signals that may induce neurogenic processes following ES cell transplantation (Bentz 2010). However, it is also likely that different cell types will have different transplantation time points for optimal survival, migration, and proliferation. For example, bone marrow-derived stem cells (BMSCs) have been suggested to engraft better as damage progresses in injured tissue (Prockop 2002).

4. Inflammatory response

Inflammatory leukocytic recruitment and diffuse neuronal degeneration are pathological processes resulting from TBI. While the normal brain is generally considered to be a relatively immunologically privileged organ, the injured brain is certainly not. The presence of an inflammatory response in the injured parenchyma may increase damage of the host brain in the early post-traumatic phase, while becoming more beneficial in the chronic phase (Allan 2001, Lenzlinger 2001). The effects of this inflammatory reaction on the graft survival following TX are still poorly understood. Our working group examined the time dependent fate of ES cells following ipsi- and contralateral implantation into rat brains injured via FP injury. Double-staining for GFP and macrophage antigens revealed stem cells phagocytosed by infiltrated and activated macrophages, indicating the loss of implanted stem cells was due to an early posttraumatic inflammatory response. Macrophage infiltration was shown to be less pronounced when stem cells were implanted into completely intact healthy brains. We therefore suggested that the massive macrophage infiltration at graft sites might be ascribed to the combined stimulus exhibited by the FP brain injury and the cell implantation (Molcanyi 2009).

5. Tumor formation

It has to be taken into consideration that the highly proliferative characteristics (selfrenewal potential) of ES cells combined with the ability to differentiate into all embryonic germinal layers (pluripotency) present a potential threat of tumor development (teratoma, teratocarcinoma) when they are transplanted into the adult CNS. However, tumorigenesis has been observed after implantation of undifferentiated human ESCs into healthy rat brains, giving rise to teratomas and malignant teratocarcinomas (Thomson 1998). Accordingly, Erdö compared the tumorigenic outcome after implantation of D3, clone BAC-7 ESCs in a homologous (mouse to mouse) vs. xenogeneic (mouse to rat) stroke model. In injured and healthy mouse brains, both transplanted undifferentiated and pre-differentiated murine ESCs produced highly malignant teratomas, while mouse ESCs xenotransplanted into injured rat brain migrated towards the lesion and differentiated into neurons at the border zone of the ischemic infarct, suggesting that tumorigenesis may be related to the host animal rather than to the differentiation status of the implanted cells (Erdo 2003)

Our scientific group has repeated an analogous experiment in the model of traumatic brain injury vs. healthy rat brain. After TBI we observed a scarce tumor formation, but in healthy rat brains the above mentioned cell line lead to formation of malignant teratocarcinomas in the majority of engrafted animals. The absence of tumor formation in animals suffering from traumatic brain injury was linked to the already described inflammatory response. As we believe, the tumorigenic fraction of implanted graft may have been scavenged by activaed macrophages, alongside with concomitant survival of stem cells turning into healthy neural phenotypes. Great caution is needed when stem cells are implanted in experimental settings of diseases associated with inflammatory response (such as stroke or traumatic brain injury) as the tumorigenic threat may stay unveiled (in case tumorigenic fraction is being removed by activated immune cells) (Molcanyi 2009).

6. Functional outcome

Neurobehavioral assessment of outcome has always played an integral part in traumatic brain injury (TBI) research (Hamm 2001).

After transplantation of the C17.2 Neural Stem Cells intracerebrally in the acute period after TBI we were able to show, that these cells survive, differentiate, and attenuate posttraumatic neurological deficits in the chronic postinjury period. Brain-injured mice that received NSC transplants showed significantly improved performance in the rotating pole test during the 8-week observation period. In addition, brain-injured animals that received NSCs in the ipsilateral hemisphere exhibited improved performance in the rotarod test during the 12-week observation period. Similar effects were achieved by, transplantation of undifferentiated ES cells following experimental traumatic brain injury. Hereby the TX significantly attenuates the impairment of motor function. Performance on the rotarod test and sensorimotor scores improved significantly when brain injured animals received ES cells. This is in accordance with previous brain injury studies that also reported recovery of function following cell transplantation. Improved behavioural outcome on sensorimotor and locomotor tests in brain injured animals was also demonstrated following transplantation of pre-differentiated ES cells, or minced fetal cortical grafts (E16), respectively (Hoane 2004).

But, the effects on neurobehavior outcome can be influenced by the medication the animals receive. Cyclosporin A (CsA) is widely used in clinical situations to attenuate graft rejection following organ and central nervous system transplantation. Therefore, we evaluated the influence of post-injury CsA administration on behavioral recovery after TBI. Hereby we found that, injured animals treated with CsA showed a significant improvement in motor function and in sensorimotor function, when compared to vehicle treated, injured animals. In conclusion, the treatment with CsA improves certain aspects of motor and sensorimotor function following experimental TBI. Therefore in animal studies analyzing functional outcome, these effects have to be controlled for (Riess 2001).

However, all the mechanisms of how stem cells attenuate a neurologic impairment are not completely clarified, yet. In the light of the recent data, the true cell replacement of lost/injured tissue seems to be highly unlikely. Humoral/trophic mechanisms accompanied by cell-cell interactions have been proposed to play a central role of stem cell grafting effect. Our scientific group has investigated the behaviour of stem cells in-vitro after conditioning with the cerebral extract derived from healthy and injured rat brain. Stimulated cells have produced statistically significant amounts of various neurotrophic factors, proving this phenomena to be possibly one of the regenarative mechanisms following stem cell transplantation (Bentz 2007).

7. Conclusion and further prospectives

In the present review we have summarized our experience with in-vivo and in-vitro set-ups of cell therapy in TBI. Our results demonstrate the influence of brain microenvironment on alterations on stem cell fate, the significance of appropriate timepoint for TX and the effect of the time dependent inflammatory response following TBI. However, despite the promising results concerning functional improvement following ES cell TX following TBI the clinical use of ES cells is complicated due to ethical and immunological concerns (Molcanyi 2007, Riess 2007, Erdö 2003). These concerns might be overcome by using autologous pluripotent stem cells derived from a patient's own somatic cells by ectopic expression of pluripotency factors (Hochedlinger 2006). These, induced pluripotent stem (iPS) cells are widely recognised as a powerful alternative to ES cells as potential therapeutic agents, with unique advantages. Like ES cells, they are pluripotent and can be used to obtain tissue-specific cells or progenitors of therapeutic interest (such as neurons and their progenitors). But on the other side, iPS cells are likely to carry a higher risk of tumorigenicity than ES cells, due to the inappropriate reprogramming of these somatic cells, the activation of exogenous transcription factors, or other reasons (Tsuji 2010). In a most recent study it has been shown that the transplantation of iPS-derived neurospheres into the spinal cord directly after contusive injury in mice resulted in cell differentiation into all three neural lineages without forming teratomas or other tumors. These cells also participated in remyelination and induced the axonal regrowth of host 5HT-positive serotonergic fibers, promoting locomotor function recovery. These findings suggest that iPS cell-derived neurospheres may be a promising cell source for therapy of spinal cord injury (Tsuji 2010). It is required, that the therapeutic potency of these cell-source will also be evaluated in models of TBI to prove their effectiveness and safety as a clinical therapy for human after TBI.

8. References

Allan S, Rothwell N (2001). Cytokines and acute neurodegeneration. Nat. Rev. Neurosci. 2, 734–744.

Bentz K, Molcanyi M, Riess P, et al. (2007). Embryonic stem cells produce neurotrophins in response to cerebral tissue extract: Cell line-dependent differences.J Neurosci Res. 2007 85(5):1057-64.

Breitbach M, Bostani T et al. (2007). Potential risks of bone marrow cell transplantation into infarcted hearts. Blood. 15;110(4):1362-9. Epub 2007 May 4.

Bruns J, Hauser W (2003) The epidemiology of traumatic brain injury: a review. Epilepsia. 44 Suppl10:2–10.

Cao, Q, Benton, R et al. (2002).Stem cell repair of central nervous system injury. J. Neurosci. Res. 68, 501–510.

Erdö, T, Buhrle C et al. (2003). Host-dependent tumorigenesis of embryonic stem cell transplantation in experimental stroke. J. Cereb. Blood Flow Metab. 23,780–785.

Hamm R (2001) Neurobehavioral assessment of outcome following traumatic brain injury in rats: an evaluation of selected measures. Journal of Neurotrauma. 18(11): 1207-1216.

Han S, Kang D et al. (2003). Grafted lineage-restricted precursors differentiate exclusively into neurons in the adult spinal cord. Exp Neurol. 2002 Oct;177(2):360-75. Erratum in: Exp Neurol. 180(2):172.

Hochedlinger K (2006), Jaenisch R. Nuclear reprogramming and pluripotency. *Nature.* 441:1061-67

Hoehn M, Kustermann E et al (2002). Monitoring of implanted stem cell migration *in vivo*: a highly resolved *in vivo* magnetic resonance imaging investigation of experimental stroke in rat. Proc. Natl. Acad. Sci. USA 99,16267–16272.

Hofstetter C, Schwarz E et al.(2002). Marrow stromal cells form guiding strands in the injured spinal cord and promote recovery. Proc. Natl. Acad. Sci. USA 99, 2199–2204.

Lenzlinger P, Morganti-Kossmann M et al. (2001). The duality of the inflammatory response to traumatic brain injury.Mol. Neurobiol. 24, 169–181.

Lu D, Li Y et al. (2001). Intraarterial administration of marrow stromal cells in a rat model of traumatic brain injury. J. Neurotrauma 18, 813–9.

Lu D, Mahmood A et al. (2001). Adult bone marrow stromal cells administered intravenously to rats after traumatic brain injury migrate into brain and improve neurological outcome. NeuroReport 12, 559–63.

Maegele M, Engel D, et al. (2007). Incidence and outcome of traumatic brain injury in an urban area in Western Europe over 10 years. Eur Surg Res. 39:372–9.

Mahmood A, Lu D et al.(2001b). Intracranial bone marrow transplantation after traumatic brain injury improves functional outcome in adult rats. J. Neurosurg. 94, 589–95.

Martinez-Serrano A, Bjorklund A (1998).*Ex vivo* nerve growth factor gene transfer to the basal forebrain in presymptomatic middle-aged rats prevents the development of cholinergic neuron atrophy and cognitive impairment during aging. Proc. Natl. Acad. Sci. USA 95, 1858–1863.

Martinez-Serrano A, Fischer W et al. (1996). Long-term functional recovery from age-induced spatial memory impairments by nerve growth factor gene transfer to the rat basal forebrain. Proc. Natl. Acad. Sci. USA 93,6355–6360.

Miyazono M, Lee V et al. (1995). Proliferation, cell death, and neuronal differentiation in transplanted human embryonal carcinoma (NTera2) cells depend on the graft site in nude and severe combined immunodeficient mice. Lab Invest. 73(2):273-83.

Molcanyi M, Riess P et al. (2007). Trauma-associated inflammatory response impairs embryonic stem cell survival and integration after implantation into injured rat brain. J Neutotrauma. 24:625-37.

Molcanyi M, Riess P et al. (2009) Developmental potential of the murine embryonic stem cells transplanted into the healthy rat brain--novel insights into tumorigenesis.Cell Physiol Biochem. 24:87-94. Epub 2009 Jul 1.

Muir J, Raghupathi R et al. (1999). Terminally differentiated human neurons survive and integrate following transplantation into the traumatically injured rat brain. J. Neurotrauma 16, 403–414.

Philips M, Mattiasson G et al.(2001). Neuroprotective and behavioral efficacy of nerve growth factor-transfected hippocampal progenitor cells transplants after experimental traumatic brain injury. J. Neurosurg. 94, 765–774.

Philips, M, Muir J et al. (1999). Survival and integration of transplanted postmitotic human neurons following experimental brain injury in immunocompetent rats. J. Neurosurg. 90, 116–124.

Prockop D (2002). Adult stem cells gradually come of age. Nat. Biotechnol. 20, 791–792.

Riess P, Bareyre Fet al. (2001). Effects of chronic, post-injury cyclosporin A administration on motor and sensorimotor function following severe, experimental traumatic brain injury. Restor. Neurol. Neurosci. 18, 1–8.

Riess P, Molcanyi M et al. (2007). Embryonic stem cell transplantation after experimental traumatic brain injury dramatically improves neurological outcome, but may cause tumors.J Neurotrauma. 24(1):216-25.

Riess P, Zhang C et al. (2002). Transplanted neural stem cells survive, differentiate, and improve neurological motor function after experimental traumatic brain injury. Neurosurgery 51, 1043–1054.

Ryder E, Snyder E et al. (1990). Establishment and characterization of multipotent neural cell lines using retrovirus vector-mediated oncogene transfer. J.Neurobiol. 21, 356–375.

Schouten J, Fulp C et al. (2004). A review and rationale for the use of cellular transplantation as a therapeutic strategy for traumatic brain injury. J Neurotrauma. 21:1501-38.

Soares HD, McIntosh TK. (1991) J Neural Transplant Plast.2(3-4):207-20.)

Sofroniew MV, Vinters HV. Astrocytes: biology and pathology Acta Neuropathol. 2010 Jan;119(1):7-35. Epub 2009 Dec 10.

Sosin D, Sacks J et al. (1996). Pediatric head injuries and deaths from bicycling in the United States. Pediatrics. 98:868–70.).

Thomson JA, Itskovitz-Eldor J et al. (1998) Embryonic stem cell lines derived from human blastocysts. Science. 282:1145–7.

Trojanowski J, Kleppner S et al. (1997). Transfectable and transplantable postmitotic human neurons: a potential "platform" for gene therapy of nervous system diseases. Exp. Neurol. 144, 92–97.

Trojanowski J, Mantione J et al.(1993). Neurons derived from a human teratocarcinoma cell line establish molecular and structural polarity following transplantation into the rodent brain. Exp. Neurol. 122,283–294.

Tsuji O, Miura K et al. (2010). Therapeutic potential of appropriately evaluated safe-induced pluripotent stem cells for spinal cord injury. Proc Natl Acad Sci U S A. 13;107(28):12704-9.

Watson D, Lonhi L et al. (2003). Genetically modified NT2N human neuronal cells mediate long-term gene expression as CNS grafts *in vivo* and improve functional cognitive outcome following experimental traumatic brain injury. J. Neuropathol. Exp. Neurol. 62,368–380.

24

Autologous Stem Cell Infusion for Treatment of Pulmonary Disease

Neal M. Patel and Charles D. Burger
Mayo Clinic Florida
USA

1. Introduction

Over the past several decades stem cell therapy has been characterized as having the potential to dramatically change the treatment of human disease. Impact of such therapy on pulmonary disorders remains unknown. A lack of evidence for a function of exogenous cells in lung repair limits our ability to proceed into clinical trials. Therefore we must rely on personal experience and case series to better understand the clinical effects of stem cell therapy in pulmonary disease. In this report, we discuss four patients with common pulmonary diseases who were treated with the direct infusion of peripherally harvested autologous stem cells into the pulmonary arteries. Although these four patients were followed by Mayo Clinic, the procedure was performed by Regenocyte Therapeutic, a privately owned and operated Florida-based enterprise. The procedure is not approved by the United States Food and Drug Administration. In addition, it was not performed in the context of an approved clinical trial. Each of the patients in this series independently pursued and personally paid for this experimental and controversial treatment. Since the practice is not approved in the United States, the stem cell infusion was performed at the Centro Otorrino Laringologia Hospital in Santo Domingo, Dominican Republic. This report summarizes the clinical experience of these patients based on unrelated clinical assessments at the Mayo Clinic performed both before and after the procedure..

2. Case 1

2.1 Clinical history and course

A 44 year-old Caucasian man and former smoker (38 pack years, quit 2004) presented for evaluation in December 2006 for symptoms of progressive dyspnea (World Health Organization (WHO) functional class III), fatigue, and presyncopal spells. His past medical history was significant for mild chronic obstructive pulmonary disease (COPD) and pulmonary hypertension (PH) diagnosed in 2004, with right heart catheterization (RHC) revealing a mean pulmonary artery pressure (MPAP) of 37 mmHg. Additional PH risk factors included obesity (BMI 33.2), and mild obstructive sleep apnea. The patient had recently started sildenafil (25mg dose) on an as needed basis and reported some reduction in dyspnea after each use.

His laboratory evaluation at Mayo Clinic was normal except for an elevated hemoglobin (17.7 mg/dL). He was hypoxic on room air (pH 7.4, PaO2 49, PaCO2 33, SaO2 90%), and his pulmonary function test (PFT) revealed mild obstruction with a reduction in diffusing capacity for carbon monoxide (DLCO) (Table 1). High-resolution computed tomography scan of the chest showed diffuse emphysema. The electrocardiogram was normal and the contrast echocardiogram (ECHO) demonstrated mild PH with a right ventricular systolic pressure (RVSP) of 49 mmHg and MPAP 31 mmHg. There was mild left-ventricular hypertrophy but right-ventricular (RV) size and function were normal with an estimated right atrial pressure (RAP) of 5 mmHg. There was no shunt identified (Table 2). A ventilation-perfusion (V/Q) lung scan excluded chronic thromboembolic disease (CTED). The patient achieved 519 meters in the six-minute walk test (6MWT) performed on 2 liters per minute (lpm) oxygen with a nadir saturation of 84%. Overnight oximetry on room air demonstrated a low baseline saturation (mean of 87%) and no significant sleep-disordered breathing.

The patient underwent a repeat RHC with cardiopulmonary exercise testing (CPX). The RHC revealed mild PH at rest that worsened significantly with exercise (MPAP increased from 30 to 58 mmHg) (Table 3). The maximal oxygen consumption was 15.7 mlO2/kg/min with an inadequate increase in stroke volume with exercise. Pulmonary arterial vasodilator trial revealed no significant response to intravenous epoprostenol.

His PH was classified as WHO diagnostic group 3 (Simmoneau 2009); therefore, regular use of adequate oxygen was recommended. He also started tiotropium inhaler. In addition he was instructed to take sildenafil regularly at 20mg three times daily (Alp 2006). Over the next 9 months the patient reported some improvement; however, he remained dyspneic with minimal exertion. Objective measures of his PH remained stable (Table 2). Sildenafil use had ceased and he remained non-compliant with oxygen, using it only at night or when significantly dyspneic. He was evaluated for possible lung transplantation but was not felt to be a good candidate due to noncompliance and his body mass index. The recommendations were for weight loss and improved compliance with his medical regimen.

In November 2007, he was seen in follow-up and was again counseled to be more compliant with oxygen and to restart sildenafil at 20mg three times daily. The patient had been seen by a cardiologist outside our institution for intermittent chest pain and palpitations and had been initiated on diltiazem 180mg daily.

After exploring alternative therapies at a private facility, the patient had blood collected for possible stem cell therapy. On March 4, 2008 he traveled to the Dominican Republic where he received multiple infusions, of what was reported to be the harvested stem cells, directly into the pulmonary arteries.

Shortly after the infusion, he reported feeling like "Superman," running 1 to 2 hours in the first few days. By his report, his room air saturations rose to 99% at 2 weeks, but deteriorated subsequently. Other perceived benefits included decreased fatigue, chest pain, palpitations, and leg pain and even improved vision and mental acuity. His symptomatic improvements persisted at subsequent visits despite becoming more sedentary and gaining 4.5 kg. Notably, he had become more compliant with the use of oxygen to about 3 lpm by nasal cannula at rest.

Objectively, there were also improvements noted in his 6MWT and in the echocardiographic estimates of his pulmonary pressures. (Table 2) These improvements were sustained at repeated follow-ups. However, the oxygen requirement remained 8 to 15 lpm with exertion.

During his evaluation in March 2010, the patient reported that many of the initial subjective improvements had waned. Overall his objective measures remained relatively stable. (Table 2) He had a repeat RHC demonstrating mild resting but severe exercise induced PH (Table 3).

3. Case 2

3.1 Clinical history and course

A 78 year-old Caucasian man presented with dyspnea on exertion (WHO functional class II). He had smoked cigars for 19 years and quit 20 years ago. He was diagnosed with moderate COPD and was treated with inhaled bronchodilators. A few months prior to his presentation at Mayo Clinic, an extensive evaluation including a RHC, a V/Q scan and a pulmonary arteriogram demonstrated PH due to CTED. Those studies were reviewed and felt to be conclusive. His medications included hydrochlorothiazide, furosemide, and warfarin. A PFT demonstrated moderate obstruction, air trapping, a reduced DLCO, and oxygen desaturation with exercise (Table 1). The 6MWT and ECHO results are reported in Table 2. Arterial blood gas on room air was: pH 7.47, PaO2 67, PaCO2 37; and the BNP was 178 pg/mL. Cardiac MRI revealed RV enlargement and hypertrophy with a reduced RV ejection fraction to 27%. A repeat RHC was performed, and it was felt the PH was multi-factorial secondary to COPD, CTED, and mild pulmonary venous hypertension (Table 3). Given the patient's age, minimal symptoms, multiple mechanisms for his PH, and his co-morbidities, he was considered a high risk for pulmonary endarterectomy, and the patient elected for conservative therapy. An inferior vena cava filter was placed and life-long anti-coagulation was recommended. In addition, supplemental oxygen with exercise and sildenafil 20 mg three times daily (Reichenberger 2007) were initiated.
Six-months later, there was a slight improvement in the 6MWT (Table 2). He was now requiring increasing oxygen to maintain saturations at rest (1.5 lpm) and during exertion. Given persistently elevated right-heart pressures and RV enlargement and hypokinesis, bosentan was added to his regimen. Eight months after adding bosentan, he continued to show improvement in the 6MWT but the ECHO was unchanged. Oxygen needs continued to increase, now requiring 2 lpm at rest and 15 lpm for the 6MWT (Table 2).
Three months later, he underwent adult stem infusion. This date corresponded to 18 months on sildenafil 20 mg three times daily and 12 months on bosentan at full dose. One week after the harvest, he traveled to Santo Domingo for the infusion into his pulmonary arteries via a pulmonary artery catheter at Centro Otorrino Laringologia Hospital. Repeat ECHO one month later showed a RVSP of 55 mmHg which was exactly the same as his baseline value one month prior to the infusion. Another ECHO performed by Regenocyte Therapeutics three months after the stem cell infusion demonstrated a RVSP 41 mmHg. A PFT four months after stem cell infusion was essentially unchanged (Table 1).
The patient was seen in follow up at Mayo Clinic approximately 5 and 12 months after his stem cell infusion. He stated that he felt neither better nor worse after his stem cell treatment and was on 3 lpm oxygen to maintain an adequate resting saturation. His objective assessment showed some benefit at 5 months. That benefit had waned one year after stem cell infusion. His treatment regimen included diuretics, warfarin, sildenafil 20 mg three times daily, bosentan 125 mg twice daily (Jais 2008) and oxygen. He has also been regularly participating in a formal pulmonary rehabilitation program. No changes were made in his medical regimen.

Date	PFT Results FEV1	FEV1/FVC	TLC	RV	DLCO	Pulse Oximetry Rest/Exercise	Six-Minute Walk
	L (%*)	%	L (%*)	L (%*)	ml/min/mmHg (%)	%	meters
Case 1							
Dec-06	3.31 (73)	73	6.68 (89)	2.06(119)	16.3 (49)	93/86	530
Mar-08	STEM	CELL	INFUSION				
Sep-08	3.13 (71)	71	6.98 (95)	2.26(127)	16.3 (50)	94/87	536
Case 2							
Oct-07	1.32 (54)	44	6.50 (111)	3.60 (138)	11.8 (64)	94/87	***
Mar-09	STEM	CELL	INFUSION				
Jul-09	1.33 (55)	39	6.8 (115)	3.40 (132)	11.3 (62)	91**	***
Case 3							
Jan 09	1.81 (49)	59	5.03 (71)	1.55 (70)	17.2 (62)	96/92	N/A
Jun-09	1.84 (49)	58	5.29 (73)	1.98 (89)	17.2 (62)	95/92	N/A
Jan-10	STEM	CELL	INFUSION				
Jun-10	1.74 (46)	57	5.39 (75)	2.19 (98)	17.7 (65)	97/94	N/A
Case 4							
Jun-10	0.45 (12)	31	6.91 (116)	5.40 (369)	11.1 (33)	86	255**
Jul-10	STEM	CELL	INFUSION				
Aug-10	0.45 (12)	33	6.72 (113)	5.06 (346)	10.7 (32)	92	341**

Legend: Pulmonary function test results, oxygenation and modified Borg dyspnea score data on four patients who had pulmonary artery stem cell infusion. Key: FEV1 – Forced Expiratory Volume in One Second, FVC – Forced Vital Capacity, TLC – Total Lung Capacity, RV – Residual Volume, DLCO – Diffusing Capacity of Carbon Monoxide

* Percent Predicted. **Test performed on oxygen. ***See Table 2. N/A – Not Available

Table 1. Pulmonary Function and Six-Minute Walk Testing- Pre and Post Stem Cell Infusion

			6 Minute Walk			Echo Data					
DATE	**BNP pg/ml**	**WHO Functional Class**	**Distance (meters)**	**Sa02(%)/ FIO_2 (L)**	**Borg Score**	**RAP**	**TR Jet Velocity**	**RVSP**	**MPAP**	**RV Size**	**RV Function**
CASE 1											
Dec-06	20	3	530	84/8	1	5	3.3	49	31	Normal	Normal
Apr-07	25	3	550	88/6	3	5	3.5	54	45	Mild	Normal
Oct-07	17	3	470	91/10	3	10	3.5	59	43	Normal	Normal
Nov-07	31	2	596	88/10	3	5	3.6	57		Mild	Normal
Feb-08	54	3	512	91/10	2	5	3.6	57	41	Normal	Normal
Mar-08	**STEM**	**CELL**									
May-08	26	2	556	91/8	1	5	3.1	43	28	Normal	Normal
Nov-08	16	3	580	86/8	3	5	3.2	46	33	Normal	Normal
Feb-09	39	3	600	93/15	4	5	3.3	49	29	Mild	Normal
May-09	57	3	619	90/10	3	5	3.2	46	N/A*	Mild	Normal
Mar-10	19	3	586	92/15	5	5	3.3	49	36	Normal	Normal
CASE 2											
Aug-07	178	2	291	91/4	5	5	4.3	79	47	Moderate	Moderate
Apr-08	119	2	339	89/15	5	5	4.3	79	47	Moderate	Moderate
Dec-08	38	2	384	90/15	4	15	4.0	79	58	Moderate	Moderate
Mar-09	**STEM**	**CELL**									
Aug-09	40	2	401	94/15	3	N/A*	N/A*	N/A*	N/A*	Mild	Normal
May-10	87	3	373	96/15	4	16	4.0	80	58	Moderate Severe	Moderate Severe

Legend: Brain natriuretic peptide levels, WHO functional class (1 to 4 scale), six-minute walk distances with modified Borg dyspnea score and echocardiography results on Cases 1 and 2 who had pulmonary artery stem cell infusion. Key: BNP – Brain Natriuretic Peptide, WHO – World Health Organization, Borg Score – Modified Borg Dyspnea Score on a scale of 0 to 10, Sa02 – Oxygen Saturation, FIO2 – Fraction of Inspired Oxygen, RAP – Right Atrial Pressure, TR – Tricuspid Regurgitant (peak), RVSP – Right Ventricular Systolic Pressure, MPAP – Mean Pulmonary Artery Pressure, RV – Right Ventricle, Normal and Mild describe degree of enlargement and ventricular dysfunction

* Unable to estimate due to incomplete tricuspid regurgitant jet

Table 2. Cardiopulmonary Data for Cases 1 and 2

		RAP (mmHG)	PAP (mmHG)	MPAP (mmHg)	PAOP (mmHg)	CO/CI (L/min, L/min/m²)	PVR (dyne sec cm⁻⁵)	Vasodilator Response	MPAP During Exercise (mmHg)
Case 1	Baseline 1/2007	4	45/25	30	12	5.1/2.1	282	No	58
	Follow-up 3/2010	6	44/20	28	14	8.3/3.3	135	No	70
Case 2	Baseline 10/2007	11	85/34	51	16	5.8/3.2	478	No	N/A *
	Follow-up	NOT	DONE						

Legend: Hemodynamic data from right heart catheterization on Cases 1 and 2. Key: RAP - Right Atrial Pressure, PAP - Pulmonary Artery Pressure, MPAP - Mean Pulmonary Artery Pressure, PAOP - Pulmonary Artery Occlusion Pressure, CO - Cardiac Output, CI - Cardiac Index, PVR - Pulmonary Vascular Resistance

* Exercise study not performed

Table 3. Baseline and Post Stem Cell Transfusion Right Heart Catheter Data

4. Case 3

4.1 Clinical history and course

A 66-year-old Caucasian man presented in January of 2009 with symptoms of worsening dyspnea on exertion. Patient had a past medical history of hypertension and transient ischemic attacks. His pulmonary history dates back to 2002 at which time he was told to have scarring of his lungs. He had been followed by outside pulmonologist who had reported stable "scarring" of his lungs. The patient endorses a history of asbestos exposure, as well as a history of being in industrial chemical salesman. He had a less than 5 pack year history of smoking and denied any exposure to pets, birds, hot tubs, or other antigens suggestive for hypersensitivity pneumonitis.

Patient underwent extensive testing including high resolution CT scan, which revealed extensive upper and lower lobe fibrosis, focal atelectasis and pleural thickening at the left base, and extensive linear calcifications in both lungs associated with fibrosis. Comparing this study to a previous scan performed in 2002 the findings had progressed slightly. Pulmonary function testing was performed with spirometry showing a mixed pattern of obstruction and restriction, lung volumes consistent with mild restriction, and a mildly reduced DLCO (Table 1). Echocardiogram revealed normal left ventricular function, normal diastolic function, normal RV size and function, and indeterminate pulmonary pressures secondary to incomplete tricuspid regurgitant jet. Laboratory work was notable for negative hypersensitivity panel, fungal serologies, and normal blood counts. Patient underwent bronchoscopy for bronchoalveolar lavage and transbronchial biopsies. Cultures were negative and pathology of transbronchial biopsy revealed elastic fibrous tissue. Referral to cardiothoracic surgery was made for possible open lung biopsy however it was decided that the yield would be low given the extensive fibrotic changes and outweigh potential benefits. Patient was given a diagnosis of idiopathic pulmonary fibrosis and obstructive lung disease. Follow-up chest imaging and pulmonary function testing was planned. Bronchodilator therapy was offered, but refused by patient.

In June of 2009 patient returned with no significant change in his symptoms. A repeat CT scan showed stable diffuse fibrotic changes. Pulmonary function testing was performed and found to be essentially unchanged (Table 1). Given the relative stability of patient's symptoms and pulmonary function testing further follow-up was recommended. At this time, the patient inquired about treatment of his pulmonary fibrosis with stem cell therapy. Given the lack of safety and efficacy data in the use of stem cell therapy for the patient's disease, treatment with this modality was discouraged.

In January 2010, under the oversight of Regenocyte Therapeutic, the patient underwent a bone marrow derived autologous stem cell infusion in the Dominican Republic. He reported that stem cells were infused into three different areas of each lung via a pulmonary artery catheter. Patient returned to Mayo Clinic in June 2010 for follow-up at which time he reported a significant subjective improvement in his symptoms of dyspnea on exertion. More specifically he states that immediately after the infusion he noted no change in Approximately 3 months after stem cell infusion, he professed to less dyspnea with exercise. For example, in the past it would take him approximately one hour to walk 2 miles and several months following stem cell therapy he was able to do so in one half hour. The patient's subjective improvements were unfortunately not correlated with any improvement in his pulmonary function testing (Table 1). CT imaging also remained unchanged showing stable diffuse fibrotic lung disease.

5. Case 4

5.1 Clinical history and course

A 30-year-old man, originally from Albania who has lived in the United States for the past 25 years, presented in June 2010 with a primary complaint of recurrent pneumonias. Patient had a history of repeated pneumonias since the age of 2. He had been hospitalized numerous times throughout his life secondary to pneumonia. He described outside evaluation as including bronchoscopy and nasal endoscopy with biopsy. Suspected diagnosis at time of consultation was primary ciliary dyskinesia. Medications at time of evaluation included fluticasone/salmeterol 250/50, tiotropium, and oxygen use nightly. Past medical history was otherwise negative. The patient has been a lifelong nonsmoker and denied any pulmonary family history.

At Mayo Clinic, he underwent testing to further evaluate his history and suspected diagnosis of primary ciliary dyskinesia. CT Scan showed widespread hyperinflation with mosaic attenuation, bronchial wall thickening, and bronchiectasis. The distribution of this was diffuse and not lobar predominant. Pulmonary function testing revealed severe obstruction, hyperinflation, reduction in DLCO, and resting saturation of 86%. The patient was able to ambulate 255 meters on 6 minute walk with end-of-walk oxygen saturation 97% on 2 L nasal cannula (Table 1). Echocardiogram performed at time of evaluation revealed preserved left ventricular systolic and diastolic function, moderate to severe RV enlargement as well as decrease in RV systolic function. Right ventricular systolic pressure was estimated at 63 mmHg, with a Doppler derived MPAP of 45 mmHg. The patient was diagnosed with severe obstructive lung disease secondary to bronchiectasis, with secondary PH (WHO group 3). Primary ciliary dyskinesia was suspected to be the underlying cause. Given the severity of patient's disease he was referred to lung transplantation for evaluation.

In August of 2010 the patient returned for follow-up evaluation. About one month prior to this visit he underwent harvesting of stem cells from bone marrow and subsequent infusion into his pulmonary arteries for treatment of pulmonary hypertension under the direction of Regenocyte Therapeutic. Further details of the procedure were not available at the time of visit. He did report subjective improvements in cough, improved energy level, and decreased utilization of oxygen. He also denied any exacerbations of his bronchiectasis in the interim. Repeat pulmonary function testing showed unchanged lung volumes, persistent severe obstruction and reduction in DLCO. Interestingly the patient had an improvement in his resting oxygen saturation to 92% from previous measure of 86%. 6 minute walk also showed an improvement, with patient walking of 341 meters compared to 255 meters, with an end-of-walk oxygen saturation of 90% on 2 L nasal cannula (Table 1). Echocardiography continued to show preserved left ventricular function with slight improvement in patient's RV enlargement as well as right ventricular systolic function. Estimated RVSP was reduced to 49 mmHg with a Doppler derived MPAP of 35 mmHg.

The patient was evaluated by the lung transplantation service; however, he was not interested in proceeding given the subjective improvement he had experienced with his initial stem cell therapy.

6. Discussion

Use of stem cell therapy for the treatment of pulmonary diseases is gaining increasing attention. In the past decade, circulating bone marrow–derived cells similar to embryonal angioblasts have been identified (Asahara 1997) termed "endothelial progenitor cells (EPC)." These cells have the potential to proliferate and differentiate into mature endothelial cells. Two major types of these cells in human peripheral blood have been termed "early EPC's" and "late outgrowth EPC's," both of which appear to have potential in lung microvascular repair and restoring vascular structures of the lung. (Yoon 2005)

A number of trials are currently underway to evaluate the safety and effectiveness of the infusion of stem cells in pulmonary disease. A recently published report out of Brazil suggested both safety and improvement in quality of life in patients with severe COPD/emphysema (Ribeiro-Paes 2011). An ongoing multi-center trial is currently looking at the efficacy of PROCHYMAL™ (ex Vivo Cultured Adult Human Mesenchymal Stem Cells) intravenous infusion for treatment of moderate to severe COPD. A pilot trial of autologous EPC administration for idiopathic pulmonary arterial hypertension was conducted at Zhejiang University, Hangzhou, China. The trial was a 12 week follow-up of patients after systemic administration of autologous EPC's plus conventional therapy compared with patients receiving conventional therapy alone. (Wang 2007) The results included increased six-minute-walk capacity and improved hemodynamic variables, including MPAP, pulmonary vascular resistance, and cardiac output. No concerns with the safety of the administration of EPC's were encountered. The Pulmonary Hypertension and Cell Therapy (PHACeT) trial by the University of Toronto reported on the administration of autologous EPCs transduced to express exogenous nitric oxide synthase. Three patients in their initial panel showed a remarkable (nearly 50%) reduction in total pulmonary vascular resistance over the course of the 3-day delivery period. Subsequent study in additional patients is currently underway. (Weiss 2008)

In our report, we present four patients who had peripherally harvested autologous stem cells infused into the pulmonary arteries. The procedure was not endorsed by or performed by Mayo Clinic. In addition, it was not performed in the context of an approved clinical trial.

The procedures were performed by Regenocyte Therapeutic. The reported infusion occurred at the Centro Otorrino Laringologia Hospital in Santo Domingo, Dominican Republic, as the procedure is not approved by the United States Food and Drug Administration.

A variety of unusual challenges are illustrated by the care and interaction of these four patients at Mayo Clinic (Table 4). The inability to control for concomitant therapy complicates the interpretation of these patients' clincial course both pre and post stem cell infusion. The first patient (Case 1) reported significant subjective improvement. At the time of follow-up many of these improvements had waned; however, there may have been mild improvement in both his resting pulmonary pressures by ECHO and 6MWT. (Table 2) These reductions in MPAP appeared to be sustained up to 24 months post infusion. Unfortunately, these functional and hemodynamic improvements did not translate into any significant change in the severe hypoxemia. Case 2 did not have any change in his symptoms or WHO functional class. He continued to have increasing supplemental oxygen requirements. His six-minute walk distance had slightly improved, and his ECHO demonstrated improved RV function. Case 3 reported significant subjective improvement in exercise capacity. Objectively this did not translate into any improvement in PFT. Case 4 perhaps seemed to have the most significant improvement in objective measures including improvement in resting saturation, six minute walk distance and reduction in MPAP as estimated by ECHO.

Are any of the changes due to the stem cell infusion? It is impossible to determine in an uncontrolled setting. The role of multiple ongoing therapies and the limitations of the clinical assessments are particularly challenging to evaluate. Three of the four patients were on multiple therapies for treatment of PH and/or COPD at the time they received the stem cell infusion. Variations in oxygen requirements and supplementation may be particularly confounding since three of the four patients had obstructive lung disease with associated PH. Although validated surrogate endpoints in assessing response to therapy remain a source of ongoing discussion, measurements such as spirometry, ECHO, six minute walk distance, WHO functional class have been validated to monitor disease course, but remain imperfect. (Hoeper 2005; Snow 2007; Aduen 2009)

If there were some benefit, it is unfortunate that it will remain inconclusive. Cases 1 and 3 had subjective improvement; however, there clinical scenario was complex as outlined above. Case 4's objective improvements are interesting but may represent concomitant therapies not disclosed in this uncontrolled setting. These cases present several challenging dilemmas for the clinician in this evolving area of science. Short of lung transplantation, there are generally no cures for the underlying pulmonary disease. In Cases 1, 3, and 4, we informed the patients that the only proven therapy for COPD/IPF in the setting of hypoxia was oxygen therapy. (Nocturnal Oxygen Therapy Trial Group 1980; Medical Research Council 1981) Nonetheless, these patients had a difficult time accepting more conventional treatment and instead sought experimental therapies.

The stem cell therapy raised the issue of unproven therapies to a new level as the patients personally paid approximately $50,000 for the therapy that was administered outside the context of either a clinical or research setting. In addition, the therapy was pursued independently of our clinical recommendation or oversight. Also of interest in these particular cases, was the potential impact on the relationship between the physician and patient since the patients had pursued an unendorsed experimental therapy outside the country. Although our stance on the experimental nature of stem cell therapy for pulmonary disease has always been clear, all of these patients researched, pursued and financed the therapy on their own accord. The clinician must be prepared for any potential conflict that may ensue if the patient seeks an experimental, unapproved therapy. In our practice, we maintain that such therapy should only be administered in the context of an approved research protocol. Nonetheless, we felt committed to the ongoing management of these patients regardless of their decision to seek unapproved stem cell infusion therapy.

Patient-Centered	Provider-Centered	Knowledge-Centered
End-stage disease generates desperate interest in unproven therapies	Limited time to discuss and counsel patients regarding unproven therapies	Conventional therapies are often of limited value in severe pulmonary disease
Patients may have strong desire to receive unproven therapies regardless of consequences	Brief, simplistic cautionary advice may be insufficient to dissuade the patient from seeking unproven therapy	Knowledge gap exists by definition with unproven therapies
Inability to distinguish between for-profit endeavors and academically-based clinical trials	Assessment of ethical, legal and professional nature of extra-mural practices often complex	Provision of unproven therapy in uncontrolled (non-research) manner precludes firm conclusions from results
Patient desires to maintain relationship with primary physician and to seek unconventional and/or unproven therapies	Physician must maintain objectivity as well as primary role as patient advocate and caregiver	Cooperative interaction and research between academic centers required to advance the science

Legend: Important challenges and limitations encountered in this series of 4 patients receiving stem cell infusion independent of your care.

Table 4. Illustrative Lessons

7. Conclusion

Stem cell therapy has emerged as an experimental therapy to potentially treat a number of diseases, including pulmonary disorders. Current ongoing safety and efficacy trials will likely add significantly to our current understanding of the field, with hopes of aiding in future patient care. In a parallel fashion, patients are pursuing this form of therapy outside of controlled research trials. Such practices create significant challenges in evaluating efficacy and potentially stress the provider-patient relationship in unique ways (Table 4).

8. References

Aduen JF, Castello R, Lazano MM, et al. An alternative echocardiographic method to estimate mean pulmonary artery pressure: diagnostic and clinical implications. J Am Soc Echocardiography 2009; 22:814-19.

Alp S, Skrygan M, Schmidt WE, et al. Sildenafil improves hemodynamic parameters in COPD—an investigation of six patients. Pulm Pharmacol Ther 2006; 19:386–390.

Asahara T, Murohara T, Sullivan A, et al. Isolation of putative progenitor endothelial cells for angiogenesis. Science 1997;275:964–967.

Hoeper MM, Markevych I, Spiekerkoetter E, et al. Goal-oriented treatment and combination therapy for pulmonary arterial hypertension. Eur Respir J. 2005; 26:858-863.

Jais X, D'Armini AM, Jansa P, et al. Bosentan for treatment of inoperable chronic thromboembolic pulmonary hypertension: BENEFiT (Bosentan Effects in iNopErable Forms of chronic Thromboembolic pulmonary hypertension), a randomized, placebo-controlled trial. J Am Coll Cardiol 2008;52:2127-34.

Nocturnal Oxygen Therapy Trial Group. Continuous or nocturnal oxygen therapy in hypoxemic chronic obstructive lung disease: a clinical trial. Annals of Internal Medicine 1980;93(3):391-8.

Reichenberger F, Voswinckel R, Enke B, et al. Long-term treatment with sildenafil in chronic thromboembolic pulmonary hypertension. Eur Respir J 2007;30: 922-7

Report of the Medical Research Council Working Party. Long-term domiciliary oxygen therapy in chronic hypoxic cor pulmonale complicating chronic bronchitis and emphysema. Lancet 1981;1:681-5.

Ribeiro-Paes JT, Bilaqui A, et al. Unicentric study of cell therapy in COPD/Pulmonary Emphysema. International Journal of COPD. 2011;6:63-71.

Simmoneau G, Robbins IM, Beghetti M, et al. Updated clinical classification of pulmonary hypertension. J Am Coll Cardiol 2009; 54:S43-54.

Snow JL and Kawut SM. Surrogate end points in pulmonary arterial hypertension: assessing the response to therapy. Clin Chest Med 2007; 28:75-89.

Wang XX, Zhang FR, Shang YP, et al. Transplantation of autologous endothelial progenitor cells may be beneficial in patients with idiopathic pulmonary arterial hypertension: a pilot randomized controlled trial. J Am Coll Cardiol. 2007;49:1566–1571.

Weiss DJ, Kolls JK, Ortiz LA, Panoskaltsis-Mortari A, Prockop DJ. Stem cells and cell therapies in lung biology and lung diseases. Proc Am Thorac Soc. 2008 Jul 15;5(5):637-67.

Yoon CH, Hur J, Park KW, et al. Synergistic neovascularization by mixed transplantation of early endothelial progenitor cells and late outgrowth endothelial cells: the role of angiogenic cytokines and matrix metalloproteinases. Circulation 2005;112:1618–1627.

25

Neurologic Sequealae of Hematopoietic Stem Cell Transplantation (HSCT)

Ami J. Shah[1,2,3], Tena Rosser[2,3,4] and Fariba Goodarzian[3,5]
[1]*Division of Research Immunology/ Bone Marrow Transplantation,*
[2]*Department of Pediatrics, Keck School of Medicine,*
[3]*The Saban Research Institute of Childrens Hospital Los Angeles,*
[4]*Division of Neurology,*
[5]*Department of Pediatric Radiology*
USA

1. Introduction

Hematopoietic stem cell transplantation (HSCT) has been used to cure many pediatric patients with both malignant and non-malignant diseases in whom conventional therapy has failed. HSCT is generally performed after the administration of high sublethal doses of chemotherapy or chemoradiotherapy to achieve myeloablation, immunosuppression and eradication of abnormal cells. The use of high dose cytotoxic therapy before HSCT places children at risk for acute and long-term side effects which may differ or be more severe than those seen in children who did not receive HSCT. One important organ that may be damaged from cytotoxic therapies used to treat recipients of HSCT has been the brain.

Survival rates for childhood ALL improved after the introduction of treatments to prevent the development of central nervous system (CNS) leukemia. During the past 25 years, CNS prophylaxis for leukemia has consisted of either intrathecal chemotherapy alone or a combination of both chemotherapy and cranial irradiation. There is a wide range of deficits associated with this CNS prophylaxis. Neurological sequelae include severe leukoencephalopathy to more subtle and mild dysfunction, such as learning disorders and abnormalities that can be seen on CT scan or MRI (Packer R 1986, Brouwers P 1985).

In contrast, patients with malignant brain tumors often present with neurological symptoms of their primary disease such as a) increased intracranial pressure b) problems with balance and cranial nerve findings, which are associated with infratentorial tumors and c) seizures, which are often seen in those with supratentorial tumors. The treatment of malignant brain tumors involves chemoradiotherpay in addition to neurosurgery. The cranial irradiation that these children receive is often more than double that received by patients with acute leukemias. Therefore, the neurological sequelae of patients with this particular malignancy are greater than that seen for patients with a hematological malignancy.

The neurocognitive sequelae following cranial irradiation and chemotherapy for childhood acute leukemia and brain tumors have been well documented (Fletcher et al 1988, Smibert et al 1996, Anderson et al 1997, Stehbens et al 1991, Butler and Copeland 1993, Moleski 2000, Ris and Noll 1994, Mulhern and Butler 2004). However, information regarding the impact of HSCT and the neurocognitive function of HSCT recipients is still emerging, especially among patients with genetic diseases and those who have a primary immune deficiency.

The different types of transplants and their complications can add to the neurological complications that follow HSCT. In general, there are three different types of HSCT: 1) autologous, which is used for the treatment of solid tumors and lymphomas, 2) syngeneic and 3) allogeneic, which is used for the treatment of leukemias, genetic diseases and immunodeficiencies.

In this chapter we review both the acute neurological complications that can occur following HSCT and the long-term clinical significance of these complications, especially the neurocognitive outcomes of this therapy.

2. Neurological sequelae following HSCT

Neurological complications following HSCT have not been studied well, especially in the pediatric population. The majority of studies have focused upon adult patients A prospective study of 180 adult HSCT patient autopsies (177 allogeneic and 3 autologous) detected neuropathologic abnormalities in over 90% of the subjects (Bleggi-Torres et al, 2000). The most frequent findings in this study were subarachnoid hemorrhage (32%), intraparenchymal hemorrhage (27%) and fungal infections (9%). There were no abnormal findings in 9% of the autopsies. A smaller series of autopsy studies showed similar findings (Mohrmann et al, 1990). Although there were a high number of patients with nervous system involvement following HSCT, we do not know if these findings led to clinically significant neurological complications.

The incidence of clinical neurologic symptoms varies from 10-40%. Weber et al evaluated 165 pediatric patients following HSCT. Of these patients, 67% were allogeneic and 33% were autologous. Of these 165 patients, 24% developed neurologic complications. The cause of the neurological complications included: infections (n=10), drug toxicity (n= 5), cerebrovascular events (n=2), and CNS relapse (n=2). Among the infections, aspergillus and toxoplasmosis were the most frequent pathogens isolated (Weber et al, 2008).

A study of patients with Hodgkin Disease who received an autologous HSCT showed early symptoms of neurological complications such as encephalopathy, seizures, cerebral hemorrhage and psychiatric symptoms (Snider et al, 1994). A separate study of a similar population showed the incidence of clinically significant neurological complications as being 3.2%. (Guerrero et al, 1999).

The incidence of neurological symptoms varies depending upon 1) the disease and the therapies used to treat that disease prior to HSCT, 2) type of transplant (autologous, syngeneic and allogeneic and the complications associated with the specific type of transplant 3) conditioning therapy (chemotherapy alone versus chemoradiotherapy, 4) immunosuppressive therapies and 5) the infections that may occur from being immunosuppressed.

3. Impact of disease

There are various reasons for patients to receive HSCT. Among the common diseases include a) hematological malignancies such as acute leukemia, b) malignant solid tumors such as brain tumors and c) nonmalignant diseases such as genetic diseases, immune deficiencies and hemoglobinopathies. All of these diseases can have an impact on the neurological outcome of patients, even without a transplant.

Leukemia patients who are either high-risk patients or have CNS disease, often receive between 1800 or 2400 cGy cranial irradiation for the prophylaxis or treatment of CNS leukemia. Many believe that the threshold between 1400 and 1800 cGy does not have an impact upon neurocognitive function. However, it is known that therapies greater than 1800 cGy appear to affect the neurocognitive function of leukemia patients. In contrast, patients with brain tumors, often receive cranial radiation that can be more than twice that traditionally given in the treatment of ALL. Despite differences in the therapy, there are similar declines that are seen in the neurocognitive function of both groups of patients, especially those who are young at diagnosis.

For patients with non-malignant diseases, the pre-transplant neurological sequelae are often related to the disease. For example, patients with sickle cell disease often have a history of strokes prior to HSCT, this can impact a patient's neurological status prior to coming to transplant. In addition, patients with hemoglobinopathies who are on hypertransfusion protocols, may have evidence of iron overload that may involve the brain. Patients with genetic diseases such as adrenoleukodystrophy (ALD) and Krabbe's disease often have disease specific neurological findings prior to HSCT. These patients have lesions in their CNS that can affect both the motor and mental development of these patients. Although there may be some improvement following HSCT, in general these changes do not resolve completely.

Patients with immunodeficiency, such as severe combined immunodeficiency (SCID) or Wiskott Aldrich Syndrome, often have had serious life threatening infections prior to their diagnosis, which may impact their overall neurological function. Therefore, it is important to assess patients both before and after HSCT to really understand the impact of HSCT upon neurological function.

4. Impact of type of transplant

Patients who receive an allogeneic transplant are thought to have increased neurologic side effects compared to those receiving an autologous transplant. However, a retrospective study of 425 (mostly adult patients) found that the incidence of neurologic complications was similar in both autologous and allogeneic patients (Saiz et al , 2004). The complications that were seen varied depending upon the type of disease and type of transplant. For example, patients with Acute Myelogenous Leukemia (AML) receiving an autologous transplant developed more subdural hematomas due to prolonged thrombocytopenia; whereas, patients with central nervous system (CNS) infections were more common in those receiving an allogeneic transplant due to prolonged immunosuppression.

Allogeneic transplant recipients are vulnerable to developing graft versus host disease (GVHD). The risk is greatest among patients receiving an unrelated transplant despite the stem cell source. There has been one report of possible central nervous system GVHD in two patients. Two patients developed hemiparesis, seizure, encephalopathy and MRI findings of hyperintense white matter lesions on T-2 weighted images. A brain biopsy on one patient and an autopsy finding on the other revealed perivascular lymphocytic infiltrates of primarily donor origin. It is difficult to make an unequivocal diagnosis of this entity, since there are a number of other etiologies that may explain these findings. However, physicians should be made aware of this possibility (Kamble, et al, 2007).

The source of stem cells may also affect the neurologic outcomes. There are three different sources of stem cells: bone marrow, cord blood and peripheralb blood stem cells. Patients who receive peripheral blood stem cells often have a delay in platelet recovery, thus increasing the risk of subdural hematomas. Alternatively, patients with cord blood transplants can have delayed engraftment, thus increasing the period of neutropenia or thrombocytopenia. This can either increase the risk of infections or increase the chances of CNS bleeding, both of which can increase neurologic complications.

Neurologic complications have also been reported following the infusion of stem cells. In one report of 179 consecutive patients, 3 developed complications following the infusion of cryopreserved stem cells (Hoyt et al, 2000). The symptoms were global amnesia (n=1), cerebral infarction (n=2). The patients who developed infarcts showed evidence of a cerebral hemorrhage due to thrombocytopenia and the other showed cerebral aspergillus on autopsy. Therefore, although there are sporadic reports of complications due to the infusion of cells, the risk of the stem cell infusion itself is very low.

5. Impact of conditioning therapy

Ther pretransplant conditioning therapy consists of either high doses of chemotherapy or total body irradiation (TBI). The high dose chemotherapy often crosses the blood brain barrier and the TBI has a direct cytotoxic effect upon the brain. This pretransplant conditioning regimen and the therapy used for the prophylaxis against graft versus host disease (GVHD) are thought to be responsible for the majority of neurological complications following HSCT. The use of various chemotherapy regimens and the use of TBI vary depending upon the type of disease that is being transplanted. For example, patients with nonmalignant disorders are often treated with high dose busulfan while most patients with acute lymphoblastic leukemia are treated with TBI in addition to a chemotherapeutic agent. All agents can produce some acute neurotoxicities and may even have some late onset neurotoxicities. The most common agents that are used in transplant conditioning regimens and their side effects are shown in Table 1.

The impact of the conditioning regimen generally affects the patients during the first 30 days post-HSCT, or the acute post-transplant period. Complications during this time period include pancytopenia, metabolic abnormalities, coagulopathies, drug toxicities and infections due to neutropenia. During this period, subdural hematomas are common due to thrombocytopenia and /or platelet refractoriness. Many patients also develop endothelial cell damage from alkylators such as cytoxan. This may also increase patient's risk of developing subdural hematomas. Intraparenchymal hemorrhages are less common, but may occur as well. These are usually associated with myeloid leukemias.

During the acute period, patients are severely neutropenic. As a result, HSCT recipients are at risk of developing a variety of infections they include bacterial, viral and fungal infections. Table 2 details the most common neurologic complications that occur post-HSCT.

Drug	**Neurologic Toxicities**
Busulfan	Reversible encephalopathy with somnolence, confusion, decreased alertness, myoclonus, headaches and hallucinations. Seizures in approximately 10% of patients. Prophylactic use of antiepileptics is recommended. (De La Camara et al 1991 and Vassal et al 1990)
Carboplatin	Ototoxicity in over 80% of children with neuroblastoma (Parsons et al 1998)
Carmustine	Optic disc and retinal microvaculopathy with variable degrees of visual loss. Very high doses 9> 1,500 mg/ m2) can be associated with severe irreversible encephalopathy (Johnson et al 1999, Burger et al 1981)
Cytarabine arabinoside	Pancerebellar syndrome+/- diffuse encephalopathy with lethary, confusion, and seizures (Baker et al 1991)
Etoposide	Acute confusional state, somnolence, and seizures, which resolve spontaneously (Leff et al 1988)
Ifosphamide	Encephalopathy with lethargy, confusion and seizures in 10-40% of patients. Visual or auditory hallucinations, myoclonus, or muscle rigidity have been reported, which is usually self limited, but have been reported to progress to a coma. (Pratt et al 1990, DiMaggio et al 1994)
Thiotepa	Chronic encephalopathy with progressive declines in cognitive and behaviorial function

Table 1. Neurologic Toxicity of Commonly Used Drugs in HSCT Conditioning Regimens

One of the major complications of CNS therapies is the development of methotrexate induced leukoencephalopathy. The dose of intravenous methotrexate (not intrathecal methotrexate) correlates with the development of leukoencephalopathy. The incidence of this phenomenon is 16-69%. Patients may have no neurological symptoms. Methotrexate induced Leukoencephalopathy does not progress to progressive multifocal leukoencephalopathy ((Knaap et al 2005, Pande et al 2006, Reddick et al 2005).

	< 30 days	30-100 days	> 100 days
Infections	Fungus (aspergillus,candida, etc) → →	→→→→	
	Virus (HHV6 and CMV)		Herpes Zoster
		Toxoplasmosis	PML (JCV)
	Gram negative bacteria	Gram + bacteria	Encapsulated bacterial infections
Drug induced neurotoxicity	1) Cyclosporine or tracrolimus induced encephalopathy (PRES) 2) Methotrexate encephalopathy 3) Chemotherapy related side effects	→	Multifocal leukoencephalopathy
Vascular complications	1) Subdural hemorrhage, 2) intracranial hemorrhage, 3) thrombotic disease		
Post-HSCT			1) Post-transplant lymphoproliferative disease, 2) relapse or 3) secondary malignancy
Causative issue	Neutropenia	Acute GVHD and deficiency of cellular immunity	Chronic GVHD and deficiency of both cellular and humoral immunity

Table 2. CNS complications following HSCT

Progressive multifocal leukoencephalopathy is a more devastating complication that is differentiated from methotrexate induced leukoencephalopathy by the presence of more extensive areas of white matter disease. Approximately 7% of patients who receive cranial irradiation and/ or intrathecal chemotherapy prior to HSCT and intrathecal methotrexate following HSCT develop progressive leukoencephalopathy 4-5 months post-HSCT. Clinical manifestations of this problem include dysarthria, confusion, seizures, upper motor neuron weakness and coma. Leukoencephalopathy is usually chronic and often fatal. (Bleyer, 1981, Thompson et al, 1986). Leukoencephalopathy with severe neurological sequelae and death have also been attributed to amphotericin B following TBI in transplant recipients (Devinsky et al, 1987). Figure 1 shows the presence of progressive multifocal leukoencephalopathy in an 18 year old patient with ALL in 3rd remission 4 months following an unrelated BMT. The patient had received 1800 cGy of cranial irradiation prior to HSCT for the treatment of his ALL and received 1360 cGy of TBI as part of his HSCT conditioning regimen. The patient developed seizures and confusion and subsequently went into a coma. Lumbar puncture did not reveal any infectious etiology. The MRI shows extensive white matter changes consistent with progressive multifocal leukoencephalopathy.

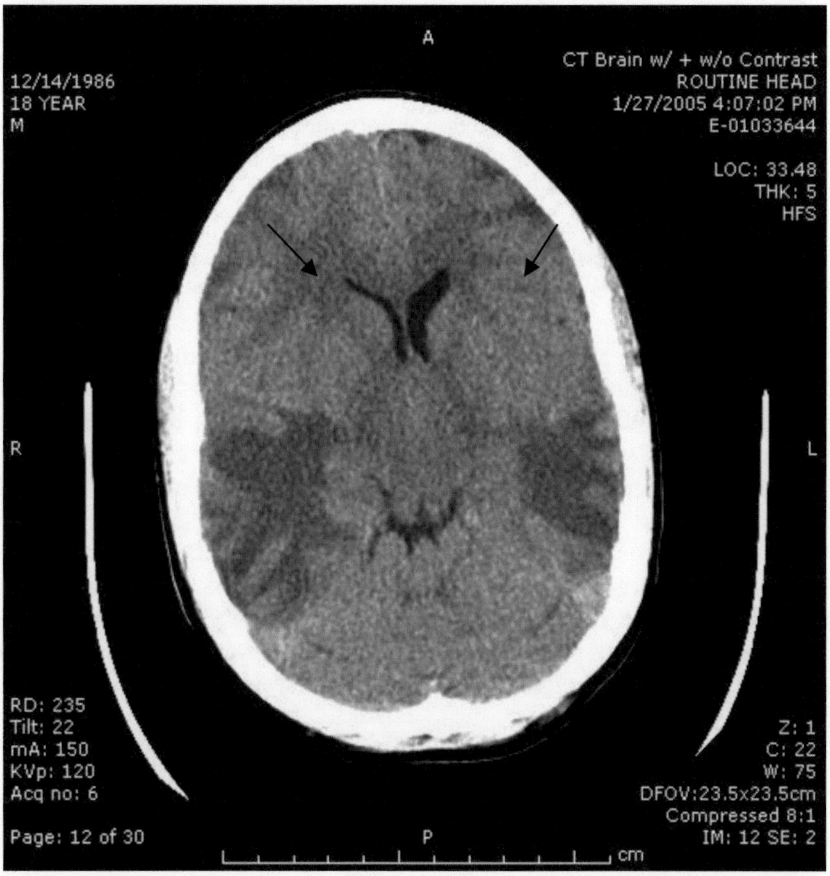

Fig. 1. An 18 year old male with ALL in 3rd remission following an unrelated bone marrow transplant. His conditioning regimen consisted of TBI 1360cGy. Four months post-HSCT he developed seizures and a coma.. An MRI of his head showed bilateral lesions in the deep subcortical white matter of the temporal lobes. The MRI findings are consistent with progressive multifocal leukoencephalopathy

Many patients are unable to tolerate the intensive chemotherapies that are used in myeloablative transplants. Thus many centers have begun to employ the use of reduced intensity conditioning regimens (RIC). RIC often results in increased GVHD, which may also increase the neurological sequelae and this may negate the beneficial long term effects that would be seen of decreased drug toxicity. Although it is expected that a RIC will have fewer neurological complications, there are no studies that have been completed in this area.

6. Impact of immunosuppression

Patients are often very immunosupressed following HSCT. A variety of complications can occur due to immunosuppression. The second period of transplantation occurs during the time between 30-100 days. During this period, patients may develop GVHD, due to alloreactive donor derived T cells from the graft attacking nonshared recipient alloantigens in target tissues such as the skin, liver and GI tract. During this time, radiation given during the conditioning period damages the host tissue from the mucosa, liver and other tissues. Activated cells from the damaged recipient tissues secrete inflammatory cytokines, which upregulate expression of recipient antigens. Presentation of recipient antigens to donor T cells occurs and subsequent proliferation and differentiation of these activated T cells occurs during this time. Activated CD4+ T cells expand and generate cytokines promote differentiation of CD8+ T cells to cytotoxic T cells, which kill recipient cells and cause further tissue damage.

In order to prevent GVHD, patients are often treated with either Cyclosporin A (CSA) or tacrolimus. Both of these agents can cause neurotoxicity. The most frequent complications seen in patients receiving these agents are tremors (40%), paresthesias (11%) [Walker et al 1988]. Approximately 5% of patients develop isolated seizures. Both of these agents can also cause posterior reversible encephalopathy (PRES syndrome). The pathogenesis of PRES is unclear. However, both of these agents disrupt the blood brain barrier or interfere with cerbrovascular autoregulation, resulting in vasogenic edema. This process can rarely progress to irreversible cytotoxic edema (Hinchey et al 1996, Schwarz et al 1995, Mckinney et al 2007). Patients who develop this syndrome develop headache, confusion, blurry vision, visual hallucinations, seizures and motor dysfunction. MRI findings include white matter edema which usually affects the parietal-occipital regions symmetrically and is best seen on fluid- attenuated inversion recovery (FLAIR) sequences. PRES can occur despite normal drug levels. The symptoms of this condition resolve after discontinuation of the drug. The development of PRES does not preclude the safe use of the alternate agent and patients have frequently switched back and forth.

Many HSCT recipients are on corticosteroid therapy, which can also lead to neurological complications. One of the major side effects of corticosteroid therapy includes steroid-induced myopathy, which can occur early in therapy. Steroid-induced myopathy is often characterized by proximal muscle weakness, usually in the hip and neck flexors. Discontinuation of corticosteroids will improve the condition, but that may take many months to achieve. Corticosteroids are also well known to cause a number of psychiatric symptoms such a hallucinations, insomnia, anxiety, mood swings and memory problems.

After one hundred days post-HSCT, chronic GVHD may occur. The immune system generally tends to recover, however if chronic GVHD occurs, there tends to be a prolongation of both cellular and humoral recovery. Patients are at high risk of developing infections with a variety of infections that can include bacterial, viral and fungal infections. Since HSCT recipients are on anti-inflammatory medications such as corticosteroids, it may be difficult for patients to display the classic signs of CNS infections such as fever and meningismus. In addition, many patients have other ongoing complications from their transplant that CNS infections are not often considered.

Patients with CNS infections can be grouped into those who have nonfocal generalized CNS dysfunction such as headaches and meningismus or into those with focal findings suggestive of a mass lesion or stroke. In the first month following HSCT, most patients are neutropenic and are at risk of bacterial infections. In the first month following transplant, gram negative bacterial infections are more predominant, but in the later period (30-100 days) of HSCT, gram positive infections predominate. Methicillin-resistant *Staphylococcus aureus or epidermidis* (MRSA/ MRSE) and *Listeria monocytogenes* may cause meningitis, encephalitis and ventriculitis.

Viral infections post-HSCT include those who develop adenovirus or Coxsackie virus B4 can also produce encephalitis (Krouwer et al 2003, Cree et al 2003). Acquired viral infections such as with human herpes virus 6 (HHV6) and cytomegalovirus (CMV) infections are also common during this period. Most children develop primary HHV6 infection between 6 months and 5 years of age (Yoshikawa et al 2003, Gorniak et al 2006). After the primary infection, this virus remains latent in the body and reactivates upon immunosuppression of the host. HHV6 often reactivates concurrently with GVHD and can develop in approximately 28%-78% of HSCT recipients (Yoshikawa et al 2003, Gorniak et al 2006). Symptoms that have been associated with CNS HHV6 infections include short-term memory loss, mental status changes, fever and headache. The mortality of HHV6 infections is greater than 50%. Figure 2 shows the MRI of a 14 year old female patient with ALL and chronic GVHD. 2 months following HSCT she developed fevers, headaches, seizures and confusion. A CT scan was read as normal, however the MRI showed hyperintensity in the limbic system. A lumbar puncture revealed HHV6 infection and she was treated with foscarnet. This patient recovered from her HHV6 infection, however the mortality of this infection is more than 50% (Yoshikawa et al 2003, Gorniak et al 2006).

Most pediatric patients have indwelling central venous catheters during their transplant. These patients are at risk of developing fungal infections due to Candida. Candidemia can become disseminated and involve the CNS with septic emboli. Aspergillus infections are the most common fungal infection following HSCT. These often occur when patients are either neutropenic or when their T cells have not recovered (day 0-100). MRI findings of patients with Aspergillus often show ring enhancing lesions that may have hemorrhage surrounding the lesion. Aspergillus preferentially occurs in the basal ganglia, cerebral hemispheres, and the corpus callosum. Unfortunately, a biopsy of this area is very difficult and treatment is often started empirically. The treatment is usually liposomal amphotericin, which may add to neurological complications post-HSCT. There are newer agents that are also available for the treatment of aspergillus. The mortality of cerebral aspergillosis approaches 85-100% depending upon the degree of neurological compromise at presentation (Guermazi et al 2003, Miaux et al 1995).

The most common parasitic infection following HSCT is *Toxoplasma gondii*. Patients present with focal symptoms and mild to severe encephalopathy. MRI findings on these patients show the presence of multiple mass lesions affecting the basal ganglia with minimal to no enhancement. Patients who have not received prophylactic antibiotics with trimethoprim sulfamethoxazole are at risk of developing this infection.

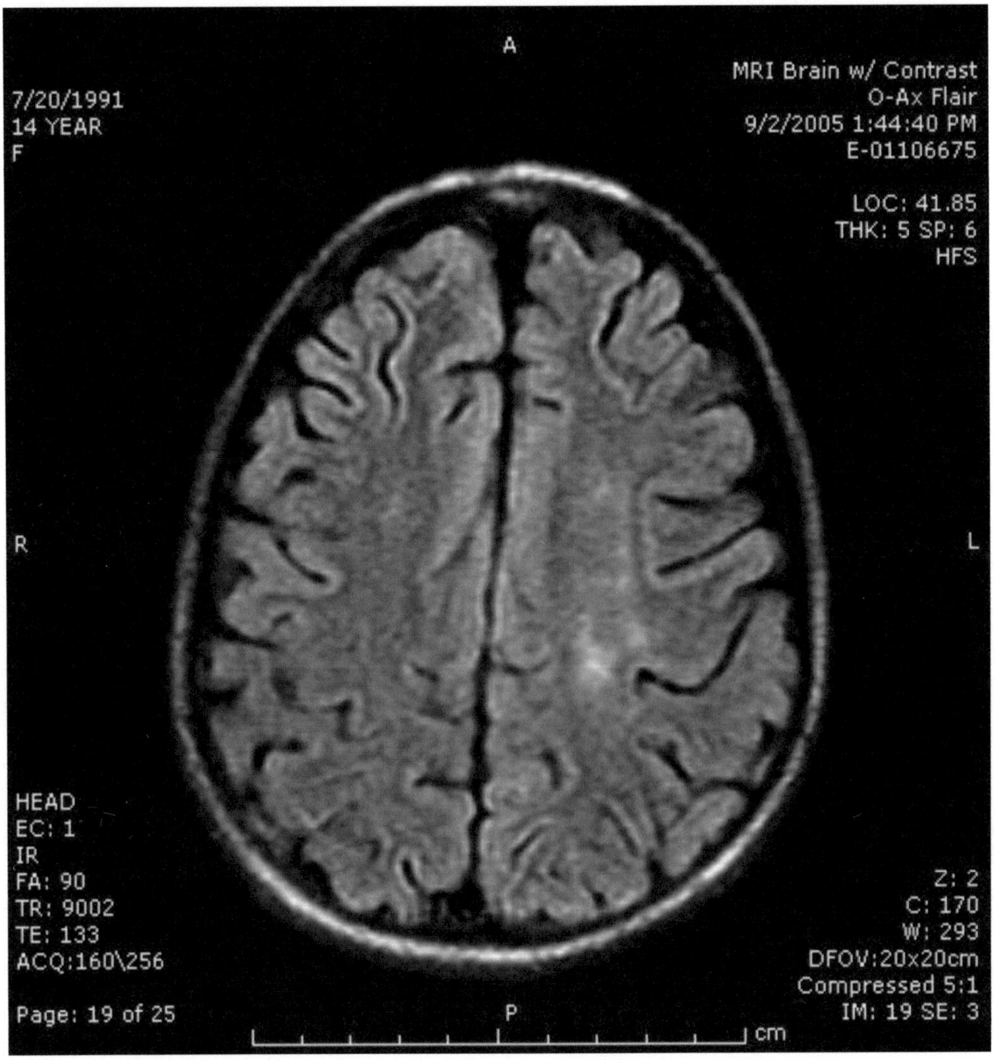

Fig. 2. An MRI of a 14 year old female with ALL in 3rd remission following an unrelated cord blood transplantation. She developed severe graft versus host disease fever, confusion and seizures. A lumbar puncture revealed the presence of HHV6. Flair T2 demonstrates biparietal white matter increased signal intensity more on the left side. This is consistent with HHV6 lesion.

Finally, post-transplant lymphoproliferative disease (PTLD) covers a spectrum that involves benign polyclonal lymphoid hyperplasia to malignant monoclonal lymphoma. Often *Ebstein Barr Virus* (EBV) is the infectious etiology of the malignant lymphoma. In children, the risk is related to a primary EBV infection in a previously seronegative younger patient (Newell et al, 1996). These patients are usually treated with either reduced immunosuppression or chemotherapy. When PTLD does involve the CNS, the prognosis is poor.

7. Neurocognitive function following HSCT

The potential impact of the disease, intensive therapy, social isolation and altered family dynamics in a young child is uncertain and cannot be studied well in pediatric patients. Pot-Mees (1989) documented the profound psychosocial effects of HSCT on patients and families. Recipients of HSCT, despite the underlying disease for which they are transplanted, often face prolonged periods of time in the hospital. Many do not attend school on a regular basis. During the time that they are receiving therapy, they often have some malnutrition due to nausea and vomiting from the therapy to treat their underlying disease. Even after their therapy is completed, some children have an adaptive school program. After HSCT, many children do not have a normal immune system for a long period after HSCT. As a result, they are not involved with extracurricular activities like other age-matched peers.

One would expect that HSCT would increase the neurocognitive deficits seen in pediatric leukemia survivors. The current studies of neurocognitive functioning upon HSCT recipients have yielded mixed results. Most studies of neurocognitive function upon HSCT recipients contain patients with all diseases (malignant and non-malignant), varying types of transplants and conditioning regimens and different age groups. This makes it difficult to interpret the impact of HSCT upon specific disease groups.

Some studies have shown no significant differences in intellectual functioning in patients who are pre-HSCT versus those who are one-year post-HSCT evaluations, except for those patients who have been younger at diagnosis (Kaleita et al 1989, Kramer et al 1992, Simms et al 1998, Phipps et al 1995, Simms et al 2002, Phipps et al 2000, Chou et al 1996, Phipps et al 2008). Whereas, other studies have reported that those patients who receive cranial irradiation have declining neurocognitive function in terms of IQ, achievement, memory and fine motor skills (Cool et al 1996, Smedler et al 1995, Said et al , Kramer et al 1997, Shah et al 2008).

Consistently, all of these previous studies have shown that a younger age at diagnosis and treatment have resulted in significant declines in neurocognitive function In addition, deficits in fine motor skills appear to be more pronounced in HSCT recipients who received cranial irradiation at a younger age than those who received cranial irradiation at an older age (Phipps 2000, Cool 1996, Smedler 1990). Since cranial irradiation appears to affect the information processing, memory and attention mechanisms of the brain, patients who have received cranial irradiation (brain tumor patients and those who had CNS leukemia or received propylaxis for CNS leukemia) are at particular risk of developing long-term cognitive deficits.

Phipps et al conducted the largest pediatric trial relating to the neurocognitive function of patients post-HSCT. This prospective study had 102 patients, who received a full battery of tests for up to 3 years post HSCT. In this study, the neurocognitive function (cognitive and academic functioning) was not impacted by HSCT in patients greater than 6 years of age. There was no impact of TBI upon neurocognitive function. The greatest decline in neurocognitive function occurred in those < 3 years of age. A follow up study done in 2008, increased the population being evaluated to 268 patients. Although the majority of patients had a malignant disease, there were minimal neurocognitive delays. There were subgroups of patients that were increased risk of neurocognitive delays. These subgroups included those who received TBI, recipients of unrelated transplants, and those who developed graft versus host disease (Phipps et al 2008).

For patients who received HSCT for genetic diseases of immunodeficiencies, the information regarding neurocognitive function post-HSCT is sparser. Most patients with genetic disease have problems due to their underlying disease. Following HSCT, the developmental outcomes vary depending upon the disease that they were transplanted for and their pre-HSCT performance scores. Prasad et al transplanted 159 pediatric patients with a variety of inherited metabolic disorders. Among patients with MPS I (Hurler's Disease), most have shown continued increases in their neurocognitive skills, with 81% of them being in age appropriate classes. However, for patients with MPS III (Krabbe's disease), there was less of an impact of HSCT upon neurocognitive function. Most children with Krabbe's disease have severe neurological complications of this disease. Of the surviving patients with Tay-Sachs, there was no impact of HSCT upon their neurocognitive function and the surviving patients are severely debilitated (Prasad 2008).

There have not been any formal studies upon the neurocognitive function of patients with hemoglobinopathies following HSCT. Walters et al recently reported the outcomes of patients receiving HSCT for sickle cell disease. In that study of 55 long-term survivors of histocompatible HSCT, 19 patients developed neurological complications post HSCT, which consisted primarily of seizures. One patient had graft rejection and subsequently developed a second thrombotic stroke. However, no formal neurocognitive testing was done on these patients to determine the long-term clinical impact of transplant on this group of patients (Walters et al, 2009).

Many patients with immunodeficiencies, present to a doctor only after they have had a life-threatening illness. Since these children are very young at diagnosis, they are usually socially isolated after the diagnosis is made. They are not exposed to their age-matched peers. In addition, they often have a history of chronic infections, especially their respiratory systems.

For patients with immunodeficiencies, there have been only 2 major studies that have focused on this population alone. First, Titman et al reported the results of formalized neurocognitive testing upon 106 patients with immunodeficiencies. This study showed that the underlying genetic disease, presence of ADA deficient SCID and consanguinity were associate with a worse outcome. Age at transplantation ($\leq$ 3 years of age versus > 3 years of age), length of stay in the hospital, and chemotherapy conditioning regimen were not significant factors (Titman, 2008).

Lin et al performed a prospective study on the neurocognitive function of 16 patients with SCID. Although there were significant decreases in the formal neurocognitive test scores, the results showed that these children did learn and acquire developmental milestones, but at a slower rate than their age matched normal peers. There was no impact of age ($\leq$ 8 months versus > 8 months of age), presence of serious infections prior to HSCT, or length of stay in the hospital (Lin, 2009).

8. Conclusion

Patients who receive HSCT are at great risk of developing neurological sequelae due to either their primary disease, the therapies used to treat the disease and the complications of these treatments. The complications can be hard to diagnose. These complications can have serious implications for the long-term survivor of HSCT. Formal testing of HSCT recipients neurocognitive function should be required for patients

9. References

[1] Anderson V, Godber T, Smibert E and Ekert H. (1997) Neurobehaviorial sequlae following cranial irradiation and chemotherapy in children: an analysis of risk factors. Pediatr Rehabil 1: 63-76.

[2] Baker WJ, Royer GL JR, Weiss RB: (1991) Cytarabine and neurologic toxicity. J Clin Oncol 9: 679-693.

[3] Bleggi-Torres LF, de Medeiros BC, Werner B, et al. (2000) Neuropathological findings after bone marrow transplantation: An autopsy study of 180 cases. Bone Marrow Transplant 25: 301-307.

[4] Bleyer WA (1981). Neurologic sequelae of methotrexate and ionizing radiation: a new classification. Cancer TreatmentReports 65 (Suppl) 89-98.

[5] Brouwers P, Riccardi R, Fedio P Poplack DG (1985). Long term neuropsychological sequelae of childhood leukemia. Correlation with CT brain scan abnormalities. J Pediatr 106: 723-728.

[6] Brouwers P, Riccardi R, Poplack D, et al (1984). Attentional deficits in long term survivors of childhood acute lymphoblastic leukemia (ALL). J Clin Neuropsych 6: 325-336.

[7] Burger PC, Kamenar E, Schold SC, et al (1981). Encephalomyelopathy following high-dose BCNU therapy. Cancer 48:1318-1327.

[8] Butler RW and Copeland DR. (1993) Neuropsychological effects of central nervous system prophylactic treatment in childhood leukemia: methodological considerations. J Ped Psych, 18:319-338.

[9] Chou RH, Wong GB, Kramer JH et al. (1986) Toxicities of total body irradiation for pediatric bone marrow transplantation. Int J Radiat Oncol Biol Phys 34: 843-851.

[10] Cool VA. Long-term neuropsychological risks in pediatric bone marrow transplant: what do we know? Bone Marrow Transplant (1996) 18 Suppl 3: S45-S49.

[11] Cree, BC, Bernardin GL, Hays AP et al. (2003) A fatal case of coxsackie virus B4 meningoencephalitis. Arch Neurol 60: 107-117.

[12] De La Camara R, Tomas JF, Figuera A et al. (1991) High dose busulfan and seizures. Bone Marrow Transplant 7: 363-4.

[13] Devinsky O, Lemann W, Evans AC, et al. (1987) Akinetic mutism in bone marrow transplant recipient following total body irradiation and amphotericin B chemoprophylaxis. A positron emission tomographic and neuropathologic study. Arch Neurol, 44: 414-7.

[14] DiMaggio JR, Brown R, Baile WF, et al. (1984) Hallucinations and ifosfamide indued neurotoxicity. Cancer 73: 1509-1514.

[15] Fletcher JM and Copeland DR (1988) Neurobehavioral effects of central nervous system prophylactic treatment of cancer in children. J Clin Exp Neuropsychology 10: 495-538.

[16] Gorniak RJT, Young GS, Wiese DE, Marty FM, Schwarz RB. (2006) MR imaging of human herpes virus –6-associated encephalitis in 4 patients with anterograde amnesia after allogeneic hematopoietic stem cell transplantation. Am J Neuroradiol 27:887-891.

[17] Guermazi A, Gluckman E, Tabti B, Myaux Y (2003) Invasive central nervous system aspergillosis in bone marrow transplant recipients: an overview. Eur Radio 13:1480-1485.

[18] Guerrero A, Perez-Simon JA, Gutierrez N et al (1999) Neurological complications after autologous stem cell transplantation. Eur Neurol 41:48-50.
[19] Hinchey J, Chaves C, Appignani B, et al. (1996) A reversible posterior leukoencephalopathy syndrome. N Eng J Med 334:494-500.
[20] Hoyt R, Szer J, Grigg A. (2000) Neurological events associated with the infusion of cryopreserved bone marrow and / or peripheral blood progenitor cells. Bone Marrow Transplant 25: 1285-1287.
[21] Johnson DW, Cagnoni PJ, Schossau TM et al (1999) Optic disc and retinal microvasculopathy after high-dose chemotherapy and autologous hematopoietic progenitor cell support. Bone Marrow Transplant 24: 785-792.
[22] Kaleita TA, Shilds D, Tesler A, Feig SA (1989) Normal neurodevelopment in four young children treated with bone marrow transplantation for acute leukemia or aplastic anemia. Pediatrics 83: 753-757.
[23] Kamble RT, Chang CC, Sanchez S, Carrum G (2007) Central nervous system graft versus host disease: a report of two cases and literature review. BMT 39: 49-52.
[24] Kramer JH, Crittendon MR, Halberg FE, et al. (1992) A prospective study of cognitive functioning following low-dose cranial radiation for bone marrow transplantation. Pediatrics 90: 447-450.
[25] Krouwer HG, Wijdicks EF. (2003) Neurologic complications of bone marrow transplantation. Neurol Clin 21: 319-352.
[26] Leff RS, Thompson JM, Daly MB, et al. (1988)Acute neurologic dysfunction after high-dose therapy for malignant glioma. Cancer 62:32-35.
[27] Lin M, Epport K, Azen C et al (2009) Long term neurocognitive function of pediatric patients with severe combined immune deficiency (SCID): pre and post hematopoietic stem cell transplant. J Clin Immunol 29: 231-237.
[28] Mckinney AM, Short J, Truwit CL et al. (2007) Posterior reversible encephalopathy syndrome: incidence of atypical regions of involvement and imaging findings. Am J Radiol 189:904-912.
[29] Miaux Y, Willimas M, Guermazi A et al. (1995) MR of cerebral aspergillosis in patients who have had bone marrow transplantation Am J Neurradiol 16:555-562.
[30] Mohrmann RL, Mah V Vintners HV: (1990) Neuropathologic findings after bone marrow transplantation: An autopsy sudy. Hum Pathol 21:630-639.
[31] Moleski M. (2000) Neuropsychological, neuroanatomical, and neurophysiolical consequences of CNS chemotherapy for acute lymphoblastic leukemia. Arch Clin Neuropsychol 5: 603-630.
[32] Mulhern RK and Butler RW (2004). Neurocognitive sequelae of childhood cancers and their treatment. 7: 1-14.
[33] Newell KA, Alonso EM, Whitington PF et al. (1996) Posttransplant Lymphoproliferative disease in pediatric liver transplantation. Interplay between primary Ebstein -Barr virus infection and immunosuppression. Transplantation. 62:370-375, 1996.
[34] Packer RJ, Zimmerman RA, Bilaniuk LT (1986) Magnetic resonance imaging in the evaluation of treatment-related central nervous system damage. Cancer 58: 635-640.
[35] Parsons SK, Neault MW, Lehman LE, et al (1998) Severe ototoxicity following carboplatin-containing conditioning regimen for autologous marrow transplantation for neuroblastoma. Bone Marrow Transplant 22:669-674.

[36] Phipps S, Brenner M, Heslop H,. Krance R, Jayawardene M, Mulhern R (1985) Psychological effects of bone marrow transplantation on children and adolescents: a preliminary report of a longitudinal study. Bone Marrow Transplant 15:829-835.
[37] Phipps S, Dunaveant M, Srivastava DK, Bowman L, Mulhern R (2000) Cognitive and academic functioning in survivors of pediatric bone marrow transplantation. J Clin Oncol 18: 1004-1011.
[38] Phipps S, Rai SN, Leung WH, et al. (2008) Cognitive and academic consequences of stem cell transplantation in children. J Clin Onc 26: 2027-2033.
[39] Pott-Mees CC (1989): the Psychological Aspects of Bone Marrow Transplantation in Children. Delft, The Netherlands, Eburon,.
[40] Prasad VK, Mendizabal A, Parikh SH et al. (2008) Unrelated donor umbilical cord blood transplantation for inherited metabolic disorders in 159 pediatric patients from a single center: influence of cellular composition of the graft on transplantation outcomes. Blood 112: 2979-2989.
[41] Pratt CB, Goren MP, Meyer WH, et al (1990) Ifosfamide neurotoxicity is related to previous cisplatin treatment for pediatric solid tumors. J Clin Oncol 8: 1399-1401.
[42] Ris MD, and Noll RB. (1994) Long-term neurobehaviorial outcome in pediatric brain tumor patients: review and methodological critique. J Clin Exp Neurolopsych 16: 21-42.
[43] Saiz A, Graus F (2004): Neurological complications of hematopoietic cell transplantation. Semin Neurol 24: 427-434.
[44] Schwarz RB, Bravo SM, Klufas RA et al. (1995) Cyclosporine neurotoxicity and its relationship to hypertensive encephalopathy: Ct and MR findings in 16 cases. Am J Radiol 165:627-631.
[45] Shah AJ, Epport K, Azen C et al (2008). Progressive declines in neurocognitive function among pediatric recipients of hematopoietic stem cell transplantation for hematological malignancies. J Pediatr Hematol Oncol. 30:411-418.
[46] Simms S, Kazak AE, Gannon T, Goldwein J, Bunin N (1998) Neuropsychologival ouitcome of children undergoing bone marrow transplantation. Bone Marrow Transplant 22: 181-184.
[47] Simms S, Kazak, AE, Golomb V, Goldwein J. Bunin N (2002) Cognitive behaviorial and social outcome in survivors of childhood stem cell transplantation. J Ped Hem Onc 25: 5-7.
[48] Smedler AC, ringden K Bergman H, Bolme P (1990): Sensory-motor and cognitive functioning in children who have undergone bone amrrow transplantation. Acta Paedr Scan 79: 613-621.
[49] Smibert E, Anderson V, Godber T, Ekert H (1996) Risk factors for intellectual and educational sequelae of cranial irradiation in childhood acute lymphoblastic leukemia. Br J Cancer 73: 825-30.
[50] Snider S, Bashir R, Bierman P (1994) Neurologic complications after high dose chemotherapy and autologous bone marrow transplantation for Hodgkins disease. Neurology 44:681-684.
[51] Stehbens JA, Kaleita TA, Noll RB et al. (1991) CNS prophylaxis of childhood leukemia: What are the long-term neurological, neuropsychological and behaviorial effects? Neuropsycholga Review 2: 1147-177.

[52] Thompson CB, Sanders JE, Flournoy N et al (1986) The risks of central nervous system relapse and leukoencephalopathy in patients receiving marrow transplants for acute leukemia. Blood 67, 195-9.

[53] Titman P, Pink E, Skucek E, O'Hanlon K et al. (2008) Cognitive and behavioral abnormalities in children after hematopoietic stem cell transplantation for severe congenital immunodeficiencies. Blood 112: 3907-3913.

[54] Vassal G, Deroussent A, Hartmann O, et al (1990). Dose-dependent neurotoxicity of high-dose busulfan in children: A clinical pharmacological study. Cancer Res 50:6203-6207.

[55] Walker RW, Brochstein JA (1988). Neurologic complications of immunosuppressive agents. Neurol Clin 6:261-278.

[56] Walters MC, Hardy K, Edwards S et al (2009). Pulmonary, gonadal and central nervous system status after bone marrow transplant for sickle cell disease. Biol Blood Marrow Transplant. 2009 Oct 9. [Epub ahead of print]

[57] Weber C, Schaper J, Tibussek D et al (2008) Diagnostic and therapeutic implications of neurological complications following paediatric haematopoietic stem cell transplantation. BMT 42: 253-259.

[58] Yoshikawa T (2003) Hujman herpesvirus 6 and -7 infections in transplantation. Pediatr Transplantation 7:11-17.

Adenoviral Infection – Common Complication Following Hematopoietic Stem Cell Transplantation

Iwona Bil-Lula[1,2], Marek Ussowicz[3] and Mieczysław Woźniak[1,4]
[1]*Department of Clinical Chemistry, University of Medicine, Wrocław*
[2]*Laboratory of Molecular Biology, Central University Hospital in Wrocław*
[3]*Department of Paediatric Bone Marrow Transplantation, Oncology, and Haematology, University of Medicine in Wrocław*
[4]*Department of Pharmacology, University of Saskatchewan*
[1,2,3]*Poland*
[4]*Canada*

1. Introduction

Allogeneic stem cell transplantations (allo-SCT) increase in recipients the risk of life-threatening opportunistic infections caused by different pathogens. In recent years, adenoviruses (AdV) have been recognized as an emerging pathogen causing serious post-transplant complications with very high mortality (Fowler et al., 2010). Hematopoietic stem cell transplantation is a standard treatment in malignant and non-malignant disorders of bone marrow, solid tumors, immunodeficiencies, and autoimmune disorders (Gyurkocza et al., 2010; Ljungman et al., 2010). Intensive immunosuppressive treatment for prevention both of graft rejection and graft-versus-host disease (GvHD) and the therapy of the latter one, makes the patients at risk of developing opportunistic infections. The immune reconstitution period following hematopoietic transplantation is therefore accompanied by high incidence of adenoviral infections due to profound immunodeficiency (Echavarri´a, 2008; Gyurkocza et al., 2010). Lack of immunological competence caused by invasive cancer therapy and conditioning procedures results mainly in lowering the number of CD3+ lymphocytes in peripheral blood. Lymphopenia increases patient`s risk for de novo infection or reactivation of a latent virus that mainly occurs during the early post-transplantation period and usually leads to disseminated disease. The time of the infection onset is also vital, because AdV infections occurring after day +100 from the transplantation do not seem to be associated with life-threatening disease (Lion et al., 2003). According to different retrospective and prospective studies, adenovirus has been found to infect from just a few percent to a several dozen percent of patient post-HSCT (Akiyama et al., 2001; Arnberg et al., 2002; Benko et al., 2000; Ebner et al., 2005; Howard et al., 1999; Mahr & Gooding, 1999; Lauer et al., 2003; Robin et al., 2007; Rutala et al., 2006; Stone et al., 2003; Walls et al., 2003; Watzinger et al., 2004) with mortality in disseminated disease reaching even 80% (Akiyama et al., 2001; Mori et al., 2003; Russell, 2000; Walls et al., 2003; Watzinger et al., 2004). Adenoviruses are commonly detected in stool, pharyngeal swabs, urine and whole blood in HSCT recipients usually causing limited infections of the respiratory tract, gastrointestinal system and urinary tract but sometimes progress to disseminated disease affecting many organs and whole systems, too (Watcharananan et al., 2010). The diagnosis of adenoviral infection can be difficult due to complexity of the multiorgan disease non-specific symptoms resembling other infections or acute graft versus host disease (aGvHD). This review describes adenoviral infections in recipients of hematopoietic stem cell transplantations and focuses attention on infection course, risk factors, possible complications, diagnostic methods for adenovirus identification and therapy strategies.

2. Human adenoviruses-general description

The original virus was isolated in 1953 from surgically removed adenoids from a child by Rowe et al. (Huebner et al., 1954; Rowe et al., 1953), thus it has been called adenovirus. Adenoviruses are ubiquitous in the environment contaminated with human feces or sewage. To date, 54 antigenic types of human adenoviruses belonging to genus *Mastadenovirus* and family *Adenoviridae* have been described, and over half of them have been recognized as pathogenic for humans. They have been divided into seven species (from A to G) on the basis of their morphological, hemagglutinating and oncogeneic properties as well as their genome size, DNA sequence and electrophoretic mobility of virion polypeptides (Table 1). Adenoviruses from species B and D were further subdivided into subspecies: B1, B2 and D1, D2 and D3, respectively. In addition to this classification several AdV genotypes can be identified within previously mentioned serotypes (Echavarri´a, 2008, 2009; Ebner et al. 2005; Stone et al., 2003; Wadell et al., 1980).

Adenoviruses are non-enveloped, 70-90 nm in diameter particles with icosahedral symmetry. The capsid is composed of 252 capsomers: 240 hexamers and 12 pentons (Russell, 2009; Rux et al., 2003). Several different proteins in AdV particle can be distinguished, but three major proteins such as hexon, penton base and fiber protein create the main capsid structure. The hexon is a homotrimer of three identical polypeptide chains (pII) formulating a triangular vertex of surface loops. It is the most abundant protein in adenoviral particle (more than 80% of AdV2 capsid protein) formulating hexon molecules symmetrically distributed in the capsid (Burnett, 1999). The hexon possess the group− and type−specific determinants which are utilized in diagnostic procedures such as ELISA test, hemagglutination inhibition and serum neutralization. Occurrence of these specific antigen determinants on a viral surface is determined by the presence of variable and hypervariable regions (HVRs) in nucleotide sequence of the hexon gene. Since today, nine hipervariable regions (HVR1-HVR9) have been distinguished (Crawford-Miksza & Schnurr, 1996; Rux et al., 2003). They are situated within the loops at the top of the hexon molecule, hence can be utilized in diagnostic tests. The hexon protein is also an immunodominant T-cell target, however other structural components of the virion may also be immunogenic. Twelve penton bases (pIII) creating capsid vertex provide the basis for radially projecting trimeric fibers (pIV) of different lengths according to the AdV serotype (10-37 nm). The external end of the fiber is terminated with a C-terminal knob which posses several antigen determinants participating in cell receptors binding, hemagglutination in vitro and virus internalization into a host cell (Burmeister et al., 2004; Cusack, 2005; Echavarri´a, 2008). Genetic information of adenoviruses is encoded by about 34-36 kb double stranded linear DNA, containing more than 50 coding regions (Echavarri´a, 2008).

Species	Serotypes	Features#
A	12, 18, 31	Highly oncogenic GC%: 47-49 FL: 28-31nm HA: rat (incomplete)
B	B1: 3, 7, 16, 21, 50† B2: 11, 14, 34, 35	Poorly oncogenic GC%: 48-52 FL: 9-11 nm HA: Rhesus
C	1, 2, 5, 6	Nononcogenic GC%: 55-59 FL: 23-31nm HA: rat (incomplete)
D	D1: 9, 19 D2: 15, 22 D3: 8, 10, 13, 17, 20, 23-30, 32, 33, 36-39, 42-49, 51†, 53, 54*	Nononcogenic GC%: 54-59 FL: 12-13 nm HA: rat, mouse, human, dog, pig, monkey
E	4	Nononcogenic GC%: 57 FL: 17 nm HA: Rat (incomplete)
F	40, 41	Nononcogenic GC%: 51-59 FL: 29 nm HA: rat (atypic)
G	52	ND

Data adopted from references: (Echavarria, 2006; Hierholzer, 1992; Kaneko et al., 2009; Wadell et al., 1980; Walsh et al., 2010; and from National Center for Biotechnology Information database (http://www.ncbi.nlm.nih.gov);

† according to Ebner et al. (2005) serotype 50 was reclassified into species D and serotype 51 into species B;

* firstly described in year 2009, (proposal) (Kaneko et al., 2006);

ND- not determined, FL-fiber length, HA- hemaglutination;

Table 1. Adenovirus classification and general features.

3. Virus transmission in stem cell recipients

The most common source of adenoviral infection after haematopoietic stem cell transplantation is reactivation of a latent virus persisting in lymphocytes of peripheral blood, in adenoids, intestines and kidneys through low-grade replication. AdV1, AdV2 and AdV5 (C) are the most common serotypes remaining latent after primary exposure in the childhood (Lee et al., 2010; Malekshahi et al., 2010). They also play the significant role in development of AdV infections after SCT, therefore confirming the role of virus reactivation during an early post-transplant period. The next source of AdV infection in stem cell recipients might be the virus transmission from the positive donor to the negative recipient.

There are many reports confirming more than 4-fold increased risk of primary adenoviral infections in HSCT recipients grafted from previously infected (seropositive) donors in comparison to uninfected donors (Runde et al., 2001; Walls et al., 2003).
Adenoviruses can spread directly through respiratory droplets hence upper respiratory tract infections occur very often. They are able to replicate in the environment of the gastrointestinal tract due to their resistance to low pH of a stomach and proteolytic enzymes in gastrointestinal secretions, thus allowing the virus to achieve a high viral load in the gut. Because viral loads in stool samples of infected patients achieve the highest levels in comparison to other clinical materials, it can be suggested that fecal-oral transmission is the most common route of infection spread (Baldwin et al., 2000; Chakrabarti et al., 2002; Hale et al., 1999; Howard et al., 1999; Jeulin et al., 2011; Lion et al., 2010; Rutala et al, 2006).
There are also some reports suggesting possible virus spread via unsterile instruments and medical equipment, direct staff-to-patient transmission or inadequate air conditioning in the post-transplantation ward. Adenoviruses are strongly resistant to chemical or physical agents and adverse pH conditions and are able to remain infectious for a long period outside of the body, but this route of infection can be prevented by disinfection procedures in transplantation units (Artieda et al., 2010; Gerber et al., 2001; Gray, 2006; Leruez-Ville et al., 2006; Lessa et al., 2007; Romero et al., 2010).

4. Pathomechanism and clinical manifestations of adenoviral infections

Individual serotypes of *Adenoviridae* family show varied tissue tropism. They infect post-mitotic cells of even highly differentiated tissues such as skeletal muscles, lungs, brain and heart muscle (Russell, 2009; Zhang & Bergelson, 2005). The infection cycle is divided into two stages involving virus adsorption and internalization (Figure 1). Attachment to the host cell receptor and virus entry occurs via the fibre protein and penton base. Majority of adenoviruses (mainly those from species A, C, D, E and F) use the coxackie-adenovirus receptor (CAR) which is a member of the immunoglobulin family and is situated on a surface of most human tissues such as heart muscle, brain, liver, pancreas, intestines, lungs and kidneys (Dechecchi et al, 2001; Raschperger et al., 2006; Russell, 2000, 2009; Rux et al., 2003; Zhang & Bergelson, 2005). In human cells lacking CAR receptor, other cell molecules are used for virus attachment. For example, AdV2 and AdV5 are able to bind the α-domain of MHC class I (*Major Histocompatibility Complex Class I)* molecule or heparan sulfate glycosaminoglycans (HS GAG). Additionally, these serotypes are capable to recognize and interact with vascular cell adhesion molecule-1 (VCAM-1) which is situated on atherosclerotic lesions of vascular endothelium. More rarely, α(2-3)- sialic acid, sBAR and sB2AR and CD46, CD80, CD86 are used (mainly serotypes: 3, 7, 8, 11, 14, 16, 19, 21, 37, 34 , 35 and 50) (Benko et al., 2000; Burmeister et al., 2004; Chu et al., 2001; Sharma et al., 2009; Wang et al, 2007; Zhang & Bergelson, 2005). Virus binding to the cell receptor and interaction between penton base and integrins on the target cell induce internalization of the viral particle into the host cell by clathrin-coated vesicles and endosomes. Several proteins and intra-endosomal mechanisms lead to disintegration of virus particle and subsequently to viral core release. Viral DNA is transported to the cell nucleus by microtubule-mediated transport and undergoes DNA transcription and replication (Cusack, 2005; Kelkar et al., 2004; Russell, 2000). Replication process is divided into an early and late phase. Sometimes, an intermediate phase is also marked out. The early phase (6-8 hours) includes virus penetration into the host cell, transport of viral DNA to cell nucleus and transcription/translation of the "early genes" such as E1A, E1B, E2, E3, E4. The products of the early genes act as regulatory factors. They change cell functions to facilitate further replication of the virus and transactivate other early transcription units. Second phase of AdV replication leads to transcription and translation of the "late genes" (L1-L5) encoding mainly structural proteins of the virion shell. Assembly and maturation of AdV particles take place in the nuclei of infected cells. The whole process usually takes only 4-6 hours (Greber et al. 1997, Gonçalves & de Vries, 2006; Matthews & Russell, 1998; Russell, 2000).

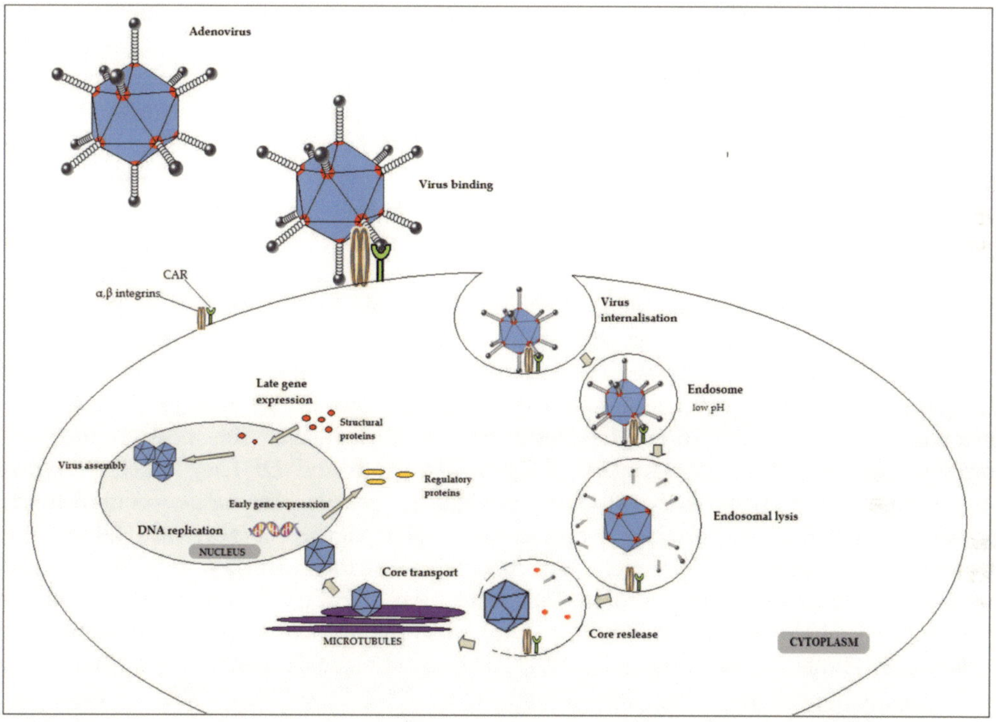

Fig. 1. AdV infection cycle. An interaction between penton base and α,β integrins induces internalization of viral particle into the host cell. Intra-endosomal mechanisms lead to disintegration of virion shell. Viral DNA is transported into the cell nucleus providing material for DNA replication. The early phase of replication includes expression of the "early genes" (E1-E4) acting as regulatory factors and facilitating virus replication. Second phase of AdV replication leads to expression of L1-L4 encoding mainly structural proteins determining virus assembly and maturation.

Adenoviral infection and active virus replication in the host cell can exert three different cytopathological effects. In most cases AdV infection leads to cell lysis. AdV genome contains at least four different coding regions (E1a, E4ORF4, E4ORF6/7 i ADP, which products can induce cell death and lysis (Braithwaite & Russell, 2001). Alternatively, after acute phase of the infection, the virus might remain in latent phase. Human adenoviruses, especially AdV1, AdV2, AdV5 from species C, possess the ability to prolonged persistence in adenoids, intestines, renal parenchyma and lymphocytes in peripheral blood with periodic shedding of virus in feaces and respiratory secretions (Akiyama et al., 2001; Braithwaite & Russell, 2001; Echavarri´a, 2008; Watzinger et al., 2004). The third possible outcome of adenoviral infection is an oncogenic transformation of infected cells, that is observed in animal models (Russell, 2009).

Primary adenoviral infections affect predominantly pediatric population with more than 50% of children infected before the age of five (Cooper et al., 2003). Children are often infected with adenovirus types 1, 2, 3, 4, 5, 7 and 30 which are the most common causes of tonsillopharyngitis and pneumonia (Carballal et al., 2002; Faden et al., 2005). These antigenic types are also the most frequent in recipients of SCT indicating virus reactivation or droplets transmission as most reliable routes of virus spread. AdV infections in immunocompetent humans are usually asymptomatic, localized and tend to be self-limited. They are restricted to the respiratory system, genitourinary and gastrointestinal tract infections, occasionally affecting conjunctiva and cornea. Recipients of HSCT and other imunocompromised patients present much broader spectrum of clinical manifestations. According to different studies, the estimated rate of AdV infection after HSCT ranges from 3–47% (Akiyama et al., 2001; Arnberg et al., 2002; Ebner et al., 2005; Howard et al, 1999; Lauer et al., 2004; Mahr & Gooding, 1999; Rutala et al., 2006; Stone et al., 2003; Walls et al, 2003; Watcharananan et al, 2010; Watzinger et al., 2004;) with mortality reaching up to 80% in patients with disseminated disease (Akiyama et al., 2001; Kaneko et al., 2009; Russell, 2000; Walls et al., 2003; Watzinger et al., 2004). The wide range in reported AdV incidence can results from different diagnostic methods, variety of body sites analyzed as well as demographic differences and lack of strict criteria for defining adenovirus infection or disease. AdV infections in immunocompromised patients tend to be invasive. The most common are infections of AdVs from subgroup C, followed by A, B and D. They cause exudative pharyngitis, acute respiratory disease epidemics, febrile laryngitis, conjunctivitis, keratoconjunctivitis, necrotizing enterocolitis, pharyngeal–conjunctival fever and hemorrhagic cystitis. Less frequent are testitis, nephritis, arthritis, myocarditis and pericarditis. Infections of the gastrointestinal tract are especially frequent in young children and include gastroenteritis, mesenteric lymphadenitis, intussusception, hepatitis, and appendicitis (Table 2) (Akiyama et al., 2001; Arnberg et al., 2002; Benko, 2000; Chakrabarti et al., 2000; Chmielewicz et al., 2005; Ebner et al., 2005; Echavarria, 2009; Ephros et al., 2009; Howard et al., 1999; Ison, 2006; Jones et al., 2007; Kaneko et al., 2009; Lim et al., 2005; Mori et al., 2003; Robin et al., 2007; Runde et al., 2001; Venard et al., 2000; Walsh et al, 2009). Most adenoviral infections in patients following HSCT occur during an early post-transplantation period, within 2-3 months (Ephros et al., 2009; Lion et al., 2010; Watcharananan et al., 2010). Adenovirus infections in patients after stem cell transplantations can occur throughout the year but there are some reports suggesting seasonal fluctuations of adenoviral infections indicating greater incidence in autumn and winter, 52-70% (Bil-Lula et al., 2010).

An intensity of virus replication depends on a site of infection. The highest viral loads are observed in gastrointestinal tract infections (even more than 10^8 copies/g of stool) and in upper respiratory tract infections (10^4 -10^6 copies/ml). Whole blood viraemia remains at a moderate level and usually does not exceed 10^3 – 10^4 copies per ml, whereas in plasma samples only trace amounts of AdV particles can be detected (< 10^2 copies/ml). Some of patients with moderate viraemia are able to self-eliminate of the virus within 2-4 weeks. However most patients presenting disseminated infections (> 10^7 copies/ml) die due to adenoviral infection within short time (Ephros et al., 2009; Heim et al., 2003; Lion et al., 2003; Walls et al., 2005).

Clinical symptoms	Involved serotypes
Acute follicular conjunctivitis and keratoconjunctivitis	1, 3, 4, 5, 7, 8, 9, 11, 16, 19, 26, 27, 30, 37, 53, 54
Acute hemorrhagic cystitis (HC)	3, 7, 11, 21, 34, 35
Acute respiratory disease	1, 2, 3, 4, 5, 6, 7, 14, 21, 30
Exudative pharyngitis	1, 2, 3, 5, 7
Gastroenteritis	1, 2, 12, 16, 18, 31, 40, 41, 52
Hepatitis, appendicitis	1, 2, 3, 5, 7
Intussusception	1, 2, 5
Meningitis, encephalitis	2, 3, 7, 12, 32
Myocarditis	7, 21
Nephritis and kidney damage	11, 14, 34, 35
Pertussis–like syndrome	5
Pharyngitis	1, 2, 3, 5, 7
Pharyngoconjunctival fever (PCF)	3, 4, 7, 14
Pneumonia	1, 2, 3, 4, 7, 11, 14, 21, 30, 34, 35
Prolonged diarrhea	40, 41, 52
Ulcerative changes in the female genital organs, cervicitis	19, 37
Urethritis	1, 3, 4, 7, 14, 16, 19, 34, 35, 37, 50

Table 2. Clinical manifestations of adenoviral infection in stem cell transplant recipients (summary table).

5. Anti-viral response and viral immunoavoidance

Virus penetration into human organism induces both specific and non-specific response to infection. Firstly, virus by binding with the cell receptors activates congenital mechanisms of anti-viral defense. An induction of the inflammatory response stimulates interferon production by both immunocompetent and non-immunocompetent cells. Macrophages, natural killers (NK cells) and complement proteins are also involved in anti-adenovirus defense indirectly by killing infected cells or due to production of cytokins such as Il-1, Il-6 and TNF-α. Additionally, Smith et al. (2008) confirmed the role of the epithelial defensins which coat the capsid vertex of adenoviral particle and inhibit virus uncoating in cytoplasm of infected cell (Smith & Nemerow, 2008). Another way of virus elimination from the organism is induction of pro-apoptotic mechanisms in infected cells due to e.g. p53, TNFα and Bax proteins (Randall & Goodbourn, 2008; Russell, 2000, 2009; Smith & Nemerow, 2008). Very important role in viral clearance is played by acquired immunity mechanisms. The cellular response is limited to CD3+ cells activation (both helper and cytotoxic lymphocytes) which also produce TNFα and INFγ. Activation of CD8+ cells due to presentation of hexon determinants in MHC I complex leads to eradication of infected cells by means of cell perforation and lyses. This mechanism prevents virus replication and further spread of viral particles. The humoral response is focused on production of

neutralizing antibodies. The neutralizing antibodies recognize HVRs and the fiber determinants leading to agglutination of virus particles and interrupting infection of new cells. The presence of anti-adenoviral antibodies in patient serum grants permanent immunity against AdV (Baldwin et al., 2000; Crawford-Miksza & Schnurr, 1994; Leen et al, 2008; Russel, 2009; Rux et al., 2003; Schilham et al., 2002; Walls et al., 2003).
Intensity and length of immunosuppressive therapy are tailored to the risk of graft-rejection or graft versus host disease.
The prevention and therapy of GvHD is crucial in allogeneic HSCT recipients, which explains its intensity and diversity by combining drugs (cyclosporine, tacrolimus, mycophenolate mofetil) with biological agents (OKT-3, alemtuzumab, ATG) or graft engineering methods like T-cell depletion/CD34+ positive selection. Immunosupression determinates stable engraftment and prevents from GvHD, but increases risk of disease relapse or opportunistic infections, due to depletion of donor lymphocytes necessary for graft-versus-tumor (GVT) effects and delayed immune reconstitution (Gyurkocza et al., 2010).
Immune defense against adenoviral infections is hampered by viral abilities for avoidance of both humoral and cellular host immune response. By means of E1A (early region 1A) and VARNAs (virally associated RNAs) they developed the ability to inhibit human interferons α and β. The product of E1B inhibits apoptosis of infected cells. Moreover, product of E3 coding region can inhibit a transport of MHC I particles to the cell membrane disturbing viral antigens presentation to the cytotoxic lymphocytes (CLS) (Braithwaite & Russell, 2001; Lauer et al., 2004; Mahr & Gooding, 1999; Russell, 2000; Stone et al., 2003). Some of E3 products interfere with pro-inflammatory and cytolytic activity of TNF or remove Fas and TRAIL receptors from cell surface (Echavarria, 2009). The fact, that these proteins are expressed during early stage of transcription, protects infected cells from immune surveillance.

6. Risk factors for development of adenoviral infection after HSCT

The hematopoietic stem cell transplantation recipients constitute a group of patients with an extremely high risk of death due to opportunistic infections, and among them the AdVs are accounted to most challenging pathogens responsible for fatal outcome. The problem of AdV infections after HSCT is a consequence of different factors affecting recipient`s immunity. Development of graft versus host disease and its prophylaxis or therapy with potent immunosuppressants are the reasons for numerous opportunistic infections. The anti-leukemic efficacy and early complications of HSCT result from both intensity of the conditioning regimen and the graft-versus-leukaemia (GVL) effect. However, conditioning strategies with high doses of cytotoxic and immunosupressive drugs lead to tissue damage, cytokine storm and profound impairment of patients immunity, thus increasing patient`s risk of de novo infection or reactivation of latent adenovirus during a post-transplantation period. Reduced-intensity conditioning (RIC) regimens lower the risk for AdV infection due decreased organ damage and lower proinflammatory cytokines secretion, hence protecting the patient from aGvHD but maintaining an effect of GVL (Couriel et al., 2004; Gyurkocza et al., 2010; Hill& Ferrara, 2000; Pérez-Simón et al., 2002, 2005). The clinical course of AdV infection after HSCT is different from the one observed in immunocompetent individuals. AdVs usually cause permanent and stubborn infections in those patients because immunological response to adenoviruses following SCT is very poor. In rare situations, AdV can be responsible for graft rejection or delayed implantation of stem cells (Hierholzer, 1992; Ison, 2006).

Risk factors affecting patient`s susceptibility to adenovirus infections after stem cells transplantation are well known (Figure 2). Many reports indicate increased morbidity among recipients of allogeneic grafts from matched unrelated donors (MUD) in comparison to partially matched family donors (PMFD) (Baldwin et al. 2000; Bruno et al., 2003; Ebner et al., 2005; Gu et al., 2003; Hale et al., 1999; Howard et al., 1999; Ison, 2006; Legrand et al., 2001; Lion et al., 2003). An incidence of AdV infections and patient`s mortality due to infection in this group reach about 5-47% and 60%, respectively (Hierholzer, 1992; Howard et al., 1999; Leruez-Ville et al., 2006; Lim et al., 2005; Robin et al., 2007; Runde et al., 2001; Venard et al., 2000; Walls et al., 2005). In recipients of autologous transplantations AdV infections prevalence is much lower, about 1-14% (Baldwin et al., 2000; Bruno et al., 2003; Hale et al., 1999; Howard et al., 1999; Teramura et al., 2004). Significant differences in occurrence of adenoviral infections are believed to result from more aggressive and longer immunosuppressive therapy after allogeneic transplantations (Walls et al., 2003). Lack of endogenous T-cell immunity makes patient more predisposing to development of adenoviral infection and disseminated disease with fatal outcome (Bil-Lula et al., 2010; Feuchtinger et al., 2008; Watcharananan et al., 2010). Therefore, delayed immune recovery after stem cell transplantation has critical impact on progression of adenoviral infections. Moreover, second allotransplantation also increases patient`s susceptibility to AdV infection (Bruno et al., 2003).

Another risk factor for development of adenoviral infections in HSCT recipients is patient`s age. There are many reports confirming that children and adolescents are at greater risk of adenovirus infection than adults (Robin et al., 2007; Walls et al., 2005). Greater susceptibility of younger patients may results from immaturity and worse efficiency of immune system in comparison to older recipients. Moreover, persistent infections are predominantly caused by AdV from species C, which are common etiological factor of adenoviral infections in a childhood. Combination of immunosupression and young age of graft recipients can lead to intensified replication of adenoviruses and to greater viral load in clinical samples (more than 10^5 copies /ml) in comparison to adult recipient (viral load usually not exceeds 10^3 copies /ml) (Baldwin et al., 2000; Bruno et al., 2003; Chakrabarti et al., 2002; Howard et al., 1999; Heim et al., 2003; Ison, 2006; Robin et al., 2007; Walls et al., 2003).

The role of T-cells in controlling the AdV infection is unquestionable. The graft engineering methods involving ex-vivo removal of T-cells by CD34+ positive selection or by lymphocyte depletion are very effective in respect of GvHD reduction but cause profound long-term T-cell deficiency and function impairment. Adenoviral infections due to reactivation of latent virus are also reported more frequently (more than 70%) in recipients of manipulated grafts in comparison to those who received graft containing donor`s lymphocytes (25%). Moreover, some reports suggested delayed recovery and lower overall survival in patients undergoing T-cells depletion (Chakrabarti et al, 2002; Ison, 2006; Lion et al., 2003; Venard et al., 2000; Walls et al., 2003, 2005).

The development and severity of graft versus host disease are typically identified independent risk factors for progression of adenoviral infection after stem cell transplantation. The mechanism of GvHD induction is multifactorial and despite intensive research only moderately understood, the incompatibilities in HLA and in minor histocompatibility antigens (mHA) are generally recognized causes for its occurrence

(Gyurkocza et al., 2010). It was reported that moderate to severe aGvHD and its therapy may facilitate virus replication in recipients of SCT (Baldwin et al., 2000; Bil-Lula et al, 2010; Bruno et al., 2003; Ison, 2006; Robin et al., 2007; Runde et al., 2001; Watcharananan et al., 2010). GvHD requires immunosuppressive therapy which lowers the number of lymphocytes and hampers the immunity of the recipient, thus enabling adenoviral infections. A low lymphocyte count due to non-specific T-cell depletion or delayed T-cell reconstitution is considered as important predictor for detection of adenoviral infections. An inadequate humoral response due to B lymphocyte impairment contributes to greater susceptibility to AdV infection (Chakrabarti et al., 2000, 2002; Heemskerk et al., 2005).

Some authors emphasize that presence of adenovirus in peripheral blood or plasma is an indicator of impending symptomatic adenoviral disease and multiorgan failure in patients undergoing HSCT. Chakrabarti et al. (2002) and others confirmed a strong correlation between the presence of adenovirus in whole blood/plasma and an increased mortality due to infection (9-86%) (Chakrabarti et al., 2002; Ebner et al., 2005; Echavarria et al, 2001; Lion

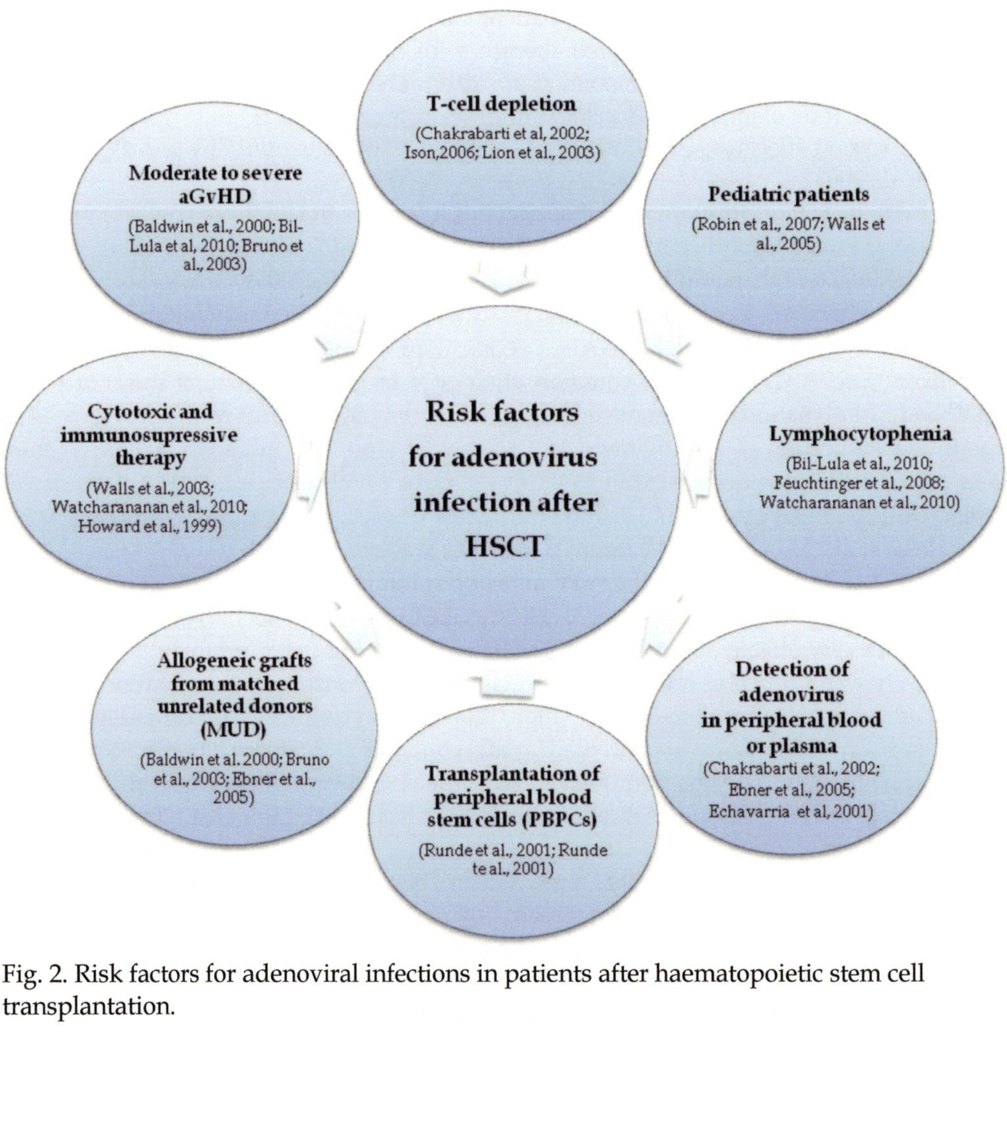

Fig. 2. Risk factors for adenoviral infections in patients after haematopoietic stem cell transplantation.

et al., 2003; Robin et al, 2007; Schilham et al., 2002; Watcharananan et al., 2010). High viral load in plasma samples (10^6-10^7 copies/ml) should be considered as a risk factor for serious post-transplantation complications (Ison, 2006; Claas et al., 2005), but the issue is still controversial because other authors did not confirm the relationship between viral load and severity of infection (Walls et al., 2005). In some studies were reported patients who despite high viral load in blood recovered from adenoviral infection without any treatment. It suggests that not each adenoviral infection needs antiviral treatment (Lankester et al., 2002; Schilham et al., 2002).

It needs to be mentioned that the source of progenitor cells also plays a significant role in development of adenoviral infection. It was confirmed that transplantation of peripheral blood progenitor cells (PBPCs) increases an incidence of AdV infections in hematological patients in comparison to bone marrow transplantations (BMT) due to high number of lymphocytes transferred with the graft. Adenovirus latency in peripheral lymphocytes may be a source of infection in those patients (Runde et al., 2001). On the other hand, unrelated cord blood transplantations (UCB) are suggested to be an independent risk factor for AdV infections (Robin et al., 2007) due to lack of mature lymphocytes in CB regarded as the major component of antiviral defense. Therefore it is still controversial issue.

7. Co-infections in patients undergoing stem cell transplantation

Recipients of stem cell transplantations with AdV infection are prone to life-threatening multiple opportunistic infections. Simultaneous infections of adenoviruses and CMV (4,3-73%), EBV (2,8-34%), polyoma BK (BKV) (1,7-20%), HSV (6,7-26,6%) or RSV (3,8-13%) are most frequently reported (Baldwin et al., 2000; Bruno et al., 2003; Chakrabarti et al., 2002; Leruez-Ville et al., 2006; Lion et al., 2003; Myers et al., 2005; Watcharananan et al., 2010). Co-infections of AdV and BKV are usually detected in urinary tract of immunocompromised patients leading to intensification of hemorrhagic cystitis sympthoms (Akiyama et al., 2001). It was also found that CMV viraemia leads to more than 4-fold increase in patient`s risk for development of AdV infection (Baldwin et al., 2000; Watcharananan et al., 2010). In rare cases, co-infections of AdV and other viruses such as: HSV1, RSV, EBV and rotaviruses were observed (Legrand et al., 2001; Robin et al., 2007). Many reports suggested that infection of one AdV serotype increases patient`s risk for co-infection with second type of human adenovirus leading to extensive organ and whole system involvement (Echavarria et al., 2006; Kroes et al., 2007; Watcharananan et al., 2010).

8. Diagnostic methods for adenovirus detection and identification

Permanently growing number of hematopoietic transplantations in recent years and high risk of viral complications following this procedure demands an implementation of better and more effective diagnostics methods for detection and monitoring of viral infections. There are several methods commonly used for adenovirus detection and identification in clinical samples. Serological tests, virus isolation in cell culture, microscopic techniques and molecular methods found their way into clinical practice.

Virus isolation in cell culture is still considered a "gold standard" for detection of adenoviral infection. Different cell lines like A549, Graham 293 and HEp-2 can be used for virus isolation from clinical samples such as stool, throat swabs, conjunctival swabs, urine and biopsy specimens. Within several days of culturing, cythopatic effect as clumping and cell rounding can be observed (Figure 3).

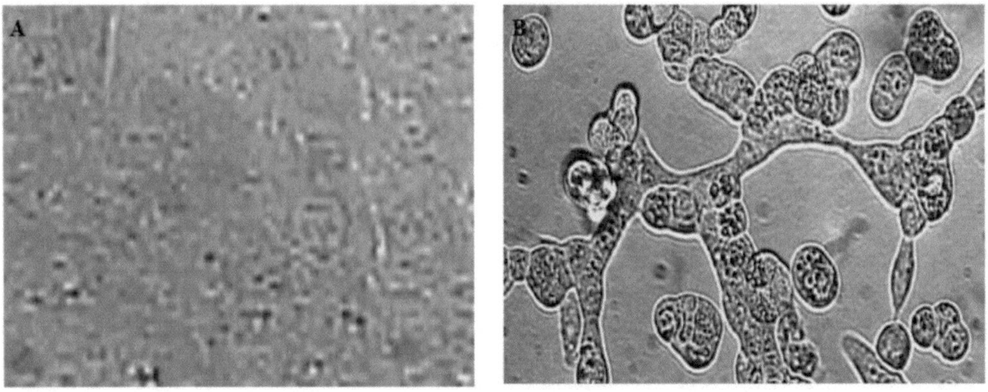

Fig. 3. Microscope image of A549 cells. (A) native culture; (B) cythopatic effect occurring within 48 hours since AdV21 infection (species B). Own collection.

This reference method provides virus identification and typing by means of serum neutralization (SN), hemagglutination inhibition (HI) and complement fixation tests. Unfortunately, the culture technique is laborious, time-consuming and lacks sensitivity to detect virus in the early phase of infection. It can also be inhibited by neutralizing antibodies or other interfering substances (Echavarria, 2009; Huang & Turchek, 2000; Raboni et al., 2003). Nowadays, due to its disadvantages it is replaced by new, more reliable methods like molecular tests.

The next conventional methods for detection of adenoviral infection are serological tests. They are used for indirect antigen detection on capsid surface (IFA, RIA) allowing for virus identification in highly concentrated samples of respiratory secretion, pharyngeal swabs and stool or to confirm AdV infection on the basis of specific antibodies detection. ELISA tests are also used for identification of AdV-antibodies in patient`s serum (Echavarria, 2009; Hierholzer et al., 1993; Madisch et al., 2005; Raboni et al., 2003). Although serological tests are commonly used in routine diagnostics of viral infections they are not recommended in adenoviral infections due to their insensitivity, low specificity, false negative results during the window period and difficulties arising from virus variability (Chirmule et al., 1999; Chmielewicz et al., 2005; Crawford-Miksza & Schnurr, 1994; Echavarria, 2009; Raboni et al., 2003). Furthermore, serological tests are useless in recipients of SCT due to profound immunosuppression and using of anti-viral drugs.

The characteristic icosahedral morphology of adenoviruses can be utilized in virus detection by use of electron microscopy (EM) and other microscopic techniques. Infected cell are characterized by presence of enlarged nuclei containing crystalline inclusions of adenoviral particles. Nonetheless, this method is limited by access to unique equipment and EM requires a large number of viral particles in clinical sample hence it can be mainly used in diagnostics of acute gastroenteritis and upper respiratory tract infections (Chirmule et al., 1999; Roingeard, 2008).

Numerous limitations of conventional methods make these techniques impractical in routine diagnostics. The culture collection, serological tests and microscopic techniques are laborious, time-consuming, and insensitive especially in an early phase of the infection. Hence in recent years, molecular methods dominate virus detection and identification in clinical practice. Identification of adenovirus by polymerase chain reaction (PCR) facilitates accurate and rapid diagnostics and surveillance. PCR is currently the most widespread method used for detection of microbial infections. In comparison to conventional techniques it offers high sensitivity and specificity as well as possibility to obtain a reliable result within several hours. This method allows detecting even small number of viral particles in largely cell-free fluids such as plasma, serum, urine (centrifuged), cerebrospinal fluids and others, preferably before clinical manifestation and tissue damage (Ephros et al., 2009). It can also be implemented for detection of AdV serotypes which are potentially considered as not growing in routine cell culture. Moreover, polymerase chain reaction may be employed for detection of variance and mutations in virus genome (Powledge, 2004). In combination with other methods such as sequencing of hyper variation regions (HVRs) or restriction fragment length analysis, PCR may be used for AdV typing. During last years, the quantitative real-time PCR (RQ-PCR) technique has been successfully implemented more often in viral diagnostics (Bil-Lula et al., 2010; Claas et al., 2005; Echavarria et al., 1999, 2001; Gu et al., 2003; Heim et al., 2003; Hierholzer et al., 1993; Lankester et al., 2002; Lion et al., 2003; Watzinger et al., 2004). Use of highly specific probes allows sensitive detection and quantification of viral equivalents in clinical samples. Sensitive and reliable monitoring of viral replication in immunocompromised patient is extremely important due to possibility for dissemination of infection and poor outcome. It has also a prognostic significance for the patient. Moreover, monitoring of active replication and increasing viral load in clinical samples constitutes the basis for differentiation of active and latent infection in recipients of stem cell transplantations. On the other hand, an early detection of adenovirus due to highly sensitive RQ-PCR creates the opportunity for reduction of immunosuppressive therapy or early initiation of preemptive anti-viral agents before onset of fulminant disease, simultaneously determining therapy effectiveness (Mori et al., 2003). It was confirmed that weekly surveillance of samples for AdVs and early intervention with anti-AdV agents results in significant reduction in the disseminated disease rate and fatal outcomes (Sivaprakasam et al., 2007; Yusuf et al., 2006).

9. Treatment strategies for adenoviral infections

There are many reports describing treatment trials of adenoviral infections in recent years. However, treatment of AdV infections remains a serious problem due to lack of unequivocally proven effectiveness of used drugs. Therefore adenoviral infections in immunocompromised patients should be considered as a serious, life-threatening complication.

Due to numerous toxicities and limitations, administration of anti-AdV drugs should be tailored to the patients situation and risk of dismal outcome. In severe AdV infections usually cidofovir (CDV) or ganciclovir are administered (Bordigoni et al., 2001; Hoffman et al., 2001; Legrand et al., 2001). Other anti-viral drugs such as vidarabine and ribavirine may be implemented (Bordigoni et al., 2001; Miyamura et al., 2000). It was demonstrated that ribavirin (usually used in treatment of HCV and RSV infections) is highly effective in treatment of localized AdV infections in urinary tract but is of limited efficacy in disseminated infections (Bordigoni et al., 2001; Gavin & Katz, 2002; Hoffman et al., 2000; Lankester et al., 2004). Some studies also confirmed partially the ability to virus elimination from urine samples after cidofovir administration in patients suffering from hemorrhagic cystitis. Unfortunately, CDV shows numerous side effects such as renal toxicity, carcinogenicity and toxic injury of muscles (Bordigoni et al., 2001; Feuchtinger et al., 2006,

2008). Despite all these side effects, lack of other more effective medicaments dictates the application of these drugs in therapy of adenoviral infections of urinary tract, gastroenterititis and pneumonia (Bruno et al., 2003; Ebner et al., 2006; Heim et al., 2003; Hoffman et al., 2001; Legrand et al., 2001, Uchio et al., 2007). New antiviral agents are under development. CMX001 is a new formulation of cidofovir. It is oral lipid conjugate of CDV, potentially highly effective against all dsDNA viruses, including adenoviruses (Fowler et al., 2010; Randhawa et al., 2006). Novel form of cidofovir is characterized by convenient oral supply, decreased dose and lack of affinity to kidney tissue. Due to insufficient effectiveness of previously used drugs, new clinical trials are still carrying on. New candidates for treatment of adenoviral infections are also zalcitabine and stavudine, commonly used as anti-HIV agents (Inoue et al., 2009; Romanowski et al., 2009). Zalcitabine is a reverse transcriptase inhibitor which may be used in treatment of AdV infections caused by serotypes: 2, 3, 4, 8, 19 and 37. Unfortunately, only low dosed of zalcitabine may be administrated because of its mitochondrial toxicity and possibility of peripheral neuropathy and esophageal varices. Stavudine, in turn, is an analogue of pyrimidine nucleoside which inhibits the activity of reverse transcription. It may be potentially used in treatment of AdV3 and AdV4 infections (Inoue et al., 2009; Romanowski et al., 2009; Uchio et al., 2007).

Timing of therapeutical intervention plays important role, too. There are many reports confirming the significant role of early detection of viraemia before the appearance of clinical symptoms (Legrand et al, 2001; Teramura et al., 2004; Watzinger et al., 2004). Initiation of antiviral treatment before the onset of invasive disease is beneficial. It was documented that even intravenous supplementation of antiviral agent during active virus replication does not result in recovery in some patients (Chakrabarti et al, 1999). Therefore, an early diagnosis of infection and adequately early implementation of antiviral therapy can significantly reduce the effects of AdV infections (Chakrabarti et al., 2002; Gavin & Katz, 2002; Sivaprakasam et al., 2007; Walls et al., 2003). Drug cytotoxicity and its limited efficacy in adenoviral infections should encourage clinicians to focus on an earlier diagnostics and monitoring of adenoviral infections.

In recent years, immunomodulation has been most frequently proposed as a new therapeutic strategy. Adoptive transfer of T-cell immunity from the donor to the recipient (DLI) and infusion of donor immunoglobulins (IVIG) combined with reduced immunosupression has become a new treatment option for patients with an insufficient number of AdV-specific T-cells. It was demonstrated that infusion of donor AdV-specific lymphocytes T is well tolerated and may lead to partially reconstruction of recipient`s immune system leading to prophylactics or treatment of adenoviral infections (Amrolia et al., 2006; Bordigoni et al., 2001; Feuchtinger et al., 2008; Fujita et al., 2008; Ison, 2006; Lion et al., 2010). Successful reduction of AdV replication is achieved mainly due to infusion of selected AdV-specific lymphocytes. Infusion of unselected donor lymphocytes is associated with higher risk of aGvHD due to high alloreactivity which requires more aggressive immunosupresive therapy conducing development of viral infections (Walter et al., 1995). The sufficient immune response is achieved mainly due to infusion of combined CD4+/CD8+ (60%) or solely CD4+ (40%) cells (Feuchtinger et al., 2008). However, heterogeneity of *Adenoviridae* family in the context of immunomodulation is questionable. Human immune system recognizes the surface determinants of AdV capsid which are highly variable between AdV serotypes. Hence, the success of T-cells administration may be dependent on infecting serotype (Ebner et al., 2006). On the other hand, human adenoviruses exhibit high cross-reactivity against different clinically relevant species of AdV and therefore there is a large chance that successful DLI may become a new therapeutic strategy for treatment of adenoviral infections after SCT, as antiviral therapy has revealed limited success (Bordigoni et al, 2001; Feuchtinger et al., 2008; Lion et al., 2010). Lion et al. (2010) suggested that initiation of cidofovir/adoptive transfer of AdV-specific T cells may reduce proliferation of adenovirus until recovery of host immune system. These considerations suggest that simultaneous detection and treatment of adenoviral infection at early stage might prevent life-treatening disseminated infections in recipients of hematopoietic stem cell.

Anti-adenovirus treatment needs careful consideration in many clinical aspects. The knowledge on AdV pathogeneicity needs more elucidation. The retrospective studies of Wall et al. (2005), Hale et al. (2003) and others proved that there is a possibility for elimination of high viraemia without any treatment due to sufficient host T-cell response (Ephros et al., 2009). The opportunity for 'watch and wait' strategy is a new unexplored clinical option, but the identification of patients with benign clinical course is impossible yet.

10. Conclusion

Taking into account that more than 50000 of hematopoietic stem cell transplantations are performed every year and nearly 10000 originate from unrelated donors, the number of patients who may be affected from adenoviral infections seems to be significant. In view of worldwide distribution of adenoviruses, numerous routes of infection and limited effectiveness of anti-AdV treatment, adenoviral infections in patients undergoing hematopoietic stem cell transplantations should be always considered as serious, life-threatening post-transplant complication which demands rapid and unequivocal diagnosis defining patient`s outcome.

11. References

Akiyama, H., Kurosu, T., Sakashita, C., Inoue, T., Mori, Si., Ohashi, K., Tanikawa, S., Sakamaki, H., Onozawa, Y., Chen, Q., Zheng, H. & Kitamura T. (2001). Adenovirus is a key pathogen in hemorrhagic cystitis associated with bone marrow transplantation. *Clin. Infect. Dis.*,Vol.32, No.9, pp.1325–1330.

Amrolia, P,.J., Muccioli-Casadei, G., Huls, H., Adams, S., Durett, A., Gee, A., Yvon, E., Weiss, H., Cobbold, M., Gaspar, H.B., Rooney, C., Kuehnle, I., Ghetie, V., Schindler, J., Krance, R., Heslop, H.E., Veys, P., Vitetta, E. & Brenner M.K. (2006). Adoptive immunotherapy with allodepleted donor T-cells improves immune reconstitution after haploidentical stem cell transplantation. *Blood*, Vol.108, No.6, pp.1797-1808.

Arnberg, N., Pring-Akerblom, P. & Wadell, G. (2002). Adenovirus type 37 uses sialic acid as a cellular receptor on Chang C cells. *J. Virol.*, Vol.76, No,17, pp.8834–8841.

Artieda, J., Montes, M., Vicente, D., Martínez, C., Piñeiro, L. & Mendiola, J. (2010). Outbreak of follicular conjunctivitis caused by adenovirus in a geriatric centre. *Enferm. Infecc. Microbiol. Clin.*, Vol.28, No.10, pp.690-693.

Baldwin, A., Kingman, H., Darville, M., Foot, A.B., Grier, D., Cornish, J.M., Goulden, N., Oakhill, A., Pamphilon, D.H., Steward, C.G. & Marks D.I. (2000). Outcome and clinical course of 100 patients with adenovirus infection following bone marrow transplantation. *Bone Marrow Transplant.*, Vol.26, No,12, pp.1333-1338.

Benko, M., Harrach, B. & Russell, W.C. (2000). Family adenoviridae, In: *Virus taxonomy*, van Regenmortel, M. V. H., Fauquet, C.M. & Bishop D.H.L. (Ed.), pp. 227-238, Academic Press, New York.

Bil-Lula, I., Ussowicz, M., Rybka, B., Wendycz-Domalewska, D., Ryczan, R., Gorczyńska, E., Kałwak, K. & Woźniak, M. (2010a). PCR diagnostics and monitoring of adenoviral infections in hematopoietic stem cell transplantation recipients. *Arch. Virol.*, Vol.155, No.12, pp.2007-2015.

Bil-Lula, I., Ussowicz, M., Rybka, B., Wendycz-Domalewska, D., Ryczan, R., Gorczyńska, E., Kałwak, K. & Woźniak, M. (2010b). Hematuria due to adenoviral infection in bone marrow transplant recipients. *Transplant Proc.* Vol.42, No.9, pp.3729-3734.

Bordigoni, P., Carret, A.S., Vanard, V., Witz, F. & Le Faou, A. (2001). Treatment of adenovirus infections in patients undergoing allogeneic haematopoietic stem cell transplantation. *Clin. Infect. Dis.*,Vol.32, No.9, pp.1290–1297.

Braithwaite, A.W. & Russell, I.A. (2001). Induction of cell death by adenoviruses. *Apoptosis*, Vol.6, No.5, pp. 359–370.

Bruno, B., Gooley, T., Hackman, R.C., Davis, C., Corey, L. & Boeckh, M. (2003). Adenovirus infection in hematopoietic stem cell transplantation: effect of ganciclovir and impact on survival. *Biol. Blood and Marrow Transplant.*,Vol.9, No.5, pp.341-352.

Burmeister, W.P., Guilligay, D., Cusack, S., Wadell, G. & Arnberg, N.(2004) Crystal structure of species D adenovirus fiber knobs and their sialic acid binding sites. *J Virol*, Vol.78, No.14, pp. 7727-7736.

Burnett, R. M. (1999). The structure of adenovirus, In: Structural biology of viruses, Chiu W, Burnett R.M. & Garcea, R.L. (Ed.), pp. 209–238. Oxford University Press, New York.

Carballal, G., Videla, C., Misirlian, A., Requeijo, P.V. & Aguilar, M. C. (2002). Adenovirus type 7 associated with severe and fatal acute lower respiratory infections in Argentine children. *BMC Pediatr.*, Vol.2:6.

Chakrabarti, S., Collingham, K.E., Fegan, C.D. & Milligan, D.W. (1999). Fulminant adenovirus hepatitis following unrelated bone marrow transplantation: failure of intravenous ribavirin therapy. *Bone Marrow Transplant.*, Vol. 23,pp. 1209–1211.

Chakrabarti, S., Collingham, K.E., Fegan, C.D., Pillay, D. & Milligan, D.W. (2000). Adenovirus infections following haematopoietic cell transplantation: is there a role for adoptive immunotherapy? *Bone Marrow Transplant.* , Vol.26, No.3, pp.305–307.

Chakrabarti, S., Collingham, K.E., Stevens, R.H., Pillay, D., Fegan, C.D. & Milligan, D.W. (2000). Isolation of viruses from stools in stem cell transplant recipients: a prospective surveillance study. *Bone Marrow Transplant.*, Vol.25, No.3, pp.277–282.

Chakrabarti, S., Mautner, V., Osman, H., Collingham, K.E., Fegan, C.D., Klapper, P.E., Moss, P.A. & Milligan D.W. (2002). Adenovirus infections following allogeneic stem cell transplantation: incidence and outcome in relation to graft manipulation, immunosuppression, and immune recovery. *Blood*, Vol.100, No.5, pp.1619-1627.

Chirmule, N., Propert, K., Magosin, S., Qian, Y., Qian, R. & Wilsonn, J. (1999). Immune responses to adenovirus and adeno-associated virus in humans. *Gene Ther.*, Vol.6, No.9, pp.1574–1583.

Chmielewicz, B., Nitsche, A., Schweiger, B. & Ellerbrok, H. (2005). Development of a PCR-based assay for detection, quantification and genotyping of human adenoviruses. *Clin. Chem.*,Vol.51, No.8, pp.1365-1373.

Chu, Y., Heistad, D.D., Cybulsky, M.I. & Davidson, B.L. (2001). Vascular cell adhesion molecule-1 augments adenovirus-mediated gene transfer. *Arterioscler. Thromb. Vasc. Biol.*, Vol.21, No.2, pp.238-242.

Claas, E.C., Schilham, M.W., Brouwer, C.S., Hubacek, P., Echavarria, M., Lankester, A.C., Tol, M.J. & Kroes, M.C. (2005). Internally controlled real-time PCR monitoring of adenovirus DNA load in serum or plasma of transplant recipients. *J. Clin. Microbiol.*, Vol.43, No.4, pp.1738–1744.

Cooper, R.J., Hallett, R., Tullo, A.B. & Klapper, P.E. (2000). The epidemiology of adenovirus infections in Greater Manchester, UK 1982-96. *Epidemiol. Infect.*, Vol.125, No.2, pp.333-345.

Couriel, D.R., Saliba, R.M., Giralt, S., Khouri, I., Andersson, B., de Lima, M., Hosing, C., Anderlini, P., Donato, M., Cleary, K., Gajewski, J., Neumann, J., Ippoliti, C., Rondon, G., Cohen, A. & Champlin, R. (2004). Acute and chronic graft-versus-host disease after ablative and nonmyeloablative conditioning for allogeneic hematopoietic transplantation. *Biol. Blood Marrow Transplant.*, Vol.1, No.3, pp.178-185.

Crawford-Miksza, L.K. & Schnurr, D.P. (1994). Quantitative colorimetric microneutralization assay for characterization of adenoviruses. *J. Clin. Microbiol.*, Vol.32, No.9, pp.2331–2334.

Crawford-Miksza, L. & Schnurr, D.P. (1996). Analysis of 15 adenovirus hexon proteins reveals the location and structure of seven hypervariable regions containing serotype-specific residues. *J. Virol.*, Vol.70, No.3, pp.1836–1844.

Cusack S. (2005). Adenovirus complex structures. *Curr. Opin. Cell Biol.*, Vol.15, No.2, pp.237-243.

Dechecchi, M.C., Melotti, P., Bonizzato, A., Santacatterina, M., Chilosi, M. & Cabrini, G. (2001). Heparan sulfate glycosaminoglycans are receptors sufficient to mediate the initial binding of adenovirus types 2 and 5. *J. Virol.*, Vol.75, No.18, pp.8772–8780.

Ebner, K., Pinsker, W. & Lion, T. (2005). Comparative sequence analysis of the hexon gene in the entire spectrum of human adenovirus serotypes: phylogenetic, taxonomic, and clinical implications. *J. Virol.*,Vol.79, No.20, pp. 12635-12642.

Ebner, K., Rauch, M., Preuner, S. & Lion, T. (2006). Typing of human adenoviruses in specimens from immunosuppressed patients by PCR-fragment length analysis and real-time quantitative PCR. *J. Clin. Microbiol.*, Vol.44, No.8, pp.2808–2815.

Ebner, K., Suda, M., Watzinger, F. & Lion, T. (2005). Molecular detection and quantitative analysis of the entire spectrum of human adenoviruses by a two-reaction real-time PCR assay. *J. Clin. Microbiol.*, Vol.43, No.7, pp.3049-3053.

Echavarri´a, M. (2008). Adenoviruses in immunocompromised hosts. *Clin. Microbiol. Rev.*, Vol.21, No.4, pp.704–715.

Echavarria, M. (2009). Adenoviruses. In: Principles and practice of clinical virology, Zuckerman, A.J., Banatvala, J.E., Griffiths, P.D., Schoub, B. & Mortimer, P. (Ed.),pp.463- 488, Willey-Blackwell, UK.

Echavarria, M., Forman, M., van Tol, M.J., Vossen, J.M., Charache, P. & Kroes, A.C. (2001). Prediction of severe disseminated adenovirus infection by serum PCR. *Lancet*, Vol.358, No. 9279, pp.384-385.

Echavarria, M., Maldonado, D., Elbert, G., Videla, C., Rappaport, R. & Carballal, G. (2006). Use of PCR to demonstrate presence of adenovirus species B, C, or F as well as coinfection with two adenovirus species in children with flu-like symptoms. *J. Clin. Microbiol.*, Vol.44, No.2, pp.625–627.

Echavarria, M.S., Ray, S.C., Ambinder, R., Dumler, J.S. & Charache, P. (1999). PCR detection of adenovirus in a bone marrow transplant recipient: hemorrhagic cystitis as a presenting manifestation of disseminated disease. *J. Clin. Microbiol.*, Vol.37, No.3, pp.686-689.

Ephros, M., Friedman, B.C., Elhasid, R., Kra-Oz, Z., Shaked-Mishan, P., Sattinger, J. & Kassis, I. (2009). Incidence and clinical manifestations of adenoviral infection among children undergoing allogeneic stem cell transplantation. *Isr. Med. Assoc. J.*, Vol.11, No.12, pp.744-748.

Faden, H.S., Wynn, R.J., Campagna, L. & Ryan, R.M. (2005). Outbreak of adenovirus type 30 in a neonatal intensive care unit. *J. Pediatr.* Vol.146, No.4, pp.523-527.

Feuchtinger, T., Matthes-Martin, S., Richard, C., Lion, T., Fuhrer, M., Hamprecht, K., Handgretinger, R., Peters, C., Schuster, F.R., Beck, R., Schumm, M., Lotfi, R., Jahn, G. & Lang, P. (2006). Safe adoptive transfer of virus-specific T-cell immunity for the treatment of systemic adenovirus infection after allogeneic stem cell transplantation. *Br. J. Haematol.*, Vol.134, No.1, pp.64-76.

Feuchtinger, T., Richard, C., Joachim, S., Scheible, M.H., Schumm, M., Hamprecht, K., Martin, D., Jahn, G., Handgretinger, R. & Lang, P. (2008). Clinical grade generation of hexon-specific T cells for adoptive T-cell transfer as a treatment of adenovirus infection after allogeneic stem cell transplantation. *J. Immunother.*, Vol. 31, No.2, pp.199-206.

Fowler, C.J., Dunlap, J., Troyer, D., Stenzel, P., Epner, E. & Maziarz, R.T. (2010.) Life-threatening adenovirus infections in the setting of the immunocompromised allogeneic stem cell transplant patients. *Adv. Hematol.*, doi:10.1155/2010/601548. Epub 2010 Jun 16.

Fujita, Y., Rooney, C.M. & Heslop, H.E. (2008). Adoptive cellular immunotherapy for viral diseases. *Bone Marrow Transplant.*, Vol.41, No.2, pp.193–198.

Garnett, C.T., Erdman, D., Xu, W. & Gooding, L.R. (2002). Prevalence and quantitation of species C adenovirus DNA in human mucosal lymphocytes. *J. Virol.*, Vol.76, No.21, pp.10608–10616.

Gavin, P.J. & Katz, B.Z. (2002). Intravenous ribavirin treatment for severe adenovirus disease in immunocompromised children. *Paediatrics*, Vol.110, 1 Pt 1: e9.

Gerber, S.I., Erdman, D.D., Pur, S.L., Diaz, P.S., Segreti, J., Kajon, A.E., Belkengren, R.P. & Jones, R.C. Outbreak of adenovirus genome type 7d2 infection in a pediatric chronic-care facility and tertiary-care hospital. *Clin. Infect. Dis.*, Vol.32, No.5, pp.694-700.

Gonçalves, M.A. & de Vries, A.A. (2006). Adenovirus: from foe to friend. *Rev. Med. Virol.*, Vol.16, No.3, pp.167-186.

Gratwohl, A., Baldomero, H., Aljurf, M., Pasquini, M.C., Bouzas, L.F., Yoshimi, A., Szer, J., Lipton, J., Schwendener, A., Gratwohl ,M., Frauendorfer, K., Niederwieser, D., Horowitz, M. & Kodera, Y. (2010). Hematopoietic stem cell transplantation: a global perspective. *JAMA*. Vol.303, No.16, pp.1617-1624.

Gray, G.C. (2006). Adenovirus Transmission—Worthy of Our Attention. *JID,* Vol.94, No.7, pp.871-873.

Greber, U.F., Suomalainen, M., Stidwell, R.P., Boucke, K., Ebersold, M.W. & Helenius, A. (1997). The role of the nuclear complex in adenovirus DNA entry. *EMBO J.*, Vol.16, No.19, pp.5998–6007.

Gu, Z., Belzer, S.W., Gibson, C.S., Bankowski, M.J. & Hayden. R.T. (2003). Multiplexed, real-time PCR for quantitative detection of human adenovirus. *J. Clin. Microbiol.*, Vol.41, No.10, pp.4636- 4641.

Gyurkocza, B., Rezvani, A. & Storb, R.F. (2010). Allogeneic hematopoietic cell transplantation: the state of the art. *Expert Rev Hematol.*, Vol.3, No.3, pp.285-299.

Hale, G.A., Heslop, H.E., Krance, R.A., Brenner, M.A., Jayawardene, D., Srivastava, D.K. & Patrick, C.C. (1999). Adenovirus infection after pediatric bone marrow transplantation. *Bone Marrow Transplant.*, Vol.23, No.3, pp. 277- 282.

Hale, G.A., Rochester, R.J., Heslop, H.E., Krance, R.A., Gingrich, J.R., Benaim, E., Horwitz, E.M., Cunningham, J.M., Tong, X., Srivastava, D.K., Leung, W.H., Woodard, P., Bowman, L.C. & Handgretinger, R. (2003). Hemorrhagic cystitis after allogeneic bone marrow transplantation in children: clinical characteristics and outcome. *Biol. Blood Marrow Transplant.*, Vol.9, No.11, pp.698-705.

Heemskerk, B., Lankester, A.C., van Vreeswijk, T., Beersma, M.F., Claas, E.C., Veltrop-Duits, L.A., Kroes, A.C., Vossen, J.M., Schilham, M.W. & van Tol, M.J. (2005). Immune reconstitution and clearance of human adenovirus viremia in pediatric stem-cell recipients. *J. Infect. Dis.*, Vol.191, No.4, pp.520–530.

Heim, A., Ebnet, C., Harste, G. & Pring-Akerblom, P. (2003). Rapid and quantitative detection of human adenovirus DNA by real-time PCR. *J. Med. Virol.*, Vol.70, No.2, pp.228-239.

Hierholzer, J.C. (1992). Adenoviruses in the immunocompromised host. *Clin. Microbiol. Rev.*, Vol.5, No.3, pp.262-274.

Hierholzer, J.C., Halonen, P.E., Dahlen, P.O., Bingham, P.G. & McDonough, M.M. (1993). Detection of adenovirus in clinical specimens by polymerase chain reaction and liquid-phase hybridization quantitated by time-resolved fluorometry. *J. Clin. Microbiol.*,Vol.31, No.7, pp.1886-1891.

Hill, G.R. & Ferrara, J.L. (2000). The primacy of the gastrointestinal tract as a target organ of acute graft-versus-host disease: rationale for the use of cytokine shields in allogeneic bone marrow transplantation. *Blood.*, Vol.95, No.9, pp.2754-2759.

Hoffman, J.A., Shah, A.J., Ross, L.A. & Kapoor, N. (2001). Adenoviral infections and a prospective trial of cidofovir in pediatric hematopoietic stem cell transplantation. *Biol. Blood Marrow Transplant.*, Vol.7, No.7, pp.388-394.

Howard, D.S., Phillips, II G.L., Reece, D.E., Munn, R.K., Henslee-Downey, J., Pittard, M., Barker, M. & Pomeroy, C. (1999). Adenovirus infections in hematopoietic stem cell transplant recipients. *Clin. Infect. Dis.*, Vol.29, No.6, pp.1494–1501.

Huang, Y.T. & Turchek, B.M. (2000). Mink lung cells and mixed mink lung and A549 cells for rapid detection of influenza virus and other respiratory viruses. *J. Clin. Microbiol.*, Vol.38, No.1, pp.422-423.

Huebner, R.J., Rowe, W.P., Ward, T.G., Parrott, R.H. & Bell, J.A. (1954). Adenoidal-pharyngoconjunctival agents: a newly recognized group of common viruses of the respiratory system. *N. Engl. J. Med.*, Vol.251, No.27, pp.1077–1086.

Inoue, H., Sonoda, K.H., Ishikawa, M., Kadonosono, K. & Uchio, E. (2009). Clinical evaluation of local ocular toxicity in candidate anti-adenoviral agents in vivo. *Ophthalmologica.*, Vol.223, No.4, pp.233-238.

Ishiko, H., Shimada, Y., Konno, T., Hayashi, A., Ohguchi, T., Tagawa, Y., Aoki, K., Ohno, S. & Yamazaki, S. (2008). Novel human adenovirus causing nosocomial epidemic keratoconjunctivitis. *J. Clin. Microbiol., Vol.*46, No.6, pp.2002-2008.

Ison, M.G. (2006). Adenovirus infections in transplant recipients. *Clin. Infect. Dis.*, Vol.43, No.3, pp.331–339.

Jeulin, H., Salmon, A., Bordigoni, P. & Venard, V. (2011). Diagnostic value of quantitative PCR for adenovirus detection in stool samples as compared with antigen detection and cell culture in haematopoietic stem cell transplant recipients. *Clin. Microbiol. Infect.* doi: 10.1111/j.1469-0691.2011.03488.x. [Epub ahead of print] Jones, M.S. 2nd, Harrach, B., Ganac, R.D., Gozum, M.M., Dela, Cruz, W.P., Riedel, B., Pan, C., Delwart, E.L. & Schnurr, D.P. (2007). New adenovirus species found in a patient presenting with gastroenteritis. *J. Virol.*, Vol.81, no.11, pp.5978–5984.

Kaneko, H., Iida, T., Ishiko, H., Ohguchi, T., Ariga, T., Tagawa, Y., Aoki, K., Ohno, S. & Suzutani, T. (2009). Analysis of the complete genome sequence of epidemickeratoconjunctivitis-related human adenovirus type 8, 19, 37 and a novel serotype. *J. Gen. Virol.*, Vol.90, (Pt 6), pp.1471-1476.

Kelkar, S.A., Pfister, K.K., Crystal, R.G. & Leopold, P.L. (2004). Cytoplasmic dynein mediates adenovirus binding to microtubules. *J. Virol.*, Vol.78, No.18, pp.10122–10132.

Kroes, A.C., de Klerk, E.P., Lankester, A.C., Malipaard, C., de Brouwer ,C.S., Claas, E.C., Jol-van der Zijde, E.C. & van Tol, M.J. (2007). Sequential emergence of multiple adenovirus serotypes after pediatric stem cell transplantation. *J. Clin. Virol.*, Vol.38, No.4, pp.341-347.

Lankester, A.C., Heemskerk, B., Claas, E.C., Schilham, M.W., Beersma, M.F., Bredius, R.G., van Tol, M.J. & Kroes, A.C. (2004). Effect of ribavirin on the plasma viral DNA load in patients with disseminating adenovirus infection. *Clin. Infect. Dis.*, Vol.38, No.11, pp.1521-1525.

Lankester, A.C., van Tol, M.J., Claas, E.C., Vossen, J.M. & Kroes, A.C. (2002). Quantification of adenovirus DNA in plasma for management of infection in stem cell graft recipients. *Clin. Infect. Dis.*, Vol.34, No.6, pp.864–867.

Lauer, K.P., Llorente, I., Blair, E., Seto, J., Krasnov, V., Purkayastha, A., Ditty, S.E., Hadfield, T.L., Buck, C., Tibbetts, C. & Seto, D. (2004). Natural variation among human adenoviruses: genome sequence and annotation of human adenovirus serotype 1. *J. Gen. Virol.*, Vol.85, (Pt 9), pp.2615-2625.

Lee, J., Choi, E.H. & Lee, H.J. (2010). Clinical severity of respiratory adenoviral infection by serotypes in Korean children over 17 consecutive years (1991-2007). *J. Clin. Virol.*, Vol.49, No.2, pp.115-120.

Leen, A.M., Christin, A., Khalil, M., Weiss, H., Gee, A.P., Brenner, M.K., Heslop, H.E., Rooney, C.M. & Bollard, C.M. (2008). Idenification of hexon-specific CD4 and CD8 T-cell epitopes for vaccine and immunotherapy. *J. Virol.*, Vol.82, No.1, pp.546-554.

Legrand, F., Berrebi, D., Houhou, N., Freymuth, F., Faye, A., Duval, M., Mougenot, J.F., Peuchmaur, M. & Vilmer, E. (2001). Early dignosis of adenovirus infection and treatment with cydofovir after bone marrow transplantation in children. *Bone Marrow Transplant.*, Vol.27, No.6, pp.621-626.

Leruez-Ville, M., Chardin-Ouachee, M., Neven, B., Picard, C., Le Guinche, I., Fischer ,A., Rouzioux, C. & Blanche, S. (2006). Description of an adenovirus A31 outbreak in a paediatric haematology unit. *Bone Marrow Transplant.*, Vol.38, No.1, pp.23-28.

Lessa, F.C., Gould, P.L., Pascoe, N., Erdman, D.D., Lu, X., Bunning, M.L., Marconi, V.C., Lott, L., Widdowson, M.A., Anderson, L.J. & Srinivasan, A. (2009). Health care transmission of a newly emergent adenovirus serotype in health care personnel at a military hospital in Texas, 2007. *J. Infect. Dis.*, Vol.200, No.11, pp.1759-1765.

Lim, A.K., Parsons, S. & Ierino, F. (2005). Adenovirus tubulointerstitial nephritis presenting as a renal allograft space occupying lesion. *Am. J. Transplant.*, Vol.5, No.8, pp. 2062–2066.

Lion, T., Baumgartinger, R., Watzinger, F., Matthes-Martin, S., Suda, M., Preuner, S., Futterknecht, B., Lawitschka, A., Peters, C., Potschger, U. & Gadner, H. (2003).

Molecular monitoring of adenovirus in peripheral blood after allogeneic bone marrow transplantation permits early diagnosis of disseminated disease. *Blood*, Vol.102, No.3, pp.1114-1120.

Lion, T., Kosulin, K., Landlinger, C., Rauch, M., Preuner, S., Jugovic, D., Pötschger, U., Lawitschka, A., Peters, C., Fritsch, G. & Matthes-Martin, S. (2010). Monitoring of adenovirus load in stool by real-time PCR permits early detection of impending invasive infection in patients after allogeneic stem cell transplantation. *Leukemia*, Vol.24, No.4, pp.706-714.

Ljungman, P., Bregni, M., Brune, M., Cornelissen, J., de Witte, T., Dini, G., Einsele, H., Gaspar, H.B., Gratwohl, A., Passweg, J., Peters, C., Rocha, V., Saccardi, R., Schouten, H., Sureda, A., Tichelli, A., Velardi, A. & Niederwieser, D. (2010). Allogeneic and autologous transplantation for haematological diseases, solid tumors and immune disorders: current practice in Europe 2009. *Bone Marrow Transplant.*, Vol.45, No.2, pp.219–234.

Madisch, I., Harste, G., Pommer, H. & Heim, A. (2005). Phylogenetic analysis of the main neutralization and hemagglutination determinants of all human adenovirus prototypes as a basis for molecular classification and taxonomy. *J. Virol.*, Vol.79, No.24, pp.15265-15276.

Mahr, J. A. & Gooding, L.R.(1999). Immune evasion by adenoviruses. *Immunol. Rev.*,Vol.168, pp.121–130.

Malekshahi, S.S., Azad, T.M., Yavarian, J., Shahmahmoodi, S., Naseri, M. & Rezaei, F.(2010). Molecular detection of respiratory viruses in clinical specimens from children with acute respiratory disease in Iran. *Pediatr. Infect. Dis. J.*, Vol.29, No.10, pp.931-933.

Matthews, D.A. & Russell, W.C. (1998). Adenovirus core protein V is delivered by the invading virus to the nucleus of the infected cell and later in infection is associated with nucleoli. *J. Gen. Virol.*, Vol.79, (Pt 7), pp.1671–1675.

Miyamura, K., Hamaguchi, M., Taji, H., Kanie, T., Kohno, A., Tanimoto, M., Saito, H., Kojima, S., Matsuyama, T., Kitaori, K., Nagafuji, K., Sato & T., Kodera, Y. (2000). Successful ribavirin therapy for severe adenovirus hemorrhagic cystitis after allogeneic marrow transplant from close HLA donors rather than distant donors. *Bone Marrow Transplant.*, Vol.25, No.5, pp.545-548.

Mori, K., Yoshihara, T., Nishimura, Y., Uchida, M., Katsura, K., Kawase, K., Hatano, I., Ishida, H., Chiyonobu, T., Kasubuchi, Y., Morimoto, A., Teramura, T. & Imashuku, S. (2003). Acute renal failure due to adenovirus-associated obstructive uropathy and necrotizing tubulointerstitial nephritis in a bone marrow transplant recipient. *Bone Marrow Transplant.*, Vol.31, No.12, pp.1173–1176.

Myers, G.D., Krance, R.A., Weiss, H., Kuehnle, I., Demmler, G., Heslop, H.E. & Bollard, C.M. (2005). Adenovirus infection rates in pediatric recipients of alternate donor allogeneic bone marrow transplants receiving either antithymocyte globulin (ATG) or alemtuzumab (Campath). *Bone Marrow Transplant.*, Vol. 36, No.11, pp.1001-1008.

Pérez-Simón, J.A., Caballero, D., Diez-Campelo, M., Lopez-Pérez, R., Mateos, G., Cañizo, C., Vazquez, L., Vidriales, B., Mateos, M.V., Gonzalez, M. & San Miguel, J.F. (2002). Chimerism and minimal residual disease monitoring after reduced intensity conditioning (RIC) allogeneic transplantation. *Leukemia*, Vol.16, No.8, pp.1423-1431.

Pérez-Simón, J.A., Díez-Campelo, M., Martino, R., Brunet, S., Urbano, A., Caballero, M.D., de León, A., Valcárcel, D., Carreras, E., del Cañizo, M.C., López-Fidalgo, J., Sierra, J. & San Miguel, J.F. (2005). Influence of the intensity of the conditioning regimen on the characteristics of acute and chronic graft-versus-host disease after allogeneic transplantation. *Br. J. Haematol.*, Vol.130, No.3, pp.394-403.

Powledge, T.M. (2004). The polymerase chain reaction. *Adv. Physiol. Educ.,* Vol.8, No.1-4, pp.44-50.

Raboni, S.M., Siqueira, M.M., Portes, S.R. & Pasquini, R. (2003). Comparison of PCR, enzyme immunoassay and conventional culture for adenovirus detection in bone marrow transplant patients with hemorrhagic cystitis. *J. Clin. Virol.*, Vol.27, No.3, pp.270-275.

Randall, R.E. & Goodbourn, S. (2008). Interferons and viruses: an interplay between induction, signaling, antiviral responses and virus countermeasures. *J. Gen. Virol.*,Vol.89, (Pt 1), pp.1-47.

Randhawa, P., Farasati, N.A., Shapiro, R. & Hostetler, K.Y. (2006). Ether lipid ester derivatives of cidofovir inhibit polyomavirus BK replication in vitro. *Antimicrob Agents Chemother.*, Vol.50, No.4, pp.1564-1566.

Raschperger, E., Thyberg, J., Pettersson, S., Philipson, L., Fuxe, J. & Pettersson, R.F. (2006). The coxsackie- and adenovirus receptor (CAR) is an in vivo marker for epithelial tight junctions, with a potential role in regulating permeability and tissue homeostasis. *Exp. Cell Res., Vol.*312, No.9, pp.1566–1580.

Robin, M., Marque-Juillet, S., Scieux, C., Peffault de Latour, R., Ferry, C., Rocha, V., Molina, J.M., Bergeron, A., Devergie, A., Gluckman, E., Ribaud, P. & Socié, G. (2007). Disseminated adenovirus infections after allogeneic hematopoietic stem cell transplantation: incidence, risk factors and outcome. *Haematologica,*Vol.92, No.9, pp.1254-1257.

Roingeard, P. (2009). Viral detection by electron microscopy: past, present and future. *Biol. Cell,* Vol.100, No.8, pp.491–501.

Romanowski, E.G., Yates, K.A. & Gordon, Y.J. (2009). The in vitro and in vivo evaluation of ddC as a topical antiviral for ocular adenovirus infections. *Invest. Ophthalmol. Vis. Sci.*, Vol.50, No.11, pp.5295-5299.

Romero, C.P., Vogel, G.M., Solanes, F. & Luchsinger, F.V. (2010). Adenovirus keratoconjunctivitis acquired in an ophthalmology clinic. *Rev Chilena Infectol.*, Vol.27, No.2, pp.148-152.

Rowe, W.P., Huebner, R.J., Gilmore, L.K., Parrott, R.H. & Ward, T.G. (1953). Isolation of a cytopathogenic agent from human adenoids undergoing spontaneous degeneration in tissue culture. *Proc. Soc. Exp. Biol. Med.*, Vol.84, No.3, pp.570-573.

Runde, V., Ross, S., Trenschel, R., Lagemann, E., Basu, O., Renzing-Köhler, K., Schaefer, U.W., Roggendorf, M. & Holler, E. (2001). Adenoviral infection after allogeneic stem cell transplantation (SCT): report on 130 patients from a single SCT unit involved in a prospective multi center surveillance study. *Bone Marrow Transplant.*,Vol.28, No.1, pp.51–57.

Russell, W.C. (2000). Update on adenovirus and its vectors. *J. Gen. Virol.*, Vol.81, (Pt 11), pp.2573–2604.

Russell, W.C. (2009). Adenoviruses: update on structure and function. *J. Gen. Virol.*, Vol.90, (Pt 1), pp.1–20.

Rutala, W.A, Peacock, J.E., Gergen, M.F., Sobsey, M.D. & Weber, D.J. (2006). Efficacy of hospital germicides against adenovirus 8, a common cause of epidemic keratoconjunctivitis in health care facilities. *Antimicrob. Agents Chemother.*, Vol.50, No.4, pp.1419–1424.

Rux, J.J., Kuser, P.R. & Burnett, R.M. (2003). Structural and phylogenetic analysis of adenovirus hexons by use of high-resolution x-ray crystallographic, molecular modeling, and sequence-based methods. *J. Virol.*, Vol.77, No.17, pp.9553-9566.

Schilham, M.W., Claas, E.C., van Zaane, W., Heemskerk, B., Vossen, J.M., Lankester, A.C., Toes, R.E., Echavarria, M., Kroes, A.C. & van Tol, M.J. (2002). High levels of adenovirus DNA in serum correlate with fatal outcome of adenovirus infection in children after allogeneic stem-cell transplantation. *Clin. Infect. Dis.*, Vol.35, No.5, pp.526-532.

Segerman, A., Arnberg, N., Erikson, A., Lindman, K. & Wadell, G. (2003). There are two different species B adenovirus receptors: sBAR, common to species B1 and B2 adenoviruses, and sB2AR exclusively used by species B2 adenoviruses. *J. Virol.*, Vol.77, No.2, pp.1157-1162.

Sharma, A., Li, X., Bangari, D.S, & Mittal, S.K. (2009). Adenovirus receptors and their implications in gene delivery. *Virus Res.*, Vol.143, No.2, pp.184-194.

Sivaprakasam, P., Carr, T.F., Coussons, M., Khalid, T., Bailey, A.S., Guiver, M., Mutton, K.J., Turner, A.J., Grainger, J.D. & Wynn, R.F. (2007). Improved outcome from invasive adenovirus infection in pediatric patients after hemopoietic stem cell transplantation using intensive clinical surveillance and early intervention. *J. Pediatr. Hematol. Oncol.*, Vol.29, No.2, pp.81-85.

Smith, J.G. & Nemerow, G.R. (2008). Mechanism of adenovirus neutralization by human α-defensins. *Cell Host Microb.*, Vol.3, No.1, pp.11-19.

Stone, D., Furthmann ,A., Sandig, V. & Liebera, A. (2003). The complete nucleotide sequence, genome organization, and origin of human adenovirus type 11. *Virology*, Vol.309, No.1, pp.152-165.

Teramura, T., Naya, M., Yoshihara, T., Kanoh, G., Morimoto, A. & Imashuku, S. (2004). Adenoviral infection in hematopoietic stem cell transplantation: early diagnosis with quantitative detection of the viral genome in serum and urine. *Bone Marrow Transplant.*, Vol.33, No.1, pp.87-92.

Uchio, E., Fuchigami, A., Kadonosono, K., Hayashi, A., Ishiko, H., Aoki, K. & Ohno, S. (2007). Anti-adenoviral effect of anti-HIV agents in vitro in serotypes inducing keratoconjunctivitis. *Graefes. Arch. Clin. Exp. Ophthalmol.*, Vol.245, No.9, pp.1319-1325.

Venard, V., Carret, A.S., Corsaro, D., Bordigoni, P. & Le Faou, A. (2000). Genotyping of adenoviruses isolated in an outbreak in a bone marrow transplant unit shows that diverse strains are involved. *J. Hosp. Infect.*, Vol.44, No.1, pp.71-74.

Wadell, G., Hammarskjöld, M.L., Winberg, G., Varsanyi, T.M. & Sundell, G. (1980). Genetic variability of adenoviruses. *Ann. N. Y. Acad. Sci.*, Vol.354, pp.16-42.

Walls, T., Shankar, A.G. & Shingadia, D. (2003). Adenovirus: an increasingly important pathogen in paediatric bone marrow transplant patients. *Lancet Infect. Dis.*,Vol.3, No.2, pp.79-86.

Walls, T., Hawrami, K., Ushiro-Lumb, I., Shingadia, D., Saha, V. & Shankar, A.G. (2005). Adenovirus infection after pediatric bone marrow transplantation: is treatment always necessary? *Clin. Infect. Dis.*, Vol.40, No.9, pp.1244-1249.

Walsh, M.P., Chintakuntlawar, A., Robinson, C.M., Madisch, I., Harrach, B., Hudson, N.R., Schnurr, D., Heim, A., Chodosh, J., Seto, D. & Jones, M.S. (2009). Evidence of molecular evolution driven by recombination events influencing tropism in a novel human adenovirus that causes epidemic keratoconjunctivitis. *PLoS One*, Vol.4, No.6, e5635.

Walsh, M.P., Seto, J., Tirado, D., Chodosh, J., Schnurr, D., Seto, D. & Jones, M.S. (2010). Computational analysis of human adenovirus serotype 18. *Virology*, Vol.404, No.2, pp.284-289.

Walter, E.A., Greenberg, P.D., Gilbert, M.J., Finch, R.J., Watanabe, K.S., Thomas, E.D. & Riddell, S.R. (1995). Reconstitution of cellular immunity against cytomegalovirus in recipients of allogeneic bone marrow by transfer of T-cell clones from the donor. *N. Engl. J. Med.*, Vol.333, No.16, pp.1038-1044.

Wang, H., Liaw, Y.C., Stone, D., Kalyuzhniy, O., Amiraslanov, I., Tuve, S., Verlinde, C.M., Shayakhmetov, D., Stehle, T., Roffler, S. & Lieber, A. (2007). Identification of CD46 binding sites within the adenovirus serotype 35 fiber knob. *J. Virol.*, Vol.81, No.23, pp.12785–12792.

Watcharananan, S.P. , Kiertiburanakul, S., Piyatuctsanawong, W., Anurathapan, U., Sungkanuparph, S., Pakakasama, S., Chantratita, W. & Hongen, S. (2010). Cytomegalovirus, adenovirus, and polyomavirus co-infection among pediatric recipients of allogeneic stem cell transplantation: characteristics and outcome. *Pediatr. Transplant.*, Vol.14, No.5, pp.675–681.

Watzinger, F., Suda, M. , Preuner, S., Baumgartinger, R., Ebner, K., Baskova, L., Niesters, H.G., Lawitschka, A. & Lion, T. (2004). Real-time quantitative PCR assays for detection and monitoring of pathogenic human viruses in immunosuppressed pediatric patients. *J. Clin. Microbiol.*, Vol.42, No.11, pp.5189-5198.

World Health Organization (WHO). *Water Recreation and Disease. Plausibility of Associated Infections: Acute Effects, Sequelae and Mortality* by Kathy Pond. Published by IWA Publishing, London, UK. ISBN: 1843390663

Zhang, Y. & Bergelson, J.M. (2005). Adenovirus receptors. *J. Virol.*, Vol.79, No.19, pp.12125–12131.

Yea, C., Dembowy, J., Pacione, L. & Brown, M. (2007). Microtubule-mediated and microtubule-independent transport of adenovirus type 5 in HEK293 cells. *J. Virol.*, Vol.81, No.13, pp.6899–6908.

Yusuf, U., Hale, G.A., Carr, J., Gu, Z., Benaim, E., Woodard, P., Kasow, K.A., Horwitz, E.M., Leung, W., Srivastava, D.K., Handgretinger, R. & Hayden, R.T. (2006). Cidofovir for the treatment of adenoviral infection in pediatric hematopoietic stem cell transplant patients. *Transplantation*, Vol.81, No.10, pp.1398-1404.

A Systematic Review of Nonpharmacological Exercise-Based Rehabilitative Interventions in Adults Undergoing Allogeneic Hematopoietic Stem Cell Transplantation

M. Jarden
Department of Haematology, The University Hospital of Copenhagen,
The University Hospital's Centre for Nursing and Care Research,
Copenhagen
Denmark

1. Introduction

Allogeneic hematopoietic stem cell transplantation (allo-HSCT) is an established treatment. More than 15,000 procedures are performed worldwide each year for a number of hematological malignancies such as acute myeloid and lymphoid leukemia, and bone marrow failure syndromes (Gratwohl et al. 2002; Frassoni, 2004). It is predicted that transplantation rates for allo-HSCT will continue at the same or higher level in the immediate future (Gratwohl et al. 2002). Despite clinical cure in 20-70% of all patients, depending on disease presentation, long term sequelae of immunosuppression, chemotherapy toxicities and graft-versus-host disease (GvHD) debilitate a large number of patients (Gratwohl et al. 2002). Moderate to severe GvHD develops in 40-50% of patients undergoing allo-HSCT (Bearman et al. 1988; Weisdorf et al. 1990; Roy et al. 1992; Hings et al. 1994). Factors limiting the efficacy of this treatment are death due to recurrence or treatment-related death due to infection or organ failure in the cytopenia and later immunosuppressed phase immediately post-HSCT. Over the last decades, the cumulative effects of improvements in supportive care, drug dosing, stem cell technology and prophylaxis of GvHD have led to an increased number of complete remissions (Devergie, 2004). However, with the increasing number of transplants performed and the growing number of survivors, a shift in clinical focus from not only improving survival but also improving short and long-term quality of life has emerged (Andrykowski et al. 1995). Patients in the treatment and recovery phase of HSCT commonly experience adverse physical and emotional reactions. Fatigue and muscle weakness can limit ability to accomplish activities of daily living. Additionally, depression, anxiety, fear, and frustration add to the difficulties of recovering from HSCT (Syrjala et al. 1993; Andrykowski et al. 1995). Several studies confirm that high levels of physical and psychological stress have been observed in patients prior to and at the start of HSCT and during follow-up periods (Baker et al. 1997; Molassiotis & Morris, 1997; McQuellon et al. 1998; Fife et al. 2000). The mechanisms are not fully known, but it is assumed that several factors such as total body irradiation (TBI), chemotherapy, GvHD, infections, long-term inactivity or bed rest and side-effects from medication can contribute to the physical and emotional weakening of the patient. Recipients of allo-HSCT with low Vitamin D levels and low bone mineral density were likely to have received corticosteroids, have experienced GVHD and an elevated parathyroid hormone level (Sproat et al. 2011; Massenkeil et al. 2001). Vitamin D insufficiency and deficiency can cause osteomalacia, bone pain, muscle weakness, musculoskeletal pain, headache, fatigue, and may precipitate or exacerbate osteopenia and osteoporosis and increase risk of skeletal fracture (Knutsen et al. 2010; Sproat et al. 2011).

Patients that have undergone HSCT experience treatment-related symptoms during and after treatment that can affect health related quality of life (HR-QOL). Patients experience multiple somatic, affective and cognitive symptoms during and after aggressive cancer treatment, where eleven to thirteen simultaneously occurring symptoms have been reported (Portenoy et al. 1994). During hospitalization for standard allo-HSCT, patients are typically on prolonged bed rest, and experience complications from the myeloablative treatment, ie. acute GvHD, side effects from medications (immunosuppression & steroids), frequent infections and psychological reactions that can be debilitating. It is reported that HR-QOL is lowest during inpatient time (Grulke et al. 2011). The most commonly reported symptoms are fatigue, diarrhea, insomnia, poor appetite, diminished concentration, mouth dryness, dyspnoea, loss of hair and poor body image perception (Jarden et al. 2009; Molassitis et al.1997; Larsen et al. 2007). After HSCT, fatigue, dyspnoea and insomnia remain at elevated levels (Grulke et al. 2011). Psychosocial wellbeing after transplant is influenced by mucositis toxicity, and other side effects, and psychological factors as anxiety, distress and social support have a significant impact on how severely patients experience mouth pain (Schulz-Kindermann et al. 2002) Fatigue is one of the most frequent and distressful side effects reported by patients who have undergone HSCT (Jarden et al. 2009; So et al. 2003), and it has been shown that physical activity decreased and this decline coincided with diminished physical, emotional, role and cognitive functioning during the initial post transplantation period (Danaher et al. 2006). Bevans et al. found that patients experienced multiple symptoms and high symptom distress after allo-HSCT conditioning (Bevans et al. 2008) Further, fatigue was the main symptom interfering with daily life in 79% of patients (Molassiotis & Morris, 1999), and in 11% of patients at 100 days post allo-HSCT (Bevans et al. 2008). Loss of physical strength seem to be more pronounced in patients on corticosteroid treatment, and the causes of an impairment of physical performance are not fully understood, though low activity levels have been suggested to be a substantial contributor (Carlson et al. 2006). One study in patients undergoing HSCT showed a correlation between the number of symptoms experienced and poor functional status and general health (Larsen et al. 2007) and in another study, changes in HRQoL could be explained entirely by changes in functional limitations and somatic symptoms (Broers et al. 2000). Further, symptom bother from GVHD had a direct effect on functional performance (Mitchell et al. 2010). A Danish study found patients prior to allo-HSCT to have lower VO_2 max scores and elevated fatigue levels than the normal population, and these scores were unchanged six months after transplantation (Kalo et al. 2007). Furthermore, persons diagnosed with hematological disease have difficulty returning to the work force (deBoer et al. 2008) and have an increased risk for early retirement (Carlsen et al. 2008), while unemployed leukaemia patients, especially those with lower social support have significantly elevated levels of stress, anxiety, and depression (deBoer et al. 2008). In HSCT, predictors of slower return to work include physical dysfunction and female gender (Kirchhoff AC et al. 2010). Maintaining daily function and reducing fatigue and treatment-related symptoms can be important goals and there is therefore, a continued need for nonpharmacological strategies that address the specific impairments experienced by patients undergoing allo-HSCT.

There is a rapidly increasing literature on the effects of exercise on cancer rehabilitation, especially for breast cancer patients, on whom the majority of research has been conducted (Courneya et al. 2011, Conn et al. 2006). Despite that physical exercise showed positive effects on cardiorespiratory fitness, treatment-related symptoms and physiological effects, the extent of these positive results still need to be established. A qualitative and quantitative review and meta-analysis found only small to moderate effect of physical activity interventions on these outcomes (Schmitz et al. 2005; Conn et al. 2006). Physical activity is reported as being well tolerated in cancer survivors during and after treatment, however, conclusions about adverse effects are inconclusive (Schmitz et al. 2005). Recent guidelines for exercise prescription for cancer survivors from the American College of Sports Medicine (Schmitz et al. 2010) report no contraindication for starting an exercise program in patients undergoing either autologous or allogeneic HSCT - however, issues regarding, the ideal time for starting a program safely and effectively, type of program, frequency, intensity and duration is not confirmed, especially in relation to the HSCT treatment trajectory. Exercise has been proposed as a nonpharmacologic adjuvant therapy to combat the physiological and psychological symptoms of HSCT (Wiskemann & Huber, 2008). However, little work exists in utilizing exercise interventions specifically in the allo-HSCT setting. It is documented that there is a decline in exercise levels in cancer patients from prediagnosis to postdiagnosis (Courneya et al.1997) and more specifically, a low level of "naturally-occurring" exercise amongst patients undergoing HSCT is reported, suggesting that a structured intervention may be necessary in order to promote exercise in this population (Courneya et al. 2000). The majority of the earlier research done in adult patients with hematological disease is in the context of high dose chemotherapy with stem cell support (autologous HSCT or HD-SCS). To date, there are eight published studies that incorporated exercise regimes in the high dose chemotherapy-stem cell support context (HD- SCS) (Coleman et al. 2003; Decker et al 1989; Hayes et al. 2003, Dimeo et al. 1996, 1997, 1999, 2003, including one on-going study (Peerson et al. 2010), and five studies in mixed HD-SCS and allo-HSCT populations (Dimeo et al. 1999, Baumann et al. 2005; Wilson et al. 2005, Knols et al. 2010, Danaher Hacker et al. 2011). There are however fundamental and important differences in the two types of treatment (HD-SCS and allo-HSCT) including the conditioning regimes, i.e. total body irradiation in allo-HSCT, and origin of stem cells, i.e. patients undergoing HD-SCS are supported with their own stem cells, and therefore the donor related challenges (GvHD) in allo-HSCT are not present, and lastly, the overall duration of hospitalization for HD-SCS is much shorter compared to allo-HSCT. One 3 year prospective study, found that patients undergoing HD-SCS had better self-rated physical function (symptomotology, physical status and energy level) as compared to the allo-HSCT group (Prieto et al 2005). The HD-SCS and mixed group (HD-SCS and allo-HSCT, and age) exercise-based study sizes ranged between 12-70 participants. There are five randomized trials (Peerson et al. 2010; Baumann et al. 2005; Coleman et al. 2003; Hayes et al. 2004; Dimeo et al. 1996), one study used a minimization procedure (Knols et al. 2010) and six single-group trials (Danaher Hacker et al. 2011; Wilson et al. 2005; Decker et al 1989; Dimeo et al.

1997, 1999, 2003). Exercise was tested during hospitalization in three studies (Baumann et al. 2005; Dimeo et al. 1996; 2003), one study during hospitalization and continued to out-patient (Dimeo 1999), and eight studies after hospital discharge (outpatient and home based programs) (Wilson et al. 2005; Decker et al 1989; Coleman et al. 2003; Hayes et al. 2004; Dimeo et al. 1997, Knols et al. 2010; Peerson et al. 2010; Danaher Hacker et al. 2011). Feasibility was established, no adverse events registered and beneficial effects were reported on aerobic capacity (Hayes et al. 2004; Dimeo et al. 2003; Dimeo et al. 1996); muscle strength (Hayes et al. 2004; Knols et al. 2010), body composition (Coleman et al. 2003; Hayes et al. 2003), immunological function (Dimeo et al. 1997; Hayes et al. 2003; Knols et al. 2010), treatment-related symptoms i.e. fatigue (Dimeo et al. 1999; Wilson et al. 2005) and HRQoL (Wilson et al. 2005; Hayes et al. 2004). These positive and potentially important results for HD-SCS and mixed groups are encouraging, but we need to look exclusively at the allo-HSCT adult patient group in order to evaluate the role and impact of exercise rehabilitation in this treatment context alone.

The primary objective of the systematic review is to summarize the exercise-based rehabilitative interventions in adults with haematological disease undergoing allogeneic hematopoetic stem cell transplantation (allo-HSCT) on feasibility and safety, and effectiveness related to physical and functional capacity, health related quality of life, treatment-related symptoms and medical related outcomes.

2. Method

The systematic literature search is based on PRISMA guidelines (preferred reporting items for systematic reviews and metaanalyses) developed from Cochrane Collaboration (Moher D et al. 2009).This review includes 1) a systematic literature search with the identification of all intervention trials in adult patients in the allo-HSCT setting during the past 25 years, 2) a uniform presentation of all trials and a synthesis of the characteristics and findings, 3) an appraisal of the methodological quality of the trials, 4) a summary and 5) conclusions and future research.

2.1 Search strategy and data extraction

The systematic literature review covers 25 years of research: 1986 - 2011 (Table 1). Searches were carried out in MEDLINE and EMBASE using search items bone marrow cell transplantation, bone marrow transplantation, stem cell transplantation, physical activity, physical fitness and exercise. Identified articles were searched for additional references. The search was limited to include randomized controlled trials (RCT), controlled clinical trials, adults and English articles. Eligibility criteria are shown in Table 2. Only studies that included patients from the allogeneic setting are included in this review. Studies that included patients in either the auto-HSCT / HD-SCS setting or mixed settings (both autologous and allogeneic) were excluded, though compiled in the literature search for background and reference purposes. Full articles were obtained for remaining abstracts and information was extracted from identified articles and organized under the following headings: authors, sample (n, type of treatment), age, design and study period, exercise-based intervention, duration, frequency/intensity, and results (Table 3). All articles were independently reviewed and appraised for rigor of method and analysis.

Date	Search	Limits	Database	Results
1/7/2011	"Bone Marrow Cell Transplantation" "Bone Marrow Transplantation" "Stem Cell Transplantation" Allogeneic[tw] Autologous[tw] "Physical activity" "Physical fitness" "Exercise"	English Publication date from 1986/01/01 to 2011/07/01	Medline (NLM)	90
("Bone Marrow Cell Transplantation" OR "Bone Marrow Transplantation" OR "Stem Cell Transplantation") AND (Allogeneic OR Autologous[AND ("Physical activity" OR "Physical fitness" OR Exercise) AND (English[lang] AND ("1986/01/01" : "2011/07/01")				
Date	**Search**	**Limits**	**Database**	**Results**
1/7/2011	"Bone Marrow Cell Transplantation". "Bone Marrow Transplantation". "Stem Cell Transplantation". "Allogeneic" "Autologous". "Physical activity". "Physical fitness" "Exercise"	English language Publication year: 1986-Current	EMBASE (OvidSP)	90
(("Bone Marrow Cell Transplantation" or "Bone Marrow Transplantation" or "Stem Cell Transplantation") and ("Allogeneic" or " Autologous") and ("Physical activity" or "Physical fitness" or "Exercise")), (English[lang] AND ("1986/01/01" : "2011/07/01")				

Table 1. Results of a systematic literature search with keywords

Inclusion criteria	Exclusion criteria
• Journal articles reporting primary research of exercise-based intervention studies prior to, during or after allo-HSCT • Randomized clinical trials or controlled studies • Participants ≥18 years • Recipients of allo-HSCT for a hematologic disease • Published in English between 1986 and 2011	• Studies that include auto-HSCT/HD-SCT • Studies that include a mixed population of auto-HSCT and allo-HSCT • Review articles or case study • Clinical reports • Dissertations • Conference abstracts • Editorials or letters to the editor

Allo-HSCT indicates allogeneic stem cell transplantation; auto-SCT, autologous stem cell transplantation; HD-SCT, high dose chemotherapy with stem cell support

Table 2. Eligibility criteria

Authors	Sample	Age	Design and study period	Exercise-based Intervention	Duration	Frequency/ Intensity	Results
Cunningham et al (1986)	N=40 Attrition 25% 30 completers: IG1=10 IG2=10 CG=10 **Allo-HSCT** Hematopoietic diagnoses: AML, ALL - all receiving TPN	25 years* (range 14-41) IG1: 20.8 (range 14-33) IG2: 26.0 (15-38) CG: 22.5 (15-41)	RCT - three groups IG1=resistive exercise IG2=resistive exercise CG= usual care/ no exercise **In-patient** During hospitalization, after last dose of cyclophosphamide	Supervised IG1 and IG2: strength resistance training **Contraindication for exercise:** Platelets <10,000/mm Bleeding Cardiac arrhythmia Temp: >39.5C	5 weeks	IG1 3 days/week IG2 5 days/week 30 min per session Resistive exercise with 15 repetitions of 8 different exercises	**Program compliance:** not reported. **Safety**: muscle and joint stiffness reported in CG (5 of 5 patients). Independent mean active days for IG1 19.8, IG2 19.5 and CG 18.7). *Pre-post* No significant intergroup differences in body composition CG ↓ creatinine excretion (p=0.039) IG1and IG2 → creatinine excretion
Mello M, et al. (2003)	N= 32 Attrition 44% 18 completers: IG=9 CG=9 **Allogeneic HSCT** Hematopoietic diagnoses: CML, AML, SAA, NHL, MDS	29 years* (range 18-44) IG: 27.9 (18-39) CG: 30.2 (18-44)	RCT - two groups During HSCT IG=exercise program CG=usual care **In-patient/Outpatient** Exercise initiated after neutrophil engraftment and continued in the outpatient facility	Supervised IG: active ROM exercise (5 exercises), muscle stretching (3 muscle groups) and a modified treadmill walking program **Contraindication for exercise:** Hemoglobin<10mg/dl Platelets<20,000/mg/dl	6 weeks	5 days /week 40 min per session Aerobic training: 5 sets of 3 min walking at a comfortable pace w/ 3 min rest between sets Progression to 2 sets of 10 min walking at comfortable pace alternated with 20 min walking at accelerated pace with HR no higher than 70% of MHR. Mild to moderate intensity	**Program compliance:** not reported. **Safety:** No adverse events **Muscle strength:** *Pre-post* No significant intergroup differences Significant Intragroup differences: CG ↓ Shoulder (abductors & flexors) CG↓ Elbow (excl. DM extensors) CG ↓ and IG ↓ Knee (flexors) CG↓ Ankle (flexors)

Authors	Sample	Age	Design and study period	Exercise-based Intervention	Duration	Frequency/ Intensity	Results
Kim et al. (2005, 2006)	N= 42 Attrition 17% 35 completers: IG=18 CG=17 **Allo-HSCT** Hematopoietic diagnoses: AML, ALL, SAA	33.6 years* (range 20-48) IG: 32.9 ($\pm$7.0) CG: 34.3($\pm$7.8)	RCT – two groups IG: bed exercises CG: usual care **In-patient** During entire hospitalization	Supervised IG: Bed exercises: Preliminary exercise 10 min Relaxation breathing 10 min ROM, stretching and relaxation 10 min	6 weeks	30 minutes daily Low intensity	**Program compliance:** not reported **Safety**: not reported **Hematological Outcomes:** *Pre-post* IG: ↑ lymphocyte count (not significant, however interaction between groups and times significant (p=0.031). CG: ↓ lymphocyte count (p<0.05) **Psychological Measures:***Pre-post* Intergroup significant changes IG ↓ Depression: BDI (p=0.0001) IG ↓ Anxiety: STAI (p=0.0001)
Carlson et al. (2006)	N= 12 Attrition 0% (12 completers) **Allo-HSCT** Hematopoietic diagnoses: CLL,CML,AML,NHL, FL, MDS	47 (range 28-55)	QE (No control group) **Outpatient** After allo-HSCT: 39 months (range 9-92 months)	Supervised Individualized endurance training on cycle ergometer	12 weeks	3 days/week 1st workout 30 min (RPE 2: light to moderate) 2nd workout 15 min (RPE 6: hard to very hard) 3rd workout 20 min (RPE 4: somewhat hard)	**Program compliance:** 89% **Safety:** No adverse events **Physiological Outcomes:** *Pre-post* ↑ St.V (p<0.005), ↓HR (p<0.005) and ↓ RPE (p<0.005) **Psychological Measures:***Pre-post* ↓ Fatigue: BFI (p<0.001) ↓ Fatigue: FACT-F (p<0.001) ↑ Vigor: POMS (p<0.001)
DeFor et al. 2007	N=100 Attrition 15% 85 completers: IG=43 CG=42	47 years* (range 18-68) IG: 46 (18-68)	RCT – two groups IG: walking regime CG: no formal exercise	Not supervised – compliance was encouraged three times /wk during hospitalization	14 weeks	Daily In-hospital 15 min twice a day on a treadmill	**Program compliance**: 100% for 10 patients (24%). 28 (68%) and 24 (62%) patients exercised at least 5 times/wk for at least 15 min. during hospitalization, and

Authors	Sample	Age	Design and study period	Exercise-based Intervention	Duration	Frequency/ Intensity	Results
	Allo-HSCT Hematopoietic diagnoses: AML, ALL,CML, NHL, HD, MDS	CG: 49 (22-64)	**In-patient/home-based** From day of admission to 100 days	Inpatient: endurance training - walking regime on treadmill Home-based: endurance training - walking regime		After discharge 30 min walk once a day at comfortable speed	after discharge, respectively. **Safety:** Not reported **Performance:** *Pre-post* KPS: No significant intergroup differences KPS: IG less ↓ than CG - ´significant for subgroup of nonmyeloablative conditioned patients (p=0.04) **Physical and Emotional wellbeing:** *Pre-discharge-100 days* IG ↑ than CG - significant at discharge for subgroup of nonmyeloablative conditioned patients (physical p<0.01 and emotional p=0.02), and after 100 days (physical p<0.01)
Jarden et al. 2007, 2009, 2009	N= 42 patients Attrition 19% 34 completers IG=17 CG=17 **Allo-HSCT** Hematopoietic diagnoses: CML, AML, ALL, AA, MDS, WM, PNH, MF	40.8 years* (range 18-60) IG: 45 (18-60) CG: 38 (18-55)	RCT - two groups IG: exercise-based intervention CG: standard care (PT) start after stem cell infusion 1-3 days/wk. **In-patient** Initiated on first day of admission to discharge	Supervised Aerobic exercise on cycle ergometer, resistance training with hand and ankle weights (9 muscle groups), ROM and stretching (6 exercises), relaxation training and psycho-education **Contraindications for intervention:** Platelets <20 x 9[10]/l Hemoglobin <5g/dl Bleeding, petecchiae Temp: >38C	6 weeks	5 days/week 1 h $\pm$ 10 min Aerobic exercise on cycle ergometer no higher than 75% MHR, for 15-30 min daily. ROM and stretching daily, resistance training 3/wk, All exercises 2 sets, 12 reps. Progressive relaxation twice weekly, psychoeducation daily. RPE up to 13 (somewhat hard)	**Program compliance**: 90% (67-100%) **Safety:** No adverse reactions or injuries **Physiological Outcomes** *Pre-post* IG → VO_{2max}, muscle strength and functional capacity. Highly significant intergroup differences for all outcomes favouring IG **PRO for HR-QoL** *Pre-post-3 & 6 mo* No significant difference in QOL between groups

Authors	Sample	Age	Design and study period	Exercise-based Intervention	Duration	Frequency/ Intensity	Results
							Clinical Outcomes Significant intergroup difference: IG fewer days receiving TPN (p=0.019) **PRO for treatment related Symptoms** *Pre-post-3 mo- 6 mo* Significant intergroup differences: IG ↓ diarrhea (p=0.014) EORTC IG ↓ symptom prevalence: diminished concentration, memory problems, nausea, nervousness (p<0.01) SCT-SAS IG ↓ symptom severity: fatigue, loss of appetite, diminished concentration, sleep difficulties, nausea (p<0.05) SCT-SAS IG ↓ symptom cluster severity: gastrointestinal, cognitive, functional and mucositis (p<0.01)
Shelton et al. 2009	N=61 Attrition 13% 53 completers: IG1=26 IG2=27 **Allo-HSCT** Hematopoietic diagnoses: AML,ALL,CML, CLL, NHL, HD, Lymphoma	46.3 years* (range 22-70) IG1: 43.7 (22-68) IG2: 48.9 (29-70)	RCT – two groups IG1: Supervised IG2: Self directed **Out-patient** After allo-HSCT and within the previous 6 months	Supervised IG1: Aerobic (upper and lower extremity ergometer cycle, treadmill) and resistive exercises (weight machines, 6 exercises). Self-directed IG2: Received oral, written and practical instruction in 8 resistance exercises using elastic bands of differing resistance	4 weeks	3 days/week 20-30 min aerobic training + resistance exercise IG1: Aerobic intensity on cycle or treadmill 60-75% and resistance exercise 1-3 sets of 10 rep. IG2: 8 resistance exercises 1-3 sets of 10-15 rep. Patient was instructed to increase	**Program compliance:** IG1: 75% IG2: Not reported **Safety:** Not reported **Performance test results:** *Pre-post* No significant intergroup differences Significant Intragroup differences: IG1↑ 6MWD 12% (p<0.05) and ↑ 50 foot walk time (p=0.05), → remaining performance tests.

Authors	Sample	Age	Design and study period	Exercise-based Intervention	Duration	Frequency/ Intensity	Results
				and a walking program. After 4 weeks and post testing, patients were offered the supervised program. All participants received information about exercising safely in the context of their medical status.		walking up to 30 consecutive min. 3x/wk.	IG2 ↑ 6MWD 9.8% ($p<0.05$) **Fatigue** No significant difference in fatigue levels between groups, however there was a 20 and 10% decrease in fatigue scores in the supervised and unsupervised groups, respectively.
Inoue et al. 2010	N= 26 Attrition 0% 26 completers: IG1=13 IG2=13 **Allo-HSCT** Hematopoietic diagnoses: AML, ALL, CML, NHL, AA, MDS	48.5 years* (range 20-62) IG1: 43 (20-55) CG: 54 (27-62)	Convenience sample – two different treatment groups receiving same intervention IG1: myeloablative IG2: nonmyeloablative **In-patient** Exercise initiated after neutrophil engraftment until discharge	Supervised Aerobic exercise on cycle ergometer, and walking in a corridor, muscle strength (3 large muscle groups) and stretching (5 exercises)	IG1 approx. 12 weeks IG2 approx. 8 weeks Exact duration not reported: Program started after neutrophil engraftment IG1: median 17 days (median hosp. days 101) IG2: median 14 days (median hosp. days 71)	5 days/week 20-40 min Aerobic training on ergometer cycle 60% MHR, strength and stretching exercises modified according to patients activity level and condition	**Program compliance:** Not reported **Safety:** Not reported, however toxicities due to treatment did not affect mean daily steps. IG1: The degree of physical activity had negative correlation with the duration of hospitalization ($r=-.71$; $p=0.0071$).
Wiskemann et.al. 2010	N= 105 Attrition 24% 80 completers: IG1=40 CG=40	48.8 years* (range 18-71) IG:47.6 (18-70)	Allocation through minimization method - multicenter IG: Partly supervised exercise program	Partly supervised (in-patient) Self-directed (at home) Introduction to program and training	16 weeks *Prior HSCT* IG 3 weeks CG 2 weeks *During HSCT*	5 sessions/week (3- 5 endurance and 2 resistance training sessions) Light aerobic activity	**Program compliance:** 87.3% (79% of exercise logs were returned) Prior HSCT: 87.5% During HSCT:83%

Authors	Sample	Age	Design and study period	Exercise-based Intervention	Duration	Frequency/ Intensity	Results
	Allo-HSCT Hematopoietic diagnoses: AML, secondary AML, ALL, CML, CLL, AA, MDS, MPS, MM, other lymphomas	CG: 50 (20-71)	CG: Standard care - PT 3 days/week 30 min./session during hospitalization, step counters to record physical activity **At-home** (prior HSCT) **In-patient** (during HSCT) **At-home** (after HSCT)	manual inkl. exercise DVD. Individualized exercise program includes aerobic warm-up, endurance and strength training. **Contraindictions for Intervention:** Platelets <10-20 x 9^{10}/l Hemoglobin <8 g/dl Temp: >38C Pain, nausea and dizziness	IG 6 weeks CG 6 weeks *After HSCT* IG 7 weeks CG 7.5 weeks	and stretching Endurance training: walking/cycle 15-40 min Strength training: stretch bands 8-20 rep., 2-3 sets). Three different protocols were applied 1) focus on extremities, 2) entire body and 3) bed exercises (in-patient) Endurance RPE 12-14 Resistance RPE 14-16	After HSCT: 91.3% **Safety:** Not reported **Physiological Outcomes** *Pre-post* Significant intergroup difference in favor of IG: IG → and CG ↓ 6MWD (p=0.02) **PRO for HR-QoL** *Pre-post* Significant intergroup differences in favor of IG: IG↓ and CG↑ general and physical fatigue (p=0.009, p=0.01) MFI IG ↓ and CG↑ fatigue (p=0.01) POMS IG↓ and CG↓ physical functioning (p=0.03) EORTC IG↓ and CG↑ depression (p=0.05) POMS Significant intergroup difference in favour of CG: IG↑ and CG ↓ anxiety (p=0.01) HADS
Baumann et al. 2011	N= 47 patients Attrition 30% 33 completers IG=17 CG=16 **Allo-HSCT** Hematopoietic diagnoses: CML, AML, ALL, CLL, MPS, MDS, CMML, MM, PID	42.1 years* (range 28-57) IG: 41.4 (±11.78) CG: 42.8 (±14.04)	RCT - two exercise groups IG1: Aerobic endurance and ADL **In-patient** Initiated 6 days prior to stem cell infusion CG: standard care (PT) **In-patient** Initiated after stem cell infusion	Supervised IG: Aerobic endurance cycle ergometer ADL including strength, coordination stretching, walking and stair climbing CG: Mobilization 10 min, (gymnastics and coordination training), and stretching 5 min.	IG 4.5 weeks CG 3.8 weeks	5 days/week IG: 2 sessions/ day aerobic 20-30 min/day interval training, RPE: 80% of achieved watt load in WHO-test. ADL: 20 min/day RPE: slightly strenuous CG: 20 min/day	**Program compliance:** Not reported **Safety:** **Aerobic endurance, Strength, Lung function :** *Pre-post* No significant differences between exercise groups, however CG ↓ aerobic endurance (p=0.009)

Authors	Sample	Age	Design and study period	Exercise-based Intervention	Duration	Frequency/ Intensity	Results
				Contraindictions for exercise: Platelets <10-20 x 9^{10}/1 Hemoglobin <8 g/dl Temp: >38C Strong pain, infection, restricted consciousness, somnolence, confusion, dizziness, nausea and vomiting. Training was interrupted during cardio- and nephrotoxic chemotherapy.		RPE: Low intensity: not strenuous	CG ↓ muscle strength (p=0.022) **PRO for HR-QoL** *Pre-post* No significant differences between exercise groups, however CG ↑ fatigue (p=0.046) EORTC IG and CG ↓ physical functioning (p=0.005 and p=0.002) IG ↑emotional state (p=0.028)

*median age

AA indicates aplastic anemia, AML, acute myelogenous leukemia**;** ALL, acute lymphocytic or lymphoblastic leukemia ; CLL, chronic lymphocytic leukemia; CML, chronic myelogenous leukemia; CMML, chronic myelomonocytic leukemia; HD, Hodgkins disease; MDS, myelodysplastic syndrome; MF, myelofibrosis; MM, multiple myeloma; MPS, myeloproliferative syndrome; NHL, Non-Hodgkin lymphoma; PNH, paroxysmal nocturnal hemoglobinuria; WM, waldenstrom macroglobulinemia
RPE, Rate of Perceived Exertion; St.v, Stroke Volume; HR, Heart Rate; IG, Intervention Group; IG2, Intervention Group 2; CG, Control Group; TPN, Total Parenteral Nutrition;
MHR, Maximal Heart Rate; DM and NDM, Dominant Non Dominant; KPS, Karnofsky Score; PT, Physical therapy; PRO, patient reported outcome; HR-QoL, health related Quality of Life; TR-symptoms, treatment related symptoms; SCT-SAS, Stem cell transplantation-Symptom Assessment Scale; EORTC-QLQ-C30 European Organization for Research and Treatment of Cancer Quality of Life Questionnaire; HADS, Hospital Anxiety and Depression Scale; FACT-An or FACT-F, Functional Assessment of Cancer therapy – Anemia or Fatigue scale; POMS, Profile of Mood States; MFI, Multidimensional Fatigue Inventory; BFI, Brief Fatigue Inventory; BDI, Beck Depression Inventory; STAI, State Trait Anxiety Inventory; 6MWD, 6 minute walk distance , → no change; ↑ increase; ↓ decrease..

Table 3. Physical exercise based studies in allogeneic HSCT on aerobic capacity, muscle strength, health-related quality of life and treatment related symptoms

3. Results

3.1 Exercise-based studies in the allo-HSCT setting

In this review, 10 studies met the inclusion criteria (Baumann et al. 2011; Carlson et al. 2006; Cunningham et al. 1986; Defor et al. 2007; Inoue et al., 2010; Jarden et al., 2009; Kim and Kim 2006; Mello et al. 2003; Shelton et al. 2009; Wiskemann et al. 2011). Of these, three were from the USA and two from Germany, and respectively, one from Brazil, Canada, Denmark, Japan and South Korea. Cunningham et al. carried out the very first exercise training trial for the allo-HSCT population in 1986 and although the participants included children and adults (range 14 - 41 years), this study is included in the review because of its focus being in the allo-HSCT setting only. Jarden et al. published three articles and Kim et al. two articles, each based on one trial, however each article has a different focus and purpose.

All studies were designed as prospective intervention trials that tested an exercise-based program. The primary and secondary outcomes were study feasibility and safety; physiological outcomes i.e. aerobic, muscle strength and function; psychosocial outcomes i.e. health-related QoL, emotional wellbeing; treatment-related symptoms i.e. fatigue; and hospital or disease-related outcomes i.e. days of hospitalization, creatinine excretion, lymphocyte counts. Baseline to post assessment ranged between 4 - 16 weeks [mean 7.3] and one study had follow-up tests to 6 months (Jarden et. al. 2009).

3.2 Sample characteristics

In all, 406 patients with different haematological diseases (AA,, AML, ALL, CLL , CML, CMML, FL, HD, MDS, MF, MM, MPS, NHL, PNH, WM, other lymphomas)[1] across 10

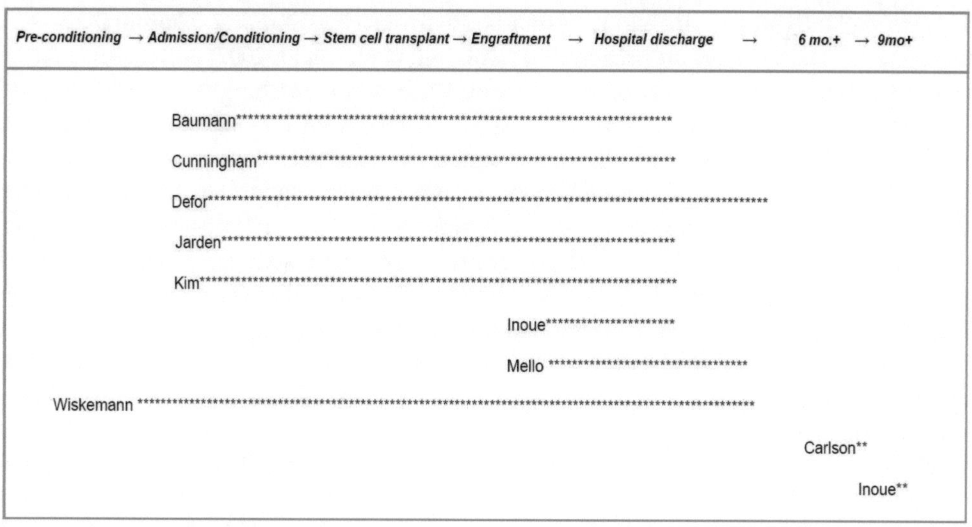

Table 4. Intervention phase

[1] AA indicates aplastic anemia, AML, acute myelogenous leukemia; ALL, acute lymphocytic or lymphoblastic leukemia ; CLL, chronic lymphocytic leukemia; CML, chronic myelogenous leukemia; CMML, chronic myelomonocytic leukemia; HD, Hodgkins disease; MDS, myelodysplastic syndrome; MF, myelofibrosis; MM, multiple myeloma; MPS, myeloproliferative syndrome; NHL, Non-Hodgkin lymphoma; PNH, paroxysmal nocturnal hemoglobinuria; WM, waldenstrom macroglobulinemia

studies are included in this review. The sample size of the studies ranged from 12 to 105 [Mean 50.7] and the patients were of mixed gender between 14 and 71 years [mean 42.6]. Cunningham was the only study that included patients less than 18 years of age. Four studies were initiated prior to conditioning and throughout the hospitalization period (Baumann et al. 2011; Cunningham et al. 1986; Defor et al. 2007; Jarden et al. 2009; Kim and Kim 2006; Wiskemann et al. 2011), two studies after marrow engraftment and throughout hospitalization (Mello et al., 2003; Inoue et al., 2010), and three of these continued post allo-HSCT (Defor et al. 2007; Mello et al. 2003; Wiskemann et al. 2011). Two studies were initiated post allo-HSCT in the out-patient or home setting, within 6 and 39 months (range 9-92), respectively (Shelton et al. 2009; Carlson et al. 2006). The approximate start and endpoint of each intervention is illustrated in Table 4.

3.3 Type of exercise-based interventions

The duration of the exercise-based interventions ranged between 4-16 weeks [mean 7.3]. For interventions initiated prior to and during hospitalization, 6 of the studies were supervised (Cunningham et al. 1986; Mello et al. 2003; Kim and Kim 2006; Jarden et al. 2009; Inoue et al. 2010; Baumann et al. 2011) and 2 were partly supervised (DeFor et al. 2007, Wiskemann et al. 2011). In the out-patient context, Carlson's study was fully supervised, while Shelton's study had one supervised and one self-directed study arm. The frequency of all interventions ranged between 3 and 5 days/week or daily. When reported, the intensity of training in general was between low/mild and comfortable to moderate but not exceeding 70-75% of maximum heart rate or in Rate of Perceived Exertion (RPE) - being somewhat hard.

In the in-patient context, one study tested strength resistive training (Cunningham et al. 1986), one a walking program (treadmill or walking) (Defor et al. 2007) and another mixed-type low intensity bed exercises of stretching and relaxation breathing (Kim and Kim 2006), and 5 studies instituted mixed-type exercise up to moderate intensity (Mello et al. 2003, Jarden et al. 2009, Inoue et al. 2010, Wiskemann et al. 2011, Baumann et al. 2011) by combining aerobic training (treadmill, cycle, walking or stair climbing) with one or more other moderate intensity exercise (range of motion or ADL (activities of daily living), coordination exercises, muscle stretching, resistive exercises with free weights or elastic bands) and low intensity progressive relaxation training and education (Jarden et al. 2009). Three studies (Mello et al. 2003; Wiskemann et al. 2011, Defor et al. 2007) continued the programs after hospital discharge. In the out-patient only context, one study tested an ergometer cycle program (Carlson et al. 2006), and the other, aerobic (cycle or treadmill) and resistive exercises (weight machines) vs. selfdirected walking and resistive exercises including patient information regarding exercise safety (Shelton et al. 2009). All in all, three studies were unidimensional (one exercise component), of which, two were aerobic training (Carlson et al. 2006; DeFor et al. 2007), and one resistance training (Cunningham et al. 1986), and seven studies had mixed type training (Baumann et al. 2011; Inoue et al. 2010; Jarden et al. 2009; Kim and Kim 2006; Mello et al. 2003; Shelton et al. 2009; Wiskemann 2011), of which, one was of low intensity (Kim and Kim 2006), and one study included both low and moderate intensity components (Jarden et al. 2009). Only two studies incorporated educational (Shelton et al. 2009) or psychoeducational (Jarden et al. 2009) elements in the program.

3.4 Feasibility and safety

No adverse events, reactions or injuries were reported, though not all studies reported safety outcomes. The overall attrition rates ranged between 0 and 44% (mean 18.7). Program compliance was reported by five studies (Carlson et al. 2006; DeFor et al. 2007; Jarden et al. 2009; Shelton et al. 2009; Wiskemann et al. 2011). Carlson et al. reported an overall 89% compliance. Defor et al. reported that 24% of all patients had 100% compliance and that 62% of the study's population exercised at least 5 times/wk for at least 15 min during hospitalization and after discharge, respectively. Jarden et al. reported 90% intervention compliance (range 67-100), and 83% in-hospital and 87% at-home compliance (Wiskemann et al. 2011), while Shelton et al. reported 75% for the supervised intervention, though did not report for the self-directed intervention. Some studies had safety screening parameters, in which contraindication for exercise included platelet counts <10 and 20 x 10^9/l (Cunningham et al. 1986, Mello et al. 2003, Jarden et al. 2009; Wiskemann et al. 2011, Baumann et al. 2011), and haemoglobin <5 g/dl (Jarden et al. 2009), <8 (Wiskemann et al. 2011, Baumann et al. 2011), and <10 (Mello et al. 2003); temp>38°C (Cunningham et al. 1986; Jarden et al. 2009; Wiskemann et al. 2011, Baumann et al. 2011), and adverse symptoms as bleeding, petecchiae, pain, nausea, dizziness. Baumann et al. interrupted training during cardio and nephrotoxic chemotherapy.

3.5 Physiological outcomes - aerobic/endurance

A mixed-type exercise program 5 days/week at moderate intensity during the entire hospitalization was able to maintain aerobic capacity, while the control group decreased, and this difference was highly significant ($p<0.01$) (Jarden et al. 2009). Another similar mixed-exercise program during hospitalization showed a significant decrease in aerobic endurance ($p=0.009$) in the control group, but no significant difference between intervention and control groups (Baumann et al. 2011). A 12 wk. aerobic training program on cycle ergometer initiated between 9 and 92 months after HSCT showed cardiovascular effects through increased stroke volume ($p<0.005$) and decreased heart rate ($p<0.005$), with a decreased RPE ($p<0.005$) (Carlson et al. 2006).

3.6 Muscle strength

Mello et al. combined aerobic (treadmill) and ROM exercise program initiated during hospitalization and continued into the outpatient facility over 6 weeks showed significant decrease in upper and lower extremity strength in the control group, however differences between intervention and control groups were not significant.

During hospitalization, four muscle strength tests (chest press 1 RM, leg extension 1 RM, elbow flexor and knee extensor Newton) decreased by 2 - 4 % in the intervention group, while decreasing between 19 - 25 % in the control group. This difference was significant ($p<0.01$) (Jarden et al. 2009). Baumann et al., however, did not find a significant difference between groups for muscle strength, but the pre to post muscle strength scores decreased significantly in the control group ($p=0.022$).

3.7 Functional capacity

Wiskemann et al. maintained the 6 minute walk distance (6MWD) in the intervention group, while the control group decreased and the intergroup difference was significant ($p=0.02$). The 2 minute stair climb test in the intervention group decreased by 14 % and for the control group 38%, and the difference between groups was significant ($p<0.01$) (Jarden et al. 2009). A walking program did not show significant differences between groups for the Karnofsky score, though a subgroup analysis of the nonmyeloablative conditioned patients showed that this patient group decreased significantly less than the control group ($p=0.04$) (DeFor et al. 2007). There were no significant differences for the 6MWD between a supervised and self directed intervention, still both groups significantly improved by 12% ($p<0.05$) and 9.8% ($p<0.05$), respectively (Shelton et al. 2009). Also, the supervised group improved the 50 foot walk time ($p=0.05$) and maintained other performance tests, though no significance between groups.

3.8 Health-related quality of life

A bed exercise study with relaxation breathing, ROM and stretching during hospitalization decreased depression (BDI) $p=0.0001$ and anxiety (STAI) $p=0.0001$ as compared to the control group (Kim and Kim 2006). An in- to out-patient walking program showed improvements in physical and emotional wellbeing on a self reported score from 1 - 10, with 1 being very poor and 10 being very good. At discharge, physical wellbeing was better in the exercise group ($p<0.01$). Among the nonmyeloablative group, emotional wellbeing was better in the exercise group ($p=0.02$) at discharge and at 100 days, physical wellbeing was superior in the exercise group ($p<0.01$) (DeFor et al. 2007). A mixed exercise intervention during the entire hospitalization showed no difference in QoL and emotional wellbeing between groups (EORTC-QLQ-C30, FACT-An, HADS), though HRQOL was maintained in the intervention group at post testing, and there was significant improvement in emotional wellbeing at 3 and 6 months, $p=0.045$ and $p=0.012$, respectively (FACT-An) and significant decrease in anxiety at 3 and 6 months, $p=0.021$ and $p<0.0001$, respectively (HADS). The control group significantly decreased overall HRQOL $p=0.0005$ (FACT-An) at post testing, and significantly reduced physical functioning ($p=0.004$) and worsened three gastrointestinal symptoms (nausea and vomiting ($p=0.048$), appetite loss ($p=0.004$) and diarrhea ($p=0.011$) (EORTC QLQ C-30) (Jarden et al. 2009). Wiskemann et al. study beyond discharge showed between group differences in favor of the intervention group in physical functioning ($p=0.03$)(EORTC-QLQ-C30) and decreased depression ($p=0.05$), though showed a significant increase in anxiety ($p=0.01$) in the intervention group (HADS). Baumann et al. found no significant difference between groups on the EORTC-QLQ-C30, though the pre-post differences in the intervention and control group decreased significantly for physical functioning ($p=0.005$ and $p=0.002$). Intervention group improved emotional state ($p=0.028$), but again, no differences between groups (Baumann et al. 2011). Carlson et al.'s out-patient endurance program significantly improved intergroup vigor scores ($p<0.001$) on POMS.

3.9 Treatment-related symptoms

There was a decreased symptom prevalence in diminished concentration, memory loss, nausea and nervousness ($p<0.01$) and decreased symptom severity in fatigue, loss of appetite, sleep difficulties and nausea ($p<0.05$) on the SCT-SAS scale (Jarden et al. 2009). Further, diarrhea was significantly decreased (EORTC-QLQ-C30) ($p=0.014$) (Jarden et al. 2009). Symptom cluster analyses revealed a significant decrease in symptom severity in gastrointestinal, cognitive, functional and mucositis clusters over time and up to 6 months after allo-HSCT ($p<0.01$). Wiskemann et al. found a significant decrease between groups in both general and physical fatigue ($p=0.009$, $p=0.01$) MFI and a significant decrease in fatigue

($p=0.01$) (POMS). Baumann et al. was unable to show between group differences in fatigue scores, but the control group increased fatigue at post testing ($p=0.046$). Carlson et al. showed a significant decrease in fatigue ($p<0.001$) (BFI and FACT-F).

3.10 Medical related outcomes

Cunningham et al. did not find significant changes in body composition, though a decreased creatinine excretion in the intervention group (Cunningham et al. 1986). Kim et al. tested a bed exercise effect on lymphocyte counts, and reported no significant differences between groups, however there was an interaction between groups and times ($p=0.031$), there was also a decrease in lymphocyte count in the control group ($p<0.05$). Jarden et al. reported the intervention group as receiving fewer days of TPN ($p=0.019$) with no changes in BMI between groups at post testing. There were no other differences between groups regarding hospitalization days, bone marrow engraftment days, days with fever, and though there was a 19% difference in the incidence of GvHD favoring the intervention group, this was not statistically different. Defor et al. and Inoue et al. showed no difference in days of hospitalization between groups, but Inoue et al. showed that the degree of physical activity had a negative correlation with the duration of hospitalization ($r=-.71$; $p=0.0071$).

3.11 Methodological quality of the studies

Seven studies were designed as randomized trials (Cunningham et al. 1986; Mello et al. 2003; Kim and Kim 2006; Defor et al. 2007; Jarden et al. 2009; Shelton et al. 2009; Baumann et al. 2011), one study allocated through the minimization method (Wiskeman et al. 2011), one study was a convenience sample studying the effect of the same exercise intervention on two different allo-HSCT treatment groups (myeloablative or nonmyeloablative conditioning regime) (Inoue et al. 2010) and one study did not have a control group (Carlson et al. 2006). Further, Shelton et al. studied the effect of two different interventions (supervised vs. self directed) in the outpatient/home setting. The control group was described in most studies as receiving standard or usual care, including either no formal training or the hospital units' standard physical therapy (PT). Generally, standard PT was described as being introduced later during hospitalization, i.e. after stem cell infusion, and at less frequent intervals and lower intensity levels. The study arms were not similar at baseline in two (Kim and Kim 2006; Baumann et al. 2011) of the eight groups in which there was an intervention and control group present. The control group in Kim et al. had a significant higher lymphocyte level at baseline compared to the intervention group and Baumann et al. had twice as many males as females in the exercise group as compared to the control group at baseline. All studies reported the eligibility criteria to which the study population was chosen. Only one study stated that the outcome assessor was blinded (DeFor et al. 2007). In the relevant studies, none reported blinding of the exercise trainer or the patients. Two studies reported performing intention-to-treat analyses (Jarden et al. 2009; Wiskemann et al. 2011).

4. Summary

This is the first literature review of exercise-based interventions in the allo-HSCT context. The purpose of this systematic review was to summarize the exercise-based rehabilitative interventions in adults with haematological disease undergoing allogeneic hematopoetic stem cell transplantation (allo-HSCT) on feasibility and safety, and effectiveness related to physical and functional capacity, health related quality of life, treatment-related symptoms and medical related outcomes. To date, 10 intervention studies and 13 articles have been published that incorporated exercise-based regimes in the allo-HSCT context, and though we have found encouraging and important results, making direct trial comparisons can be a challenge due to the small sample sizes, the wide range of different primary and secondary outcomes and measurements, varying types of interventions, and different start and end points, duration, frequency and intensity of the different exercise components. Most of the studies in this review were randomized trials, however, control groups received varying standard care regimes, and there was a lack of outcome assessor blinding, trainer blinding and patient blinding which decreases the general methodological quality of the studies. Taking these methodological limitations into consideration, this review however finds important results pertinent to the allo-HSCT clinical setting.

The results suggest that exercise interventions are feasible and safe. No study reported adverse events as a direct result of testing or exercising, though not all studies reported safety. Five studies reported safety screening parameters, which may have contributed to patients being able to exercise safely. Of the five studies that reported compliance rates, it would suggest that patients are capable of participating adequately in a daily exercise program during and after allo-HSCT. Two mixed type exercise studies implemented during the entire hospitalization suggest a stabilization in aerobic endurance during hospitalization (Jarden et al. 2009; Baumann et al. 2011), and one 12 week out-patient endurance study found significant positive cardiovascular effects (Carlson et al. 2006). In regards to muscle strength, mixed type exercise during hospitalization (Baumann et al. 2011; Jarden et al. 2009) and continuing in the outpatient context (Mello et al. 2003) showed significant muscle strength decreases in control group, but only one study found significant differences between groups (Jarden et al. 2009) suggesting that the loss of muscle strength was minimized. Mixed type exercise during hospitalization (Jarden et al. 2009) and continued after discharge (Wiskemann et al. 2011) significantly decreased loss of function (2 minute stair climb) (Jarden et al. 2009) and significantly maintained function (6MWT) (Wiskemann et al. 2011). DeFor et al.'s walking program during and after discharge did not show significant differences between groups for Karnofsky score, though a subgroup analysis of the nonmyeloablative group showed a significant reduction in loss of performance in the intervention group. (DeFor et al. 2007). Both supervised and self directed mixed type exercise in the post HSCT was shown to improve function significantly on the 6MWD and for the supervised group improvement was reported for the 50 foot walk time (Shelton et al. 2009). A mixed type low intensity exercise program during hospitalization significantly decreased depression (BDI) and anxiety as compared to the control group (Kim and Kim 2006), while two mixed-type moderate intensity exercise programs during hospitalization did not show a significant effect on HR-QoL (Baumann et al. 2011, Jarden et al. 2009), but Wiskemann et al.'s mixed type exercise that continued after discharge showed a significant improvement in physical functioning and decreased depression, but also a significant increase in anxiety (Wiskemann et al. 2011). DeFor's in- to out-patient walking program reported significant improvements in emotional wellbeing among the nonmyeloablative patient group at hospital discharge and by 100 days, physical wellbeing was significantly improved. Carlson et al.'s 12 week out-patient endurance program significantly increased vigor. There was a significant longitudinal decrease in prevalence and intensity of several symptoms and symptom clusters, including fatigue up to 6 months (Jarden et al. 2009) and in two studies exercising after discharge, a significant decrease in fatigue scores was reported (Wiskemann et al. 2011, Carlson et al. 2006). Hospital or treatment-related outcomes as body composition or immunological and infectious parameters i.e. lymphocyte counts, days to bone marrow engraftment, days with fever, incidence of GvHD., number of transfusions received) was not affected by exercise. One exercising group received significantly fewer days of TPN as compared to the control group (Jarden et al. 2009). There was no effect on duration of hospitalization (Inoue et al. 2010, Jarden et al. 2009, Defor et al. 2007), though Inoue et al. reported that the level of physical activity had a negative correlation with the number of hospitalization days (Inoue et al. 2010).

It is suggested in recent literature that the optimal training program for persons with cancer combine both aerobic and muscle strength training (Neiman & Courneya, 2006, Courneya & Friedenreich, 2011). A review from 2008 of 15 exercise trials in the in-patient and out-patient HSCT setting, included patients in the allo-HSCT and HD-SCS context and resulted in tentative recommendations (Wiskemann & Huber, 2008): mixed exercise: endurance (up to daily) and resistance training (2-3 x/wk)), from 10 - 30 minutes at moderate intensity (BORG scale 12-14, 70-80% maximum HR) before, during and after hospitalization. In Liu et al. literature review from 2009 of physical exercise interventions in hæmatological cancer patients suggested it feasible to conduct exercise in this patient population, but concluded that there was a lack in methodological quality in the physical exercise studies and therefore effectiveness could not be established (Liu et al.). Recommendations based on a more recent review from 2011 of patients with mixed hematologic disease propose 'supervised' exercise during and after hematological cancer treatments 2 to 5 days per week, with adjustment for health conditions. Further, a combination of aerobic and resistance exercise is suggested, with a varying intensity between 40-70% of maximum heart rate and full body resistance exercises at 8-12 reps and 2-3 sets with slow progression over time. Health and neutropenic screening for exercise participation is also recommended according to Jones et al. (Battaglini 2011).

Exercise recommendations, however for patients during and after hospitalization for specifically allo-HSCT have not been developed, and in order to determine the appropriate and optimal exercise prescription / intervention for patients undergoing allo-HSCT, trials in the allo-HSCT context were examined in this review for type, duration, frequency and intensity. Taken the methodological limitations into consideration, a partly to fully supervised and daily, mixed-type exercise (aerobic and resistant exercises, also ROM and stretching) at moderate intensity (70-75% MHR) started at least prior to conditioning and carried out during allo-HSCT is feasible and can maintain or decrease loss of aerobic capacity, muscle strength and function at hospital discharge. However, continuing the program after hospital discharge had further physical, functional and symptom related benefits (decrease fatigue, reduced depression). Attaining positive results require a relatively high compliance rate, over 85%. Therefore, screening parameters and contraindication for training criteria should be instituted to not only keep the patient safe, but also enable adequate participation in the program. Based on this review, it can be suggested that postponement or modification of training include: platelets < 20 x 10^9/l, haemoglobin < 5-8 g/l, temp > 38°C and adverse symptoms as bleeding, petecchiae, pain, nausea, dizziness. It is not clear from this review, whether training during conditioning inkl. cardio and nephrotoxic chemotherapy should be postponed, and to what extent, that is, during active infusion or hours/days following chemotherapy. We may also consider incorporation of low intensity exercise as relaxation breathing or progressive relaxation in the program, especially when a higher exercise intensity level is not possible (Kim and Kim, 2006; Jarden et al. 2009). Psychosocial, educational and motivational approaches may be integrated to maintain compliance levels, support exercise motivation and efficacy and increase independent activity and lifestyle changes.

5. Conclusion and future research

This chapter is a systematic review of the rehabilitation research carried out in the allo-HSCT context during the past 25 years. These findings, despite a number of methodological issues, indicate positive physiological and emotional benefits from exercise in patients during and after allo-HSCT. Exercise during treatment may help patients decrease loss of or maintain aerobic and functional capacity, and muscle strength, and when exercise is continued or instituted in the outpatient/home context there are improvements in aerobic and functional capacity. These results also indicate that exercise-based interventions have multidimensional benefits, including maintaining or improving HRQOL and reducing the most persistent treatment-related symptoms, especially fatigue. This review provides general guidelines for exercise in the allo-HSCT context. However, with improved methodological approaches, future research may provide clinicians with more specific rehabilitation guidelines.

5.1 Future research

Future studies are encouraged to institute certain methodological stringencies, including inclusion of homogenous groups (same diagnosis or treatment group), larger patient populations, perhaps, multi-institutional studies, randomized designs that clearly describe treatment allocation and stratification methods, as well as details regarding the control group. Further, point estimates for primary outcomes, effect size calculations, as well as Intention-to-Treat analyses are recommended. When and if possible, an effort to blind outcome assessors, data entry keyers and statisticians would improve methodological quality. It is also important that future studies justify the chosen intervention and clearly describe the individual exercise components duration, intensity and frequency as well as screening parameters for intervention participation, and lastly, documentation of compliance rates in order to properly evaluate the effect of the intervention. Decisions regarding test time-points and the most relevant and comprehensive outcome measurements need to be associated with challenges directly related to allo-HSCT. Therefore, GvHD, treatment related symptoms i.e. fatigue, insomnia, pain, gastrointestinal complaints, skin and bodily changes, poor physical, functional and muscle capacity, low levels of physical activity and decreased bone density, low Vitamin D levels and reduced HRQOL including problems with sexuality, body image, social wellbeing and job function need to be considered in this treatment group. More studies are needed that examine the entire treatment trajectory and continue well into the out-patient and home setting. Research that aims to promote multidimensional benefits may consider multimodal interventional designs that combine physical exercise with other psychosocial and educational approaches. Additionally the role of vitamin D and exposure to sunlight in combination with physical exercise on bone health, function and general wellbeing may be warranted. Further, there is a need to study the patients' own experience as a participant in an exercise-based intervention in order to develop programs that are tailored to fit the needs of patients during and after allo-HSCT. Finally, translational intervention studies are needed that support and improve the ability of patients to cope with their life situation during treatment, as well as to function optimally after allo-HSCT in daily life, including return to employment.

6. Conflict of interest

The author declares no conflict of interest.

7. Acknowledgements

I thank Anders Larsen (Information specialist - The University Hospital's Centre for Nursing and Care Research, Copenhagen Denmark) for verifying and assisting with the literature search strategy. I would also like to thank Kristine Marie Jarden (Research assistant - University of Southern Denmark) for her assistance in manuscript preparation.

8. References

Andrykowski MA, Breiner CB, Altmaier EM. (1995) Quality of life following bone marrow transplantation: findings from a multicenter study. Br J Cancer; 71:1322-9.

Baker F, Marcellus D, Zabora J, Polland A, Jodrey D. (1997) Psychological distress among adult patients being evaluated for bone marrow transplantation. Psychosomatics 1997;38:10-19

Battaglini CL. (2011) Physical Activity and hematological Cancer Survivorship. *Physical Activity and Cancer*, Recent Results in Cancer Research 186, DOI: 10.1007/978-3-642-04321-7_12, Springer-Verlag Berlin Heidelberg, 275-304.

Baumann FT, Zopf EM, Nykamp E, Kraut L, Schüle K, Elter T, Fauser AA, Bloch W. (2011) Physical activity for patients undergoing an allogenic hematopoietic stem cell transplantation: benefits of a moderate exercise intervention. European Journal of Haematology; accepted article: 10.1111/j.1600-0609.2011.01640.x

Bearman SI, Applebaum FR, Buckner CD. (1988) Regimen-related toxicity in patients undergoing bone marrow transplantation. J Clin Oncol;6:1562-68.

Bevans MF, Mitchell SA, Marden S. (2008) The symptom experience in the first 100 days following allogeneic hematopoietic stem cell transplantation (HSCT).Support Care Cancer;16:1243-54.

Broers S, Kaptein AA, Le Cessie S, Fibbe W, Hengeveld MW. (2000) Psychological functioning and quality of life following bone marrow transplantation: a 3-year follow-up study. J Psychosom Res;48:11-21.

Carlsen K, Oksbjerg Dalton S, Frederiksen K, Diderichsen F, Johansen C. (2008) Cancer and the risk for taking early retirement pension: a Danish cohort study. Scand J Public Health;36(2):117-25.

Carlson LE, Smith D, Russell J, Fibich C, Whittaker T. (2006) Individualized exercise program for the treatment of severe fatigue in patients after allogeneic hematopoietic stem-cell transplant: a pilot study. Bone Marrow Transplant;37:945-54.

Coleman EA, Coon S, Hall-Barrow J, Richards K, Gaylor D, Stewart B. (2003) Feasibility of exercise during treatment for multiple myeloma. Cancer Nurs 2003;26:410-9.

Conn V, Hafdahl A, Porock D, McDaniel R, Nielsen P. (2006) A meta-analysis of exercise interventions among people treated for cancer. Support Care Cancer; 14:699-712.

Courneya KS, Friedenreich (2011) *Physical Activity and Cancer*, Recent Results in Cancer Research 186, DOI: 10.1007/978-3-642-04321-7_12, Springer-Verlag Berlin Heidelberg.

Courneya KS, Keats MR, Turner AR. (2000) Physical exercise and quality of life in cancer patients following high dose chemotherapy and autologous bone marrow transplantation. Psychooncology;9:127-36.

Cunningham BA, Morris G, Cheney CL, Buergel N, Aker SN, Lenssen P. (1986) Effects of resistive exercise on skeletal muscle in marrow transplant recipients receiving total parenteral nutrition. J Parenter Enteral Nutr;10:558-563.

Danaher Hacker E, Larson JL, Peace D. (2011) Exercise in Patients Recieving Hematopoietic Stem Cell Transplantation: Lessons Learned and Results From a Feasibility Study. Oncology Nursing Forum;38:216-223.

Danaher Hacker E, Ferrans C, Verlen E, Ravandi F, Van Besien K, Gelms J, Dieterle N. (2006) Fatigue and Physical activity in patients undergoing hematopoietic stem cell transplant. Oncol Nurs Forum;33:615-24.

Decker WA, Turner-McGlade J, Fehir KM. (1989) Psyhcological aspects and the physiological effects of a cardiopulmonary exericse program in patients undergoing bone marrow transplantation for acute leukemia. Transplant Proc 21;21(1Pt3):3068-3069.

de Boer AG, Verbeek JH, Spelten ER, Uitterhoeve AL, Ansink AC, de Reijke TM, Kammeijer M, Sprangers MA, van Dijk FJ. (2008) Work ability and return-to-work in cancer patients. Br J Cancer;98(8):1342-7.

DeFor TE, Burns LJ, Gold EM, Weisdorf DJ. (2007) A randomized trial of the effect of a walking regimen on the functional status of 100 adult allogeneic donor hematopoietic cell transplant patients. Biol Blood Marrow Transplant;13:948-55.

Devergie A. Graft vs. host disease. (2004) In: Apperley J, Carrerras E, Gluckman E, Gratwohl A, Masszi T, eds. The EBMT Handbook 2004 revised edition: haemopoietic stem cell transplantation. Maastricht: European Group for Blood and Marrow Transplantation; 63-168.

Dimeo F, Bertz H, Finke J, Fetscher S, Mertelsmann R, Keul J. (1996) An aerobic exercise program for patients with haematological malignancies after bone marrow transplantation. Bone Marrow Transplant;18:1157-60.

Dimeo F, Fetscher S, Lange W, Mertelsmann R, Keul J. (1997) Effects of aerobic exercise on the physical performance and incidence of treatment related complications after high dose chemotherapy. Blood;90:3390-4.

Dimeo F, Schwartz S, Fietz T, Wanjura T, Boning D, Thiel E. (2003) Effects of endurance training on the physical performance of patients with hematological malignancies during chemotherapy. Support Care Cancer;11:623-628.

Dimeo F, Stieglitz RD, Novelli-Fischer U, Fetscher S, Keul J. (1999) Effects of physical activity on the fatigue and psychologic status of cancer patients during chemotherapy. Cancer;85:2273-7.

Fife BL, Huster GA, Cornetta KG. (2000) Longitudinal study of adaptation to the stress of bone marrow transplantation. J Clin Oncol;29:1539-49.

Frassoni F. Stem Cell Transplantation in adults. (2004) In: Apperley J, Carrerras E, Gluckman E, Gratwohl A, Masszi T, eds. The EBMT Handbook 2004 revised edition: haemopoietic stem cell transplantation. Maastricht: European Group for Blood and Marrow Transplantation; 2004. p. 239-54.

Gratwohl A, Baldomero H, Horisberger B, Schmid C, Passweg J, Urbano-Ispizua A. (2002) Current trends in hematopoietic stem cell transplantation in Europe. Blood; 100:2374-86.

Grulke N, Albani C, Bailer H. (2011) Quality of Life in patients before and after haematopoietic stem cell transplantation measured with the European Organization for Research and Treatment of Cancer (EORTC) Quality of Life Core Questionnaire QLQ-C30. Bone Marrow Transplant, accepted 28 March 2011;1-10.

Hayes S, Davies PS, Parker T, Bashford J. (2003) Total energy expenditure and body composition changes following peripheral blood stem cell transplantation and participation in an exercise program. Bone Marrow Transplant;31:331-8.

Hayes S, Davies PS, Parker T, Bashford J, Newman B. (2004) Quality of life changes following peripheral blood stem cell transplantation and participation in a mixed-type, moderate-intensity, exercise program. Bone Marrow Transplant; 33:553-8.

Hayes SC, Rowbottom D, Davies PS, Parker TW, Bashford J. (2004) Immunological changes after cancer treatment and participation in an exercise program. Med Sci Sports Exerc; 35:2-9.

Hayes SC, Davies PS, Parker TW, Bashford J, Green A. (2004) Role of a mixed-type, moderate intensity exercise programme after peripheral blood stem cell transplantation. Br J Sports Med;38:304-9.

Hings IM, Severeson R, Filipovich AH. (1994) Treatment of moderate and severe acute graft-versus host disease after allogeneic bone marrow transplantation. Transplantation; 58:437-42.

Inoue J, Ono R, Okamura A, Matsui T, Taekoshi H, Miwa M, Kurosaka M, Saura R, Shimada T. (2010) The impact of Early Rehabilitation on the Duration of Hospitalization in Patients After Allogenic Hematopoietic Stem Cell Transplantation. Transplantation Proceedings; 42: 2740-2744.

Jarden M, Baadsgaard MT, Hovgaard DJ, Boesen E, Adamsen L. (2009) A randomized trial on the effect of a multimodal intervention on physical capacity, functional performance and quality of life in adult patients undergoing allogeneic SCT. Bone Marrow Transplantation;43(9):725-737.

Jarden M, Nelausen K, Hovgaard D, Boesen E, Adamsen L. (2009) The effect of a multimodal intervention on treatment-related symptoms in patients undergoing hematopoietic stem cell transplantation: A randomized controlled trial. Journal of pain and symptom management;38(2):174-190.

Jarden M, Hovgaard D, Boesen E, Quist M, Adamsen L. (2007) Pilot study of a multimodal intervention: mixed-type exercise and psychoeducation in patients undergoing allogeneic stem cell transplantation. Bone Marrow Transplantation; 40:793-800.

Kalo J, Agergård J, Aadahl M, Visby R, Halkjær Kristensen J, Jacobsen N. (2007) Fatigue in hematological patients treated with allogeneic stem cell transplantation. J Danish Med Assoc;169:3194-7.

Kim SD, Kim HS. (2006) A series of bed exercises to improve lymphocyte count in allogeneic bone marrow transplantation patients. Eur J Cancer Care (Engl); 15:453-7.

Kim S-D, Kim H-S. (2005) Effects of a relaxation breathing exercise on anxiety, depression, and leukocyte in hemopoietic stem cell transplantation patients. Cancer Nursing; 28(1):79-83.

Kirchhoff AC, Leisenring W, Syrjala KL. (2010) Prospective predictors of return to work in the 5 years after hematopoietic cell transplantation. J Cancer Surviv;4(1)33-44.

Knutsen KV, Brekke M, Gjelstad S, Lagerløv P. (2010) Vitamin D status in patients with musculoskeletal pain, fatigue and headache: a cross-sectional descriptive study in a multi-ethnic general practice in Norway.Scand J Prim Health Care. 28(3):166-71.

Larsen J, Nordstrom G, Ljungman P, Gardulf A. (2007) Factors associated with poor general health after stem-cell transplantation. Support Care Cancer;15:849-57.

Massenkeil G, Fiene C, Rosen O, Michael R, Reisinger W, Arnold R. (2001) Loss of bone mass and vitamin D deficiency after hematopoietic stem cell transplantation: standard prophylactic measures fail to prevent osteoporosis Leukemia;15(11):1701-5

McQuellon RP, Russell GB, Rambo TD, Craven BL, Radford Perry JJ, Cruz J, Hurd DD. (1998) Quality of life and psychological distress of bone marrow transplant recipients: the 'time trajectory' to recovery over the first year. Bone Marrow Transplant; 21:477-86.

Mello M, Tanaka C, Dulley FL. (2003) Effects of an exercise program on muscle performance in patients undergoing allogeneic bone marrow transplantation. Bone Marrow Transplant; 32:723-8.

Mitchell SA, Leidy NK, Mooney KH, Dudley WN, Beck SL, LaStayo PC, Cowen EW, Palit P, Comis LE, Krumlauf MC, Avila DN, Atlam N, Fowler DH, Pavletic SZ. (2010) Determinants of functional performance in long-term survivors of allogeneic hematopoietic stem cell transplantation with chronic graft-versus-host disease (cGVHD). Bone Marrow Transplant;45(4)Epub 2009.

Moher D, Liberati A, Tetzlaff J, Altman DG, The PRISMA Group (2009) Preferred Reporting Items for Systematic Reviews and Meta-Analyses: The PRISMA Statement. PLoS Med 6(7): e1000097. doi:10.1371/journal.pmed.1000097

Molassiotis A, Morris PJ. (1997) Suicide and suicidal ideation after marrow transplantation. Bone Marrow Transplant;19:87-90.

Molassiotis A, Morris PJ. (1999) Quality of life in patients with chronic myeloid leukemia after unrelated donor bone marrow transplantation. Cancer Nurs; 22:340-9.

Nieman DC, Courneya KS. (2006) Immunological conditions. In: American College of Sports Medicine. ACSM's Resource Manual For Guidelines For Exercise Testing and Prescription. 5th ed. Philadelphia: Lippincott Williams & Wilkins; 2006. p. 528-542.

Persoon S, Kersten MJ, ChinAPaw M JM, Buffart LM, Burghout H, Schep G, Brug J, Nollet F. (2010) Design of the Exercise Intervention after Stem cell Transplantation (EXIST) study: a randomized controlled trial to evaluate the effectiveness and cost-effectiveness of an individualized high intensity physical exercise program on fitness and fatigue in patients with multiple myeloma or (non-) Hodgkins lymphoma treated with high dose chemotherapy and autologous stem cell transplantation. BMC; 10:671
http://www.biomedcentral.com/1471-2407/10/671.

Portenoy RK, Thaler HT, Kornblith AB, Lepore JM, Friedlander-Klar H, Coyle N, Smart-Curley T, Kemeny N, Norton L, Hoskins W, et al. (1994) Symptom prevalence, characteristics and distress in a cancer population. Qual Life Res;3:183-9.

Prieto JM, Atala J, Blanch J, Carreras E, Rovira M, Cierera e, Gasto C. (2005) Patient-rated emotional and physical functioning among hematologic cancer patients during hospitalization for stem-cell transplantation. Bone Marrow Transplant; 35(3):307-14.

Roy J, McGlave PB, Filipovich AH. (1992) Acute graft-versus-host disease following unrelated donor marrow transplantation: failure of conventional therapy. Bone Marrow Transplant;10:77-82.

Schulz-Kinderman F, Hennings U, Ramm G. (2002) The role of biomedical and psychosocial factors for the prediction of pain and distress in patients undergoing high-dose therapy and BMT/PBSCT. Bone Marrow Transplant;29:341-51.

Schmitz KH, Holtzmann J, Courneya K, Masse LC, Duval S, Kane R. (2005) Controlled physical activity trials in cancer survivors: a systematic review and meta-analysis. Cancer Epidemiol Biomarkers Prev;14:1588-1595.

Schmitz KH, Courneya KS, Matthews C, Demark-Wahnefried W. Galvao DA, Pinto BM, Irwin ML, Wolin KY, Segal RJ, Lucia A, Schneider CM, von Gruenigen VE, Schwartz AL. (2010) American College of Sports Medicine Roundtable on Exercise Guidelines for Cancer Survivors: Special Communications - Roundtable Consensus Statement, Medicine & Science in Sports & Exercise, DOI:10.1249/ MSS.0b013e3181e0c112;1409-1423

Shelton ML, Lee JQ, Morris GS, Massey PR, Kendall DG, Munsell MF, Anderson KO, Simmonds MJ, Giralt SA. (2010) A randomized control trial of a supervised versus a self-directed exercise program for allogeneic stem cell transplant patients. Psycho-Oncology;18: 353-359.

So WK, Dodson J, Tai JW. (2003) Fatigue and quality of life among Chinese patients with hematologic malignancy after bone marrow transplantation. Cancer Nurs; 26:211-9.

Sproat L, Bolwell B, Rybicki L, Dean R, Sobecks R, Pohlman B, Andresen S, Sweetenham J, Copelan E, Kalaycio M. (2011) Vitamin d level after allogeneic hematopoietic stem cell transplant. Biol Blood Marrow Transplant. 17(7):1079-83.

Syrjala KL, Chapko MK, Vitaliano PP, Cummings C, Sullivan KM. (1993) Recovery after allogeneic marrow transplantation: prospective study of predictors of long-term physical and psychosocial functioning. Bone Marrow Transplant;11:319-27.

Weisdorf D, Haake R, Blazar B. (1990) Treatment of moderate/severe acute graft-versus-host disease after allogeneic bone marrow transplant: an analysis of clinical risk features and outcome. Blood; 75:1024-30.

Wilson RW, Jacobsen PB, Fields KK. (2005) Pilot study of a home-based aerobic exercise program for sedentary cancer survivors treated with hematopoietic stem cell transplantation. Bone Marrow Transplant;35:721-7.

Wiskemann J, Huber G. (2008) Physical exercise as adjuvant therapy for patients undergoing hematopoietic stem cell transplantation. Bone Marrow Transplant; 41:321-9.

Wiskemann J, Dreger P, Schwerdtfeger R, Bondong A, Huber G, Kleindienst N, Ulrich CM, Bohus M. (2010) Effects of a partly self-administered exercise program before, during and after allogeneic stem cell transplantation. Blood; 117(9):2604-2613.